TS 155.6 .D48 Desrochers A. (A. Model: control automate

Modeling and Control
of
Automated Manufacturing Systems

Modeling and Control of Automated Manufacturing Systems

Alan A. Desrochers

IEEE Computer Society Press Tutorial

Modeling and Control of Automated Manufacturing Systems

ALAN A. DESROCHERS

Electrical, Computer, and Systems Engineering Department
Rensselaer Polytechnic Institute

IEEE Computer Society Press

Washington ● Los Alamitos ● Brussels ● Tokyo

Published by

IEEE Computer Society Press
1730 Massachusetts Avenue, N.W.
Washington, D.C. 20036-1903

Printed in the United States of America

Copyright © 1990 by IEEE, Inc.

Copyright and Reprint Permissions: Abstracting is permitted with credit to the source. Libraries are permitted to photocopy beyond the limits of U.S. copyright law for private use of patrons those articles in this volume that carry a code at the bottom of the first page, provided the per-copy fee indicated in the code is paid through the Copyright Clearance Center, 29 Congress Street, Salem, MA 01970. Instructors are permitted to photocopy isolated articles for noncommercial classroom use without fee. For other copying, reprint or republication permission, write to Director, Publishing Services, IEEE, 345 East 47th Street, New York, NY 10017. All rights reserved. Copyright © 1990 by The Institute of Electrical and Electronics Engineers, Inc.

IEEE Computer Society Order Number 1916
Library of Congress Number 89-45825
IEEE Catalog Number EH0294-9
ISBN 0-8186-8916-1 (casebound)
ISBN 0-8186-5916-5 (microfiche)

Additional copies may be ordered from:

| IEEE Computer Society
Order Department
10662 Los Vaqueros Circle
Los Alamitos, CA 90720-2578 | IEEE Service Center
445 Hoes Lane
P.O. Box 1331
Piscataway, NJ 08855-1331 | IEEE Computer Society
13, Avenue de l'Aquilon
B-1200 Brussels
BELGIUM | IEEE Computer Society
Ooshima Building
2-19-1 Minami-Aoyama,
Minato-Ku
Tokyo 107 JAPAN |

THE INSTITUTE OF ELECTRICAL AND ELECTRONICS ENGINEERS, INC.

Preface

This tutorial surveys and summarizes the present research directions related to the systems issues in automation and flexible manufacturing. Chapter 1 establishes the theme of the tutorial which is the modeling and control of automated manufacturing systems when there is uncertainty in machine availability and product demand. Modeling methods that deal directly with these uncertainties are presented in Chapter 2.

The control methods presented in this tutorial are emerging mathematical techniques that have demonstrated potential for changing the way in which modern manufacturing systems are operated. The first approach uses the concepts and system principles from modern dynamic control theory. These are the control theoretic methods contained in Chapter 3. Chapter 4 presents a new approach called perturbation analysis. This technique is related to linearization of discrete-event dynamic systems and enables the very efficient optimization of manufacturing system parameters. Supervisory control is the subject of Chapter 5 in which Petri nets are presented as a method to coordinate the various manufacturing subsystems. The Petri net is also useful for modeling the asynchronous and concurrent events that occur in an automated manufacturing system. In addition, they can be used to evaluate the performance of the system. These modeling and control methods are presented as an alternative to the traditional queueing approach.

Chapter 6 is a collection of various mathematical techniques that have been used to measure and improve the performance of these automated systems. Finally, the last chapter provides concluding remarks and a discussion of simulation languages and artificial intelligence in manufacturing.

The tutorial is at the level of a first year graduate student in engineering. Although electrical engineers will probably feel most comfortable with this material, it is also suitable for mechanical, industrial, and chemical engineers who have a background in basic control theory and optimization methods. No prior knowledge of manufacturing is required. In addition, the chapters are not considered prerequisites for each other, although Chapter 1 should be read first.

The target audience consists of new researchers to this field and practicing manufacturing engineers. Educating themselves is complicated since the manufacturing literature is scattered throughout many books and journals. This tutorial attempts to sort the new and important topics. This should benefit those who are beginning to research this field. In addition, the material on perturbation analysis and Petri nets should be valuable to the practicing manufacturing engineer since applications have already been demonstrated in these new areas.

The initial work on this tutorial was done while I was on sabbatical at the Laboratory for Information and Decision Systems at M.I.T. I would especially like to thank Dr. Stanley Gerschwin for extending an invitation to me and for his insightful discussions, seminars, and lectures on automated manufacturing systems.

I would also like to acknowledge the support of the Computer Integrated Manufacturing Research Program and the Flexible Manufacturing Systems Research Program administered through Rensselaer's Center for Manufacturing Productivity and Technology Transfer, as well as the support of the IBM Manufacturing Systems Engineering Curriculum Grant and the Sloan Foundation.

I am also grateful to Margaret Brown at the IEEE Computer Society for her detailed editorial assistance, to Professor Jon Butler at the Naval Postgraduate School for obtaining prompt and thorough reviews, and to Ruth Houston and Charmaine Darmetko for typing the manuscript.

Alan A. Desrochers

Table of Contents

Preface v

Chapter 1: Introduction to Modeling and Control Issues in Automated Manufacturing ... 1

Production: A Dynamic Challenge 12
    M.E. Merchant (*IEEE Spectrum*, May 1983, pages 36-39)
Factory Automation: An Automatic Assembly Line for the Manufacture of Printers 16
    H. Tanimoto (*Computer*, December 1984, pages 50-68)
A Control Perspective on Recent Trends in Manufacturing Systems 35
    S.B. Gershwin, R.R. Hildebrant, R. Suri and S.K. Mitter
    (*IEEE Control Systems Magazine*, April 1986, pages 3-15)

Chapter 2: Modeling Methods for Manufacturing Systems 49

Markovian Modelling of Manufacturing Systems 60
    R.P. Davis and W.J. Kennedy, Jr. (*International Journal of Production Research*,
    March 1987, pages 337-351)
Analysis of Transfer Lines Consisting of Two Unreliable Machines with Random
Processing Times and Finite Storage Buffers 74
    S.B. Gershwin and O. Berman (*AIIE Transactions*,
    March 1981, pages 2-11)
Models for Understanding Flexible Manufacturing Systems 84
    J.A. Buzacott and J.G. Shanthikumar (*AIIE Transactions*,
    December 1980, pages 339-349)
Modeling FMS by Closed Queuing Network Analysis Methods 96
    G. Menga, G. Bruno, R. Conterno, and M.A. Dato (*IEEE Transactions on Components,
    Hybrids, and Manufacturing Technology*, September 1984, pages 241-248)
"Optimal" Operating Rules for Automated Manufacturing Systems 104
    J.A. Buzacott (*IEEE Transactions on Automatic Control*, February 1982, pages 80-86)

Chapter 3: Control-Theoretic Methods 111

An Algorithm for the Computer Control of a Flexible Manufacturing System 119
    J. Kimemia and S.B. Gershwin (*IIE Transactions*,
    December 1983, pages 353-362)
Performance of Hierarchical Production Scheduling Policy 129
    R. Akella, Y. Choong, and S.B. Gershwin (*IEEE Transactions on Components,
    Hybrids, and Manufacturing Technology*, September 1984, pages 225-240)
Optimal Control of Production Rate in a Failure-Prone Manufacturing System 145
    R. Akella and P.R. Kumar (*IEEE Transactions on Automatic Control*,
    February 1986, pages 116-126)
Production Control of a Manufacturing System with Multiple Machine States 156
    A. Sharifnia (*IEEE Transactions on Automatic Control*,
    July 1988, pages 620-625)
A Dynamic Optimization Model for Integrated Production Planning:
Computational Aspects 162
    W.A. Gruver and S.L. Narasimhan (*IEEE Transactions on Systems, Man,
    and Cybernetics*, November 1979, pages 689-695)

Chapter 4: Discrete Event Dynamic Systems 169

A Gradient Technique for General Buffer Storage Design in a Production Line 177
    Y.C. Ho, M.A. Eyler, and T.T. Chien (*International Journal of Production Research*, December 1979, pages 557-580)

Perturbation Analysis Explained 201
    Y.C. Ho (*IEEE Transactions on Automatic Control*, August 1988, pages 761-763)

Parametric Sensitivity of a Statistical Experiment 204
    Y.C. Ho (*IEEE Transactions on Automatic Control*, December 1979, pages 982-983)

A New Approach to the Analysis of Discrete Event Dynamic Systems 206
    Y.C. Ho and C. Cassandras (*Automatica*, March 1983, pages 149-167)

An Event Domain Formalism for Simple Path Perturbation Analysis of
Discrete Event Dynamic Systems 225
    C. Cassandras and Y.C. Ho (*IEEE Transactions on Automatic Control*, December 1985, pages 1217-1221)

Performance Sensitivity to Routing Changes in Queueing Networks
and Flexible Manufacturing Systems Using Perturbation Analysis 230
    Y.C. Ho and X.R. Cao (*IEEE Journal of Robotics and Automation*, December 1985, pages 165-172)

Chapter 5: Modeling and Control Using Petri Nets 239

Putting Petri Nets to Work 252
    T. Agerwala (*Computer*, December 1979, pages 85-94)

Applications of Petri Net Based Models in the Modelling and Analysis of
Flexible Manufacturing Systems 262
    M. Kamath and N. Viswanadham (*Proceedings of the 1986 IEEE International Conference on Robotics and Automation*, 1986, pages 312-317)

Performance Analysis Using Stochastic Petri Nets 268
    M.K. Molloy (*IEEE Transactions on Computers*, September 1982, pages 913-917)

Performance Evaluation of Asynchronous Concurrent Systems Using Petri Nets 273
    C.V. Ramamoorthy and G.S. Ho (*IEEE Transactions on Software Engineering*, September 1980, pages 440-449)

Generalized Petri Net Reduction Method 283
    L.-K. Hyung, J. Favrel, and P. Baptiste (*IEEE Transactions on Systems, Man, and Cybernetics*, March/April 1987, pages 297-303)

An Autonomous, Decentralized Control System for Factory Automation 290
    N. Komoda, K. Kera, and T. Kubo (*Computer*, December 1984, pages 73-83)

A Petri Net-Based Controller for Flexible and Maintainable Sequence Control
and Its Applications in Factory Automation 301
    T. Murata, N. Komoda, K. Matsumoto, and K. Haruna (*IEEE Transactions on Industrial Electronics*, February 1986, pages 1-8)

Chapter 6: Mathematical Methods for Improving Performance 309

Optimal Control of a Queueing System with Two Heterogeneous Servers 319
    W. Lin and P.R. Kumar (*IEEE Transactions on Automatic Control*, August 1984, pages 696-703)

Cooperation among Flexible Manufacturing Systems 327
    O. Berman and O. Maimon (*IEEE Journal of Robotics and Automation*, March 1986, pages 24-30)

Measuring Decision Flexibility in Production Planning 334
    J.B. Lasserre and F. Roubellat (*IEEE Transactions on Automatic Control*, May 1985, pages 447-452)

Optimal Scheduling for Load Balance of Two-Machine Production Lines 340
    S. Mitsumori (*IEEE Transactions on Systems, Man, and Cybernetics,*
    June 1981, pages 400-409)

Concurrent Routing, Sequencing, and Setups for a Two-Machine
Flexible Manufacturing Cell.. 350
    E.-J. Lee and P.B. Mirchandani (*IEEE Journal of Robotics and Automation,*
    June 1988, pages 256-264)

Selection of Process Plans in Automated Manufacturing Systems 359
    A. Kusiak and G. Finke (*IEEE Journal of Robotics and Automation,*
    August 1988, pages 397-402)

Chapter 7: Additional Readings in Manufacturing Systems Engineering 365

Author Index 373

Chapter 1: Introduction to Modeling and Control Issues in Automated Manufacturing

1.1: Introduction

The major purpose of this tutorial is to summarize the present research directions related to the systems issues in manufacturing. The increased flexibility of automated manufacturing systems has encouraged researchers to view these complex systems as dynamic ones. As a result, modeling and control theory tools have emerged as likely candidates to investigate the underlying dynamics of automated manufacturing systems. This tutorial presents several of the most promising techniques.

At first glance, it may appear that this tutorial covers a wide range of topics. This is because the study of automated manufacturing systems has not yet been unified. By examining several of the most promising approaches, we hope to get closer to the goal of a unified theory of manufacturing automation.

This chapter emphasizes that an automated manufacturing system is a dynamic system. It introduces some of the physical components that are found in a manufacturing system, some measures of performance, the hierarchical nature of the control problem, and a detailed explanation of how a specific product is manufactured. Finally, the control theory perspective is outlined.

1.2: Manufacturing a Product

This section introduces some manufacturing concepts and some common terminology. Additional details can be found in the books by Groover [1,2], Ranky [3,4], Hartley [5], O'Grady [6], and an excellent handbook by Kief and Olling [7]. These are mainly surveys of manufacturing components, their capabilities and limitations, as well as some of the integration issues. Here, the discussion centers around the problems that arise and the decisions that must be made to manufacture a product efficiently. Specifically, what are some possible ways to manufacture a product? What problems and decisions might you encounter along the way?

Consider the following manufacturing problem. We wish to manufacture a part type that requires four different operations A, B, C, and D. The demand for this part type requires that we produce N finished parts. Figure 1.1 shows one possible manufacturing scenario.

In this case, each machine is dedicated to one operation. In addition, each machine relies on the one before it and the one after it. Since individual machines are likely to have different production rates, this will lead to machines that are starved or blocked. A *starved* machine is one that must remain idle because the preceding machine cannot supply it with parts. Similarly, a *blocked* machine is forced to be idle because it has no place to dispatch its finished parts. Machine failures and random repair times would also lead to the same situation.

Based on the orders for the part type, you would soon come to grips with the meaning of *system capacity*. Excess capacity (too many machines representing too large an investment) would result in idle machines and workers. Not enough capacity would result in unsatisfied demand that translates into unsatisfied customers. Clearly, a certain degree of production planning is required.

1.2.1: Buffers

To help smooth out variations in production, you would next be likely to add temporary storage between the machines as shown in Figure 1.2.

In this case, if machine C fails, B does not have to be blocked, and D is not necessarily starved. In general, *buffers* allow machines to continue working while another machine undergoes scheduled maintenance, tool changes, or repairs. However, now we must determine what size the buffer should be. In general, a large buffer size is not desirable. This will be addressed in later chapters.

1.2.2: Machining Centers

Now suppose that we are willing to make a larger investment in the form of machining centers. *Machining centers* have a tool magazine that permits automatic tool changing that makes them capable of performing more than one operation. Figure 1.3 illustrates this manufacturing strategy.

This setup has an increased *flexibility* regarding the variations in part types that it can produce. It is suitable for low-volume custom orders. Also note that failures and repairs are decoupled in the sense that if one machine fails it does not affect the production of the others. By adding a part transportation system to these machines, the performance of the system can be improved, which leads to a flexible manufacturing system.

1.2.3: Flexible Manufacturing Systems

Figure 1.4 shows the essential features of a *flexible manufacturing system* (FMS). In addition, the parts transportation system, the fixtures, and the machines are assumed to be under the control of a common computer (system). Several surveys have appeared that outline the benefits of FMS and the new control problems that arise [12-15].

An FMS has the ability to make different part types and to make rapid changes from one part type to another. The part transportation system allows alternate routes that can be used to compensate for machine failures. Although not

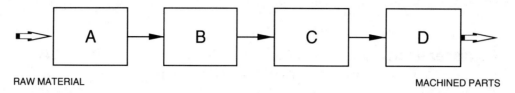

Figure 1.1: Serial Production

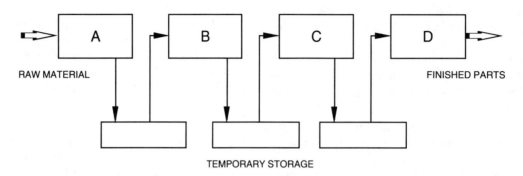

Figure 1.2: Buffers Can Be Used to Smooth Variations in Production

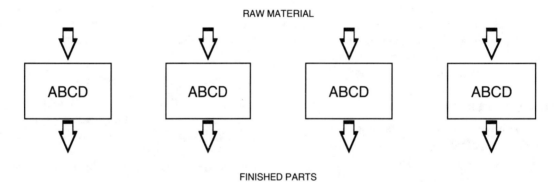

Figure 1.3: Manufacturing with Machining Centers

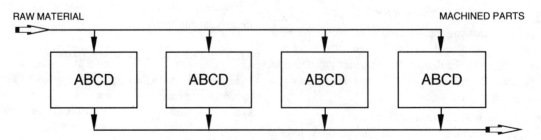

Figure 1.4: A Flexible Manufacturing System

shown in Figure 1.4, buffers may still be included. In fact, a central buffer is a new possibility.

The FMS raises new questions when it comes to controlling this type of system. For example, how are jobs sequenced to minimize completion time? How are production rates maximized? How are jobs dispatched to individual machines? Recall that the parts transportation system gave us flexibility and economy, but it also gave us numerous routings which give the control problem a combinatorial flavor. Combinatorial optimization problems are often very difficult to solve. As a result, we view these automated manufacturing systems as dynamic systems and concentrate on answers that are based on the underlying structure of the system. Techniques for achieving this goal are contained in this tutorial.

1.2.4: Job Shops, Flow Shops, and FMS

A more formal definition of an FMS is possible by considering two standard manufacturing layouts: the job shop and the flow shop [10,21].

In the *job shop* (Figure 1.5), different parts follow different paths through the machines. There is no common pattern for the flow of jobs through the shop. The flexibility of the job shop allows it to process a wide variety of small-volume orders. The disadvantage occurs with the long in-process times and the corresponding large in-process inventories.

In the *flow shop*, shown in Figure 1.6, jobs undergo the same sequence of operations. Volumes are large and production is very efficient. Highly automated versions of the flow shop result in *transfer lines* (or assembly lines). The drawback of this production scheme lies in the fact that the conversion to another product is a time consuming task.

An FMS attempts to achieve the advantages of both the job shop and the flow shop. This can be summarized as follows [19]:

> A Flexible Manufacturing System (FMS) is an integrated, computer-controlled complex of automated material handling devices and numerically controlled (NC) machine tools that can simultaneously process medium-sized volumes of a variety of part types. This new production technology has been designed to attain the efficiency of well-balanced, machine-paced transfer lines, while utilizing the flexibility that job shops have to simultaneously machine multiple part types.

There is still some uncertainty about what constitutes an FMS. Browne et al. [19] have made further classifications in an attempt to clarify the meaning of flexibility. Figure 1.7 illustrates the degree of flexibility in the context of three manufacturing scenarios [7].

The design and operation of an FMS involves the solution of several subproblems [59]. These include,

- An economic justification
- Selection of parts to be manufactured in the FMS (parts grouping)
- Selection of machine tools
- Workcell design
- Selection of a storage system (local buffers or central storage)
- Design of a material handling system
- Selection of fixtures and pallets
- Design of computer systems and communication networks
- Layout and integration of the machines, storage system, parts handling system, and computer systems

Detailed discussions of these issues can be found in [9,59].

1.3: The Manufacturing System Hierarchy

The management of many manufacturing processes can be decomposed into a *hierarchy*. Figure 1.8 shows a typical hierarchical approach to production planning and scheduling. Several variations of this structure have been adopted [16-18, 22-24].

Each level has a different function and is concerned with the manufacturing system on a different time scale. Note that the *time scales* range from years and months to minutes and seconds on the shop floor.

This wide range of time scales is an asset to the solution of the control problem. This can be seen by observing the dynamics (or relative dynamics) above and below the K-th level of the hierarchy. First, let the time scales be separated by, say, an order of magnitude. Then, an observer at the K-th level sees an approximately constant (non-dynamic) system when looking above the K-th level. Variables above the K-th level look like setpoint commands for the K-th level. Similarly, the observer sees a dynamic system when looking below the K-th level. These concepts will be important for the development of the control theoretic methods in Chapter 3.

The function of each level is closely related to its corresponding time scale. The reprints in this tutorial are mainly concerned with the lower three levels. At the shop loading level, raw materials are ordered and lot sizes are determined for the next one or two months. The *lot sizes* (orders) are then translated into a detailed scheduled at the next level. This *schedule* determines the number of each part type to be made but says nothing about which specific machine will actually do the job. Thus, the scheduler provides aggregate information. The detailed allocation of a part type to a specific machine is the function of the real-time *dispatching* level. This level decides which job will be done next and on which machine.

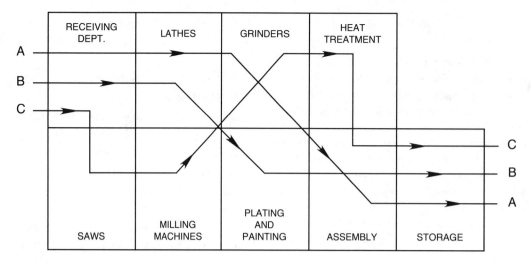

Figure 1.5: The Job Shop

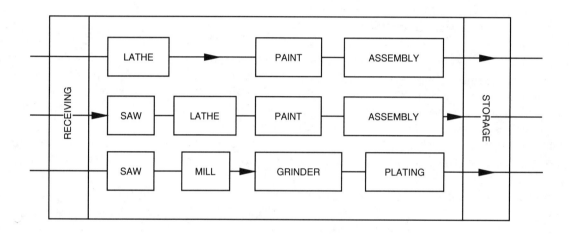

Figure 1.6: The Flow Shop

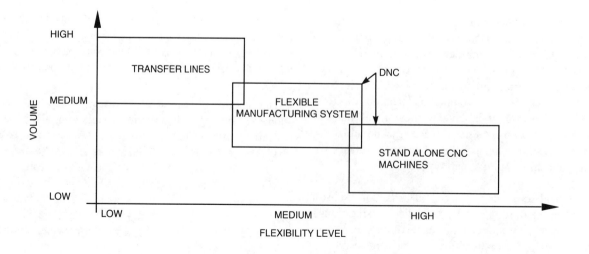

Figure 1.7: Manufacturing Strategy and Flexibility

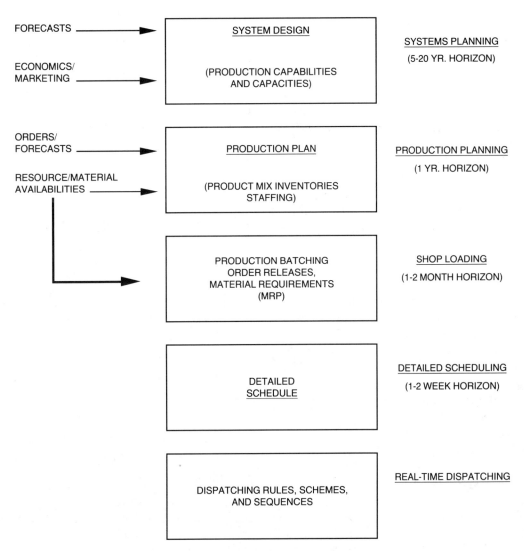

Figure 1.8: The Production Control Hierarchy

1.3.1: The Automated Manufacturing Research Facility at the National Institute of Standards and Technology

The Automated Manufacturing Research Facility (AMRF) is a national research laboratory that has implemented many of the most recent ideas in automated manufacturing [16-18]. Albus [16] and McLean [18] have been strong proponents of the hierarchical formulation. Since the production hierarchy is a concept on which nearly all researchers agree, in this tutorial, we adopt the terminology [18] originated by the National Bureau of Standards (now the National Institute of Standards and Technology) to describe each level of the hierarchy. In descending order, the levels of the hierarchy are

- *Facility:* highest level of control composed of manufacturing engineering, information management, and production management
- *Shop level control:* responsible for real-time management of jobs and resources on the shop floor
- *Cell level controller:* sequence batch jobs of similar parts
- *Workstation controller:* directs and coordinates the equipment within a workstation
- *Machine controller:* controls an individual piece of equipment (e.g., a robot control system or a machine tool controller).

The AMRF papers [16-18] are a good source of information in regard to real-life problems that are encountered in setting up an automated manufacturing system. In addition, the AMRF is periodically demonstrated to the public and is available to researchers as a test bed.

1.4: Some Unavoidable Modeling Issues

There are several key modeling issues that complicate the control of an automated manufacturing facility. Any strategy for operating such a system must be able to handle:

- Uncertainty in product demand knowledge (at all levels of the production hierarchy)
- Finite manufacturing capacity
- Random machine failures and repair rates.

In a mathematical sense, these are all very tough to deal with, yet they are very realistic.

1.4.1: Product Demand Knowledge

Total product demand knowledge is based on actual orders that are only known for a specific period into the future plus some forecasted value obtained from prior experience as well as seasonal and cyclic variations. Canceled orders also add to this uncertainty.

Uncertainty in product demand makes it difficult to set manufacturing capacity. How many machines should be purchased? How many factories should be built? This uncertainty in demand can occur at all levels of the production control hierarchy.

1.4.2: Finite Manufacturing Capacity

If manufacturing capacity were excessive, the control problem would become trivial. There would be plenty of machines to do any job at any desired time. In reality, one would like to get away with the least number of machines (capital investment) while still satisfying demand.

Capacity is not just the number of machines on the shop floor. Instead, the true capacity is related to the sources of uncertainty in a manufacturing system. One source of uncertainty is the reliability of the machines. Machines are prone to fail at random times, and the time to repair is also a random variable. This is strongly related to the plant maintenance program and procedures.

Other sources of uncertainty that affect capacity are worker and material absence; variations in the quality of the raw materials, which may affect process yield; variation in tool wear; rework; and variation in machine processing times because of quality of raw materials or perhaps operator experience. These types of uncertainties must be taken into account when estimating system capacity.

Because there are so many uncertainties, it is difficult to calculate (or even define) capacity. Consequently, many assumptions are often made to facilitate the modeling process. Nevertheless, capacity is a central concept for modeling and controlling automated manufacturing systems.

1.4.3: Additional Modeling Issues

Models and control algorithms should also be able to handle:

- setup times (possibly stochastic)
- finite buffer sizes

When a machine switches from making one part to another, new fixtures and tools are generally required. Thus, the setup of the machine must be physically changed. This takes a certain amount of time to accomplish and is called the *setup time*.

The setup time could be a random variable, perhaps because of different machine operators or the physical location of the new fixtures and tools. Other modeling issues include (but are not limited to) maintenance scheduling, the routing and sequencing of parts, interactions between manufacturing subsystems, and process plan selection. These topics are treated in Chapters 2 and 6.

The major point is that the *same system* may require different models to study different issues. Once again, this is related to the lack of a unified body of knowledge in this area.

This text strives to emphasize emerging techniques that address all of these issues.

1.5: Some Performance Measures

There are several performance criteria that are used to determine control policies for operating an automated manufacturing system. Among these are

- Minimize the total time required to complete all the jobs (i.e., minimize the *makespan*)
- Minimize the setup costs
- Meet the due date
- Minimize the mean time in the shop (mean flow time)
- Minimize machine idle time
- Minimize mean number of jobs in the system
- Minimize percentage of jobs late
- Minimize mean lateness of jobs
- Minimize mean queue time.

This is only a partial list of the criteria that one can expect to encounter in the literature. Next, several of the criteria on this list are defined and discussed.

1.5.1: Minimize the Makespan

One possible goal is to minimize the total time required to complete all of the jobs. This requires the determination of the proper sequence of jobs. Furthermore, the sequence is the solution to a combinatorial optimization problem. Because of the large dimensionality of manufacturing systems, we wish to avoid this problem all together. In fact, there is no general optimal solution to this problem for $M > 3$, where M is the number of machines [10]. In addition, the problem formulation does not take into account the modeling issues described in Section 1.4.

The benefit of minimizing the makespan is well documented by Merchant (the first reprint). One advantage is that parts spend less time in the shop. This decreases the cost of carrying an inventory of unfinished parts on the shop floor (i.e., the *work in process* is reduced). Also, there is a decrease in the inventory of finished parts that are waiting for other parts to be processed so that the two can be assembled.

1.5.2: Minimization of Setup Costs

A paint-making company uses the same machine for mixing different paint colors [10]. The preparation of the machine depends on the color that was processed previously. Note that the sum of the processing times for all jobs is constant, while the setup time depends on the job sequence. The problem is to find the job sequence that minimizes the sum of the setup times.

The solution to this problem is identical to the one for the traveling salesman problem. A salesman starting in one city wishes to visit each of n-1 other cities, once and only once, and return to the starting city. Thus, he seeks the optimal order of visiting the cities that minimizes the total distance traveled. The distance between cities is analogous to the setup times of the jobs. This is a combinatorial optimization problem that has an optimal solution [10]. In general, a well-defined combinatorial problem has an optimal solution but it may not be easy to obtain. However, the optimal solution of most traveling salesman problems can be obtained in a reasonable amount of time.

1.5.3: Meet the Due Date

In some cases, like military suppliers, delivering the product on time is of utmost importance.

Meeting the due date is a function of the job sequence and may incur additional labor costs, inventory carrying costs, or machine tool costs, but these are of secondary importance for this class of problems.

1.6: The Operations Research Perspective

Approaches to the planning, control, and scheduling of automated manufacturing systems originates from at least two distinct disciplines: operations research and control theory. From a general perspective, operations research and control theory have some common characteristics. First, both are concerned with decision making based upon some analytical model. Second, these models represent the system behavior with a limited degree of accuracy, and they depend heavily on the issues being studied. It is entirely possible to employ several models for the same system, each representing a different aspect of the system's behavior. Buzacott and Yao [58] emphasize this point in their review of analytical models for flexible manufacturing systems. The operations research approach will be summarized in this section, while the remainder of this tutorial focuses on modeling and control techniques from the control theory perspective.

The problem of planning and scheduling activities in a manufacturing facility has received considerable attention from the operations research and industrial engineering communities over the last 25 years. A concise summary of simulation models, queueing theory approaches, integer programing models, heuristic algorithms, and a variety of other techniques, along with an extensive list of references, has been compiled by O'Grady and Menon [12]. Kusiak [59] has also done an extensive review of operational research models and scheduling issues for flexible manufacturing systems.

1.6.1: Queueing Theory Models

Queueing theory has been a popular approach to the modeling of automated manufacturing systems since the classic work of Jackson [30]. Later, Solberg [31] was one of the first to formulate a queueing network model of a flexible manufacturing system. This resulted in the CAN-Q software, which generates approximate performance results, quickly. These can be used as a rough cut or preliminary solution. The weakness of this approach is in the assumptions that are made in the queueing theory derivations (e.g., infinite buffers and exponential distributions).

A review of queueing network results as applied to flexible manufacturing systems has been done by Buzacott and Yao [32]. One problem with queueing networks is the difficulty in obtaining solutions for general networks. This has led to the use of approximation schemes [33,34]. Finally, O'Grady and Menon [12] point out that, "while the approach is impressive in its mathematical treatment and is useful in providing approximate guidelines at the preliminary system design stage, it is generally of limited value in the operational planning and control function."

1.6.2: Mathematical Programming Models

Production planning and scheduling problems have been approached through a variety of mathematical programming techniques. Usually, the problem is formulated as the minimization of an objective function subject to a set of constraints. The decision variables might be the production rates, inventory levels, or resources required over the planning horizon. Typical constraints include production capacity, precedence of jobs, workforce, or customer orders. Possible objective functions are tardiness, lateness, throughput, and flowtime. Depending upon the mathematical technique employed, a variety of assumptions are made regarding the uncertainty in job processing times, release dates, manufacturing capacity, machine reliability, etc.

Integer programming models have been used to solve certain production planning problems [12]. Stecke [48] solved several production planning problems for the flexible manufacturing system at Caterpillar Tractor, Peoria, Illinois. The tool assignment problem has also been approached [49] by using integer programming methods. Other applications have been done by O'Grady and Menon [50,51].

1.6.3: Scheduling [52]

There are primarily two types of scheduling approaches for conventional manufacturing systems: deterministic and stochastic. In the deterministic case, there is a list of jobs to be processed by a set of machines. The parameters of the jobs, such as processing times and due dates, are assumed to be known in advance. In the stochastic case, the information on the parameters of the jobs is only known through their distribution functions. A comprehensive review of models for stochastic scheduling problems can be found in Pinedo and Schrage [53] and Weiss [37].

The operations research approach in scheduling theory is succinctly presented in four important works. Conway et al. [57] present a collection of results pertaining to scheduling theory up to the early 1960s. Baker [46] presents a good introductory exposition of scheduling theory. Coffman [47] provides a compilation of papers on scheduling that deals with computational complexity of scheduling problems. Complexity theory has provided the research community with problem classifications that allow the identification of problems that are computationally tractable and non-tractable. Lawler et al. [36] characterize and classify deterministic scheduling problems according to their machine configuration, job characteristics and objective criteria. They discuss the computational complexity of each problem within a three-field problem classification. Graves [35] and most recently Vasquez-Marquez [52] have also surveyed the operations research approach to scheduling.

In these surveys, known solutions to single machine, parallel machine, and job shop problems are categorized. Typical of the results found from this body of work are the sequencing rules for the two-machine flow shop case. Here, it is necessary to specify a list of jobs that have to be processed by two machines in order. Johnson's algorithm [see Section 6.6.3] sequences the n jobs based on processing time to minimize the maximum flow time.

1.6.4: Dispatching Rules

The application of optimization-based scheduling algorithms has fallen far short of the optimality goal in practice. In many cases, heuristic dispatching rules are used to order and schedule jobs. These rules represent a commonly used mechanism for releasing and sequencing jobs through the shop. They indicate which job should have priority, given that a machine is available. Unlike global optimization schemes, dispatch rules use local information (e.g., the state of the upstream or downstream queue).

Typical dispatching rules include shortest processing time, work in next queue, last in first out (LIFO), and first in first out (FIFO) priority rules. These simple rules have significant importance since they provide an event-driven feedback-control law for shop floor sequencing. As events occur, such as operation complete or new job release, these rules offer a mechanism to dictate the required action without recomputing an expensive optimization-based scheduling algorithm.

Although dispatching rules are not usually based on optimization theory, for some very simple shop configurations, some heuristic rules have been shown to yield optimal results. An example is the shortest processing time rule with dynamic release dates to minimize flowtime in the one-machine case. A thorough simulation-based comparison of several common dispatching rules has been reported on by Dar-El [60]. An extensive survey of known dispatching rules has been compiled by Panwalkar and Iskander [54].

1.6.5: Heuristics

Production planning and scheduling problems are very complex and even simple models yield optimal solutions for only a few special cases. This level of complexity encourages the use of heuristic schemes for increasing the scope of feasibility and application. McMillan [55] has reviewed the application of heuristic algorithms for resource allocation and assembly line balancing. Heuristics for job shop scheduling can be found in Gere [56] and in Panwalkar and Iskander [54].

1.6.6: Optimal Design Issues

The goal of any modeling technique is to achieve improved performance of the manufacturing system. There is an impressive amount of work in the operations research literature that addresses this problem. This is highlighted in [40–45].

1.6.7: Summary

This brief section has attempted to point out that the operations research approach to production planning, scheduling, and control is a well established and mature discipline with many proven methodologies applicable to manufacturing systems. Specifically, the operations research and industrial engineering communities have made significant contributions in queueing networks, stochastic control of queueing systems, deterministic and stochastic scheduling, complexity theory, and heuristic approaches to the solution of manufacturing control problems.

1.7: The Control Theory Perspective

Numerous mathematical techniques for modeling and controlling automated manufacturing systems have been emerging from the systems and control-theory communities. The methods presented in this tutorial are techniques that have demonstrated potential for changing the way in which modern manufacturing systems are operated. The first approach uses the concepts and system principles from modern dynamic control theory. These are the control theoretic methods discussed in Chapter 3. Chapter 4 presents a new approach called perturbation analysis. This technique is related to linearization of dynamic systems and enables the very efficient optimization of manufacturing system parameters. Supervisory control is the subject of

Chapter 5 where Petri nets are presented as a method to coordinate the various manufacturing subsystems. The Petri net is also useful for modeling the asynchronous and concurrent events that occur in an automated manufacturing system. In addition, they can be used to evaluate the performance of the system. These modeling and control methods are presented as an alternative to the traditional queueing approach. By studying and examining these approaches, we hope to get closer to uncovering the underlying principles of manufacturing.

1.8: Reprints

The first reprint, "Production: A Dynamic Challenge" by Merchant, provides several real-life manufacturing examples that illustrate the benefits of automated manufacturing. These examples also introduce several physical components of a manufacturing system (e.g., computer numerically controlled machines, pallets, and automatically guided vehicles). The paper also identifies the performance measures of a manufacturing system, which in turn identify the control issues. Important terminology is also introduced.

The next reprint, "Factory Automation: an Automatic Assembly Line for the Manufacture of Printers" by Tanimoto, explains all the details that must be considered to manufacture a product. Good figures of more manufacturing equipment are included. Figures 16–18 show the extent of the control problem. Additional case studies can be found in [25–27].

The final reprint, "A Control Perspective on Recent Trends in Manufacturing Systems" by Gershwin et al., establishes the major theme of the tutorial which is to summarize the present research directions related to the systems issues in manufacturing. This specific paper presents the hierarchical framework from a control-theory point of view and surveys some present practical methods. A more succinct and general overview of research needs in manufacturing systems is presented in [28].

1.9: References

1.9.1: Books

1. M.P. Groover, *Automation, Production Systems, and Computer-Aided Manufacturing*, Prentice Hall, Englewood Cliffs, N.J., 1980.
2. M.P. Groover, *CAD/CAM: Computer-Aided Design and Manufacturing*, Prentice Hall, Englewood Cliffs, N.J., 1984.
3. P. Ranky, *The Design and Operation of FMS*, North Holland Publishing Company, New York, 1983.
4. P. Ranky, *Computer Integrated Manufacturing, An Introduction with Case Studies,* Prentice Hall, Englewood Cliffs, N.J., 1985.
5. J. Hartley, *FMS at Work*, North Holland Publishing Company, New York, 1984.
6. P.J. O'Grady, *Controlling Automated Manufacturing Systems*, Kogan Page, London 1986.
7. H.B. Kief and G. Olling, *Flexible Automation, The International Guidebook on Computer Integrated Manufacturing,* Becker Publishing Company, Naples, Fla., 1985.
8. Draper Lab, *The Flexible Manufacturing Systems Handbook*, The Charles Stark Draper Laboratory, Cambridge, Mass., 1982.
9. K.E. Stecke and R. Suri (editors), *Proceedings of the Second ORSA/TIMS Conference on Flexible Manufacturing Systems: Operations Research Models and Applications,* Elsevier, New York, 1986.
10. E.A. Elsayed and T.O. Boucher, *Analysis and Control of Production Systems,* Prentice Hall, Englewood Cliffs, N.J., 1985.
11. R.K. Jurgen (editor), *Computers and Manufacturing Productivity*, IEEE Press, New York, 1987.

1.9.2: Survey Papers

12. P.J. O'Grady and U. Menon, "A Concise Review of Flexible Manufacturing Systems and FMS Literature," *Computers in Industry*, Vol. 7, 1986, pp. 155–167.
13. C. Dupont-Gatelmand, "A Survey of Flexible Manufacturing Systems," *Journal of Manufacturing Systems,* Vol. 1, No. 1, 1982, pp. 1–15.
14. W. Eversheim and P. Herrmann, "Recent Trends in Flexible Automated Manufacturing," *Journal of Manufacturing Systems,* Vol. 1, No. 2, pp. 139–147.
15. A. Kusiak, "Flexible Manufacturing Systems: A Structured Approach," *International Journal of Production Research*, Vol. 23, No. 6, 1985, pp. 1057–1073.

1.9.3: The AMRF

16. J.A. Simpson, R.J. Hocken, and J.S. Albus, "The Automated Manufacturing Research Facility of the National Bureau of Standards," *Journal of Manufacturing Systems*, Vol. 1, No. 1, 1982, pp. 17–31.
17. A.T. Jones and C.R. McLean, "A Proposed Hierarchical Control Model for Automated Manufacturing Systems," *Journal of Manufacturing Systems*, Vol. 5, No. 1, 1986, pp. 15–25.

18. C. McLean, M. Mitchell, and E. Barkmeyer, "A Computer Architecture for Small Batch Manufacturing," *IEEE Spectrum*, May 1983, pp. 59–64.

1.9.4: General References

19. J. Browne, D. Dubois, K. Rathmill, S. Sethi, and K.E. Stecke, "Classification of Flexible Manufacturing Systems," *The FMS Magazine*, April 1984, pp. 114–117.
20. K.E. Stecke, "Planning and Control Models to Analyze Problems of Flexible Manufacturing," *Proceedings of the IEEE 23rd Conference on Decision and Control*, IEEE Press, New York, 1984, pp. 851–854.
21. J.T. Black, "Cellular Manufacturing Systems Reduce Setup Time, Make Small Lot Production Economical," *Industrial Engineering*, Nov. 1983, pp. 36–48.
22. D.A. Bourne and M.S. Fox, "Autonomous Manufacturing: Automating the Job-Shop," *IEEE Computer*, Sept. 1984, pp. 76–86.
23. R.K. Jurgen (editor), Special issue on "Data Driven Automation: Toward a Smarter Enterprise," *IEEE Spectrum*, May 1983.
24. K. Gardiner (editor), Special issue on "Aspects of Manufacturing Systems and Their Integration," *IEEE Transactions on Components, Hybrids, and Manufacturing Technology*, Vol. CHMT-7, No. 3, Sept. 1984.

1.9.5: Applications

25. B.G. Boehlert and W.J. Trybula, "Successful Factory Automation," *IEEE Transactions on Components, Hybrids, and Manufacturing Technology*, Vol. CHMT-7, No. 3, Sept. 1984, pp. 218–224.
26. S. Inaba, "An Experience and Effect of FMS in Machine Factory," *IEEE Control Systems Magazine*, June 1982, pp. 3–9.
27. G.A. Olig, "An Overview of an FMS System," *IEEE Transactions on Industry Applications*, Vol. IA-21, No. 2, March-April 1985, pp. 318–323.

1.9.6: Research Needs

28. J.F. Cassidy, T.Z. Chu, M. Kutcher, S.B. Gershwin, and Y.C. Ho, "Research Needs in Manufacturing Systems," *IEEE Control Systems Magazine*, Aug. 1985, pp. 11–13.

1.9.7 Operations Research/Industrial Engineering Perspective

29. J.A. Buzacott and D.D. Yao, "Flexible Manufacturing Systems: A Review of Analytical Models," *Management Science*, Vol. 32, 1986, pp. 890–895.
30. J.R. Jackson, "Jobshop Like Queueing Systems," *Management Science*, Vol. 10, No. 1, 1963, pp. 131–142.
31. J.J. Solberg, "A Mathematical Model of Computerized Manufacturing Systems," *Proceedings of the Fourth International Conference on Production Research*, Tokyo, Japan, 1977.
32. J.A. Buzacott and D.D. Yao, "On Queueing Network Models of Flexible Manufacturing Systems," *Queueing Systems*, Vol. 1, 1986, pp. 5–27.
33. G.R. Bitran and D. Tirupati, "Multiproduct Queueing Networks with Deterministic Routing: Decomposition Approach and the Notion of Interference," *Management Science*, Vol. 34, No. 1, 1988, pp. 75–100.
34. W. Whitt, "The Queueing Network Analyzer," *Bell System Technical Journal*, Vol. 62, No. 9, 1983, pp. 2779–2815.
35. S.C. Graves, "A Review of Production Scheduling," *Operations Research*, Vol. 29, No. 4, 1981, pp. 646–675.
36. E.L. Lawler, J.K. Lenstra, and A.H.G. Rinnoy Kan, "Recent Developments in Deterministic Sequencing and Scheduling: A Survey," in M.A.H. Dempster et al., editors, *Deterministic and Stochastic Scheduling*, D. Reidel Publishing Company, Boston, Mass., 1982, pp. 35–73.
37. G. Weiss, "Multiserver Stochastic Scheduling," in M.A.H. Dempster et al., editors, *Deterministic and Stochastic Scheduling*, D. Reidel Publishing Company, Boston, Mass., 1982.
38. J.M. Harrison and L.M. Wein, "Scheduling Networks of Queues: Heavy Traffic Analysis of a Two Station Closed Network," *Operations Research*, Vol. 37, 1989.
39. L.M. Wein, "Scheduling Networks of Queues: Heavy Traffic Analysis of a Two Station Network with Controllable Inputs," *Operations Research*, Vol. 37, 1989.
40. J.G. Shanthikumar and D.D. Yao, "On Server Allocation in Multiple Center Manufacturing Systems," *Operations Research*, Vol. 36, No. 2, 1988, pp. 333–343.
41. K.E. Stecke and J.J. Solberg, "The Optimality of Unbalancing Both Workloads and Machine Group Sizes in Closed Queueing Networks of Multiserver Queues," *Operations Research*, Vol. 33, 1985, pp. 882–910.
42. B. Vinod and J.J. Solberg, "Optimal Design of Flexible Manufacturing Systems," *International Journal of Production Research*, Vol. 23, 1985, pp. 1141–1151.

43. G.R. Bitran and D. Tirupati, "Trade-Off Curves, Targeting and Balancing in Manufacturing Networks," *IC² Working Paper WP #87-08-05*, IC² Institute, The University of Texas, Austin, and in *Operations Research*, 1989.
44. R. Conway, W. Maxwell, J.O. McClain, and L.J. Thomas, "The Role of Work-In-Process Inventory in Serial Production Lines," *Operations Research*, Vol. 36, No. 2, 1988, pp. 229–241.
45. J.G. Shanthikumar and D.D. Yao, "Optimal Server Allocation in a System of Multiserver Stations," *Management Science*, Vol. 33, 1987, pp. 1173–1180.
46. K.R. Baker, *Introduction to Sequencing and Scheduling*, John Wiley and Sons, Inc., New York, 1974.
47. E.G. Coffman, Jr. (Ed.), *Computer and Job Shop Scheduling Theory*, John Wiley and Sons, Inc., New York, 1976.
48. K.E. Stecke, "Formulation and Solution of Nonlinear Integer Production Planning Problems for Flexible Manufacturing Systems," *Management Science*, Vol. 29, No. 3, 1983, p. 273.
49. E. Canuto, G. Menga, and G. Bruno, "Analysis of Flexible Manufacturing Systems," in *Efficiency of Manufacturing Systems*, B. Wilson, C.C. Berg and D. French (eds.), New York, Plenum Press, 1983, p. 189.
50. P. O'Grady, "The Application of Discrete Modern Control Theory to the Problem of Production Planning and Control," Ph.D. Thesis, University of Nottingham, Nottingham, England.
51. P.J. O'Grady and U. Menon, "A Multiple Criteria Approach for Production Planning of Automated Manufacturing," *Engineering Optimization*, Vol. 8, No. 3, 1985, p. 161.
52. A. Vasquez-Marquez, "Concurrent Resource Scheduling for Flexible Manufacturing Systems," Ph.D. Thesis, Rensselaer Polytechnic Institute, Troy, New York, May 1988.
53. M. Pinedo and L. Schrage, "Stochastic Shop Scheduling: A Survey," in *Deterministic and Stochastic Scheduling*, M.A.H. Dempster et al., editors, D. Reidel Publishing Company, Boston, Mass., 1982.
54. S.S. Panwalkar and W. Iskander, "A Survey of Scheduling Rules," *Operations Research*, Vol. 25, No. 1, 1977, pp. 45–61.
55. C. McMillan, Jr., *Mathematical Programming: An Introduction to the Design and Application of Optimal Decision Machines*, John Wiley, New York, 1970.
56. K.S. Gere, Jr., "Heuristics in Job Shop Scheduling," *Management Science*, Vol. 13, No. 3, p. 167.
57. R. Conway, W. Maxwell, and L.W. Miller, *Theory of Scheduling*, Addison-Wesley, Reading, Mass., 1967.
58. J.A. Buzacott and D.D. Yao, "Flexible Manufacturing Systems: A Review of Analytical Models," *Management Science*, Vol. 32, 1986, pp. 890–895.
59. A. Kusiak, "Application of Operational Research Models and Techniques in Flexible Manufacturing Systems," *European Journal of Operational Research*, Vol. 24, 1986, pp. 336–345.
60. E.M. Dar-El and S.C. Sarin, "Scheduling Parts in FMS to Achieve Maximum Machine Utilization," *Proceedings of the First ORSA/TIMS Conference on Flexible Manufacturing Systems*, University of Michigan, Ann Arbor, Mich., 1984, pp. 300–306.

Production: a dynamic challenge

What's needed is flexibility—the ability to manufacture efficiently a constantly changing variety of products. The computer makes it possible

Most manufacturing is not mass production. If it were, hard automation—the inflexible turning out of hundreds of thousands or even millions of identical parts—could handle most needs. But even in the United States, the world's largest center of mass production, 75 percent of the parts produced by metalworking, for example, are in lots of less than 50 pieces. What is needed here and in most manufacturing is flexible automation—automation that can handle a large and constantly changing variety of produced items.

The on-line availability of computers makes flexible automation possible. Computers also provide a second capability: moment-by-moment optimization of manufacturing processes and decision making. These two capabilities work hand in hand with a third—systems integration—to yield the ultimate for most needs: a computer-integrated manufacturing system [Fig. 1]. Such a system can be applied to both the hard components of manufacturing—the machinery and other equipment—and the soft components, like information flow and data bases.

Social and economic factors are hastening the trend toward flexible automation. On the social side, workers increasingly are shunning jobs in manufacturing; machinery must fill the gap. At the same time employers and governments are endeavoring to improve working conditions in factories, and automation is providing many improvements [see "Social attitudes spur flexible automation," p. 39]. As for economic factors, manufacturers are turning to automation to enhance profits both by speeding the processing of goods and getting more work from costly equipment [see "Economic needs foster flexible automation," p. 38].

But full systems of computer-integrated automation do not yet exist. Instead microcosms of at least a portion of such systems are now operating in the form of flexible manufacturing systems (FMS). Such systems generally consist of groups of machine tools served by automated tool-and-workpiece transport and handling equipment, all operating under integrated hierarchical computer control to produce broad families of machined parts. The results already obtained from these systems indicate that full computer-integrated automation is feasible.

The FMS now operating at Messerschmitt-Bölkow-Blohm (MBB) in Augsburg, West Germany, is one of the most sophisticated in the world today. At Augsburg, MBB builds the center section—the wing box and its related structure—for the swing-wing Tornado fighter plane. Ten Tornados a month are being produced against a total initial order of 800.

In 1975, when plans were made to produce the plane at Augsburg, company officials decided to use computer-integrated manufacturing as fully as possible. This decision was motivated in part by the fact that 40 percent of the manufacturing cost of the Tornado would be in its machined parts, even with use of the most advanced computer automation.

MBB officials then developed a long-range master plan that they called computer-integrated and automated manufacturing (CIAM). The goal was to reduce manufacturing costs by integrating the design, production planning and control, and production operations into a single computer-automated system. This was deemed essential to keep the company competitive in the world aircraft market in the face of West Germany's high-valued currency and high wages.

The major initial development in the CIAM program was the FMS for machining titanium parts of the Tornado's center section. The cost of the system [Fig. 2] was in excess of $50 million. Its basic elements are 28 numerically controlled (NC) machining centers and multispindle gantry and traveling column machines averaging $1 million each; a fully automated tool flow and changing system; an automated workpiece-transport system; and coordinated computer control of all these elements.

Benefits of FMS are high

The system has demonstrated remarkable efficiencies. Machines cut metal at least 75 percent of the time. The lead time for production of a Tornado is only 18 months, compared with 30 months for planes produced by noncomputerized means. The number of NC machines required was reduced by 44 percent compared with what would have been needed by using standalone NC machines. Required personnel were cut by 44 percent, floor space by 30 percent, parts-flow time by 25 percent, and capital investment costs by 9 percent.

The reduced capital investment costs alone clearly demonstrate that the increased machine utilization obtained by use of full computer-integrated automation sets free more than enough money to purchase the additional sophisticated capital equipment necessary for implementing flexible manufacturing systems. The increased productivity these figures represent is impressive.

Nonquantified benefits experienced with the system include increased quality in the form of higher reproducibility, lower reworking costs, and lower scrap rates; lower quality-assurance costs; improved adherence to production schedules; decreased paper flow; and improved working conditions resulting from decreased risk of accidents and relief from heavy physical labor and monotonous work.

In another case history, in Japan, where FMS implementations are already advancing rapidly, efforts are being directed toward developing the capability for such systems to run unattended at night. One of the first systems to partly realize this objective has been developed by the Toshiba Tungaloy Co. at its Kawasaki plant. It was put into service in August 1980 and produces milling cutters and single-point cutting tool bodies.

The Kawasaki installation consists of one NC lathe, one ver-

M. Eugene Merchant   Cincinnati Milacron Inc.

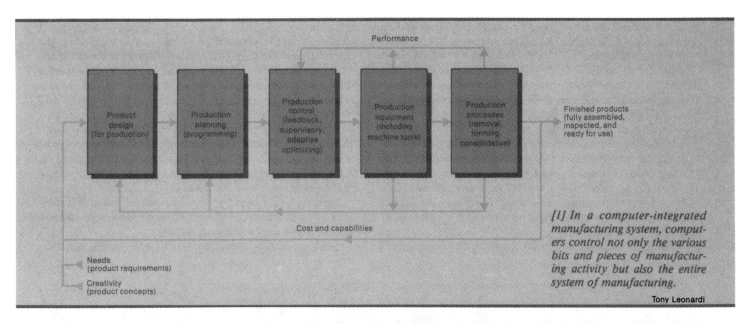

[1] In a computer-integrated manufacturing system, computers control not only the various bits and pieces of manufacturing activity but also the entire system of manufacturing.

Tony Leonardi

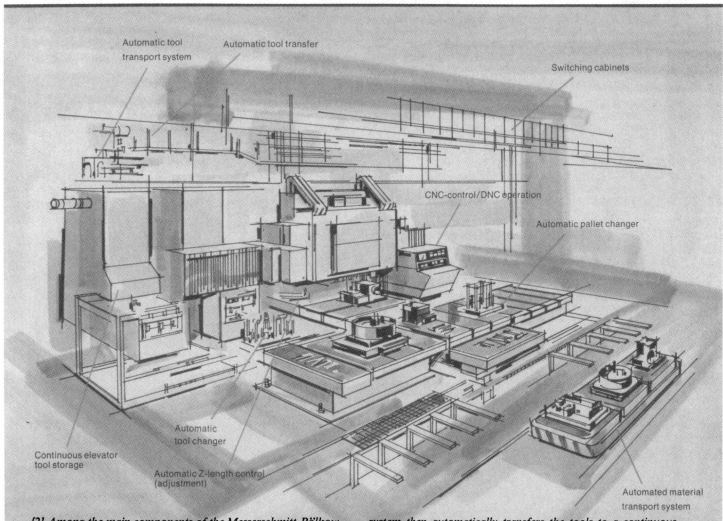

[2] Among the main components of the Messerschmitt-Bölkow-Blohm flexible manufacturing system is one of the 28 NC machine tools at the center. The automated workpiece transfer system brings workpieces to and from each machine tool by means of computer-controlled carts (lower right). The fully automated tool-transport and tool-changing system brings tools to each machine via an overhead transport system (top). The system then automatically transfers the tools to a continuous elevator tool-storage facility (left). This facility in turn interfaces with the automatic tool-changing mechanism of the machine tool. All three subsystems—the machine tools, the work-transfer system, and the tool-transfer system—are coordinated, controlled, and automated by a hierarchical distributed computer system.

tical machining center, three five-axis machining centers, and one multispindle NC grinder. All machines operate under hierarchical computer control. The FMS can handle 4000 different types of parts weighing up to about 200 kilograms and can produce parts to an accuracy of within 0.0001 inch. A special feature of the system is that the three five-axis machining centers can run unattended at night from about 9 p.m. to 8 a.m.

During that period they machine a variety of parts automatically. The parameters monitored during unattended operation are spindle motor current and length measurement of small drills. Whenever a tool breaks or becomes worn, the machine in which that tool is installed shuts down automatically. Normally, however, tools are changed automatically after a preprogrammed amount of cutting time, before they have experienced excessive wear.

The results obtained with this system in the first year and a half of operation are striking. They were confirmed by comparing the Kawasaki system with another Tungaloy factory producing a similar range of cutting tools with a mix of conventional and NC stand-alone machines. The Kawasaki plant reduced the number of machine tools required by 88 percent, from 50 to 6. It reduced personnel by 77 percent, from 79 to 16; increased product yield from 95 to 98 percent; increased the average use of the machines from 20 to 73 percent; reduced required floor space 76 percent, from 1500 to 360 square meters; reduced the average number of operations per part by 47 percent, from 15 to 8; and reduced the total average processing time from raw stock to finished product by 77 percent, from 18.6 to 4.2 days.

Obviously very large benefits of increased productivity, decreased costs, increased use of capital equipment, and reduction of work in process are being realized.

Another example from Japan is an FMS operating at the

Economic needs foster flexible automation

A variety of long-term economic factors are generating a major need for flexible automation. Two stand out: the tremendous economic waste when huge quantities of work are tied up in processing and the equally great economic loss resulting from very low utilization of capital equipment in noncomputerized factories. The significance of these two factors is illustrated by a simple example of machined parts production.

The average workpiece in a conventional metalworking factory spends only about 5 percent of its time on machine tools. Of that 5 percent, only about 30 percent (or 1.5 percent overall) is productive time in removing metal [see figure, below].

This example pinpoints the two main areas where the greatest improvements can be made today in the economy and productivity of metalworking manufacturing. The first is reducing the time during which parts are processed in a shop. Reducing this time decreases the inventory of unfinished parts on the shop floor and also the inventory of finished parts waiting for the processing of other parts so product assembly can proceed. The second area for improvement is in machine use.

The capital lying idle in parts being processed is often as great as all the capital invested in the plant and its equipment—enough to build another entire plant. The application of flexible automation has already demonstrated major potential to free most of the idle capital.

The 30-percent machine use indicated in the accompanying figure must be viewed in light of the fact that the average machine spends approximately 50 percent of its time waiting for parts to work on. As a result the average machine tool in a batch type of shop is being used productively to cut metal only about 15 percent of the time it is manned. Yet it is seldom manned 24 hours a day, so that actual use may be lower than 5 percent. This represents only about 5-percent use of the capital invested in these machine tools. The capital lying idle here is more than enough to equip the plant with the much more fully flexible machinery made possible by flexible automation.

The key long-term socioeconomic force at work today in industrialized countries, as documented by the Society of Manufacturing Engineers in a position paper, is the continuing need to reduce the cost of creating "real," or tangible, wealth.

Reducing the cost of creating real wealth results in improvements in living standards, quality of life, employment, and the general economic well-being of a country. And the prime way to reduce the cost is by increasing manufacturing productivity through improvement of manufacturing technology.
—M.E.M.

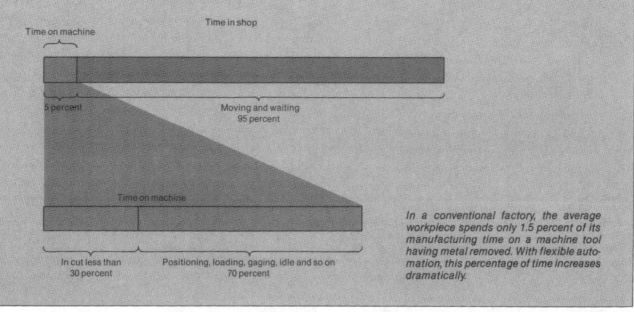

In a conventional factory, the average workpiece spends only 1.5 percent of its manufacturing time on a machine tool having metal removed. With flexible automation, this percentage of time increases dramatically.

Niigata Engineering Co.'s internal combustion engine plant in Niigata. The company experimented for many years with running individual computer-controlled machine tools unattended at night. Success here led to the development of the first full FMS capable of running unattended for nearly a full day. It was put into service in 1981.

The Niigata FMS machines 30 different types of cylinder heads in lot sizes ranging from 6 to 30 parts. The system runs 21 hours a day. Unattended nightime operation includes automatic mounting and demounting of some of the many different workpieces on pallets by a robot.

The number of machines required has been reduced by 81 percent, from 31 (including 6 NC machines) to 5; the operator force has been cut 87 percent, from 31 to 4; the production time during which parts are being turned out has been increased from 9 to 21 hours per day; and the lead time for these parts has been reduced from 16 to 4 days.

A major factor in Niigata's success is the fact that the "operator" (actually the supervisor) of each machine tool has total responsibility and authority for its success or failure. The supervisor does all the NC programming and all the NC tape tryout on first-run parts; specifies the tooling, including required uniformity; carries out final inspection of all tools, rejecting any that are unacceptable; has final say on acceptance of the metallurgical properties of the workpiece; and controls and supervises all maintenance on the system. If a machine or the system shuts down automatically at night before it has finished its supply of parts, the supervisor is held responsible. But this almost never happens.

Robots produced in unattended plant

Yet another case history: Fujitsu Fanuc has recently put into operation an entire factory for producing robots, small machining centers, and wire-cut electrodischarge machines. The machining section of the factory operates almost wholly unattended on the night shift.

There are 29 machining cells (22 machining centers served by automatic pallet changers and seven robot-served cells), each consisting of one or more NC machine tools and a robot or pallet changer to keep the machine or machines loaded with parts during the night. Workpieces on pallets are transported to and from the cells and to and from computer-controlled automatic stacker cranes by computer-controlled, wire-guided carts.

During the day there are 19 workers on the machining floor. On the night shift no one is on the machining floor, and there is only one man in the control room. In a 24-hour period, machine availability is running close to 100 percent and machine use is averaging 65 to 70 percent. As a result, this 2050-square-meter factory is producing 100 robots, 75 machining centers, and 75 electrodischarge machines each month.

A number of other Japanese companies are in various stages of implementation of computer-controlled unattended nightshift operations. With that capability, many of these companies are now beginning to add to the 24-hour operation so-called "just-in-time" (*kanban*) production. Here they let the requirements of the assembly floor "pull" the needed parts through the factory rather than depending on scheduling to "push" them through. They do this by loading the highly flexible manufacturing cells and systems with just the right mix of parts needed on the assembly floor in the next 24 hours. Thus work in process is reduced virtually to zero, and tremendous amounts of idle capital are available for other purposes.

These case histories illustrate dramatically the benefits to be derived from flexible automation. Indirectly they also send a clear message to other manufacturing companies throughout the industrialized world: If high-technology manufacturing is not in the company's future plans, it may be a company with no future.

About the author

M. Eugene Merchant is principal scientist, responsible for manufacturing research, for Cincinnati Milacron Inc. in Cincinnati, Ohio, with which he has been affiliated since 1936. He has done basic and applied research on manufacturing processes, equipment, and systems and the future of manufacturing technology. He has presented and published numerous papers on these subjects in the United States and abroad. Dr. Merchant holds the bachelor of science degree in mechanical engineering from the University of Vermont, a Ph.D. in science from the University of Cincinnati, and honorary doctorates from the University of Vermont and the University of Salford, England. He was president of the Society of Manufacturing Engineers (1967-77), vice president of ASME (1973-75), and president of the International Institution for Production Engineering Research (1968-69). He is a Fellow or member of numerous technical organizations and has received many awards including the Richards Memorial Award of the ASME (1959), the George Schlesinger Prize of the City of Berlin (1980), and the Otto Benedikt Prize of the Computer and Automation Institute of Hungary (1981).

Social attitudes spur flexible automation

Three long-range social factors are creating problems in manufacturing today, and flexible automation is helping to solve all of them. First, in times of prosperity, workers are increasingly reluctant to work in manufacturing. Second, in response to this, employers are increasing their efforts to satisfy workers and to improve working environments. And third, governments are moving from a somewhat passive role to a more active one in requiring improved working conditions.

In all industrialized countries of the world, there is a growing shortage of willing, capable manufacturing workers. This shortage is generating a major need for flexible automation to fill the gap.

At the same time many employers now recognize clearly the workers' needs for job satisfaction and freedom from potentially dangerous or unhealthy conditions. F. Herzburg, writing in the *Harvard Business Review* (January 1968, pp. 53-62), noted that while the so-called hygiene factors of a job—like company policy and administration, supervision, work conditions, and salaries—can cause dissatisfaction if they are inadequate, seeing that they are adequate will not alone ensure ongoing job satisfaction. Instead, Mr. Herzburg said, such satisfaction derives from the adequacy of the so-called motivator factors of a job—like opportunity for achievement, recognition, responsibility, advancement, and growth. The major means for achieving this type of satisfaction is through participation in a company's decision making.

Here, too, flexible automation is beneficial. It can give workers, through the use of interactive software and hardware and distributed control, access to the total manufacturing data base.

Flexible automation can also free workers from unpleasant and potentially dangerous or unhealthy conditions on the job by getting them off the factory floor. In most industrialized countries, governments are now requiring that technology be developed to improve working conditions. For example, in the United States, the relatively new Occupational Safety and Health Act will, in the long run, require manufacturers to eliminate the need for workers to place their hands, arms, or any other parts of their bodies into a dangerous machine area—such as a press—and to keep the average noise level in a factory over an eight-hour period below 90 or 85 dBa. —*M.E.M.*

Factory Automation: An Automatic Assembly Line for the Manufacture of Printers

Reprinted from *Computer*, December 1984, pages 50-68. Copyright © 1984 by The Institute of Electrical and Electronics Engineers, Inc. All rights reserved.

Hiromu Tanimoto
Oki Electric Industry Company, Ltd.

Flexible enough to cope with model changes and diversified users' requirements, this robotics system is the company's first step toward totally unmanned operation.

Because of recent qualitative and quantitative changes in market demands, many information-processing products on today's production floor will soon become obsolete and will be replaced by products with more sophisticated and diversified functions. To meet these market demands and maintain steady growth, information-processing equipment manufacturers must have the ability to manufacture a variety of high-quality products with low costs and short production cycles. They can satisfy this requirement by eliminating the uncertainty intrinsic to manual operations, by employing machinery for stable processes, and by automating the manufacturing process. This is the major reason why manufacturers seek automated assembly lines.

The Takasaki Plant of Oki Electric Industry Company, Ltd., in Japan performs the integrated operations of developing, designing, and manufacturing over 1200 types of information-processing equipment, encompassing not only individual products, such as computers, various on-line terminals, peripherals, and input and output devices, but complex system products as well. Aiming at complete unmanned production, we have been automating the Takasaki plant through a number of stages: the "point" stage for automation of a single product, the "line" stage for automation of subassembly lines for various products, and the "plane" stage for automation of the entire factory floor. We have now developed and implemented an automatic production line for 18-pin and 24-pin dot impact printers, and we believe this line will lead to our target of total plant automation.

Preliminary considerations

Automated systems must implement total production control in the most efficient manner so that sales, design, manufacturing, service, and any other activities are properly coordinated and integrated to meet user needs for product performance, quality, price, and delivery.

General system requirements. The system must be able to respond to diverse user needs by accommodating quality assurance, introduction of new products, quick product delivery, and low production costs. From the viewpoint of production, the system must have unified information control, simple production control, and flexible manufacturing processes, yet be efficient in all respects. Thus the system must

(1) be provided with automatic system control facilities—production control, quality control, and operations control;

(2) be hierarchical, allowing individual functions to be performed independently and yet also allowing the

total system to work in an integrated manner;

(3) be versatile, responding to diverse user needs;

(4) help improve product quality, reduce costs, and shorten production cycle time; and

(5) contain integrated manufacturing lines and feature simplified material handling, minimized in-process storage, and balanced load over the line.

The manufacturing process. In designing an automation system, we must consider not only the basic system characteristics discussed above, but also the following characteristics of the manufacturing process to be automated:

(1) Components of products must suit automatic assembly without loss of performance and quality. Recomposition may be necessary, including combining, integrating, dividing, or deleting the present components.

(2) Motion analysis of individual work processes—their work area, speed, mode, modularity, flexibility, etc.

(3) Balance among different processes in the same line, among different lines, and in total work load within the system—between warehouse and line, between line and postprocesses, between outside operation and line, etc.

(4) Volumes of materials and information related to production control. The basic unit of transport and the unit of decentralized processing must be determined for the total system to work efficiently.

(3) The production cycle time from material entry to product output must be no greater than 24 hours.

(4) Stock control, material-handling control, and workstation control must be performed by a hierarchical central control system containing microprocessors and IF800 personal computers (our products) for decentralized processing.

Analysis of work assignments. A key to successful implementation of an automated system is optimization of work assignments between robots and people. Before appropriate assignments can be made, assembly processes must be broken down into individual work elements according to their motion characteristics. Product variety, assembly processes, and production processes must be analyzed in order to accomplish variety reduction for efficient automation.

Economic factors are another essential consideration in the assignment of work. Initial investment, running costs, and personnel expenses of the system to be developed must be evaluated with reference to the existing production technology.

In deciding whether a particular job could be automated, we considered the means used to accomplish it and the work elements underlying it. We classified wiring and making adjustments, for example, as human work from economic and technological viewpoints. We analyzed the assembly processes, dividing them into elements such as "material supply," "insertion," "jointing," "transport," and "control," and we assigned an automation tool to each element.

Material handling. When developing the material-handling system, we considered the following factors with the goals of unified information control and short lead time:

(1) Tact time, as determined by production volume within a given period. Tact time is the time necessary to complete a process and send the work to the next workstation.

(2) Products to be assembled.

(3) Operating efficiency.

(4) Material supply.

(5) Transport unit.

In addition, we evaluated the following items on the basis of the line configuration determined as a result of the foregoing process analysis:

(1) Time for transport between lines.

(2) Material volumes to be distributed and stored.

(3) Transport method.

(4) Warehousing capacity.

(5) Storage within processes.

We designed the material-handling tools (automated warehouse, unmanned carrier, conveyor line, etc.) according to the results of the above evaluation.

Production control. Our automatic assembly line features material control, timely transport between processes, and schedule control based on information and production control, the core feature of the system. Automatic material-handling was imple-

System implementation

We established the following requirements when we planned our system implementation:

(1) The system must cover all processes from material entry to product shipment, as shown in Figure 1.

(2) Automated processes must be flexible to allow for future changes.

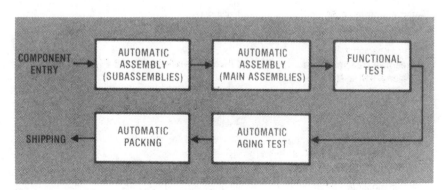

Figure 1. General flow of automated production process.

mented mainly by introducing automated warehouses and unmanned carriers.

To shorten production time, reduce inventory levels, and lower managerial and material-handling expenses, we

(1) simplified and computerized the production control system;

(2) minimized in-process storage;

(3) adopted the shortest path between processes;

(4) eliminated the need to reload materials; and

(5) standardized and aligned component pallets.

System design targets. Finally, on the basis of the preliminary considerations described above, we set the following specific design targets for our automated system:

(1) Total system control based on information accumulated in the central control unit (systematic control from production planning to individual workstation).

(2) Independent, yet integrated, functions achieved through a combination of decentralized and centralized processing.

(3) Automatic control of input (number of components) and output (product turnover).

(4) On-line monitoring of operations and collection and analysis of quality data.

(5) Priority control of material transport.

(6) Efficient random transport.

(7) Automatic production model changes on the line.

(8) Extensive use of robots for flexible automation.

(9) Separation and integration of subunit and main unit assembly processes.

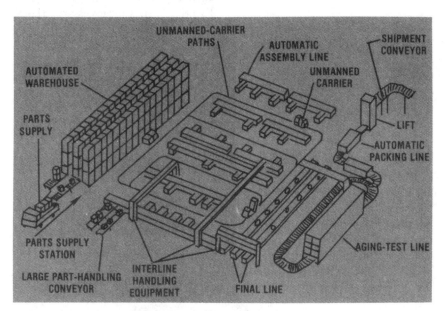

Figure 2. Automatic assembly line.

Description of the automatic assembly line

As shown in Table 1 and Figure 2, the automatic printer assembly line consists of a parts entry line, an automated warehouse, unmanned carriers, automatic assembly lines, final processing lines, a rack-type aging-test line, an automatic packing line, and a product-shipping line. All processes from component loading to product shipment are integrated into a single line. Components are first subassembled into units, which are then assembled by a main assembly line (Figure 3).

The individual lines are described here in the order of actual production flow:

Parts entry line. All components supplied from outside the line are appropriately arranged (to facilitate automatic assembly) in the magazine

Table 1. Overall system for automatic manufacture of printers.

ITEM	SPECIFICATION
Printers assembled	ET-5300, ET-8300, ET-8400, etc.
Number of workstations	80 max.
Tact time	30 seconds
Number of lines	12
Automatic assembly	10
Final assembly	1
Aging test	1
Automated warehouse	1 unit
Unmanned carrier	4 units
Main component cells	
Free-flow base machines	11 units (overall length: 198 m)
Material flow system	
Carry-in	100 m (round-trip)
Carry-out	230 m (round-trip)
Pick and place hands	4 units
Assembly robots	
Medium size (4 axes)	28
Medium size (6 axes)	21
Parts supply traverser	53 units
Automatic screwdrivers	21 units

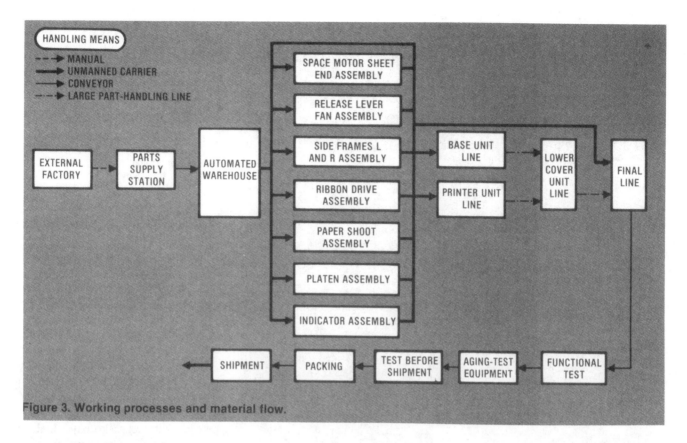

Figure 3. Working processes and material flow.

of a component pallet. Components are loaded into the production line by the component-loading line, which consists of a roller conveyor with a round-trip travel length of 100 meters. Large components—the lower cover, base frame, and upper cover of the printer—are directly conveyed into the assembly line through a dedicated loading line.

The loading line is designed to allow unattended loading of components and their direct entry to the production line, standardized arrangement of components being carried in, elimination of the need for controlling the number of components being loaded, and quality assurance in component transport.

All component pallets have a standardized size of 450 × 540 millimeters, and differences in component shape and size are accommodated by the height to which components are stacked on the pallet. For this purpose, stacking is classified into three-, five-, and seven-layer stacking.

Automated warehouse. The pallet from the component-loading line arrives at the entrance station of the automated warehouse, where bar codes on the pallet are read with a laser scanner. Components are thus identified, and they enter the automated warehouse. Specifications of the automated warehouse are given in Table 2.

The purpose of the automated warehouse is to allow timely and stable loading of components onto the production line and to automate component identification, model selection, and component control in response to different models. To serve this purpose, we implemented decentralized processing control by incorporating Oki IF800 personal computers.

Unattended transport. When a station in the assembly line requests access to a pallet, the pallet exits the automated warehouse and moves to the requesting station on one of four railless, unmanned carriers. The specifications of the unmanned carrier are given in Table 3.

At present, the production line has 71 magazine transfer stations, and access from these stations occurs about 240 times a day. At each access, 300 components on an average are transported to the assembly line. The average transport time per access is about 2.5 minutes.

We have developed a random-transport technique to move the four

Table 2. Automated warehouse specifications.

ITEM	SPECIFICATION
Number of cells	344 cells
Pallet storage capacity	1720 pallets maximum
Delivery time	3 minutes minimum
	10 minutes maximum
Parts storage capacity	Parts necessary for one day's work for each type
Storage method	By each component block

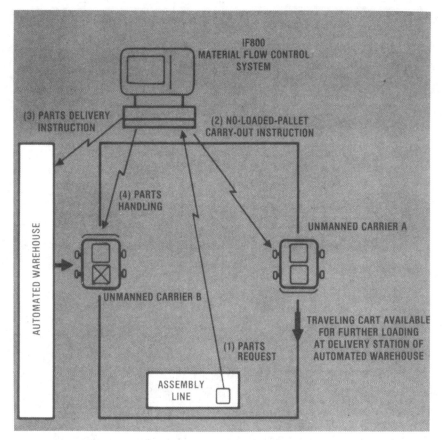

Figure 4. Unmanned random-transport handling system.

Table 3. Unmanned-carrier specifications.

ITEM	SPECIFICATION
Number of units	4
Running speed	65 m/min.
Travel method	Roller drive
Travel pattern	Random
Number of loading/ unloading stations	71

unmanned carriers to the 71 stations efficiently. Figure 4 illustrates the technique. Unlike the conventional, fixed-pattern system, which allows only a predetermined sequence of jobs, our handling system makes it possible to give running carriers new instructions if necessary and to assign additional jobs even during transport. As a result, one carrier can simultaneously transport pallets destined for two different stations. This has raised material distribution efficiency by 30 percent with respect to pallet loading and unloading operations. What is more, the random-transport technique has brought about timely supply of components, flexible transport, implementation of priority control on unmanned carriers, efficient material-handling control, better maintainability (utilization of carrier path space), and automation of model selection.

Figure 5 shows the automatic assembly line and an unmanned carrier.

Automatic assembly line. The automatic assembly line consists of seven subassembly lines and three main assembly lines. To each workstation on these lines an assembly robot developed by Oki is allocated; it can respond to different models and specification changes. Figure 6 is a general view of the automatic assembly line from the central control room on the second floor.

As shown in Figure 7, the basic components of the automatic assembly line are "free-flow" base machines, component-loading traversers, assembling robots, and auxiliary equipment (screwdriving robots, etc.).

Component-loading traverser. We have developed a component-loading

Figure 5. The automatic assembly line and an unmanned carrier.

traverser to receive pallets with different heights, or stack levels, from unmanned carriers (Figure 8). The line contains 53 component-loading traversers, located at the entrances to the assembly line. The traverser lifts stacked pallets one by one to separate them and loads them onto the picking area of the robot, beginning with the pallet at the bottom of the stack. In the picking area, a positioning pin is raised to lift the magazine in the pallet and fix it in position. Once the components to be assembled are loaded in position, the robot takes them out of the pallet and places them on the assembling jigs running on the base machine.

The empty pallet is ejected through the bottom of the traverser and returns to the original position on the separator (Figure 9). When all pallets on the traverser become empty, they are collected by an unmanned carrier, and stacked pallets full of components are supplied so that assembly can proceed without interruption. (A patent is pending for the traverser techniques.)

Robot. The automatic assembly line contains 49 numerical-control robots and 21 screwdriving robots. Many NC robots introduced in the basic assembly processes are intended to enhance flexibility for different models and design changes. As shown in Figure 10, the processes for assembling various dot printers were analyzed to

Figure 6. The automatic assembly line seen from the central control room.

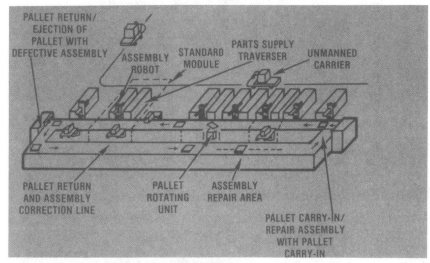

Figure 7. Standard line configuration.

Figure 8. Traverser loading parts onto assembly line.

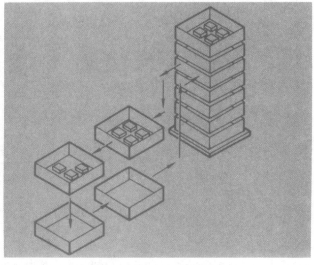

Figure 9. Pallet flow in the traverser.

determine the required performance (such as operating range and load capacity) of the robot. The performance characteristics and the appearance of the robot are shown in Table 4 and Figure 11, respectively.

The assembling robot uses a cylindrical coordinate system, which satisfies needs related to work and operation elements, assures a wide operation range because of its structural characteristics, and allows positioning accuracy to be controlled easily. The number of control axes is six: three for the cylindrical coordinate system and three for attitude control.

Figure 12 shows the free-flow base machine on the automatic assembly line, and Figure 13 shows a robot performing assembly.

Multipurpose hand. The assembling robot has a multipurpose hand (Figure 14) developed to hold different types of components. By use of this hand, the robot performs various complex assembly operations when combined with Oki's NC equipment, which performs auxiliary functions. Components loaded by two parts feeders can be assembled by one robot, and high flexibility is available to solve tooling-related problems, which often raise difficulties in switching from one model to another.

Ejection from the line. When an assembling error is found in the line, the pallet containing the faulty unit passes through all subsequent workstations and is fed to the pallet-out line, which goes backward from the line end to the line top. On the pallet-out line, only recovery from the assembling error is performed. Then the pallet reenters the assembly line, passes through the stations where processing was already done, and enters the processes which have not been finished, flowing toward the line end (see Figure 7).

Final line. Unfinished products are fed from the automatic assembly line through the interline transport equipment to the final line for completion of assembly. On the final line, wiring, adjustments, and certain other functions are performed manually because of economic and technical factors.

Free-flow base machines are used on a total of 11 automatic assembly and final lines. The overall length of the lines amounts to 198 meters.

Automatic aging-test line. Completely assembled products are tested for functionality and fed to the aging-test line, where burn-in and continuous-printing tests are performed. "Aging" refers here to storing and testing a device until its characteristics become constant, and this line, consisting of two rows with four levels each (a total of eight lines), is a rack-type self-loading and unloading warehouse.

Automatic packing and shipping line. After passing the aging tests, products are subjected to shipping tests, packed automatically, and transported on a roller conveyor (228 meters over the forward and backward paths)

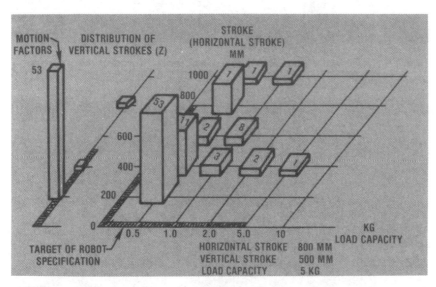

Figure 10. Determination of necessary robot performance.

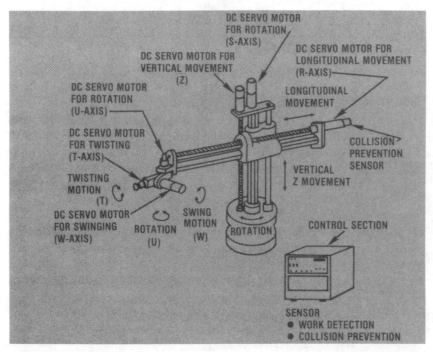

Figure 11. Construction of NC robot. (R), (S), (T), (U), (W), and (Z) represent axes of robot motion.

Table 4. Robot performance (model M-2).

MECHANICAL SECTION			CONTROL SECTION	
Degree of freedom	4	6	Control method	Numerical
Construction	Cylindrical coordinates		Control function	Point-to-point positioning
Working range				
Z-axis	300 mm	300 mm	Input system	Absolute
R-axis	450 mm	450 mm		
S-axis	270°	270°	Control system	DC servo motor
T-axis	N/A	410°		
U-axis	270°	270°	Program language	Motion statement level
W-axis	N/A	100°		
Max. speed	1 m/sec	1 m/sec	External interface	Operation command; type selection command; error message
Load capacity	5 kg	2 kg		
Repeatability	±0.02 mm	±0.05 mm	Teaching	Sequence program selection; job input

Figure 12. The free-flow base machine on the assembly line.

Figure 13. A robot performing assembly.

Table 5. Scale of software.

SOFTWARE	CAPACITY
Robot program	
Control	64 KB
Teaching	
Memory capacity	64 KB
Number of steps	2400
System program	
Host control	90 KB
Handling control	110 KB
+ Monitor	38 KB
Total line control	46 KB
+ Monitor	80 KB
Line control	8 KB
Station control	32 KB

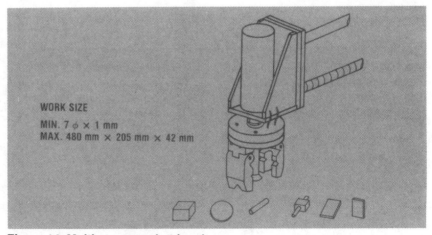

Figure 14. Multipurpose robot hand.

Figure 15. An assembled ET-8300 printer.

to the shipping center outside the factory. Figure 15 shows an assembled ET-8300 printer.

System control

The automatic assembly system is controlled as shown in Figures 16 and 17. The component devices of the control system are 10 IF800/M30 personal computers, 100 8085 microprocessors, and 49 8086 and 8087 microprocessors. The system software capacity is itemized in Table 5. The features of the system are as follows:

(1) Both hardware and software are of hierarchical structure.

(2) Under total system control, each process functions independently through decentralized processing.

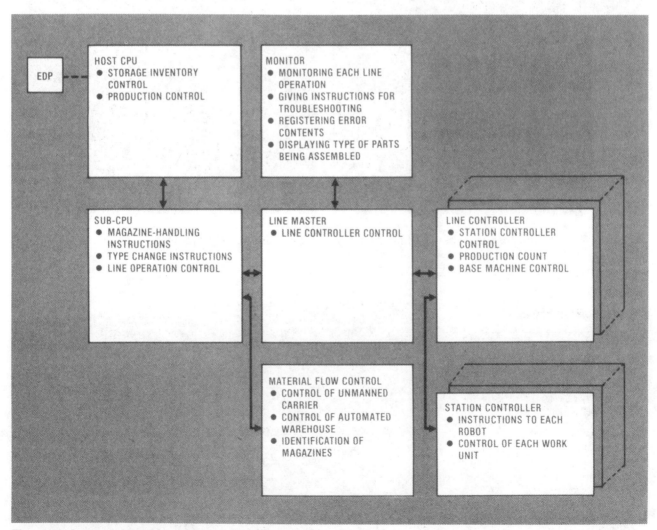

Figure 16. System control functions.

(3) Equipment configuration at each hierarchical level—NC robot, screwdriving robot, component-loading equipment, base machine, microprocessor-based control equipment, etc.—is standardized so that systematization can be easily implemented.

(4) Software development periods are remarkably short because of the newly developed operating system featuring a timing-chart entry technique. Unlike the conventional assembler-level coding method, our technique allows timing charts to be entered directly through conversation with the system so that programs can be prepared automatically.

(5) Control programs for modules at each hierarchical level are standardized to the basic structure shown in Figure 18 so that they can be handled through a shared database. The control program for each module consists of a high-level interface unit, a low-level interface unit, a job control unit, a low-level control unit, a special-processing unit, a data table, and program control and service elements. The function of each module is de-

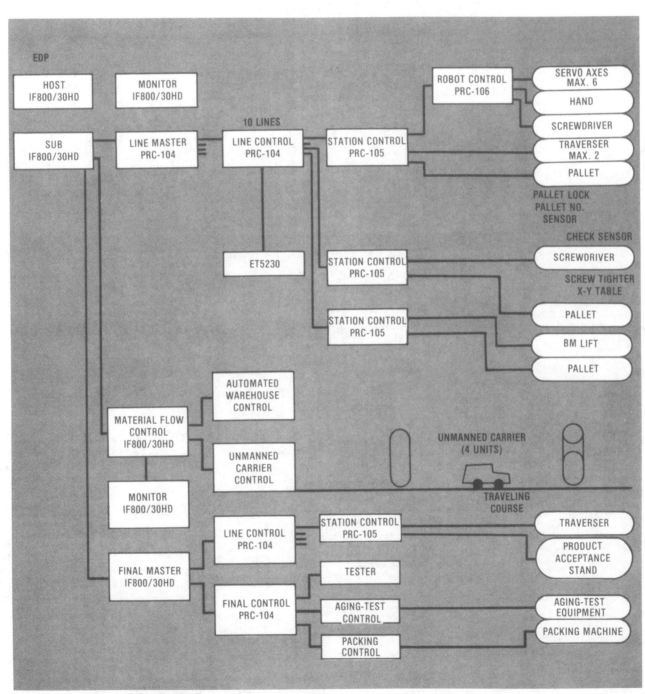

Figure 17. System control flowchart.

fined in the data table, job control unit, and low-level control unit so that flexibility over system functions is assured.

(6) Line operations and material handling are monitored on the CRT of the IF800 personal computer and in the central control room on a real-time basis. What is more, records about trouble on the line and in component transport are filed for use in computing the utilization rate of each function, mean time between failures, mean time to repair, and other maintenance-related factors, as well as for troubleshooting.

System functions. As shown in Figure 16, the following system functions are performed through decentralized processing.

(1) Host CPU functions: production plan entry, production control, component stock control, output of

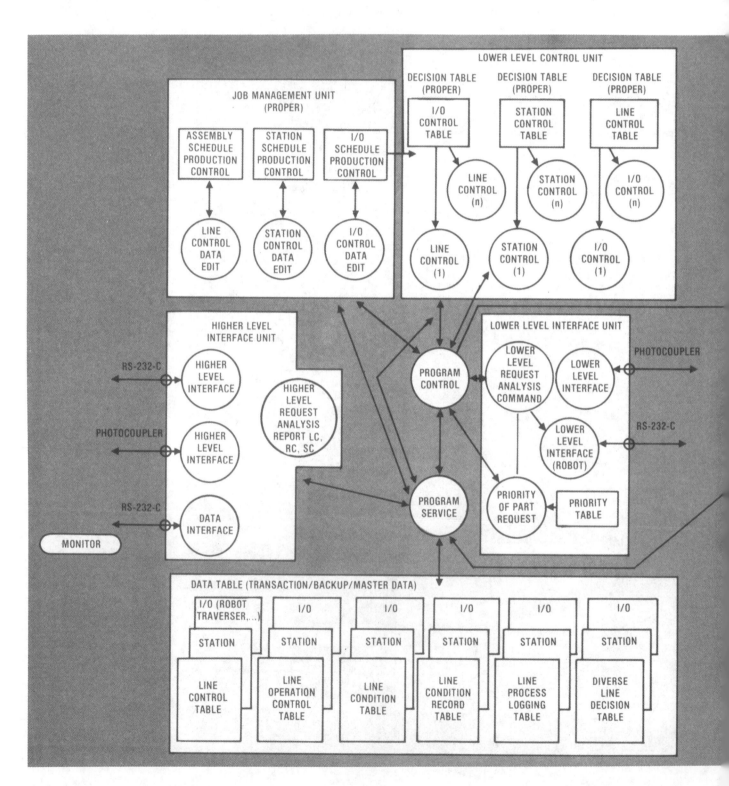

information about missing components.

(2) Sub-CPU functions: production instructions, job assignment to modules according to line status and other information, transport priority assignment (16 steps) in response to component requests from assembly line.

(3) Material control functions: control of entry to and exit from the automated warehouse and random-transport control of unmanned carriers.

(4) Line master functions: production job assignment to individual assembly lines and status control for each line.

(5) Line controller functions: instructions to individual workstation controllers and control of production from the line.

(6) Station controller functions: operation control for each workstation and directions to robots.

(7) Monitor functions: material-handling monitor, line operation monitor, and error data logging.

Information flow. We outline processing in the control system of the automatic printer assembly line with emphasis on information flow (see Figure 19):

Production plan. The system production plan is prepared according to the planning data supplied by the electronic data-processing system positioned at a higher level than the automatic production line, and is entered to the host computer.

Starting production. The computer processes the entered production plan and sends production models, lot numbers, and other information to the subcomputer. Next, the subcomputer categorizes the received information, and sends appropriate data (robot operation mode, production model, etc.) to the line master controller, the material distribution controller, etc. The line master controller sends operation instructions to the station controller (which controls the operation of each workstation) through each corresponding line controller. The line master controller also sends the operation mode and instructions to the robot controller.

Line operation control. At each workstation on the assembly line, the assembling robot performs according to the operation sequence it was taught in advance. The robot controller's internal RAM (backed up by a battery) can store up to 30 operation sequences. When the model is changed, the operation pattern is also changed automatically according to the operation mode command from the higher level controller.

To perform assembly, the robot mounts components in sequence onto assembly pallet jigs running on the free-flow base machine. The work-

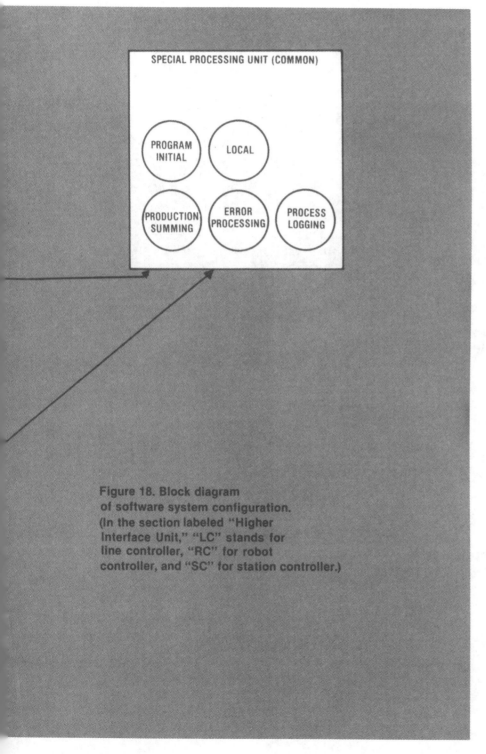

Figure 18. Block diagram of software system configuration. (In the section labeled "Higher Interface Unit," "LC" stands for line controller, "RC" for robot controller, and "SC" for station controller.)

station controller counts the assembled units. When the count reaches the value prescribed by the line controller (higher level computer)—i.e., when the planned number of items has been assembled—the station controller notifies the line controller that production has been completed.

When an assembly error occurs, the station controller transmits error information to the line controller and turns on the alarm lamps on the corresponding workstations. Upon receiving error information, the line controller outputs it to the printer installed on the line and transmits it to the line monitor via the line master controller and subcomputer. Assembly errors are detected by various sensors (optical, magnetic, or mechanical) mounted on the hands of the robots.

The line controller has another error-handling function: When a retouch error occurs, the line controller notifies other station controllers of the number of the jig pallet where the error occurred and commands them to skip that pallet.

Request for components. Each workstation on the automatic assembly line sends a component request to the material-handling controller before the supplied components have been used up. The request is assigned a priority by the subcomputer so that the material-handling controller can process component requests in order. The material-handling controller di-

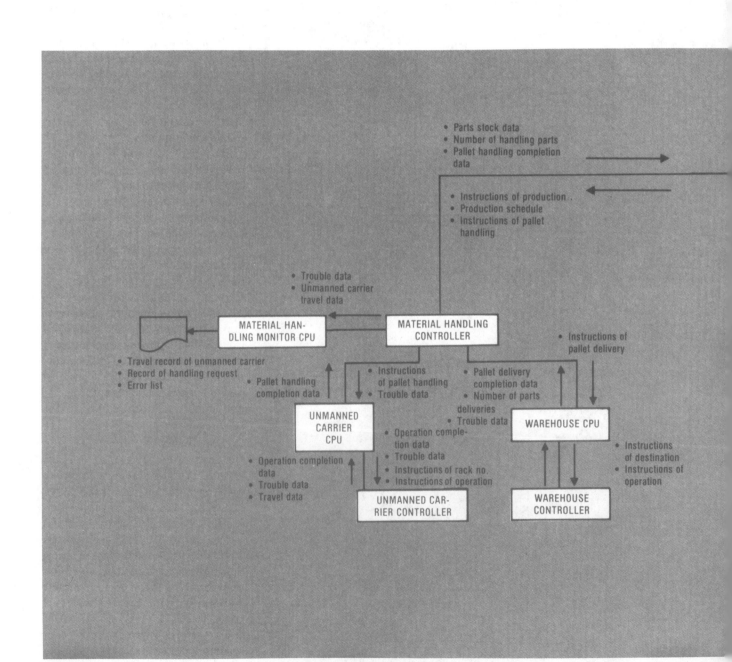

Figure 19. Information flow on the automatic assembly line.

rects the automated-warehouse controller to fetch requested components and commands the unmanned-carrier controller to transport them.

Upon receiving a fetch command, the warehouse controller checks the requested components. If they are not present in the warehouse, the warehouse controller informs the host computer of missing components. The host computer immediately displays missing component information on the CRT. To prevent the warehouse from running short of components, the host computer always inquires about the stock in the automated warehouse. When the stock drops below a specific level, the host computer displays a warning message on the CRT.

The unmanned-carrier controllers are always aware of what the carriers are doing and where they are running. Upon receiving a transport command, the unmanned-carrier controller sends the command to the available carrier nearest the starting position of the requested transport. The carrier is equipped with a radio unit and communicates with the carrier controller through a communication line buried under the floor along the carrier's guideline.

Production control and model switching. The host computer acquires production information from each automatic assembly line. When a line completes the planned production, the host computer informs the line about

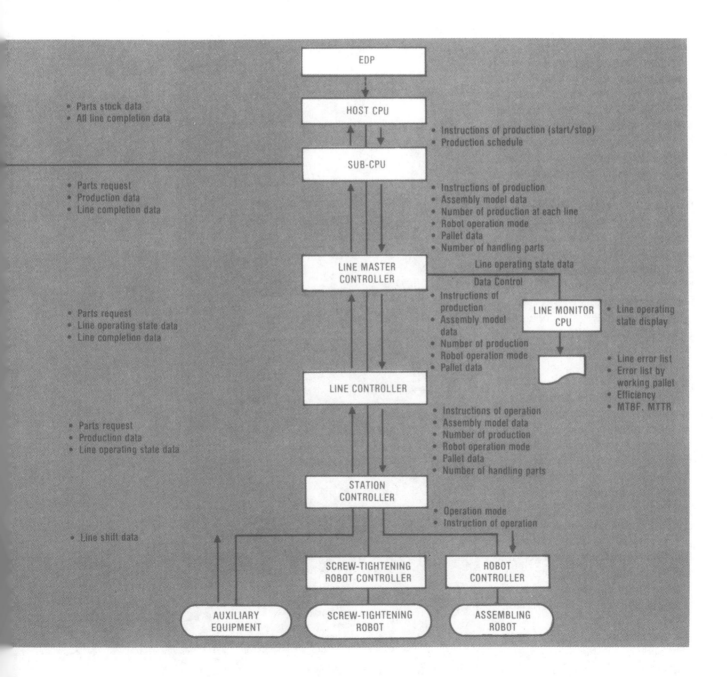

the model and parameters of the next production. If the next model differs from the present model, the subcomputer notifies the material-handling controller and line controller that the model must be changed. The line controller, thus notified, sends the new information necessary for the next production to the station controller and directs changes of component pallets, robot operation mode, and other mechanical arrangements.

The material-handling controller commands the automated warehouse to provide components needed to produce the next model and directs the unmanned carriers to change component pallets.

Robot control system. The robot system configuration is shown in Figure 20, and the robot control system is shown in Figure 21. Robot operation sequence programs are prepared by the robot-programming system IF800/M30, a Z-80-based computer. Robot operations are described in an Oki-developed robot language. Programs are translated by the robot language compiler into an intermediate code. Standard patterns of robot operation sequences are filed in advance and can be invoked for preparation of a robot operation sequence program. This greatly facilitates programming procedures.

The intermediate code produced by compilation is transmitted to the robot controller and stored in the battery-supported memory area designated by the robot controller.

The robot controller has the following features: In on-line mode, the robot operation mode is specified by the higher level controller. The robot controller can accept two types of operation sequence programs at the same time. The order of the two can be specified. As directed by the higher level controller, the robot controller can command the robot to start or stop each operation step and perform error-handling processing. This means that the robot can be freely manipulated through the higher level controller.

A teaching box is used to teach the robot operating points and to operate the robot as taught. (The assembly line operator uses switches on the teaching box to produce step-by-step robot motion. The robot then "memorizes" the sequence of switching operations, which become part of its control program. See Figure 20.) It is also possible to teach the robot manually for movements along each axis of the arm, linear movements, and wrist attitude control. Operation sequences can be performed stepwise or continuously.

Operation monitor. Operations on the line and material distribution can be monitored on the IF800 CRT (information is classified by colors corresponding to status). Also, real-time monitoring is possible in the central control room. The monitor functions and displayed information are shown here:

Line monitor

(1) Line status display
 • Line operation status
 • Line error information
 • Model being produced
(2) RS-232-C interface (between main program and line master)
(3) Line operation data file management
 • Management of error files, each provided for a line station
(4) Floppy disk drive backup
(5) List output

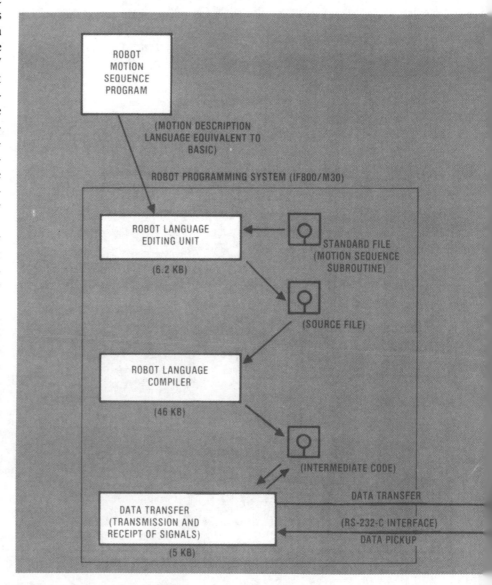

Figure 21. Robot control system.

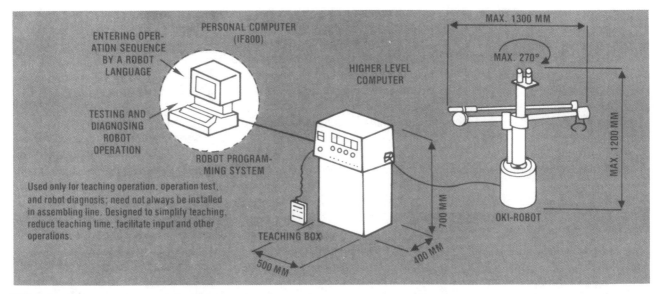

Figure 20. Robot system configuration.

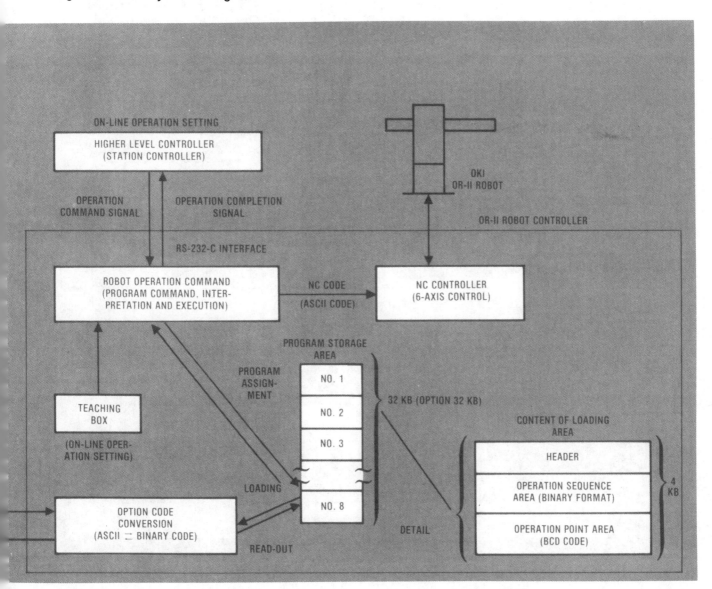

- Line operation error list (per day and month, by model)
- Jig error list (per day and month, by model)
- Quality control list (per day and month, by line and station)

(6) Quality data management (for each line and each station)
- Operating ratio
- MTBF (mean time between failures)
- MTTR (mean time to repair)

Material-handling monitor

(1) Unmanned-carrier status display
- Station assignment
- Transport request
- Carrier position

(2) Display of rack master error in automated warehouse
- Error display for each model

(3) RS-232-C interface (between main program and carrier master)

(4) Material-handling record file management
- Carrier transport record file management
- Warehouse rack master file management

(5) Floppy disk drive data backup

(6) List output
- Carrier operation record list
- Transport request data list
- Warehouse operation list
- Carrier error list

Figure 22 gives an example of material-handling monitor display, and Figures 23 and 24 give examples of line monitor display.

Results of system implementation

In order to adapt our printer products to the automatic assembly line, we have conducted extensive variety reductions (VRs). Some of the components were combined into a single piece and others were divided into two or more pieces for ease of mechanical handling. Also, changes in shapes and reductions in the number of parts have been made. In the case of the ET-8300, for example, the number of mechanical parts was reduced by 31 percent to 269 pieces. Further, of these 269

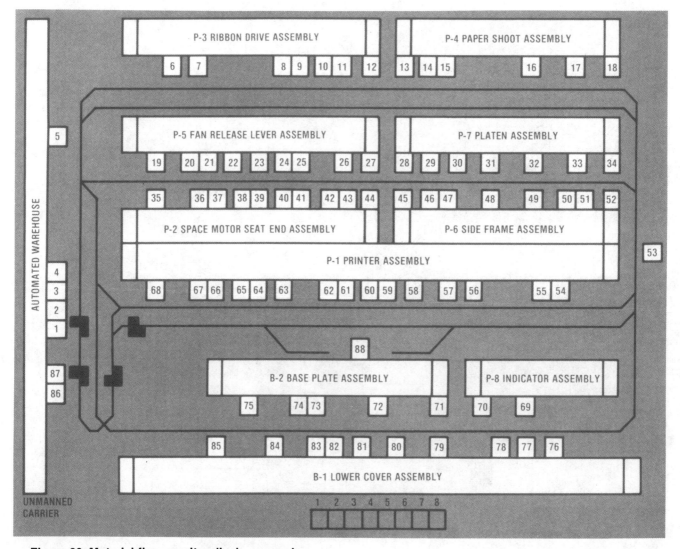

Figure 22. Material flow monitor display example.

pieces, 219 pieces (81.4 percent) are now assembled by robots. Consequently, at the monthly production rate of 10,000 units, assembly laborers were reduced by 75 percent to 27 persons, and production time from receiving components to shipment of the finished product was reduced by 60 percent to 24 hours.

In addition, the failure rate of the product in the customer site and the assembly cost exclusive of material cost both have been reduced to one tenth of what they were when these products were manually assembled.

Figure 25 shows the production record since November 1983, when the automatic printer assembly line went into operation.

We developed the automatic assembly line by exploiting Oki's technologies in a wide range of fields, including systems engineering, robotics, and material handling. It is a total system, assuming the form of a plant and integrating all processes comprising

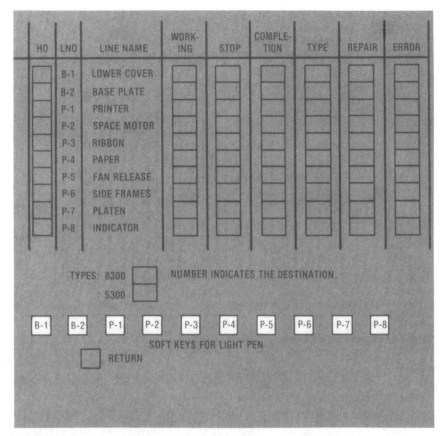

Figure 23. Line monitor display example.

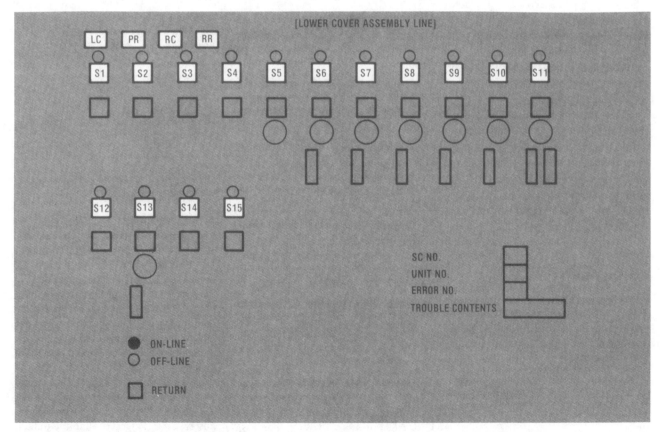

Figure 24. Line monitor display example.

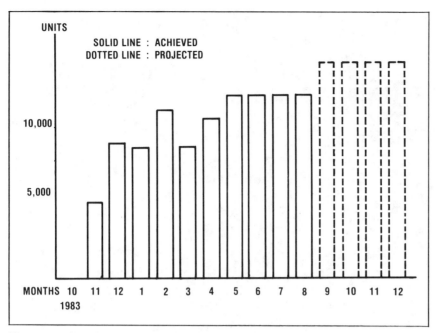

Figure 25. Automatic assembly line production.

planning, component loading, and product shipping. Thoroughly system-oriented design, intensive utilization of mechanical and electronic technologies, and systematic combination of electronics equipment and mechanical facilities have produced an integrated effect of reliable automation in our plant. This has not only improved productivity but has also encouraged improvements in product quality and in all processes from material procurement to product shipment.

There still remain subjects to be further examined for advancing factory automation; these include component identification techniques, integration of component-processing steps into the system, component arranging techniques, and enhancement of functions for checking assembly quality. □

Bibliography*

F. Horiguchi, "Development and Introduction of Assembly Robot (Oki-Robot)," 21st Lecture on Industrial Robot Application Techniques, Japan Industrial Association, Dec. 1982.

F. Horiguchi, "Printer Automatic Assembly Factory," 118th Lecture at Kansai

Robotization: Its Implications for Management is available in Japanese and English. The other works listed are all in Japanese.

Branch of Japan Soc. Mechanical Engineers, Jan. 1984.

H. Tanimoto, "Case Study of Enterprises Coping with FA," *Industrial Daily News*, Factory Management, Tokyo, special issue, Vol. 28, No. 14, Dec. 1982, pp. 152-160.

H. Tanimoto, "Printer Automatic Assembly Line," *J. Japan Soc. Mechanical Engineers*, Vol. 86, No. 779, Oct. 1983, pp. 80-86.

H. Tanimoto, "Small Printer Assembly Line with Robots," *Electronic Material*, Industrial Research Publ. Co., Vol. 22, No. 6, June 1983, pp. 48-56.

H. Tanimoto, "Small Printer Automatic Assembly Line," *Proc. 51st Sem. Automatic Assembly Committee of Japan Soc. of Precision Engineering*, Jan. 1983, pp. 29-53.

H. Tanimoto, "TPS Production Method of Oki-Takasaki Meets FA Age Requirements," *Industrial Daily News*, Factory Management, Tokyo, Vol. 28, No. 10, Sept. 1982, pp. 85-98.

"New Factory of Tohoku Oki Electric Industry Co.," photo reportage, *Production and Handling*, New Technology Publishing Co., Tokyo, Vol. 23, No. 11, Nov. 1982, pp. 33-36.

"APL for Printer Assembly," *Robotization: Its Implications for Management*, Japan Management Association Research Inst., ed., Fuji Corp., Tokyo, June 1983, pp. 205-220.

"Chinese Character Printer Automatic Assembly Line at Oki-Takasaki," *Nikkei Mechanical*, Nikkei McGraw-Hill, Tokyo, No. 159, Jan. 30, 1984, pp. 66-75.

"Full Automatic and Unmanned Production of Chinese Character Printer at Oki-Takasaki," photo reportage, *Unmanned Production Technique*, New Technology Publishing Co., Tokyo, Vol. 25, No. 3, Mar. 1984, pp. 49-56.

Hiromu Tanimoto is a corporate director and general manager of the Takasaki Plant, in the Data-Processing Group of Oki Electric Industry Company, Ltd. He joined Oki in 1958 and has spent most of the years since in the production area, most recently applying automated production techniques to the company's various information-processing products.

Tanimoto has contributed a number of articles on automated production to various industry periodicals. He received the BS degree at Chiba University, Japan.

Tanimoto's address is Oki Electric Industry Co., Takasaki Plant, Gumma, Japan.

A Control Perspective on Recent Trends in Manufacturing Systems

Stanley B. Gershwin, Richard R. Hildebrant, Rajan Suri, and Sanjoy K. Mitter

INTRODUCTION

While the technology of manufacturing (including processes and computer hardware and software) is improving rapidly, a basic understanding of the *systems issues* remains incomplete. These issues include production planning, scheduling, and control of work in process. They are complicated by randomness in the manufacturing environment (particularly due to machine failures and uncertainty and variability in production requirements), large data requirements, multiple-level hierarchies, and other issues that control engineers and systems engineers have studied in other contexts.

The purpose of this paper is to present an interpretation of recent progress in manufacturing systems from the perspective of control. We believe that this community has a vocabulary and a view of systems that can be helpful in this area. However, in order for this group to make that impact, it is essential that they learn the problems and terminology and become familiar with recent research directions. This paper is intended to present certain issues in manufacturing management in a way that will facilitate this.

We will establish a framework for manufacturing systems issues that is heavily influenced by control and systems thinking. We will then summarize current practice and current research, and critique them from the point of view of that framework.

GENERAL PERSPECTIVE

The purpose of manufacturing system control is not different in essence from many other control problems: it is to ensure that a complex system behave in a desirable way. Many notions from control theory are relevant here, although their specific realization is quite different from more traditional application areas. The standard control theory techniques do not apply: we have not yet seen a manufacturing system that can be usefully represented by a linear system with quadratic objectives. This is not surprising; standard techniques have been developed for what have been standard problems. Manufacturing systems can be an important area for the future of control; new standard techniques will be developed.

Some central issues in manufacturing systems include *complexity, hierarchy, discipline, capacity, uncertainty,* and *feedback*. Important notions of control theory include state and control variables, the objective function, the dynamics or plant model, and constraints. It would be premature to try to identify these with all the issues outlined in this paper; it would even go against the purpose of the paper, which is to stimulate such modeling activity. In this section, we describe the relevance of these notions to the manufacturing context for readers whose primary background is in control and systems.

Complexity

Manufacturing systems are large-scale systems. Enormous volumes of data are required to describe them. Optimization is impossible; suboptimal strategies for planning based on hierarchical decomposition are the only ones that have any hope of being practical.

Hierarchy

There are many time scales over which planning and scheduling decisions must be made. The longest term decisions involve capital expenditure or redeployment. The shortest involve the times to load individual parts, or even robot arm trajectories. While these decisions are made separately, they are related. In particular, each long-term decision presents an assignment to the next shorter term decision maker. The decision must be made in a way that takes the resources—i.e., the capacity—explicitly into account. The definition of the capacity depends on the time scale. For example, short time-scale capacity is a function of the set of machines operational at any instant. Long time-scale capacity is an average of short time-scale capacity.

Machine-Level Control

At the very shortest time scale is the machine-level control. This includes the calculation and implementation of optimal robot arm trajectories; the design of "ladder diagrams" for relays, microswitches, motors, and hydraulics in machine tools; and the control of furnaces and other steps in the fabrication of semiconductors. Other short scale issues include the detailed control of a cutting tool: in particular, adaptive machining.

There is no rule that determines exactly what this shortest time scale is. A robot arm movement can take seconds while a semiconductor oxidation step can take hours.

The issue at this time scale is the optimization of each individual operation. Here, one can focus on minimizing the time or other cost of each separate movement or transformation of material. One can also treat the detailed relationships among operations. An example of this is the *line balancing* problem. Here, a large set of operations is grouped into tasks to be performed at stations along a production line. The objective is to minimize the maximum time at a station, which results in maximum production rate.

Other control problems at this time scale include the detection of wear and breakage of machine tools, the control of temperatures and partial pressures in furnaces, the automatic control of the insertion of electronic components into printed circuit boards, and a vast variety of others.

Cell Level

At the next time scale, one must consider the interactions of a small number of machines. This is cell-level control and includes the operation of small, flexible manufacturing systems. The important issues include routing and scheduling. The control problem is ensuring that the specified volumes are actually produced. At this level, the detailed specifications of the operations are taken as given. In fact, for many purposes, the opera-

---

Presented at the 1984 Conference on Decision and Control, Las Vegas, NV, December 12-14, 1984. Stanley B. Gershwin and Sanjoy K. Mitter are with the Laboratory for Information and Decision Systems, Massachusetts Institute of Technology, Cambridge, MA 02139. Richard R. Hildebrant is with the Charles Stark Draper Laboratory, Cambridge, MA 02139. Rajan Suri was with the Division of Applied Science, Harvard University, when this work was done. He is currently at the Department of Industrial Engineering, University of Wisconsin-Madison, Madison, WI 53706.

tions themselves may be treated as black boxes.

The issue here is to move parts to machines in a way that reduces unnecessary idle time of both parts and machines. The loading problem is choosing the times at which the parts are loaded into the system or subsystem. The routing problem is to choose the sequence of machines the parts visit, and the scheduling problem is to choose the times at which the parts visit the machines.

The important considerations in routing include the set of machines available that can do the required tasks. It is often not desirable to use a flexible machine to do a job that can be done by a dedicated one, since the flexible machine may be able to do jobs for which there are no dedicated machines.

In scheduling, one must guarantee that parts visit their required machines while also guaranteeing that production requirements are met. At this level, the issue is allocating system resources in an efficient way. These resources include machines, transportation elements, and storage space.

A control problem at this level is to limit the effect of disruptions on factory operations. Disruptions are due to machine failures, operator absences, material unavailability, surges in demand, or other effects that may not be specified in advance but which are inevitable. This problem may be viewed as analogous to the problem of making an airplane robust to sudden wind gusts, or even to loss of power in one of three or more engines.

Factory Level

At each higher level, the time scale lengthens and the area under concern grows. At the next higher level, one must treat several cells. For example, in printed circuit fabrication, the first stage is a set of operations that prepares the boards. Metal is removed, and holes are drilled. At the next stage, components are inserted. The next stage is the soldering operation. Later, the boards are tested and reworked if necessary. Still later, they are assembled into the product. This process takes much time and a good deal of floor space.

Issues of routing and scheduling remain important here. However, setup times become crucial. That is, after a machine or cell completes work on one set of parts of the same or a smaller number of types, it is often necessary to change the system configuration in some way. For example, one may have to change the cutters in a machine tool. In printed circuit assembly, one must remove the remaining components from the insertion machines and replace them with a new set for the next set of part types to be made. The scheduling problem is now one of choosing the times at which these major setups must take place. This is often called the *tooling problem*.

Other issues are important at still longer time scales. One is to integrate new production demands with production already scheduled in a way that does not disrupt the system. Another class of decisions is that pertaining to medium-term capacity, such as the number of shifts to operate and the number of contract employees to hire for the next few months. Another decision, at a still longer time scale, is the expansion of the capital equipment of the factory. At this time scale, one must consider such *strategic goals* as market share, sales, product quality, and responsiveness to customers.

Discipline

Specified operating rules are required for complex systems. Manufacturing, communication, transportation, and other large systems degenerate into chaos when these rules are disregarded or when the rules are inadequate. In the manufacturing context, all participants must be bound by the operating discipline. This includes the shop-floor workers, who must perform tasks when required, and managers, who must not demand more than the system can produce. It is essential that constraints on allowable control actions be imposed on all levels of the hierarchy. These constraints must allow sufficient freedom for the decision makers at each level so that choices that are good for the system as a whole can be made, but they must not be allowed to disrupt its orderly operation.

Capacity

An important element in the discipline of a system is its capacity. Demands must be within capacity or excessive queuing will occur, leading to excessive costs and, possibly, to reduced effective capacity. High-level managers must not be allowed to make requirements that exceed their capacity on their subordinates; subordinates must be obliged to accurately report their capacities to those higher up.

All operations at machines take a finite amount of time. This implies that the rate at which parts can be introduced into the system is limited. Otherwise, parts would be introduced into the system faster than they could be processed. These parts would then be stored in buffers (or worse, in the transportation system) while waiting for the machines to become available, resulting in undesirably large work in process and reduced effective capacity. The effect is that throughput (parts actually produced) may drop with increasing loading rate when loading rate is beyond capacity. Thus, defining the capacity of the system carefully is a very important first step for on-line scheduling.

An additional complication is that manufacturing systems involve people. It is harder to *measure* human capacity than machine capacity, particularly when the work has creative aspects. Human capacity may be harder to *define* as well, since it can depend on circumstances such as whether the environment is undergoing rapid changes.

Defining, measuring, and respecting capacity are important at *all* levels of the hierarchy. No system can produce outside its capacity, and it is futile, at best, and damaging, at worst, to try. On the other hand, it may be possible to expand the capacity of a given system by a learning process. This is a goal of the Japanese *just-in-time* (JIT) approach, which takes place over a relatively long time scale.

It is essential, therefore, to determine what capacity is, then to develop a discipline for staying within it, and finally to expand it.

Uncertainty

All real systems are subject to random disturbances. The precise time or extent of such disturbances may not be known, but some statistical measures are often available. For a system to function properly, some means must be found to desensitize it to these phenomena.

Control theorists often distinguish between random events and unknown parameters, and different methods have been developed to treat them. In a manufacturing system, machine failures, operator absences, material shortages, and changing demands are examples of random events. Machine reliabilities are examples of parameters that are often unknown. Desensitization to uncertainties is one of the functions of the operating discipline. In particular, the system's capacity must be computed while taking disturbances into account, and the discipline must restrict requirements to within that capacity. The kinds of disturbances that must be treated differ at different levels of the time-scale hierarchy: at the shortest time scale, a machine failure influences which part is loaded next; at the longest scale, economic trends and technological changes influence marketing decisions and capital investments.

It is our belief that such disturbances can have a major effect on the operation of a plant. Scheduling and planning must take these events into account, in spite of the evident difficulty in doing so.

Feedback

In order to make good decisions under uncertainty, it is necessary to know something about the current state of the system and to use this information effectively. At the shortest time scale, this includes the conditions of the machines and the amount of material already processed. Control engineers know that designing good feedback strategies is generally a hard problem. It is essential, especially at the short time scale, that these decisions are calculated quickly and be relevant to long-term goals. The trade-off between optimality and computation gives rise to many interesting research directions.

SURVEY AND CRITIQUE OF PRACTICAL METHODS

The manufacturing environment is one of the richest sources of important and challenging control problems of which we are aware. Until recently, however, the classical and modern control community has not been attracted to this opportunity. One reason is undoubtedly that the manufacturing area has never been perceived as needing the help. Extreme competition from overseas manufacturers has, more than anything else, changed this perception.

Another reason the manufacturing area has not enjoyed the attention of control theorists is that the area has not been, and some argue is still not, amenable to their techniques. This is because, in part, modeling large complex systems is difficult. Also, there has not been sufficient information available for feedback control that is current or even correct. Control theory has, to a large extent, implicitly assumed a plant that is automatically controlled; manufacturing systems are run largely on manual effort. All this is beginning to change, however, due to the availability of inexpensive computation, the installation of more fully automated systems, and the additional requirements of flexibility, quality, etc., that are placed on these systems.

A wide variety of methods that deal with scheduling and planning are available to industry. The purpose of this section is to survey these methods and to critique them according to the outline of the previous section. A representative survey of current practice in controlling manufacturing systems is provided in this section. The intention is to give the reader perspective on the current state of manufacturing control.

Factory-Level Control

Traditional Framework

The manufacturing community is accustomed to thinking about production control within a particular mature framework. All the functions necessary for planning and executing manufacturing activities in order to make products most efficient have been grouped into a few large areas. These areas and the general interrelationships among them are shown in Fig. 1.

This diagram shows a tremendous amount of interaction, where information is fed forward and back, among the different areas. Also, the diagram deals mainly with the resource allocation aspect of the production control problem. Other important traditional areas that are integral to a successful control system are receiving, cost planning and control, and such financial functions as accounts receivable, accounts payable, etc.

Function Descriptions Brief descriptions of the functions performed within the major areas are given below:

Forecasting: Demand is projected over time horizons of various lengths. Different forecast models are maintained by this function.

Master Scheduling provides "rough-cut" capacity requirements analysis in order to determine the impact of production plans on

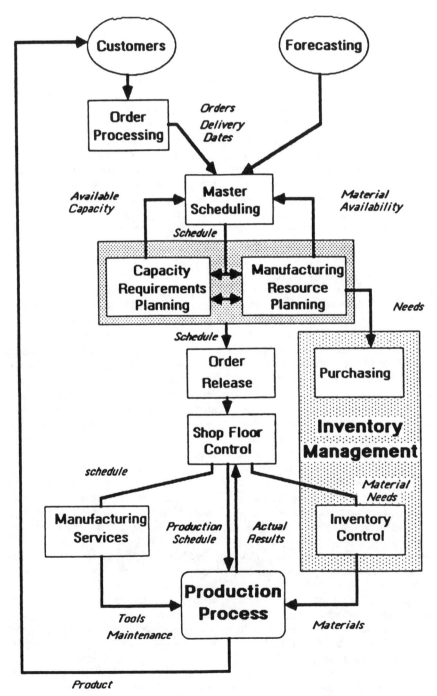

Fig. 1. Traditional framework in production control.

plant capacity. Comparisons are made between the forecast and actual sales order rate, sales orders and production, and, finally, scheduled and actual production.

Material Requirements Planning (MRP) determines quantity and timing of each item required—both manufactured and purchased. For each end-item, the quantity of all components and subassemblies is determined, and, by working backward from the date of final assembly, MRP determines when production or ordering of these subassemblies should occur. A more detailed capacity requirements analysis is made, and operation sequencing is determined. Also, lot-sizing is performed at this stage. While MRP tends to be highly detailed, there is no mechanism to take random events and unknowns into account; so it can lead to excess computer usage, misleading precision, delays in providing schedules, and rigidity.

Capacity Requirements Planning forecasts workcenter load for both released and planned orders and compares this figure with available capacity. The user specifies the number and duration of time periods over which the analysis is performed.

Order Release is the connection between manufacturing planning and execution. When an order is scheduled for release, this function creates the documentation required for initiating production.

Shop Floor Control is a lower level, "real-time" control function that is responsible for carrying out the production plan. This function performs priority dispatching and tracking of the product, as well as ancillary material and tooling. Data are collected on the disposition of the product and the performance of workcenters (utilization, efficiency, and productivity).

Inventory Control performs general accounting and valuation functions, as well as controlling the storage location of materials. It also often supports priority allocation of material to products or orders, and aids in filling order requisitions.

Planning Procedure These functions have always been performed. Many companies are organized according to these areas. For example, separate dedicated groups of people are often given the responsibility of controlling inventory, planning master schedules, etc.

The advent of the computer age brought software products that mirror almost exactly the functional framework outlined above. It is possible to buy software that addresses each functional area. In fact, the software is usually modularized so that the system may be acquired piecemeal.

The difficulty of the planning process is illustrated with the experiences of a large manufacturer of bed linens. They fabricate a basic set of products: sheets, pillow cases, and accessories. However, because bed linens have almost become high-fashion items, they come in a wide variety of sizes, prints, and styles. The force that drives the manufacturing process is initially the long-term forecast, but what is important is the ability to satisfy customer demand in the short to medium term. Customer demand manifests itself as an order for a particular set of products to be delivered at a particular time.

The basic problem they have is that orders are not being satisfied even though the overall level of finished inventory is extremely high. The product is being made, but not the right product at the right time!

Considering the large number of different products produced, the fact that production resources must be shared among those products, the limits on production capacity, and the difficulty of modeling manufacturing system dynamics, it is not surprising that the bed linen manufacturer has problems. Many companies do. The manufacturers that are most successful at controlling their operations generally make a small number of products and are able to forecast customer demand fairly accurately.

Recent Trends in Production Control

Finite-Capacity Material Requirements Planning Traditional MRP has offered little more than a computerized method of keeping voluminous records on material, and the resulting resource, requirements. There has never been an attempt, in any but the most superficial way, to account for the actual resource capacity in production planning and control. It has always been handled in an iterative, ad hoc, manual fashion. The manual approach is often a frustrating and impossible task.

There is growing interest in devising better factory-level models that integrate actual resource capacity with production requirements. In fact, one or two products that claim this capability have come onto the market within the past few years.

Products that attempt to perform finite-capacity planning often meet mixed reviews because their treatment cannot be comprehensive. A model formulation and its associated optimization procedure can be specified in a relatively straightforward way, but solving the problem with finite computational resources is impossible. Practical approaches must reduce the problem by making, what often turn out to be, limited assumptions.

Also, factory dynamics for different industries, while often similar, can be quite different. Consider the domestic manufacturer of candies and confection that became enamored over the dazzling performance reports of a particular finite-capacity scheduler. This scheduler can be quite adept at modeling and controlling factories where discrete parts are manufactured, but the candy manufacturer's process was continuous! No amount of hammering could bend the finite-capacity scheduler into a shape that would solve their problems.

The Just-in-Time or Kanban Approach

The just-in-time (JIT), or kanban (Kb), approach to manufacturing control is a Japanese refinement to the approach discussed above. The objective of this recent trend in material control is to reduce the need for large, expensive inventories of materials and subassemblies. By requiring that external and internal suppliers deliver just the right items, at just the right place, at just the right time, this objective may be met.

Kanban is a particular control implementation for forcing a just-in-time philosophy. A *kanban* is a job ticket that accompanies a part through the assembly process. When the part is actually installed in an assembly or subassembly, the kanban is sent back to its source to trigger the production of a new part. The control variable is the number of kanban tickets in the system.

The high risks of interrupted production due to low inventories are somewhat mitigated by imposing a great degree of discipline on all facets of manufacturing. Maintenance procedures and scheduling must be tightened up, lest the flow of parts that are needed downstream stop. Outside suppliers must ensure high quality in order to reduce the need for elaborate, and inventory-producing, inspections. Also, very good predictability of transportation times and strong communication ties are required of suppliers that participate in a JIT program. The long-term benefits of this discipline can lead to productivity increases beyond the simple reduction of inventory carrying costs [63], [32].

Implementing the JIT philosophy usually results in smaller and more frequent deliveries of materials. This can exacerbate the still necessary task of inventory management. Although zero inventory is an appealing goal, it should be moderated to the extent that costs required to achieve it increases.

The JIT philosophy for production control is most applicable where production requirements are known and fixed far in advance, and where buffering is not required to smooth the unavoidable effects of process time variations. This last point is illustrated by the material flow associated with flexible manufacturing systems, job shops, or any other system where a variety of parts with wide variations in process times share the same resources. Even without machine failures, buffers are required to reap the maximum production.

The JIT approach works best in applications such as the assembly process for products with predictable sales (refrigerators, automobiles, etc.). The uncertainties in these applications are not high enough to require intermediate buffering in order to achieve the maximum production rate possible.

Cell-Level Control

Traditional Approach

There have been very few successful approaches to scheduling the activities in a cell. Simulation is one that is widely used to determine scheduling strategies, floor layout, and other planning problems. However, it is expensive in both human and computer time since simulations, to be credible, tend to be complex and require a great deal of data. Many simulation runs are required to make a decision; the decision parameter must be "tuned" until optimal, or at least satisfactory, behavior is found.

Recent Developments in Cell Design and Control

Recent developments in automation and new constraints on the "flexibility" of the manufacturing process are beginning to alter the traditional concept of a cell and how that cell is to be controlled. One direction of development, called *group-technology cells*, *flowlines*, or *cellular manufacturing*, was stimulated by reports of Japanese successes. A family of products with very similar operation sequences is manufactured from start to finish in a single cell. This is intended to lead to a simplification of product flow and scheduling, tighter coupling of operations, less inventory, and greater worker coordination.

A second, stimulated by advances in automation and control technology, is the *flexible manufacturing system*, which is described below. A good overview of cellular manufacturing concepts can be found in Black [6] or Schonberger [64].

Flexible Manufacturing System Control

A modern example of a cell is a flexible manufacturing system (FMS), which consists of several machines and associated storage elements, connected by an automated materials handling system. It is controlled by a computer or a network of computers. The purpose of the flexibility and versatility of the configuration is to meet production targets for a variety of part types in the face of disruptions, such as demand variations and machine failures.

In an FMS, individual part processing is practical because of two factors: the automated transportation system and the setup or changeover time (the time required to change a machine from doing one operation to doing another), which is small in comparison with operation times. The combination of these features enables the FMS to rapidly redistribute its capacity among different parts. Thus, a properly scheduled FMS can cope effectively with a variety of dynamically changing situations.

The size of these systems ranges from approximately 5 to more than 25 machines. They are also specifically designed for the concurrent processing of a number of different parts (5 to 10 unique parts types is not unusual), each of which may require a variety of processing (milling, drilling, boring, etc.). An FMS of average size, built for a large manufacturer of agricultural equipment, is shown in Fig. 2. There are, altogether, 16 machines that are serviced by automatically guided vehicles (AGVs). An area for loading and unloading parts on and off the AGVs is set aside to one end of the system. Parts enter the system here, proceed through their respective process plan, and exit.

A flexible manufacturing system is a simple cell whose main objective is to meet a

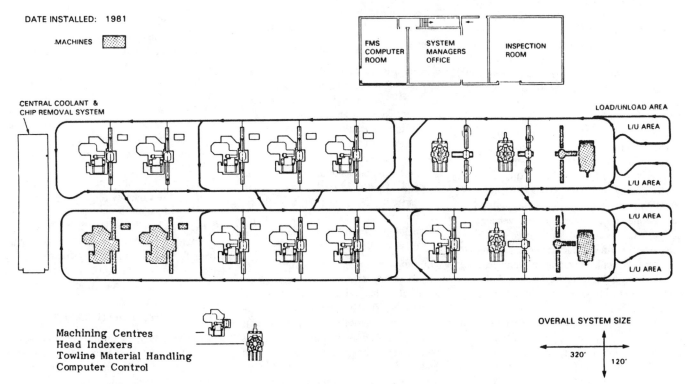

Fig. 2. Flexible manufacturing system of average size.

predefined master production schedule. The operational decisions that must be made include:

- Allocation of operations (and tools) to machines, such that the following, often conflicting, subobjectives must be met:
  — Workload requirements are evenly balanced among the machines and material handling systems.
  — Machine failures have a minimum effect on other machines' work availability.
  — Work-in-process requirements are minimized.
  — Processing redundancy (duplicate tooling) is maximized.
- Reallocation of operations and tools to machines when machines fail, such that, in addition to those objectives listed above, tool-changing effort is minimized.
- Real-time allocation of resources for processing pieceparts, such that:
  — Workload requirements are evenly balanced among the machines and material handling systems.
  — Quality of processed parts is maximized.

The first two areas are generally not well addressed by the vendor community and, in fact, often cause the users major operational problems when trying to run an FMS. Because of the difficulty in juggling the conflicting objectives under sometimes severe constraints (limits on the number of tool pockets per machine and on the weight the tool chain may bear), it is very difficult to manually allocate processing to resources. Some recent strides have been taken in solving this problem, but the capability is not yet widespread and has not yet been integrated into the operating software that controls FMSs.

The real-time scheduling of parts to machines, however, is addressed directly by the vendors that supply "turn-key" FMSs. Each vendor usually takes a unqiue approach to the scheduling problem (this is motivated, in part, by the unique aspects of each vendor's design) and, because of a perceived proprietary edge, is often reluctant to divulge the details of its implementation. Nevertheless, after analyzing the behavior of many FMSs over a period of time, we can make the following observations.

The decision classes, or control variables, for scheduling the activity in an FMS are listed below. In principle, one can construct a detailed schedule before the fact. In practice, however, the complexity of the problem (the large number of possible decision choices and uncertainty in material and resource availability) prevents this.

The general approach taken by the practitioners of FMS control is that of *dispatch scheduling*. Decisions are made as they are needed. Very little information is considered when making these decisions. The criteria and constraints for a variety of questions related to dispatch scheduling are:

- *Part sequence into FMS:* Since an FMS can process a number of different parts and since these parts are required in certain ratios relative to one another, active control of the part input sequence is required.
- *Sequencing of fixturings:* Many parts must make a number of passes through the system in order to process different sides. The sequence for these separate passes could be chosen to enhance the performance of the system.
- *Sequencing of operations:* Once in the system, a part must often visit a number of different machines before processing is complete. The sequence of these separate machine visits could be chosen to enhance the performance of the system.
- *Machine choice:* Often a particular operation may be performed at more than one machine. When this is true, a choice must be made among the possibilities.
- *Cart choice:* Many FMSs employ a number of separate carts for transporting parts from machine to machine. When the need arises for transporting a part, a choice among the carts of the system must be made.
  — *cart movement:* Carts are always moving, except while undergoing load/unload operation or while queuing at an occupied node. Shortest routes are chosen when there is a destination. Deadlocks are checked for periodically.
  — *requests for carts:* Intervals are computed and parts are introduced. Backlogs of parts are tracked. The closest cart with the correct pallet is chosen for loading parts. The closest empty pallet is chosen for each part coming off a shuttle.
- *Operation and frequency selection for quality check:* Many FMSs being built are equipped with a coordinate measuring machine (CMM). The purpose of this machine is to monitor the quality of the parts being processed as well as the processes themselves. Through the measurement of part dimensions, the nature of process errors (tool wear, machine misalignment, fixture misalignment, etc.) may be inferred. Because the CMM resource is limited, the intelligent selection of operations to measure and the frequency with which to measure them is required in order to ensure that quality standards are satisfied and that processing errors are quickly identified.

Machine-Level Control

The bottom tier of the manufacturing structure is comprised of individual workstations, which may be actual machinery or even lone workers (as is the case with manual assembly systems). Control at the machine level does not really include material flow, scheduling, or other logistical considerations. These issues have been accounted for at the cell and factory levels.

The problems encountered at this level are sometimes more in line with those that have been traditionally treated within the classical and modern control framework. The domain is often continuous, rather than discrete, and there is often opportunity for instrumenting the machinery for full automatic control and feedback.

The Traditional Approach

In the beginning, there were just hand tools. All control and feedback was accomplished through eye-hand coordination. This continued to be the case, for the most part, up until recently (1950s). The tools (lathes, drill presses, etc.) became larger and more complex, but the principle remained the same. Then computers were applied and numerically controlled (NC) machinery was the result. Here the position, feed, and speed of the tool relative to the part is controlled through standard feedback techniques. In addition, the different operations a part required could be programmed to occur automatically on one machine in the proper sequence.

Operation sequencing is generally performed open loop: there has not been sufficient reason to alter the sequence. This is changing in some environments where there is full automation. It may happen that a tool breaks part way through a "tape segment." If the part has to leave the machine and come back for any reason (quality-control check, extract broken tool, etc.), it is difficult to pick up where the processing left off.

Recent Developments

Until recently, the position of the tool, and its feed rate and speed relative to the part, has been controlled in an entirely open-loop manner. Regardless of what was happening (wearing of tools, anomalies in casting dimensions and quality, etc.), these variables would remain constant. This is beginning to change. By monitoring the power requirements of a particular cut, the condition of the casting/tooling combination can be determined and adjustments made.

Electronic vision is another means by which feedback is being used in control at the machine level. These systems check for the presence or absence of tools in the spindle. Other techniques for measuring the wear on these tools are also being employed.

SURVEY AND CRITIQUE OF RECENT RESEARCH

This section reviews recent developments in analysis and optimization of manufacturing systems. It is intended for control engineers who wish to become familiar with research in this field. We should emphasize at the outset that there is a large body of literature available on traditional approaches to manufacturing. For example, the area of production and operations management (POM) occupies a significant place in most business schools, and many textbooks exist for this well-developed area [85]. Here we will restrict ourselves to the *systems* aspects of manufacturing problems, to areas relevant to the framework as discussed earlier, and to recent developments in these areas, which we believe could significantly affect the progress of the field.

We use the time-scale hierarchy mentioned, rather than the framework of practical methods. The former is more appealing from the control point of view, and perhaps more amenable to rigorous development.

We adopt the distinction, as proposed in Suri [78], between *generative* and *evaluative* techniques or models. A generative technique is one that takes a set of criteria and constraints, and generates a set of decisions. An evaluative technique is one that takes a set of decisions and predicts (evaluates) the performance of a system under those decisions. (The terms *prescriptive* or *normative*, and *descriptive*, are also used for these two categories.)

While we concentrate on recent developments, it is appropriate to comment briefly on early research in manufacturing systems that can be found in the management science and operations research literature. Much of this was directed at production planning and scheduling problems.

In particular, a great deal of the work on generative techniques for production scheduling and planning was concerned with the mathematical problem of fitting together the production requirements of a large number of discrete, distinct parts [24]. Such combinatorial optimization problems are very difficult in the sense that they often require an impractical amount of computer time. Furthermore, they are limited to deterministic problems so that random effects, including machine failures and demand uncertainties, cannot be analyzed. An excellent review of production scheduling methods, including the use of heuristics and hierarchical approaches to solve large problems, can be found in Graves [29].

The early work on evaluative models was mainly an attempt to represent the random nature of the production process by using queuing-theoretic models, such as the classic M/M/1 and M/G/1 queues [52].

The applicability of queuing theory to manufacturing was considerably enhanced by the development of network-of-queues theory [46], [28]. However, only the more recent development of efficient computational algorithms and good approximation methods has enabled the implementation of reasonable "first-cut" evaluative models of fairly complex manufacturing systems, as described later.

Another early development in the area of evaluative models was the use of computer-based simulation methods, which employ a "Monte Carlo" approach to system evaluation. With the growing accessibility of computing power, the development of easy-to-use simulation packages, and the advancement of simulation theory, this area has made major strides forward recently.

Long-Term Decisions

In this section, we consider decisions that involve considerable investment in plant, equipment, or new manufacturing methods. Typically, such decisions may take over a year to implement and may have an operational lifetime of 5 to 20 years during which they are expected to pay back.

Generative Techniques

Traditional systems-based approaches for generating long-term decisions include the production planning and hierarchical approaches mentioned above, as well as strategic planning, forecasting, decision analysis, and location analysis. We do not deal with these here, but an overview and literature survey can be found in [77].

In the context of automated manufacturing, mathematical programming techniques (LP and IP) have been applied to selection of equipment and of production strategies [30], [73], [87]. However, the constraints involved in the mathematical programming problem formulations can be very complex. Whitney [86] has proposed *sequential decisions*, which is a new framework for developing heuristic algorithms for solving these complex optimization problems. It has been successfully applied to the problem of selecting parts and equipment for manufacturing in a very large organization.

Some recent approaches, which should be of interest to the control community, use dynamic investment models for long-term decision making. The decision to invest in alternative manufacturing strategies (and equipment), over a period of time, is formulated as an optimal control problem [7], [25], [53]. Such models offer qualitative insight to help decision makers faced with the complex set of investment alternatives that modern manufacturing systems involve. However, practical application and use of these models remains to be seen.

Evaluative Techniques

The evaluation of long-term effects of a decision on an enterprise is a particularly difficult problem, and evaluative models for long-term planning deal primarily with strategic and accounting issues [45], [47], [53]. Strategic issues involve such questions as how improved product quality or response time to orders will affect the market share. Accounting issues require models to trade off current expenditures with future (uncertain) revenue streams. Neither of these areas is of primary interest to the current audience. However, we should mention two factors. The first is that evaluative models for strategic and accounting issues are currently undergoing radical changes, in the face of the (relative) failure of U.S. industry to make prudent investments [1], [57].

The second important point, which is often missed by those undertaking modeling/analysis studies, is that the long-term decisions are influenced to a large extent by these strategic and accounting issues. Even though we do not cover them here, it is important for analysts working in this general area not to lose sight of the forest for the trees. Many modeling and analysis efforts fail to be useful to the manufacturing community because they focus on minor technical points and do not provide the overall insight that is needed for this stage of the planning process. Professor Milton Smith (of Texas Technological University) said at the recent First ORSA/TIMS Conference on Flexible Manufacturing Systems that around a hundred man-years had been expended on solving minimum makespan scheduling problems, but he did not know of a single company that used minimum makespan to schedule their shop-floor operations!

Medium-Term Decisions

Here we are concerned with a time period ranging anywhere from a day to a year, and the scope of the decisions involves primarily trade-offs between different modes of operation, but with only minor investments in new equipment/resources.

Traditionally, such decisions have been the domain of master scheduling, MRP, and inventory management systems. These systems generally do not account for uncertainty in a direct manner, but rather, in an indirect way through the use of "safety" values, whether in stocks, lead times, or other quantities. Master scheduling and MRP systems work to a deterministic plan, which gets updated periodically (say once a week or once a month); see [51] and [84] for insights into these points.

Inventory theory models and analyzes the effects of uncertainty to derive optimal stock policies. Inventory stocking policies assume that each item stocked has an exogenous demand, modeled by some stochastic process, and attempt to find the best stocking policy for each item.

The fact that the demand on inventory comes as a result of the master scheduling and MRP decisions is ignored, and thus it is clear that much useful information for decision making is being thrown away. Of course, it is the size and complexity of a manufacturing system that makes it very difficult to solve the entire problem simultaneously. Nevertheless, we feel that suitable structures can be developed to make the decision making more coherent across these components. Some attempts in this direction are described in this section.

Generative Techniques

We begin by reviewing generative techniques for this level of decision making. Control theorists are familiar with the concept of time-scale decomposition and hierarchical control, and should therefore readily understand the idea behind hierarchical production planning. It partitions the problem into a hierarchy of subproblems, with successively shorter time scales. The solution of each subproblem imposes constraints on lower subproblems. The advantages of the hierarchical approach are many: in addition to computational savings, this approach requires less detailed data, and it mimics the actual organizational structure [58].

The original ideas for this approach based the hierarchical structure on intuitive and heuristic arguments. However, control theorists should find it interesting that Graves [30] showed, by the use of Lagrange multipliers, that the Hax-Meal hierarchy could be derived as a natural decomposition of a primal optimization problem. An alternative hierarchy, based not on optimality but rather feasibility considerations, is derived by Suri [74], also using multiplier methods.

These hierarchical approaches assume that demand and capacity are known and deterministic over a period of time, and then re-solve the planning problem periodically. Recent developments in manufacturing systems have sought to represent uncertainty explicitly in the problem formulation. This uncertainty includes not just demand but also equipment failures (hence, randomly varying capacity). Since this usually leads to an intractable problem, the contribution of the new approaches is primarily in ways that they propose to formulate the problem or approximate the solution.

Hildebrant and Suri [36] proposed a hierarchical procedure where the hierarchy is derived from heuristic arguments based on tractability considerations, but the interaction between the levels is based on a mathematical programming problem. To get around the difficulty of solving a stochastic optimization problem, they proposed an "open-loop feedback" policy where the dynamics of the system between failure states is replaced by a static average of the time spent in each failure state (or each capacity condition). The technique showed reasonable improvement over existing heuristics [35].

Kimemia [49] and Kimemia and Gershwin [50] have derived an alternative, closed-loop solution to this problem. Their approach has also been to separate the relatively long-term issues (the response to machine failures and to production backlogs and surpluses) from the short-term problem of part dispatching. The long-term problem is modeled as a continuous dynamic programming problem. A feedback control law, which determines the next part to be loaded and when it should be loaded as a function of current machine state and current production surplus, is sought.

This formulation, which reflects the disruptive nature of machine failures, had previously been proposed by Olsder and Suri [61], but they had concluded that it was too hard to solve exactly. The contribution of Kimemia and Gershwin has been to find a good approximation to the exact solution. Essentially, this involves two steps in a dynamic programming framework: separating the top-level problem (the solution to a Bellman equation) into a number of subproblems, obtained formally through a constraint-relaxation procedure, and then approximating the value function for each subproblem by a quadratic.

More recent work by Gershwin, Akella, and Choong [27] has further simplified the computational effort. Simulation results indicate that the behavior of a manufacturing system is highly insensitive to errors in the cost-to-go function, so the Bellman equation can be replaced by a far simpler procedure.

Evaluative Models

Evaluative models for this decision-making level involve both analytic approaches and simulation. Important features are the ability to represent production uncertainties (such as machine failures) and limited buffer stocks, in order to trade off between the two. For large systems, this is, again, analytically intractable. The earliest work in this field is surveyed by Koenigsberg (1959). Notable contributions were made by Buzacott [8]–[10], who looked at various approximate analytic models that gave insight into these issues.

Most of these analytic studies are based on the Markov models of transfer lines and other production systems. An appreciation for the difficulty of the problem is seen from the fact that the largest general model for which an exact solution is available involves three machines with two buffers between them [26].

A promising, recent development has come out of a technique for decomposing a production line into a set of two-machine one-buffer subsystems [26]. This idea had been previously proposed by Zimmern [90] and Sevastyanov [68], but an efficient and accurate method had not been developed. Gershwin's procedure for solving this system is analogous to the idea of solving two-point boundary value problems. Numerical results indicate that the method is very accurate, and, what is more, fairly large problems (20 machines) can be solved in reasonable time. The technique is, however, currently restricted to the case where the cycle times of the machines are deterministic and equal. Altiok [3] has recently developed methods for systems with more general phase-type processing time distributions.

Other evaluative techniques include queuing network models, and simulation. Both of these methods can be used for short-term decision making as well. However, we feel that queuing network models are best suited to more aggregated decision making, while simulation is more suited to detailed decisions. Therefore, we discuss the former here and the latter in the next subsection, although the particular application may suggest the use of one or the other technique for either of these levels.

A fairly recent development in (analytic) evaluative models of manufacturing systems has been the growing use of *queuing network* models for system planning and operation. A simple-minded, static, capacity allocation model does not take into account the system dynamics, interactions, and uncertainties inherent in manufacturing systems. Queuing network models are able to incorporate some of these features, albeit with some restrictions, and thus enable more refined evaluation of decisions for manufacturing systems. The increased use of such models stems primarily from the advances made in the computational algorithms available to solve queuing networks, both exactly and approximately.

Buzen's algorithm [14] made the solution of these systems tractable. Solberg [71] applied this to capacity planning for FMS, and Stecke [72] used it for solving production planning problems. Shanthikumar [70] developed a number of approximate queuing models for manufacturing. The development of the *mean value analysis* (MVA) technique for solving these networks (Reiser and Lavenberg) opened up a host of new extensions and approximations. Various approximate MVA algorithms have been developed [66], [4], which enable fast and accurate solution of very large networks. Hildebrant and Suri [36] and Hildebrant [35] applied MVA techniques to both design and real-time operation problems in FMS.

An extension [81] enabled efficient solution of systems with machine groups. Another recent extension, called *priority mean value analysis* (PMVA) [69], allows a wide variety of operational features to be modeled.

An important reason for the increasing popularity of such models in manufacturing is that they have proved their usefulness in the area of computer/communication systems modeling, in terms of giving reasonable performance predictions. Recent analysis has given a basis to the robustness of queuing network models for use in practical situations [76].

The disadvantages of queuing network models are that they model many aspects of the system in an aggregate way, and they fail to represent certain other features, such as limited buffer space. (Some recent developments, e.g., Buzacott and Yao [13] and Suri and Diehl [80] do allow limited buffer sizes.) The output measures they produce are average values, based on a steady-state operation of the system. Thus, they are not good for modeling transient effects due to infrequent but severe disruptions such as machine failures.

However, the models tend to give reasonable estimates of system performance, and they are very efficient: that is, they require relatively little input data, and do not use much computer time. A typical FMS model [81] might require 20 to 40 items of data to be input, and run in 1 to 10 sec on a microcomputer, in contrast to the much larger numbers for simulation. Thus, these models can be used interactively to quickly arrive at preliminary decisions. More detailed models can then refine these decisions.

Queuing network models suggest themselves for use in the middle level of a hierarchy. Development of queuing network models along with suitable control aspects to tie in the lower and higher levels of the hierarchy could be a useful topic of research.

Lasserre [55] and Lasserre and Roubellat [56] represent the medium-term production planning problem as a linear program of special structure, and develop an efficient solution technique for it.

Short-Term Decisions

Decisions at this level typically have a time frame of from a few minutes up to about a month. Traditional generative models have included those for lot sizing and scheduling, using both exact approaches as well as heuristics or rules. There have been a number of recent, interesting developments in this area, which are now described.

Traditional lot-sizing models traded off the cost of setting up a machine with the cost of holding inventory, on an individual product basis. The Japanese (just-in-time and kanban) approaches have challenged these concepts as being narrow-minded and myopic in terms of the long-term goals of the organization. They advocate operation with minimal or no inventory, claiming that this not only saves inventory carrying costs but also gives rise to a learning process that leads to more balanced production in the long run [63], [32].

This thesis is becoming more widely accepted in U.S. industries as well. However, reduced inventory leads to line stoppages and inefficiencies in the short term. It is, therefore, logical to ask what is the optimal rate of reducing inventory so that short-term losses are traded off against long-term gains due to increased learning. There has been some preliminary investigation of this point [79]. It is a problem that would fit naturally into an optimal control framework, and further investigation would be useful.

There have also been some recent studies indicating that lot sizing in a multi-item environment ought to be treated as a vector optimization problem. The idea is that the lot size of each product affects the production rate of other products, primarily through the queuing of each lot of parts waiting for other lots to be done at each machine. Therefore, one ought to consider the joint problem of simultaneously optimizing all the lot sizes. This integer programming problem would normally be computationally intractable for any realistic manufacturing system. However, by modeling the system as a queuing network and then solving a resulting nonlinear program, some recent results have been obtained [48]. This is a promising development that needs further exploration.

Hitz [37], [38] studied the detailed, deterministic scheduling of a special class of flexible manufacturing systems: flexible flow shops. In these systems, parts follow a common path from machine to machine. He found that by grouping parts appropriately, he could design a periodic sequence of loading times. This substantially reduced the combinatorial optimization problem.

Erschler, Roubellat, and Thomas [21] describe a deterministic, combinatorial scheduling technique that searches for a *class* of optimal decisions. Rather than deciding which part to send into the system next, it presents to the user a set of candidate choices. This flexibility is intended as a response to the random events such as machine failures that are difficult to represent explicitly in a scheduling model.

Perhaps the most widely used evaluative tool for manufacturing systems today is *simulation* [65]. The term "simulation" in this context refers specifically to computer-based discrete event simulation. Such a model mimics the detailed operation of the manufacturing system, through a computer program that effectively steps through each event that would occur in the system (or to be more precise, each event that we wish to model).

In principle, simulation models can be made very accurate—the price is the programming time to create the model, the input time to generate detailed data sets, and the computer time each time the model is run. In addition, the more phenomena that the analyst tries to represent, the more complex the code, and the more likely there are errors, some of which may never be found.

It is sometimes forgotten that the accuracy of a simulation is limited by the judgment and skill of the programmer. Detail and complexity are not necessarily synonymous with accuracy, if major classes of phenomena are left out. (While simulation can be used at any of the levels of the decision-making process, we choose to describe it here since it can examine the most detailed operation of a manufacturing system.)

Two reasons for the recent popularity of simulation are the number of software tools that have been developed to make simulation more accessible to manufacturing designers, and the decrease in computing costs and the availability of microcomputers. These factors make it well worth an organization's effort to use simulation before making large investments. In addition, there have been many developments in the design and analysis of simulation experiments, which have contributed to the acceptance of simulation as a valid and scientific methodology in this field.

Recent developments in software tools for simulation can be categorized into simulation languages, "canned packages," interactive model development (or graphical input), and animation (or output graphics). The two input/output graphics features will not be discussed further here, but good examples are SIMAN (for input) and SEE WHY (for output).

Although simulation languages have been around for a while, the last five years have seen the development of many powerful languages, such as GPSS/H, SIMSCRIPT II.5, and SLAM II, as well as the development of languages specially tailored to the manufacturing user (e.g., SIMAN and MAP/1). Also, most languages are now available on microcomputers as well (e.g., GPSS/PC, SIMSCRIPT II.5, SIMAN, MICRONET).

Another development, specially geared to the manufacturing designer, has been the development of canned packages, which do not require programming skills, but are completely data driven (e.g., GCMS, GFMS, SPEED). Of course, they have a number of structural assumptions built into them, in terms of how the manufacturing system operates, but can be useful for very quick analysis of a system. At the other end of the spectrum, for very detailed simulation, it may be necessary to resort to a programming language such as FORTRAN or PASCAL. See [5] for a discussion of the trade-offs among these options.

It should be noted that simulation is useful for planning, as well as off-line analysis of operating strategies. However, its structure and computation-time requirements make it currently unsuitable for on-line decision making. Even though simulation is perhaps the most widely used computer-based performance evaluation tool for manufacturing systems, we would recommend greater use of analytic and queuing network models prior to conducting the more expensive simulation studies — in comparison to the numbers quoted for queuing network models, a simulation model might require 100 to 1000 data items and 15 to 10,000 sec to run on a microcomputer.

In the area of simulation design and analysis, there have been several developments that should be of interest to the control community. The analysis of simulation outputs — which involves parameter identification, confidence interval generation, detection of bias and initial transients, and run length control — has used many techniques from time-series analysis and spectral methods (see [54] for a survey). Parameter optimization in simulations involves stochastic approximation techniques [40], [59], [83], which again are familiar ground to our community.

A recent development, called *perturbation analysis of discrete event systems*, enables very efficient optimization of parameters in simulations (see [44] for a survey). This technique is related to linearization of dynamic systems, and, again, has parallels with conventional dynamic systems [39], [41]. Essentially, it enables the gradient vector of system output with respect to a number of parameters to be estimated by observing only one sample path. In this sense, it is an evaluative and "semigenerative" tool, since it not only evaluates decisions but also suggests directions for improving the decisions.

While much of the original work on perturbation analysis relied on experimental results to demonstrate its accuracy [42], [43], recent analyses have given it a more rigorous foundation [75] and also proved that it is probabilistically correct for certain systems [17], [40], [83], as well as better than repeated simulation [16], [89].

Another recent, interesting development has been the application of the Petri net theory to the performance analysis of manufacturing systems [23]. In the past, the main use of Petri nets (in computer science) was to answer such qualitative questions as: Will there be any deadlocks? However, there have been some important recent advances in the theory of timed Petri nets.

Following some work by Cunningham-Greene [20], Cohen et al. [18], [19] have developed a linear systems theoretic view of production processes. This enables efficient answers to some complex performance questions. It also gives rise to a parallel set of control-theoretical concepts for discrete events systems, e.g., transfer functions, controllability, observability. The main disadvantages it has currently are that it can only deal with completely deterministic situations and that it is only evaluative, not generative. However, it is a promising new development.

One of the most useful areas requiring more research is that of real-time control of manufacturing systems, at a detailed level. Little theoretical research has been done on this, apart from the large body of heuristics that exists for scheduling [29]. A few researchers have treated the issues in a formal way [11], [2], [15]. This seems to be an area for control theorists to apply their expertise.

Indeed, Ho et al. [44] have coined the term DEDS, for discrete event *dynamic* system, to emphasize that manufacturing systems are a class of dynamic systems, and that there are concepts from dynamic system theory that need to be developed or applied for DEDS as well. In the past, we have seen DEDS analyzed either by purely probabilistic approaches (e.g., Markov chains, queues) or by purely deterministic approaches (scheduling and other combinatorial methods). The work by Cohen et al., as well as the perturbation analysis approach, have shown the use of a dynamic systems view of the world.

As an example, Suri and Zazanis [89] have used perturbation analysis combined with stochastic approximation to adaptively optimize a queuing system. This could be used, for example, for improving the choice of lot sizes for a number of different parts, while a facility is operating — the approach is simple to implement and has obvious applications in real systems. However, many interesting questions of convergence, etc., remain to be answered for this adaptive method.

CONCLUSION

We have described a framework for many of the important problems in manufacturing systems that need the attention of people trained in control and systems theory. We have shown how existing practical methods solve those problems, and where they fall short. We have also shown how recent and on-going research fits into that framework. An important goal of this effort has been to encourage control theorists to make the modeling and analysis efforts that will lead to substantial progress in this very important field.

References

[1] R. A. Abbott and E. A. Ring, "The MAPI Method — Its Effects on Productivity: An Alternative Is Needed," vol. 2, no. 1, pp. 15–30, 1983.

[2] R. Akella, Y. Choong, and S. B. Gershwin, "Performance of Hierarchical Production Scheduling Policy," *IEEE Trans. on Compon., Hybr., and Manuf. Tech.*, Sept. 1984.

[3] T. Altiok, "Approximate Analysis of Exponential Tandem Queues with Blocking," *Eur. J. Oper. Res.*, vol. 11, pp. 390–398, 1982.

[4] Y. Bard, "Some Extensions to Multiclass Queuing Network Analysis," in *Performance of Computer Systems* (M. Arato, ed.), North Holland, Amsterdam, 1979.

[5] J. P. Bevans, "First, Choose an FMS Simulator," *American Machinist*, pp. 143–145, May 1982.

[6] J. T. Black, "An Overview of Cellular Manufacturing Systems and Comparison to Conventional Systems," *Industrial Engineering*, vol. 15, no. 11, pp. 36–40, Nov. 1983.

[7] M. C. Burstein and M. Talbi, "Economic Justification for the Introduction of Flexible Manufacturing Technology: Traditional Procedures Versus a Dynamics Based Approach," *Annals of Oper. Res.*, 1985.

[8] J. A. Buzacott, "Automatic Transfer Lines with Buffer Stocks," *Int. J. Prod. Res.*, vol. 5, no. 3, pp. 183–200, 1967.

[9] J. A. Buzacott, "The Role of Inventory Banks in Flow-Line Production Systems," *Int. J. Prod. Res.*, vol. 9, no. 4, pp. 425–436, 1971.

[10] J. A. Buzacott, "The Production Capacity of Job Shops with Limited Storage Space," *Int. J. Prod. Res.*, vol. 14, no. 5, pp. 597–605, 1976.

[11] J. A. Buzacott, "Optimal Operating Rules for Automated Manufacturing Systems," *IEEE Trans. on Auto. Contr.*, vol. AC-27, no. 1, pp. 80–86, Feb. 1982.

[12] J. A. Buzacott and J. G. Shanthikumar, "Models for Understanding Flexible Manufacturing Systems," *AIIE Trans.*, pp. 339–350, Dec. 1980.

[13] J. A. Buzacott and D. D. W. Yao, "Flexible Manufacturing Systems: A Review of Models," Working Paper No. 7, Dept. of IE, Univ. of Toronto, Mar. 1982.

[14] J. P. Buzen, "Computational Algorithms for Closed-Queuing Networks with Exponential Servers," *C. ACM*, vol. 16, no. 9, pp. 527–531, Sept. 1973.

[15] C. Cassandras, "A Hierarchical Routing Control Scheme for Material Handling Systems," *Annals of Oper. Res.*, 1985.

[16] X. R. Cao, "Convergence of Parameter Sensitivity Estimates in a Stochastic Experiment," *Proc. IEEE Conf. Dec. and Contr.*, Dec. 1984.

[17] X. R. Cao, "Sample Function Analysis of Queuing Networks," *Annals of Oper. Res.*, 1985.

[18] G. Cohen, D. Dubois, J. P. Quadrat, and M. Viot, "A Linear System-Theoretic View of Discrete-Event Processes," *Proc. 22nd IEEE Conf. Dec. and Contr.*, Dec. 1983.

[19] G. Cohen, D. Moller, J. P. Quadrat, and M. Viot, "Linear System Theory for Discrete Event Systems," *Proc. 23rd IEEE Conf. Dec. and Contr.*, Dec. 1984.

[20] R. A. Cunningham-Greene, "Describing Industrial Processes and Approximating Their Steady-State Behaviour," *Op. Res. Q*, vol. 13, pp. 95–100, 1962.

[21] J. Erschler, F. Roubellat, and V. Thomas, "Aide à la Décision dans L'Ordancement d'Atelier en Temps Réel," 3è me Journées Scientifiques et Techniques Automatisée, ADEPA, Toulouse, June 1981.

[22] Draper Lab., *The Flexible Manufacturing Systems Handbook*, The Charles Stark Draper Laboratory, Cambridge, Massachusetts, 1982.

[23] D. Dubois and K. E. Stecke, "Using Petri Nets to Represent Production Processes," *Proc. 22nd IEEE Conf. Dec. and Contr.*, Dec. 1983.

[24] B. P. Dzielinski and R. E. Gomory, "Optimal Programming of Lot Sizes, Inventory, and Labor Allocations," *Manag. Sci.*, vol. 11, no. 9, pp. 874–890, July 1965.

[25] C. Gaimon, "The Dynamic, Optimal Mix of Labor and Automation," *Annals of Oper. Res.*, 1985.

[26] S. B. Gershwin, "An Efficient Decomposition Method for the Approximate Evaluation of Production Lines with Finite Storage Space," MIT Laboratory for Information and Decision Systems, Rep. LIDS-R-1309, Dec. 1983.

[27] S. B. Gershwin, R. Akella, and Y. C. Choong, "Short-Term Production Scheduling of an Automated Manufacturing Facility," MIT Laboratory for Information and Decision Systems, Rep. LIDS-FR-1356, Feb. 1984.

[28] W. J. Gordon and G. F. Newell, "Closed-Queuing Systems with Exponential Servers," *Op. Res.*, vol. 15, no. 2, pp. 254–265, Apr. 1967.

[29] S. C. Graves, "A Review of Production Scheduling," *Op. Res.*, vol. 29, no. 4, pp. 646–675, 1981.

[30] S. C. Graves, "Using Lagrangian Techniques to Solve Hierarchical Production Planning Problems," *Manag. Sci.*, vol. 28, no. 3, pp. 260–275, Mar. 1982.

[31] S. C. Graves and B. W. Lamar, "A Mathematical Programming Procedure for Manufacturing System Design and Evaluation," *Proc. IEEE Int. Conf. on Cir. and Compu.*, 1980.

[32] R. W. Hall, *Zero Inventories*, Dow-Jones Irwin, 1983.

[33] A. C. Hax and H. C. Meal, "Hierarchical Integration of Production Planning and Scheduling," in *Logistics* (M. A. Geisler, ed.), North Holland, 1975.

[34] P. Heidelberger and P. D. Welch, "A Spectral Method for Confidence Interval Generation and Run Length Control in Simulations," *C. ACM*, vol. 24, no. 4, pp. 233–245, Apr. 1981.

[35] R. R. Hildebrant, "Scheduling Flexible Machining Systems When Machines Are Prone to Failure," Ph.D. Thesis, MIT, 1980.

[36] R. R. Hildebrant and R. Suri, "Methodology and Multilevel Algorithm Structure for Scheduling and Real-Time Control of Flexible Manufacturing Systems," *Proc. 3rd Int. Sym. on Large Engin. Sys.*, Memorial Univ. of Newfoundland, pp. 239–244, July 1980.

[37] K. L. Hitz, "Scheduling of Flexible Flow Shops," MIT Laboratory for Information and Decision Systems, Rep. LIDS-R-879, Jan. 1979.

[38] K. L. Hitz, "Scheduling of Flexible Flow Shops—II," MIT Laboratory for Information and Decision Systems, Rep. LIDS-R-1049, Oct. 1980.

[39] Y. C. Ho, "A Survey of the Perturbation Analysis of Discrete Event Dynamic Systems," *Annals of Oper. Res.*, 1985.

[40] Y. C. Ho and X. R. Cao, "Perturbation Analysis and Optimization of Queuing Networks," *JOTA*, vol. 40, no. 4, pp. 559–582, 1983.

[41] Y. C. Ho and C. Cassandras, "Computing Costate Variables for Discrete Event Systems," *Proc. 19th IEEE Conf. Dec. and Contr.*, Dec. 1980.

[42] Y. C. Ho, M. A. Eyler, and T. T. Chien, "A New Approach to Determine Parameter Sensitivities of Transfer Lines," *Manag. Sci.*, vol. 29, no. 6, pp. 700–714, 1983.

[43] Y. C. Ho, M. A. Eyler, and T. T. Chien, "A Gradient Technique for General Buffer Storage Design in a Serial Production Line," *Int. J. Prod. Res.*, vol. 17, no. 6, pp. 557–580, 1979.

[44] Y. C. Ho, R. Suri, X. R. Cao, G. W. Diehl, J. W. Dille, and M. A. Zazanis, "Optimization of Large Multiclass (Nonproduct Form) Queuing Networks Using Perturbation Analysis," *Large-Scale Systems*, 1984.

[45] G. K. Hutchinson and J. R. Holland, "The Economic Value of Flexible Automation," *J. of Manuf. Sys.*, vol. 1, no. 2, pp. 215–228, 1982.

[46] J. R. Jackson, "Jobshoplike Queuing Systems," *Manag. Sci.*, vol. 10, no. 1, pp. 131–142, Oct. 1963.

[47] R. Jaikumar, "Flexible Manufacturing Systems: A Managerial Perspective," Harvard Business School, Working Paper, 1984.

[48] U. S. Karmarkar, S. Kekre, and S. Kekre, "Lotsizing in Multi-Item Multi-Machine Job Shops," Univ. of Rochester, Grad. School of Mgt., Working Paper QM8402, Mar. 1984.

[49] J. G. Kimemia, "Hierarchical Control of Production in Flexible Manufacturing Systems," MIT Laboratory for Information and Decision Systems, Rep. LIDS-TH-1215.

[50] J. G. Kimemia and S. B. Gershwin, "An Algorithm for the Computer Control of Production in Flexible Manufacturing Systems," *IEE Trans.*, vol. 15, no. 4, pp. 353–362, Dec. 1983.

[51] O. Kimura and H. Terada, "Design and Analysis of Pull System: A Method of Multistage Production Control," *Int. J. Prod.*

Res., vol. 19, no. 3, pp. 241–253, 1981.
[52] L. Kleinrock, *Queuing Systems,* John Wiley, 1975.
[53] N. Kulatilaka, "A Managerial Decision Support System to Evaluate Investments in FMSs," *Annals of Oper. Res.*, 1985.
[54] A. M. Law, "Statistical Analysis of Simulation Output Data," *Oper. Res.*, vol. 31, no. 6, pp. 983–1029, Dec. 1983.
[55] J. B. Lasserre, "Étude de la Planification à Moyen Terme d'une Unite de Fabrication," thesis presented to Université Paul Sabatier for the degree of Docteur Ingenieur, 1978.
[56] J. B. Lasserre and F. Roubellat, "Une Methode Rapide de Resolution de Certaines Programmes Lineaires a Structure en Escalier," *RAIRO Oper. Res.*, vol. 14, no. 2, pp. 171–191, May 1980.
[57] L. C. Leung and J. M. A. Tanchoco, "Replacement Decision Based on Productivity Analysis—An Alternative to the MAPI Method," *J. Manuf. Sys.*, vol. 2, no. 2, pp. 175–188, 1983.
[58] H. C. Meal, "Putting Production Decisions Where They Belong," *Harvard Business R.*, vol. 62, no. 2, pp. 102–111, Mar. 1984.
[59] M. S. Meketon, "Optimization in Simulation: A Tutorial," presented at Winter Simulation Conf., WSC83, Dec. 1983.
[60] G. J. Olsder and R. Suri, "Time-Optimal Control of Flexible Manufacturing Systems with Failure Prone Machines," *Proc. 19th IEEE Conf. Dec. and Contr.*, Dec. 1980.
[61] *Operations Research,* Special Issue on Simulation, vol. 31, no. 6, 1983.
[62] M. Reiser and S. S. Lavenberg, "Mean Value Analysis of Closed Multichain Networks," *J. ACM*, vol. 27, pp. 313–323, Apr. 1980.
[63] R. J. Schonberger, *Japanese Manufacturing Techniques,* Free Press, 1982.
[64] R. J. Schonberger, "Integration of Cellular Manufacturing and Just-In-Time Production," *Industrial Engineering*, vol. 15, no. 11, pp. 66–71, Nov. 1983.
[65] T. Schriber, "The Use of GPSS/H in Modeling a Typical Flexible Manufacturing System," *Annals of Oper. Res.*, 1985.
[66] P. J. Schweitzer, "Approximate Analysis of Multiclass Closed Networks of Queues," presented at Int. Conf. on Stochastic Contr. and Optimization, Amsterdam, 1979.
[67] A. Seidman and P. J. Schweitzer, *Real-Time On-Line Control of a FMS Cell,* Working Paper QM8217, Grad. School of Manag., Univ. of Rochester, New York, 1982.
[68] B. A. Sevastyanov, "Influence of Storage Bin Capacity on the Average Standstill of a Production Line," *Theory Prob. Appl.*, pp. 429–438, 1962.
[69] S. Shalev-Oren, A. Seidman, and P. J. Schweitzer, "Analysis of Flexible Manufacturing Systems with Priority Scheduling: PMVA," *Annals of Oper. Res.*, 1985.
[70] J. G. Shanthikumar, "Approximate Queuing Models of Dynamic Job Shops," Ph.D. Thesis, Dept. of IE, Univ. of Toronto, 1979.
[71] J. J. Solberg, *CAN-Q User's Guide,* Report No. 9 (Revised), School of IE, Purdue Univ., W. Lafayette, Indiana, 1980.
[72] K. E. Stecke, *Production Planning Problems for Flexible Manufacturing Systems,* Ph.D. Dissertation, School of IE, Purdue Univ., W. Lafayette, Indiana, 1981.
[73] K. E. Stecke, "Formulation and Solution of Nonlinear Integer Production Problems for Flexible Manufacturing Systems," *Manag. Sci.*, vol. 29, no. 3, pp. 273–288, Mar. 1983.
[74] R. Suri, *Resource Management Concepts for Large Systems,* Pergamon Press, 1981.
[75] R. Suri, "Robustness of Queuing Network Formulae," *J. ACM*, vol. 30, no. 3, pp. 564–594, July 1983a.
[76] R. Suri, "Infinitesimal Perturbation Analysis of Discrete Event Dynamic Systems: A General Theory," *Proc. 22nd IEEE Conf. Dec. and Contr.*, Dec. 1983b.
[77] R. Suri, "Quantitative Techniques for Robotic Systems Analysis," Chapter in *Handbook of Industrial Robotics* (S. Y. Nof, ed.), Wiley, 1984b.
[78] R. Suri, "An Overview of Evaluative Models for Flexible Manufacturing Systems," *Annals of Oper. Res.*, 1985.
[79] R. Suri and S. DeTreville, "Getting from Just-In-Case to Just-In-Time: Insights from a Simple Model," *J. Oper. Manag.*, 1985.
[80] R. Suri and G. W. Diehl, "A Variable Buffer Size Model and Its Use in Analyzing Closed Queuing Networks with Blocking," *Manag. Sci.*, 1985.
[81] R. Suri and R. R. Hildebrant, "Modeling Flexible Manufacturing Systems Using Mean Value Analysis," *J. Manuf. Sys.*, vol. 3, no. 1, pp. 27–38, 1984.
[82] R. Suri and C. K. Whitney, "Decision Support Requirements in Flexible Manufacturing," *J. Manuf. Sys.*, vol. 3, no. 1, pp. 61–69, 1984.
[83] R. Suri and M. A. Zazanis, "Perturbation Analysis Gives Strongly Consistent Estimates for the M/G/1 Queue," *Manag. Sci.*, to appear in 1986.
[84] T. Tabe, R. Muramatsu, and Y. Tanaka, "Analysis of Production Ordering Quantities and Inventory Variations in a Multistage Ordering System," *Int. J. of Prod. Res.*, vol. 18, no. 2, pp. 245–257, 1980.
[85] R. J. Tersine, *Production and Operations Management,* North Holland, 1980.
[86] C. K. Whitney, "Control Principles in Flexible Manufacturing," submitted to *J. Manuf. Sys.*, 1984.
[87] C. K. Whitney and R. Suri, "Algorithms for Part and Machine Selection in Flexible Manufacturing Systems," *Annals of Oper. Res.*, 1985.
[88] D. D. W. Yao and J. A. Buzacott, "Modeling a Class of Flexible Manufacturing Systems with Reversible Routing," Tech. Rep. 8302, Dept. of IE, Univ. of Toronto, Aug. 1983.
[89] M. A. Zazanis and R. Suri, "Comparison of Perturbation Analysis with Conventional Sensitivity Estimates for Regenerative Stochastic Systems," submitted to *Oper. Res.*, 1985.
[90] B. Zimmern, "Etudes de la Planification des arrets aleatoires dans les chaines de production," *Revue de statistique appliquee,* vol. 4, pp. 85–104, 1956.

Stanley B. Gershwin received the B.S. degree in engineering mathematics from Columbia University, New York, in 1966, and the M.A. and Ph.D. degrees in applied mathematics from Harvard University, Cambridge, Massachusetts, in 1967 and 1971. He is a Principal Research Scientist at the MIT Laboratory for Information and Decision Systems and a Lecturer in the MIT Department of Electrical Engineering and Computer Science. He is also Assistant Director of the MIT Laboratory for Information and Decision Systems, and President of Technical Support Software, Inc. (TSSI). In 1970–1971, he was employed by the Bell Telephone Laboratories in Holmdel, New Jersey, where he studied telephone hardware capacity estimation. At the Charles Stark Draper Laboratory in Cambridge, Massachusetts, from 1971 to 1975, Dr. Gershwin investigated problems in manufacturing and in transportation. His interest in these areas, as well as control, optimization and estimation, continues at MIT and at TSSI. He has studied traffic assignment, the measurement of traffic-flow parameters, and related areas in transportation. He has investigated routing optimization in networks of machine tools and the effect of limited buffer storage space on transfer lines and assembly networks. During June 1981, Dr. Gershwin was the guest of the French scientific agency (CNRS) and visited laboratories in Toulouse, Bordeaux, Paris, and Valenciennes to observe their research in manufacturing systems.

Richard R. Hildebrant is a member of the research staff at the Charles Stark Draper Laboratory in Cambridge, Massachusetts, where he is Chief of the Industrial Operations Methods Section. His background in modern and classical control theory has been applied to the design and analysis of aircraft, navigation, and other aerospace systems. More recently, his professional interests have focused on the application of operations research, AI, computer science, manufacturing engineering, and related disciplines to the development of scheduling and control software for manufacturing systems. His expertise, ranging from the design, analysis, and specification of complex manufacturing and assembly systems to the development and implementation of real-time scheduling and control software, has been employed by numerous major corporations undergoing both large and small CAM projects. Dr. Hildebrant received his Ph.D. from the Massachusetts Institute of Technology in 1980.

Rajan Suri is currently Associate Professor of Industrial Engineering at the University of Wisconsin-Madison. He received his bachelor's degree (1974) from Cambridge University (England) and his M.S. (1975) and Ph.D. (1978) from Harvard University. In 1981, he received the IEEE Donald P. Eckman Award for outstanding contributions in his field. His current interests are in modeling and decision support for manufacturing systems, specializing in flexible manufacturing systems. He is the author of many journal publications and several books, and has chaired international conferences on this subject. He is a consultant to major industrial corporations and also a principal of Network Dynamics, Inc., a Cambridge Massachusetts, firm specializing in software for modeling manufacturing systems.

Sanjoy Mitter was born in Calcutta, India, and educated in India and England. He received a Ph.D. in automatic control from Imperial College of Science and Technology, University of London, in 1965. He has taught at Case Western Reserve University and has been at MIT since 1969, where he is currently Professor of Electrical Engineering and Director, Laboratory for Information and Decision Systems. He has held visiting appointments at numerous institutions in Asia, Europe, and the USA, including the School of Mathematics, Tata Institute of Fundamental Research, Bombay, India; Scuola Normale Superiore, Pisa, Italy; Imperial College, London; and INRIA, France. He is a Fellow of the IEEE. Dr. Mitter has made extensive contributions in research in the broad area of systems and control, statistical signal processing, and mathematical problems related to the above areas.

Chapter 2: Modeling Methods for Manufacturing Systems

2.1: Introduction

Models for automated manufacturing systems should be able to represent finite buffer capacity, random machine failures and repair rates, and should be useful for calculating the various performance measures described in Chapter 1. This chapter reviews several well established methods for this modeling problem. It concentrates on the Markov chain approach and models based on queueing theory. More recent research in manufacturing system modeling has focused on perturbation analysis and the Petri net approach. These topics will be treated separately in Chapters 4 and 5.

The modeling process can be approached from many different mathematical points of view. It is not appropriate to say that one model is better than another without first asking: What are the issues being studied? It is quite acceptable to use many different models for the same system to study different aspects of that system. In fact, it is quite likely that at one time or another, we will need models from all of the chapters of this text.

2.2: "Markovian Modeling of Manufacturing Systems" (R.P. Davis and W.J. Kennedy, Jr.)

This section gives a brief review of discrete-state discrete-transition *Markov chains*. It is intended to provide the necessary background for reading the reprint by Kennedy and Davis. Additional details about Markov processes can be found in [1–4].

This reprint has been included because it illustrates a very basic approach to the modeling of manufacturing systems. It also helps us appreciate the more complex methods presented in this chapter and it is essential for understanding the Petri net methods in Chapter 5.

2.2.1: System State

A Makov model consists of a set of discrete states. This set is exhaustive and describes all of the possible states in which the system can be. *Transitions* from state i to j occur with probability p_{ij}.

Example: The state of a manufacturing system can be defined by considering the simple transfer line that consists of two machines (M_1 and M_2) and one buffer as shown in Figure 2.1.

Each machine is described by a processing time, a time to fail, and a time to repair. These quantities could be deterministic in which case we obtain a discrete-state discrete-transission Markov chain (or a discrete-time Markov chain). They could also be random variables that would lead to a *continuous-time Markov process*. This section is concerned only with Markov chains leaving the continuous-time case for the next section.

The first problem is to model this simple transfer line as a Markov chain. We assume that the first machine is never starved and that the last machine is never blocked.

We can indicate the basic idea of a Markov chain by modeling just one of the machines. In this case, the state of a machine is described by its status, α_i, where

$$\alpha_i = \begin{cases} 1 & \text{machine } i \text{ is up} \\ 0 & \text{machine } i \text{ is under repair} \end{cases}$$

This leads to the simple Markov chain shown in Figure 2.2. Note that the transition from one state to another is related to the probability of failure (or repair).

The state of the transfer line consists of the status of each machine plus the amount of material present in the system. Specifically, the state of the transfer line is

$$s = (n, \alpha_1, \alpha_2)$$

where

$$n = \text{\# pieces in the buffer} + \text{\# pieces in machine 2}$$
$$0 \leq n \leq N$$

Since, n, α_1, and α_2 are all integers, the system is described by a set of mutually exclusive and collectively exhaustive states $s_1, s_2, \dots s_m$. The system can only be in one of these states at any given instant of time.

The system may undergo a change of state (or a *state transition*) at discrete instants of time according to a set of probabilities. Let

$$P[s_i(k)] = \text{probability the system is in state } s_i \text{ at time } k$$

Now we can say that a change of state will occur with probability,

Figure 2.1: Transfer Line

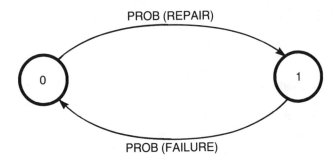

Figure 2.2: Markov Model for a Simple Machine

$$P[s_j(k) \mid s_a(k-1), s_b(k-2), s_c(k-3) \ldots]$$
$$1 \leq j,a,b,c, \leq m$$
$$k = 1,2,3,\ldots$$

which is referred to as the transition probability.

2.2.2: The Markov Condition

If $P[s_j(k) \mid s_a(k-1), s_b(k-2), s_c(k-3) \ldots] = P[s_j(k) \mid s_a(k-1)]$ for all $k,j,a,b,c\ldots$, then the system is a discrete-state discrete-transition Markov process. The implication of this condition is that the history of the system prior to its arrival at a has no affect on the transition to j. In a sense, the system has no memory.

2.2.3: State Transition Probabilities

For a Markov chain, we define the *state transition probabilities* as

$$p_{ij} = P[s_j(k) \mid s_i(k-1)] \quad 1 \leq i,j \leq m$$

and each p_{ij} is independent of k. These probabilities can be included in a *state transition matrix*,

$$P = \begin{pmatrix} p_{11} & p_{12} & \cdots & p_{1m} \\ p_{21} & p_{22} & \cdots & p_{2m} \\ \cdot & \cdot & & \cdot \\ \cdot & \cdot & & \cdot \\ \cdot & \cdot & & \cdot \\ p_{m1} & p_{m2} & \cdots & p_{mm} \end{pmatrix}$$

Also note that the state transition probabilities satisfy

$$0 \leq p_{ij} \leq 1$$

and

$$\sum_{j=1}^{m} p_{ij} = 1 \quad i = 1,2,\ldots m$$

because the set of states is mutually exclusive and collectively exhaustive.

The matrix, P, provides a complete description of the Markov process and it will be used to answer numerous questions about the system. For example,

Q1. What is the probability that the system will occupy s_i after k transitions if its state at $k = 0$ is known?

We can answer this question by finding a difference equation that relates the next state to the present state. We start with,

$$P[s_j(k+1)] = P[s_1(k)] p_{1j} + P[s_2(k)] p_{2j} + \ldots P[s_m(k)] p_{mj}$$

where

$P[s_j(k+1)] =$ probability that the system is in state s_j at time $k+1$.

In the equation above, $j = 1,2,\ldots m$. If we write out all m of these, we can put them in matrix form and obtain

$$(P[s_1(k+1)] \; P[s_2(k+1)] \ldots P[s_m(k+1)]) =$$

$$(P[s_1(k)] \; P[s_2(k)] \ldots P[s_m(k)]) \cdot$$

$$\begin{pmatrix} p_{11} & p_{12} & \cdots & p_{1m} \\ p_{21} & p_{22} & \cdots & p_{2m} \\ \cdot & \cdot & & \cdot \\ \cdot & \cdot & & \cdot \\ \cdot & \cdot & & \cdot \\ p_{m1} & p_{m2} & \cdots & p_{mm} \end{pmatrix}$$

or

$$P(k+1) = P(k)P \quad k = 0,1,2,\ldots$$

To answer question Q1, we solve this matrix difference equation by induction,

$$P(1) = P(0)P$$
$$P(2) = P(1)P = P(0)P^2$$
$$\cdot$$
$$\cdot$$
$$\cdot$$

which results in

$$P(k) = P(0)P^k \quad k = 0,1,2,\ldots$$

Example: Find $P(k)$ for the Markov chain in Figure 2.3.
Solution: From the state diagram,

$$P = \begin{pmatrix} .5 & .5 \\ .4 & .6 \end{pmatrix}$$

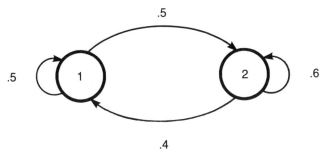

Figure 2.3: A Two-State Markov Chain

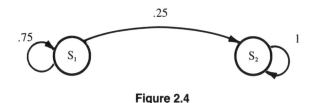

Figure 2.4

To find $P(k)$, we need P^k, which can be found from the Cayley-Hamilton [2] technique as

$$P^k = \frac{(.1)^{k-1} - 1}{9} I + \frac{10 - (.1)^{k-1}}{9} P$$

where I is the 2×2 identity matrix. This can be evaluated for a specific k, then

$$P(k) = P(0)P^k .$$

One case of particular interest is the one in which k goes to infinity. For this example,

$$P^k = \begin{pmatrix} 4/9 & 5/9 \\ 4/9 & 5/9 \end{pmatrix} \text{ as } k \longrightarrow \infty$$

Cleary, P^k approaches a constant, this also leads to another important property that can be seen by letting $P(0) = (1,0)$. Then $P(k) = (4/9, 5/9)$ as k goes to infinity. Similarly, if $P(0) = (0,1)$, then $P(k)$ remains the same. Thus, the *limiting state probabilities* are independent of the initial state. Many Markov processes exhibit this property and are referred to as *ergodic processes*.

2.2.4: State Classification

Limiting state probabilities: if as $k \longrightarrow \infty$, $P(k)$ approaches a constant then the limiting state probabilities exist and are independent of the initial conditions.

Let π be $1 \times m$ vector defined as

$$\lim_{k \to \infty} P(k) = \pi$$

then π must satisfy

$$\pi = \pi P$$

For the previous example, this equation can be solved along with the constraint $\sum_i^m \pi_i = 1$ to obtain

$$(\pi_1 \quad \pi_2) = (4/9 \quad 5/9)$$

Transient State: s_i is a *transient state* if you can leave the state but never return to it. Consider Figure 2.4. In this figure, s_1 is a transient state. Note that its limiting state probability is 0.

Trapping State: s_i is a *trapping state* if the system enters the state and remains there. In Figure 2.4, s_2 is a trapping state and its limiting state probability is 1.

Recurrent Chain: A *recurrent chain* is a set of states from which the system cannot exit. The system makes jumps within this set indefinitely.

Finally, we summarize some useful Markov chain facts:

- Every Markov process must have at least one recurrent chain
- A recurrent chain is a generalized trapping state
- A Markov process that has one recurrent chain must be completely ergodic since wherever the process starts it will end up in the recurrent chain.
- If a process has two or more recurrent chains then the ergodic property no longer holds.
- A transient state is a state the system occupies before commiting itself to a recurrent chain.

These are all of the concepts necessary to duplicate the results in the Davis and Kennedy reprint. How true to life are these models? What are their advantages and disadvantages?

2.3: Analysis of Transfer Lines

This section introduces the concepts and tools that are prerequisites for Gershwin and Berman's "Analysis of Transfer Lines Consisting of Two Unreliable Machines with Random Processing Times and Finite Storage Buffers."

2.3.1: Continuous-Time Markov Chains

The Markov models of Section 2.2 are discrete time. As a result, the transitions occur at uniformly spaced intervals of time; hence, the name discrete-transition Markov chains. Since in many cases, processing, failure, and repair rates are random variables, we would like to be able to handle transitions that occur at random time intervals. This requires the theory of continuous time Markov chains, which is used extensively in the Gershwin and Berman reprint.

2.3.2: Transition Rates and the State Probabilities

In continuous-time Markov chains, we deal with *transition rates* as opposed to state transition probabilities.

Definition: a_{ij} is the transition rate of a process from state i to state j, $i \neq j$. A is the transition rate matrix.

Definition: Given that the present state is s_i, the conditional probability that $s_i \rightarrow s_j$ during the time interval dt is $a_{ij} \, dt$. (Note that a_{ij} has units of sec^{-1} and so a_{ij}^{-1} = expected time between transitions.)

Let,

$P[s_i(t)]$ = probability that the system occupies state i at time t.

Find the probability of being in state j at time $t + dt$.

First we note that

$P[s_j(t + dt)] = P[s_j(t)]$ (probability of making no transition from j)
$+ P[s_1(t)] a_{1j} \, dt + P[s_2(t)] a_{2j} \, dt \ldots$
$\ldots + P[s_{j-1}(t)] a_{j-1,j} \, dt + P[s_{j+1}(t)] a_{j+1,j} \, dt$
$\ldots + P[s_m(t)] a_{mj} \, dt$

Next, the probability of making no transition from j = 1 − probability of making a transition from j

$$= 1 - \sum_{i \neq j}^{m} a_{ji} \, dt$$

Substituting and rewriting yields,

$$P[s_j(t + dt)] = P[s_j(t)] \left(1 - \sum_{i \neq j}^{m} a_{ji} \, dt\right)$$
$$\sum_{i \neq j}^{m} P[s_i(t)] a_{ij} \, dt$$
$$j = 1, 2, \ldots m$$

The goal is to find a solution to this equation. First, define the diagonal elements of the *transition-rate matrix* A by,

$$a_{jj} = - \sum_{i \neq j}^{m} a_{ji}$$

then

$$P[s_j(t + dt)] = P[s_j(t)](1 + a_{jj} dt) + \sum_{i \neq j}^{m} P[s_i(t)] a_{ij} \, dt$$

or

$$P[s_j(t + dt)] - P[s_j(t)] = \sum_{i=1}^{m} P[s_i(t)] a_{ij} \, dt$$

Dividing by dt and taking the limit leads to

$$\frac{d}{dt} P[s_j(t)] = \sum_{i=1}^{m} P[s_i(t)] a_{ij} \qquad j = 1, 2 \ldots m$$

Next, define

$$P(t) = (P[s_1(t)] \quad P[s_2(t)] \ldots P[s_m(t)])$$

which results in

$$\frac{d}{dt} P(t) = P(t) A$$

which has solution

$$P(t) = P(0) e^{At} \qquad (2.1)$$

Example: Transition rates and the exponential distribution.

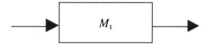

M_1 may be either working (state 1) or not working (state 2). If it is working, there is a probability $5dt$ that it will break down in a short interval dt; if it is not working, there is a probability $4dt$ that it will be repaired in dt.

Find the probability that the machine will be operating at time t, if it is operating at $t = 0$.

Remark: $5dt$ and $4dt$ are equivalent to saying that the operating time between breakdowns is an exponentially distributed random variable with mean 1/5, while the time for repair is exponentially distributed with mean 1/4. This can be seen by examining the properties of *the exponential distribution*, which is given by

$$p(\tau) = \frac{1}{\bar{\tau}} e^{-\tau/\bar{\tau}}$$

For this distribution, $E(\tau) = \bar{\tau}$, which leads to the following important points:

- The random variable in the exponential distribution is the time interval between adjacent events
- The average time interval between adjacent events is $\bar{\tau}$
- The rate at which events occur is $1/\bar{\tau}$

Solution: Let the time between failures (time under repair) be an exponentially distributed random variable with rate $p_i(r_i)$. M_1 can then be modeled as shown in Figure 2.5. This process has the state transition rate matrix of

$$A = \begin{pmatrix} -5 & 5 \\ 4 & -4 \end{pmatrix}$$

Next, solve $P(t) = P(0) e^{-At}$ where $P(0)$ has been given as $(1 \quad 0)$. This leads to the solution where

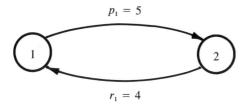

Figure 2.5

$$P(t) = (4/9 \quad 5/9) + e^{-9t}(5/9 \quad -5/9)$$

and so

$$P(\infty) = (4/9 \quad 5/9)$$

2.3.3: "Analysis of Transfer Lines Consisting of Two Unreliable Machines with Random Processing Times and Finite Storage Buffers" (S.B. Gershwin and O. Berman)

In the Gershwin and Berman article, the machines are described by processing, failure, and repair rates, which are exponentially distributed random variables with rates μ_i, p_i, and r_i. The states are described by $s(n,\alpha_1,\alpha_2)$ and it is desired to calculate $p(n,\alpha_1,\alpha_2)$, the probability of being in state $s(n,\alpha_1,\alpha_2)$. From these probabilities, performance measures, such as system production rate, machine efficiency, average in-process inventory, probability machine 1 is blocked, etc., can be calculated. These performance measures require steady-state probabilities that can be obtained from the balance equations.

2.3.4: The Balance Equations

The *balance equations* are a very important result obtained from the steady-state solution to equation (2.1). In steady state

$$0 = \dot{P}(t) A = \pi A$$

This leads to the conclusion that (in steady state) the rate of leaving state (n,α_1,α_2) = rate of entering it

This can be easily verified for the previous example. The balance equations are used to establish equation 4.2 in the Gershwin and Berman reprint. Once the balance equation has been set up, one would like to find a closed-form expression for $p(n,\alpha_1,\alpha_2)$. In general, this is a very difficult problem. However for the two machine one-buffer transfer line, the authors are able to obtain such an explicit expression.

2.3.5: Solving for the Steady-State Probabilities

Finding a closed-form expression for $p(n,\alpha_1,\alpha_2)$ involves using the balance equations. A good "guess" for the solution is to try

$$p(n,\alpha_1,\alpha_2) = p(n)\, p(\alpha_1)\, p(\alpha_2)$$

that is, take the joint distribution as the product of the individual distributions where

$$p(n) = c_0 X^n$$
$$p(\alpha_1) = c_1 Y_1^{\alpha_1}$$
$$p(\alpha_2) = c_2 Y_2^{\alpha_2}$$

or

$$p(n,\alpha_1,\alpha_2) = C\, X^n\, Y_1^{\alpha_1} Y_2^{\alpha_2}.$$

To find C, X, Y_1, and Y_2 we need some definitions. The states $s(n,\alpha_1,\alpha_2)$ are defined for $0 \leq n \leq N$. The states corresponding to $n = 0, N$ are called *Boundary states* and those described by $1 \leq n \leq N - 1$ are internal states. These can be used to classify the balance equation into *internal equations* (balance equations that contain only internal states) and *boundary equations* (balance equations that contain only boundary states). The internal equations are used to find X, Y_1, Y_2 while the boundary equations are used to find C. This results in a closed-form solution that is then used to study the effect of varying μ_i, p_i, and r_i on several performance measures.

2.3.6: "Summary of the Analysis of Transfer Lines Consisting of Two Unreliable Machines with Random Processing Times and Finite Storage Buffers" (S.B. Gershwin and O. Berman)

The authors provide a very thorough analysis of a two-machine transfer line, which is a very fundamental component of a manufacturing system. The reason for including this paper is because the authors obtain a closed-form solution that provides a great deal of insight into the operation of this system. Numerous graphical results are presented for this purpose.

2.4: "Models for Understanding Flexible Manufacturing Systems" (J.A. Buzacott and J.G. Shanthikumar)

Transfer lines are examples of fixed automation since only one part path is usually allowed. In a flexible manufacturing system, material handling is possible. This allows a path to follow several possible routes. This feature, along with complete computer control of the system, gives rise to the name FMS.

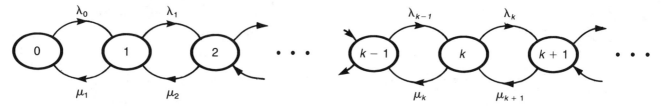

Figure 2.6: The Birth-Death Process

Buzacott and Shanthikumar survey models based on Markov chains and *queueing theory*. Many of the theoretical concepts of Section 2.2 are applicable here (e.g., exponential distributions). However, some additional mathematical background should be helpful.

2.4.1: Modeling Assumptions

The complete system consists of the FMS and a queue of jobs waiting for processing. One of the assumptions is that jobs arrive at the queue according to a *Poisson process* with parameter λ.

The Poisson process: The Poisson process can be derived from the continuous-time *birth-death process* [3], which is a special case of a Markov process in which transitions from state E_k are permitted only to neighboring states E_{k+1} or E_{k-1} as shown in Figure 2.6. Note that this is a continuous-time process in a discrete-state space. At this point, the reader should accept the fact that the birth-death process is capable of providing a lot of important and interesting results in queueing theory [1-3].

The simplest case to consider is the pure birth system where we assume $\mu_k = 0$ for all k. In terms of our modeling problem, this corresponds to the arrival of jobs at some FMS (like the birth of new members into a population). We now wish to find the probability that k arrivals occur during the time interval $(0,t)$.

Referring to Figure 2.6, we again use the fact that the

$$\text{rate of leaving state } E_k = \text{rate of entering it}$$

which leads to

$$\lambda_{k-1} P_{k-1}(t) = \lambda_k P_k(t) \qquad (2.3.1)$$

where $P_k(t)$ is the probability of being in state k at time t. The difference between the two flow rates is the probability flow rate into this state, or

$$\frac{dP_k(t)}{dt} = \lambda_{k-1} P_{k-1}(t) - \lambda_k P_k(t) \qquad (2.3.2)$$

To simplify the solution to (2.3.2), we further assume that

$$\lambda_k = \lambda$$

or

$$\frac{dP_k(t)}{dt} = -\lambda P_k(t) + \lambda P_{k-1}(t), k \geq 0 \qquad (2.3.3)$$

$$\frac{dP_o(t)}{dt} = -\lambda P_o(t), \quad k = 0$$

Assuming that the system has no jobs at time 0, then

$$P_k(0) = \begin{cases} 1 & k = 0 \\ 0 & k \neq 0 \end{cases}$$

We can now solve for $P_0(t)$ to obtain

$$P_0(t) = e^{-\lambda t}$$

Inserting this into (2.3.3) and setting $k = 1$,

$$\frac{dP_1(t)}{dt} = -\lambda P_1(t) + \lambda e^{-\lambda t}$$

which has the solution

$$P_1(t) = \lambda t e^{-\lambda t}$$

Repeating these substitutions for $P_2(t), P_3(t), \ldots$ leads to the solution of (2.3.3) as

$$P_k(t) = \frac{(\lambda t)^k}{k!} e^{-\lambda t}, k \geq 0, t \geq 0 \qquad (2.3.4)$$

This is the *Poisson distribution* which is central to queueing theory. The Poisson process appears often in nature especially where there is an aggregate effect of independent individuals (or jobs, customers, particles, automobiles, etc.) under observation.

$P_k(t)$ gives the probability that k arrivals occur during the time interval $(0,t)$. An important feature of the Poisson process is that the average number of arrivals in an interval of length t is λt, that is,

$$E[K] = \sum_{k=0}^{\infty} k P_k(t) = \lambda t$$

where K is the number of arrivals in the interval of length t. This is also an intuitive result since the average arrival rate is λ per second, then the average number of arrivals in an interval of length t must be λt. It can also be shown that the variance of the Poisson process is equal to λt.

Exponentially distributed processing times: Another assumption is that the processing times are all exponentially distributed with mean $1/\mu_i$. As it turns out, there is a relationship between the Poisson process and the exponential distribution. In fact, for a Poisson arrival process, the time *between* arrivals is exponentially distributed (with parameter λ).

Several other results from queueing theory will be presented here to facilitate the understanding of the models.

2.4.2: The A/B/m/K/M Queue

Many types of queues exist depending on

A— the distribution of the interarrival times of jobs
B— the distribution of the processing time for each job
m— the number of machines (servers)
K— the buffer size (storage capacity)
M— the available number of jobs (customer population)

For example, the M/M/1 queue indicates there is one machine where jobs arrive according to a Poisson process and where they are processed according to an exponential distribution (signified by M). Also, the notation indicates infinite storage space (since K is missing) and that infinite (no M) jobs are available, meaning that the machine is never starved. It has the state space diagram shown in Figure 2.6.

Analysis of the M/M/1 queue: Let $\lambda_k = \lambda$ and $\mu_k = \mu$. When using the flow rate concept, the equations for the steady state probability distribution are

$$p(k)\lambda = p(k+1)\mu \quad k = 0,1,\ldots N-1$$
$$\sum_{k=0}^{N} p(k) = 1 \quad (2.3.5)$$

We will later let N go to infinity and obtain the average number of jobs in the system.

Solution: Assume $p(k) = c\rho^n$, and substitute it into (2.3.5) and solve for ρ to obtain,

$$\rho = \frac{\lambda}{\mu}$$

To find c, use the second equation in (2.3.5) to obtain

$$c = (\rho - 1)/(\rho^{N+1} - 1)$$

The average buffer level can be found from

$$\bar{n} = \sum_{k=0}^{N} kp(k)$$

After much algebra, this leads to

$$\bar{n} = \frac{\rho}{\rho - 1} \frac{N\rho^{N+1} - (N+1)\rho^N + 1}{\rho^{N+1} - 1} \quad (2.3.6)$$

If $\rho < 1$, then the limit of (2.3.6) as N goes to infinity is

$$\frac{\rho}{1 - \rho}$$

Buzacott and Shanthikumar use this in the section of their paper entitled: "Determining inventory levels and waiting times." They also make use of Little's Law [3].

Little's Law: This law states that the average number of customers in a queueing system ($\bar{N}$) is equal to the average arrival rate of customers to that system (λ) times the average time spent in that system (T), that is,

$$\bar{N} = \lambda T$$

This is used in the reprint to calculate mean-inventory levels and mean-waiting times.

Remark: It is worthwhile to review the assumptions of the model. First, it assumes infinite buffer capacity. Second, it doesn't take into account random-machine failures and repairs. These models seem to be best suited for a first approximation to the manufacturing system behavior.

2.4.3: "Summary of Models for Understanding Flexible Manufacturing Systems" (J.A. Buzacott and J.G. Shanthikumar)

This reprint presents simple models for understanding the relative merits of several manufacturing configurations. It introduces balanced systems, conventional job shops, single-class job shops, flow shops, pure job shops, and a two machine-two class job shop. Common modeling assumptions are introduced. The reprint discusses the importance of job routing, implications of common buffers, local buffers, and the effects of breakdowns. In general, the paper is a comprehensive treatment of the detailed aspects of modeling that one needs to understand before considering the control issues.

In a more recent paper, Buzacott and Yao [36] provide a qualitative review of several analytical models for flexible manufacturing systems. The authors point out that it is difficult for one single model to describe fully the behavior of a flexible manufacturing system. Instead models, which address a specific set of issues, are devised. The work of several research groups are assessed for their strengths and weaknesses.

2.5: "Modeling FMS by Closed Queueing Network Analysis Methods" (G. Menga, G. Bruno, R. Conterno, and M. Actis Dato)

Another technique for modeling automated-manufacturing systems is based on the theory of networks of queues. The results developed in this paper are based on these ideas. As a result, we will summarize this important background material here; details and derivations can be found in Kleinrock [3].

2.5.1: Networks of Queues

In a network of queues, jobs enter the system at various points, queue for service, and upon departing from a machine, the job proceeds to another machine for additional service (or processing). In these problems, the topology of the network becomes important because it describes the permissible transitions between the machines. Also, the routes taken by the individual jobs now must be described.

Manufacturing systems can be modeled as either open or closed queueing networks. In an *open queueing system*, jobs arrive at the system with a certain rate and eventually exit from the network, never to return again. In a *closed network*, it is assumed that there is a fixed and finite number of customers in the system and that they are trapped there in the sense that no other jobs may enter or leave. A special case of these systems is a *cyclic queue*, where the last stage is connected back to the first.

2.5.2: Jackson Networks

Jackson [42] considered an arbitrary (open) network of queues. His system consisted of N nodes where the i-th node consisted of m_i exponential servers (more than one machine permitted at a node) each with parameter μ_i. In addition, the i-th node receives arrivals from outside the system as a Poisson process at rate γ_i. After leaving the i-th node, a customer (job) proceeds to the j-th node with probability r_{ij}. The problem is to calculate the total average-arrival rate of jobs at a given node. What is significant about Jackson's work, is that he was able to show that each node in the network behaves as if it were an independent M/M/m system with a Poisson input rate λ_i.

Jackson was also able to show some interesting results in regard to the steady-state probability distribution of the number of customers at each of the N nodes. First, define the state of this N-node system as the vector $(k_1, k_2, \ldots k_N)$ where k_i is the number of customers at the i-th node. Jackson then showed that the steady-state probability distribution associated with this i state vector is

$$p(k_1, k_2, \ldots k_N) = p_1(k_1) p_2(k_2) \ldots p_N(k_N)$$

where $p_i(k_i)$ is the marginal distribution of finding k_i customers at the i-th node. Furthermore, $p_i(k_i)$ turns out to be the solution to the classical M/M/m system. This result is usually called *Jackson's Theorem*. Note the similarity of this result with the approach taken by Gershwin and Berman in Section 2.3.3. Both solutions are the so-called *product form* solution for Markovian queues in steady state.

2.5.3: Mean Value Analysis

Menga et al. modify the mean-value analysis algorithm of Reiser [43], who in his work has outlined a procedure for performance evaluation of multiclass single-server closed-queueing networks with general service times and a first-come first-served service rule. His method is summarized here while additional details can be found in [43].

We start with a simple closed chain consisting of N servers in series with the output of the last server connecting to the input of the first server. Then for each queue i, $i = 1, 2, \ldots N$, we can define the steady state quantities:

W—total number of customers in the chain

τ_i—mean service time

$n_i(W)$—mean queue length (including customers in service)

$t_i(W)$—mean queueing time (including customers in service)

$\lambda(W)$—throughput of the chain

From these definitions, the following relations can be intuitively obtained,

$$t_i(W) = \tau_i + \tau_i^\times \text{ (mean queue length of queue } i \text{ upon arrival)} \quad (2.4.1)$$

This states that an arriving job stays in the queue for its own service time plus the time it takes to remove the existing backlog upon arrival.

The throughput can then be calculated as,

$$\lambda(W) = W / \sum_{i=1}^{N} t_i(W) \quad (2.4.2)$$

which leads to the mean queue length of queue i,

$$n_i(W) = \lambda(W) t_i(W) \quad (2.4.3)$$

Equation (2.4.2) is simply Little's Law (Section 2.4) applied to the entire chain while (2.4.3) represents Little's Law applied to each queue.

2.5.4: Theorem

In a closed multi-chain queueing network with product form solution, an arriving job observes the steady state solution of the queueing network with one less customer in the arriving job's chain.

The proof can be found in [47]. The emphasis in this theorem is on the multi-chain case. Using the theorem, (2.4.1) becomes

$$t_i(W) = \tau_i + \tau_i n_i(W-1) \qquad (2.4.4)$$

The idea is to solve (2.4.2) through (2.4.4) recursively starting with $n_i(0) = 0, i = 1,2,...N$.

2.5.5: Related Research

Cavaille and Dubois [45] have also used mean-value analysis for the performance evaluation of flexible manufacturing systems. It allows a departure from the usual assumption of exponential processing times on machines.

Suri and Hildebrant [44] use mean-value analysis to model flexible manufacturing systems. Their paper also includes a brief tutorial on the theory of mean-value analysis.

Finally, Hildebrant [46] uses mean-value analysis to schedule flexible machining systems that are prone to failure.

2.6: "'Optimal' Operating Rules for Automated Manufacturing Systems" (J.A. Buzacott)

In this reprint, Buzacott uses a variety of modeling techniques, including dynamic programming and Markov processes [1-2], to answer a number of important questions about the control policies for operating transfer lines and flexible manufacturing systems. Markov chain models are used and solutions are obtained for the optimal operating policy for a transfer line, the optimum single repairman strategy and optimal operating sequences. In addition, the advantages of coordinated control and the effect of system status information on input control are considered.

The reader should compare the similarity between Buzacott's inventory levels and the buffer hedging points, which will be presented in Chapter 3.

The background material for this paper has been presented in Sections 2.2 and 2.3. The solution to the state-transition equations is obtained in a manner similar to Gershwin and Berman's solution (second reprint in this chapter).

2.7: References

2.7.1: Markov Processes

1. R.A. Howard, *Dynamic Programming and Markov Processes*, M.I.T. Press, Cambridge, Mass., 1960.
2. R.A. Howard, *Dynamic Probabilistic Systems*, Vol. 1 (Markov Models) and Vol. II (Semi-Markov and Decision Processes), John Wiley and Sons, Inc., New York, 1971.
3. L. Kleinrock, *Queueing Systems, Volume 1: Theory*, John Wiley and Sons, Inc., New York, 1975.
4. S.M. Ross, *Introduction to Probability Models*, 2nd edition, Academic Press, Orlando, Fla., 1980.
5. J.W. Foster, III and A. Garcia–Diaz, "Markovian Models for Investigating Failure and Repair Characteristics of Production Systems," *IIE Transactions*, Vol. 15, No. 3, Sep. 1983, pp. 202–209.
6. S.A. El-Asfouri, B.C. McInnis, and A.S. Kapadia, "Estimation of Markov Processes with Application to the Flow of Jobs in a Computer System," *IEEE Transactions on Systems, Man, and Cybernetics*, Vol. SMC-11, No. 10, Oct. 1981, pp. 718–723.
7. M. Alam and D. Gupta, "Performance Modeling and Evaluation of Computerized Manufacturing Systems Using Semi-Markov Approach," *Proceedings of the First ORSA/TIMS Conference on Flexible Manufacturing Systems*, Ann Arbor, Mich., 1984, pp. 142–153.

2.7.2: Transfer Lines (Also See Buffer Storage Allocation, Chapter 6)

8. S.B. Gershwin and I.C. Schick, "Modeling and Analysis of Three-Stage Transfer Lines with Unreliable Machines and Finite Buffers," *Operations Research*, Vol. 31, No. 2, March–April 1983, pp. 354–380.
9. S.B. Gershwin, "An Efficient Decomposition Method for the Approximate Evaluation of Tandem Queues with Finite Storage Space and Blocking," *Operations Research*, Vol. 35, No. 2, March–April 1987, pp. 354–380.
10. Y.F. Choong and S.B. Gershwin, "A Decomposition Method for the Approximate Evaluation of Capacitated Transfer Lines with Unreliable Machines and Random Processing Times," *IIE Transactions*, Vol. 19, No. 2, June 1987, pp. 150–159.
11. E. Canuto, A. Villa, and S. Rossetto, "Transfer Lines: A Deterministic Model for Buffer Capacity Selection," *ASME Journal of Engineering for Industry*, Vol. 104, May 1982, pp. 132–138.
12. T.M. Altiok and S. Stidham, Jr., "A Note on Transfer Lines with Unreliable Machines, Random Processing Times, and Finite Buffers," *IIE Transactions*, Vol. 14, No. 2, June 1982, pp. 125–127.
13. P. Caseau and G. Pujolle, "Throughput Capacity of a Sequence of Queues with Blocking Due to Finite Waiting Room," *IEEE Transactions on Software Engineering*, Vol. SE-5, No. 6, Nov. 1979, pp. 631–642.
14. A Villa, B. Fassino, and S. Rossetto, "Buffer Size Planning Versus Transfer Line Efficiency," *ASME Journal of Engineering for Industry*, Vol. 108, May 1986, pp. 105–112.
15. D.A. Castanon, B.C. Levy, and S.B. Gershwin, "Diffusion Approximations for Three Stage Transfer Lines with Unreliable Machines and Finite Buffers," *Proceedings of the 21st IEEE Conference on Decision and*

Control, Orlando, Fla., Dec. 1982, pp. 1066–1067, and *Report No. LIDS-P-1183*, M.I.T., Cambridge, Mass., March 1982.

16. W. Eva, "Dynamic Behavior of N-Stage Transfer Lines with Unreliable Machines and Finite Buffers," *Proceedings of the 1986 IEEE International Conference on Robotics and Automation*, San Francisco, Calif., April 1986, pp. 973–976.

17. B.R. Sarker, "Some Comparative and Design Aspects of Series Production Systems," *IIE Transactions*, Vol. 16, No. 3, Sept. 1984, pp. 229–239.

18. E.A. Elsayed and C.C. Hwang, "Analysis of Two-Stage Manufacturing Systems with Buffer Storage and Redundant Machines," *International Journal of Production Research*, Vol. 24, No. 1, 1986, pp. 187–201.

19. J.G. Shanthikumar, "On the Production Capacity of Automatic Transfer Lines with Unlimited Buffer Space," *AIIE Transactions*, Vol. 12, No. 3, Sept. 1980, pp. 273–274.

20. K. Okamura and H. Yamashina, "Justification for Installing Buffer Stocks in Unbalanced Two Stage Automatic Transfer Lines," *AIIE Transactions*, Vol. 11, No. 4, 1979, pp. 308–312.

21. M.H. Ammar and S.B. Gershwin, "A Partially Formulated Method for Solving Three Machine Transfer Lines," *Laboratory for Information and Decision Systems Report No. LIDS-TM-1046*, M.I.T., Cambridge, Mass., Sept. 1980.

22. M.M. Ibrahim, "Modeling and Analysis of Automated Manufacturing Systems with Focus in Equivalence and Computational Complexity," *Laboratory for Information and Decision Systems Report No. LIDS-TH-1190*, M.I.T., Cambridge, Mass., Dec. 1981.

23. S.B. Gershwin, "Representation and Analysis of Transfer Lines with Machines That Have Different Processing Rates," *Annals of Operations Research*, Vol. 9, 1987, pp. 511–530.

2.7.3: Models for Flexible Manufacturing Systems

24. K.E. Stecke, "Planning and Control Models to Analyze Problems of Flexible Manufacturing," *Proceedings of the 23rd IEEE Conference on Decision and Control*, Las Vegas, Nev., Dec. 1984, pp. 851–854.

25. C. Buyukkoc, "An Approximation Method for Feedforward Queueing Networks with Finite Buffers, A Manufacturing Perspective," *Proceedings of the 1986 IEEE International Conference on Robotics and Automation*, San Francisco, Calif., April 1986, pp. 965–972.

26. J.T. Lim and S.M. Meerkov, "Asympotic Analysis of a Simple Model of an Assembly Line," *Proceedings of the 24th Conference on Decision and Control*, Ft. Lauderdale, Fla., Dec. 1985, pp. 2012–2015.

27. K. Hitomi, M. Nakajima, and Y. Osaka, "Analysis of the Flow-Type Manufacturing Systems Using the Cyclic Queueing Theory," *ASME Journal of Engineering for Industry*, Vol. 100, Nov. 1978, pp. 468–474.

28. R.A. Marie, "An Approximate Analytical Method for General Queueing Networks," *IEEE Transactions on Software Engineering*, Vol. SE-5, No. 5, Sept. 1979, pp. 530–538.

29. J.A. Buzacott and D. Gupta, "Impact of Flexible Machines on Automated Manufacturing Systems," *Proceedings of the Second ORSA/TIMS Conference on Flexible Manufacturing Systems*, (eds. K.E. Stecke and R. Suri), Elsevier, New York, 1986, pp. 257–268.

30. J.J. Solberg and S.Y. Nof, "Analysis of Flow Control in Alternative Manufacturing Configurations," *ASME Journal of Dynamic Systems, Measurement, and Control*, Vol. 102, Sept. 1980, pp. 141–147.

31. S.B. Gershwin and M. Ammar, "Reliability in Flexible Manufacturing Systems," *Proceedings of the 1979 IEEE Conference on Decision and Control*, IEEE Press, New York, pp. 540–545.

32. D.D. Yao and J.A. Buzacott, "Models of Flexible Manufacturing Systems with Limited Local Buffers," *International Journal of Production Research*, Vol. 24, No. 1, 1986, pp. 107–118.

33. R. Suri, "An Overview of Evaluation Models for Flexible Manufacturing Systems," *Proceedings of the First ORSA/TIMS Conference on Flexible Manufacturing Systems*, Ann Arbor, Mich., 1984, pp. 8–15.

34. M.V. Kalkunte and R.G. Sargent, "Modelling Flexible Manufacturing Systems with Event Graphs," *Proceedings of the First ORSA/TIMS Conference on Flexible Manufacturing Systems*, Ann Arbor, Mich., 1984, pp. 183–198.

35. H.C. Co and R.A. Wysk, "The Robustness of CAN-Q in Modelling Automated Manufacturing Systems," *International Journal of Production Research*, Vol. 24, No. 6, 1986, pp. 1485–1503.

36. J.A. Buzacott and D.D. Yao, "Flexible Manufacturing Systems: A Review of Analytical Models," *Management Science*, Vol. 32, No. 7, 1982, pp. 890–895.

2.7.4: Queueing Network Models

37. D.D. Yao and J.A. Buzacott, "Modelling the Performance of Flexible Manufacturing Systems," *International Journal of Production Research*, Vol. 23, No. 5, 1985, pp. 945–959.

38. A. Seidmann, S. Shalev-Oren, and P.J. Schweitzer, "An Analytical Review of Several Computerized Closed Queueing Network Models of FMS," *Proceedings of the Second ORSA/TIMS Conference on Flexible Manufacturing Systems,* (eds. K.E. Stecke and R. Suri), Elsevier, New York, 1986, pp. 369–380.

39. Y. Dallery, "A Queueing Network Model of Flexible Manufacturing Systems Consisting of Cells," *Proceedings of the 1986 International Conference on Robotics and Automation*, San Francisco, Calif., April 1986, pp. 951–958.

40. R. Conterno, G. Menga, and S. Quaglino, "Performance Evaluation of FMS by Heuristic Queueing Network Analysis," *Proceedings of the 1986 International Conference on Robotics and Automation,* San Francisco, Calif., April 1986, pp. 959–964.

41. M. Ammar and S.B. Gershwin, "Equivalent Relations in Queueing Models of Assembly/Disassembly Networks," *Technical Report GIT-ICS-87/45*, Georgia Institute of Technology, Atlanta, Ga., Dec. 1987.

42. J.R. Jackson, "Networks of Waiting Lines," *Operations Research*, Vol. 5, 1957, pp. 518–521.

2.7.5: Mean Value Analysis

43. M. Reiser, "A Queueing Network Analysis of Computer Communication Networks with Window Flow Control," *IEEE Transactions on Communications,* Vol. COM-27, No. 8, Aug. 1978, pp. 1199–1209.

44. R. Suri and R. Hildebrant, "Modelling Flexible Manufacturing Systems Using Mean Value Analysis," *Journal of Manufacturing Systems*, Vol. 3, No. 1, 1984, pp. 27–38.

45. J.B. Cavaille and D. Dubois, "Heuristic Methods Based on Mean Value Analysis for Flexible Manufacturing Systems Performance Analysis," *Proceedings of the 21st IEEE Conference on Decision and Control*, Orlando, Fla. Dec. 1982, pp. 1061–1065.

46. R. Hildebrant, "Scheduling Flexible Machining Systems Using Mean Value Analysis," *Proceedings of the 1980 IEEE Conference on Decision and Control*, IEEE Press, New York, Dec. 1980, pp. 701–706.

47. M. Reiser and S.S. Lavenberg, "Mean Value Analysis of Closed Multichain Queueing Networks," *IBM Research Report RC 7023*, Yorktown Heights, N.Y., 1978.

Markovian modelling of manufacturing systems

ROBERT P. DAVIS† and W. J. KENNEDY Jr†

Successively more complex Markov models are used to describe an easily understood production process. As the models become more complex, they become more useful, in particular including more cost categories and more types of data. Rework and work-in-process inventories are modelled, together with inspections and scrap rates, and a method is shown for taking tool wear into account. The scheduling impacts of rate-dependent tool wear are also analysed using the Markov models developed.

Introduction

An important problem that confronts a manager of any manufacturing facility is to develop an understanding of the manufacturing operations in sufficient depth to make operating and planning decisions. The modelling of manufacturing systems has proven to be a valuable means for gaining this understanding. The system characteristics which form the focal point of these modelling efforts range from aggregate capacity planning issues to detailed, and time dependent, sequencing and routing decisions based on the probabilistic behaviour of the system and its individual components. This paper uses a first order Markov model to illustrate the relationships of design and control issues in production systems modelling.

Primitive, probabilistic models such as the models described here play an important role in: (a) providing insight to the decision maker of the sensitivity of a production system organization to fundamental variations in system performance; and (b) assisting in making the transition from deterministic models to more realistic, stochastic models and the essential data required to capture the most sensitive probabilistic behaviour of the system. The analysis presented in this paper also shows clearly the interactions between design and control decisions and gives a systematic way to evaluate the relative costs and benefits of collecting data on system performance. This analysis is also valuable in showing the place of models in the design and control of manufacturing systems. In particular, problems associated with production flow and capacity planning are illustrated.

Numerous approaches to modelling production systems have been reported in the literature and span the spectrum from deterministic linear programs to rather sophisticated (and problem specific) simulation models (Engesser *et al.* 1975, Inoue and Eslick 1975, Pritsker 1984, Pegden 1982). Production systems have been described in terms of queues by Buzacott (1968), Buzacott and Shanthikumer (1980), Stecke and Solberg (1985) and many others, but queueing models for production systems become complicated very quickly as such systems approach a realistic size. DuBois and Stecke (1983) have proposed the use of Petri nets for such processes but suggest that their use still requires substantial development. Markov models have also been used to describe failure and repair characteristics of

---

Revision received June 1986.

† Department of Industrial Engineering, Clemson University, 104 Freeman Hall, Clemson, South Carolina 29634-0920, U.S.A.

production systems, by Foster and Garcia-Diaz (1983) and Lie *et al.* (1977). None of the cited papers, however, use Markov analysis to show how a production system can be described in a way clear to the production planner, with a variable degree of complexity, and with costs, scrap rates, and data needs taken into account.

Markov models in general have a number of advantages. First, the data to be collected can be explained easily, since each element of a Markov transition matrix can be thought of as the relative frequency with which a transition takes place from one named state to another. Second, the computations involve well-known and easily implemented matrix operations. Expanding the size of a problem does not present computational difficulties. Third, the quantities calculated from a Markov analysis have meaning in the context of most problems, so that the results of a Markov analysis are easy to interpret.

The disadvantages of first-order Markov models may be significant in a particular problem. First-order Markov models rely on single step transition matrices, and the description of some systems requires that more steps be retained (Kennedy 1978). Further, the transition probabilities may change with time, complicating the analysis considerably. Finally, many people are not familiar with Markov modelling and analysis techniques, and the language, if not interpreted in the context of a problem, may be an obstacle.

In the literature, Greene (1984) gives one page to this technique in his chapter on decision making but gives little detail. Egbelu *et al.* (1982) give a procedure for selecting a casting/machining strategy based on the economic tradeoffs, including costs of casting, turning, boring, and facing. They give a relationship between casting precision and the amount of machining. Their procedure assumes a threshold value for casting precision, below which machining is not required. Davis and Miller (1979) developed a parametric Markov model to provide information on the relationships between machine characteristics and machine requirements, but their paper did not extend to rework, nor did it take scrap and control issues into account. This paper extends previous work by listing and describing costs and production planning decisions associated with successively more realistic and complex Markov models for an easily understood production process.

Most production planning decisions include one or more of the following specific problems:

(*a*) determining the number of machines required for a given production rate;
(*b*) calculating the effect of scrap and rework on production rates and costs;
(*c*) calculating the real production rates, taking into account the delays caused by maintenance, tool changing, and the like; and
(*d*) determining the effect of rate-dependent maintenance on possible scheduling options and on equipment requirements.

The objective of this paper is to show how models of production systems—in this case, Markov models—assist in resolving these decision issues. The paper shows how such models can be developed in an evolutionary process, with each step in the evolution requiring more data and yielding additional useful information.

The process of evolutionary model development is exemplified below in four scenarios. The first scenario is a three operation serial system with one category of scrap. General cost categories are used, and estimates can be obtained for production rates and the number of machines required. This model is extended in scenario 2 by the addition of inspection and rework. More detailed costs can be incorporated, the number of machines can be estimated with more confidence, and the benefits from quality control can be quantified. Scenario 3 extends the model to problems where production times and tool change intervals depend upon production rates. The result is a more accurate estimate of equipment requirements, production rates, and scrap and rework amounts. In the final scenario, imbalance of production rates is taken into account by allowing for work-in-process inventories and overtime.

Scenario 1. A serial system

Scenario 1 gives an example of a system described in gross detail. This system, shown in Fig. 1, is designed to produce a shaft in three operations—turn, mill, and drill. An incoming inspection is made on bar stock, and unacceptable material is scrapped. After each operation, a cursory inspection is performed to determine whether the machined dimension is within specifications. If specifications are not met, the machined part is scrapped.

The data needed to describe this system include only scrap rates and operating rates as shown in Table 1. If these data are accurate, operating rates are based on standard times and include time for tool changes and maintenance. As presented, however, it is difficult to know what the rates include, and only an approximation can be made to the number of machines needed to produce the prescribed number of finished shafts per day. Costs are also given in gross categories as shown below:

(a) Costs—equipment, incoming material less scrap revenue, and processing.
(b) Data needed—operating rates (3 machines) and scrap rates.

A model for this system would enable the user to estimate the costs and production consequences of changing the production rates of scrap rates. Such a model can be developed from the operations needed to process raw data as in Table 2.

An alternative to the above calculation is to create and use a model derived from the network shown in Fig. 2. In this model, the nodes in the network correspond to the processes in Fig. 1 and the percentages correspond to the scrap and acceptable material rates (1—scrap) in Table 1. A from–to matrix is developed from these

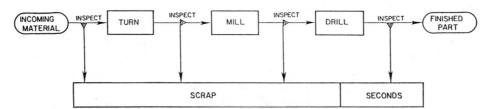

Figure 1. Illustrative serial system.

Turn	Drill	Mill
25 parts/hr.	25 parts/hr.	25 parts/hr.
3% scrap rate	5% scrap rate	2% scrap rate

Incoming material: scrap rate = 0·2%.

Table 1. Operating and scrap rates.

Process	Number of parts processed	Processing rate × 8 hours	Number of machines
Incoming material	554·78	—	—
Scrap	1·11	—	—
Turn	553·67	200	(2·77) 3
Scrap	16·61	—	—
Mill	537·06	200	(2·69) 3
Scrap	26·86	—	—
Drill	510·20	200	(2·55) 3
Scrap	10·20	—	—
Finished parts	500·00		

Table 2. Calculations for scenario 1.

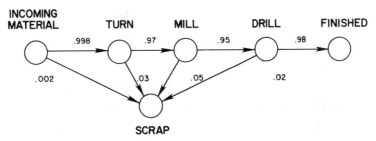

Figure 2. Network diagram for scenario 1.

figures and given in Fig. 3. In this matrix, the rows correspond to the 'from' nodes—or 'states'—and the columns to the 'to' nodes—or 'states'. The number in the interaction between a given row and column gives the probability that a unit in the state given by the row will go on the next step to the state given by the column and indicates the data that need to be obtained for a description of the system. In Fig. 3, for example, a number for p23 is needed. Data indicates that the appropriate value is 0·97, corresponding to the percentage of parts from turn (node 2) that go to mill (node 3), as shown in Fig. 4. A brief examination of Figs. 1 and 2 shows that ultimately everything ends up as either finished product, node 5, or scrap, node 6. Scrap and finished product are called 'absorbing' states, and the probability of absorption in either state from a given state shows the percentage that start in the given state and finish in the absorbing state. The absorbing probabilities are given in Table 3.

From State	meaning	To	1	2	3	4	5	6
1	Incoming material		p_{11}	p_{12}	p_{13}	p_{14}	p_{15}	p_{16}
2	Turn machine centre		p_{21}	p_{22}	p_{23}	p_{24}	p_{25}	p_{26}
3	Mill machine centre		p_{31}	p_{32}	p_{33}	p_{34}	p_{35}	p_{36}
4	Drill machine centre		p_{41}	p_{42}	p_{43}	p_{44}	p_{45}	p_{46}
5	Finished product		p_{51}	p_{52}	p_{53}	p_{54}	p_{55}	p_{56}
6	Scrap		p_{61}	p_{62}	p_{63}	p_{64}	p_{65}	p_{66}

Figure 3. General Markov transition matrix for scenario 1.

Markovian modelling of manufacturing systems

To From	1	2	3	4	5	6
1	—	0·998	—	—	—	0·002
2	—	—	0·97	—	—	0·03
3	—	—	—	0·95	—	0·05
4	—	—	—	—	0·98	0·02
5	—	—	—	—	1·00	—
6	—	—	—	—	—	1·00

Figure 4. Markov transition matrix for data of scenario 1.

From state no.	By state no.	Probability
1	5	0·90126386
1	6	0·09873614
2	5	0·90307
2	6	0·09693
3	5	0·931
3	6	0·069
4	5	0·98
4	6	0·02

Table 3. Absorption probabilities of each absorbing state.

The states of the system, together with the probability matrix that describes the transitions between states, make up a Markov model for the system. Once the Markov model is programmed into a computer, with the data of Fig. 4, the model yields the total percentage of scrap, the percentage of good parts from the turning operation that become good finished product, or any one of a number of other variables. For example, to calculate the number of parts processed as incoming material in order to yield 500 finished parts, divide 500 by the probability of absorption from state 1 (incoming material) into state 5 (finished material). Absorption probabilities are easily calculated (Foster and Garcia-Diaz 1983). These computations can then be incorporated into a cost model for the actual system, and system design studies can be performed.

Scenario 2. Serial system with rework

Scenario 1 assumed that parts were either good or bad. In general, dimensions, surface characteristics, and material composition follow probability distributions such as those given in Fig. 5. When such variation is known to occur, it is often advantageous to have rework as an alternative to scrapping. Figure 6 gives a diagram showing the use of rework in such a situation. In this case, an inspector is used to decide which units to scrap and which to rework. Parts for rework are machined on either a precision lathe or a precision milling machine and inspected again, with some of the reworked parts going to scrap and some to the next process.

The information needed for this scenario includes the data from scenario 1 with the addition of processing rates and scrap rates from the precision turn and precision mill operations. Additional data is also needed on unacceptable output from turn and drill operations to determine how much is reworkable and how much is scrap.

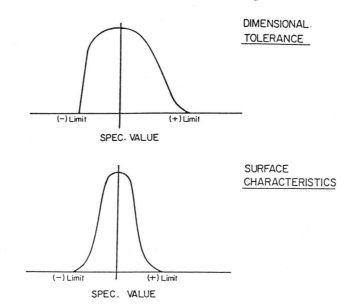

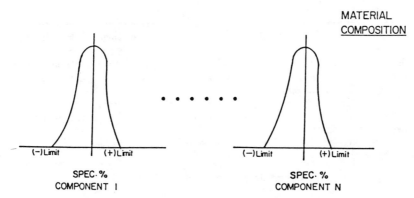

Figure 5. Probability distributions of material states.

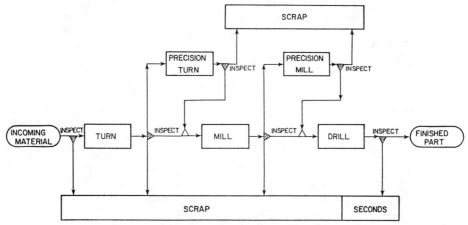

Figure 6. Illustrative serial system with rework stages.

Additional costs defined include those of inspection and those associated with processing of rework. Equipment costs will also increase to include a precision lathe and a precision mill. Table 4 shows the data and costs needed for this scenario.

A model for scenario 2 enables additional analyses not possible with that for scenario 1. In particular, many of the quality costs can be evaluated in terms of their benefits. As an example consider Table 5. In this table note that the amount of incoming material is reduced from 554·78 units to 532·60 units while requiring two additional machines. Whether this is worthwhile depends upon the magnitude of the costs involved.

The network model and the Markov matrix for the network are shown in Figs. 7 and 8, respectively, where the states are defined to correspond to the nodes of the network, and the nodes of the network correspond to the situation modelled, as before. In this new scenario, there are three absorbing states—two scrap states and one finished product state. In a more complex problem there could be many more states, possibly with several output states corresponding to each production process. The resulting probabilities are given in Table 6.

Operating and scrap rates

Incoming material: scrap rate = 0·2%.

Turn	Drill	Mill	Precision turn	Precision drill
25 parts/hr	25 parts/hr	25 parts/hr	10 parts/hr	10 parts/hr
2% scrap rate	2% scrap rate	2% scrap rate	1% scrap rate	2% scrap rate
1% rework rate	3% rework rate			

(a) Costs—equipment; incoming material less scrap revenue; processing (machining and rework); and inspection.

(b) Data Needed—Operating rates (5 machines); scrap percentages; and rework percentages.

Table 4. Operating and scrap rates, costs, and data.

Process	Number of parts processed	Processing rate × 8 hours	Number of machines
Incoming material	532·60	—	—
Scrap	1·06	—	—
Turn	531·54	200	(2·66) 3
Scrap	10·63	—	—
Precision Turn	5·31	80	(0·07) 1
Scrap	0·05	—	—
Mill	520·85	200	(2·60) 3
Scrap	10·41	—	—
Precision Mill	15·62	80	(0·20) 1
Scrap	0·31	—	—
Drill	510·13	200	(2·55) 3
Scrap	10·20	—	—
Finished parts	500·00	—	—

Table 5. Calculations for scenario 2.

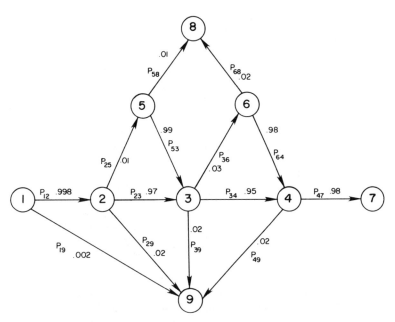

Figure 7. Network diagram for scenario 2.

To From	1	2	3	4	5	6	7	8	9
1	—	0·998	—	—	—	—	—	—	0·002
2	—	—	0·97	—	0·01	—	—	—	0·02
3	—	—	—	0·95	—	0·03	—	—	0·02
4	—	—	—	—	0·98	—	—	—	0·02
5	—	—	0·99	—	—	—	—	0·01	—
6	—	—	—	0·98	—	—	—	0·02	—
7	—	—	—	—	—	—	1·00	—	—
8	—	—	—	—	—	—	—	1·00	—
9	—	—	—	—	—	—	—	—	1·00

Figure 8. Markov transition matrix for data of scenario 2.

Scenario 3. Scenario 2 with times included

No model of a manufacturing system is complete without some consideration of times. The time categories usually include tool replacement and calibration, inspection time on each part, maintenance time on equipment, and repair time.

Tool replacement times may depend upon production rates. Suppose that these tool replacement and calibration intervals hold for the production rates given in Fig. 3: turn—70 parts; mill—60 parts; drill—150 parts; precision turn—80 parts; and precision mill—60 parts. Table 7 gives additional necessary time data. With this data, computations such as that in Table 9 can be performed. These calculations give more realistic machine requirements than those shown in Tables 2 and 5, with the assumption that each machine centre (turn, mill, drill) must produce the same number of finished equivalent units (i.e. taking scrap into account) per day.

As well as giving better estimates of the number of machines needed, scenario 3 also gives more realistic estimates of costs than those based on scenario 2, and a production system based on this model will also be more realistic. The additional

From state no.	By state no.	Probability
1	7	0·93863874
1	8	0·00068656
1	9	0·06067470
2	7	0·94051978
2	8	0·00068794
2	9	0·05879228
3	7	0·95981200
3	8	0·00060000
3	9	0·03958800
4	7	0·98
4	8	0
4	9	0·02
5	7	0·95021388
5	8	0·01059400
5	9	0·03919212
6	7	0·9604
6	8	0·02
6	9	0·0196

Table 6. Absorption probabilities for scenario 2.

Tool replacement/calibration time = 4 minutes each time

Inspection time—minutes/part:

Incoming material	1·0 min	Turn	1·5 min
Precision turn	3·0 min	Mill	2·0 min
Precision mill	4·0 min	Drill	4·0 min

Maintenance time:	Lathe	Mill	Drill
	15 min	15 min	10 min

Precision lathe: 25 min   Precision mill: 30 min

Repair time:	Lathe	Mill	Drill
	3 hrs	4 hrs	2 hrs

Precision lathe: 4 hrs   Precision mill: 6 hrs

Table 7. Times for scenario 3.

	Turn	Mill	Drill	Precision turn	Precision mill
Parts per change	70	60	150	80	60
Changes per day	7·6	8·68	3·4	—	—
Change time per day	30·4 min	34·7 min	13·6 min	—	—
PM/change	15 min	20 min	10 min	25 min	30 min
PM/day	114 min	173·6 min	34 min	—	—
Available time per machine	5·59 hrs	4·53 hrs	7·21 hrs	8 hrs	8 hrs
No. machines	(3·8)4	(4·6)5	(2·83)3	1	1

Table 8. Machine requirements with available time.

realism has its cost, however—the necessity to collect the additional data shown below.

(a) Costs—equipment; incoming material less scrap revenue; processing (machining and rework); and inspection.
(b) Data needed—operating rates (5 machines); scrap and rework percentages; maintenance and repair times and frequencies; and tool change times.

Scenario 4. Fewer machines, more overtime, more inventory

The idea of using more time with fewer machines is usually worth considering. If equipment is limited to three machines in each of the three primary machining centres, Table 9 shows that the cost is 1·50 hours of overtime for each lathe or turning machine and 2·41 hours on each milling machine. It is also necessary to plan for a work-in-process inventory of 70·5 parts between the mill and drill operations. The data needed and the components of a cost model for this situation is shown below:

(a) Costs—equipment; incoming material less scrap revenue; processing (machining and rework); inspection; inventories (value added at each process, holding cost and stockout cost); and labour (straight time and overtime).
(b) Data needed—operating rates (5 machines); scrap and rework percentages; maintenance and repair times and frequencies; tool change times; and inventory space requirements between processes.

Rate-dependent scenarios

In the above development, it is assumed that the production rates for each machine centre are fixed at the rates given in Table 4. Relaxing this assumption can lead to the consideration of different alternatives, some of which may be better than those derived above. In general, it could be assumed that scrap rates, the need for maintenance, and the need for tool changes will all depend upon the production rates. Typical curves relating these characteristics to production rates are shown in Figs. 9–12.

As an example of the effect of different operating rates, consider the possible operating rates shown in Table 10. One question is whether the operating rates can be

	Turn	Mill	Drill	Precision turn	Precision mill
Parts	531·54	520·85	510·13	5·31	15·62
No. machines	3	3	3	1	1
Capacity (parts/hr)	75	75	75	10	10
Operating time (hrs)	7·09	6·94	6·80	0·53	1·56
Tool change and maintenance time (hrs)	2·41	3·47	0·79		
Total time (hrs)	9·50	10·41	7·59		

Note: Difference in processing time of 2·82 hours at 25 parts per hour requires inventory capacity of 70·5 parts at drilling machine.

Table 9. Machine requirement calculations with overtime.

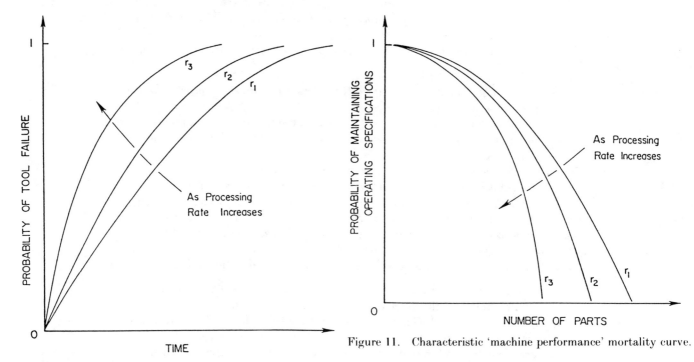

Figure 9. Characteristic 'tool failure' mortality curve.

Figure 11. Characteristic 'machine performance' mortality curve.

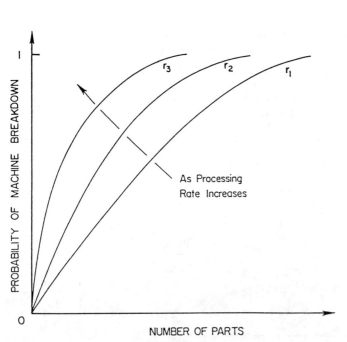

Figure 10. Characteristic 'machine failure' mortality curve.

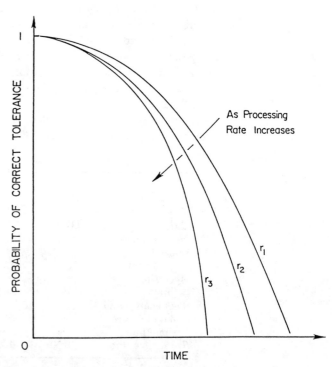

Figure 12. Characteristic 'tool performance' mortality curve.

Turn parts/hr	Drill parts/hr	Mill parts/hr	Precision turn parts/hr	Precision drill parts/hr
20	20	20	10	10
25	25	25	20	20
30	30	30		

Tool replacement calibration time = 4 minutes.

	Turn pts	Mill pts	Drill pts	Precision turn pts	Precision mill pts
Rate	20–100	20–80	20–200	10–80	10–60
	25– 70	25–60	25–150	20–50	15–40
	30– 40	30–40	30– 80		

Table 10. Effects of different operating rates.

changed to balance production in each cell with a limited amount of overtime, say two hours, and no more than three machines. Before this question can be answered it is necessary to calculate the number of machine centre hours needed per productive hour at a given rate, including maintenance and calibration. For turning at 20 parts/hour per machine, maintenance and calibration take up, per hour, $3 \times (20/100) \times (4+5)$ minutes, i.e. 19 hours, giving 1·19 total hours per hour of productive time. The corresponding rate at 25 parts/hour is 1·34 hours per productive hour, and other rates are calculated similarly. These calculations are necessary since the total number of working hours is to be no greater than 10 hours per day. The daily production must be met by the chosen combination of rates, imposing another constraint. These constraints are shown in Table 11. Note that at the rate of 20 parts/hour/machine, the time required in the drill machine centre is 9·09 hours so that all the drill centre machines can be run at the slower rate.

In the above scenario, the percentages of scrap and rework will probably change, and a new network like Fig. 7 can be developed with the corresponding Markov model.

The costs for the above model are those of previous scenarios with an additional requirement of collecting the data shown in Table 10.

Applicability and limitations of the Markov models

Each of the above scenarios gave a production situation that was easily described and analysed using Markov techniques. The output from each model included values that would be of use to production planners. These techniques could also be extended to more complicated situations, and the results could be used as inputs in the development of cost models for equipment selection, replacement, and maintenance.

Although generally helpful, the Markov models might not be applicable to some permutations of the situations described above. In the above scenarios, it was assumed that machines did not interfere with each other. Including machine interference would require a time-dynamic model, probably a simulation. Other real conditions whose analysis would profit from such modelling would include machines

(a) Problem: more machines or more overtime cost money.

(b) Possible solution: run machines at different rates.

(c) Analysis:
Turn: $1{\cdot}19\,t(T1) + 1{\cdot}34\,t(T2) \leq 10{\cdot}00$
        $60\,t(T1) + 75\,t(T2) = 531{\cdot}34$
        where $t(T1)$ = number of hours at 20 parts per hour
        $t(T2)$ = number of hours at 25 parts per hour

Mill: $1{\cdot}50\,m(T1) + 1{\cdot}71\,m(T2) \leq 10{\cdot}00$
        $75\,m(T1) + 90\,m(T2) = 520{\cdot}85$
        where $m(T1)$ = number of hours at 25 parts per hour
        $m(T2)$ = number of hours at 30 parts per hour

Drill: $1{\cdot}07\,d(T1) + 1{\cdot}17\,d(T2) \leq 10{\cdot}00$
        $60\,d(T1) + 75\,d(T2) = 510{\cdot}13$
        where $d(T1)$ = number of hours at 20 parts per hour
        $d(T2)$ = number of hours at 25 parts per hour

(d) Analysis results:
Turn: 4·26 hours @ 20 parts/hr
        3·61 hours @ 30 parts/hr
        Total time = 10 hours/day for 3 machines

Mill: 1·38 hours @ 25 parts/hr
        4·64 hours @ 30 parts/hr
        Total time = 10 hours/day for 3 machines

Drill: 9·09 hours @ 20 parts/hr
        Total time = 9·09 hours/day for 3 drilling machines

Table 11. Effects of varying processing rates.

that break down, production demands that change with time, learning, and complex costs. For all of these situations, data is necessary. A more realistic model will be more useful and will require more data. The steps in the development of such a model can follow the evolutionary process illustrated above.

Des modèles Markov de plus en plus complexes servent à décrire un procédé de production favile à appréhender. Au fur et à mesure qu'ils s'élaborent, les modèles s'avèrent plus utiles, comprenant notamment plus de catégories de coûts et un plus grand nombre de types de données. Des inventaires de refaçon et de travaux en cours sont modélisés ainsi que des contrôles et des taux de mise au rebut. Une méthode permettant de tenir compte de l'usure des outils est indiquée. Les effets sur la programmation de l'usure d'outils qui est fonction des taux sont également analysés au moyen des modèles Markov développés.

Einige zunehmend komplizierter werdende Markowsche Modelle werden nacheinander für die Beschreibung eines leicht verständlichen Fertigungsverfahrens eingesetzt. In gleichem Maße wie die Modelle komplizierter werden, steigt auch ihre Nützlichkeit, besonders wenn sie eine größere Zahl von Kostenklassen und Datenarten beinhalten. Die Nachbearbeitung, Halbfertigerzeugnisse, Fertigungskontrollen und Ausschußquoten werden nachgebildet, und es wird ein Verfahren gezeigt, das sogar den Gesamtverschleiß berücksichtigt. Die Auswirkungen, die der von der Mengenleistung abhängige Werkzeugverbrauch auf die Ablaufplanung hat, werden ebenfalls mit Hilfe der entwickelten Markowschen Modelle analysiert.

References

BUZACOTT, J. A., 1968, Prediction of the efficiency of production systems without internal storage. *The International Journal of Production Research*, **6**, 173.

BUZACOTT, J. A., and SHANTHIKUMER, J. G., 1980, Models for understanding flexible manufacturing systems. *A.I.I.E. Transactions*, **12**, 339.

DAVIS, R. P., and MILLER, D. M., 1979, Analysis of the machine requirements problem through a parametric Markovian model. SME technical paper MS79-960.

DU BOIS, D., and STECKE, K., 1983, Using Petri nets to represent production processes. *Proceedings of the 22nd IEEE Conference on Decision & Control*, December.

EGBELU, P. J., DAVIS, R. P., WYSK, R. A., and TANCHOCO, J. M. A., 1982, An economic model for the machining of cast parts. *Journal of Manufacturing Systems*, **1**, 207.

ENGESSER, W. F., INOUE, M. S., and MERCER, W. G., 1975, Production optimization through resource planning and management networks. *Proceedings of the 1975 Systems Engineering Conference*, American Institute of Industrial Engineers.

FOSTER, J. W., III, and GARCIA-DIAZ, A., 1983, Markovian models for investigating failure and repair characteristics of production systems. *I.I.E. Transactions*, **15**, 202.

GREENE, J. H., 1984, *Operations Management: Productivity and Profit* (Reston, Virginia: Reston Publishing Company).

INOUE, M. S., and ESLICK, P. O., 1975, Application of RPMS methodology to a goal programming problem in a wood product industry. *Proceedings of the 1975 Systems Engineering Conference*, American Institute of Industrial Engineers.

KENNEDY, W. J., JR., 1978, The model that didn't work. *Interfaces*, **8**, 54.

LIE, C. H., HWANG, C. L., and TILLMAN, F. A., 1977, Availability of maintained systems: a state-of-the-art survey. *A.I.I.E. Transactions*, **9**, 247.

PEGDEN, G. D., 1982, *Introduction to SIMAN* (State College, Pennsylvania: Systems Modeling Corporation).

PRITSKER, A. A. B., 1984, *Introduction to Simulation and SLAM II*, second edition (West LaFayette, Indiana: Systems Publishing Corporation).

STECKE, K., and SOLBERG, J. J., 1985, The optimality of unbalancing both workloads and machine group sizes in closed queueing networks of multi-server queues. *Operations Research*, **33**, 882.

Analysis of Transfer Lines Consisting of Two Unreliable Machines with Random Processing Times and Finite Storage Buffers

STANLEY B. GERSHWIN

Laboratory for Information and Decision Systems
Massachusetts Institute of Technology
Cambridge, Massachusetts 02139

ODED BERMAN

Faculty of Management
University of Calgary
Calgary, Canada T2N 1N4

Abstract: A Markov process model of a transfer line is presented in which there are two machines and a single finite buffer. The machines have exponential service, failure, and repair processes. The movement of discrete parts is represented. The model is analyzed and a compact solution is obtained. Limiting behavior is investigated and numerical results are discussed.

Reprinted with permission from *AIIE Transactions*, Volume 13, Number 1, March 1981, pages 2-11. Copyright Institute of Industrial Engineers, 25 Technology Park/Atlanta, Norcross, GA 30092.

Introduction

1.1 Problem Description

■ Production systems are often composed of many processing stages through which material must pass. The performance of such a system is impaired by variations in the behavior of the stages, whether due to failures or processing time fluctuations.

The effects of these variations can often be mitigated by the use of storage buffers between processors, which temporarily hold in-process inventory. When one stage is under repair or taking an unusually long time to process a piece, a buffer may enable work to continue elsewhere.

The purpose of this paper is to introduce a model of the production system illustrated in Fig. 1, the two-stage flow shop or transfer line. This model is shown to be analytically tractable, and its behavior is demonstrated (by theoretical and numerical argument) to agree with intuition. Because numerical results on production rate and average in-process inventory are so easily obtained, this model can be helpful in the design of production systems. That is, it can be used to choose an optimal buffer size or to choose among different processing options for performing the operations at stage 1 and stage 2.

In Figure 1, workpieces enter the first machine (or stage, or processor), and an operation takes place. The pieces are then moved to the buffer and they later proceed to the second machine. When processing is complete there, workpieces leave the system.

Simple systems of the form of Fig. 1 have been extensively studied. The version discussed here is a new model which accounts for random machine failures, variations in processing times, and a finite buffer.

Each machine is described by three exponentially distributed random variables: the processing time, the time to fail, and the time to repair. The machines, in general, may differ.

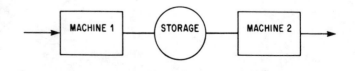

Fig. 1. Two-machine transfer line.

Received January 1979; revised December 1979, June 1980. Paper was handled by Applied Probability and Statistics Department.

1.2 Comparison with Earlier Work

In this section we describe related studies. Because the work reported here is analytic, we refer only to other analytic investigations and not to simulations. We also restrict our attention to systems with finite queues between machines. Other surveys can be found in Koenigsberg (1959), Buchan and Koenigsberg (1963), and Buzacott and Hanifin (1978a and b).

The related literature can be divided into four classes. The first consists of papers on systems without failures. The motion of discrete parts is modeled, and the processing times are random variables. Several authors [Hunt (1956), Hildebrand (1967 and 1968), Neuts (1968 and 1970), Reiser and Konheim (1974), and Lavenberg (1975)] allow an infinite input queue upstream of the first machine and have parts arriving at random. Others [Hillier and Boling (1966 and 1967), Hatcher (1969), Knott (1970), Kraemer and Love (1970), Muth (1973 and 1977), and Rao (1975)] assume, as we do, that there is an inexhaustible supply of parts to the first machine. A related assumption (which is always made, often tacitly) is that there is always space available into which the last machine can discharge a part. We make this assumption below.

Stages are assumed to have exponentially distributed processing times by Hunt (1956), Patterson (1964), Hillier and Boling (1966), Gordon and Newell (1967), Hildebrand (1968), Hatcher (1969), and Kraemer and Love (1970). Hillier and Boling (1967) extend this to Erlang distributions. More general distributions are treated by Muth (1973 and 1977) and Rao (1975). It should be noted that where the probability distribution $p(n)$ of the number of pieces n in a single storage is explicitly calculated [Gordon and Newell (1967), Hatcher (1969), Kraemer and Love (1970), Hitomi et al. (1977)], it is of the form

$$p(n) = cX^n, \qquad (1.1)$$

where c and X are parameters.

A second class of papers include those that assume deterministic processing times, but that allow random failures. These include Buzacott (1967a, 1967b, 1969) Gershwin (1973), Sheskin (1974, 1976), Soyster and Toof (1976), Artamonov (1977), Okamura and Yamashina (1977), Ignall and Silver (1977), Buzacott and Hanifin (1978a), Schick and Gershwin (1978), Gershwin and Schick (1979a, 1979b, and 1980). All models assume an unlimited supply of pieces to the first machine, an unlimited amount of room for the last machine to dispose of its pieces, and use a Markov chain approach. All assume geometrically distributed times between failures and times to repair except Buzacott, who generalizes these assumptions.

In all the two-machine papers where up- and down-times are geometrically distributed and where probability distributions are calculated explicitly [Buzacott (1967a), Gershwin (1973), Sheskin (1974), Artamonov (1977), Schick and Gershwin (1978), Gershwin and Schick (1980)], the following statement is true. Let n be the number of pieces in the buffer and α_i ($i=1, 2$) be a binary variable indicating whether machine i is operational ($\alpha_i=1$) or under repair ($\alpha_i=0$). Let $p(n, \alpha_1, \alpha_2)$ be the probability of that state. Then

$$p(n, \alpha_1, \alpha_2) = X^n f(\alpha_1, \alpha_2), \qquad (1.2)$$

where X is a parameter and f is a function of machine states. (This equation is valid everywhere except for certain states in the vicinity of the boundaries, i.e., where the buffer is empty or full.) In particular Gershwin and Schick (1980) find

$$p(n, \alpha_1, \alpha_2) = cX^n Y_1^{\alpha_1} Y_2^{\alpha_2}, \qquad (1.3)$$

where c, X, Y_1, and Y_2 are parameters [see also Schick and Gershwin (1978)]. Equation (1.3) appears below in a slightly different form.

A third class of papers analyzes models in which parts are not treated as discrete items. Instead, the material to be processed is treated as a continuous fluid. Failures, repairs, and finite buffers are studied. This includes the work of Finch, Vladzievski, and Sevast'yanov (1962), described in Koenigsberg (1959) and Buzacott and Hanifin (1978a). See also Murphy (1975 and 1978), Schick and Gershwin (1978), and Wijngaard (1979).

The fourth class is one in which individual parts are represented, machine operation times are random, and failure and repair times are random. As far as these authors are aware only Buzacott (1972) makes these assumptions. The research reported below also falls into this class.

1.3 Contribution of this Paper

As indicated, this is evidently only the second paper that treats systems with discrete workpieces, finite buffers, random processing times, and random failures. In both this paper and the previous [Buzacott (1972)] a continuous time Markov chain is used to represent a two-stage production line. However, both the models and the solution techniques differ significantly. Model differences are described in Section 2.

In this paper, an efficient analytic technique is devised to calculate the steady state probability distribution of the Markov chain. This distribution is used to calculate such performance measures as system production rate, machine efficiency or utilization, and average in-process inventory. The results are easy to calculate. That is, a computer program which is short and efficient suffices for the calculation of the probability distribution and of the performance measures. In fact, a program has been written for the TI-59 pocket calculator to evaluate the performance measures [Ward (1980)]. When cost information is available, these measures are valuable in the optimal design or modification of a two-stage production system.

Theoretical results are obtained in this paper concerning conservation of pieces, limiting behavior as one machine becomes much more or much less productive then the other, and other matters. While such results as conservation of pieces are not surprising, they show that in certain important areas, the model behaves reasonably and agrees with intuition.

This paper is intended as a starting point for the study of more general systems. The approach follows closely a study of two-machine systems with deterministic processing times [Gershwin (1973), Schick and Gershwin (1978)] which has been extended to three-machine lines [Gershwin and Schick (1979a, 1979b, 1980)]. Other generalizations of this model have already been considered: an analysis of three-machine lines and assembly merges appears in Ammar (1980), and a study of two-machine lines with Erlang processing times appears in Gershwin and Berman (1978) and Berman (1979).

1.4 Outline

The assumptions of the model are presented in the next section, where the differences between this model and Buzacott's (1972) are made explicit. Formulas for such performance measures as efficiencies, production rate, and average in-process inventory are presented in Section 3. Section 4 contains the detailed balance equations which define the Markov chain. Theoretical results, based on these equations, appear in Section 5 and 6. The latter describes limiting behavior, as one machine becomes much more or much less productive than the other. Section 7 contains a closed form representation of the probability distribution. Numerical experience is described in Section 8 and conclusions and future research directions can be found in Section 9.

2. Model Description and Assumptions

The system consists of two machines that are separated by a finite storage buffer (Fig. 1). Workpieces enter machine 1 from the outside. Each piece is operated upon in machine 1, then passes to the buffer and then proceeds to machine 2. After being operated on in machine 2, the piece leaves the system.

A machine can be in two possible states: *operational* or *under repair*. The binary variable α_i is defined to be 1 when machine i is operational and 0 when machine i is under repair. When a machine goes from state 1 to state 0, it is said to *fail*. A *repair* takes place when the transition from $\alpha_i = 0$ to $\alpha_i = 1$ occurs. When a machine is in state 1, it may be processing workpieces. It processes no pieces while it is under repair.

Even if a machine is operational, it cannot process pieces if none are available to it or if there is no room in which to put processed pieces. In the former condition, the machine is said to be *starved*; in the latter it is *blocked*. It is assumed here that the first machine is never starved and the second is never blocked.

Blocked or starved machines, because they are not operating, are not vulnerable to failure. This assumption differs from Wijngaard (1979), who allows failures of idle machines.

Service, failure, and repair times for machine i are assumed to be exponential random variables with parameters μ_i, p_i, r_i; $i = 1, 2$, respectively. These quantities are the *processing rate*, *failure rate*, and *repair rate*, respectively. When a machine is under repair, it remains in that state for a period of time which is exponentially distributed with mean r_i^{-1}. This period is unaffected by the states of the other machine or of the storage.

When a machine is up, it operates on a piece if it is not starved or blocked. It continues operating until either it completes the piece or a failure occurs, whichever happens first. Either event can happen during the time interval $(t, t+\delta t)$ with probability approximately $\mu_i \delta t$ or $p_i \delta t$ respectively, for small δt.

The repair and workpiece completion models are similar to those of Buzacott (1972). However, in Buzacott's failure model, the probability of a failure before a completion is independent of the time spent on the piece. Here, instead, the longer an operation takes, the more likely it is that a failure occurs before the work is complete. Buzacott's model would seem, therefore, to be appropriate where the predominant cause of failure is the transfer mechanism, clamping, or some other action that takes place exactly once during an operation. The model presented here would seem to better represent failures in mechanisms that are vulnerable during an entire operation, such as tools or drivers.

The amount of material in storage is represented by the integer n,

$$0 \leqslant n \leqslant N. \tag{2.1}$$

This is the number of pieces in the buffer plus the number of pieces in machine 2. Most frequently, there is one piece in machine 2. There are no pieces in that machine if the buffer becomes empty and machine 2 completes the processing of the last remaining piece.

To summarize, seven numbers are required to completely characterize a two-machine line: p_i, r_i, μ_i, $i = 1, 2$, and N.

The state of the system is denoted by

$$s = (n, \alpha_1, \alpha_2). \tag{2.2}$$

The probability that the system is in this state is written $p(n, \alpha_1, \alpha_2)$. The use of this distribution to calculate system performance measures is demonstrated in the following section.

3. Measures of Performance

The *isolated efficiency* e_i of machine i is defined as

$$e_i = \frac{r_i}{r_i + p_i} \tag{3.1}$$

This is the fraction of time machine i would be producing pieces if it were in isolation, that is, if it had an endless supply of raw workpieces and unlimited reservoir in which to store processed pieces.

The *isolated production rate*, ρ_i, is given by

$$\rho_i = \mu_i e_i . \quad (3.2)$$

This is the rate at which machine i would process parts in isolation.

To characterize machines in systems the *efficiency* E_i of machine i is defined as the probability that the ith machine is operating on a piece, or as the fraction of time in which the ith machine produces pieces. We can express the efficiencies in terms of the steady state probabilities as:

$$E_1 = \sum_{n=0}^{N-1} \sum_{\alpha_2=0}^{1} p(n,1,\alpha_2) , \quad (3.3)$$

$$E_2 = \sum_{n=1}^{N} \sum_{\alpha_1=0}^{1} p(n,\alpha_1,1) . \quad (3.4)$$

It is important to distinguish E_i from e_i. The former is affected by the other machines and the storage while the latter is a characteristic of machine i only.

In Lemma 5 below we show that

$$\mu_1 E_1 = \mu_2 E_2 . \quad (3.5)$$

The quantity $\mu_i E_i$ can be interpreted as the rate at which pieces emerge from machine i. Equation (3.5) is then a conservation of flow law, and we can define

$$P = \mu_i E_i . \quad (3.6)$$

This is the *production rate* of the system.

Another important measure of system performance is the expected in-process inventory. This can be written

$$\bar{n} = \sum_{n=0}^{N} \sum_{\alpha_1=0}^{1} \sum_{\alpha_2=0}^{1} np(n,\alpha_1,\alpha_2) . \quad (3.7)$$

4. The Markov Process Model Equations

In this section we present the balance equations satisfied by the probability distribution $p(N, \alpha_1, \alpha_2)$ defined in Section 2. These equations represent the model in steady state.

The complete set of equations can be written in compact form if some definitions are made. Let $I_i(-1) = I_i(N+1) = 0$,

$$\left.\begin{array}{l} I_1(N) = 0 \\ I_1(n) = 1 , n \leq N-1 \\ I_2(0) = 0 \\ I_2(n) = 1 , n \geq 1 \end{array}\right\} \quad (4.1)$$

That is, $I_i(n) = 0$ if machine i is starved or blocked and 1 otherwise. Then the probability equations are

$$p(n, \alpha_1, \alpha_2) \sum_{i=1}^{2} [r_i(1-\alpha_i) + (\mu_i+p_i)\alpha_i I_i(n)]$$

$$= p(n-1, 1, \alpha_2) \mu_1 \alpha_1 I_1(n-1)$$

$$+ p(n+1, \alpha_1, 1) \mu_2 \alpha_2 I_2(n+1)$$

$$+ p(n, 1-\alpha_1, \alpha_2)[r_1\alpha_1 + p_1(1-\alpha_1)I_1(n)]$$

$$+ p(n, \alpha_1, 1-\alpha_2)[r_2\alpha_2 + p_2(1-\alpha_2)I_2(n)] ,$$

$$0 \leq n \leq N ; \alpha_1 = 0,1; \alpha_2 = 0,1. \quad (4.2)$$

The left side of this equation represents the rate of leaving state (n, α_1, α_2) and the right side is the rate of entering it. To understand the left side, note that the first term in the brackets is the rate of leaving the state by repair, which is only possible if $\alpha_i=0$. The second is the rate of leaving the state by completion of a piece or by failure, which are only possible if $\alpha_i=1$ and $I_i(n)=1$. The right side can be understood similarly.

5. Theoretical Results

In this section we present some theoretical results. These results are important in providing insight into the model as well as a basis for our discussions in the following sections. Detailed proofs can be found in Gershwin and Berman (1978).

Lemma 1 identifies certain states as transient. This information is part of the closed form expressions for the probability distribution described in Section 7.

LEMMA 1

$$p(0,0,0) = p(0,1,0) = p(N,0,0) = p(N,0,1) = 0 . \quad (5.1)$$

Lemma 2 asserts that the rate of transitions from the set of states in which machine 1 is under repair to the set of states in which machine 1 is operational is equal to the rate of transitions in the opposite direction.

LEMMA 2

$$r_1 \underbrace{\sum_{n=0}^{N} \sum_{\alpha_2=0}^{1} p(n,0,\alpha_2)}_{\text{Probability that machine 1 is under repair}} = p_1 \underbrace{\sum_{n=0}^{N-1} \sum_{\alpha_2=0}^{1} p(n,1,\alpha_2)}_{\text{Probability that machine 1 can operate}} \quad (5.2)$$

This is proved by adding Eq. (4.2) for $0 \leq n \leq N$, $\alpha_1 = 0$, $\alpha_2 = 0,1$. Lemma 3 establishes a corresponding result for machine 2.

LEMMA 3

$$r_2 \sum_{n=0}^{N} \sum_{\alpha_1=0}^{1} p(n,\alpha_1,0) = p_2 \sum_{n=1}^{N} \sum_{\alpha_1=0}^{1} p(n,\alpha_1,1) \quad (5.3)$$

$\underbrace{\phantom{r_2 \sum_{n=0}^{N} \sum_{\alpha_1=0}^{1} p(n,\alpha_1,0)}}_{\text{Probability that machine 2 is under repair}} \quad \underbrace{\phantom{p_2 \sum_{n=1}^{N} \sum_{\alpha_1=0}^{1} p(n,\alpha_1,1)}}_{\text{Probability that machine 2 can operate}}$

This is proved by adding Eq. (4.2) for $0 \leq n \leq N$, $\alpha_1 = 0, 1$, $\alpha_2 = 0$.

Lemma 4 asserts that the rate of transitions from the set of states with n pieces in storage to the set of states with $n+1$ pieces in storage is equal to the rate of transitions in the opposite direction

LEMMA 4

$$\mu_1 \sum_{\alpha_2=0}^{1} p(n,1,\alpha_2) = \mu_2 \sum_{\alpha_1=0}^{1} p(n+1,\alpha_1,1), \quad 0 \leq n \leq N-1. \quad (5.4)$$

Probability that machine 1 is operational with n pieces in storage   Probability that machine 2 is operational with $n+1$ pieces in storage

This can be proved by induction. For $n=0$, Eq. (4.2) is added for $n=0$, $\alpha_1=0, 1$, $\alpha_2=0,1$. The result is established for $n>0$ by adding (4.2) for $n>0$ fixed and $\alpha_1=0,1$, $\alpha_2=0,1$.

Lemma 5 shows that the rate at which pieces emerge from machine 1 is equal to the rate at which they emerge from machine 2. This establishes a relationship between the efficiencies E_i of the machines and defines the production rate P of the system. This is discussed in Section 3 [(Eqs. (3.5) and (3.6)].

LEMMA 5

$$\mu_1 \sum_{n=0}^{N-1} \sum_{\alpha_2=0}^{1} p(n,1,\alpha_2) = \mu_2 \sum_{n=1}^{N} \sum_{\alpha_1=0}^{1} p(n,\alpha_1,1) \quad (5.5)$$

Probability that machine 1 can produce a piece   Probability that machine 2 can produce a piece

PROOF: Equation (5.5) is simply Eq. (5.4) summed for $n = 0, \ldots, N-1$.

Lemma 6 establishes a conditional independence among machines. It shows that if the storage is not full, machine 1 is operational exactly as often as it would be if it were not connected to another machine and a storage. Similarly, when the storage is not empty, machine 2 is operational just as often as it would be in isolation. For this lemma, prob (A) is the probability of event A and prob (A|B) is the conditional probability of event A given event B. The quantities e_i and E_i are defined in Section 3.

LEMMA 6

$$\text{prob}\,(\alpha_1 = 1 \mid n \neq N) = e_1 \quad (5.6)$$

$$\text{prob}\,(\alpha_2 = 1 \mid n \neq 0) = e_2. \quad (5.7)$$

PROOF: Equation (5.6) only is proved. The same method applies to (5.7). By definition of conditional probability,

$$\text{prob}\,(\alpha_1 = 1 \mid n \neq N) = \frac{\text{prob}\,(\alpha_1 = 1, n \neq N)}{\text{prob}\,(n \neq N)}. \quad (5.8)$$

This can be written

$$\text{prob}\,(\alpha_1 = 1 \mid n \neq N) = \frac{E_1}{E_1 + D_1}, \quad (5.9)$$

where E_1 is defined in Eq. (3.3) and D_1 is defined as

$$D_1 = \text{prob}(\alpha_1 = 0, n \neq N) = \sum_{n=0}^{N-1} \sum_{\alpha_2=0}^{1} p(n,0,\alpha_2). \quad (5.10)$$

Lemma 1 and Lemma 2 imply that

$$r_1 D_1 = p_1 E_1 \quad (5.11)$$

and (5.6) follows from (5.9), (5.11), and (3.1). A similar proof suffices for (5.7).

Note that these results can also be written

$$E_1 = e_1 \text{ prob } (n \neq N), \quad (5.12)$$

$$E_2 = e_2 \text{ prob } (n \neq 0). \quad (5.13)$$

6. Limiting Cases

In this section, we consider the effects of letting machine parameters be extremely large or extremely small. In statement 1 of Lemma 7, we examine the effects of letting $\rho_i \to 0$; that is, the isolated production rate of machine i is made very small. This can happen if $\mu_i \to 0$ or $r_i \to 0$ or $p_i \to \infty$, or some combination. It is interesting to observe that the results in statement 1 hold true whatever the cause of ρ_i being small. In statement 2, $\rho_i \to \infty$. Machine i becomes very productive, which happens if $\mu_i \to \infty$.

The results presented here are borne out by the graphs in Section 8.

LEMMA 7

1. If $\rho_i \to 0$, then $P \to 0$, $E_i \to e_i$, and $E_j \to 0, j \neq i$.

2. If $\rho_i \to \infty$, then $P \to \rho_j$, $E_i \to 0$, and $E_j \to e_j, j \neq i$.

PROOF

By combining (3.5), (3.6), (5.12), and (5.13) we obtain

$$P = \rho_1 \text{ prob}(n \neq N) = \rho_2 \text{ Prob}(n \neq 0). \quad (6.1)$$

To prove statement 1, note that (6.1) implies that as $\rho_1 \to 0$, $P \to 0$ and prob $(n \neq 0) \to 0$. By (5.13), $E_2 \to 0$. We also have prob $(n=0) \to 1$ so that prob $(n=N) \to 0$ or prob $(n \neq N) \to 1$. Consequently, (5.12) implies $E_1 \to e_1$. Statement 1 can be completed in the same manner by allowing $\rho_2 \to 0$. Statement 2 can be proved similarly.

7. Closed Form Expressions for the Steady State Probabilities

In this section, a set of explicit expressions for the probability distribution $p(n, \alpha_1, \alpha_2)$ is found. Details of these derivations are in Gershwin and Berman (1978).

7.1 Analysis of Internal Equations

We define *internal states* as states (n, α_1, α_2) where $1 \leq n \leq N-1$.[1] All other states are boundary states. *Internal equations* are the balance equations that do not include any boundary states. The rest are called *boundary equations*. Following the analysis in Gershwin and Schick (1980), we propose a solution for the internal equations of the form

$$P(n, \alpha_1, \alpha_2) = c X^n Y_1^{\alpha_1} Y_2^{\alpha_2}, \quad 1 \leq n \leq N-1, \quad (7.1)$$

where c, X, Y_1, Y_2 are parameters to be determined.

By substituting (7.1) into the internal equations we find that those equations are satisfied if X, Y_1, Y_2 satisfy the following three nonlinear equations:

$$\sum_{i=1}^{2} (p_i Y_i - r_i) = 0, \quad (7.2)$$

$$\mu_1 \left(\frac{1}{X} - 1\right) = (p_1 Y_1 - r_1)\left(1 + \frac{1}{Y_1}\right), \quad (7.3)$$

$$\mu_2 (X-1) = (p_2 Y_2 - r_2)\left(1 + \frac{1}{Y_2}\right). \quad (7.4)$$

Equations (7.2) - (7.4) can be reduced to the following fourth degree polynomial in Y_1:

$$(p_1 Y_1 - r_1)(Y_1^3 + s Y_1^2 + t Y_1 + u) = 0, \quad (7.5)$$

where s, t and u are given by

$$s = \frac{1}{p_1}(-\mu_2 - 2r_1 - r_2 - p_2 + \mu_1 + p_1), \quad (7.6)$$

$$t = \frac{1}{p_1^2}(\mu_2 r_1 - r_2 \mu_1 - r_1 \mu_1 - p_2 \mu_1 - \mu_2 p_1$$

---

[1] This is in contrast to the deterministic processing time case (Schick and Gershwin, 1978) in which the internal states are those in which $2 \leq n \leq N-2$.

$$-2 p_1 r_1 - p_1 r_2 - p_1 p_2 + \mu_2 r_2 + r_1^2$$
$$+ r_1 r_2 + r_1 p_2), \quad (7.7)$$

$$u = \frac{1}{p_1^2}[\mu_2(r_1 + r_2) + r_1 r_2 + r_1^2 + r_1 p_2]. \quad (7.8)$$

Equations (7.2) - (7.4) imply that to the solution

$$Y_{11} = r_1/p_1 \quad (7.9)$$

of (7.5) there corresponds

$$Y_{21} = r_2/p_2 \quad (7.10)$$

$$X_1 = 1. \quad (7.11)$$

The other three values of Y_1 (Y_{1j}, $j = 2,3,4$) may be obtained by the standard solution to a cubic equation [Chemical Rubber (1959)] and the corresponding values of Y_{2j} and X_j ($j = 2,3,4$) from (7.2) and (7.3).

We have shown that (7.1) satisfies the internal equations of Section 4 when (X, Y_1, Y_2) is one of the four triples defined by (7.5), (7.2), (7.3). Since the internal equations are linear, any linear combination also satisfies them. We therefore conclude that, for internal states,

$$p(n, \alpha_1, \alpha_2) = \sum_{j=1}^{4} c_j X_j^n Y_{1j}^{\alpha_1} Y_{2j}^{\alpha_2}, \quad 1 \leq n \leq N-1 \quad (7.12)$$

where c_j, $j=1, 2, 3, 4$, are parameters to be determined.

7.2. Analysis of the Boundary Equations

There are a total of eight boundary states. The probabilities of four of them are specified by Lemma 1. The other four are characterized in the next lemma.

LEMMA 8

$$p(0,0,1) = \frac{1}{r_1} \sum_{j=1}^{4} c_j(p_1 Y_{1j} Y_{2j} + \mu_2 X_j Y_{2j}) \quad (7.13)$$

$$p(0,1,1) = \sum_{j=1}^{4} c_j Y_{1j} Y_{2j} \quad (7.14)$$

$$p(N,1,0) = \frac{1}{r_2} \sum_{j=1}^{4} c_j X_j^{N-1}(\mu_1 Y_{1j} + p_2 X_j Y_{1j} Y_{2j}) \quad (7.15)$$

$$p(N,1,1) = \sum_{j=1}^{4} c_j X_j^N Y_{1j} Y_{2j} \quad (7.16)$$

Note that (7.14) and (7.16) are in the internal (7.12) form. Furthermore, the coefficients c_j satisfy

$$c_1 = 0, \qquad (7.17)$$

$$\sum_{j=2}^{4} c_j Y_{1j} = 0, \qquad (7.18)$$

$$\sum_{j=2}^{4} c_j X_j^N Y_{2j} = 0, \qquad (7.19)$$

and the normalization equation,

$$\sum_{n=0}^{N} \sum_{\alpha_1=0}^{1} \sum_{\alpha_2=0}^{1} p(n,\alpha_1,\alpha_2) = 1. \qquad (7.20)$$

This is proved by observing that (7.12) - (7.16) satisfy all boundary equations except those in which $(n,\alpha_1,\alpha_2) = (N\text{-}1, 0,1), (1,1,0), (0,1,1),$ and $(N,1,1)$. Considerable manipulation is required to show that these equations are equivalent to (7.17) - (7.19). It should be noted that two cases must be considered: $\rho_1 = \rho_2$ and $\rho_1 \neq \rho_2$. In the former case, Y_{11} is a *double* root of (7.5).

7.3 The Algorithm

All the steady state probabilities, as well as the efficiencies, the production rate, the in-process inventory, and any other measure of interest can be obtained from the following.

STEP 1
Find $Y_{1j}, Y_{2j}, X_j; j = 2,3,4$ using the cubic formula [Chemical Rubber (1959)], (7.2), and (7.3).

STEP 2
Solve Eqs. (7.18) - (7.20) to obtain $c_j; j = 2,3,4$.

STEP 3.
Use Lemma 1 and Eqs. (7.12) - (7.16) to evaluate all probabilities. These probabilities can be used to evaluate the measures of performance of Section 3. This procedure can be made somewhat more efficient by performing the necessary sums analytically. This obviates the evaluation of the internal states. A program based on this approach has been written for the TI-59 pocket calculator [Ward (1980)].

8. Computational Experience

In this section we describe the results of a set of numerical experiments. These experiments demonstrate that the model behaves reasonably, that is, as one would expect an actual transfer line to behave as its parameters are varied. Table 1 summarizes the cases that were considered. Seven cases were

Table 1: Summary of numerical experiments.

Case	Varied Parameter	Range	Graph in Fig.
1	μ_1	.1 - 1000	2
2	μ_2	.1 - 1000	2
3	p_1	.1 - 1000	3
4	p_2	.1 - 1000	3
5	r_1	.1 - 1000	4
6	r_2	.1 - 1000	4
7	N	2 - 20	5

studied because in each, one parameter was varied over the range indicated, and all others were held constant. The standard parameter values were: $\mu_1=1, \mu_2=2, p_3=3, p_4=4, r_1=5, r_2=6, N=4$.

The graphs of the production rates ($P^{(1)}$ and $P^{(2)}$) and average in-process inventories ($\bar{n}^{(1)}$ and $\bar{n}^{(2)}$) of the first two cases are plotted in Fig. 2. The superscripts refer to case numbers.

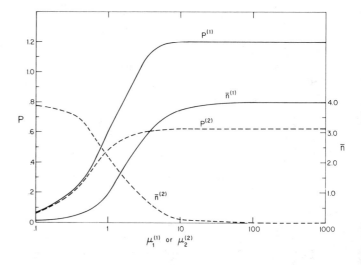

Fig. 2. Effect of machine speed on production rate and average in-process inventory.

In case 1, as μ_1, the rate of service for machine 1, increases, the production rate P increases to a limit of $\rho_2 = 1.2$ as predicted by Lemma 7. That is, there is a saturation effect, and no amount of increase in the speed of machine 1 can improve the productivity of the system. Note that as the first machine is speeded up, the average amount of material in the storage, $\bar{n}$, increases.

In case 2, in which μ_2 is varied, the production rate P increases to $\rho_1 = .625$ as μ_2 becomes large. This is again consistent with Lemma 7. When the second machine is very fast, it frequently empties the storage. Consequently $\bar{n}$ decreases and machine 2 is often starved.

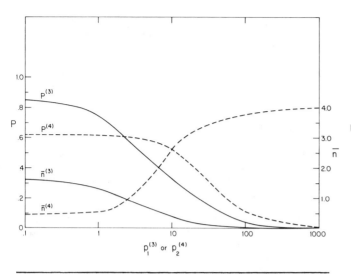

Fig. 3. Effect of machine failure rate on production rate and average in-process inventory.

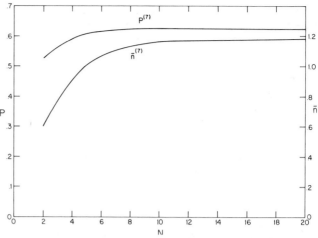

Fig. 5. Effect of storage size on production rate and average in-process inventory.

In Cases 3 and 4 failure rates p_i are varied. System production rates $P^{(3)}$ and $P^{(4)}$ and average in-process inventories are plotted in Fig. 3.

In both cases, as p_i increases, production rate decreases. As before, when the second machine is more productive, $\bar{n}$ is small, and when the first machine is more productive, $\bar{n}$ is large.

Figure 4 contains the graphs of production rates and average in-process inventories for Cases 5 and 6, in which r_1 and r_2 are varied. Again, as a machine becomes more productive (r_i increases) the system's production rate increases. Note however that Lemma 7 does not apply when $r_i \to \infty$ because ρ_i has a finite limit. As in the previous cases, when the first machine becomes more productive, $\bar{n}$ increases and when the second becomes more productive, $\bar{n}$ decreases.

The model's behavior, in these six cases, have the following characteristics in common: as any machine becomes more productive, due to μ_i or r_i increasing or p_i decreasing, the system's production rate increases. The average in-process inventory increases when the first machine becomes more productive, and it decreases when the second machine becomes more productive.

In Fig. 5 are plotted the production rate and average in-process inventory for Case 7, in which the storage size, N, is varied from 2 to 20.

As N increases, the production rate appears to increase to a limit. This limit seems to be the production rate in isolation of the least productive machine (ρ_1). [See Buzacott (1967b)]. Note also that as $N \to \infty$, $\bar{n}$ approaches a finite limit. This is evidently because machine 2 is the more productive stage in this system.

The purpose of these numerical experiments is the same as the purpose of some of the lemmas: to demonstrate that the model behaves reasonably. Because production rate and mean in-process inventory are easy to calculate using the procedure of Section 7, the model should be a useful tool for manufacturing engineers to use in evaluating alternative configurations of two-stage transfer lines.

9. Conclusions and Future Research

We have calculated the steady state probabilities for the two-machine transfer line subject to failures and exponentially distributed processing times. These probabilities are used in the calculation of efficiencies, the production rate, and the average in-process inventory. Theoretical and computational results demonstrate that the model behaves in a manner consistent with intuition.

Future research includes various generalizations of these results. Lines and networks of three or more machines should be investigated. The results of Gershwin and Schick (1980), Gershwin and Ammar (1979), Ammar (1980), indicate that this is feasible, although somewhat more complicated than the results presented here. The replacement of the exponential probability distribution with Erlang distributions [Gershwin and Berman (1978), Berman (1979)] would extend the applicability of this model. Further

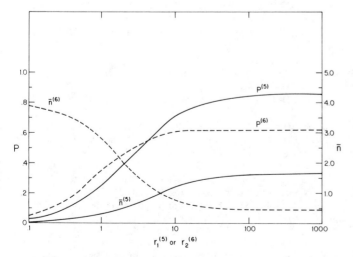

Fig. 4. Effect of machine repair rate on production rate and average in-process inventory.

numerical experience with these results should be obtained, particularly to investigate the differences between the exponential and deterministic processing time systems. If the differences are small, it may be possible to bypass the Erlang case altogether.

It should be relatively easy to develop computer programs for transfer line layout. That is, given cost data (capital cost for machines and buffers; cost rate for in-process inventory; cost for delaying demand), profit rate and demand data, it is straightforward to evaluate each possible configuration and then to search for the best.

Acknowledgments

Thanks are due to Mr. Rami Mangoubi for his help in debugging and running the computer program and to Ms. Ellen Hahne and Mr. Mostafa Ammar for their valuable suggestions. We are also grateful for the advice of two anonymous reviewers.

The MACSYMA language is supported in part by the U.S. Department of Energy (formerly the Energy Research and Development Administration) under Contract E(11-1)-3070 and the National Aeronautics and Space Administration under Grant NSG 1323.

This research was carried out in the MIT Laboratory for Information and Decision Systems (formerly the Electronics Systems Laboratory) with support extended by National Science Foundation Grants NSF/RANN APR76-12036 and NSF DAR78-17826.

References

Ammar, M. H. (1980), "Modelling and Analysis of Unreliable Manufacturing Assembly Networks with Finite Storages," MIT Laboratory for Information and Decision Systems Report LIDS-TH-1004.

Artamonov, G. T. (1977), "Productivity of a Two-Instrument Discrete Processing Line in the Presence of Failures," *Cybernetics* **12**, 3, 464-468.

Berman, O. (1979), "Efficiency and Production Rate of a Transfer Line with Two Machines and a Finite Storage Buffer," MIT Laboratory for Information and Decision Systems, Report LIDS-R-899.

Buchan, J. and E. Koenigsberg (1963), *Scientific Inventory Management,* Prentice Hall, Englewood Cliffs, N.J.

Buzacott, J. A. (1967a), "Markov Chain Analysis of Automatic Transfer Line with Buffer Stock," unpublished PhD Thesis, Department of Engineering Production, University of Birmingham.

Buzacott, J. A. (1967b), "Automatic Transfer Lines with Buffer Stocks," *Int. J. of Prod. Res.* **5**, 3, 183-200.

Buzacott, J. A. (1969), "Methods of Reliability Analysis of Production Systems Subject to Breakdown," in *Operations Research and Reliability,* Daniel Grouchko, ed., Proc. of a NATO Conf., Turin, Italy, 211-232.

Buzacott, J. A., (1972), "The Effect of Station Breakdown and Random Processing Times on the Capacity of Flow Lines with In-Process Storage," *AIIE Transactions,* **4**, 4, 308-312.

Buzacott, J. A., and L. E. Hanifin (1978a), "Models of Automatic Transfer Lines with Inventory Banks – A Review and Comparison," *AIIE Transactions* **10**, 2, 197-207.

Buzacott, J. A., and L. E. Hanifin (1978b), "Transfer Line Design and Analysis – An Overview," Fall IE Conference.

Chemical Rubbert Company (1959), *Handbook of Chemistry and Physics,* Forty-First Edition.

Gershwin, S. B. (1973), "Reliability and Storage Size, Parts I and II," Unpublished Charles Stark Draper Laboratory Memoranda FM47000-107A, FM44700-110.

Gershwin, S. B., and M. H. Ammar (1979), "Reliability in Flexible Manufacturing Systems," *Proceedings of the 1979 Conference on Decision and Control,* Ft. Lauderdale, Florida.

Gershwin, S. B. and O. Berman (1978), "Analysis of Transfer Lines Consisting of Two Unreliable Machines with Random Processing Times and a Finite Storage Buffers," MIT Laboratory for Information and Decision Systems Report ESL-FR-834-7.

Gershwin, S. B., and I. C. Schick (1978a), "Analysis of Transfer Lines Consisting of Three Unreliable Machines and Two Finite Storage Buffers," MIT Laboratory for Information and Decision Systems Report ESL-FR-834-9.

Gershwin, S. B., and I. C. Schick (1979b), "Analytic Methods for Calculating Performance Measures of Production Lines with Buffer Storages," *Proceedings of the 1978 Conference on Decision and Control,* 618-624.

Gershwin, S. B. and I. C. Shick (1980), "Modelling and Analysis of Two- and Three-Stage Transfer Lines with Unreliable Machines and Finite Buffers, MIT Laboratory for Information and Decision Systems Report LIDS-R-979.

Gordon, W. J. and G. F. Newell (1967), "Cyclic Queuing Systems with Restricted Length Queues," *Operations Research* **15**, 2, 266-277.

Hatcher, J. M. (1969), "The Effect of Internal Storage on the Production Rate of a Series of Stages Having Exponential Service Times," *AIIE Transactions* **1**, 2, 150-156.

Hildebrand, D. K. (1967), "Stability of Finite Queue, Tandem Server Systems," *Journal of Applied Probability* **4**, 571-583.

Hildebrand, D. K. (1968), On the Capacity of Tandem Server, Finite Queue, Service Systems," *Operations Research* **16**, 72-82.

Hillier, F. S., and R. Boling (1969), "The Effect of Some Design Factors on the Efficiency of Production Lines with Variable Operations Times," *Journal of Industrial Engineering* **17**, 12, 651-658.

Hillier, F. S. and R. Boling (1967), "Finite Queues in Series with Exponential or Erlang Service Times – A Numerical Approach," *Operations Research* **15**, 2, 286-303.

Hitomi, K., M. Nakajima, and Y. Osaka (1977), "Analysis of the Flow-Type Manufacturing Systems Using the Cyclic Queuing Theory," Transactions of the ASME, *Journal of Engineering for Industry,* paper No. 77-WA/PROD-32.

Hunt, G. C. (1956), "Sequential Arrays of Waiting Lines," *Operations Research* **4**, 6, 674-683.

Ignall, E., and A. Silver (1977), "The Output of a Two-Stage System with Unreliable Machines and Limited Storage," *AIIE Transactions* **9**, 2, 183-188.

Koenigsberg, E. (1959), "Production Lines and Internal Storage – A Review," *Management Science* **5**, 410-433.

Knott, A. D. (1970), Letter to the Editor, *AIIE Transactions* **2**, 3, 273.

Kraemer, S. A. and R. F. Love (1970), "A Model for Optimizing the Buffer Inventory Storage Size in a Sequential Production System," *AIIE Transactions* **2**, 1, 64-69.

Lavenberg, S. J. (1975), "Stability and Maximum Departure Rate of Certain Open Queueing Networks Having Finite Capacity Constraints," IBM Research Report, RJ 1625.

Murphy, R. A. (1975), "The Effect of Surge on System Availability," *AIIE Transactions* **7**, 4, 439-443.

Murphy, R. A. (1978), "Estimating the Output of a Series Production System," *AIIE Transactions* **10**, 2, 139-148.

Muth, E. J. (1973), The Production Rate of a Series of Work Systems with Variable Service Times," *Int. J. of Prod. Res.* **11**, 2, 155-169.

Muth, E. J. (1977), "Numerical Methods Applicable to a Production Line with Stochastic Servers," *TIMS Studies in the Management Sciences* **7**, 143-159.

Neuts, M. F. (1968), "Two Queues in Series with a Finite, Intermediate Waitingroom," *J. of Applied Probability* **5**, 123-142

Neuts, M. F. (1970), "Two Servers in Series, Studied in Terms of a Markov Renewal Branching Process," *Advances in Applied Probability* **2**, 110-149.

Okamura, K. and Yamashina, H. (1977), "Analysis of the Effect of Buffer Storage Capacity in Transfer Line Systems," *AIIE Transactions* **9**, 2, 127-135.

Patterson, R. L. (1964), "Markov Processes Occurring in the Theory of Traffic Flow through an N-Stage Stochastic Service System," *J. of Ind. Eng.* **15**, 4, 188-193.

Rao, N. P. (1975), "Two-Stage Production Systems with Intermediate Storage," *AIIE Transactions* **7**, 4, 414-421.

Reiser, M. and A. G. Konheim (1974), "Blocking in a Queuing Network with Two Exponential Servers," IBM Research Report RJ 1360.

Schick, I. C. and S. B. Gershwin (1978), "Modeling and Analysis of Unreliable Transfer Lines with Finite Interstage Buffers," MIT Laboratory for Information and Decision Systems, Report ESL-FR-834-6.

Sevast'yanov, B. A. (1962), "Influence of Stage Bin Capacity on the Average Standstill Time of a Production Line," *Theory of Probability and Its Applications* **7**, 4, 429-438.

Sheskin, T. J. (1974), "Allocations of Interstage Storage Along an Automatic Transfer Production Line with Discrete Flow," unpublished PhD Thesis, Department of Industrial and Management Systems Engineering, Pennsylvania State University.

Sheskin, T. J. (1976), "Allocation of Interstage Storage Along an Automatic Production Line," *AIIE Transactions* **8**, 1, 146-155.

Soyster, A. L. and D. I. Toof (1976), "Some Comparative and Design Aspects of Fixed Cycle Production Systems," *Naval Research Logistics Quarterly* **23**, 3, 437-454.

Ward, J. E. (1980), "TI-59 Calculator Programs for Three Two-Machine One-Buffer Transfer Line Models," MIT Laboratory for Information and Decision Systems Report LIDS-R-1009.

Wijngaard, J. (1979), "The Effect of Interstage Buffer Storage on the Output of Two Unreliable Production Units in Series, with Different Production Rates," *AIIE Transactions* **11**, 1, 42-47).

Stanley B. Gershwin received the BS degree in Engineering Mathematics from Columbia University in 1966 and the MA and PhD degrees in Applied Mathematics from Harvard University in 1967 and 1971, respectively. He was employed by the Bell Telephone Laboratories in 1970-71, where he studied the estimation of the capacity of telephone switching equipment. At the Charles Stark Draper Laboratory he studied problems in manufacturing and in transportation. His interests in these areas continues at the MIT Laboratory for Information and Decision Systems, where he is a Principal Research Scientist. His current responsibilities include the role of Principal Investigator of a project in the modeling and analysis of automated manufacturing and material handling systems. He has investigated routing optimization in networks of machine tools and the effects of reliability and limited buffer storage space. He is a member of the IEEE Control Systems Society, the American Association for the Advancement of Science, and ORSA.

Oded Berman is an Associate Professor at the Faculty of Management, University of Calgary. He received his PhD (1978) in Operations Research from MIT, and a BA in Economics and Statistics from Tel Aviv University. He taught at MIT and was also a research staff member at the Electronic Systems Laboratory of MIT. Berman's papers were accepted for publication (or published) in the *European Journal of Operations Research, INFOR. Transportation Science,* and in the *Proceedings of the IEEE on Systems, Man, and Cybernetics.*

Models for Understanding Flexible Manufacturing Systems

J. A. BUZACOTT

Department of Industrial Engineering
University of Toronto
Toronto M5S 1A4, Canada

J. G. SHANTHIKUMAR

Department of Industrial Engineering
and Operations Research
Syracuse University
Syracuse, NY 13210

Abstract: The basic features of flexible manufacturing systems are reviewed and models for determining the production capacity of such systems are developed. These models show the desirability of a balanced work load, the benefit of diversity in job routing if there is adequate control of the release of jobs (a job shop can be better than a flow shop), and the superiority of common storage for the system over local storage at machines. The models are extended to allow for material handling delays between machines and for unreliable machines. It is also shown that production capacity models can be used to develop good approximations to the mean number of jobs in the system for given job arrival rates and machine utilizations.

■ A flexible manufacturing system (FMS) consists of machines where production operations are performed, linked by a material handling system and all under central computer control. In contrast to a transfer line where all parts follow the same sequence of operations, the material handling system permits the parts to follow a variety of different routings. Flexible manufacturing systems are sometimes also called computerized manufacturing systems and they are essentially automated job shops [2, 16, 17]. The FMS should be able to process a large total volume of small to medium batches of parts.

Flexible routings can be achieved by

1. providing separate paths between each pair of machines where part movement might occur; or

2. using a common material handling device through which all parts pass and which connects all machines.

(For example, a loop conveyor passing all machines [8] or a rail mounted transfer car [16].)

Received May 1979; revised July 1980. Paper was handled by Facilities Design/Location Analysis Department. Work supported by National Research Council of Canada under grant A4372 and by Canadian Commonwealth Scholarship and Fellowship Committee. An earlier version of this paper was presented at the ORSA/TIMS meeting, Los Angeles, November 1978.

Because of the linkage between machines created by the material handling system it is necessary to analyze them using a systems perspective. The objective of this paper is to show how some relatively simple models can be used to understand the inherent relative merits of some alternative systems designs.

Storage in Flexible Manufacturing Systems

In the classical job shop it is inevitable that, whether planned or not, a substantial amount of space is occupied at each machine by work waiting to be processed at the machine or waiting to be moved to the next machine on its route. With FMS it is necessary to plan for the storage space either in front of individual machines or for the system as a whole.

If the machines are linked by a material handling system that permits no storage and a machine breaks down, all other machines that process parts either going to or coming from the failed machine may be forced down. The disruption may eventually affect all machines in the system. Also, as the utilization of the machines increases, it is possible that no part movement between machines can occur until the longest operation anywhere in the system is completed. Furthermore, the diversity of routes taken by different parts may result in machines being blocked, because a part which has completed its operation at the machine is unable to move on to its next machine which is still busy. Therefore it is desirable to provide some storage to reduce the effect of breakdowns, variability in operation times, and the blocking of machines due to the diversity of part routing.

There are two basic alternatives which can be used in providing storage, local storage at machines (see [11] or the Hüller system in [14]), or a common storage for the system which is accessible by all machines. There are a number of ways suggested for providing the common store. One method of achieving flexible routing is to link each machine to an AS/RS system or a set of pigeon holes [18] in which parts in process are stored. On completion of an operation at a machine the part goes to some location in the AS/RS system. When the next machine on the part's route becomes available, the part can be retrieved and sent to the machine (see Fig. 1A). Alternatively, if a loop conveyor is used to move parts between machines the parts on the loop can be considered as being in the common storage, that is, the size of the common storage is the number of spaces for parts on the loop conveyor (see Fig. 1B). Reference [1] describes an application of the loop conveyor system in the clothing industry.

Some systems use both common and local storage [8]. For example, in loop conveyor systems a small number of storage spaces may be provided in front of and after each machine. The purpose of these local storage spaces is to reduce the time the machine is idle waiting for the conveyor to bring the next part.

In the models which follow, transit times between machines or between machines and store will initially be ignored. However, in a subsequent section the inclusion of transit times in the models will be discussed.

Control of Flexible Manufacturing Systems

Because of the complexity of flexible manufacturing systems, the potential diversity of part routing and the variability in operation times, it is necessary to give careful consideration to the control of the system at three levels: (1) pre-release planning, (2) release or input control, and (3) operational control.

1. At the pre-release phase, the parts which are to be manufactured by the system are decided, constraints on the operation sequence identified, and operation durations estimated.

2. The purpose of input control is to determine the sequence and timing of the release of jobs to the system.

3. At the operational control level the movement of parts between machines must be ensured. If a number of alternatives exist as to which part should go on a machine, the conflict must be resolved. Furthermore, if machines should break down it is necessary for the system to respond so that the consequent disruption is minimized.

Because of the inherent complexity of operational control it is often considered advantageous to transfer certain decisions from the operational level to the pre-release level. For example, if the operation sequence constraints permit alternate routings, some specific routing may be chosen at the pre-release level in order to remove this decision from operational control. In non-automated systems the usual philosophy is to make operational control as simple as possible.

Input control and operational control can also be simplified by reducing the amount, both in frequency and in detail, of information collected about what is happening in the system. For example, it may be preferable to decide which job to release next to the system on the basis of detailed information on the work load at each machine and how it will change (consequent on the routing of jobs) once the present operations at the machines have been completed. However, such information may be costly to collect and use as the basis for decision. Instead the decision on releasing a job might be based solely on the number of jobs in the system and the availability of a space for the job in the system if it were released.

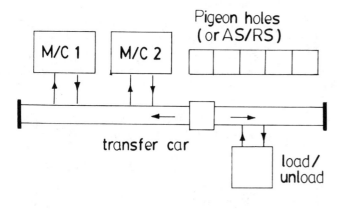

Fig. 1A. Flexible manufacturing system with AS/RS or pigeon holes.

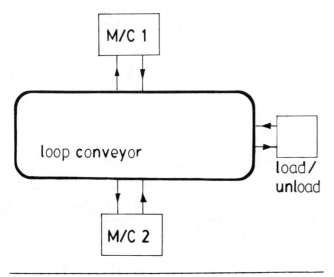

Fig. 1B. Flexible manufacturing system with loop conveyor.

It must be appreciated that, if control costs can be ignored, any restriction on the options available at the operational level is disadvantageous. Ashby's Principle of Requisite Variety states that the greater the control options available, i.e., the more variety available to the controller, the less will be the effect of disturbances on system performance. Thus, insofar as it is technically and economically possible, advantage should be taken of the flexibility inherent in the FMS.

Unfortunately, the more complex the information structure the more complex is the model required to describe the system and the less likely simple solutions can be found. Thus none of the models to be considered in this paper permits control based on knowing in advance the processing times of a specific job.

Modelling the System

The purpose of the models is to develop a general understanding of the way in which the performance of a flexible manufacturing system depends on its configuration. Thus, some features are simplified so that results can be obtained using queueing models with analytic solutions.

The system modelled consists of the flexible manufacturing system and the queue of jobs waiting to enter it. This queue will be called the initial queue.

The general assumptions are (cf. [15]):

(i) There are M machines, each of which can only process one job at a time.

(ii) Jobs arrive at the system and join the initial queue according to a Poisson process with parameter λ.

(iii) There are R classes of jobs processed by the system. The probability that an arriving job is class r and has first operation on machine i is q_{ir}.

(iv) Processing times are all exponentially distributed with mean $1/\mu_i$ at machine i, independent of job class.

(v) Within each job class the routing of jobs is determined by a probability transition matrix $P_{ij,r}$ for class r. $1 - \sum_{j=1}^{M} P_{ij,r}$ is the probability a class r job leaves the system after processing at machine i.

(vi) At machine i there is storage space N_i, including the space for the job on the machine. This storage space is only used by jobs whose next operation is on machine i.

(vii) Within the flexible manufacturing system itself there is a common storage space for z jobs. This space is accessible by all machines.

(viii) The first K jobs in the initial queue are stored in such a way that the class and machine for the first operation of each of these jobs are known and any of these K jobs can be selected for release to the flexible manufacturing system. $K = 1$ corresponds to FCFS queue discipline.

If, when a machine finishes processing of a job, there is no available space to which the job can be moved, then the machine will be *blocked* and unable to work on the next job.

One parameter of interest is e_{ir}, the expected number of times an arriving job will visit machine i with class r. It can be found by solving the M equations

$$e_{ir} = q_{ir} + \sum_{j=1}^{M} P_{ji,r} e_{jr} \ (i = 1, \ldots, M).$$

If the arrival rate of jobs at the system is λ, U_i, the utilization of machine i, is given by

$$U_i = \lambda a_i$$

where $a_i = e_i/\mu_i$ and $e_i = \sum_{r=1}^{R} e_{ir}$.

There are several types of system which will be considered:

(1) a balanced system: $a_i = a \ (\forall_i)$

(2) a conventional job shop: $e_i = 1, P_{ii} = 0 \ (\forall_i)$

(3) a single class job shop: $e_i = 1, P_{ii} = 0, (\forall_i), R = 1$.

In the special case of a two machine single class job shop in which $q_1 = q$ it follows that $q_2 = 1 - q$, $P_{12} = q$, $P_{21} = 1 - q$.

(4) a flow shop: a single class job shop in which $q_1 = 1, q_i = 0$ $(i \neq 1), P_{i,i+1} = 1 \ (i = 1, 2, \ldots, M-1)$ and all other $P_{ij} = 0$.

(5) a pure job shop: a single class job shop in which

$$q_i = 1/M \ (\forall_i), P_{ij} = 1/M \quad (\forall_i, \forall_{j \neq i}).$$

(6) a two machine two class job shop: This is a job shop where, for class 1, $P_{12,1} = 1$ and all other $P_{ij,1}$ are zero, while, for class 2, $P_{21,2} = 1$ and all other $P_{ij,2}$ are zero. Also $q_{11} = q$, $q_{12} = 0$, $q_{21} = 0$, $q_{22} = 1 - q$. Note that $e_1 = e_2 = 1$.

In any balanced system for which $e_i = 1 \ (\forall_i)$ it follows that $\mu_i = \mu \ (\forall_i)$.

Calculating system performance

The simplest measure of system performance which can be used to compare alternatives is the production capacity, denoted by PC. There are several approaches to calculating the production capacity.

(1) Calculate the departure rate from the system when the initial queue is infinite.

(2) Find the maximum arrival rate for which the system is stable by setting up the system state equations and finding the conditions under which the equations have a solution.

(3) Consider the closed system equivalent to the original open system of flexible manufacturing system and at least the first K positions in the initial queue. In the closed system there is a fixed number of jobs L and a job leaving the flexible manufacturing system is immediately fed back to join the initial queue with probability q_{ir} that the job is of class r and has its first operation on machine i. Then, in the closed system calculate the rate at which jobs are fed back, $TH(L)$. Then the production capacity is $\lim_{L \to \infty} TH(L)$.

It can be shown that (2) and (3) are equivalent (see Schweitzer [15]). Method (1) can give a higher value of capacity if, in applying it, some assumption is made which implies either that there is infinite storage space in the system or that K is infinite (see Schweitzer's example). However, with finite storage space in the system and K finite it appears method (1) will agree with methods (2) and (3). Method (3) has been used to obtain most of the results which follow; those results obtained in [7] using method (1) have been checked to see that they would be unchanged if method (3) were used.

The Models

I. Systems With Local Storage Only

IA. Infinite Local Storage —
(i) All jobs released on arrival.

If the queue discipline at each machine is FCFS then the basic models are those of Jackson [9] for a single class of jobs and Basket et al [4] for multiple job classes. They showed that the system can be decomposed and the machine queues treated separately. From their results

$$PC = 1/a_{max}$$

where

$$a_{max} = \max_i a_i$$

It follows that it is desirable for the system to be so designed that all machines have approximately the same utilization.

(ii) A maximum of C jobs allowed in the flexible manufacturing system. FCFS release.

This corresponds to the case where only C pallets on which jobs can be located are available. FCFS is assumed at the machines and FCFS is assumed for release of jobs when pallets become available.

Schweitzer shows that

$$PC = q(C-1)/q(C)$$

$$= \frac{\sum_{n_1+n_2+\cdots=C-1} \prod (a_i)^{n_i}}{\sum_{n_1+n_2+\cdots=C} \prod (a_i)^{n_i}}$$

where n_i is the number of jobs in the queue at machine i;
In a balanced system of M machines

$$PC = \frac{C}{a(C+M-1)}.$$

Note that in both (i) and (ii) the capacity is independent of the routing of the jobs, i.e., all routings giving the same set of e_i will give the same capacity.

Solberg [16] describes the use of this model for a computerized manufacturing system with a single class of jobs.

(iii) A maximum of C jobs allowed in the flexible manufacturing system. Idle machine rule release.

The idle machine rule for release of jobs is to only release a job to the shop when a machine is idle. The job to be released must be one whose first operation is on the idle machine. The production capacity will depend on the routing of jobs and K, the number of jobs in the initial queue which are accessible for release to the shop.

In a balanced two machine pure job shop it can be shown that, for large K and FCFS at the machines,

$$PC = \mu(1 - [½]^C)$$

as compared to $PC = \mu C/(C+1)$ in case (ii). There is thus a significant improvement in capacity.

The dependence on job routing can be seen from a balanced two machine single class job shop where $C = 2$. For large K it can be shown that (see appendix 1)

$$PC(q) = PC(1-q)$$

and

$$\frac{PC(q)}{\mu} = \frac{2-q}{3-2q} \qquad 0 \leq q < ½.$$

That is, the capacity is a maximum for a pure job shop while, as would be expected, there is no difference between FCFS and the idle machine rule in a flow shop. Figure 2 shows a plot of $PC(q)$ over the range of q from 0 to 1.

The dependence of the capacity on K for a balanced two machine pure job shop with $C = 2$ can be shown to be

$$PC(K) = \mu \frac{3K+1}{4K+2} \qquad K \geq 1$$

$K = 1$ implies FCFS release.

While no closed form solutions exist for a three machine pure job shop the idle machine rule is also effective in increasing the production capacity. Table 1 indicates the improvement which can be obtained in a balanced system.

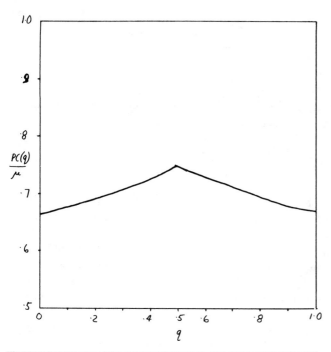

Fig. 2. Production capacity of balanced job shop using idle machine rule for release (C=2).

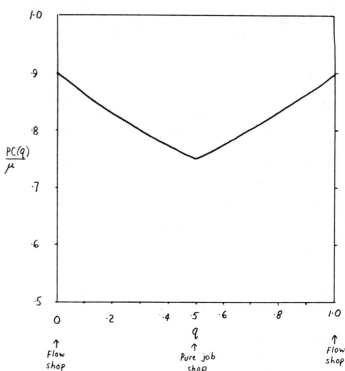

Fig. 3. Production capacity of a balanced job shop with local storage and jobs released when space available (N=8).

Table 1: PC/μ for three machine pure job shop.

C	2	3	4	5
FCFS	.500	.600	.667	.714
Idle machine	.533	.667	.762	.840

IB. Finite Local Storage —

(i) Jobs released to system whenever space is available.

Since blocking of machines can occur it is necessary to define a priority for what happens when a space becomes available to which a blocked machine can transfer its jobs. It is assumed that this transfer has priority over release of jobs from the initial queue. The capacity depends on the routing of jobs, K and N.

For a balanced two machine single class job shop in which $N_1 = N_2 = N$ it can be shown that for large K

$$PC(q) = \mu \frac{(1-q^2)(1+q)^{N-1} - (1-2q)}{(1+q-2q^2)(1+q)^{N-1} - (1-2q)} \quad 0 < q < \tfrac{1}{2}$$

$$PC(0) = \mu (N+1)/(N+2)$$

$$PC(\tfrac{1}{2}) = \mu 3/4$$

$$PC(q) = PC(1-q) \qquad \tfrac{1}{2} < q \leq 1.$$

It can also be shown for a balanced two machine two class job shop when $q = \tfrac{1}{2}$ that the capacity is also $3\mu/4$.

Figure 3 shows a plot of $PC(q)$ as a function of q. Maximum capacity is obtained for a flow shop while the capacity of a pure job shop is independent of the storage space provided at the machines. This result indicates why poorly controlled job shops have inherently low production capacity and high work in process (cf. [12]).

(ii) A maximum of C jobs allowed in the flexible manufacturing system ($C \leq \Sigma N_i$).

If $C \leq \min N_i$ then FCFS release can be used and the results of IA(ii) apply. Alternatively, the idle machine rule can be used and the results of IA(iii) apply.

When $C = \min N_i + 1$, blocking can occur on one machine. However, there will be no job in the flexible manufacturing system waiting for the blocked machine so the blocked machine can be looked on as providing an extra storage space. So again the results of IA(ii) or IA(iii) apply as appropriate.

However, once $C > \min_i N_i + 1$, blocking will occur and the results of IA(ii) or IA(iii) are no longer valid.

Consider two machine systems where $N_1 = N_2 = N$, and $C \geq N+1$. Suppose FCFS is used for release of jobs unless the job at the head of the initial queue is for a machine where there is no storage space available. In this case a job whose first operation is on the other machine is selected for release so that the number of jobs in the system is kept at C.

Then in a balanced two machine pure job shop it can be shown that, for large K, (see [7]),

$$PC(C) = \mu \frac{4N - 2C + 3}{4N - 2C + 4} \qquad N+1 \leq C \leq 2N.$$

The maximum value of the capacity occurs when $C = N+1$ and it is then

$$PC(N+1) = \mu(2N+1)/(2N+2).$$

However, for a balanced two machine flow shop

$$PC(C) = \mu(N+1)/(N+2) \qquad N+1 \leqslant C \leqslant 2N.$$

Thus it can be seen that greater capacity can be achieved with a pure job shop than a flow shop. Even greater capacity can be achieved if the idle machine rule is used in a pure job shop.

II. Common Storage Only

A maximum of C jobs allowed in the system.

In a two machine system the maximum value of C possible is $z + 2$ if the common store has z units capacity. For $C \leqslant z + 1$ no blocking will occur, while for $C = z + 2$ blocking of one machine can occur but no job in the store would then be waiting to go on that machine. The machine is then being used only as a storage space.

It follows that the results obtained in IA(ii) and IA(iii) apply for all possible values of C with maximum production capacity obtained when $C = z + 2$.

Alternatively, the results in IB(ii) for $C = N + 1 = z + 2$ will also apply, i.e.

$$PC = \mu(2z+3)/(2z+4)$$

in a balanced two machine pure job shop with FCFS release except when a machine would otherwise be idle. It can be seen that the common store automatically restricts the number of jobs in the system to the optimum value.

In a system with more machines the maximum value of C is $z + M$. Again for $C \leqslant z + 2$ the results of IA(ii) apply. However, while no results are available for $z + 2 < C \leqslant z + M$, it would be expected that capacity will increase over this range.

Implications of Models

There are three general conclusions which can be drawn from the models.

(1) The Importance of Balance

This was pointed out in IA(i). If the job routing is determined at the pre-production planning level this means that they should be chosen so as to minimize the maximum a_i subject to the technological constraints on operation sequence. This problem can be formulated as an LP.

When the number of jobs in the system is restricted then strict balance is not necessarily optimum. The determination at the pre-production level of the optimum routing becomes a non-linear programming problem (see [10]). However, the difference in capacity between strict balance and the optimum routing is generally quite small unless the number of jobs is restricted to a number comparable to the number of machines.

Achieving balance will be easiest when there is a fair degree of flexibility in assigning jobs to machines or all machines are capable of a wide variety of operations.

(2) The Importance of Diversity in Job Routing

In addition to a balanced workload on the machines, it is desirable to have a diversity of jobs with different routings since, if the release of jobs to the flexible manufacturing system can be controlled, higher production capacity can be achieved. The two machine models show that the effectiveness of a given storage space or restricted number of pallets in a pure job shop is double that in a flow shop. Even greater improvement can be obtained using the idle machine rule.

For a flexible manufacturing system to perform at its best, there should be jobs having diverse routes available. Alternatively, for jobs with some flexibility in the sequence in which operations are performed, it is better *not* to fix the sequence at the pre-production planning level. The flexibility in sequence can be used to create diverse routings.

(3) The Superiority of Common Storage over Local Storage

The models show that common storage is superior to local storage. Local storage can be used, but only if there is close control over the release of jobs to the system so that little blocking occurs. A common storage space automatically achieves control over the number of jobs in the system (with two machines at the optimum level). The only reason for providing local storage is to reduce delays while the material handling system moves job between machines or between machines and the central store. Thus it is likely that space for only one or two jobs is required.

Some Extensions of the Models

(a) *Allowing for material handling system delays*

Some of the models can be extended to allow for finite transit times between machines and delays in the material handling system. In particular, the Posner and Bernholtz model [13] of closed queueing systems with time lags can be used to find $TH(L)$ provided there is infinite local storage at machines, FCFS queue discipline and no restriction in the number of jobs in transit at any instant (i.e., no congestion can occur in the material handling system). The model permits the transit times to have general distributions.

The key result of the Posner and Bernholtz model is that a closed queueing network with time lags and M stations is equivalent to an $M+1$ station system with no time lags in which the $M+1th$ station has L parallel channels each having service time $b = \bar{H}$ where $\bar{H}$ is the mean transit time of a job between stations. Thus the transit times create an extra service centre which, using Schweitzer's terminology, is IS (infinitely many parallel servers). Thus Schweitzer's results can be used to determine the production capacity of a flexible manufacturing system in which a maximum of C jobs are allowed in the system. We have

$$PC = q(C-1)/q(C)$$

where

$$q(K) = \sum_{S+T=K} \left\{ \sum_{\substack{n_1, n_2, \ldots, n_M \geq 0 \\ n_1 + n_2 + \ldots + n_M = S}} \prod_i (a_i)^{n_i} \right\} b^T/T!$$

In the case of a balanced system

$$q(K) = \sum_{S+T=K} \left\{ a^S \prod_{u=1}^{M-1} (S+u) \right\} b^T/T!$$

In the limit, as $C \to \infty$, Schweitzer shows that

$$PC = 1/a_{\max}$$

i.e., the production capacity is then not affected by finite transit times.

In general, transit times have a negligible effect on the production capacity in the situations where the Posner and Bernholtz model is valid. The effect increases with diminishing C and increasing ratio of b/a (mean transit time / mean processing time). For example, with $M = 2$, $C = 4$, and $b/a = .2$, the production capacity is reduced by less than 1% from its value with zero transit times. It is only when transit times become comparable with processing times that the effect would be significant at low values of C.

If there is a single material handling device jobs could be delayed, waiting for the device to complete other moves. The Posner and Bernholtz model is then not valid. Solberg [16] assumed that such a device could be treated as a further server in a closed Jackson-type queueing network. Such a model assumes that there is infinite storage for jobs waiting to be served by the material handling device. This means that when a machine completes a job the machine is not then blocked, waiting for the material handling device to move the job. That is, there must be local storage at the machine for completed jobs so the machines can begin the next job as soon as the previous job is completed.

(b) *Determining inventory levels and waiting times*

Although the production capacity is probably the most important measure of system performance, it is also desirable to develop measures of performance for a given arrival rate λ. The most useful are the mean inventory levels and the mean waiting times, specifically:

L_T = expected total number of jobs in the flexible manufacturing system and in the initial queue

L_F = expected number of jobs in the flexible manufacturing system

W_T = expected time between arrival of a job at the initial queue and departure from the system

W_F = expected time between release of a job to the flexible manufacturing system and its departure.

Little's formula gives

$$L_T = \lambda W_T \quad \text{and} \quad L_F = \lambda W_F .$$

The exact calculation of L_T and L_F requires setting up and solving the system state equations. This approach is usually too complex to give meaningful solutions. However, a number of alternative approximations can be used.

The simplest is to regard the system as being equivalent to an $M/M/1$ queue in which the service rate equals PC.
Hence

$$L_T^{①} = \rho/(1-\rho) \quad \text{and} \quad L_F^{①} = \sum_{n=0}^{C-1} n\rho^n(1-\rho) + C\rho^C$$

where $\rho = \lambda/PC$ and C is the maximum number of jobs allowed in the flexible manufacturing system.

A better approximation is an approach first suggested by Avi Itzhak and Hayman [3] for modelling central server computer systems and developed by Buzacott [6] for modelling job shops with limited storage space and control of input. In this approach the system is regarded as equivalent to a single server with state dependent exponentially distributed service time.

The approximation technique is

(i) Using the closed queueing model determine $TH(L)$ for each $L \geq 1$.

(ii) Consider an open system with single server and service rate when there are L customers in the system, μ_L given by $\mu_L = TH(L)$.

Then the probability of u jobs in the system is given by

$$p(u) = \prod_{L=1}^{u} (\lambda/\mu_L) \; p(0)$$

and

$$\sum_{u=0}^{\infty} p(u) = 1 .$$

(iii) Determine $L_T^{②} = \sum_{u=0}^{\infty} u \; p(u)$

$$L_F^{②} = \sum_{u=0}^{\infty} n_u p(u)$$

where n_u is the expected number of jobs in the flexible manufacturing system given that there are u jobs in the system as a whole.

Table 2: Comparison of approximate and simulation results for L_T and L_F when $N_1 = N_2 = 4$.

u	C	Approximation			Simulation		
		L_T^1	L_T^2	L_F^2	Observed Utilization	L_T	L_F
.4	2	1.500	1.667	1.067	.401	1.654 ± .083	1.069
	4	1.000	1.364	1.272	.397	1.368 ± .068	1.257
	5	0.800	1.335	1.294	.408	1.398 ± .068	1.345
	8	1.143	1.351	1.330	.403	1.365 ± .070	1.340
.5	2	3.000	3.200	1.400	.499	3.308 ± .165	1.405
	4	1.667	2.136	1.796	.501	2.104 ± .105	1.788
	5	1.250	2.012	1.850	.500	2.063 ± .086	1.884
	8	2.000	2.113	2.004	.493	2.109 ± .086	1.979
.6	2	9.000	9.231	1.754	.601	9.236 ± .907	1.761
	3	4.000	4.421	2.147	.601	4.672 ± .275	2.156
	4	3.000	3.577	2.423	.600	3.661 ± .183	2.430
	5	2.000	3.064	2.524	.602	3.213 ± .151	2.538
	6	2.182	3.188	2.730	.606	3.400 ± .183	2.794
	7	2.571	3.322	2.893	.601	3.499 ± .262	2.938
	8	4.000	3.686	3.082	.596	3.595 ± .243	3.063
.7	2	←-------------------------- UNSTABLE --------------------------→					
	3	14.000	14.487	2.703	.697	16.478 ± 3.88	2.705
	4	7.000	7.686	3.157	.697	7.866 ± .611	3.157
	5	3.500	5.016	3.308	.698	5.374 ± .280	3.355
	8	14.000	11.468	5.241	.698	13.932 ± 2.361	5.428

While in many situations $n_u = u$ for $u < C$ and $n_u = C$ for $u \geqslant C$, with limited storage space at the machines n_u can be less than the above values. For example, with FCFS release, all queue spaces may be full at the machine to which the first job in the initial queue should go. n_u can be calculated from the closed queueing model used to determine $TH(u)$.

Simulation experiments have been carried out to test the validity of the approximations. Table 2 shows a comparison of simulation and approximation in the case where there is finite local storage and a maximum of C jobs allowed in the flexible manufacturing system [case IB (ii) above]. The results given are for a balanced two machine pure job shop in which $N_1 = N_2 = 4$. The simulation used the regenerative technique and for L_T the 90% confidence limits are given. Some results for $TH(L)$ needed in obtaining the approximate values of $L_T^{(2)}$ are given in appendix 2.

The state dependent queueing model gives very good predictions for both L_T and L_F. The simple queueing model gives reasonably good results for L_T when the arrival rate approaches the production capacity but otherwise they are relatively poor. The L_F predictions are poor and they are not shown on the table. In one case, $C = 2$, it is possible to get exact solutions for L_T and L_F. For L_T it can be shown that

$$\text{exact } L_T = \frac{2U(1+U^2/12)}{(1+U/2)(1-3U/2)}$$

$$\text{approx } L_T^{(2)} = \frac{2U}{(1+U/2)(1-3U/2)}$$

where $U = \lambda a$.

Since the maximum value of U for which the system is stable is $U = 2/3$ it can be seen that the maximum percentage error between exact and approximate model is 4%.

The simulation model was used to investigate some other sequencing rules. In a balanced two machine system the lowest value of L_T was obtained with the shortest processing time queue discipline at the machines and least total processing time for release from the initial queue. However, these rules do not appear to increase the production capacity.

(c) *Effect of breakdown*

It is possible to adapt a model of a transfer line with breakdowns to represent a particular flexible manufacturing system where machines break down.

In a Markov model of a two stage transfer line the system state is described by the set of states of the stages, i.e., whether the stage is working or failed, and by the level x of the inventory bank where x ranges from O to z, the bank capacity.

Now consider a two machine flexible manufacturing system with identical constant processing times at the machines. Suppose the job routing defines a two class balanced job shop, i.e., jobs can either follow sequence 1→2 or sequence 2→1. In a Markov model the system can be described by the set of states of the machines and by another variable x ($-z \leqslant x \leqslant z$), where $x > 0$ implies that the bank has x jobs which have been to machine 1 only, and $x < 0$ implies that the bank contains $-x$ jobs which have been to machine 2 only.

Suppose the jobs are following sequence 1→2 and machine 1 breaks down. Machine 2 will first process the jobs in the bank until it is emptied. Then it will switch to the other class of jobs, refilling the bank with jobs which have been to machine 2 only. Then, when machine 1 is repaired it will process the jobs in the bank. The transitions between states of a flexible manufacturing system operated according to this policy are identical with the transitions between the states of an ordinary transfer line with bank capacity $2z$. That is, the performance of the flexible manufacturing system with a bank of capacity z is identical to the performance of a transfer line with a bank of $2z$ and having station characteristics (breakdown rate and repair rate) identical to the characteristics of the machines of the flexible manufacturing system (cf. [5]).

This result is analogous to that obtained on p. 344 comparing the production capacity of a flow line and a flexible manufacturing system without breakdowns; the effectiveness of a given bank capacity in a flexible manufacturing system is double what it is in a flow line.

Considering the effect of breakdowns in the case of identical machines, the production rate PR of the flexible manufacturing system is given by

$$\frac{PR}{PC} = \frac{2 + 2L(1+x)}{2(1+2x) + 2L(1+x)^2}$$

where $x = p\tau$, $L = z/\tau$, p is the breakdown rate of a machine per cycle, τ is its repair time in cycles, and PC is the production capacity without breakdowns. PC equals $1/a$ for identical constant processing (cycle) times of a minutes.

Conclusions

The general implication of all the models considered is that flexible manufacturing systems have the potential for significantly higher capacity than the equivalent flow line system. In two machine systems a general rule is that the effectiveness of a given storage capacity is doubled if there is sufficient diversity in job routing available. The flexible manufacturing system makes use of the flexibility afforded by the diversity in routing to be less affected by disturbances due to variability in job processing times or machine breakdowns.

However, it must be noted that the improvement in production capacity is dependent on sufficient control capability being provided. If the bare minimum control is used, e.g., release all jobs to the system as soon as there is space available for them, then the performance can be worse than the equivalent flow line. More control is required than the flow line (where such elementary rules can be safely used) and so it is to be expected that the flexible manufacturing system are more susceptible to disturbances in control. This is probably why their development has been contingent on the use of computer control and automated data collection on job progress. However, it is possible that the superior performance of semi-autonomous groups occasionally observed in manufacturing has been due to the group intuitively developing means of responding to disturbances by using any inherent flexibility in the sequence of task performance or work assignment.

In addition to the high initial cost of flexible manufacturing systems and the newness of the concept, there are other barriers to their widespread use. To achieve a reasonably balanced utilization of machines, to make use of any diversity of routing of different jobs or to use alternative routings for a given job type requires careful overall planning. Justification for the purchase of a system will occur when a company has a relatively fixed manufacturing program encompassing a variety of models and a diversity of parts. Thus the degree to which balance and diversity can be achieved will be identified before making the decision to install the system.

Comparing general purpose and special purpose machines, and observing the differences between ease of achieving balance and amount of material handling, we expect that most flexible manufacturing systems will be initiated with general purpose machines integrated by a material handling system and common store to achieve a balanced load, economy in space utilization, and the ability to easily respond to individual machine breakdowns or stoppages.

It is quite likely that special purpose machines will be used in flexible manufacturing systems only when most jobs follow the same sequence of machines. Such systems would be an alternative to transfer lines where the transfer line cycle time is relatively long and the transfer time between machines need not be minimized. The common store and computer controlled material handling system would permit machines to operate with different cycle times (particularly useful if their breakdown rates are different), and would enable jobs to be fed back to an earlier stage if an automatic inspection station reveals the need for rework or adjustment. They also permit a modular approach to system design and installation so that individual machines could be added to or removed from the system if a change in workload or mix of operations is required.

Appendix 1. Derivation of Results

Results in IA(iii) Idle machine rule in a two machine single class job shop with $C = 2$.

In the closed system consisting of the flexible manufacturing system and part of the initial queue take $L = K + C$. It can be shown that $TH(L)$ is independent of L for $L \geq K + C$.

Write the state of the system as $(ij, k\ell, u)$ where i and j are the number of jobs at machines 1 and 2 respectively, k and ℓ are the number of jobs among the first K in the initial queue which have first operation at machines 1 and 2 respectively, and u is the number of jobs in the initial queue not in the first K. With $L = K + C$, the only states in which u is non-zero are $(10, K0, 1)$ and $(01, 0K, 1)$.

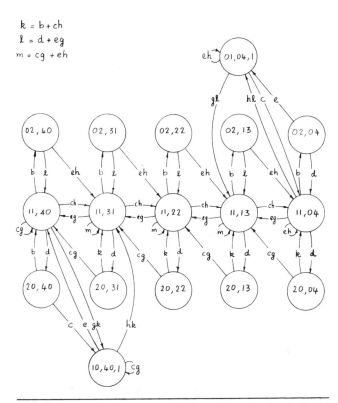

Fig. 4. State transition diagram with idle machine rule and $C=2$, $K=4$.

The system state transition diagram is shown in Fig. 4 (for $K=4$), where the following symbols have been used.

$b = \mu_1 P_{12}$ $\qquad d = \mu_2 P_{21}$

$c = \mu_1(1-P_{11}-P_{12})$ $\qquad e = \mu_2(1-P_{21}-P_{22})$

$g = q_1$ $\qquad h = q_2$

If P_{11} or P_{22} were non-zero then associated with every state would also be a loop corresponding to a job returning to the same machine when its operation is completed. These transitions are not shown.

In the particular case of a balanced two machine single class job shop where $q_1 = q$, and $\mu_1 = \mu_2 = \mu$ it can be shown that the state probabilities are related by

$\Pr(20, K\text{-}r\ r, 0) = (1-q)\Pr(11, K\text{-}r\ r, 0)\ (0 \leq r \leq K)$

$\Pr(02, K\text{-}r\ r, 0) = q \Pr(11, K\text{-}r\ r, 0)\ \ (0 \leq r \leq K)$

$\Pr(10, K0, 1) = \Pr(11, K0, 0)$

$\Pr(01, 0K, 1) = \Pr(11, 0K, 1)$

$\Pr(11, K0, 0) = \dfrac{q}{2(1-q)} \Pr(11, K\text{-}1\ 1, 0)$

$\Pr(11, K\text{-}r\ r, 0) = \dfrac{q}{1-q} \Pr(11, K\text{-}r+1\ r+1, 0)$
$\qquad\qquad\qquad\qquad (1 \leq r \leq K-2)$

$\Pr(11, 1K\text{-}1, 0) = \dfrac{2q}{(1-q)} \Pr(11, 0K, 0)$

and

$TH(L) = \mu \displaystyle\sum_{r=0}^{K} \{(1-q)\Pr(20, K\text{-}r\ r, 0)$

$\qquad + \Pr(11, K\text{-}r\ r, 0) + q \Pr(02, K\text{-}r\ r, 0)\}$

$\qquad + (1-q)\Pr(10, K0, 1) + q \Pr(01, 0K, 1)$

$\dfrac{PC(q)}{\mu} = \dfrac{2-q-(1+q)[q/(1-q)]^K}{3-2q-(1+2q)[q/(1-q)]^K}$ $\qquad q \neq 1/2$

$\dfrac{PC}{\mu} = \dfrac{3K+1}{4K+2}$ $\qquad q = 1/2$

for K large

$\dfrac{PC(q)}{\mu} = \dfrac{2-q}{3-2q}$ $\qquad 0 \leq q < 1/2$

$\dfrac{PC(q)}{\mu} = \dfrac{1+q}{1+2q}$ $\qquad 1/2 < q \leq 1$

Appendix 2. Details of Approximate Model for $L_T^{(2)}$

The state dependent queueing model for approximating L_T and L_F requires that $TH(L)$ be determined for all values of $L \geq 1$.

In case 1B(ii), i.e., a two machine balanced pure job shop, it can be shown that, with $N_1 = N_2 = N$,

(a) When $C \leq N$

$\qquad aTH(L) = L/(L+1) \qquad L < C$

$\qquad aTH(L) = C/(C+1) \qquad L \geq C$

Hence in the approximate model with $\lambda a = U$

$\qquad p(u) = (u+1)U^u\, p(0) \qquad u < C$

$\qquad p(u) = (C+1)U^u [(C+1)/C]^{u-C} p(0) \qquad u \geq C$

(b) When $C > N$

Closed form solutions do not exist for $TH(L)$ for all L, C, N combinations. The three cases where closed form solutions exist are:

(i) $L \leq N$

$\qquad aTH(L) = L/(L+1) \qquad$ for all C

(ii) $C = 2N$, $L \geqslant 2N$

$$aTH(L) = \frac{3(L-2N) + 8 - (1/2)^{N-3}}{4(L-2N) + 10 - (1/2)^{N-3}} \quad N \geqslant 3$$

$$aTH(L) = 3/4$$

(iii) $N < C < 2N$, $L = C$

$$aTH(C) = \frac{2N-C+4 - (1/2)^{C-N-2}}{2N-C+5 - (1/2)^{C-N-2}}$$

For other values of L, N, C it was conjectured, by analogy with the $C = 2N$ result above, that for a given N

$$aTH(L) = \frac{e_c + f_c(L-C)}{g_c + h_c(L-C)} \quad L \geqslant C$$

and, since

$$aTH(\infty) = f_c/h_c$$

and

$$aTH(C) = e_c/g_c$$

a further relationship was required to find $TH(L)$.

$TH(C+1)$ can be found for each N, C combination although no general formula exists and, since

$$aTH(C+1) = \frac{e_c + f_c}{g_c + h_c}$$

$aTH(L)$ can be found.

For example, when $N = 4$, $C = 5$, this procedure gives [6]

$$aTH(L) = \frac{15 + 9(L-5)}{18 + 10(L-5)} \quad L \geqslant 5$$

Figure 5 shows a plot of $aTH(L)$ when $N = 4$ for $C = 4, 5,$ and 8. The decrease of $TH(L)$ for $C = 8$ indicates the effect of blocking.

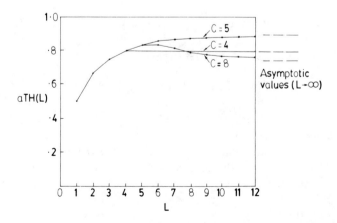

Fig. 5. $aTH(L)$ in a two machine system with $N_1 = N_2 = 4$ for different values of C.

References

[1] Arnold, I. and Feldman, J., "The Eton 2000 production and workplace system," 399-403 in *Proceedings AIIE 1978 Spring Annual Conference,* Toronto, May 23-26 (1978).

[2] Athans, M., Cook, N. H., et al,,"Complex materials handling and assembly system," Report No. ESL - IR - 771, Electronic Systems Laboratory, M.I.T. (1977).

[3] Avi-Itzhak, B. and Heyman, D. P., "Approximate queueing models for multiprogramming computer systems," *Operations Research,* **21**, 1212-1230 (1973).

[4] Basket, F., Chandy, K. M., Muntz, R. R. and Palacios, F. G., "Open, closed and mixed networks of queues with different classes of customers," *J.A.C.M.,* **22**, 248-260 (1975).

[5] Buzacott, J. A., "Reliability of systems with in-service repair," PhD Thesis, University of Birmingham (1967).

[6] Buzacott, J. A., "On the optimal control of input to a job shop," Working Paper #74-004, Dept. of Industrial Engineering, University of Toronto. (Presented at ORSA-TIMS meeting, Boston,1974.)

[7] Buzacott, J. A., "The production capacity of job shops with limited storage space," *Int. J. Prod. Res.,* **14**, 5, 597-605 (1976).

[8] Cook, N. H., "Computer managed parts manufacture," *Scientific American,* 22-29 (Feb. 1975).

[9] Jackson, J. R., "Jobshop-like queueing systems," *Management Science,* **10**, 131-142 (1963).

[10] Kimemia, J. and Gershwin, S. B., "Network flow optimization in flexible manufacturing systems," Paper LIDS-P-866 Laboratory for Information and Decision Systems, M.I.T. (1978).

[11] Perry, C. B., "Variable-mission manufacturing systems," 409-433 in "NC: 1971 the opening door to productivity and profits," ed. by M. A. DeVries, Numerical Control Society Inc. (1971).

[12] Plossl, G. W. and Wight, O. W., "Production and inventory control," Prentice-Hall, 251-253 (1967).

[13] Posner, M. and Bernholtz, B., "Closed finite queueing networks with time lags," *Operations Research,* **16**, 962-976 (1968).

[14] Scharf, P. and Schulz, E., "Integrierte flexible Fertigungssysteme," *wt. Z. ind. Fertigung,* **63**, 130-136 and 199-206, (1973).

[15] Schweitzer, P. J., "Maximum throughput in finite-capacity open queueing networks with product-form solutions," *Management Science,* **24**, 217-223 (1977).

[16] Solberg, J. J., "A mathematical model of computerized manufacturing systems," Paper presented at 4th International Conference on Production Research, Tokyo, 22-30 August (1977).

[17] Warnecke, H. J. and Scharf, P., "Some criteria for the development of integrated manufacturing systems," 401-408 in "Development of production systems," ed.C. H. Gudnason and E. N. Corlett, Taylor & Francis (1974).

[18] Williamson, D.T.N., "The pattern of batch manufacture and its influence on machine tool design," *Proc. Instn. Mech. Engrs.,* **182**, pt. 1, 870-875 (1968).

Dr. John Buzacott is a professor of Industrial Engineering at the University of Toronto. His research interests include the development of models of production systems, the dynamics of technological change and the evaluation of system reliability. He holds a BE degree from the University of Sydney, Australia, and an MS and PhD from the University of Birmingham, England. He is a member of the Editorial Board of the International Journal of Production Research.

Dr. J. G. Shanthikumar is an Assistant Professor of Industrial Engineering and Operations Research at Syracuse University. His research interests are in production systems modelling and analysis, queueing theory, scheduling, and simulation. He holds a BSc degree from the University of Sri Lanka, Sri Lanka, and an MASc and PhD from the University of Tornoto, Toronto, Canada. He is a member of AIIE, IEEE, ORSA, TIMS, and Canadian Operational Research Society.

Modeling FMS by Closed Queuing Network Analysis Methods

GIUSEPPE MENGA, GIORGIO BRUNO, RENATO CONTERNO, AND MASSIMO ACTIS DATO

Abstract—The problem of scheduling concurrent lots of different components on a flexible manufacturing system (FMS) structured with a closed transportation network is approached. The production model and hence the decision/control process is seen here decomposed in two hierarchical levels. 1) At the lowest (microscopic) level, with a heuristic closed queuing network analysis, the average throughputs of the different lots contemporaneously present in the system are estimated. Routing of parts has the objective of balancing machine utilization. 2) At the highest (macroscopic) level, given the throughput estimates, lots are scheduled in order to optimize a production performance index and to satisfy release and due date constraints. Time-phased orders for materials and decisions for sequencing lots on the FMS are generated.

I. INTRODUCTION

A. FMS Architecture

FLEXIBLE manufacturing systems (FMS) are receiving a growing interest as structures suitable for increasing machine utilization, reducing manufacturing lead time and in-process inventory, and providing flexibility in small batch manufacturing. The typical structure of the FMS considered here consists of a set of workstations interconnected by a closed palletized transportation system (Fig. 1). Workstations can be either machining centers or service facilities where operations such as assembly, testing, washing, drying, and loading/unloading of parts are performed. A computer allows the plant to operate automatically and to process different component types simultaneously by assigning operations to the machines, by mounting proper tools and by controlling the advancing and the routing of parts; dedicated fixtures are used to accomodate parts on pallets. Such an FMS organization, which was originated for metal-cutting applications, can now be found as well in the assembly of printed circuit boards [1], integrated circuit fabrication [9], and related manufacturing processes in the electronic industry.

The FMS architecture makes it not only possible but also requires, in order to obtain a balanced utilization level of the resources (machines, tools, and fixtures), that a relatively high number of different types of properly matched parts be contemporaneously present in the system.

B. FMS Planning and Control Problem

Two markedly different approaches can be applied in controlling FMS's. In the first, production is viewed as a continuous process and the product flow mixing is controlled

Manuscript received September 1984.
G. Menga, G. Bruno, and R. Conterno are with Dipartimento di Automatica e Informatica, Politecnico di Torino, Corso Duca degli Abruzzi 24, 10129 Torino, Italy.
M. Actis Dato is with COMAU S.p.A., Direzione Automazione.

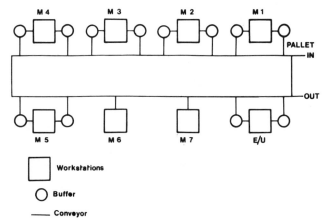

Fig. 1. FMS with closed transportation network.

in time according to the demand. In the second, production is viewed as a discrete process where predefined orders must be scheduled in a given time horizon. The work of Akella, Choong, and Gershwin in this special section is an example of the first approach, while the present paper is mostly related to the second.

Operation research literature has directed a great deal of attention in the last years to production scheduling [5]. The typical production problem related to FMS in the discrete case is the "job shop" problem. It can be stated as follows. R jobs must be processed on a system of I machines. Each job r is composed of M_r different tasks, each task j, orderly or not, can be freely performed on a subset $\mathcal{J}_r$ of the available machines.

Assigned release times, due dates, and job characteristics, a scheduling problem can be stated, and eventually solved, that optimizes a production performance index (tardiness, flow time, etc.) over a given time horizon. Considering that parts are grouped in lots (of identical or similar parts) the "job shop" problem can be observed from two extreme points of view: 1) by considering lots as indivisible entities, i.e., jobs (batch production); 2) by considering elemental parts as entities (single orders).

In FMS as proposed here, in neither of the interpretations can the classical available job-shop scheduling techniques be directly applied; in fact,

1) the previously mentioned reason of economy and efficiency and the closed network nature of the transportation system forbid a pure batch processing of lots on the FMS as a whole (one lot at a time);
2) conversely, because of the stochastic environment of the

production system, characterized by random machine failures, uncertainties on working times and on availability times of the materials, it does not make sense to plan elemental operations or parts on the machines beyond a very short time horizon.

C. A Two-Level Hierarchical Approach to Production Scheduling

A careful analysis of the planning/control structure of a FMS that operates on different lots of small size reveals that both scheduling aspects at the lot and part level coexist and interact. On one side, the groups of lots contemporaneously sharing the FMS resources must be scheduled. On the other, single parts and operations must be assigned to the machines. This structure lends itself to a two-level hierarchical decomposition of the problem.

1) At the lowest level, parts and elemental operations are considered. Service initiation and termination at the machines are indicated here as "micro-events." The resulting control actions are loading, routing through the machines, repetitively unloading parts of active lots until the respective lot completions and substitutions. In fact, we assume that, because of refixturing times, preemption of lots on the FMS is not allowed.

The routing is performed by a real time "dispatcher" with the aim being to balance the machine loads and to cope with uncertainties (machine or tool breakdowns), in order to maintain the production level even in degraded conditions. The stochastic behavior of the process, including the "dispatcher," is described by using closed-queuing network analysis techniques. By ignoring transients due to substitutions of lots and for a given configuration (of lots), average throughputs for each type of part, average queue lengths, and average machine utilizations are computed.

2) At the highest level the lots are considered. Lot introduction and completion in the FMS are indicated here as "macro-events." Based on the average data (throughputs and machine utilizations) resulting from the lowest level a deterministic scheduling problem for optimally sequencing lots on the FMS is approached. As a result, time-phased orders of materials and requests for setting fixtures on the pallets are delivered.

Hildebrand and Suri [7] have also approached the control of FMS with the introduction of closed queuing network analysis. However their work is related to our microscopic level, and at this level we offer a different solution for the dispatching algorithm.

In Section II the scheduling problem will be formulated more exactly; the queuing network model will be stated in Section III, while the implementation of the described algorithms will be outlined in Section IV, where a few numerical examples will be given and compared with the simulation results.

II. THE SCHEDULING PROBLEM

A. Macroscopic Model

A scheduling problem does not take into account the so-called microscopic events, which are likely to occur at a high rate during the working process but, on the contrary, it is concerned with the decisions in response to macroscopic events, such as,

- ready times and completion times of a lot
- significant changes in the availability of workstations
- temporary unavailability of fixtures because of refixturing.

A lot r is characterized by a ready time ρ_r, a due date σ_r, and a size S_r. To simplify the formulation only the first type of macro-events will be explicitly assumed to occur and the decisions will be constrained by the assumptions that

- a lot cannot be interrupted while being processed
- the decision on the next lot to be worked is taken at a completion time.

Now the lot scheduling appears as the solution of an N-stages decision problem, with $N \geq R$, R being the number of lots to be sequenced. The macroscopic FMS model traces the evolution of the macroscopic state $x(i) = \{y(i), v(i)\}$, $i = 0, 1, \cdots, N - 1$; $x(i)$, $y(i)$, $v(i)$ are r-dimensioned vectors whose rth component refers to lot r; $y_r(i)$ indicates the number of backlogged parts of lot r at the beginning of decision stage i, the variable $v_r(i)$ may assume four different values according to whether lot r is

a) not yet released
b) ready to be processed
c) under operation
d) terminated.

Let Ω_i be the set of micro-events which are likely to occur during stage i according to some probability measure. The components of $y(i)$ obviously satisfy

$$y_r(i+1) = y_r(i) - \lambda_r(v(i), \Omega_i)\Delta_i$$
$$y_r(0) = s_r \qquad (1)$$

where λ_r is the average production rate during stage i for lot r and Δ_i is the duration of stage i. The sequence of the decision times t_i is given by

$$t_{i+1} = t_i + \Delta_i$$

and Δ_i satisfies

$$\Delta_i = \min_r \{y_r(i)/\lambda_r(i), \max(\rho_r - t_i, 0)\}.$$

B. Lot Scheduling in FMS

A usual criterion for sequencing lots is to complete each lot before its due date, because of the extra costs of backlogging. This constraint can be included in a smooth performance index $C_r(t_r)$ of the waiting time $\tau_r = t_r - \rho_r$ (t_r being the instant at which the lot is released by the machine).

A possible shape of $C_r(t_r)$ is given by

a) $C_r(t) = t - \rho_r$ until the due date is reached
b) $C_r(t) = \sigma_r - \rho_r + \gamma_r(t - \sigma_r)$, with $\gamma_r > 1$ to take into account backlogging ($t > \sigma_r$)
c) $C_r(t) = 0$ prior to release time ($t \leq \rho_r$).

If T is the time horizon on which a set of lots $r = 1, \cdots, R$ should be completed, the performance index to be minimized is

$$J = \Sigma_r C_r(t_r), \qquad t_r \leq T. \tag{2}$$

To further pursue this idea we introduce the cost of a stage i (between two subsequent completion times) $L_i(x(i), u(i), \Omega_i)$, which depends on the macroscopic state vector $x(i)$, the decision $u(i)$, and the set Ω_i. L_i is simply expressed using the stage duration Δ_i as

$$L_i = \Sigma_r g_r[\Delta_i(x(i), u(i), \Omega_i)]$$

where the summation is extended to lots which are ready and not yet completed, and g_r is a function depending on the due date and defined by the shape of $C_r(t)$.

Then the performance index can be expressed as the sum

$$J = \sum_{i=0}^{N-1} L_i(x(i), u(i), \Omega_i) \tag{3}$$

and the lot scheduling is defined as the sequence of decisions $u^*(0), \cdots, u^*(i), \cdots$, which minimizes the expected value of J in respect to the sequence of event sets Ω_i and subject to the constrained resources of the plant (machines and fixtures availability) and to zero backlog ($x(N) = 0$).

III. HEURISTIC CLOSED QUEUING NETWORK ANALYSIS

Between two contiguous macro-events the plant behaves as a queuing network with fixed number of pallets in operation and fixed multiple part classes. Closed queuing networks are widely used in computer performance evaluation [3]. In recent years the application of similar techniques has also been proposed for production systems [2], [6], [12].

It is common knowledge that the exact solution of closed queuing networks imposes several practical limitations referring to

1) the service time distributions that must be exponential;
2) the queuing policy that when of the first in/first out (FIFO) type does not allow multiple classes;
3) the routing strategy that must be random.

In the next section the assumed routing policy or "dispatching function" in the FMS is described. The queuing network model will be stated in Section III-B.

A. The Dispatcher

The FMS dispatcher is a software module in charge of the real-time routing of the parts through the available machines. The dispatcher is promptly informed when a machine goes out of service, so that it can be cancelled from the available machines list.

The effect of machine and tool breakdowns and repair times have been discussed by several authors, for example, [7], [8]. This paper is more concerned with the interaction of concurrent lots in an FMS, so the limited resource availability is just roughly represented by slightly underloading the machines.

Once the production schedule has been stated, the dispatcher is supposed to balance the machine loads, as much as possible. In practice this is simply obtained by routing each part to the available machine, among those admissible for that particular operation, with minimum unfinished work.

We now provide an algorithm that computes the optimal routing coefficients iteratively. Such procedure well fits the iterative queuing network algorithm that will be presented in Section III-B so that both computations can be carried out at the same time.

Let us consider R jobs composed of a single task that can be performed arbitrarily on a subset $\mathcal{J}_r$ of the available I machines. This case easily encompasses more general situations by simply splitting complex jobs in their sequence of elemental tasks. Let $1 \leq r \leq R$ and $1 \leq i \leq I$ be the indices of lots and machines, respectively, and I_r the dimension of the subset $\mathcal{J}_r$. Let τ_i^r be the average service time and λ^r the average throughput, hence $q_i^r = \lambda^r \tau_i^r$ the percent utilization or load, at machine i for lot r. θ^r is the $I_r \times 1$ routing vector of lot r, with elements

$$\theta_i^r \geq 0, \qquad \text{for all } i, \sum_{i=1}^{I_r} \theta_i^r = 1.$$

θ^r represents the percent of parts of lot r addressed to each machine. q_i^0 is a fixed load present on machine i.

In matrix form the total utilization of the machines can be written as

$$l = q^0 + \sum_{r=1}^{R} A^r \theta^r, \quad l = l(\theta),$$

$$\theta = \{\theta^r, r = 1, 2, \cdots, R\} \tag{4}$$

where $l = (l_1, \cdots, l_I)^T$ and $q^0 = (q_1^0, \cdots, q_I^0)^T$ are $I \times 1$ vectors; A^r is a $I \times I_r$ matrix with only one element different from zero in each column located in position (i, j) and assuming the value $q_i^r, i \in \mathcal{J}_r$.

The problem of balancing the utilization is here approached by determining the vector $\theta = \{\theta^r\}$ that minimizes a performance index given by the difference between maximum and minimum utilization of the machines, i.e.,

$$\min \; (l_M(\theta) - l_m(\theta)),$$

$$l_M = \max_i l_i, \quad l_m = \min_i l_i.$$

We first prove that the minimum of the performance index with respect to an arbitrary θ^r (θ^j, for all $j \neq r$ are kept constant) is simply obtained by solving an auxiliary constrained quadratic programming problem.

Then we show that the recursive minimization with respect to $\theta^j, j = 1, 2, \cdots /\text{mod } r$ generates a sequence of monotone decreasing values of the performance that converges in the minimum.

Theorem 1 (Balancing Utilization with One Routing Vector): Given the machine utilization by

$$\tilde{l} = \tilde{q}^0 + \tilde{A}\tilde{\theta}$$

where $\tilde{l}$, $\tilde{q}^0$, $\tilde{\theta}$ are vectors of dimension $\tilde{I} \times 1$, $\tilde{A} = \text{diag}(a_i)$ is a $\tilde{I} \times \tilde{I}$ diagonal matrix and $\tilde{\theta}$ a routing vector, the solution of the quadratic programming problem

$$\min_{\tilde{\theta}} \|\tilde{q}^0 + \tilde{A}\tilde{\theta}\|_{\tilde{A}^{-1}}^2$$

where

$$\|\tilde{l}\|_{\tilde{A}^{-1}}^2 = \tilde{l}^T \tilde{A}^{-1} \tilde{l}$$

with inequality constraints

$$\tilde{\theta}_i \geq 0, \quad i = 1, 2, \cdots, \tilde{I}$$

and equality constraints

$$\Sigma_{i=1}^{\tilde{I}} \tilde{\theta}_i = 1$$

solves also the problem

$$\min_{\tilde{\theta}} (\tilde{l}_M(\tilde{\theta}) - \tilde{l}_m(\tilde{\theta})).$$

The proof of the theorem is given in the Appendix.

Theorem 2 (Convergence of the Iterative Minimization): In the minimum each $\tilde{\theta}^r$, $r = 1, 2, \cdots, R$ satisfies Theorem 1.

Proof: The proof derives immediately from the additive form of the utilization vector (4).

B. Closed Queuing Network Model

A recent new family of iterative methods for the approximate analysis of closed queuing networks (mean value analysis (MVA)) offers numerically efficient techniques that can easily incorporate heuristics to extend the range of application [4]. Iterative methods are based on a set of nonlinear equations that relates, approximately, unknown parameters of the network, and iteration is used for their solution. The approach developed here is based on the well-known iterative MVA equations of Reiser [11]. Our original contribution is to incorporate in Reiser's equations the behavior of the real-time dispatcher. The basic equations derived from [11] are given in the Nomenclature.

We shall also use the notation $n_i^j(r-)$, $q_i^j(r-)$, etc., to denote mean values upon arrival of a part of class r. The results are

$$\lambda^r = W^r / (\Sigma_{i \in Q(r)} \theta_i^r t_i^r) \tag{5}$$

$$n_i^r = \lambda^r \theta_i^r t_i^r \tag{6}$$

$$t^r = \tau_i^r + c_i^r \{\Sigma_{j \in R(i)} \tau_i^j [n_i^j(r-) - d_i^j \theta_i^j q_i^j(r-)]\}. \tag{7}$$

With reference to previous results the inclusion of the dispatcher into the network is obtained operating as follows:

1) by adapting recursively in the iterative solution of (5), (6), and (7) the routing coefficients θ_i^r with the technique described in the previous section;
2) by recognizing that mean queue lengths and hence mean waiting times experimented by one part are shorter when routing is done by a "dispatcher" than when routing is random. The empirical coefficients c_i^r account, in fact, for this "speed-up."

Experiments performed in several practical cases and described in [8] show that the tuning of d_i^r and c_i^r is not critical and that the approximation error of the results lies below 15 percent. Nevertheless an empirical rule has been formulated to avoid simulation and tuning in the special case when any of the lots can be executed on any of m_i equal parallel machines at work stage i, $i = 1, 2, \cdots,$.

A dispatcher routing the incoming parts to the queue with minimum unfinished work is perfectly equivalent to a multi-server queue, in which a server is idle only if there are no waiting jobs. The mean value analysis (MVA) has already been extended to handle multiple server stations when the jobs of different classes all have the same mean service time. This is not our case; moreover, the MVA becomes computationally burdensome in the multiserver multiclass implementation. The idea of a correction coefficient has in any case been exploited.

Let stage i consist of m_i parallel machines with equal mean service times τ_i^r for lot r. For the sake of simplicity exponential service distributions are now assumed ($d_i^j = 0$). The MVA algorithm is run as if no intelligent dispatchers were in the network. We have $\theta_i^r = 1/m_i$ and (7) for any of the machines at stage i becomes

$$t_i^r = \tau_i^r + c_i \cdot \Sigma_{j \in R(i)} \tau_i^j n_i^j(r-).$$

The empirical correction coefficient c_i is computed at each iteration of the MVA algorithm and is defined by

$$c_i = 0 \quad R(i) \leq m_i$$

$$c_i = \frac{n_{\text{mult}}}{m_i n_{\text{sing}}} \quad R(i) > m_i \tag{8}$$

where $R(i)$ denotes the number of lots visiting stage i which are already in the network at the current iteration of the MVA algorithm. If $R(i) \leq m_i$, no waiting will occur. Otherwise c_i depends on n_{sing} and n_{mult}; n_{sing} is the average number of components in a $M/M/1$ queue with mean service time

$$\tau_i = \frac{1}{R(i)} \Sigma_{j \in R(i)} \tau_i^j$$

subject to the arriving rate

$$\lambda_i = (1/m_i) \Sigma_{j \in R(i)} \lambda^j$$

as computed in the current MVA iteration; n_{mult} is the corresponding parameter for a $M/M/m$ queue simulating the intelligent dispatcher. So, a single exponential class of lots is assumed for the computation of c_i, whose mean service time is given by the average of the service times of the lots visiting stage i. In this way a very simple "open loop" approximated model of the queue at stage i is established; it is quite intuitive how the ratio (8) gives a rough idea of the reduction of waiting time experimented by an incoming component when the dispatcher is used. Note that c_i now depends on the network loading.

This empirical method has proved to be accurate enough in the tested cases: relative errors below 10 percent with respect to the simulation. The important advantage is that no tuning is required.

IV. THE HIERARCHICAL PLANNING AND CONTROL SYSTEM

A. Priority Scheduling

Two modules have been described in the preceding sections, acting at different levels:

a) the scheduler, dealing with lots and macro-events at the most aggregated level;
b) the queuing network model, dealing with the micro-events.

When the system must be loaded, or, more generally, at a decision stage (end of a lot), the production plan within a time horizon of a few days must be determined. Only the initial part of a production plan is actually performed, as it is updated time by time.

The scheduling problem has been clearly defined in Section II where a cost function to be minimized has been introduced. The resulting optimization problem is extremely difficult in part due to the interactions among concurrent lots: it is a classical trend in the applications to establish some heuristic procedure. The implemented lot scheduling method relies on priorities. Lots are sequenced according to a suitable parameter and they are systematically examined for entering into production at each decision instant. Heading lots are loaded as long as the estimated machine utilizations and queue lengths lie below a given threshold. The system is assumed to perform as a steady-state closed queuing network between two macro-events. Machine utilizations, queue lengths, and completion times are efficiently estimated through the queuing network model (Section III).

A static or *a priori* sequencing may be performed according to increasing values of the parameter $(\sigma_r - \rho_r)/b_r$, where σ_r, ρ_r, and b_r are the due date, the release time, and the duration of a generic lot r. The lot duration b_r is just the sum of the service times at the machines visited by lot r which, in general, is different from the time spent by the lot in the system; this is because of the interactions with the other lots that are being processed at the same time. In this way it is clear that urgent lots are scheduled first, in agreement with the cost function defined in Section II.

A more efficient parameter, requiring some more runs of the queuing network routine, has been defined as $(\sigma_r - t_i)/b_r^*$, where t_i is the ith decision time and b_r^* is the residual duration of lot r at time t_i. This results in a "dynamic" sequencing as the lots ready for production are reordered at each decision time according to the updated available information. Both static and dynamic sequencing have been tested; a few examples will be given in the next paragraph for a comparison.

B. Numerical Results

In Fig. 2 a production plan yielded by the queuing network model is compared with the simulation for a case with 18 lots, seven machines of which four are parallel, different release times, and equal due dates; dynamic scheduling has been used. The service times for the components belonging to different lots are in the range 0.7–120 min, and the number of components for each lot varies from 10 to 100.

A significant discrepancy can just be noted for lot 5, while the two plans appear to be very similar as far as the other lots are concerned; this sort of result seems to be fairly typical.

In Figs. 3 and 4 static and dynamic scheduling are compared in an example with 35 lots. Among the possible parameters to evaluate the production plan efficiency, the following have been chosen:

$$J_1 = \Sigma_r b_r / \Sigma_r z_r$$

$$J_2 = \frac{1}{R} \sum_r (b_r/v_r)$$

$$J_3 = \frac{1}{R} \sum_r \max(0, z_r - \sigma_r)/b_r$$

where z_r is the completion time of lot r and v_r is the time spent by lot r in the system. J_1 can be interpreted as the filling coefficient of the production plan: the higher it is, the higher the degree of concurrency and planning effectiveness reached. J_2 is a measure of the interactions among the lots, or system congestion, and J_3 gives an idea of how the due dates are honored. It might also be meaningful to square the terms in J_3 to penalize big backorders.

J_1 and J_2 are not seriously affected by the two scheduling methods: $J_1 = 0.2$ and $J_2 = 0.5$ are typical values for the 35 lots example of the presented case study. Higher values are found for smaller numbers of lots. J_3 is remarkably affected if the number of lots is relevant: in the 35 lots case dynamic scheduling clearly outperforms static scheduling. However, if fewer lots must be scheduled, our experience indicates that the two methods yield roughly the same efficiency, so that the static scheduling can still be profitably used when the number of lots is of the order of 10 or 15.

NOMENCLATURE

W^r	Number of identical parts of class r.
$R(i)$	Set of classes visiting station i.
$Q(r)$	Set of stations visited by class r.
τ_i^r	Mean service time of class r at station i.
n_i^r	Mean number of class r parts waiting or being served at station i.
λ^r	Throughput of class r.
t_i^r	Mean waiting time (waiting plus service) of class r at station i.
θ_i^r	Routing factor of class r to the station i.
$q_i^r = \lambda^r \cdot \tau_i^r$	Class r utilization of station i.
d_i^r	Empirical coefficient with value between 0 and 1 that accounts for the service time distribution.
c_i^r	Empirical coefficient of "speed-up" with value between 0 and 1 accounting for the presence of a real-time dispatcher.

APPENDIX A

Proof of Theorem 1

We may assume that the vector $\tilde{\rho}$ is arranged so that the components $\tilde{q}_i^0$ of $\tilde{q}^0$ are ordered by decreasing values. Let us

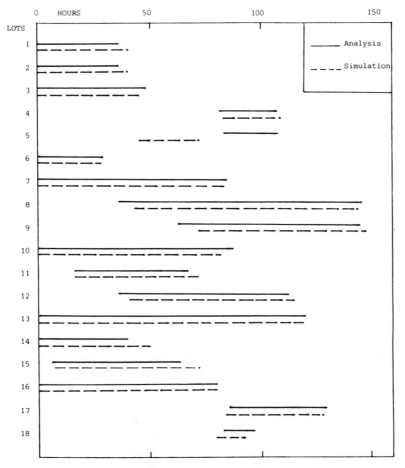

Fig. 2. Production plan: 18 lots, comparison with simulation.

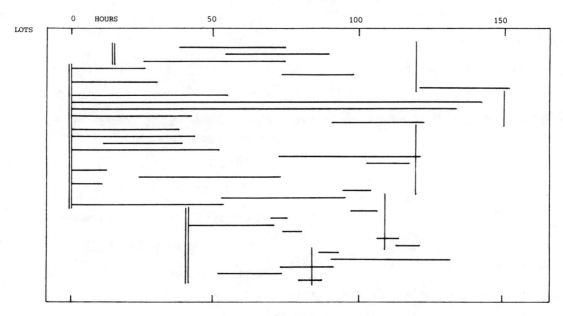

Fig. 3. Production plan: 35 lots, static routing.

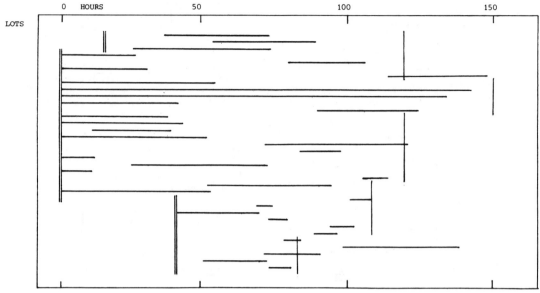

Fig. 4. Production plan: 36 lots, dynamic routing.

define the Lagrangian of the performance measure:

$$L(\theta, \lambda, g) = \frac{1}{2} \|\tilde{q}^0 + \tilde{A}\tilde{\theta}\|_{\tilde{A}^{-1}}^2 - \tilde{\theta}^T \lambda + (1 - \tilde{\theta}^T u_I)g \quad (9)$$

where

$$\sqrt{\|x\|_{\tilde{A}^{-1}}^2} = \sqrt{x^T A^{-1} x}$$

is the usual weighted Euclidean norm, $\tilde{A}$ is a diagonal positive matrix, $\lambda \geq 0$ is the $\tilde{I} \times 1$ vector of Lagrangian multipliers related to the inequality constraints, g is the scalar Lagrangian multiplier related to the equality constraint, u_T is the $\tilde{I} \times 1$ vector with all components equal to one.

Necessary conditions for a minimum are

$$\nabla_\theta L(\tilde{\theta}^*, \lambda^*, g^*) = \tilde{q}^0 + \tilde{A}\tilde{\theta}^* - \lambda^* - u_I g^* = 0 \quad (10)$$

$$\tilde{\theta}^* = \tilde{A}^{-1}(-\tilde{q}^0 + \lambda^* + u_T g^*). \quad (11)$$

If in the minimum all inequality constraints are inactive, i.e., $\lambda^* = 0$ then

$$\tilde{\theta}^* = \tilde{A}^{-1}(-\tilde{q}^0 + u_T g^*) \quad (12)$$

with

$$\tilde{\theta}_i^* \geq 0, \quad \text{for all } i$$

and

$$g^* = (1 + \Sigma_{i=j}^{\tilde{I}} a_i^{-1} \tilde{q}_i^0)/\Sigma_{i=j}^{\tilde{I}} a_i^{-1}, \quad j = 1 \quad (13)$$

satisfies the minimization problem and the utilization vector in the minimum results:

$$\tilde{l}^* = u_I g^*. \quad (14)$$

In this case we have perfect load balance and the theorem is automatically proved. Vice versa if in the minimum some of the inequality constraints are active, i.e., $\lambda_i^* > 0$ for some i, then some $\tilde{\theta}_i^*$, as given by (12) and (13), will be negative and the result will be no more correct. In this case because of the ordering of $\tilde{q}^0$, the result will certainly be $\tilde{\theta}_1^* < 0$. Then $\tilde{\theta}_1^*$ must be set equal to zero by introducing into (11) a proper $\lambda_1^* > 0$.

The remaining components $\tilde{\theta}_i^*$, $i = 2, \cdots, \tilde{I}$ are given by (12) where g^* is recomputed as in (13) with $j = 2$. If the positivity condition of the new vector $\tilde{\theta}^*$ is satisfied, then the minimization problem is solved, otherwise we repeat previous steps by setting to zero $\tilde{\theta}_2^*$ and by recomputing g^* with $j = 3$.

The procedure is ended when all $\tilde{\theta}^* \geq 0$. The utilization vector in the minimum will be in general

$$l_i^* = \begin{cases} \tilde{q}_i^0, & \text{for all } i : \tilde{\theta}_i^* = 0 \\ g^*, & \text{for all } i : \tilde{\theta}_i^* > 0. \end{cases}$$

Therefore in the minimum it results

$$l_M^* = q_1^0, \quad l_m^* = g^*.$$

We show now that the theorem is proved by giving an admissible perturbation δ to $\tilde{\theta}^*$:

$$\tilde{\theta} = \tilde{\theta}^* + \delta$$

$$\tilde{l}(\tilde{\theta}) = l^* + \tilde{A}\delta.$$

We note that δ must satisfy the condition:

$$\delta_i \geq 0, \quad \text{for all } i : \tilde{\theta}_i^* = 0.$$

Therefore it follows that

$$l_M(\tilde{\theta}) \geq l_M^*.$$

Moreover an admissible δ satisfies

$$\Sigma_{i=1}^{\tilde{I}} \delta_i = 0;$$

hence it will be either all $\delta_i = 0$ or some $\delta_i < 0$.

From this argument it is immediate to show that

$$\tilde{l}_m(\theta) = \tilde{l}_m^* - \Delta,$$

for some $\Delta > 0$, and hence that

$$\tilde{l}_m(\theta) < \tilde{l}_m^*.$$

This completes the proof of the theorem.

REFERENCES

[1] R. Akella, Y. Choong, and S. B. Gershwin, "Performance of hierarchical production scheduling policy," M.I.T. Lab. for Inf. and Dec., Systems Rep. LIDS-FR-1357, 1984.

[2] J. B. Cabaille and D. Dubois, "Heuristic methods based on mean value analysis for FMS performance evaluation," in *Proc. IEEE Conf. Decision and Control*, Orlando, FL, Dec. 1982, pp. 1061–1065.

[3] K. M. Chandy and D. Neuse, "Linearizer: A heuristic algorithm for queuing network models of computer systems," Communications of the ACM 25, 1982, pp. 126–134.

[4] E. De Souza e Silva, S. S. Lavenberg, and R. R. Muntz, "A perspective on iterative methods for the approximate analysis of closed queueing networks," Int. Workshop on Applied Mathematics and Performance/Reliability Models of Computer/Communication Systems, Univ. Pisa, Sept. 1983, pp. 191–210.

[5] S. C. Graves, "A review of production scheduling," *Operations Res.*, vol. 29, no. 4, pp. 646–675, July–Aug. 1981.

[6] R. R. Hildebrand, "Scheduling flexible manufacturing systems using mean value analysis," in *Proc. IEEE Conf. Decision and Control*, Albuquerque, NM, 1980, pp. 701–706.

[7] R. R. Hildebrand and R. Suri, "Methodology and multi-level algorithm structure for scheduling and real-time control of flexible manufacturing systems," in *Proc. 3rd Int. Symp. Large Eng. Systems*, Memorial Univ. Newfoundland, 1980, pp. 239–244.

[8] J. G. Kimemia and S. B. Gershwin, "An algorithm for the computer control of production in a flexible manufacturing system," M.I.T. Lab. for Inf. and Dec. Systems, Rep. LIDS-P-1134, 1983.

[9] D. L. Krause and D. A. Locy, "Hybrid integrated circuit manufacturing process as controlled by shop information systems," *IEEE Trans. Components, Hybrids, Manuf. Technol.*, vol. CHMT-3, no. 3, Sept. 1980.

[10] V. Minero, "Modellistica e Controllo di Sistemi Flessibili di Produzione in Italia," Dip. di Automatica e Informatica Politecnico di Torino, Dec. 1983.

[11] M. Reiser, "A queueing network analysis of computer communication networks with window flow control," *IEEE Trans. Commn.*, vol. C-27, pp. 1199–1209, Aug. 1979.

[12] J. J. Solberg, "A mathematical model of computerized manufacturing systems," presented at Proc. 4th Int. Conf. on Production Res., Tokyo, Aug. 1977.

"Optimal" Operating Rules for Automated Manufacturing Systems

J. A. BUZACOTT

Abstract — In modeling automated manufacturing systems such as transfer lines and flexible manufacturing systems it is necessary to make assumptions about how such systems will be operated. Hence, it is desirable to determine the "optimal" operating rules. In this paper a number of examples are given of how the "optimal" rules are determined so that consideration is given to the control options that can be used, the information available to the operator and the typical multilevel nature of the system's control.

Introduction

Automated manufacturing systems consist of a number of stations where production operations are carried out together with a material handling system linking the stations, with the whole system under some mixture of centralized and decentralized control and monitoring [1]. Typically, the operation of individual stations is controlled locally but the movement of parts between stations is controlled centrally. Furthermore, there will be central control of movement of parts into and out of the system with perhaps some linkage to even higher levels of control within the total manufacturing environment.

At present there are two general classes of multistation automated manufacturing systems.

One is automatic transfer lines such as are used in the automobile industry for machining cylinder blocks, transmission cases, and transaxles. These have been in relatively widespread use for about 30 years. Their distinguishing feature is that all parts processed by the line follow the same sequence of operations and part movement is synchronized, that is at fixed intervals of time (the cycle time) every part in the system is transferred to the next station.

The other class is the flexible manufacturing system (FMS) where the basic concept has been around for about 10 years or so [1],[2]. The distinguishing feature of the FMS is that provision is made for a diversity of part routings so that all parts need not visit the same sequence of machines. The FMS can be regarded as equivalent to a conventional job shop with automated material handling [3].

Manuscript received September 29, 1980, revised February 5, 1981. Paper recommended by D. G. Fisher, Past Associate Editor, Applications, Systems Evaluation, and Components Committee.

The author is with the Department of Industrial Engineering, University of Toronto, Toronto, Ont., Canada.

More recent developments in transfer line design incorporate some aspects of the FMS, for example the introduction of parallel stations or segments and the avoidance of strict synchronism between all parts of the system [4].

Because of the way in which the stations of an automated manufacturing system are linked to form an integrated system it is necessary in designing them to give careful consideration to the effect of disturbances such as breakdowns of stations or variability in the processing times of the jobs. One way in which the effect of disturbances can be reduced is through providing in process storage space. While such space is expensive to provide because of the need to maintain workpiece orientation it has been shown that it can have a significant effect on system performance (for models of the role of in process storage, see [5] and [3]).

The focus of most of the literature on transfer lines has been on the way in which the performance is determined by the characteristics of the physical equipment, such as the processing rates of the stations, the speed and capacity for simultaneous job movement of the material handling system, and the location and capacity of the in-process inventory banks.

However, the performance of the system will also be influenced by the operating policies used in its control. The purpose of this paper is to discuss a number of issues relating to the operation of transfer lines and FMS and, where possible, identify "optimal" operating policies.

Control of Automated Manufacturing Systems

It is desirable to think of the control requirements from a hierarchical perspective, with three main levels (cf. [3]).

1) Prerelease planning: Deciding which jobs are to be manufactured by the system. Identifying constraints on operation sequence.

2) Input or release control: Determining the sequence and timing of the release of jobs to the system.

3) Operational control: Ensuring movement between machines and deciding which job is to be processed next by a machine.

A further aspect of system control is the monitoring of correct performance of the operations. This aspect will not be discussed in this paper.

At each level of control the physical configuration and the decisions made at higher levels set constraints on the alternative actions. In the manufacturing environment it has always proved to be difficult to collect and use reliable detailed information about the status of the machines and the jobs being processed by the system. Thus, in defining the control problem at each level is essential to specify clearly the available information on system status (cf. [6]).

In some cases it is possible to show that a particular operating or control rule is optimum given a particular amount of information on the system status. Such results can help to identify the value of collecting more information about system status. They can also provide guidance to machine designers and provide a consistent basis for the comparison of alternative designs.

Overview of the Paper

In this paper a number of questions concerning the operation of transfer lines and FMS are considered. In particular, for transfer lines two questions are considered.
   i) Should stations be deliberately stopped to adjust inventory levels in the bank?
   ii) If there is only one repairman and two stations are down, which station should he repair?
For FMS the following questions are considered.
   iii) Should operation sequence be specified at the prerelease planning level or at the operational control level?
   iv) How does the availability of more information on job status improve input control?

I. Optimal Operating Policy for a Transfer Line

Consider a transfer line which is divided into two stages by an inventory bank of capacity z units.

There are a variety of ways in which such a line could be operated. One policy, which will be shown to be optimum, is never to deliberately stop a stage as long as the stage is capable of operating and has parts to process. However, alternative policies are known to have been used in practice.

For example, the inventory level in the bank could be kept near $z/2$. That is, after every repair of a stage (during which the bank would be used) the other stage would be stopped in order to adjust the inventory level. Such a policy would appear to overcome the blocking or starving of a stage which occurs if the other stage has a number of failures in succession.

The optimum policy can be found using dynamic programming. The behavior of the transfer line can be described by a Markov chain model where the states of the system are $\{y_1 y_2, x\}$ where $y_j = W$ means stage j is up and $y_j = R$ means stage j is under repair (it is assumed that repair can begin as soon as a stage fails). The inventory level x in the bank can take on values $0, 1, 2, \cdots, z$. The state is observed just before transfer begins.

The policy of never stopping a stage to adjust the inventory level will be assumed and then shown to be optimum.

Symbols

$v(y_1 y_2, x)$: value associated with state $y_1 y_2, x$.
g: line efficiency for the assumed policy.
a_j: failure probability per cycle of stage j.
b_j: repair probability per cycle of stage j.

$$r = a_2 b_1 / a_1 b_2.$$
$$B_1 = b_1 + b_2 - b_1 b_2 - a_1 b_2.$$
$$B_2 = b_1 + b_2 - b_1 b_2 - a_2 b_1.$$
$$u = B_1 / B_2.$$
$$A_1 = a_1 + a_2 - a_1 a_2 - b_1 a_2.$$
$$A_2 = a_1 + a_2 - a_1 a_2 - a_1 b_2.$$
$$C = A_2 B_1 / A_1 B_2.$$

Then, assuming operation dependent failures, it is known that (see Buzacott and Hanifin [5])

$$g = \frac{1 - ruC^z}{(a_1 + b_1)/b_1 - (a_2 + b_2)ruC^z/b_2} \qquad r \neq 1$$

and the value determination equations for the assumed policy are

$$v(WW, x) + g = 1 + (1 - a_1)(1 - a_2)v(WW, x) + a_1(1 - a_2)v(RW, x)$$
$$+ (1 - a_1)a_2 v(WR, x) + a_1 a_2 v(RR, x)$$
$$(0 \leq x \leq z)$$

$$v(RW, 0) + g = b_1 v(WW, 0) + (1 - b_1)v(RW, 0)$$
$$v(RW, x) + g = 1 + b_1(1 - a_2)v(WW, x - 1)$$
$$+ (1 - b_1)(1 - a_2)v(RW, x - 1)$$
$$+ b_1 a_2 v(WR, x - 1) + (1 - b_1)a_2 v(RR, x - 1)$$
$$(0 < x \leq z)$$

$$v(WR, x) + g = (1 - a_1)b_2 v(WW, x + 1) + a_1 b_2 v(RW, x + 1)$$
$$+ (1 - a_1)(1 - b_2)v(WR, x + 1)$$
$$+ a_1(1 - b_2)v(RR, x + 1) \qquad (0 \leq x < z)$$

$$v(WR, z) + g = b_2 v(WW, z) + (1 - b_2)v(WR, z)$$
$$v(RR, x) + g = b_1 b_2 v(WW, x) + (1 - b_1)b_2 v(RW, x)$$
$$+ b_1(1 - b_2)v(WR, x) + (1 - b_1)(1 - b_2)v(RR, x)$$
$$(0 \leq x \leq z).$$

Setting $v(WW, 0) = 0$ it can be shown that for $r \neq 1$ the solution to the above equations is

$$v(WW, x) = \frac{1}{D}\left\{\frac{(a_1 + b_1)x}{b_1} - \frac{C^{z-x}(1 - C^x)a_2 B_1}{a_1 b_2^2(1 - r)}\right\} \qquad (0 < x \leq z)$$

$$\left\{\begin{array}{l} v(RW, x) \\ -v(RW, 0) \end{array}\right\} = \frac{1}{D}\left\{\frac{(a_1 + b_1)x}{b_1} - \frac{C^{z-x+1}(1 - C^x)rA_1}{a_1 b_2(1 - r)}\right\} \qquad (0 < x \leq z)$$

$$\left\{\begin{array}{l} v(WR, x) \\ -v(WR, 0) \end{array}\right\} = \frac{1}{D}\left\{\frac{(a_1 + b_1)x}{b_1} - \frac{C^{z-x}(1 - C^x)A_1}{a_1 b_2(1 - r)}\right\} \qquad (0 < x \leq z)$$

$$\left\{\begin{array}{l} v(RR, x) \\ -v(RR, 0) \end{array}\right\} = \frac{1}{D}\left\{\frac{(a_1 + b_1)x}{b_1} - \frac{C^{z-x}(1 - C^x)A_1 A_2 b_1}{a_1^2 b_2(1 - r)B_2^2}\right\} \qquad (0 < x \leq z)$$

$$v(RW, 0) = -\frac{1}{b_1 D}\{1 - ruC^z\}$$

$$v(WR, 0) = -\frac{C^z(1 - r)(b_1 + b_2 - b_1 b_2)}{b_1 B_2 D}$$

$$v(RR, 0) = v(RW, 0) + \frac{b_1(1 - b_2)v(WR, 0)}{b_1 + b_2 - b_1 b_2}$$

where

$$D = (a_1 + b_1)/b_1 - (a_2 + b_2)ruC^z/b_2.$$

Now to consider the alternative policies in each state, let **1** denote the policy of deliberately stopping stage 1, and **2** denote the policy of deliberately stopping stage 2. **0** will be used to denote the policy of not stopping any stage which could operate.

Then, considering each state in turn,
WW, x **0** is better than **1** (write as **0 > 1**) if

$$v(WW, x) + g > 1 + (1 - a_2)v(WW, x - 1) + a_2 v(WR, x - 1)$$

and **0 > 2** if

$$v(WW, x) + g > (1 - a_1)v(WW, x + 1) + a_1 v(RW, x + 1).$$

These conditions reduce to the following:

$$0>1: \quad \frac{1}{D} > \frac{a_2 B_1}{b_2 A_1} \frac{C^{z-x}}{D}$$

$$0>2: \quad \frac{C^x}{D} < \frac{a_1 B_2}{b_1 A_2 D}$$

which are true for all x.

$WR, x \quad (0 \le x < z) \qquad 0>1$

if $\quad v(WR, x) + g > b_2 v(WW, x) + (1 - b_2) v(WR, x)$

or $\quad g > b_2(v(WW, x) - v(WR, x))$

which reduces to

$$1/D > C^{z-x}/D$$

which is true for $(0 \le x < z)$.

$RW, x \quad (0 < x \le z) \qquad 0>2$

if $\quad v(RW, x) + g > b_1 v(WW, x) + (1 - b_1) v(RW, x)$

or $\quad g > b_1(v(WW, x) - v(RW, x))$

which reduces to

$$1/D > C^x/D$$

which is true for $0 < x \le z$.

Hence, it follows that the policy of never deliberately stopping a stage is optimal. (This can also be shown for $r = 1$.)

The Advantage of Coordinated Control

In a transfer line transfer of workpieces from one station to the next begins simultaneously for all stations in each section of the line. However, coordination between the different sections is not necessarily provided. For example, suppose that stage 1 transfers at time $0, \tau, 2\tau, \cdots$ and stage 2 transfers at time $\delta, \delta + \tau, \delta + 2\tau, \cdots$ $(0 < \delta < \tau)$. Without coordinated control the conditions for transfer are: i) no station in stage failed, ii) space for a part in the bank following the stage, iii) availability of a part in the bank preceding the stage. As a result, stage 1 would not transfer at time $n\tau$ if the bank between stages 1 and 2 is full, even though at time $\delta + n\tau$ stage 2 will, if up, transfer and create space in the bank. Stage 1 will effectively lose a cycle of operation, and as this is equivalent to deliberately stopping it, the efficiency of the line will be reduced.

This can be overcome by providing coordinated control, that is, permitting stage 1 to delay transfer to time $n\tau + \delta + \epsilon$ if there will then be space in the bank because stage 2 has transferred. Provision would also be made for stage 2 to delay transfer if the inventory bank level is zero and it is known that stage 1 will deliver a part to the bank at time ϵ later $(0 \le \epsilon < \tau)$.

II. Optimum Single Repairman Strategy

If there is just one repairman for a two stage transfer line then it is necessary to decide which stage he should repair if both stages are simultaneously failed. Two alternative situations can be considered: a) a repair can be interrupted with no loss of time. b) a repair once begun must be completed without interruption.

In a) there is the opportunity for the repairman to begin repair on one stage and then if the other stage fails he can stop repair on that stage and switch over to work on repair of the other stage.

In b) the only situation where the repairman has to make a choice will be that in which both stages fail in the same cycle.

Dudick [7] has considered case a). He showed that the optimum repair policy was of the form: repair stage 1 if the inventory level in the bank is less than y_0, repair stage 2 if the inventory level is greater than y_0, where $0 \le y_0 \le z$. He showed that for identical stages, i.e., $a_1 = a_2$, $b_1 = b_2$ the optimum value of y_0 is $(z+1)/2$.

It is now shown that the same result applies in case b). The state description is more complex in case b) than in case a). The state of the line $\{y_1 y_2, x\}$ now permits five different values of $y_1 y_2$; both stages working (WW), one stage working and the other stage under repair (WR and RW), one stage under repair and the other stage failed and waiting repair (RF and FR).

The system state transition equations can be written down for the policy of repairing stage 1 if $x \le y_0$ and repairing stage 2 if $x > y_0$. They are

$$P(WW, 0) = (1 - a_1)(1 - a_2) P(WW, 0) + b_1(1 - a_2) P(RW, 1) + b_2 P(RW, 0)$$

$$P(WW, x) = (1 - a_1)(1 - a_2) P(WW, x) + b_1(1 - a_2) P(RW, x+1) + (1 - a_1) b_2 P(WR, x-1) \quad (0 < x < z)$$

$$P(WW, z) = (1 - a_1)(1 - a_2) P(WW, z) + (1 - a_1) b_2 P(WR, z-1) + b_2 P(WR, z)$$

$$P(WR, 0) = (1 - a_1) a_2 P(WW, 0) + b_1 a_2 P(RW, 1) + b_1 P(RF, 0)$$

$$P(WR, x) = (1 - a_1) a_2 P(WW, x) + b_1 a_2 P(RW, x+1) + (1 - a_1)(1 - b_2) P(WR, x-1) + b_1 P(RFx)$$
$$(0 < x < z)$$

$$P(WR, z) = (1 - a_1) a_2 P(WW, z) + (1 - a_1)(1 - b_2) P(WR, z-1) + (1 - b_2) P(WR, z) + b_1 P(RF, z)$$

$$P(RW, 0) = a_1(1 - a_2) P(WW, 0) + (1 - b_1)(1 - a_2) P(RW, 1) + (1 - b_1) P(RW, 0) + b_2 P(FR, 0)$$

$$P(RW, x) = a_1(1 - a_2) P(WW, x) + (1 - b_1)(1 - a_2) P(RW, x+1) + a_1 b_2 P(WR, x-1) + b_2 P(FR, x) \quad (0 < x < z)$$

$$P(RW, z) = a_1(1 - a_2) P(WW, z) + a_1 b_2 P(WR, z-1) + b_2 P(FR, z)$$

$$P(RF, x) = a_1 a_2 P(WW, x) + (1 - b_1) a_2 P(RW, x+1) + (1 - b_1) P(RF, x) \quad (0 \le x \le y)$$

$$P(RF, x) = (1 - b_1) a_2 P(RW, x+1) + (1 - b_1) P(RF, x)$$
$$(y < x < z)$$

$$P(RF, z) = 0$$

$$P(FR, 0) = 0$$

$$P(FR, x) = a_1(1 - b_2) P(WR, x-1) + (1 - b_2) P(FR, x)$$
$$(0 \le x \le y)$$

$$P(FR, z) = a_1 a_2 P(WW, z) + a_1(1 - b_2) P(WR, z-1) + (1 - b_2) P(FR, z) \quad (y < x \le z).$$

It can be shown that

$$P(WR, x) = P(RW, x+1) \quad (0 \le x < z).$$

Define

$$C_1 = \frac{(1 - a_1)(a_1 + a_2 - a_1 a_2 - a_1 b_2 + a_1 a_2 b_2)}{(1 - a_2) A_1}$$

$$C_2 = \frac{(1 - a_1) A_2}{(1 - a_2)(a_1 + a_2 - a_1 a_2 - b_1 a_2 + a_1 + a_2 b_1)}.$$

Whence it can be shown that

$$P(WR, x) = P(RW, x+1) = C_1 P(RW, x)$$
$$= C_1 P(WR, x-1) \quad (0 < x \le y)$$

$$P(WR, x) = P(RW, x+1) = C_2 P(RW, x)$$
$$= C_2 P(WR, x-1) \quad (y < x < z).$$

The reciprocal of the line efficiency can be calculated from

$$1/g = \frac{\sum_{x=0}^{z}(P(WW,x)+P(WR,x)+P(RW,x)+P(RF,x)+P(FR,x))}{\sum_{x=0}^{z}P(WW,x)+\sum_{x=1}^{z}P(RW,x)}$$

$$= 1 + \frac{a_1}{b_1} + \frac{a_2}{b_2} - a_2\left(\frac{1-C_1^{y+1}}{1-C_1} + \frac{C_1^y C_2(1-C_2^{z-1-y})}{1-C_2}\right)\bigg/D'$$

where

$$D' = (a_1 + b_2 - a_1 b_2)\frac{(1-C_1^Y)}{1-C_1} + a_2(a_2 + b_1 - a_2 b_1)$$

$$\cdot \frac{C_1^Y C_2(1-C_2^{z-1-y})}{a_1(1-C_2)} + (a_1 + a_2 - a_1 a_2)C_1^Y/a_1.$$

Now in a system consisting of two identical stages, i.e., $a_1 = a_2 = a$, $b_1 = b_2 = b$,

$$C_1 = 1/C_2 = C,$$

and

$$1/(1-C_1) = -C_2/(1-C_2) = 1/(1-C)$$

$$\begin{aligned}1/g \\ = 1 + 2a/b - \frac{a(1-C^y-C^{Y+1}+C^{2y+1-z})/(1-C)}{(a+b-ab)(1-2C^Y+C^{2Y+1-z})/(1-C)+(2-a)C^y}.\end{aligned}$$

Whence it can be shown that the optimum y is such that

$$z/2 - 1 < y < z/2$$

and if $z/2$ is even $y = z/2$.

Enforceability of Control

Even though it has been shown that the optimum choice of stage to repair has the characteristic that stage 1 is repaired first at low inventory levels and stage 2 is repaired first at high inventory levels, such a control rule could well be difficult to implement in practice unless repair was automated. This is typical of the problems in using improved operating rules or trying to transfer some of the insights gained in operating automated manufacturing systems to nonautomated systems. Control is not automatic, it requires continued management involvement to ensure that the control rules are used. Simplicity and enforceability are essential requirements of control strategies in the manufacturing environment.

III. Level for Deciding on Operation Sequence

If the sequence of operations is determined at the prerelease level for each job (taking account of any technological constraints and the capability of the machines for specific operations), then some of the flexibility of FMS is not being used. It is not surprising that the performance of an FMS can be improved by leaving the decisions on operation sequence to either the input control or operational control levels.

The nature of the improvement can be demonstrated using relatively simple models.

Prerelease Level

Consider first the determination of operation sequence at the prerelease level. Kimemia and Gershwin [8] have developed a number of mathematical programming models for this. The essential features of these models is that it is assumed that there will be FCFS (or random) release of jobs to the FMS and FCFS (or random) sequencing of jobs on the machines. Processing times on the machines are assumed to have exponential distributions and the jobs arrive in a Poisson stream.

Then if there are m machines in the system and a total of C spaces for work in process then the production capacity (the maximum arrival rate of jobs for which the system is stable) is given by (see [3])

$$PC(C) = q(C-1)/q(C)$$

where

$$q(C) = \sum_{n_1+n_2\cdots+n_m=C}\Pi(a_i)^{n_i}$$

$$a_i = \sum_r e_{ir}/\mu_i$$

and

r is the index denoting job class,
μ_i is the service rate of machine i,
e_{ir} is the expected number of visits of a job of class r to machine i.

Also,

$$PC(\infty) = 1/\max_i a_i.$$

There are a number of consequences of this formula.

1) If the e_{ir} are independent of the operation sequence, e.g., all jobs of class r must visit every machine once, then the production capacity will be the same irrespective of the operation sequence assigned at the prerelease level.

2) On the other hand, if there are alternative machines for particular operations then the production capacity will depend on the assignment of operations to machines.

For example, consider the following situation (based on an illustration in [8]).

There are two job classes 1 and 2 and two machines A and B. All jobs in class 2 must go first to machine A and then to machine B. However, jobs in class 1 require only one operation which can be performed either at machine A or machine B.

Then the prerelease planning decision is to decide what fraction of class 1 jobs should be assigned to machine A. The remaining $1-q$ will be assigned to machine B.

Suppose one-third of the jobs are class 1 and two-thirds of the jobs are class 2. Then

$$a_A = (2+q)/3\mu_A$$
$$a_B = (3-q)/3\mu_B$$

and the prerelease planning problem is to find

$$\max PC(q) = \frac{a_A^C - a_B^C}{a_A^{C+1} - a_B^{C+1}}$$

subject to $0 \le q \le 1$.

Suppose $\mu_A = 6$, $\mu_B = 5$.

Then if $C = \infty$ the optimum q is that which results in $a_A = a_B$, i.e., balanced load on the machines. At other values of C the optimum q does not result in balanced load, however the difference in production capacity between the optimum and balanced load is small. The following table gives the results.

C	Optimum q	PC (optimum q)	PC (balanced load)
∞	8/11	6.6	6.6
2	1	4.426	4.400
3	0.895	4.965	4.950

Input Control Level

If the operation sequence is determined at the input control level or the operational control level then it is possible to use information on the status of jobs and machines. One simple rule for deciding when to release jobs is the idle machine rule: only release a job to the FMS when the machine for the first operation of the job becomes idle and there is no

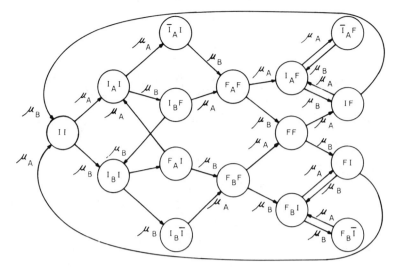

Fig. 1. State transition diagram for $C = 3$

other job already in the FMS waiting for service at the machine. When there is flexibility in operation sequence or assignment of job operations to machines this rule can be used to determine the first operation on a job. If FCFS release to the FMS is used then the job at the head of the queue of jobs waiting release to the FMS (the despatch queue) will be released as soon as any machine at which an operation is required by the job becomes idle. Once the job is in the FMS then the subsequent sequence of operations will be determined by the availability of a machine at which an operation on the job is required.

For example, consider a two machine FMS where there is a single job class requiring an operation at each machine, however the operations can be performed in sequence AB or in sequence BA. Fig. 1 shows the state transition diagram for a Markov model describing the system behavior for $C = 3$ and using the idle machine rule. The state of the system is described by whether the job at the machine is having its first (I) or second (F) operation and by which operation has been performed on the job in storage (A or B), e.g., $I_A F$ means that the job on machine A is having its first operation, the job on machine B is having its second operation and the job in storage has been to machine A. The state transition diagram is drawn for the situation where the system is operating near capacity so that there are always jobs in the input queue. $\bar{I}$ denotes that the machine is blocked.

The following table compares the production capacity when the operation sequence is determined at the prerelease level with the production capacity when the operation sequence is determined at the input control level for different values of C. It is assumed that $\mu_A = \mu_B = \mu$.

	PC/μ	
C	Prerelease Level	Input Level
2	0.667	0.750
3	0.750	0.843
4	0.800	0.885
5	0.833	0.903
∞	1	1

It can be seen that there is a significant improvement in capacity obtained if operation sequence is determined at the input control level.

The idle machine rule can be shown to be the optimum release rule for such a single class system (amongst those rules which base the decision for release on this state description, i.e., use no information on the actual processing times required by specific jobs except that the job processing times are identically and exponentially distributed) [9].

The idle machine rule can also be applied to the example discussed under (2) above. It can be shown that the production capacity is as follows.

	PC	
C	Prerelease Level	Input Level
2	4.426	4.544
3	4.965	5.213
∞	6.6	6.6.

However, the simple idle machine rule will not be the optimum release rule in this case. This is because it assumes that jobs must be released in FCFS order. As a result, machine B is sometimes idle because the first job in the despatch queue is class 2 and, even if there is a class 1 job in the despatch queue FCFS release implies that it must wait. If jobs can be released in any order from the despatch queue then the production capacity would be higher.

It is of interest to note that if there were flexibility in operation sequence for the class 2 jobs, i.e., operations could be performed either in sequence AB or sequence BA, then there would be a significant increase in production capacity using the idle machine rule. When $C = 2$ the production capacity would be 5.076 as compared to the 4.544 obtained when all class 2 jobs must follow sequence AB.

The Merit of Flow Lines?

When operation sequence is determined at the prerelease planning level then production capacity is independent of the assigned sequence when each operation must be performed at a specific machine. If all jobs are assigned the same sequence then there are obvious merits in terms of simplicity of planning and control. Laying out production in the form of a flow line results in a reduction in the distance jobs must be moved and at the same time makes it easier to justify automation of material handling. As a result operational control is very simple and input control becomes just a question of the timing of release. Thus, it is not surprising that for most of this century manufacturing system design has stressed the merits of flow lines.

However, once operation sequence is not determined at the prerelease level but left to the input or operational control levels then flow lines impose unnecessary constraints as a flow line only permits two sequences: forwards or backwards. Determining operation sequence adaptively, using information about the status of jobs and machines, is obviously more complex than the input and operational control tasks in flow lines. However, automated computer controlled systems make adaptive operation sequencing possible, indeed it is feasible with relative simple loop conveyor systems such as are used in the clothing industry.

Thus, the perceived merits of the flow line are due to the philosophy of simplifying operational and input control and transferring as much control decisions as possible to the prerelease level. This is the essence of the Taylorist philosophy of production management. Once it is recognized that it is desirable to provide more flexibility in operational control then flow lines cease to be so attractive. Of course, this conclusion assumes that technological constraints will permit flexible operation sequence. Available indications are that there is more flexibility than is conventionally assumed.

IV. Effect of System Status Information on Input Control

The performance of the FMS will be affected by the way in which it is controlled at the input level, at the operational level, and how these two levels of control interact. In particular, the control options at each level depend on the amount of information available on the status of jobs in the FMS and in the despatch queue and on the status of the machines in the FMS.

The focus of this section is on the way in which the performance of the FMS is determined by the amount of information about job and machine status used for input control.

Various amounts of information can be assumed. If there is no space limitation within the FMS then the decision on release of jobs can be made without using any information from the FMS at all. However, once there is a space limitation then it is necessary to know whether a job can enter the FMS. Beyond this minimal amount of information more information can be obtained and used up to the point where complete information on all machines and jobs is available. However, since acquiring reliable information in the manufacturing environment is often difficult and costly, it is necessary to evaluate the worth of such information in terms of improved system performance.

A number of alternative input control rules will be compared. In all cases it will be assumed that simple operational control will be used, i.e., FCFS or random sequencing of jobs at machines.

A. No Space Limitations in the FMS

Jobs can be released to the FMS as soon as they arrive. With FCFS sequencing at machines and exponentially distributed processing times the system can be modeled as a Jackson type queueing network and its performance calculated. Models for other sequencing rules at the machines are described in [10].

However, it is not necessarily advantageous to release all jobs as soon as they arrive. Jobs can be held in the despatch queue and released in batches. All jobs in a batch are processed before any job in the next batch. Priorities can be assigned to the jobs within the batch with the requirement that machines process the jobs in the sequence determined by the priorities. For example, the jobs within each batch can be sequenced in accordance with the least total processing time of all required operations.

Essentially the input control is being used to bypass the lack of operational control. While the production capacity is not affected by such input control, the release of jobs in batches can result in significant reductions in the mean flow time of jobs (or the mean number of jobs in the FMS and despatch queue) when the arrival rate of jobs is close to the production capacity. For example, if the FMS can be regarded as equivalent to a single server queue, the ratio of the arrival rate to production capacity is 0.9 and jobs are released whenever the FMS becomes empty in batches sequenced in accordance with least total processing time, then the mean flow time is about three quarters of what it would be with FCFS release (see [10]).

B. Space Limitations in FMS

Two possible FMS configurations can be considered. In one there is a common store for jobs in process so that the limitation is only on the total number of jobs in the FMS. In the other, storage space is provided at each machine so the space limitation applies to each machine. It is shown in [3] that the common store is preferable to the local store so the discussion will be restricted to the case of a common store such that the maximum number of jobs allowed in the FMS is C.

Only two machine systems will be considered and it will be assumed that jobs have random routing.

The production capacity depends on the release rule which in turn depends on what information is available from the FMS. For FCFS sequencing at the machines the following table shows the relevant results (from [11]).

	Information	Job Released to FMS	PC/μ
1)	Space available	First job i in despatch queue	$C/(C+1)$
2)	Space available and machine with least number of waiting jobs ($m/c\,i$)	First job for $m/c\,i$	$\dfrac{3\left(1-(1/2)^{C/2}\right)}{3-(1/2)^{C/2-1}}$ C even $\dfrac{4-3(1/2)^{(C-1)/2}}{4\left(1-(1/2)^{C+1/2}\right)}$ C odd
3)	Space available and machine idle	First job for idle m/c	$1-(1/2)^C$

Note that if $C=4$, PC/μ is, in case 1), 0.8; in case 2), 0.9; and in case 3), 0.937 and for all C this ranking of the rules applies.

When the arrival rate of jobs is less than the production capacity, their mean flow time can be reduced by releasing the job with the least total processing time of those meeting the required conditions concerning availability of space and the machine for its first operation. The ranking of the three rules remains unchanged.

Improved Operational Control

However, if other rules are used for operational control, e.g., choose the job with the shortest operation time, then the rules no longer have the same ranking. The idle machine rule can be worse than the other rules. The best rule obtained in simulation is shortest operation times at machines and release the job with the least total processing time as soon as space is available.

Thus, it seems that more sophisticated operational control can simplify the requirements for input control. However, the relationship between the two levels of control requires more investigation.

Conclusions

It is apparent that automated manufacturing systems are a fruitful area for developing improved control policies. These systems have considerable complexity because of the requirement for multilevel control, the inherent variability due to the possibility of machine breakdowns, the differing processing requirements of jobs (both in terms of routing and processing times) and the difficulty and cost of collecting reliable information. As a result it has proved difficult to formulate and solve adequate models of their control and operation.

Since the ability to develop exact models seems to be inherently limited by the complexity of the system, there is a need for further work in the development of good approximate models of these multilevel control systems with particular focus on the interaction between the different levels of control.

References

[1] M. P. Groover, *Automation, Production Systems and Computer-Aided Manufacturing.* Englewood Cliffs, NJ: Prentice-Hall, 1980.

[2] D. T. N. Williamson, "The pattern of batch manufacture and its influence on machine tool design," *Proc. Inst. Mech. Eng.*, vol. 182, no. 1, pp. 870–895, 1968.

[3] J. A. Buzacott and J. G. Shanthikumar, "Models for Understanding Flexible Manufacturing Systems," *AIIE Trans.*, vol. 12, no. 4, 339–350, 1980.

[4] J. A. Buzacott and L. E. Hanifin, "Transfer line design and analysis—An overview," *Proc. AIIE Fall Conf.*, Atlanta, GA, 1978, pp. 277–286.

[5] J. A. Buzacott and L. E. Hanifin, "Models of automatic transfer lines with inventory banks—A review and comparison," *AIIE Trans.*, vol. 10, no. 2, pp. 197–207, 1978.

[6] J. A. Buzacott, "On the optimal control of input to a job shop," Working Paper 74-004, Dep. Industrial Eng., Univ. of Toronto, Apr. 1974; also presented at ORSA/TIMS, Boston, MA, 1974.

[7] A. Dudick, "Fixed-cycle production systems with in-line inventory and limited repair capability," D. Eng. Sc. thesis, Columbia Univ., 1979.

[8] J. G. Kimemia and S. B. Gershwin, "Multi-commodity network flow optimization in flexible manufacturing systems," in *Complex Material Handling and Assembly Systems*, vol. II. Final Rep. Electron. Syst. Lab., M.I.T., Apr. 1980.

[9] J. A. Buzacott, "On the superiority of the production cell to job shops and flow shops," Working Paper #76-001, Dep. Ind. Eng. Univ. Toronto, Mar. 1976; presented at ORSA/TIMS scheduling Conf., Orlando, FL, 1976.

[10] J. G. Shanthikumar, "Approximate queueing models of dynamic job shops," Ph.D. thesis, Univ. of Toronto, 1979.

[11] J. A. Buzacott, "The production capacity of job shops with limited storage space," *Int. J. Prod. Res.*, vol. 14, no. 5, pp. 597–605, 1976.

Chapter 3: Control-Theoretic Methods

3.1: Introduction

The previous chapter dealt with rather detailed production models that are usually used in the lower levels of the production control hierarchy. As we move up the hierarchy, the time scale increases, and aggregate models can be used. In the control-theoretic approach, the aggregate models take the form of differential equations. Naturally, this mathematical form allows techniques from modern control theory to be used in search of a feedback control law that looks at the present inventory, machine status, and product demand to determine the present production rate.

These reprints rely extensively on modern control-theory techniques such as dynamic programming, optimal control [1-3], and stochastic control [4]. These methods have been used to formulate the production-control problem within a firm mathematical framework. The hope is that these methods will lead to feedback principles for determining production rates in an unreliable manufacturing system. This chapter presents some of the recent efforts and accomplishments toward that goal.

3.2: Aggregate-Production Models

The control theoretic methods tend to rely on aggregate models that relate part surplus levels to production rates and product demand. One common aggregate model is given by

$$\dot{x}(t) = u(t) - d(t) \qquad (3.1)$$

where $x(t) > 0$ represents a vector of part surplus levels, $u(t)$ are the production rates, and $d(t)$ represents product demand. In this model, $x(t)$ can be negative indicating that a product backlog exists.

In the control theoretic approaches, (3.1) is used in conjunction with a cost function, $g(x(t))$, that penalizes surplus and backlogs. As a result, these methods make extensive use of optimal control theory. This yields a feedback-control policy that determines production rates based on current-part surplus levels, current demand, and the present status of the machines. This is a desirable feature since decisions about production rates are made in real-time as opposed to open-loop policies that set production rates so far in advance that they are not able to respond to changes in demand or manufacturing capacity.

Aggregate models, such as (3.1), apply to time scales on the order of days or shifts. Shorter term decisions must be handled by a dispatcher. This leads to a hierarchical control structure.

3.3: Hierarchical Production Scheduling and Control of a Flexible Manufacturing System

Figure 3.1 shows the hierarchical approach to production planning that is used in the first two reprints of this chapter. This control structure makes the following assumptions:

- Part operations are on the order of seconds or minutes.
- Failure and repair times are on the order of hours.

The problem is to determine when parts should be dispatched into the flexible manufacturing system to satisfy production requirements that have been specified for a week.

At the top level of the hierarchy, machine data, part data, failure and repair rates and demand information can be found. All of these parameters take into account the manufacturing system's capacity.

The middle level computes the instantaneous production rates based on the aggregate production model of equation (3.1). These are aggregate-production rates in the sense that $u_j(t)$, an element of the production rate vector $u(t)$, only specifies the rate at which part type j should be produced. It doesn't say anything about which machine will actually do the work. This is the function of the lower level.

The dispatcher resides at the lowest level and decides when to dispatch part type j to machine i. Its decisions are made on a time scale which is on the order of minutes.

All three levels of the hierarchy work together to keep inventory as close to actual demand as possible.

3.3.1: Instantaneous-Capacity Constraints

The basic idea behind this policy is to produce parts to track the demand. Specifically, let

$D_j(T)$ = parts of type j which must be made by time T

$W_j(t)$ = total amount of type j produced by time t

Ideally, the system should satisfy

$$W_j(T) = D_j(T) \qquad (3.2)$$

This is illustrated in Figure 3.2. If we define

$$x_j(t) = W_j(t) - D_j(t) \qquad (3.3)$$

then the goal is to keep $x_j(t)$ as close to zero as possible for all t. This would not be a difficult problem except that

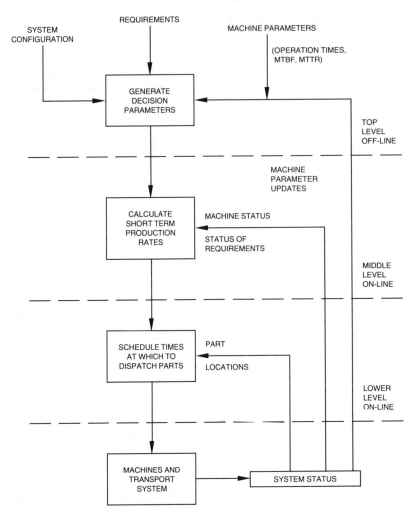

Figure 3.1: Hierarchical Approach to Production Planning

machines fail and get repaired at random time intervals. In anticipation of these failures, Gershwin et al. [13], allow the surplus to grow to a level that compensates for future production losses. This level will be called the *hedging point*, which will be chosen to keep the average value of $x_j(t)$ near zero. The key benefit to hedging point strategies is that they take into account the machine failure and repair rates through the concept of instantaneous capacity.

3.3.2: Instantaneous Capacity

The manufacturing capacity ultimately determines when parts are allowed to enter the system. If defined inaccurately, parts will end up in buffers or even in the parts transportation system. This usually happens when parts are dispatched too fast, in which case the number of parts actually produced declines.

The capacity of the system is constantly changing because of machine failures. Therefore, the instantaneous capacity of the system should be used for dispatching parts.

Assume there is only one path through a manufacturing system for each of J part types. In this case, each part type visits each machine, then

$$\tau_{i1}W_1 + \tau_{i2}W_2 + \ldots \tau_{iJ}W_J =$$
time for machine i to produce all the parts (3.4)

where

τ_{ij} = time to process part type j at machine i
W_j = number of part types to be processed at machine i during T seconds

Note that in (3.4) some of the τ_{ij}'s might be zero.

By using (3.4), parts can be processed if

$$\tau_{i1}W_1 + \tau_{i2}W_2 + \ldots \tau_{iJ}W_J \leq T_i \quad (3.5)$$

where T_i is the time available at machine i during the interval T seconds.

Next, divide (3.5) by T and let $u_j(t) = W_j/T$, which is the production rate of type j parts. Then,

$$\tau_{i1}u_1 + \tau_{i2}u_2 + \ldots \tau_{iJ}u_J \leq T_i/T \quad (3.6)$$

This directly couples the production rates with the status of the machines. In fact, if we assume small T, then T_i is either

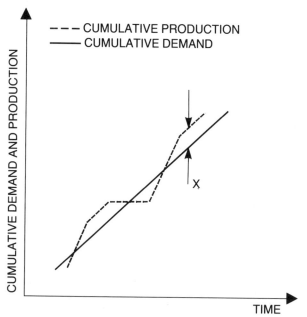

Figure 3.2: Production to Track Demand

0 or T which implies that the machine status won't change during T. In that case, we can rewrite (3.6) in terms of the machine status:

$$\alpha_i i = \begin{cases} 1 & \text{machine } i \text{ is up} \\ 0 & \text{machine } i \text{ is under repair} \end{cases}$$

then

$$\tau_{i1}u_1 + \tau_{i2}u_2 + \ldots \tau_{ij}u_j \leq \alpha_i \quad (3.7)$$

and this is the instantaneous capacity used by Gershwin et al. [13]. Note that (3.7) should be satisfied at all times for effective utilization of the manufacturing system. The hedging point strategy is designed to do this.

3.4: Hedging Point Strategies

This section presents a derivation of the hedging point that appeared in [13]. The theoretical significance of this result is explained in the first reprint "An Algorithm for the Computer Control of a Flexible Manufacturing System" by Kimemia and Gershwin.

Recall that the goal is to maintain $x_j(t)$ near a level, which we'll call $H_j(\alpha)$. This level is not zero since we anticipate machine failures. In the following derivation of $H(\alpha)$, we drop the subscript j associated with the part type.

Let

T_r = mean time to repair
T_f = mean time between failures

then the expected number of times that the machine will fail is

$$\frac{T}{T_f + T_r}$$

and the expected duration of these failures is T_r.

3.4.1: The Production Trajectory

In a manufacturing system with machines that are prone to failure, a typical history of the production will look like Figure 3.3. This trajectory captures several important points:

t_1—the machine fails, parts are removed at a rate of d, which implies that the slope between t_1 and t_3 is $-d$.

$t_3 = t_1 + T_r$—the machine is repaired.

t_3 to t_5—the machine produces at its maximum rate U, which is assumed to be greater than the demand d.

t_5—the inventory reaches the hedging point, $u(t) = d$ now.

Based on this trajectory, the problem is to find $H(\alpha)$ to minimize the weighted area above the negative (weighted by c_n) and below the positive (c_p) portions of the trajectory.

3.4.2: Area Calculation for Obtaining the Hedging Point

The area under the production trajectory is composed of

1. HT minus

2. $\dfrac{T}{T_f + T_r}$ times the area of rectangle At_1t_5C, which equals $(t_5 - t_1)H +$

3. $\dfrac{T}{T_f + T_r}$ times the areas of the triangles $At_1t_2 + t_4t_5C$, which are equal to

$$\frac{1}{2}H(t_2 - t_1), \frac{1}{2}H(t_5 - t_4), \text{ respectively,}$$

4. plus $\dfrac{T}{T_f + T_r}$ times the area of triangle t_2t_4B, which is

$$\frac{1}{2}(dT_r - H)(t_4 - t_2)$$

If we define $T' = T/(T_f + T_r)$, we weight the surplus areas by c_p, and the areas corresponding to backlogs by c_n, then

$$A = c_p(HT - T'(t_5 - t_1)H) +$$
$$c_p T'[\frac{1}{2}H(t_2 - t_1) + \frac{1}{2}H(t_5 - t_4)] +$$
$$c_n T'[\frac{1}{2}(dT_r - H)(t_4 - t_2)] \quad (3.8)$$

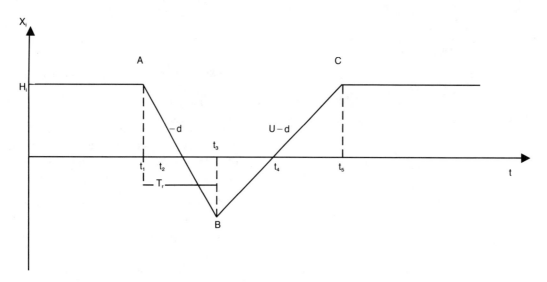

Figure 3.3: Production Trajectory

To find A, we need to calculate all of the time segments (i.e., find $(t_5 - t_1)$, $(t_2 - t_1)$, $(t_5 - t_4)$, and $(t_4 - t_2)$ in terms of T_r, T_f, H, and other known parameters).

First,

$$t_5 - t_1 = (t_5 - t_3) - (t_3 + t_1) \tag{3.9}$$

Now, the amount of parts lost during a failure equals the amount of parts built during the recovery, so we can write

$$dT_r = (t_5 - t_3)(U - d)$$

or

$$t_5 - t_3 = \frac{dT_r}{U - d} \tag{3.10}$$

The second term on the right side of (3.9) can be found directly from the trajectory:

$$t_3 - t_1 = T_r \tag{3.11}$$

Combining (3.10) and (3.11) results in

$$t_5 - t_1 = \frac{T_r U}{U - d} \tag{3.12}$$

Next, consider $t_2 - t_1$. During this time interval, the surplus drops by H at a rate of $-d$,

$$-d = \frac{H}{t_1 - t_2}$$

or

$$t_2 - t_1 = \frac{H}{d}$$

Similarly, for $t_5 - t_4$ the slope of BC is

$$\frac{H}{t_5 - t_4} = U - d$$

which yields,

$$t_5 - t_4 = \frac{H}{U - d}$$

To find $t_4 - t_2$, use

$$t_4 - t_2 = (t_4 - t_3) + (t_3 - t_2) \tag{3.13}$$

The slope in the time interval $(t_4 - t_3)$ is

$$U - d = \frac{dT_r - H}{t_4 - t_3}$$

and the slope in the interval $(t_3 - t_2)$ is

$$-d = \frac{H - dT_r}{t_3 - t_2}$$

Combining these into (3.13) gives

$$t_4 - t_2 = \frac{U(dT_r - H)}{d(U - d)}$$

Now the total weighted area can be computed as

$$A = c_p \left(HT - \frac{T}{T_f + T_r} \frac{UT_r H}{U - d} \right) +$$

$$c_p \frac{T}{T_f + T_r} \left(\frac{H^2}{2d} + \frac{H^2}{2(U - d)} \right) +$$

$$c_n \frac{T}{2(T_f + T_r)} \frac{(dT_r - H)^2}{U - d} \frac{U}{d} \tag{3.14}$$

Note that this is a quadratic in H, which satisfies the theoretical requirements developed in the reprints by Gershwin et al. Next, set

$$\frac{dA}{dH} = 0$$

and solve for H to obtain

$$H = \frac{c_n}{c_p + c_n} T_f d - \frac{c_p}{c_p + c_n} \frac{d}{U}(UT_f - d(T_f + T_r))$$

(3.15)

The hedging point is proportional to the amount of parts needed to satisfy the demand during a failure reduced by an amount proportional to the frequency of failures and the rate of production during the recovery from a failure.

The hedging point can be used to determine when to release a part into the system and how to set the production rates. Parts are released into the system whenever $x(t)$ is below the hedging point.

Production rates are determined by

$u(t) = U$ if $x(t) < H$ and the machine is up
$u(t) = d$ when $x(t) = H$
$u(t) = 0$ when $x(t) > H$

Hedging poing strategies have also been developed at the machine level by Eleftheriu and Desrochers [15]. In this case, the buffer hedging point concept has been introduced. The buffer hedging point is the level of inventory that should be maintained in the buffer in anticipation of machine failures. This results in a dispatching rule that says that parts should be dispatched from a machine whenever the buffer it feeds is below its buffer hedging point.

These hedging points are like regulators or "thermostats." The goal is to operate the entire system around these buffer hedging points. Any deviation from these setpoints causes parts to be dispatched. This results in a dispatching rule that accounts for machine failures and repairs. Buffer hedging points for several common manufacturing configurations have been derived [15].

3.4.3: "An Algorithm for the Computer Control of a Flexible Manufacturing System" (J. Kimemia and S.B. Gershwin)

In the first reprint, Kimemia and Gershwin propose a four-level hierarchical controller for a manufacturing system with failure-prone machines. Their objective is to minimize a cost function that penalizes both inventory surplus and backlogs. The first level of the controller evaluates (off-line) a cost function and generates hedging points discussed in the previous section. The second level calculates the short-term production rates based on the cost function. The third level calculates the route splits (i.e., the proportion of parts that will follow a certain path). Finally, the fourth level determines the exact dispatching times of the parts. They conclude that the optimal solution is difficult to derive and therefore propose a suboptimal policy.

In a more recent work, Maimon and Gershwin [12] clarify the routing method and extend it to the case where different machines can perform some of the same operations.

3.4.4: "Performance of Hierarchical Production Scheduling Policy" (R. Akella, Y. Choong, and S.B. Gershwin)

In the second reprint Akella et al. approximate the cost function with a quadratic function. The highlights of this approach were presented in Section 3.4 since the details of the derivation are not included in the reprint. This approximation appears to be very good, in fact, a major advantage of hedging point strategies is that they seem to be very robust to changes in the manufacturing system [15].

An appreciation for this approximation can be seen in the problem formulation.

Problem formulation: The system dynamics are described by the machine status, capacity constraints, and the production inventory model (3.1). The machine status can be modeled as an irreducible Markov chain with a finite number of states, that is,

$$P[\alpha_j(t + \delta_t) = 0 \mid \alpha_j(t) = 1] = \frac{\delta t}{T_{fj}}$$

(3.16)

where $\alpha_j(t)$ is the machine status (Section 3.3.2), $j = 1, 2, \ldots, M$ (number of machines) and T_{fj} is the time between a repair and the next failure of machine j. Similarly, on the repair side,

$$P[\alpha_j(t + \delta t) = 1 \mid \alpha_j(t) = 0] = \frac{\delta t}{T_{rj}}$$

(3.17)

where T_{rj} is the average time to repair machine j.

The capacity constraints (3.7) couple the production rate of each part type with the status of the machines given in (3.16 and 3.17).

Next, assign a cost function that penalizes both inventory surplus and backlogs of $x(t)$ as

$$g(x(t)) = \sum_{i=1}^{J} g_i(x_i(t))$$

The control problem can now be stated. Assuming initial inventory $x(t_o)$ and machine status $\alpha(t_o)$, find the controller that minimizes

$$J(x(t), \alpha(t), t_o) = E\left\{ \int_{t_o}^{t_f} g(x(t)) dt \mid x(t_o), \alpha(t_o) \right\}$$

subject to 3.1 and (3.7) and $\alpha(t)$, a Markov chain.

The exact optimal solution can be found [5] by solving

$$\min_{u(t)} \frac{\partial J_u(u^*(t), \alpha(t))}{\partial x(t)} u(t)$$

(3.18)

where $u(t)$ satisfies the capacity constraint and $J_u(u^*(t), \alpha(t))$ is defined as the cost to go function, i.e., the cost needed to finish the remaining production plan, when we are at time t and the parts level is $x(t)$, with machine status $\alpha(t)$ and $u^*(t)$ is the optimal production rate.

The large dimension of the problem and the stochastic nature of $\alpha(t)$ makes it difficult to find exact solutions. This problem is addressed in the first reprint.

In the second reprint, the cost to go function is approximated with a quadratic function which leads to the hedging points derived in Section 3.4. This reprint also includes several comparisons with other operational policies.

3.5: "Optimal Control of Production Rate in a Failure Prone Manufacturing System" (R. Akella and P.R. Kumar)

In this reprint, Akella and Kumar show that the hedging point strategy is the optimal solution for a *single-product* manufacturing system with two-machine states (up and down). They also use the same aggregate-production model and assume that the demand is constant. Let $x(t)$ be the part level at time, t, then the hedging point strategy can be summarized:

$$\text{if } x(t) \begin{cases} < \text{ hedging point, then produce at the maximum rate} \\ = \text{ hedging point, then produce to meet demand} \\ > \text{ hedging point, then produce nothing} \end{cases}$$

Their proof of optimality is based on the Hamilton-Jacobi—Bellman equation [1-3].

Bielecki and Kumar [30] have also studied this problem. They have shown that there are conditions under which a zero-inventory policy is optimal even when there is uncertainty in manufacturing capacity. Their modeling assumptions are similar to those of Akella and Kumar.

3.6: "Production Control of a Manufacturing System with Multiple Machine States" (A. Sharifnia)

Sharifnia derives equations for the steady-state probability distribution of the surplus level in a single-product-manufacturing system with multiple-machine states when the hedging point control strategy is used. This work is important because there is an arbitrary number of machine states corresponding to a failure mode of the system. This allows the consideration of many machines described by the state, $s(\alpha_1, \alpha_2, \ldots)$.

The author shows that for each machine state, the value function $J(x(t), \alpha(t))$ (equation 3.18) reaches its minimum at a hedging point. An example of a system with three-machine states is given.

3.7: "A Dynamic Optimization Model for Integrated Production Planning; Computational Aspects" (W.A. Gruver and S.L. Narasimhan)

This paper is an application of optimal-control theory to the production planning problem. The authors use a more complicated aggregate-production model, which takes into account labor, capital, and engineering expenditures. Their objective function penalizes these expenditures in addition to the backlogs and surpluses.

This work is unique in its use of optimal-control theory. By using optimal-control theory, terminal engineering costs, terminal inventory costs, and product costs can be determined as a function of inventory costs. Similarly, the cost of the product as a function of inventory and the sensitivity of inventory level to changes in the cost of engineering can both be determined. Optimal-control theory has also been used to study the effect of uncertainty in demand on determining production rates [11]. Still other applications can be found in Narasimhan and Gruver [6].

3.8: Related Research

Control theory has been applied extensively to both machine and process-control-problems. To a lesser degree, control techniques have been applied to production planning and scheduling problems. However, several research efforts have approached the production-planning problem from a control-theoretic viewpoint. A common denominator in this work is the use of some form of an aggregate-production model.

Several researchers have used the model in equation (3.1) as a basis to investigate control-theoretic approaches to the scheduling of manufacturing systems. Sethi and Thompson [27] derived exact results for the simplest of models: a one product, one-stage manufacturing system. They showed that the optimal open-loop-production trajectory was the combination of three components. The first component was a turnpike expression that attempted to keep inventory at its optimal level. The second component was a "starting correction" term accounting for initial inventory and, finally, an "ending correction" accounting for final inventory.

Bradshaw [28,29] has demonstrated a variety of well-known control techniques for cascaded, multistage manufacturing systems. They utilized both eigenvalue placement for unconstrained systems and a deadbeat-control technique for systems with bounded inputs.

Olsder and Suri [8] consider the time-optimal control of a simple two-machine two-product FMS. A two-product aggregate-production model is used to represent the FMS. Machines fail and are repaired according to a Markov process. The controls are the amount of instantaneous production rate of each product type. They must obey the production constraints such that total-system resources are not exceeded. The production-constraint set exhibits jump discontinuities based on the Markov chain model representing the machine-failure state, which consists of the parts levels and the machine-failure state. The control problem is to set the instantaneous part-production rates such that total production time is minimized while not exceeding the available production capacity. The optimal-feedback control is shown to be a function of both the machine-failure state and the inventory state. The control law is shown to be bang-bang.

For a simple example, a control switching law is computed and then shown to be the optimal.

Recently, interest has developed in controlling manufacturing systems with delays [14,21]. Lou et al. [14] consider a manufacturing system modeled as a continuous flow system with delay. This situation occurs when a workstation processes many parts at a time and the parts have average interval times that are much less than the processing time for a single part. The system can be modeled by

$$\dot{x}(t) = u(t - \tau) - d(t)$$

where $x(t)$ represents the surplus (or backlog) in the buffer, τ is the delay (processing time), $u(t)$ is the loading rate, and $d(t)$ is the deterministic and known demand rate. The authors approximate this delay system by a system of first-order differential equations without delay. The final control law is a suboptimal strategy based on a quadratic approximation to the value function.

3.9: Summary

The control theoretic approach has the potential to discover manufacturing system concepts. In control-system theory, the concepts of controllability, stability, bandwidth, disturbance rejection, etc., are well understood. The analogous concepts in manufacturing are not well understood at all.

In a manufacturing system, bottlenecks and buffer sizes growing without bounds are analogous to stability in system theory. Similarly, controllability answers the question: Can the system make the part? Bandwidth is a measure of throughput and perhaps flexibility. Disturbance rejection has already been shown [11] to be related to handling fluctuations in future product demand.

This approach is intended to expand and advance the knowledge of how these systems can be made to operate efficiently as discrete-part processors. The control-theoretic approach is likely to discover novel operating policies and manufacturing system concepts.

3.10: References

3.10.1: Books

1. D.E. Kirk, *Optimal Control Theory, An Introduction*, Prentice Hall, Englewood Cliffs, N.J., 1970.
2. A.P. Sage and C.C. White, III, *Optimum Systems Control*, Prentice Hall, Englewood Cliffs, N.J., 1977.
3. M. Athans and P.L. Falb, *Optimal Control*, McGraw-Hill, New York, 1966.
4. R.F. Stengel, *Stochastic Optimal Control*, John Wiley and Sons, Inc., New York, 1986.

3.10.2: General References

5. R. Rishel, "Dynamic Programming and Minimum Principles for Systems with Jump Markov Disturbances," *SIAM Journal on Control*, Vol. 13, No. 2, Feb. 1975, pp. 338–371.
6. S.L. Narasimhan and W.A. Gruver, "Optimal Control in Integrated R&D Production and Inventory Planning Systems," *AIIE Transactions*, Vol. 11, No. 3, Sept. 1979, pp. 198–205.
7. P.R. Kleindorfer and K. Glover, "Linear Convex Stochastic Optimal Control with Applications in Production Planning," *IEEE Transactions on Automatic Control*, Feb. 1973, pp. 56–59.
8. G.J. Olsder and R. Suri, "Time-Optimal Control of Parts-Routing in a Manufacturing System with Failure-Prone Machines," *Proceedings of the IEEE Conference on Decision and Control*, IEEE Press, New York, 1980, pp. 722-727.
9. S.B. Gershwin, R. Akella, and Y. Choong, "Short Term Production Scheduling of an Automated Manufacturing Facility," *Proceedings of the 23rd IEEE Conference on Decision and Control*, IEEE Press, New York, Dec. 1984, pp. 230–235.
10. M.F. Clifford and A.A. Desrochers, "Optimization Methods for Hybrid Manufacturing Systems," *Proceedings of the Japan-USA Symposium on Flexible Automation*, Osaka, Japan, July 1986, pp. 483–490.
11. M.F. Clifford and A.A. Desrochers, "Preview Control of Flexible Manufacturing Systems," *Proceedings of the 1987 IEEE International Conference on Robotics and Automation*, IEEE Computer Society Press, Washington, D.C., April 1987, pp. 1849–1854.
12. O.Z. Maimon and S.B. Gershwin, "Dynamic Scheduling and Routing of Flexible Manufacturing Systems that have Unreliable Machines," *Operations Research*, Vol. 36, No. 2, March–April 1988, pp. 279–292.
13. S.B. Gershwin, R. Akella, and Y.F. Choong, "Short-Term Production Scheduling of an Automated Manufacturing Facility," *IBM Journal of Research and Development*, Vol. 29, No. 4, July 1985, pp. 392–400.
14. S.X.C. Lou, G. Van Ryzin, and S.B. Gershwin, "Scheduling Job Shops with Delays," *Proceedings of the 1987 IEEE International Conference on Robotics and Automation*, IEEE Computer Society Press, Washington, D.C., April 1987, pp. 1345–1349.
15. M.N. Eleftheriu and A.A. Desrochers, "A Unified Approach to Production Planning and Scheduling," *Proceedings of the 26th IEEE Conference on Decision and Control*, Los Angeles, Calif., Dec. 1987, pp. 605–611.
16. M.N. Eleftheriu and A.A. Desrochers, "Production Planning for a Prone to Failures Manufacturing Facility with Stochastic Demand," *Proceedings of the 1988 International Conference on Computer Integrated Manufacturing*, IEEE Computer Society Press, Washington, D.C., Rensselaer Polytechnic Institute, Troy, N.Y., May 1988, pp. 74–80.

17. R. Akella, S. Rojagopal, and P. Kumar, "Part Dispatch in Multi-Image Card Lines," *Proceedings of the 1986 IEEE International Conference on Robotics and Automation*, San Francisco, Calif., April 1986, pp. 143–146.

18. S.B. Gershwin, "Stochastic Scheduling and Set-Ups in Flexible Manufacturing Systems," (eds. K.E. Stecke and R. Suri), *Proceedings of the Second ORSA/TIMS Conference on Flexible Manufacturing Systems*, Elsevier, New York, 1986.

19. V.V.S. Sarma and M. Alam, "Optimal Maintenance Policies for Machines Subject to Deterioration and Intermittent Breakdown," *IEEE Transactions on Systems, Man, and Cybernetics*, May 1975, pp. 396–398.

20. D.D. Sworder and T. Kazangey, "Optimal Control, Repair, and Inventory Strategies for a Linear Stochastic System," *IEEE Transactions on Systems, Man, and Cybernetics*, Vol. SMC-1, No. 3, July 1972, pp. 342–347.

21. G.J. Van Ryzin, "Control of Manufacturing Systems with Delay," *M.I.T., Laboratory for Information and Decision Systems, Report LIDS-TH-1676*, M.I.T., Cambridge, Mass., June 1987.

22. A. Janiak, "Time-Optimal Control in a Single Machine Problem with Resource Constraints," *Automatica*, Vol. 22, No. 6, 1986, pp. 745-747.

23. R. Conterno, E. Fiorio, G. Menga, and A. Villa, "A Large Scale System Approach to the Production Planning and Control Problem," *Proceedings of the 24th IEEE Conference on Decision and Control*, Ft. Lauderdale, Fla., Dec. 1985, pp. 2016–2021.

24. K.E. Stecke, "A Hierarchical Approach to Production Planning in Flexible Manufacturing Systems," *Proceedings of the First ORSA/TIMS Conference on Flexible Manufacturing Systems*, Ann Arbor, Mich., 1984, pp. 426–433.

25. P.J. O'Grady and M.D. Byrne, "A Continued Switching Algorithm and Linear Decision Rule Approach to Production Planning," *International Journal of Production Research*, Vol. 23, No. 2, 1985, pp. 285–296.

26. J.N. Tsitsiklis, "Optimal Dynamic Routing in an Unreliable Manufacturing System," Laboratory for Information and Decision Systems Report No. LIDS-TH-1069, M.I.T., Cambridge, Mass., Feb. 1981.

27. S.P. Sethi and G.L. Thompson, "Simple Models in Stochastic Production Planning," in *Applied Stochastic Control in Econometrics and Management Science* (eds. A. Bensoussan et al.), North Holland Publishing, New York, 1981, pp. 295–304.

28. A. Bradshaw and D. Daintith, "Synthesis of Control Policies for Cascaded Production–Inventory Systems," *International Journal of Systems Sciences*, Vol. 7, No. 9, 1976, pp. 1053–1070.

29. A. Bradshaw and Y. Erol, "Control Policies for Production—Inventory Systems with Bounded Input," *International Journal of Systems Sciences*, Vol. 11, No. 8, 1980, pp. 947–959.

30. T. Bielecki and P.R. Kumar, "Optimality of Zero-Inventory Policies for Unreliable Manufacturing Systems," *Operations Research*, Vol. 35, No. 4, July–Aug. 1988, pp. 532–541.

An Algorithm for the Computer Control of a Flexible Manufacturing System*

Reprinted with permission from *IIE Transactions*, Volume 15, Number 4, December 1983, pages 353-362. Copyright Institute of Industrial Engineers, 25 Technology Park/Atlanta, Norcross, GA 30092.

JOSEPH KIMEMIA
AT&T Information Systems Laboratories, NP 2D-108
2220 Highway 66
Neptune, New Jersey 07753

STANLEY B. GERSHWIN
MEMBER, IIE
Laboratory for Information and Decision Systems
Massachusetts Institute of Technology
35-310, 77 Massachusetts Avenue
Cambridge, Massachusetts 02139

Abstract: The problem of production management for an automated manufacturing system is described. The system consists of machines that can perform a variety of tasks on a family of parts. The machines are unreliable, and the main difficulty the control system faces is to meet production requirements while the machines fail and are repaired at random times. A multilevel hierarchical control algorithm is proposed which involves a stochastic optimal control problem at the first level. Optimal production policies are characterized, and a computational scheme is described.

■ A flexible manufacturing system (FMS) consists of a set of work stations capable of performing a number of different operations and interconnected by a transportation mechanism. The FMS produces a family of parts related by similar operational requirements or by belonging to the same final assembly. All parts in the part family are produced simultaneously. Work pieces are introduced into the system at a loading station and leave at an unloading station after undergoing a specified sequence of operations. The machines in an FMS are capable of performing operations on a random sequence of parts with negligible change over time from one part to the next. The flexibility of the system allows the choice of one or more stations for each operation. This allows production to continue even when a work station is out of service because of failure or maintenance.

The changeover time for parts in the same family is negligible in these systems because the machines are numerically controlled with a large number of tools used for each operation or because operations are performed by robots. In the first case, several tools must be selected and replaced for each operation so that no time is saved by working on two successive parts of the same type on the same machine.

In both cases, software determines the operations, and changing the software can be done nearly instantaneously compared to operation times.

The FMS concept is not limited to metal cutting; such systems can also be used for the assembly of printed circuit boards, integrated circuit fabrication, and automobile assembly lines. A survey of flexible manufacturing systems appears in Dupont-Gatelmand [3]. The ability of an FMS to produce a family of parts simultaneously results in reduced finished and in-process parts inventories and faster responses to changes in demand requirements, when compared to traditional production methods. However, careful attention must be paid to production scheduling. The high capital cost of an FMS means that efficient use of system resources is very important. In the majority of implementations, FMS's are part of a multistage manufacturing system. The input consists of parts that have undergone one or more processing stages, and the finished family of parts are assembled into different final products.

Most manufacturing systems are large and complex. It is natural, therefore, to divide the control or management into a hierarchy consisting of a number of different levels. Each level is characterized by the length of the planning

---

Received March 1982; revised June 1983 and September 1983. Paper was handled by the Manufacturing, Inventory, and Automated Production Department.

*This research was done in the Laboratory for Information and Decision Systems of the Massachusetts Institute of Technology under NSF grant numbers DAR 78-17826 and ECS 79-20834.

horizon and the kind of data required for the decision-making process. Higher levels of the hierarchy typically have long horizons and use highly aggregated data, while lower levels have shorter horizons and use more detailed information. The nature of uncertainties at each level of control also varies.

The managers of a manufacturing firm make production plans for finished products by considering forecasts of demand, sales, raw material availability, inventory levels, and plant capacity. From the resulting master plan, the requirements for the components that go into the final products can be made. The various departments responsible for the manufacture of the components schedule their activities so as to meet the requirements dictated by the master production and the materials requirements plans [4] [7].

In an FMS, the operations at the work stations and the material handling system are entirely under computer control. Decisions such as which parts should be loaded into the system and what work stations particular work pieces should visit next are taken by the FMS control computer. Human intervention is necessary only when unusual or unanticipated events take place. It is important, therefore, to develop models and algorithms which allow the FMS controller to generate production schedules which satisfy demand requirements and to exercise control over the system so that the output conforms to the schedule. The task of the controller is complicated by random failures of the work stations. A good production policy should anticipate failures and demand changes if it is to satisfy all of the objectives stated above. It is important that this policy employ feedback so as to respond to failures and to allow human operators (who can deal with a wider range of situations than envisioned by system planners) to override control decisions on rare occasions.

In systems currently operating, a number of operating policies are employed. Typically, parts are loaded into the system whenever an opportunity arises, and planned production affects few decisions internal to the FMS [9]. Such policies do not typically consider capacity or reliability and can lead to congestion and underutilization. Hutchinson [8] describes information and algorithm structures which have been implemented on an actual system. It is to be noted that a high degree of human intervention is necessary because of machine failures, human errors, maintenance, and changes in operating environment. In this paper, an FMS control policy is described. Parts are loaded into the system in a way that will not overload the system or cause congestion and yet will meet long-term production objectives. Because of the problem's complexity, the policy is organized hierarchically.

HIERARCHICAL SCHEME FOR THE OPERATIONAL CONTROL OF AN FMS

A four-level control structure specifically designed to compensate for work station failures and changes in part requirements is proposed. The hierarchy is illustrated in Fig. 1, in

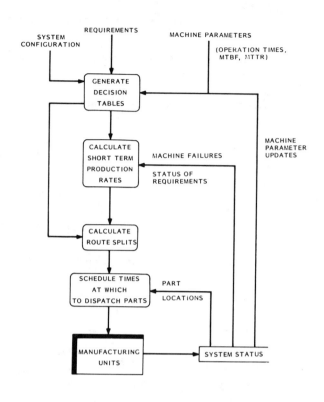

Fig. 1. The hierarchical production control scheme.

which the FMS controller is imbedded into the larger hierarchy of production management. The objective of the FMS controller is to satisfy a known, possibly time-varying demand for a family of items that is dictated by the Master Production Plan, subject to constraints imposed by the resources available.

The routing and scheduling policy described here is based on a set of assumptions on the time scales of various classes of events that occur in the operation of the flexible manufacturing system:

- The shortest time period is that of the setup when switching among the family of operations for which the machine is configured. It is assumed that these times are short compared to the following times and they may be ignored.

- The next time period is that of the typical operation. If operation times are random, then we refer to the mean of the distributions. Operation times are assumed to be orders of magnitude larger than setup times.

- The next time period is that between machine failures or repairs. Again, mean times between failures and mean times to repair (MTBF and MTTR) are considered.

- The longest time period is the planning horizon for the problem under consideration. It is assumed that demand can be specified for a time period larger than the typical

MTBF or MTTR. At this time period, the machines may be reconfigured for another part family.

The Flow Control Level (Calculates Short-Term Production Rates)

The flow control level of FMS control determines the short-term production rates of each member of the part family. The rates must be determined jointly because the parts share the time available at the work stations. In addition, the demand, the level of downstream buffer levels, and the reliability of the work stations must be taken into account.

The mix of parts being produced must be adjusted continuously so as to take into account random failures of the work stations. If, for example, a part cannot be made because a certain work station has failed, the lost production must be made up when the station is repaired. Using failure and repair statistics of the machines, the production rates should be chosen in a way that anticipates station downtime. Adequate but not excessive downstream buffer levels should be maintained so as to satisfy downstream demand.

The Routing Control Level (Calculates Route Splits)

A part entering the FMS has one or more paths it can take through the system in order to complete its processing requirements. The proportion of parts that should follow each of the available paths is chosen by the route control level of the controller. The objective is to meet the production rate dictated by the flow controller while minimizing congestion and delay within the system.

The system can be modeled as a network of queues with the stations represented as the service nodes. The arrival rate of the parts is determined by the flow control level. The flow rate on each path can then be determined by a mathematical programming technique. Alternatively, it can be calculated together with total part flow as described below.

The Sequence Controller (Schedules Times at Which to Dispatch Parts)

At the lowest level of control are scheduling algorithms that dispatch parts into the system and supervise the operations of the work stations. The objective is to maintain the flow rates chosen by the flow and route controllers.

We suggest a simple method which uses the flow rates calculated by the route controller to determine time intervals between loading parts on each path. Simulation results show that production rates and work station utilizations determined by the flow and routing levels of the controller can be achieved provided they are feasible.

Generation of Decision Tables

At the highest level of the control scheme is the off-line calculation of the control policies to be used in the flow and routing levels. In principle, this is required only when a new schedule is established. In practice, it may be prudent to include a long-term feedback loop to compile data on failure and repair rates as well as other parameters that may not be well known.

Whenever new estimates of parameters are found that are significantly different from earlier values, the calculation of control policies should be redone. While the calculation is being performed, production can continue using the previous control policy.

COMPARISON WITH OTHER WORK

The hierarchical approach to FMS planning and control has been suggested by a number of authors. Hildebrant [6] examines a three-level hierarchy that minimizes the time to produce a given number of parts. The top level calculates steady state production rates for each failure condition. Inventory levels are not considered, and a change in production requirements means that the production rates must be recomputed. The second and third levels determine loading schedules for the parts. Olsder and Suri [11] use a dynamic programming formulation for the minimum time production problem. In this case, a feedback policy results which depends on the current failure state and production levels.

Hahne [5] and Tsitsiklis [13] study the problem of maximizing throughput in a system in which parts can be routed from an upstream machine to one of two unreliable downstream machines. They show that optimal policies are piece-wise constant functions of intermediate buffer levels. Calculation of exact optimal policies for the three-machine system has large computational requirements.

The hierarchical controller described here utilizes currently available capacity while anticipating work station failures, repairs, and changes in demand requirements. It differs from Hildebrant's [6] scheme in that flow rate decisions are made on the basis of the current inventory levels as well as the current set of working machines. Part types that are backlogged will tend to be favored over part types for which a surplus exists. Therefore, this control scheme satisfies requirements for a part family without the need for large finished and in-process parts inventories.

Buzacott and Yao [2] present a survey of other research in the modeling and control of flexible manufacturing systems. A survey of routing policies within the FMS is given by Buzacott [1].

THE FLOW CONTROL LEVEL OF THE FMS CONTROLLER

Problem Formulation

The flow control level of the FMS controller determines

the production rates for the part family. The horizon is set by the FMS management and is of the order of one period of the master production plan. The routing and the sequencing levels ensure that the output of the system is the same as that set by the flow controller. For the lower levels of the hierarchy to be able to track the rates set by the flow controller, the rates must be at all times feasible for the current system configuration. It is important therefore for the flow controller to have complete and timely information on the operational status of all work stations. In addition, knowledge of the finished parts inventory is needed.

The FMS consists of M work stations. Work station m ($m = 1, 2, \ldots, M$) has L_m identical machines. The concept of work station is logical, not physical: the machines in a work station need not be located closer to one another than other machines.

A family of N part types is produced. The material flow is modeled as a continuous process. This kind of model ignores combinatorial details which are treated at the lower levels of control. Its accuracy is adequate for the time horizon treated at the flow control level which is long compared to the time needed to produce individual parts. Let $u(t) \epsilon R^N$ be the production rate for the part family, the control variable. The downstream demand rate is $d(t) \epsilon R^N$ and is known in the interval $(0, t_f)$. Finished parts are stored in downstream buffers from where the downstream demand is satisfied. Define $x(t) \epsilon R^N$ by the following differential equation:

$$\frac{dx(t)}{dt} = u(t) - d(t). \quad (1)$$

The vector $x(t)$, termed the buffer state, measures the cumulative difference between production and demand for the parts. A negative value for a component of $x(t)$ gives the backlogged demand for the corresponding part. A positive value is the size of the inventory stored in the downstream buffers. Ideally, parts in an FMS are produced as they are required, keeping the buffer state close to zero.

The state of the work stations is called the machine state and is denoted by an M-tuple of integer variables $\alpha(t)$ with the component $\alpha_m(t)$ equal to the number of operational machines at station m. Given that a machine at station m is operational, the probability of a failure in an interval of length δt is $p_m \delta t$. The probability that a failed machine is repaired during the δt time interval is given by $r_m \delta t$. The parameters p_m and r_m are the failure and repair rates for the machines at station m. The dynamics of the machine state are therefore governed by

$$P[\alpha_m(t+\delta t)=l+1 | \alpha(t)=l] = \begin{cases} (L_m-l)r_m \delta t & \text{for } 0 \leq l < L_m \\ 0 & \text{otherwise,} \end{cases} \quad (2)$$

$$P[\alpha_m(t+\delta t)=l-1 | \alpha(t)=l] = \begin{cases} l p_m \delta t & \text{for } 0 < l \leq L_m \\ 0 & \text{otherwise.} \end{cases} \quad (3)$$

Note that

$$P[\alpha_m(t+\delta t) = l_1 | \alpha_m(t) = l_2] = 0 \text{ if } |l_1 - l_2| > 1.$$

For two machine states i and j, it is convenient to define

$$\lambda_{ij} \delta t = P[\alpha(t+\delta t) = j | \alpha(t) = i] \quad \text{for } i \neq j$$

and

$$\lambda_{ii} = -\sum_j \lambda_{ij}.$$

The times between failures and to repair are thus modeled by exponentially distributed random variables with means $1/p_m$ and $1/r_m$, respectively. The machine state therefore can be modeled by an irreducible Markov chain with a finite number of states. Each state communicates with M neighbors, and transitions are due to the failure or repair of a single machine. The model assumes that machine failure rates do not depend on the part flow rate through the work stations. Where reliability does depend on the part flow rate, the failure rate then becomes a function of the production rate and part routing through the FMS. Failure and repair rates are assumed to be independent of production rates and the number of operational machines for computational and expository convenience. It is easy to extend the model and the control method of this paper to include these effects.

Exponentially distributed times between failures are suitable where machine downtime is caused by the random failure of any one of a large number of components. The exponential model is also consistent with reported field data [12]. Nonexponential distributions can be used to model failure rates that depend on the time since the last repair. However, in a practical implementation, the estimation of time-dependent failure and repair rates may prove to be difficult, whereas the mean time between failures and to repair may be readily available.

The choice of the production rate is not arbitrary. The production rate at each instant is limited by the capacity of the currently operational machines. At time t, the production rate must lie in a set $\Omega[\alpha(t)]$ which depends on the machine state and is thus subject to sudden changes.

To define $\Omega(\alpha)$, consider the machine state $\alpha = (\alpha_1, \alpha_2, \ldots, \alpha_M)$. Let y_{mn}^k be the rate at which station m performs operation k on type n parts (measured in parts per unit time interval). Let τ_{nm}^k be the time required to complete the operation. It then follows that

$$\sum_n \sum_k y_{nm}^k \tau_{nm}^k \leq \alpha_m \quad \text{for all } m. \quad (4)$$

The product $y_{nm}^k \tau_{nm}^k$ is the proportion of each unit time interval used by one or more operational machines at station m to perform operation k on type n parts. The left-hand side of Equation (4) is thus the total amount of work brought to station m per unit time by the part flow rate y_{nm}^k. The inequality follows because the amount of work brought to station m per unit time interval cannot exceed the time available at the operational machines.

Since no material is accumulated within the system, the total number of type n parts going through operation k per unit time interval is equal to the throughput of type n parts. This is expressed as

$$u_n = \sum_m y_{nm}^k \text{ for all } k \text{ and } n. \quad (5)$$

The set $\Omega(\alpha)$ is defined to be the set of all production rates $u = (u_1, u_2, \ldots, u_N)$ such that there exist feasible flow rates y_{nm}^k satisfying Equations (4) and (5). We note that $\Omega(\alpha)$ is the projection of a polyhedral set into a lower dimensional subspace. The feasible set is therefore convex and polyhedral.

This set is thus a representation of the *capacity* of the FMS. It would not be precise to denote capacity by a single number (a production rate for all the parts flowing through the system) or even a vector (a production rate for each part type). A set is required because of the sharing of resources among part types. The rate at which it is possible to manufacture one part is reduced by the production of other parts.

The flow control problem can now be stated. Given an FMS as described above, an initial buffer state $x(t_0)$ and machine state $\alpha(t_0)$, we wish to specify a production plan for $t_0 \leq t \leq t_f$ that minimizes the performance index

$$J(x, \alpha, t_0) = E\left\{\int_{t_0}^{t_f} g[x(t)] dt \mid x(t_0) = x, \alpha(t_0) = \alpha\right\} \quad (6)$$

subject to Equations (1), (2), (3) and $u(t) \in \Omega[\alpha(t)]$. The function $g[x(t)]$ penalizes the controller for failing to meet demand and for keeping an inventory of parts in the downstream buffers. The performance index $J(x, \alpha, t_0)$ is thus the expected total penalty incurred by the controller in the interval (t_0, t_f). The function $g(x)$ is given by $\Sigma_n g_n(x_n)$, where the $g_n(x_n)$ are scalar convex functions satisfying

$$\lim_{|x| \to \infty} g_n(x) = \infty$$

and $g_n(0) = 0$.

The cost function serves to enforce desired behavior on the controller. The ideal production policy would minimize the performance index by producing parts at exactly the demand rate, thereby keeping the buffer state at zero. Such a policy is impossible because of the failures of the machines.

The class of production policies to be considered consists of functions $u(x, \alpha, t)$ that satisfy for each x, α, and t:

$$u(x, \alpha, t) \in \Omega(\alpha). \quad (7)$$

The production policies are therefore feedback control laws which give a feasible production rate for each buffer and machine state in the interval (t_0, t_f).

Characterization of Optimal Production Policies

Define the "cost to go," when the production policy $u(x, \alpha, t)$ is applied, as

$$J_u(x, \alpha, t) = E\left\{\int_t^{t_f} g[x(s)] ds \mid x(t) = x, \alpha(t) = \alpha\right\}. \quad (8)$$

The cost to go is thus the expected total penalty incurred by the controller for the remaining time, given that the buffer and machine states are x and α at time t.

This function satisfies a partial differential equation. It can be derived informally by noting that, for any $\delta t > 0$,

$$J_u[x(t), \alpha(t), t] = E\left\{\int_t^{t+\delta t} g[x(s)] ds + J_u[x(t+\delta t), \alpha(t+\delta t), t+\delta t]\right\}. \quad (9)$$

For small δt, this becomes, approximately,

$$J_u[x(t), \alpha(t), t] = g[x(t)]\delta t + \sum_{\beta \neq \alpha(t)} \lambda_{\alpha\beta} \delta t J_u[x(t+\delta t), \beta, t+\delta t]$$

$$+ (1 - \lambda_{\alpha(t)\alpha(t)}\delta t)\left\{J_u[x(t), \alpha(t), t]\right.$$

$$\left. + \frac{\partial J_u[x(t), \alpha(t), t]}{\partial x} \dot{x} \delta t + \frac{\partial J_u[x(t), \alpha(t), t]}{\partial t} \delta t\right\}, \quad (10)$$

where the derivatives of J_u are evaluated at $x(t)$, $\alpha(t)$, and t. By letting δt go to zero and rearranging terms, this becomes

$$0 = g[x(t)] + \frac{\partial J_u}{\partial x}(u - d) + \frac{\partial J_u}{\partial t} + \sum_\beta \lambda_{\alpha\beta} J_u[x(t), \beta, t]. \quad (11)$$

It is also possible to show that an optimal feedback control law $u^\circ(x, \alpha, t)$ and the optimal cost to go $J_{u^\circ}(x, \alpha, t)$ satisfy

$$0 = \min_{u \in \Omega(\alpha)} \left\{g[x(t)] + \frac{\partial J_{u^\circ}}{\partial x}(u - d) + \frac{\partial J_{u^\circ}}{\partial t}\right.$$

$$\left. + \sum_\beta \lambda_{\alpha\beta} J_{u^\circ}[x(t), \beta, t]\right\}. \quad (12)$$

Note that a control u° is determined, in this equation, by

$$\min_{u \in \Omega(\alpha)} \frac{\partial J_{u^\circ}}{\partial x} u. \quad (13)$$

These results can be established formally by the techniques of Rishel [10] and Tsitsiklis [13]. We note that expression (13) is linear in u and that $\Omega(\alpha)$ is a convex polyhedral set.

An optimal policy $u°(x, \alpha, t)$ therefore takes values at extreme points of $\Omega(\alpha)$ whenever the gradient $\partial J_{u°}(x,\alpha,t)/\partial x$ exists. For each machine state α, an optimal policy divides the buffer state space into a set of regions in which the production rate is constant. Whenever the buffer state is in one of these regions, optimal production rates are constant. However, the regions do not cover the whole space. If the derivative $\partial J_{u°}(x, \alpha, t)/\partial x$ does not exist, is orthogonal to a face of $\Omega(\alpha)$, or is zero, a unique minimizing value to expression (13) does not exist. The optimal production rate in that case depends on the extreme point policies in the neighboring regions.

In the following, we consider the time-invariant case, in which $d(t) = d$, a constant, and the final time t is infinite. In that case, the criterion is the average cost, rather than the total cost, over the period $[t_o, t_f]$. As a result the cost-to-go function $J_u[x(t), \alpha(t), t]$ is time invariant, is written $J_u(x, \alpha)$, and is called the "value" function.

There are two kinds of machine states: those for which the demand rate d is feasible, i.e., those for which $d \in \Omega(\alpha)$ and those for which it is not. The former machine states are called *feasible states*; each has an associated fixed buffer. We call this buffer level x_α^H, the *hedging point*. The hedging point is so designated because if $\alpha(t)$ remains constant for a long enough period and α is a feasible state, then

$$\frac{d}{dt} J_{u°}[x(t), \alpha(t)] = \frac{\partial J_{u°}}{\partial x}(u° - d), \quad (14)$$

since $u°$ minimizes $(\partial J_{u°}/\partial x)u$ for all $u \in \Omega(\alpha)$, and since the demand rate satisfies $d \in \Omega(\alpha)$ then $(d/dt)J_{u°}[x(t), \alpha(t)]$ is negative and $J_{u°}[x(t), \alpha(t)]$ is a decreasing quantity. On the other hand, $J_{u°}$ is the integral of a positive quantity, so it can not decrease without reaching a limit. The limit is reached when $x(t)$ is equal to x_α^H (assuming that the machine state is constant for a sufficiently long time). After that time, $u°$ is set equal to d and the buffer state $x(t)$ stays constant at the hedging value.

The hedging point, which is the minimum of $J_{u°}(x, \alpha, t)$ with respect to x, is the optimal buffer level with which to hedge against future failures. When demand is close to the capacity of the system, the hedging points are at high buffer levels because failures quickly result in deficits and recovery from a deficit is slow. The gradient $\partial J_{u°}(x,\alpha,t)/\partial x$ can be regarded as a weighting on part production for an optimal control law defined by expression (13). The calculation of the optimal value function $J_{u°}$ takes into account the relative costs of backlogs and inventory storage determined by the functions $g_n(x)$. Thus a part that has a high value index and is at the same time sensitive to machine failures would have correspondingly a large weighting. The exact solution to the flow control problem requires the solution of a coupled set of differential equations (12). This is only possible for small problems. Typical flexible manufacturing systems may have ten work stations and half a dozen different part types [9]. To solve a problem of that size, a practical computational method is required.

AN ESTIMATE-BASED (EB) CONTROL SCHEME

The Approach

The optimal policy in the flow control problem is determined from the optimal value function $J_{u°}(x, \alpha, t)$ by the linear program (13). An optimal policy is a feedback law which for every machine state divides the buffer state space into regions within which the control is constant at an extreme point of the control constraint set. The hedging point x_α^H, which is the minimum of $J_{u°}(x, \alpha, t)$ with respect to x, is the optimal buffer level with which to hedge against future failures. Optimal policies cannot be computed in practice because of the large dimension of the flow control problem. We need a practical method for calculating suboptimal control laws which produce good results when used in the flow control level of the hierarchy.

Given convex functions $\psi(x, \alpha, t)$ which are estimates of the optimal value function, consider a control policy $\hat{u}(x, \alpha, t)$ determined by

$$\hat{u}(x, \alpha, t) = \underset{z \in \Omega(\alpha)}{\operatorname{argmin}} \left[\frac{\partial}{\partial x} \psi(x, \alpha, t)\right] z . \quad (15)$$

The suboptimal policy $\hat{u}(x, \alpha, t)$, like an optimal policy, divides the buffer state space into a set of regions in each of which it takes values at an extreme point of $\Omega(\alpha)$.

The estimates $\psi(x, \alpha, t)$ should exhibit the properties of the optimal value function described above. The value of the estimate should be largest for machine states with the smallest production capacity. The relative magnitudes of the components of the gradient $\partial \psi/\partial x$ should reflect both the relative value of parts and their vulnerability to machine failures. The minimum with respect to x of $\psi(x, \alpha, t)$, which determines the hedging point for the suboptimal policy, should be of a magnitude comparable to the optimal hedging buffer levels. If the estimates satisfy these criteria, we expect the suboptimal policy to perform well and to meet demand requirements when they are close to system capacity. If the optimal value of the cost index is not sensitive to the location of the region boundaries, the cost $J_{\hat{u}}(x, \alpha, t)$ corresponding to the estimate-based control policy should be close to the optimal cost.

Calculation of the Estimates

The control constraint set $\Omega(\alpha)$ is polyhedral, lies in the positive orthant, and contains the origin. Define $\bar{H}(\alpha)$ and $\underline{H}(\alpha)$ to be sets such that

$$\bar{H}(\alpha) = \{u \in R^N \mid 0 \leq u_n \leq \bar{q}_{\alpha n}\} \quad n = 1, 2, \ldots, N, \quad (16)$$

$$\underline{H}(\alpha) = \{u \in R^N \mid 0 \leq u_n \leq \underline{q}_{\alpha n}\} \quad n = 1, 2, \ldots, N, \quad (17)$$

and

$$\underline{H}(\alpha) \subseteq \Omega(\alpha) \subseteq \overline{H}(\alpha). \quad (18)$$

$\underline{H}(\alpha)$ and $\overline{H}(\alpha)$ are hypercubes, the former contained in $\Omega(\alpha)$ and the latter containing the control constraint set. For example, Fig. 2 shows the hypercubes for a sample control constraint set.

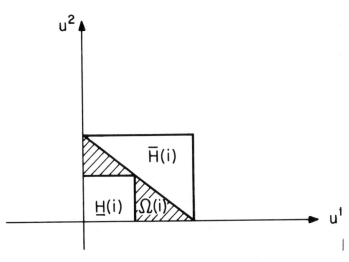

Fig. 2. An example of the hypercubes for a control constraint set.

Define $\underline{\psi}(x, \alpha, t)$ and $\overline{\psi}(x, \alpha, t)$ by the following optimization problems:

$$\overline{\psi}(x, \alpha, t) = \min_{u(s) \in \underline{H}[\alpha(s)]} E\left\{\int_t^{t_f} g[x(s)]\,ds\right\} \quad (19)$$

$$\underline{\psi}(x, \alpha, t) = \min_{u(t) \in \overline{H}[\alpha(s)]} E\left\{\int_t^{t_f} g[x(s)]\,ds\right\}, \quad (20)$$

both subject to Equations (1), (2), and (3) and with initial conditions $x(t) = x$ and $\alpha(t) = \alpha$.

From expression (8), the following holds:

$$\underline{\psi}(x, \alpha, t) \leq J_u^\circ(x, \alpha, t) \leq \overline{\psi}(x, \alpha, t). \quad (21)$$

Thus $\overline{\psi}$ and $\underline{\psi}$ are upper and lower bounds on the optimal value function. An estimate ψ of J_u° can be obtained by taking a convex combination of the lower and upper bounds.

The cost function $g(x)$ is separable. The constraints (16) and (17) affect each part separately. The optimization problems (19) and (20) are therefore decoupled and can be solved as a set of scalar problems, one for each part.

The hypercubes $\underline{H}(\alpha)$ and $\overline{H}(\alpha)$ approximate the control constraint set. If the capacity for the production of part n is small in machine state α, the corresponding limits $\underline{q}_{\alpha n}$ and $\overline{q}_{\alpha n}$ of the hypercubes are small. Likewise, if the capacity is large, the limits are also large. The calculation of the upper and lower bounds thus takes into account the relative productive capacity in all machine states; demand rates, and the value of the parts as given by the cost function $g(x)$. We expect therefore that the cost estimates satisfy the criteria above necessary for good performance by the estimate-based controller.

Implementation of the Estimate-Based Controller

There are two steps in the implementation of the EB controller. Off-line (i.e., in the top box of Fig. 1), upper and lower bounds to the optimal value function are computed by solving problems (19) and (20). In practice this is done by discretizing the problems over discrete points in time and the buffer state. The estimate $\psi(x, \alpha, t)$ is computed from the bounds and stored. On-line (in the flow control level), whenever the system enters machine state i, the control $u(t)$ is determined by the linear program (15) using the stored values of $\psi(x, \alpha, t)$.

Computational and storage costs of the EB controller grow exponentially with the number of work stations M and linearly with the number of parts N. However, the computation is done off-line, and the estimates of the optimal value function can be stored in peripheral devices. The on-line computation consists of the linear program (15) and has N variables and M constraints. Typically, N is between 5 and 10 and M between 10 and 20. The program can thus be easily solved on a small computer.

The off-line computational cost can be reduced by pruning the machine state to exclude states with low probability. With the failure and repair rates typically found in manufacturing systems, a small number of states account for over 95% of the probability. A large number of states can therefore be eliminated without substantially altering the regions and hedging points corresponding to the estimate-based control policy.

EXAMPLE

System and Analytic Results

To demonstrate the application of the hierarchical controller, consider the flexible transfer line of Fig. 3. Each station has two identical machines. Two parts are produced. The first type requires two operations, one at each station; while the second part requires a single operation which can only be performed at the first station.

The operation times and reliability data for the system are given in Tables 1 and 2. In this example, there are nine possible machine states. We will discuss only three of them, all machines operational [$\alpha = (2,2)$], one failed machine at station A [$\alpha = (1,2)$] and one station B machine failed [$\alpha = (2,1)$]. The calculation, however, must include all nine states.

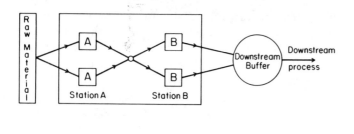

Fig. 3. A flexible transfer line.

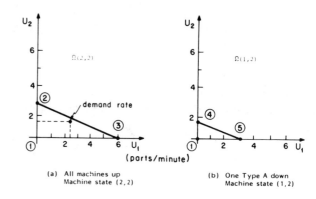

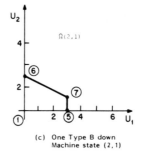

(a) All machines up Machine state (2,2)

(b) One Type A down Machine state (1,2)

(c) One Type B down Machine state (2,1)

Fig. 4. Control constraint sets.

Table 1: Processing time for the parts in minutes.

Part	Stage	
	A	B
1	0.33	0.33
2	0.67	Not required

Table 2: Reliability data in minutes.

Stage	MTBF	MTTR
A	300	30
B	300	30

The cost function is given by

$$g(x) = \sum_{n=1}^{2} |x_n|. \qquad (22)$$

Thus the system is penalized equally for being ahead or behind demand requirements.

The production constraint sets for machine states (2,2), (2,1), and (1,2) are shown in Fig. 4. The different effects of station A and station B failures are evident. The demand rate $d(t)$ for the two part types are constant at 2.5 type 1 and 1.25 type 2 parts per minute. That is, $d_1(t) = 2.5$ and $d_2(t) = 1.25$. Production can exceed demand only in machine states (2,2) and (2,1). In all other machine states, the demand rate is beyond the capacity of the system.

The control policy is characterized by the regions shown in Fig. 5. In each region, the production vector is at an extreme point of a constraint set. It is indicated by the circled numbers in Figs. 4 and 5. In finite horizon problems, the boundaries are time varying but maintain their structure. In infinite horizon problems, there is a solution only if the FMS can satisfy the long-term demand requirements. In this case a steady-state production policy exists and is characterized by constant boundaries between the regions.

The control policy was computed by evaluating an estimate of the optimal cost-to-go function and then solving Equation (15) to obtain the regions.

Also shown in Fig. 5 is the behavior of the buffer state trajectory. Initially, the system has all machines operating and the buffer state $x(0)$ is 0. (The origin of Fig. 5a.) The point $x(0)$ happens to lie on the boundary between two regions. (This is not always the case.) The production vectors in the two neighboring regions both drive the trajectory towards the boundary. The trajectory moves in the positive direction as an inventory of parts is built up as a hedge against future failures. At point (i), production equals demand and the trajectory remains constant. That is $u(x, \alpha, t) = d(t)$ and $dx/dt = 0$.

When a type A machine fails, the new production rate is found at point (ii) of Fig. 5b. Initially only type 1 parts are produced, resulting in an increase in the buffer level of type 1 parts. The buffer level of type 2 parts, as a consequence, drops. At point (iii), the trajectory meets the boundary and a mix of both parts is produced, keeping the trajectory on the boundary. After approximately 25 min, the failed machine is repaired with the buffer levels at point (iv). The production rate is found at point (v) of Fig. 5a. Type 2 parts are produced at the maximum rate to clear the backlog caused by the failure. Production of type 1 parts resumes at point (vi), and the trajectory follows the

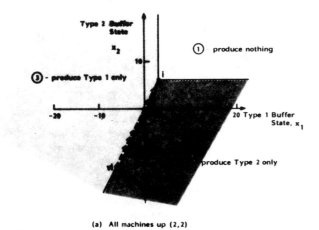

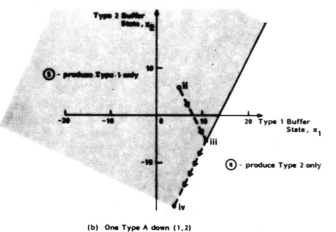

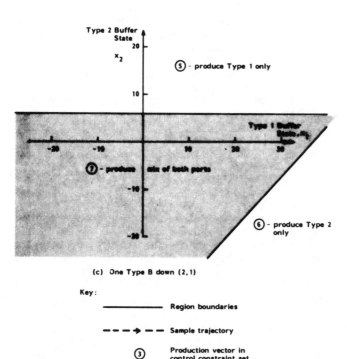

Fig. 5. Control regions corresponding to an optimal policy.

boundary to point (i), where once again production is at the demand rate. A similar set of events can be constructed for any other sequence of failures and repairs.

Simulation Results

The system of Fig. 3 was simulated with the scheduling being performed by the hierarchical controller. Each station had an internal buffer with a capacity for five pieces and a last-in-first-out discipline. The simulation model was run for an equivalent of 14 h. It should be pointed out that the buffer state $x(t)$ refers to the difference between actual and desired production levels and can therefore take on negative values, indicating backlogs. Internal storage buffers aid the sequence control level in generating schedules which maintain the production rates determined by the flow control level.

The availability and utilization of available time at each machine is given in Table 3. Station A is the system bottleneck. The controller is able to attain utilizations of 94% and 85% at the two station A machines. Station B on the other hand is lightly loaded with only 55% and 36% of the available time being used.

Table 3: Utilization and availability for the simulation.			
Stage	Machine	Availability	Utilization
A	1	0.95	0.94
A	2	0.91	0.85
B	1	0.92	0.55
B	2	0.92	0.36

Production statistics are shown in Table 4. On average, the production was 5.2 pieces behind demand for type 1 parts and 4.2 for type 2. The average in-process inventory in the system is small, 3 type 1 pieces and 1.2 type 2 pieces. At the end of the simulation, the system had produced the required number of type 2 parts and was two type 1 parts short of target. Thus the algorithm was able to track demand and at the same time keep the number of pieces inside the system small. It should be pointed out that the cost function (22) penalizes the controller equally for excess production and for backlogged demand. The preferred mode of operation is therefore to keep the buffer trajectory close to zero when all machines are operational and to clear backlogs which result when failures occur, rather than to maintain a large inventory of parts as a hedge against future failures. The behavior can be modified by penalizing backlogged demand more than excess production and by weighting the parts differently in the cost function.

Table 4: Production statistics for the simulation.				
Part	Average in-process inventory	Mean buffer state	Number of parts required	Number of parts produced
1	3.0	-5.2	2083	2081
2	1.2	-4.2	1042	1042

A portion of the buffer state plotted at one-minute intervals is depicted in Fig. 6. The flow control level implemented in the simulation calculates the production vector at one-minute intervals. This, in addition to the fact

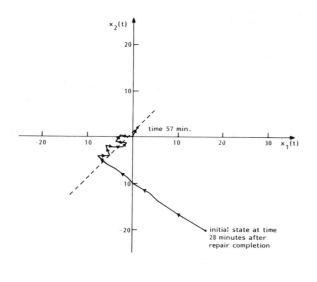

Fig. 6. A part of the buffer state trajectory for the simulation model.

that production increases by integer amounts, accounts for the chatter of the trajectory in the vicinity of the region boundaries. However, the simulated buffer state behaves in a manner very close to that predicted by theory and shown in Figs. 5a and 5b.

CONCLUSIONS

We have described a hierarchical algorithm for the control of production in an automated manufacturing system with unreliable machines. The algorithm is designed to fit into existing factory management structures. The example and the simulation results show that it is possible to accurately track demand requirements while maintaining a low in-process inventory, thereby realizing an important advantage of FMS's over traditional production methods.

The flow control level of the controller is responsible for regulating the input flow rate into the system so that production goals are met. It is important for the performance of the system that the production requirements should be feasible. The managers of the FMS should therefore have planning tools that ensure that the demand is within the capacity of the system.

This approach to short-term production planning has several desirable features. Feedback is intrinsic to the approach, so that rational responses to random events are chosen. The control policy is adapted to the whole FMS and not merely to the first machine that a part encounters. This eliminates the buildup of material inside the system and, thus, congestion. It reduces the combinatorial complexity of the scheduling problem. It explicitly takes repair and failure information into account.

Our current research is aimed at implementing this approach and reducing the off-line computational effort. It is also aimed at improving algorithm performance by better maintaining the buffer state trajectory on region boundaries, when appropriate, and by modifying the sequence control level.

REFERENCES

[1] Buzacott, J. A., " 'Optimal' Operating Rules for Automated Manufacturing Systems," *IEEE Transactions on Automatic Control* **27**, 2 (February 1982).

[2] Buzacott, J. A. and Yao, D., "Flexible Manufacturing Systems; A Review of Models," TIMS/ORSA Detroit Meeting (April 1982).

[3] Dupont-Gatelmand, C., "A Survey of Flexible Manufacturing Systems," *Journal of Manufacturing Systems* **1**, 1, 1-16 (1982).

[4] Halevi, G., *The Role of Computers in Manufacturing Processes*, John Wiley and Sons, NY (1980).

[5] Hahne, E., "Dynamic Routing in an Unreliable Manufacturing Network With Limited Resources," M.I.T. Laboratory for Information and Decision Systems, Report No. LIDS-TH-1063 (1981).

[6] Hildebrant, R., "Scheduling Flexible Machining Centers When Machines Are Prone to Failure," PhD (unpublished) thesis, M.I.T. Department of Aeronautics and Astronautics (May 1980).

[7] Hitomi, K., *Manufacturing Systems Engineering*, Taylor and Francis, London (1979).

[8] Hutchinson, G., "The Control of Flexible Manufacturing Systems: Required Information Structures," IFAC Symposium on Information-Control Problems in Manufacturing Technology, Japan (October 1977).

[9] Hutchinson, G. K. and Hughes, J. J., "A Generalized Model of Flexible Manufacturing Systems," Multi-Station Digitally Controlled Manufacturing Systems Workshop, University of Wisconsin, Milwaukee (January 1977).

[10] Rishel, R., "Dynamic Programming and Minimum Principles for Systems with Jump Markov Disturbances," *SIAM Journal on Control* **13**, 2 (February 1975).

[11] Olsder, G. J. and Suri, R., "Time Optimal Control of Parts-Routing in a Manufacturing System with Failure Prone Machines," Proceedings of the 19th IEEE Conference on Decision and Control, Albuquerque, New Mexico (1980).

[12] Schick, I. and Gershwin, S. B., "Modelling and Analysis of Unreliable Transfer Lines with Finite Interstage Buffers," M.I.T. Laboratory for Information and Decision Systems Report No. ESL-FR-834-6 (September 1978).

[13] Tsitsiklis, J., "Characterization of Optimal Policies in a Dynamic Routing Problem," M.I.T. Laboratory for Information and Decision Systems Report No. LIDS-R-1178 (February 1982).

Joseph Kimemia holds a BS degree in control systems engineering from the University of Sheffield, England, and MS and PhD degrees in electrical engineering and computer science from the Massachusetts Institute of Technology. He is a member of the technical staff at the AT&T Information Systems Laboratory. His current research interests include the design of multiprocessor systems. Dr. Kimemia is a member of the IEEE Computer Society.

Stanley B. Gershwin received the BS degree in engineering mathematics from Columbia University and the MA and PhD degrees in applied mathematics from Harvard University. He is a principal research scientist and assistant director of the MIT Laboratory for Information and Decision Systems and a lecturer in the MIT Department of Electrical Engineering and Computer Science. Dr. Gershwin's interests include traffic assignment control, routing optimization in networks of machine tools, and the effect of limited buffer storage space on transfer lines and assembly networks. He is a member of Tau Beta Pi, the IEEE Control Systems Society, AAAS, ORSA, The Society of Manufacturing Engineers Manufacturing Management Council, and IIE.

Performance of Hierarchical Production Scheduling Policy

RAMAKRISHNA AKELLA, YONG CHOONG, AND STANLEY B. GERSHWIN

Reprinted from *IEEE Transactions on Components, Hybrids, and Manufacturing Technology,* Volume CHMT-7, Number 3, September 1984, pages 225-240. Copyright © 1984 by The Institute of Electrical and Electronics Engineers, Inc. All rights reserved.

Abstract—The performance of Kimemia and Gershwin's hierarchical scheduling scheme for flexible manufacturing systems, as enhanced by Gershwin, Akella, and Choong, is described. This method calculates times at which to dispatch parts into a system in a way that limits the disruptive effects of such disturbances as machine failures. Simulation results based on a detailed model of an IBM printed circuit card assembly facility are presented. Comparisons are made with other policies and the hierarchical policy is shown to be superior.

I. INTRODUCTION

IN THIS REPORT we discuss the performance of the hierarchical production scheduling policy of Kimemia [4] and Kimemia and Gershwin [5] as it has been enhanced by Gershwin, Akella, and Choong [2]. We use a detailed simulation of an automated printed circuit card assembly line at the International Business Machines (IBM) Corporation plant at Tucson, Arizona as an experimental test bed.

We compare this with other policies for loading parts into a flexible manufacturing system. We demonstrate that the hierarchical strategy is effective in meeting production requirements (both total volume and balance among part types) while limiting average work-in-process (WIP). The purpose of this policy is to respond to disruptive events that occur as part of the production process, particularly repairs and failures. Simulation experiments described here show that the hierarchical policy allows the system to run relatively smoothly in spite of such events.

Flexible Manufacturing Systems

A flexible manufacturing system (FMS) typically consists of several production machines and associated storage elements, connected by an automated transportation system. The entire system is automatically operated by a network of computers. The purpose of the flexibility and versatility of the configuration is to meet production targets for a variety of part types in the face of disruptions such as demand variations and machine failures.

The IBM Automated Circuit Card Line is an automated assembly system for producing a variety of printed circuit cards. Workholders containing the cards move through the system from machine to machine along transportation elements which are controlled by a hierarchy of computers and microprocessors. At each of these machines electronic components are inserted into the card. Each type of card goes to a specific set of machines. The processing time of each card at any machine depends on the number and type of components that are inserted. If a machine is busy or otherwise unavailable, the workholders are stored in a buffer near the machine.

In an FMS, individual part movements are practical because of the automated transportation system. The time required to change a machine from doing one operation to doing another (the setup or changeover time) is small in comparison with operation times. These features enable the FMS to rapidly redistribute its capacity between different parts. This flexibility enables the FMS to react to potentially disruptive events such as machine failures and changes in demand.

FMS's are useful when 1) a number of related part types require operations at different machines of the same line; 2) different part types go to the same machines, but require different operations; 3) different part types go to some common machines and then to different machines; and 4) the required part-mix varies with time.

All production systems are subject to disruptive events ranging from sudden changes in demand to machine failures. Their times of occurrence cannot be predicted in advance; at best, a historical record can only provide guidelines on when they can be expected. A scheduling policy must provide for these factors. The purpose of the hierarchical policy described in this paper is to efficiently use the available information and system flexibility to anticipate and to react to disruptive events.

Hierarchical Scheduling Policy

Fig. 1 outlines the hierarchical structure of the scheduling policy. The middle level is the heart of the scheduler. It determines the short-term production rates, taking the capacity constraints of the system into account. Based on these rates the lower level determines the actual times at which parts are loaded into the system. The middle level uses machine status information and deviation from demand for its computations. It also needs certain longer term information. This is supplied by the higher level. It is computed from machine data such as failure and repair rate information, and part data such as operation times and demand.

The concept of capacity is crucial to the design of the hierarchical policy. The capacity at any instant depends on the operational states of the machines. It changes as machines fail or are repaired.

The hierarchical structure of the policy reflects the disci-

Manuscript received September 1984. This work was supported by the Manufacturing Research Center of the Thomas J. Watson Research Laboratory of the International Business Machines Corporation and by the U.S. Army Human Engineering Laboratory under Contract DAAK11-82-K-0018.

The authors are with the Laboratory for Information and Decision Systems, Massachusetts Institute of Technology, 77 Massachusetts Avenue, Cambridge, MA 02139.

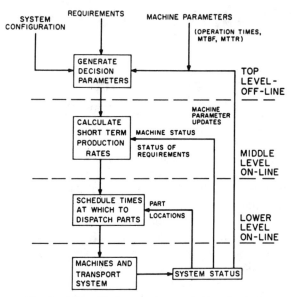

Fig. 1. Hierarchical approach to production planning.

pline that must be imposed in scheduling the FMS. If parts are loaded into the system at a rate that violates the capacity constraints, poor performance results. Material accumulates in buffers or in the transportation system, causing congestion and preventing other material from getting to the machines. Not only does the system perform below expectations, but its effective capacity is reduced.

The hierarchical policy is based on the capacity discipline. Parts are loaded into the system at rates that are within the current capacity, which is determined by the current set of operational machines. This prevents congestion from ever occurring.

In the next section we briefly describe the IBM system. In Section III we describe scheduling objectives and performance measures. The hierarchical policy and some common sense policies are described in Sections IV and V, respectively. In Section VI we compare and discuss the results, and we conclude in Section VII.

II. THE IBM AUTOMATED CARD ASSEMBLY LINE

In this section we describe a system to which the hierarchical scheduler is applicable. Our purpose in using this system is to assess the scheduler in a realistic setting.

Purpose of System

At IBM's General Products Division at Tucson, an automated card assembly line is being built up in stages, through a series of "minilines." The portion of the system of interest to us is the stage consisting of insertion machines. Printed circuit cards from a storage area upstream arrive at the loading area of the insertion stage. Each card is placed in a workholder, which is introduced into the system. It goes to the machines where the electronic components it requires are inserted. It then exits the system and goes to the downstream stages, which consist of testing and soldering machines.

There are several types of insertion machines, each of which inserts one mechanically distinct type of component. The common ones are single in-line package inserters (SIP's), dual in-line package inserters (DIP's), multiform modular inserters (MODI's) and variable center distance inserters (VCD's). By loading different components, the line can be used to assemble a variety of cards.

In order to concentrate on the operational issues of the FMS, we assume that component loading has already been determined. The changeover time is small among the family of parts producible with a given component loading. We also restrict our attention to the Miniline 1300, whose schematic is shown in Fig. 2. This consists of a DIP, a VCD, and two SIP's. Each of the machines also has an associated buffer, which can hold 30 parts.

Transportation System

The workholders are loaded at input station 301 and then move to each of the required machines. Movement is along straight or rotating elements. The straight elements are used to move parts in a single fixed direction and are represented by rectangles. The rotating elements are for 90 deg turns and are represented by circles. Representative movement times are listed in Table I.

Movement of cards in the vicinity of a work station (insertion machine, associated buffer, and transport elements) follows a common pattern. Cards arrive at a rotating element like 603 and either turn towards the insertion machines, or move straight on. The cards going to machines (e.g., 101) either wait at input elements like 605, or go into buffers like 201. After all the required components have been inserted, a similar movement takes the card out of the insertion machines and onto output element 305. After element 606 is rotated toward the work station, the card is placed on it. Element 606 is rotated back to its original position and the card is then loaded onto the next transportation element (306). Finally, after going through the entire system, the cards exit from output element 324.

Machine Parameters and Part Data

The mean time between failures (MTBF) and mean time to repair (MTTR) of the machines are listed in Table II. The average fraction of time a machine is available is the time a machine is available for production divided by the total time. This quantity, called the efficiency or availability of a machine, is also listed in Table II.

There are other random perturbations affecting the system. These include machine tool jams, which occur when a machine jams in trying to insert a component. Rather than regard this as a failure, this small but regular disturbance (approximately once every 100 insertions) can be modeled as part of the processing time.

Normally there are several part (card) types being processed in the system. We limit our experiment to only six types to better examine the hierarchical policy. Typical demand rates are listed in Table III. Also shown in Table III are the operation times required by each card type at each of the machines. These include the processing time and the time to move in and out of each machine.

Loading

Loading describes how heavily the machines in a system must be utilized to satisfy demand. The expected utilization of

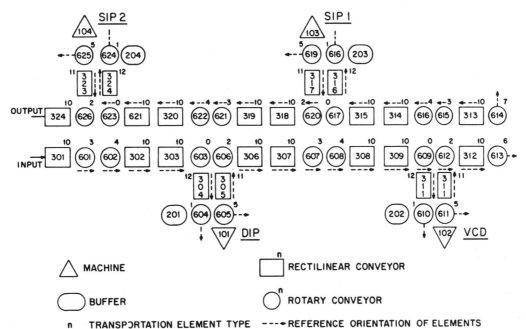

Fig. 2. IBM Miniline 1300.

TABLE I
Transportation Mechanism

Transfer Time of Card from Element to Element (straight or rotation): 1 sec.
Rotation Time: 6 sec.
Number of Movements to Transfer Card via Rotary Mode = 1 Rotation and 1 Transfer

TABLE II
Machine Parameters

MACHINE	MTBF (MINUTES)	MTTR (MINUTES)	EFFICIENCY (%)	EXPECTED(%) UTILIZATION
1	600	60	90.91	97.68
2	600	60	90.91	91.10
3	600	60	90.91	96.03
4	600	60	90.91	96.58

TABLE III
Operation Times and Demand Rates

	OPERATION TIMES (sec)					
PART TYPE MACHINE	1	2	3	4	5	6
1	40	40	0	0	20	60
2	0	0	60	30	40	40
3	0	100	0	0	70	0
4	0	0	0	80	0	80

	DEMAND RATES (parts/sec)					
PART TYPE	1	2	3	4	5	6
DEMAND RATE	.0080	.0070	.0060	.0070	.0025	.0040

a machine is the ratio of the total machine time required to the expected machine time available. The total machine time required is the product of total demand and processing time. The expected time a machine is available is its availability multiplied by the total time period. Table II displays the average utilizations for the machines in the configuration reported in the runs in Section VI. This is not IBM data; it was created to impose a heavy loading on the simulated production system. The actual utilization in any sample simulation run depends on the time history of machine failures and repairs during that run. This time sequence determines the actual amount of time a machine is available.

III. SCHEDULING OBJECTIVES AND POLICY PERFORMANCE MEASURES

Production Requirements and Scheduling Objectives

An FMS is normally only one stage of a production process, with other stages preceding and following. This necessitates coordinated production scheduling. The schedule must determine the part types and the number of each type to be produced by the FMS over a period of several days. The objective of the short term schedule is to track demand over the course of each day so as to meet the production targets set by the long-term schedule.

The production target is specified for each j as $D_j(T)$ parts of type j having to be made by time T, the production period. The cumulative production $W_j(t)$ is the total amount of material of type j produced by time t. The cumulative production must equal the total demand at time T; that is, one of the objectives is to ensure that $W_j(T)$ is equal to $D_j(T)$.

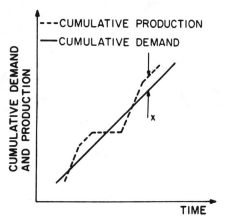

Fig. 3. Production to track demand.

It is convenient to define the demand rate

$$d_j = D_j(T)/T \tag{1}$$

and

$$D_j(t) = t d_j. \tag{2}$$

At time t, the production surplus $x_j(t)$ is the difference between the total number of parts of type j produced and the total number of parts required:

$$x_j(t) = W_j(t) - D_j(t). \tag{3}$$

Fig. 3 illustrates the cumulative demand $D_j(t)$ being tracked by the cumulative production $W_j(t)$. Our objective is to meet production targets as closely as possible at the end of time period T, or, equivalently, to keep $x_j(T)$ close to zero.

The hierarchical policy is designed to keep $x_j(t)$ as close as possible to zero for all t. It does this by allowing the production surplus to grow, when enough machines are operational, to a certain level (defined below as the hedging point). When an essential machine fails, the surplus declines and becomes negative. The level is chosen so that the average value of $x_j(t)$ is near zero.

Policy Performance Measures

The production percentage, defined as

$$P_j = W_j(T)/D_j(T) \times 100 \text{ percent}, \quad \text{for all } j \tag{4}$$

is of primary importance. This is the production of type j parts expressed as a percentage of total demand for type j. The closer this measure is to 100 percent, the better the algorithm is judged to be.

Also of interest is the average work-in-process, i.e., the average number of parts of each type present in the system. The smaller the WIP, the better the algorithm.

Finally, to compare various control policies, it is necessary to aggregate the performance measures by part type, into total performance measures. They are total production percentage

$$P = \Sigma_j W_j / \Sigma_j D_j \times 100 \text{ percent} \tag{5}$$

and total average work-in-process.

To measure the distribution of production between the various part types, we define balance as

$$B = \min_j P_j / \max_j P_j \times 100 \text{ percent}. \tag{6}$$

This is the ratio of the worst production percentage to the best percentage.

Let $T_i(\text{used})$ be the time that machine i processes parts, during the period of time T_i that it is operational. Machine utilization is given by

$$Z_i = T_i (\text{used})/T_i \times 100 \text{ percent}. \tag{7}$$

If this ratio is close to 100 percent, there is an efficient use of system resources, with very little idle time.

IV. THE HIERARCHICAL POLICY

The objective of the hierarchical scheduler is to meet production targets as closely as possible. This is to be achieved in the presence of random disturbances. Here, we treat only machine failures, although other types of uncertainties, such as random demand, will be dealt with in this framework in the future.

For efficient production, congestion in the transportation system and in internal buffers must be minimized. The hierarchical policy ensures this by respecting the system capacity constraints. The loss of production due to machine failures is compensated for by hedging, that is, by building up safety stock. We discuss these important concepts in detail below.

Capacity Constraints

All operations at machines take a finite amount of time. This implies that the rate at which parts are introduced into the system must be limited. Otherwise, parts would be introduced into the system faster than they can be processed. These parts would then be stored in buffers (or worse, in the transportation system) while waiting for the machines to be free, resulting in undesirably large work-in-process. The effect is that throughput (parts actually produced) drops with increasing loading rate, when loading rate is beyond capacity. Thus defining the capacity of the system carefully is a very important first step for on-line scheduling.

Consider a set of I machines processing J part types. Let the time to process the jth part type at machine i be τ_{ij}. Assume that W_j parts of type j must be processed at machine i during a period of T seconds.

The time required by machine i to produce all the parts is

$$\tau_{i1} W_1 + \tau_{i2} W_2 + \cdots + \tau_{iJ} W_J.$$

For the cumulative production to be feasible, this time must be less than or equal to T_i, the time available at machine i during the total time period T. (T_i is less than or equal to T. It is less when failures occur during this period.)

The parts can be processed if

$$\tau_{i1} W_1 + \tau_{i2} W_2 + \cdots + \tau_{iJ} W_J \leq T_i. \tag{8}$$

The average capacity of machine i is this limit on the number of parts that can be produced in a period of time T.

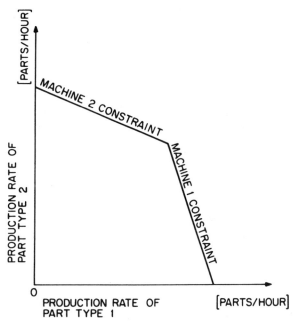

Fig. 4. Production capacity.

Because of the finite processing times, producing parts of one type implies that the time available to produce other types is reduced. The finite resource of machine availability determines, according to (8), the set of production quantities and mixes that can be produced in a given period of time. Fig. 4 describes the feasible production set of parts for a simple case.

Let $e_i = T_i/T$ be the availability of machine i. Let u_j be the average value of the production rate type j. Define the average capacity constraint set

$$\Omega = [u_j, \quad j = 1, \cdots, J$$
$$| \ \Sigma_j \tau_{ij} u_j \leqslant e_i, \quad \text{for all } i, \text{ and } u_j \geqslant 0]. \tag{9}$$

The capacity discussed so far is a long-term capacity. However it is necessary to determine at every instant whether a given part can be loaded. We must therefore find the instantaneous capacity. To do this, we first rewrite (8) as

$$\tau_{i1} u_1 + \tau_{i2} + \cdots + \tau_{iJ} u_J \leqslant T_i/T \tag{10}$$

where

$$u_j = W_j/T \tag{11}$$

is the production rate of type j parts.

For T sufficiently small, the machine operational state does not change. Depending on whether machine i is up or down, T_i is either T or 0. Denote the operational state of the machine by α_i. That is,

$$\alpha_i = \begin{cases} 0, & \text{if machine } i \text{ is down} \\ 1, & \text{if machine } i \text{ is up.} \end{cases} \tag{12}$$

Note that e_i is the average value of α_i over a long period.

The current or instantaneous capacity is then defined by

$$\tau_{i1} u_1 + \tau_{i2} u_2 + \cdots + \tau_{iJ} u_J \leqslant \alpha_i \tag{13}$$

for each i. As machines fail or are repaired, i.e., as the machine state changes, the set of feasible instantaneous production rates change. The key element of the hierarchical policy is to impose the discipline of satisfying the previous inequality at all times.

If there are several identical class i machines, α_i is a positive integer. This quantity changes as machines fail and are repaired. The machine state is defined by

$$\alpha = (\alpha_1, \alpha_2, \cdots, \alpha_I). \tag{14}$$

An instantaneous production rate is feasible only if it is a member of the capacity constraint set

$$\Omega(\alpha) = [u_j, \quad j = 1, \cdots, J$$
$$| \ \Sigma_j \tau_{ij} u_j \leqslant \alpha_i, \quad \text{for all } i, \text{ and } u_j \geqslant 0]. \tag{15}$$

Fig. 4 shows the capacity constraint set for a two part type, two machine system. Figs. 5(a) and 5(b) indicate how production rates drop to zero when one machine fails.

Fig. 5(c) describes a slightly more general situation. Here there are two part types being processed by two machines, two of which are identical ones which have been pooled together. α_1 can take the values 0, 1, 2. When one of the type 1 machines fails, the capacity set reduces to the smaller set indicated by dotted lines.

These examples indicate that when a machine fails, either some part types cannot be produced at all, or can be produced only at a reduced rate. The capacity constraint set describes precisely this notion as a function of the machine state.

To summarize, this notion of instantaneous capacity is crucial in the hierarchical policy. It describes the set of production rates one can choose from, while ensuring that queues do not build up in the system. Any choice of production rates must observe the discipline of staying within the capacity constraint set.

Hedging

Section III concludes that keeping the production surplus x_j small is an effective way of tracking demand. However failures result in a shortfall in production capacity. One compensates by building up safety stocks by overproducing when possible.

Thus rather than maintaining $x_j(t)$ at a value near zero for all t, it is reasonable to maintain it near a level $H_j(\alpha)$ while the machine state is α. We call $H_j(\alpha)$ the hedging point.

Overview of the Hierarchical Policy

The scheduler is divided into three levels, as shown in Fig. 1. The top level generates the decision parameters of the policy. These include the hedging points $H_j(\alpha)$ and other quantities. The repair and failure time data of the machines and the demand rate and processing times for each part type are required for this calculation.

The middle level computes the short-term production rates for each part type for each machine state. The lower level dispatches parts into the manufacturing system with the aim of maintaining the part loading rate equal to the computed production rate.

The top level is intended for off-line computation. It is

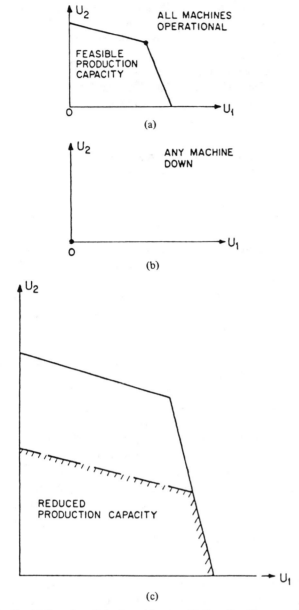

Fig. 5. (a) Capacity with both machines up. (b) Capacity with any machine down. (c) Capacity of two identical machines with one machine down.

designed to be called just once, at the start of a production run. However, if the need arises, it can be called on-line to update the decision tables.

When there is a change in machine state, i.e., when either a machine fails or is repaired, the middle level is called to compute the new values of the production rates. The resulting production surplus or buffer state trajectory is also computed. At the lowest level, parts are loaded into the system so as to follow the buffer state trajectory computed at the middle level as faithfully as possible. A detailed description of each of the levels follows.

Middle Level

This is the most important level in the hierarchy. At this level, the current production rate of each part type is determined for machine state α and buffer level x. The objective is to compute the production rates such that x approaches and then remains equal to $H(\alpha)$. This is possible only if the machine state α is such that the demand rate vector d satisfies

$$d \in \Omega(\alpha),$$

that is, only if the production rate vector u may equal or exceed d. If not, the production surplus must inevitably turn into a backlog (i.e., some components of x eventually become negative).

At the middle level, the scheduler choses production rates so that when enough capacity is present, the production surplus approaches and stays at the hedging point. If too many machines are unavailable for that, the scheduler choses from among the available production rates a set of rates to control the manner in which the production surplus declines and becomes a backlog.

Consider the situation when the machine state α is such that several part types can have production rates exceeding their demand rates. The scheduler tends to allocate manufacturing system resources to those types j whose

$$x_j - H_j(\alpha)$$

is most negative, i.e., whose production surplus is most behind its target value. It sometimes deviates from this behavior; it may allow x_j to decrease even when it is less than the hedging point so as to concentrate resources on some other part type that is farther behind or more vulnerable to future failures.

If machine state α persists for long enough, all part types k whose demands are feasible eventually have their buffer state x_j equal to the hedging point H_j. After that time, the production rate u_j is set equal to the demand rate d_j.

These desirable characteristics are the result of choosing the production rates as the solution to a certain linear programming problem. The cost coefficients are $c_1, \cdots, c_J$. They are functions of x which, along with the hedging points, are determined at the top level. Coefficient c_j tends to be negative when type j is behind or below its hedging point, and its absolute value tends to be larger for more valuable or more vulnerable parts.

The linear program minimizes a weighted sum of the production rates. It is restricted to those production rates that are currently feasible, i.e., that can be achieved by the current set of operational machines.

Linear Program: Minimize

$$c_1 u_1 + c_2 u_2 + \cdots + c_J u_J$$

subject to

$$\sum_j \tau_{ij} u_j \leq \alpha_i, \qquad \text{for all } i \tag{16}$$

$$u_j \geq 0, \qquad \text{for all } j.$$

Production rates generated according to this program automatically satisfy the instantaneous capacity constraints. This linear program is not hard to solve on-line since the number of constraints and unknowns is not large.

If the coefficients c_j are all positive, the production rates satisfying the linear program are zero. Fig. 6 shows this for a

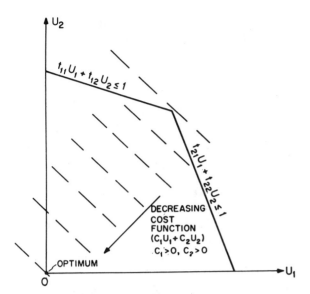

Fig. 6. Optimum production rates for all positive cost coefficients.

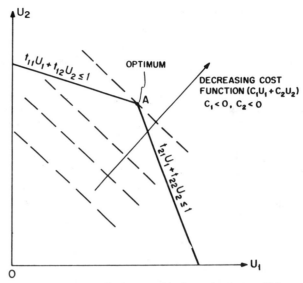

Fig. 8. Optimum production rates for all negative cost coefficients.

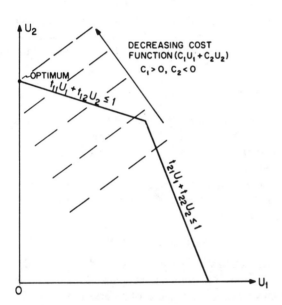

Fig. 7. Optimum production rates for positive and negative cost coefficient.

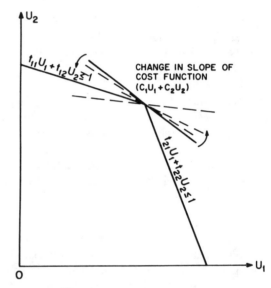

Fig. 9. Variation of cost coefficients.

simple two-machine two-part system. Fig. 7 represents the situation when one of the coefficients is negative and the others are all positive. Then the solution is such that the part type associated with the negative coefficient is produced at the maximum permissible rate. All the other production rates are set to zero.

If all the coefficients are negative, Fig. 8 shows the prevailing situation. An optimal production rate mix, corresponding to point A in the figure, is chosen. More general situations follow from these.

The cost coefficients of the linear program are given by

$$c_j(x_j) = A_j(\alpha)(x_j - H_j(\alpha)) \qquad (17)$$

where $A_j(\alpha)$ and $H_j(\alpha)$ are determined at the higher level. $A_j(\alpha)$ is a positive quantity that reflects the relative value and vulnerability of each part type.

The production surplus $x(t)$ is given by (3). It is approximately

$$x(t) = \int_0^t [u(t) - d(t)] \, dt \qquad (18)$$

since the function of the lower level is to keep the actual production rate close to the value calculated here.

As $x(t)$ changes, the coefficients of linear program change as in Fig. 9. However the production rates of the different part types remain constant, up to a point. When the coefficients change sufficiently, the production rates jump abruptly to new values.

In principle it is necessary to solve the linear program at every time instant because it is constantly changing. This was the approach followed by Kimemia [4] and Kimemia and Gershwin [5]. However this adds a computational burden which would be best to circumvent, and it leads to undersirable behavior when implemented. Gershwin, Akella, and Choong [2] discuss this behavior and a technique for eliminating it. This technique reduces much of the computational burden associated with the linear program.

To describe the behavior of the scheduling system, there are two cases to consider. The first is that the machine state is such that the demand rate is feasible; that is, that $u = d$ is a possible choice for the production rates. In this case, $x(t)$ eventually reaches the hedging point, and the cost coefficients are all zero and the linear program does not determine the value of u. Gershwin, Akella, and Choong [2] demonstrate, however, that when that happens, the solution is $u = d$ and, according to (18), $x(t)$ remains constant, at the hedging point.

When the demand is not feasible, some of the production rates must be less than the corresponding demand rates. The production surplus for these part types fall below the corresponding hedging points. The c_j coefficients then become negative and decrease. Only those part types which are feasible and at or below their hedging points are produced. The rate at which they are produced depends on the coefficients, which describe the relative deviations of the production surplus from desired values.

The system operates on a random cycle: when the machine state α is feasible, the production surplus x approaches H and then stays there. When a machine fails so that the machine state is not feasible, x moves away from H and eventually may become negative.

To complete the picture, the top level is required to determine A and H. These are functions of the relative values of the parts and of the reliabilities of the machines that they visit. The bottom level is required to choose time instants to load parts to guarantee that the production rates and production surplus calculated at the middle level are actually realized.

Lower Level

The lower level has the function of dispatching parts into the system in a way that agrees with flow rates calculated at the middle level. As described in detail in Gershwin, Akella and Choong [2], the middle level of the scheduler calculates the projected trajectory, $x^P(t)$, the best possible future behavior of $x(t)$ if no repairs or failures would occur for a long time.

The lower level treats the projected trajectory $x^P(t)$ as the value that the actual production surplus $x^A(t)$ (given by (3)) should be close to. A part of type j is loaded into the system whenever the actual production surplus $x_j^A(t)$ is less than its projected value $x_j^P(t)$. When there is a machine state change, a new projected trajectory is calculated starting at the time of the change, and the same loading process continues with the new trajectory.

A fuller description of the implementation of the loading process appears in Akella, Bevans, and Choong [1]. A qualitative description of its behavior is in Gershwin, Akella, and Choong [2].

Higher Level

The purpose of the top level of the algorithm is to provide the A_i and H_i parameters to the middle level. These quantities are used in (17) to evaluate the cost coefficients c_i of linear program (16).

Gershwin, Akella, and Choong [2] provided the following formula for the hedging point of part i, where the machine state is feasible:

$$H_i = \frac{T_r d_i (bU_i - ad_i) - T_f ad_i(U_i - d_i)}{(a+b)U_i} \quad (19)$$

where T_r is the average mean time to repair (MTTR) of all the machines part i visits, T_f is the average mean time between failures (MTBF). U_i is the average production rate of part i before x_i reaches the hedging point, and a and b are weighting parameters. The last two quantities reflect the relative penalty incurred for temporary surplus and backlog.

To further simplify the analysis, we assumed that a, b, T_r and T_f and U_i were such that

$$H_i = d_i T_r / 2. \quad (20)$$

The coefficients $A_j(\alpha)$ can be computed from the number of machines that type j parts visit. The more machines each part type visits, the more vulnerable that part type is to failures. Also the smaller the mean time between failures, the more the vulnerability. To simplify our analysis, we assumed that the mean times between failures of all the machines are the same. Thus,

$$A_j(\alpha) = \text{number of machines that type } j \text{ parts visit.} \quad (21)$$

These formulas are highly simplified, but, as the simulations show, they work very well. Further research is required to ascertain under what general conditions they can be expected to provide good results.

The reference values for the H and A parameters for the simulated system, computed according to (20) and (21), are tabulated in Table IV.

V. ALTERNATIVE POLICIES

In this section we discuss a number of simpler policies. All of them limit the number of parts in the system. The differences lie in the amount of information they use about system status and how they use this information.

There are important differences between the hierarchical policy and those described in this section. The most important is that these policies are not explicitly based on satisfying the capacity constraints. As a result, there are periods during which they load more parts than the system can process. Material accumulates in the system during those periods, leading to congestion and diminished effective capacity.

The second is that they require a fair amount of tuning to perform well. "Tuning" is the process of repeating a simulation several times in order to obtain the best values for a set of parameters. Tuning is undesirable because it is expensive. It is impractical because actual production may differ radically from tuning runs, so that good performance cannot be guaranteed.

The third difference is that the policies are not hierarchical. They do not separate the scheduling problem into a set of problems with different characteristic time scales. As a consequence they are difficult to analyze and their performance—and more importantly, the performance of any manu-

TABLE IV
REFERENCE VALUES OF CONTROL PARAMETERS

```
MACHINE STATE:  (1,1,1,1)
            A:  (1,2,1,2,3,3)
   HEDGING PT:  (15,12,15,10,6,8)

MACHINE STATE:  (0,1,1,1)
            A:  (1,2,1,2,3,3)
   HEDGING PT:  (0,0,35,31,0,0)

MACHINE STATE:  (1,0,1,1)
            A:  (1,2,1,2,3,3)
   HEDGING PT:  (34,16,0,0,0,0)

MACHINE STATE:  (1,1,0,1)
            A:  (1,2,1,2,3,3)
   HEDGING PT:  (19,0,19,16,0,13)

MACHINE STATE:  (1,1,1,0)
            A:  (1,2,1,2,3,3)
   HEDGING PT:  (19,16,19,0,12,0)

MACHINE STATE:  (any 2 machines operational)
            A:  (1,2,1,2,3,3)
   HEDGING PT:  (40,35,40,35,0,0)
```

facturing system they control—is difficult to predict other than by simulation.

These policies are based on the amount of material already loaded into the system. Cumulative production for each part type is considered to be the total number of parts loaded. It is equal to the number of parts completed ($PDONE_j$) plus the number of parts currently in the system ($PINSYS_j$). That is,

$$W_j(t) = PDONE_j(t) + PINSYS_j(t). \quad (22)$$

Also,

$$D_j(t) = d_j t \quad (23)$$

$$\begin{aligned} x_j(t) &= W_j(t) - D_j(t) \\ &= PDONE_j(t) + PINSYS_j(t) - d_j t. \end{aligned} \quad (24)$$

Simplest Policy: Policy X

This policy loads a part whose type is furthest behind or least ahead of cumulative demand. That is, it loads a type j part, where x_j is minimal.

Some limit has to be set on the total number of parts in the system in order to avoid filling up the buffers and transportation system. We define N to be the maximum permissible total number of parts in the system.

Also, buffers upstream and downstream of the FMS maybe have limited capacities, or the cost of extra inventory may be high. Thus even if production is ahead of demand, a limit E_j on excess production is useful. That is, we require that

$$x_j \leqslant E_j. \quad (25)$$

Our experience suggests that this is necessary. Production system performance is considerably degraded in the absence of this constraint.

The policy can now be described more precisely.

Policy X: At each time step,
1) do not load any part type if $\Sigma_j PINSYS_j > N$,
2) do not load a type j part if $x_j(t) > E_j$,
3) do not load a type j part if $W_j(t) > D_j(T)$, that is, if the cumulative production at time t exceeds the cumulative demand for the entire period T,
4) of the remaining part types, pick type j that minimizes $x_j(t)$, i.e., load the part type which is least ahead or furthest behind the production target.

Little's Law

Little's law [6] is useful in estimating number of parts in the system. It provides an expression for the sizes of queues of parts (N_j) in terms of the rate at which they arrive (the demand rate d_j) and the average time required for each part to be processed by the system (w_j). The expression is

$$N_j = d_j w_j. \quad (26)$$

That is,

$$\text{part in system} = \frac{(\text{demand rate}) \times (\text{average time}}{\text{required to process each part})}.$$

Note that for N_j to represent the total number of parts of type j throughout the system, w_j must include all sources of delay, including operation time, travel time, and queuing time. Queuing delay, i.e., time spent waiting in buffers or in the transportation system, is neglected for the first guess because it is difficult to calculate and because we intend to keep the number of parts in the system sufficiently small so that such delays are small.

Using this result, a first guess for N can be obtained. The expected number of parts in the system is the sum of the expected number of parts of each type, or

$$N = \Sigma_j N_j. \quad (27)$$

As the threshold limit N is increased, the following system performance is expected and is indeed confirmed by simulation runs.

1) The production rate increases—up to a limit. This limit is less than the system's capacity as calculated in Section IV.
2) The WIP increases.

In addition the balance improves. This was not expected since there is no direct connection between balance and N.

Note that an increase in the work-in-process (WIP) is particularly likely when a machine fails. The parts going to that machine can not be processed. One of these part types will soon fall furthest behind. Consequently more parts of the same type will be loaded. If N is large, the corresponding buffer eventually fill up, and the whole system becomes congested. If N is small, this problem is avoided, but the production performance will be poor, due to under-utilization of machines.

In the rest of this section, we describe other policies, which use more information than policy X to obviate some of its limitations.

More Sophisticated Policy: Policy Y

Two changes are likely to improve the performance of policy X. First, treating each part type separately should result in better balance. This is incorporated in policy Y. Considering machine operational status when loading parts is part of policy Z.

Policy Y is the same as policy X except that there is a separate threshold N_j for each part type. It can be stated as follows.

Policy Y:

1) Do not load a type j part if $PINSYS_j > N_j$.
2) Steps 2–4 are as in policy X.

The initial guesses for the N_j parameters are simple (26). While performance should improve as a result of using a policy that uses more information about the current status of the system, it comes at a price. There are more parameters to tune now, which in principle requires more computer simulation runs. We circumvented that (possibly at the price of not getting the best possible performance) by using a common scaling factor for all N_j.

Production percentage as well as balance should improve relative to policy X. This is a consequence of loading individual part types according to demand. WIP also decreases for the same reason.

Most Sophisticated Policy: Policy Z

While policy Y uses demand information for individual part types, it does not use machine failure information. When a machine fails, the flow rate of parts going to it should be set to zero. Equivalently the limit N_j should be set to zero. This ensures that the WIP does not increase due to the introduction of parts which cannot be processed. The production percentage is likely to increase as delays due to loading the wrong part types are reduced.

Policy Z:

1) Do not load a type j part if $PINSYS_j > q_j N_j$. The parameter q_j is given by

$$q_j = \begin{cases} 0, & \text{if any machine that type } j \text{ parts visit has failed} \\ 1, & \text{otherwise.} \end{cases} \quad (28)$$

2) Steps 2–4 are as in policies X and Y.

The same considerations about tuning both the N_j and the E_j parameters apply here as in policies X and Y. Note that N_j should be greater than (26) since parts should be loaded at a rate greater than d_j when their machines are operational. This means that we are making more parts of each type when we can, hedging against future machine failures.

Policy Z shares these features with the hierarchical policy. However the hierarchical policy guarantees that capacity constraints are always satisfied. Policy Z does not, so WIP can be expected to be greater. Note that the E_j parameters here are similar in their effect to the hedging points H_j in the hierarchical policy.

It is reasonable to expect that this policy behaves better than X or Y, but not quite as well as the hierarchical. Simulations confirm this.

VI. SIMULATION RESULTS

In this section we describe simulation results to evaluate the performance of the hierarchical policy. We also compare the hierarchical policy and policies X, Y, Z. We use the part and machine data of Section II. To understand the policies and not get lost in a welter of detail, a relatively small number of part types are treated.

The system is heavily loaded. That is, machines have to be used for a large percentage of the time they are operational to satisfy demand. This is the only situation in which it is meaningful to compare policies. Under lighter loading conditions, any strategy may be effective. However light loading is not generally realistic; the cost of capital equipment is such that managers will need to get the most they can from an FMS.

In these simulations, the objective is to produce a given quantity of material by the end of one shift. There is no incentive to produce more than the required amount. Consequently the maximum production of any part is 100 percent of requirements, and, because we are loading the system heavily, less than 100 percent is produced in most cases. We expect that over a longer period, such as a week, the hierarchical policy would most often fully meet the requirements imposed here.

Hierarchical Versus Policy X

Our runs correspond to an eight-hour production shift. We first examine the performance of the hierarchical policy during a given run, with different values of the hedging and A parameters. This is compared with the performance of policy X for different values of the threshold limit N on parts in the system. The highlights of the performance are summarized in Figs. 10 and 11. Tables VII–XVI contain detailed production summaries.

Fig. 10 is a plot of total production percentage versus in-process-inventory, for different parameter values of the two strategies. The reference values of the A_j and hedging points H_j are chosen as described in Section IV and tabulated in Table IV. They are varied as shown in Tables V and VI. The parameter N is chosen as described in Section V and tuned. The actual values are tabulated in Tables XII–XVI.

All the points corresponding to the hierarchical controller lie in the upper left region of the graph in Fig. 10. This indicates a high total production percentage, and a low WIP. Both high production percentage and low WIP are highly desirable, as we indicated in Section III. Simultaneously achieving these objectives demonstrates the effectiveness of the hierarchical structure.

The points corresponding to different hedging parameters are clustered close together. This shows robustness to parameter perturbations. The parameters are computed from demand, machine and part type data, which are not always known accurately. Any strategy not unduly sensitive to these is preferred. This is a very important characteristic. Not only does it imply that a great deal of data-gathering and processing

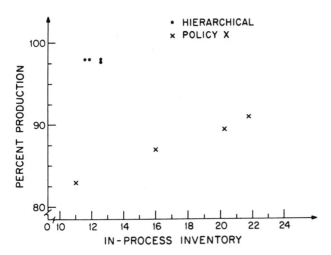

Fig. 10. Production versus in-process inventory.

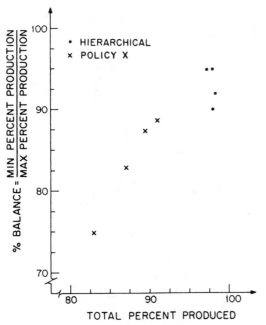

Fig. 11. Balance versus total production percentage.

TABLE V
VARIATION OF A

ORIGINAL A:	(1,2,1,1,3,3)
NEW A:	(1,2,1,5,3,8)

is not required, but it also means that the system's behavior can be expected to be stable even as its reliability drifts over time.

In contrast, the simpler policy's results are more scattered and corresponded to a combination of higher WIP and lower production percentage. The hierarchical policy far out-performs policy X.

Consider the effect of tuning policy X by increasing the threshold limit N of parts in the system. The average WIP in the system is increased in an attempt to increase the production percentage. More parts are loaded into the system and are available at the buffers so that idle time is reduced. Consequently the machines are better utilized and production percentage increases. This approach is relatively crude and

TABLE VI
VARIATION OF HEDGING POINTS

MACHINE STATE:	(1,1,1,1)
ORIGINAL HEDGING PT:	(15,12,15,10,6,8)
NEW HEDGING PT:	(15,12,15,13,6,10)
MACHINE STATE:	(0,1,1,1)
ORIGINAL HEDGING PT:	(0,0,35,31,0,0)
NEW HEDGING PT:	(0,0,35,40,0,0)
MACHINE STATE:	(1,0,1,1)
ORIGINAL HEDGING PT:	(34,16,0,0,0,0)
NEW HEDGING PT:	(34,16,0,0,0,0)
MACHINE STATE:	(1,1,0,1)
ORIGINAL HEDGING PT:	(19,9,19,16,0,13)
NEW HEDGING PT:	(19,0,19,20,0,16)
MACHINE STATE:	(1,1,1,0)
ORIGINAL HEDGING PT:	(19,16,19,0,12,0)
NEW HEDGING PT:	(19,16,19,0,12,0)

TABLE VII
HIERARCHICAL POLICY RUNS WITH VARYING CONTROL PARAMETERS

DESCRIPTION OF RUN	TOTAL PERCENT PRODUCED	BALANCE	WIP
REFERENCE	98.1	90.4	11.81
INCREASED A	98.0	95.0	12.64
INCREASED HEDGING POINTS	98.3	92.2	11.58
INCREASED A AND HEDGING POINTS	97.8	95.0	12.62

SEED = 123457.

TABLE VIII
PRODUCTION SUMMARY—HIERARCHICAL POLICY WITH REFERENCE A AND HEDGING POINTS

TYPE	REQUIREMENTS	PRODUCED	%	AVERAGE TIME SECS	AVERAGE WIP
1	230	230	100.0	193	1.54
2	201	201	100.0	435	3.04
3	172	172	100.0	179	1.07
4	201	193	96.0	410	2.77
5	71	71	100.0	523	2.29
6	115	104	90.4	573	2.08
TOTAL:	990	971	98.1	348.6	11.81

SEED = 123457.

MACHINE	UP TIME PERCENT	UTILIZATION TIME PERCENT WHEN UP
1	92.9	93.1
2	100.0	80.6
3	94.7	91.9
4	85.3	96.8

TABLE IX
PRODUCTION SUMMARY—HIERARCHICAL POLICY WITH INCREASED A

TYPE	REQUIREMENTS	PRODUCED	%	AVERAGE TIME SECS	AVERAGE WIP
1	230	226	98.3	189	1.49
2	201	198	98.5	465	3.22
3	172	172	100.0	189	1.13
4	201	191	95.0	468	3.13
5	71	71	100.0	538	1.33
6	115	112	97.4	600	2.35
TOTAL:	990	970	98.0	373.3	12.64

SEED = 123457.

MACHINE	UP TIME PERCENT	UTILIZATION TIME PERCENT WHEN UP
101	92.9	93.8
102	100	81.5
103	94.7	90.8
104	85.3	98.9

TABLE X
PRODUCTION SUMMARY—HIERARCHICAL POLICY WITH INCREASED HEDGING POINTS

TYPE	REQUIREMENTS	PRODUCED	%	AVERAGE TIME SECS	AVERAGE WIP
1	230	230	100.0	186	1.49
2	201	201	100.0	424	2.96
3	172	172	100.0	179	1.07
4	201	193	96.0	413	2.84
5	71	71	100.0	505	1.25
6	115	108	92.2	533	1.96
TOTAL:	990	973	98.3	341.1	11.58

SEED = 123457.

MACHINE	UP TIME PERCENT	UTILIZATION TIME PERCENT WHEN UP
1	92.9	93.5
2	100.0	81.1
3	94.7	91.9
4	85.3	97.6

TABLE XI
PRODUCTION SUMMARY—HIERARCHICAL POLICY WITH INCREASED A AND HEDGING POINTS

TYPE	REQUIREMENTS	PRODUCED	%	AVERAGE TIME SECS	AVERAGE WIP
1	230	226	98.3	184	1.45
2	201	198	98.5	445	3.08
3	172	172	100.0	183	1.10
4	201	191	95.0	489	3.27
5	71	71	100.0	507	1.25
6	115	111	96.5	633	2.48
TOTAL:	990	969	97.8	372.4	12.62

SEED = 123457.

MACHINE	UP TIME PERCENT	UTILIZATION TIME PERCENT WHEN UP
1	92.9	93.9
2	100.0	81.5
3	94.7	90.8
4	85.3	98.6

TABLE XII
POLICY X RUN WITH VARYING N

N	TOTAL PERCENTAGE PRODUCTION	BALANCE (%)	WIP
11	82.2	74.8	11.00
16	87.1	82.9	15.98
20	89.5	87.3	19.96
22	91.1	88.4	21.95

SEED = 123457.

disregards system capacity constraints. This is the reason that the price of increasing WIP must be paid in order to increase production percentage. In fact, if the threshold N is increased inordinately, the system gets congested.

On the other hand the hierarchical policy always satisfies the capacity constraints and is thus able to achieve low WIP. The instantaneous feedback feature, which combines system status information with hedging for future machine failures, ensures a high production percentage.

The hierarchical policy and policy X are compared with respect to balance and production percentage in Fig. 11. The hierarchical policy is superior. The total production percentage is uniformly high and robust with respect to hedging point variations. This again checks with our expectation that the exact value of the hedging point is not as important as long as it is in the right range. What matters is that the hedging should ensure that the average production surplus is close to zero.

The policy is also robust with respect to changes in A_j, though less so. While the approximation based on vulnerabil-

TABLE XIII
Production Summary—Policy X with $N = 11$

TYPE	REQUIREMENTS	PRODUCED	%	AVERAGE TIME SECS	AVERAGE WIP
1	230	199	86.5	199	1.48
2	201	172	85.6	472	2.86
3	172	146	84.9	173	0.88
4	201	174	86.6	506	3.06
5	71	46	64.8	541	0.86
6	115	87	75.7	603	1.85
TOTAL:	990	824	83.2	388.0	10.99

SEED = 123457.

MACHINE	UP TIME PERCENT	UTILIZATION TIME PERCENT WHEN UP
1	92.9	78.5
2	100.0	67.1
3	94.7	74.8
4	85.3	85.0

TABLE XIV
Production Summary—Policy X with $N = 16$

TYPE	REQUIREMENTS	PRODUCED	%	AVERAGE TIME SECS	AVERAGE WIP
1	230	205	89.1	227	1.79
2	201	177	88.1	720	4.57
3	172	153	89.0	176	0.94
4	201	181	90.0	669	4.21
5	71	53	74.6	932	1.71
6	115	93	80.9	824	2.75
TOTAL:	990	862	87.1	519.7	15.98

SEED = 123457.

MACHINE	UP TIME PERCENT	UTILIZATION TIME PERCENT WHEN UP
1	92.0	82.1
2	100.0	71.2
3	94.7	78.8
4	85.3	89.2

TABLE XV
Production Summary—Policy X with $N = 20$

TYPE	REQUIREMENTS	PRODUCED	%	AVERAGE TIME SECS	AVERAGE WIP
1	230	208	90.4	277	2.26
2	201	181	90.0	828	5.44
3	172	158	91.9	176	0.97
4	201	185	92.0	921	5.93
5	71	57	80.3	1122	2.22
6	115	97	84.3	900	3.13
TOTAL:	990	886	89.5	628.6	19.96

SEED = 123457.

MACHINE	UP TIME PERCENT	UTILIZATION TIME PERCENT WHEN UP
1	92.9	84.2
2	100.0	73.8
3	94.7	81.0
4	85.3	92.2

TABLE XVI
Production Summary—Policy X with $N = 22$

TYPE	REQUIREMENTS	PRODUCED	%	AVERAGE TIME SECS	AVERAGE WIP
1	230	211	91.7	248	2.13
2	181	90.0	90.0	984	6.55
3	172	161	93.6	184	1.03
4	201	189	94.0	1024	6.73
5	71	59	83.1	967	2.06
6	115	101	87.8	956	3.45
TOTAL:	990	902	91.1	673.2	21.95

SEED = 123457.

MACHINE	UP TIME PERCENT	UTILIZATION TIME PERCENT WHEN UP
1	92.9	85.7
2	100.0	75.6
3	94.7	81.5
4	85.3	94.4

ity to machine failures is adequate, even better balance may be possible with a more careful choice of these parameters. In any case, by redistributing available machine capacity effectively between the various part types and hedging, the hierarchical policy achieves good balance.

Policy X has lower balance, lower production percentage, and greater sensitivity to scheduling parameters (N) than the hierarchical policy. The production summaries in Tables XII–XVI show that considerably lower percentages of part types 5 and 6 are produced than are required. This is because these part types must visit more machines than the others. As a result, the likelihood of their waiting for disabled machines to be repaired is higher.

To compensate, more parts can be introduced into the system, at the expense of increased WIP. While the production of part types 5 and 6 improves, that of the other parts does not. Hence balance is better, but neither balance nor overall production percentage are as high as those achieved by the hierarchical policy.

Observe that the hierarchical policy is able to take into account these failures by hedging and building up buffer stocks (see Tables VIII–XI). The benefit of respecting capacity constraints is amply demonstrated by the much lower WIP of the hierarchical policy.

Another insight into the functioning of the hierarchical policy is provided by the machine utilization data. Under heavy loading, all the machines are scheduled to be as highly utilized as possible. Tables VIII–XI indicate that the machines that are down for the greatest periods are the ones that have the highest utilization when up. This implies that the policy is using these machines effectively. Policy X utilizes every machine much less (Tables XIII-XVI).

Comparison with Different Seeds

The same type of comparison is conducted between the hierarchical policy and policy X but for a set of different seeds. Each seed corresponds to a sequence of machine failures and repairs. That is, each seed represents a unique day. The same value of N (16) is used with each seed. The hierarchical policy is run with the same set of seeds. The results, shown in Figs. 12 and 13 and Tables XVII–XXII, are essentially similar to those seen in the previous subsection. The hierarchical policy achieves higher production percentages with lower WIP and better balance.

There is a particularly great difference between the performances of the hierarchical and policy X on certain days. The performance of the simpler policy is more variable, i.e., less predictable, from day to day. Tables XVII and XX indicate that the production percentages of the hierarchical policy stay within the range of 88.7 to 98 percent while those of policy X varies from 69 to 87.1 percent. Moreover the production balance of the hierarchical policy is in the range of 80.4 to 90.4 percent while that of policy X varies from a very low 38 to 83 percent. Table XXI shows the low percentage of type 5 parts produced for one of the runs. This variability is a serious consideration. A policy which is more predictable is more desirable to those who must make long range plans and predictions.

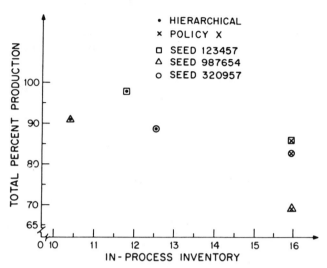

Fig. 12. Total production percentage versus in-process inventory for different seeds.

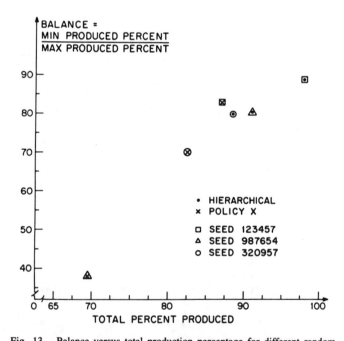

Fig. 13. Balance versus total production percentage for different random seeds.

TABLE XVII
HIERARCHICAL POLICY RESULTS WITH DIFFERENT SEQUENCES OF MACHINE REPAIRS AND FAILURES

SEED	TOTAL PRODUCTION PERCENTAGE	REFERENCE CONTROL PARAMETERS	
		BALANCE	WIP
123457	98.0	90.4	11.81
987654	91.0	80.3	10.38
320957	88.7	80.0	12.56

TABLE XVIII
PRODUCTION SUMMARY—HIERARCHICAL POLICY WITH SEED = 987654

TYPE	REQUIREMENTS	PRODUCED	%	AVERAGE TIME SECS	AVERAGE WIP
1	230	230	100	165	1.33
2	201	182	90.5	480	3.06
3	172	156	90.7	189	1.03
4	201	177	88.1	381	2.38
5	71	57	80.3	485	0.97
6	115	99	86.1	467	1.60
TOTAL:	990	901	91.0	328.6	10.38

MACHINE	UP TIME PERCENT	UTILIZATION TIME PERCENT WHEN UP
1	100.0	82.7
2	79.0	92.7
3	78.3	98.9
4	100.0	77.0

TABLE XXI
PRODUCTION SUMMARY—POLICY X WITH SEED = 987654 ($N = 16$)

TYPE	REQUIREMENTS	PRODUCED	%	AVERAGE TIME SECS	AVERAGE WIP
1	230	179	77.8	185	1.16
2	201	150	74.6	1101	5.75
3	172	123	71.5	411	1.76
4	201	149	74.1	601	3.39
5	71	21	29.6	2498	1.84
6	115	61	53.0	875	2.07
TOTAL:	990	683	69.0	650.4	15.98

MACHINE	UP TIME PERCENT	UTILIZATION TIME PERCENT WHEN UP
1	100.0	61.3
2	79.0	67.7
3	78.3	73.2
4	100.0	58.5

TABLE XIX
PRODUCTION SUMMARY—HIERARCHICAL POLICY WITH SEED = 320957

TYPE	REQUIREMENTS	PRODUCED	%	AVERAGE TIME SECS	AVERAGE WIP
1	230	192	83.5	287	2.02
2	201	169	84.1	511	3.02
3	172	166	96.5	173	0.99
4	201	201	100.0	352	2.46
5	71	58	81.7	699	2.10
6	115	92	80.0	612	1.97
TOTAL:	990	878	88.7	384.7	12.56

MACHINE	UP TIME PERCENT	UTILIZATION TIME PERCENT WHEN UP
1	74.5	99.4
2	100.0	76.6
3	100.0	72.9
4	89.7	90.7

TABLE XXII
PRODUCTION SUMMARY—POLICY X WITH SEED = 320957 ($N = 16$)

TYPE	REQUIREMENTS	PRODUCED	%	AVERAGE TIME SECS	AVERAGE WIP
1	230	200	87.0	429	3.46
2	201	171	85.1	633	4.00
3	172	141	84.3	179	0.91
4	201	172	85.6	615	3.69
5	71	44	62.0	669	1.03
6	115	86	74.8	951	2.87
TOTAL:	990	818	82.6	533.4	15.98

MACHINE	UP TIME PERCENT	UTILIZATION TIME PERCENT WHEN UP
1	74.5	98.4
2	100.0	66.7
3	100.0	70.2
4	89.7	80.2

TABLE XX
POLICY X RESULTS WITH DIFFERENT SEQUENCES OF MACHINE REPAIRS AND FAILURES WITH $N = 16$

SEED	TOTAL PRODUCTION PERCENTAGE	BALANCE	WIP
123457	87.1	83	15.98
987654	69.0	38	15.98
320957	82.6	71	15.98

Even if a policy is tuned carefully for a given run, its performance is not guaranteed to be good in runs with other seeds. This shows the impracticality of parameter tuning. Not only is tuning expensive, since it may require many simulation runs, but the parameter values determined this way may be good only for one set of repairs and failures. In contrast the hierarchical policy is not tuned for a specific failure and repair pattern.

Comparison of Hierarchical Policy and Policies Y and Z

The performance of the hierarchical policy with the reference values of the hedging parameters is also compared with that of policies Y and Z. The parameters of these policies are chosen as described in Section V. We discuss the results only for one run with a single seed.

Figs. 14 and 15 show the comparative performances of all four policies. The hierarchical strategy has the best performance. It is better than policy Z, which is better than Y, which, in turn, is better than X.

This order is a direct result of the more effective use of information. Policy X does not differentiate between part types and does not make use of machine repair state information. It performs poorly in terms of all measures. Policy Y does much better in terms of average WIP and total production percentage by differentiating among part types. Policy Z also makes use of machine state and so has lower WIP and higher balance. The implication is that effective feedback based on more information results in better performance. The series of policies culminates in the hierarchical policy, whose sophisticated information usage helps it achieve superior performance.

VII. CONCLUSION

From the simulation results, we conclude that a hierarchically structured policy designed on the basis described here and elsewhere [4], [5], [2] is very effective in scheduling a FMS. It can achieve high output with low WIP and can cope with changes and disturbances. Future research will be directed toward incorporating other kinds of uncertainties and disturbances in the hierarchical structure.

The success of the policy is a result of using feedback and adhering to the discipline of respecting system capacity constraints. Capacity limits are not just observed in the long run; they are considered as each part is considered for loading into the system. All relevant machine and system status information is fully utilized.

This approach is robust so that for a wide range of policy parameters it works very well. This obviates the need for precise machine and part data which may not always be available. It also eliminates the need to use time consuming (and thus infeasible) trial runs. Further research is needed in choosing hedging and A_j parameters for larger systems. The grouping of parts into families when there are a large number of part types is another research issue.

A variety of new problems arise when we explicitly consider the scheduling of an FMS in the context of a factory. The FMS is then one of the stages of the automated production system. It is supplied with raw material by an upstream stage. It must supply the stage which is downstream from it. Coordinated production between the different stages becomes a necessity. This influences and complicates the short term scheduling problem.

ACKNOWLEDGMENT

We are grateful for the guidance of Mr. Mike Kutcher and Dr. Chacko Abraham of IBM. Ms. Susan Moller and Mr. David White of IBM provided important advice. Mr. J.

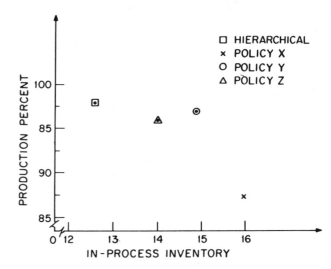

Fig. 14. Total production versus in-process inventory for various policies.

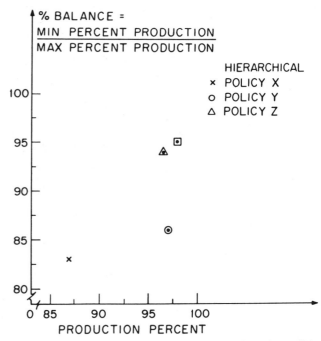

Fig. 15. Balance versus total production percentage for various policies.

Patrick Bevans of the C. S. Draper Laboratory was primarily responsible for writing the simulation, and we were assisted by M.I.T. students George Nikolau and Jean-Jacques Slotine.

REFERENCES

[1] R. Akella, J. P. Bevans, and Y. Choong, "Simulation of a flexible manufacturing system," Massachusetts Institute of Technology Laboratory for Information and Decision Systems Rep., to appear.

[2] S. B. Gershwin, R. Akella, and Y. Choong, "Short term scheduling of an automated manufacturing facility," Massachusetts Institute of Technology Laboratory for Information and Decision Systems Rep. LIDS-FR-1356, 1984.

[3] Special Issue on Data Driven Automation, IEEE Spectrum, May 1983.

[4] J. G. Kimemia, "Hierarchical control of production in flexible manufacturing systems," Massachusetts Institute of Technology Laboratory for Information and Decision Systems Rep. LIDS-TH-1215, 1982.

[5] J. G. Kimemia and S. B. Gershwin, "An algorithm for the computer control of production in flexible manufacturing systems, *IIE Trans.*, vol. 15, no. 4, pp. 353-362, Dec. 1983.

[6] J. D. C. Little, "A proof for the queuing formula: $L = \lambda W$," *Operations Res.*, vol. 9, no. 3, pp. 383-387, 1961.

Optimal Control of Production Rate in a Failure Prone Manufacturing System

RAMAKRISHNA AKELLA AND P. R. KUMAR, MEMBER, IEEE

Abstract—We address the problem of controlling the production rate of a failure prone manufacturing system so as to minimize the discounted inventory cost, where certain cost rates are specified for both positive and negative inventories, and there is a constant demand rate for the commodity produced.

The underlying theoretical problem is the optimal control of a continuous-time system with jump Markov disturbances, with an infinite horizon discounted cost criterion. We use two complementary approaches. First, proceeding informally, and using a combination of stochastic coupling, linear system arguments, stable and unstable eigenspaces, renewal theory, parametric optimization, etc., we arrive at a conjecture for the optimal policy. Then we address the previously ignored mathematical difficulties associated with differential equations with discontinuous right-hand sides, singularity of the optimal control problem, smoothness, and validity of the dynamic programming equation, etc., to give a rigorous proof of optimality of the conjectured policy. It is hoped that both approaches will find uses in other such problems also.

We obtain the complete solution and show that the optimal solution is simply characterized by a certain critical number, which we call the *optimal inventory level*. If the current inventory level exceeds the optimal, one should not produce at all; if less, one should produce at the maximum rate; while if exactly equal, one should produce exactly enough to meet demand. We also *give a simple explicit* formula for the optimal inventory level.

I. INTRODUCTION

WE consider a manufacturing system producing a single commodity. There is a constant demand rate d for the commodity, and the goal of the manufacturing system is to try to meet this demand. The manufacturing system is, however, subject to occasional breakdowns and so there are two states, a "functional" state and a "breakdown" state, in which it can be. The transitions between these two states occur as a continuous-time Markov chain, with q_1 the rate of transition from the functional to the breakdown state, and q_2 the rate of transition from the breakdown to the functional state. (Alternatively, the mean time between failures is q_1^{-1} and the mean repair time is q_2^{-1}.) When the manufacturing system is in the breakdown state it cannot produce the commodity, while if it is in the functional state it can produce at any rate u up to a maximum production rate r. We assume that $r > d > 0$.

Let $x(t)$ be the inventory of the commodity at time t, i.e., $x(t)$ = (total production up to time t) − (total demand up to time t). $x(t)$ may be negative, which corresponds to a backlog. We suppose that positive inventories incur a holding cost of c^+ per

Manuscript received December 10, 1984; revised July 18, 1985. Paper recommended by Past Associate Editor, J. Walrand. The work of the second author was supported in part by the U.S. Army Research Office under Contract DAAG-29-85-K-0094 and DAAG-29-84-K-0005 (administered through the Laboratory for Information and Decision Systems at M.I.T.). The work of the first author was supported in part by the IBM Corporation.
R. Akella is with the Laboratory for Information and Decision Systems, Massachusetts Institute of Technology, Cambridge, MA 02139.
P. R. Kumar is with the Department of Electrical and Computer Engineering and the Coordinated Science Laboratory, University of Illinois, Urbana, IL 61801.
IEEE Log Number 8406211.

unit commodity per unit time, while negative inventories incur a cost of c^-, with $c^+ > 0$, $c^- > 0$. Our goal is to control the production rate $u(t)$ at time t, so as to minimize the expected discounted cost

$$E \int_0^\infty (c^+ x^+(t) + c^- x^-(t)) e^{-\gamma t}\, dt \tag{1}$$

where $x^+ := \max(x, 0)$, $x^- := \max(-x, 0)$ and $\gamma > 0$ is the discount rate.

The problem that we address is this. When the manufacturing system is functioning, what is the optimal production rate u as a function of the inventory x? This problem has been motivated by the pioneering work of Kimemia [1] and Kimemia and Gershwin [2], where a more general problem is formulated (see Section XIII).

We obtain the complete answer to this question. The optimal solution $u(t) = \pi_{z^*}(x(t))$ is given by a critical number z^*. The optimal policy is

$$\begin{aligned}\pi_{z^*}(x(t)) &= r && \text{if } x(t) < z^* \\ &= d && \text{if } x(t) = z^* \\ &= 0 && \text{if } x(t) > z^*. \end{aligned} \tag{2}$$

Thus, whenever the manufacturing system is in the functional state, one should produce at the maximum rate r if the inventory $x(t)$ is less than z^*, one should produce exactly enough to meet demand if the inventory $x(t)$ is exactly equal to z^*, and one should not produce at all if the inventory $x(t)$ exceeds z^*. Hence, the production rate should always be controlled so as to drive the inventory level as rapidly as possible towards z^*, and once there, should maintain it at the level z^*. For this reason we shall call z^* the *optimal inventory level*.

We also obtain the following simple formula for the optimal inventory level:

$$z^* = \max\left\{0, \frac{1}{\lambda_-} \log\left[\frac{c^+}{c^+ + c^-}\cdot\left(1 + \frac{\gamma d}{q_1 d - (\gamma + q_2 + d\lambda_-)(r-d)}\right)\right]\right\} \tag{3}$$

where λ_- is the only negative eigenvalue of the matrix

$$A_1 := \begin{bmatrix} \dfrac{\gamma + q_1}{r-d} & \dfrac{-q_1}{r-d} \\ \dfrac{q_2}{d} & -\left(\dfrac{q_2+\gamma}{d}\right) \end{bmatrix}. \tag{4}$$

The motivation for the problem studied here is that it is a basic problem for manufacturing systems. The optimal policy is trivial to compute and qualitatively simple to implement, and it is hoped that these two features will render it attractive enough for use as a guideline.

From a theoretical viewpoint also, both our solution procedure and method of proof possess several interesting features. First note that the system under consideration is the following.

$$\dot{x}(t) = u(t) - d. \quad (5.\text{i})$$

$\{s(t); \; t \geq 0\}$ is a continuous-time Markov chain with state-space $\{1, 2\}$ and generator $\begin{bmatrix} -q_1 & q_1 \\ q_2 & -q_2 \end{bmatrix}$. (5.ii)

The constraint on $u(t)$ is,

$$u(t) = 0 \quad \text{if } s(t) = 2$$
$$\epsilon[0, r] \quad \text{if } s(t) = 1. \quad (5.\text{iii})$$

The goal is to minimize $E \int_0^\infty (c^+ x^+(t) + c^- x^-(t)) e^{-\gamma t} \, dt.$

$$(5.\text{iv})$$

Here $x(t)$ is the inventory at time t, $u(t)$ is the production rate at time t, and $s(t) = 1$ or 2 depending on whether the manufacturing system is in the functional state or the breakdown state, respectively. Thus, we have a continuous-time system with jump Markov disturbances and an infinite horizon discounted cost criterion. For previous work on problems of this type, we refer the reader to Rishel [3] for the case of a finite horizon cost criterion; and Krassovskii and Lidskii [4] and Lidskii [5] for a case of an infinite horizon problem. In dealing with these types of systems there are two problematical issues.

The first set of problems arises because one encounters many mathematical difficulties (see [3]) when studying the problem of optimal control for continuous-time systems with jump Markov disturbances. Consider a feedback policy $u = \pi(x)$. Then the system (5.i) satisfies

$$\dot{x} = \pi(x) - d. \quad (6)$$

However, the types of functions π we wish to consider are essentially discontinuous functions. Standard existence and uniqueness conditions for the solution of the differential equation (6) are not satisfied. Rishel [3] has considered one notion of a solution; we use another method. There are also other difficulties. Let

$V_i(x) :=$ minimum value of the cost (5.iv) when starting in the state $s(0) = i$, $x(0) = x$.

Then, informally, we have the Hamilton–Jacobi–Bellman dynamic programming equation

$$\begin{pmatrix} \min_{u \in [0, r]} (u - d) \dot{V}_1(x) \\ -d \dot{V}_2(x) \end{pmatrix} = \begin{bmatrix} \gamma + q_1 & -q_1 \\ -q_2 & \gamma + q_2 \end{bmatrix} \begin{pmatrix} V_1(x) \\ V_2(x) \end{pmatrix}$$
$$- \begin{pmatrix} 1 \\ 1 \end{pmatrix} (c^+ x^+ + c^- x^-). \quad (7)$$

It is not *a priori* clear that $V_i(\cdot)$ is a differentiable function. It is not also clear that there exists an optimal control law. Moreover, it turns out that $\dot{V}_1(x)$ does vanish for some x, and for such a value of x, the left-hand side of (7) is minimized by every u, and so the Hamilton–Jacobi–Bellman (HJB) dynamic programming equation does not prescribe the optimal u for such x. Thus, the optimal control problem is a singular one. It is this collection of mathematical problems that we shall address rigorously in the *second* half of this paper.

The second set of problems is this. Even ignoring all the mathematical difficulties mentioned above, how do we actually obtain the optimal solution? Why should we suspect that the optimal policy is of the critical number type? Why is the critical number z^* always nonnegative? Given that we want to solve the HJB equation (7), what are the appropriate boundary conditions? After determining boundary conditions how does one determine an optimal choice for z^*? It is this collection of issues dealing with the actual obtaining of the optimal policy that we address in the *first* half of the paper.

The two approaches complement each other and we hope that they will also be useful in dealing with other problems of the sort considered here.

The paper is organized as follows. In Sections II–VIII we ignore some technical questions and arrive at a conjecture for the optimal policy. Beginning with Section IX we address all the mathematical difficulties and rigorously prove the optimality of the conjectured policy.

II. Optimality of Critical Number Policy

Beginning with this section and continuing through Section VIII, we provide a sequence of informal arguments which will lead us to conjectures about the optimal policy and the optimal cost function.

In this section we give an argument to show that the optimal policy is characterized by a critical number.

Let us assume the existence of an optimal feedback policy $u(t) = \pi^*(x(t))$ and let $V_i(x)$ denote the optimal cost when starting in the state $(s(0) = i, x(0) = x)$. Fix $\{s(t, \omega); t \geq 0\}$, a realization of the continuous-time Markov chain, with $s(0, \omega) = i$.

Now we consider two different initial conditions $x_0(0)$ and $x_1(0)$ and also a convex combination $x_\alpha(0) := (1 - \alpha) x_0(0) + \alpha x_1(0)$ where $0 \leq \alpha \leq 1$. If $\pi^*(\cdot)$ is used, then the trajectories starting with the initial conditions $x_0(0)$ and $x_1(0)$ satisfy

$$\dot{x}_k(t, \omega) = \pi^*(x_k(t, \omega)) - d \quad \text{if } s(t, \omega) = 1$$
$$= -d \quad \text{if } s(t, \omega) = 2 \quad (8)$$
$$x_k(0, \omega) = x_k(0)$$

for $k = 0, 1$. Also, we have

$$V_1(x_k(0)) = E_\omega \int_0^\infty (c^+ x_k^+(t, \omega) + c^- x_k^-(t, \omega)) e^{-\gamma t} \, dt \quad (9)$$

where E_ω signifies that the expectation is taken over ω.

Suppose now that for the initial state $x_\alpha(0)$, we use the control

$$u(t, \omega) = (1 - \alpha) \pi^*(x_0(t, \omega)) + \alpha \pi^*(x_1(t, \omega)). \quad (10)$$

Note that such a control is in fact implementable because by observing $\{s(t, \omega); t \geq 0\}$ one can in fact deduce what $\{x_k(t, \omega), t \geq 0\}$ would have been for $k = 0, 1$. Such a control gives rise to the trajectory satisfying,

$$\dot{x}_\alpha(t, \omega) = (1 - \alpha) \pi^*(x_0(t, \omega)) + \alpha \pi^*(x_1(t, \omega)) - d \quad \text{if } s(t, \omega) = 1$$
$$= -d \quad \text{if } s(t, \omega) = 2$$
$$x_\alpha(0, \omega) = (1 - \alpha) x_0(0) + \alpha x_1(0). \quad (11)$$

It is easy to check from (8) and (11) that

$$x_\alpha(t, \omega) = (1 - \alpha) x_0(t, \omega) + \alpha x_1(t, \omega) \quad \text{for every } t \geq 0.$$

From the convexity of the integrand in (5.iv) it follows that

$$E_\omega \int_0^\infty (c^+ x_\alpha^+(t, \omega) + c^- x_\alpha^-(t, \omega)) e^{-\gamma t} \, dt$$
$$\leq (1 - \alpha) V_1(x_0(0)) + \alpha V_1(x_1(0)). \quad (12)$$

However, for the initial stage $x_\alpha(0)$ the control (10) is not necessarily optimal, and so

$$V_1(x_\alpha(0)) \leq E_\omega \int_0^\infty (c^+ x_\alpha^+(t, \omega) + c^- x_\alpha^-(t, \omega))e^{-\gamma t} \, dt. \quad (13)$$

From (13) and (12) we deduce that $V_1(\cdot)$ is a *convex* function. This argument is the same as that in Tsitsiklis [8]. *Assuming* that $V_1(\cdot)$ is continuously differentiable, we see that there is some z^* for which

$$\dot{V}_1(x) \leq 0 \quad \text{for } x \leq z^*$$
$$\geq 0 \quad \text{for } x \geq z^*. \quad (14)$$

From the left-hand side of the HJB equation (7), which we suspect $\{V_i(0); i = 1, 2\}$ satisfy, we see that

$$u = r \text{ minimizes } (u-d)\dot{V}_1(x) \quad \text{if } x \leq z^*$$
$$= 0 \text{ minimizes } (u-d)\dot{V}_1(x) \quad \text{if } x \geq z^*.$$

Hence, we suspect that the optimal policy $u(t) = \pi^*(x(t))$ is of the form

$$\pi^*(x) = r \quad \text{if } x < z^*$$
$$= 0 \quad \text{if } x > z^* \quad (15)$$

for some critical number z^*.

What happens at $x = z^*$? Any $u \in [0, r]$ minimizes $(u - d)\dot{V}_1(z^*)$, so we will have to determine which $u \in [0, r]$ to choose. From (15) we see that $\dot{x} > 0$ for $x < z^*$ while $\dot{x} < 0$ for $x > z^*$. Hence, when $x = z^*$, there is a "chattering" phenomenon which keeps the state x exactly at z^*. So we suspect that

$$\pi^*(x) = d \quad \text{if } x = z^* \quad (16)$$

because such a choice keeps the inventory level exactly at z^*, once it reaches z^*.

Equations (15) and (16) show that the optimal policy is characterized by the critical number z^*, which we have called the optimal inventory level.

The reader may note that the above argument relies only on the inventory cost being convex in x, and so critical number policies are likely to be optimal for more general problems also. In fact, proceeding as in the sequel one may be able to obtain explicit formulas for the optimal inventory level when the inventory cost is a convex nonnegative polynomial such as x^2 or $|x|^3$. The details will however be tedious.

III. Nonnegativity of Optimal Inventory Level

In this section we show that the optimal inventory level is nonnegative. Consider two policies $\pi^0(\cdot)$ and $\pi^z(\cdot)$, where

$$\pi^0(x) = r \quad \text{if } x < 0$$
$$= d \quad \text{if } x = 0$$
$$= 0 \quad \text{if } x > 0 \quad (17)$$

and

$$\pi^z(x) = r \quad \text{if } x < z$$
$$= d \quad \text{if } x = z$$
$$= 0 \quad \text{if } x > z. \quad (18)$$

Let $z < 0$ be some strictly negative number. Denote by $V_i^0(x)$ and $V_i^z(x)$ the costs resulting from the policies $\pi^0(\cdot)$ and $\pi^z(\cdot)$, respectively, when starting in the state $(s(0) = i, x(0) = x)$.

If $\pi^z(\cdot)$ is optimal, then from (14) we see that $V_1^z(\cdot)$ should attain a minimum at $x = z$, i.e.,

$$V_1^z(z) \leq V_1^z(x) \quad \text{for all } x. \quad (19)$$

Moreover, if $\pi^z(\cdot)$ is optimal, we should also have

$$V_1^z(x) \leq V_1^0(x) \quad \text{for all } x.$$

In particular, from the above two inequalities we should have

$$V_1^z(z) \leq V_1^0(0). \quad (20)$$

Hence, to show that $\pi^z(\cdot)$ with $z < 0$ is *not* optimal, it will suffice to show that (20) is not true.

Indeed, let $\{s(t, \omega); t \geq 0\}$ with $s(0, \omega) = 1$ be a realization, and consider the two trajectories

$$\dot{x}^0(t, \omega) = r - d \quad \text{if } x^0(t, \omega) < 0, \, s(t, \omega) = 1$$
$$= 0 \quad \text{if } x^0(t, \omega) = 0, \, s(t, \omega) = 1$$
$$= -d \quad \text{otherwise}$$
$$x^0(0, \omega) = 0$$

and

$$\dot{x}^z(t, \omega) = r - d \quad \text{if } x^z(t, \omega) < z, \, s(t, \omega) = 1$$
$$= 0 \quad \text{if } x^z(t, \omega) = z, \, s(t, \omega) = 1$$
$$= -d \quad \text{otherwise}$$
$$x^z(0, \omega) = z$$

which emanate from the initial states 0 and z, when the policies $\pi^0(\cdot)$ and $\pi^z(\cdot)$ are respectively used. It is easy to verify that

$$x^0(t, \omega) + z = x^z(t, \omega) \leq z < 0 \quad \text{for all } t \geq 0.$$

Hence,

$$c^+ x^{z+}(t, \omega) + c^- x^{z-}(t, \omega) = c^- x^{z-}(t, \omega)$$
$$= c^-[x^{0-}(t, \omega) + z^-]$$
$$= c^+ x^{0+}(t, \omega) + c^- x^{0-}(t, \omega) + c^-|z|.$$

Hence,

$$E_\omega \int_0^\infty (c^+ x^{0+}(t, \omega) + c^- x^{0-}(t, \omega))e^{-\gamma t} \, dt$$
$$< E_\omega \int_0^\infty (c^+ x^{z+}(t, \omega) + c^- x^{z-}(t, \omega))e^{-\gamma t} \, dt,$$

i.e.,

$$V_1^0(0) < V_1^z(z)$$

showing that (20) is violated, and thus that $\pi^z(\cdot)$ cannot be an optimal policy.

Hence, z^* the optimal inventory level, has to be nonnegative.

IV. The Piecewise Linear Equations for the Cost Function

Let $z \geq 0$ and denote by $V_i^z(x)$ the cost function for the policy $\pi^z(\cdot)$ defined in (18). The analog of (7) for the policy $\pi^z(\cdot)$ is

$$\begin{pmatrix} (\pi^z(x) - d)\dot{V}_1^z(x) \\ -d\dot{V}_2^z(x) \end{pmatrix} = \begin{bmatrix} \gamma + q_1 & -q_1 \\ -q_2 & \gamma + q_2 \end{bmatrix} \begin{pmatrix} V_1^z(x) \\ V_2^z(x) \end{pmatrix} - \begin{pmatrix} 1 \\ 1 \end{pmatrix}(c^+ x^+ + c^- x^-). \quad (21)$$

Denoting

$$V^z(x) := \begin{pmatrix} V_1^z(x) \\ V_2^z(x) \end{pmatrix}$$

$$A_2 := \begin{bmatrix} -\left(\dfrac{\gamma+q_1}{d}\right) & \dfrac{q_1}{d} \\ \dfrac{q_2}{d} & -\left(\dfrac{\gamma+q_2}{d}\right) \end{bmatrix}$$

$$b_1 := \begin{pmatrix} -\dfrac{1}{r-d} \\ \dfrac{1}{d} \end{pmatrix}$$

$$b_2 := \begin{pmatrix} \dfrac{1}{d} \\ \dfrac{1}{d} \end{pmatrix} \tag{22}$$

and letting A_1 be as defined in (4), it is clear that (21) can be rewritten as

$$\frac{\partial}{\partial x} V^z(x) = A_1 V^z(x) - b_1 c^- x \quad \text{for } x \leq 0$$

$$= A_1 V^z(x) + b_1 c^+ x \quad \text{for } 0 \leq x < z$$

$$= A_2 V^z(x) + b_2 c^+ x \quad \text{for } x > z. \tag{23}$$

Before we can utilize these piecewise linear equations to determine the cost function corresponding to $\pi^z(\cdot)$, we need to determine appropriate boundary conditions for (23).

V. BOUNDARY CONDITIONS

Since the vector $V^z(x)$ is two-dimensional, we need two boundary conditions for (23).

A. The First Boundary Condition

Let $\{s(t, \omega); t \geq 0\}$ be a realization. Under the policy $\pi^z(\cdot)$, the inventory is given by the differential equation

$$\dot{x}(t, \omega) = \pi^z(x(t, \omega)) - d \quad \text{if } s(t, \omega) = 1$$

$$= -d \quad \text{if } s(t, \omega) = 2.$$

In any case $|\dot{x}(t, \omega)| \leq r$ for all (t, ω), and so

$$|x(t, \omega)| \leq |x(0)| + rt \quad \text{for all } (t, \omega).$$

Hence,

$$V_i^z(x(0)) = E_\omega \left[\int_0^\infty (c^+ x^+(t, \omega) + c^- x^-(t, \omega)) e^{-\gamma t} \, dt \,\middle|\, s(0, \omega) \right.$$
$$\left. = i, \, x(0, \omega) = x(0) \right]$$

$$\leq k_1 |x(0)| + k_2 \quad \text{for some constants } k_1 \text{ and } k_2.$$

Hence, we see that

$$V^z(x) = 0(|x|) \quad \text{as } x \to \pm \infty. \tag{24}$$

Let us see how we can make (24) more usable. Solving (23) for $x < 0$ in terms of $V^z(0)$, we get

$$V^z(x) = e^{A_1 x} [V^z(0) - A_1^{-2} b_1 c^-]$$

$$+ [A_1^{-1} b_1 c^- x + A_1^{-2} b_1 c^-] \quad \text{for } x < 0. \tag{25}$$

Now note the following easily verified fact:

A_1 has one strictly positive eigenvalue, say λ_+,

and one strictly negative eigenvalue, say λ_-. (26)

Let

$$w^+ = \begin{pmatrix} 1 \\ w_2^+ \end{pmatrix} := \text{eigenvector of } A_1 \text{ corresponding to } \lambda_+. \tag{27}$$

To satisfy (24) as $x \to -\infty$, we clearly need

$$V^z(0) - A_1^{-2} b_1 c^- \in \langle w^+ \rangle \tag{28}$$

for otherwise, $V^z(x) = 0(e^{\lambda_- x})$ as $x \to -\infty$. Here, $\langle w^+ \rangle$ is the eigenspace generated by $\{w^+\}$.

Equation (28) is one boundary condition for (23).

B. The Second Boundary Condition

To obtain the second boundary condition, let us see what $V_1^z(z)$ is. Consider a system starting in state $(s(0) = 1, x(0) = z)$ and let τ, a stopping time, be the first time at which $s(\tau+) = 2$. Clearly, $x(t) = z$ for $0 \leq t < \tau$ when $\pi^z(\cdot)$ is used. Hence,

$$V_1^z(z) = E_\tau \left[\int_0^\tau e^{-\gamma t} c^+ z \, dt + e^{-\gamma \tau} V_2^z(z) \right].$$

Noting that τ is exponentially distributed with mean q_1^{-1}, by evaluating the expectation in the above equation, we get

$$V_1^z(z) = \frac{1}{q_1 + \gamma} (q_1 V_2^z(z) + c^+ z) \tag{29}$$

or, equivalently,

$$[1, 0] A_1 V^z(z) = \frac{c^+ z}{r - d}. \tag{30}$$

Equation (30) is the second boundary condition for (23).

By using the two boundary conditions (28) and (30) one can solve the piecewise linear differential equations (23) to obtain $V^z(\cdot)$, the cost function for any policy $\pi^z(\cdot)$ with $z \geq 0$.

The next question we have to face is; what is the optimal choice of z?

VI. OPTIMAL CHOICE OF z

Suppose $\pi^{z*}(\cdot)$ with $z^* > 0$ is optimal. Then,

i) $V_1^{z*}(z^*) \leq V_1^{z*}(x)$ for all x, since by (19), the optimal cost function attains a minimum at z^*.

ii) $V_1^{z*}(x) \leq V_1^z(x)$ for all x, z, since $\pi^{z*}(\cdot)$ is optimal and therefore has lower cost than any other $\pi^z(\cdot)$.

From the above, we get

$$V_1^{z*}(z^*) \leq V_1^{z*}(z) \leq V_1^z(z) \quad \text{for all } z.$$

Hence, $V_1^z(z)$ attains a minimum when $z = z^*$. Assuming now that $V_1^z(z)$ is a C^1 function of z, we see that

$$\left. \frac{d}{dz} V_1^z(z) \right|_{z = z^*} = 0. \tag{31}$$

We will call (31) the optimality condition and in the next section we will see how it can be exploited to give the optimal solution.

VII. OPTIMAL INVENTORY LEVEL

We will now utilize the piecewise linear differential equations (23), the two boundary conditions (28) and (30), and the optimality condition (31) to obtain the optimal choice for z^*.

Differentiating (29) and using (31), we get

$$\frac{d}{dz} V_2^z(z) \bigg]_{z=z^*} = -\frac{c^+}{q_1}. \quad (32)$$

However, by the chain rule

$$\frac{d}{dz} V_2^z(z) \bigg]_{z=z^*} = \frac{\partial}{\partial x} V_2^{z^*}(x) \bigg]_{x=z^*} + \frac{\partial}{\partial z} V_2^z(z^*) \bigg]_{z=z^*}.$$

Since $V_2^z(x)$, considered as a function of z, is minimized at $z = z^*$, by assuming continuous differentiability, we have

$$\frac{\partial}{\partial z} V_2^z(z^*) \bigg]_{z=z^*} = 0.$$

Hence,

$$\frac{d}{dz} V_2^z(z) \bigg]_{z=z^*} = \frac{\partial}{\partial x} V_2^{z^*}(x) \bigg]_{x=z^*} = [0, 1]\{A_2 V z^*(z^*) + b_2 c^+ z^*\} \quad (33)$$

where we have also used (23). From (32) and (33) we have

$$[0, 1]\{A_2 V z^*(z^*) + b_2 c^+ z^*\} = -\frac{c^+}{q_1}.$$

However, since $[0, 1]A_2 = [0, 1]A_1$, $[0, 1]b_2 = 1/d$, we have

$$[0, 1]A_1 V z^*(z^*) = -c^+ \left(\frac{1}{q_1} + \frac{z^*}{d}\right).$$

Combining the above equation with (30), we have

$$A_1 V z^*(z^*) = \begin{pmatrix} \dfrac{c^+ z^*}{r-d} \\ -\dfrac{c^+}{q_1} - \dfrac{c^+ z^*}{d} \end{pmatrix} = -b_1 c^+ z^* - \begin{pmatrix} 0 \\ 1 \end{pmatrix} \frac{c^+}{q_1}. \quad (34)$$

Setting (34) temporarily aside, we turn to (28). Solving (23) for $V^z(0)$ in terms of $V^z(z)$, we get

$$V^z(0) = e^{-A_1 z}[V^z(z) + A_1^{-1} b_1 c^+ z + A_1^{-2} b_1 c^+] - A_1^{-2} b_1 c^+$$

and substituting this in (28), we have

$$e^{-A_1 z}[V^z(z) + A_1^{-1} b_1 c^+ z + A_1^{-2} b_1 c^+] - A_1^{-2} b_1 (c^+ + c^-) \in \langle w^+ \rangle.$$

Since $\langle w^+ \rangle$ is invariant under $e^{A_1 z}$, we have

$$[V^z(z) + A_1^{-1} b_1 c^+ z + A_1^{-2} b_1 c^+] - e^{A_1 z} A_1^{-2} b_1 (c^+ + c^-) \in \langle w^+ \rangle. \quad (35)$$

Combining (34) and (35), and noting that $\langle w^+ \rangle$ is invariant under A_1, we get

$$-\frac{c^+}{q_1} \begin{pmatrix} 0 \\ 1 \end{pmatrix} + A_1^{-1} b_1 c^+ - e^{A_1 z^*} A_1^{-1} b_1 (c^+ + c^-) \in \langle w^+ \rangle. \quad (36)$$

This equation now gives us a "formula" to choose z^*.

Recall now from Section VI, that in obtaining (36) we made the implicit assumption that $z^* > 0$. Therefore, we now have to determine when there will be a positive solution z^* for (36).

Let us first simplify (36) a bit. Note that

$$\det A_1 = \lambda_+ \lambda_- = \frac{-\gamma(\gamma + q_1 + q_2)}{d(r-d)}$$

$$A_1^{-1} b_1 = \frac{-1}{\gamma} \begin{pmatrix} 1 \\ 1 \end{pmatrix}. \quad (37)$$

Hence, (36) simplifies to

$$e^{A_1 z^*} \begin{pmatrix} 1 \\ 1 \end{pmatrix} - \frac{c^+}{c^+ + c^-} \begin{pmatrix} 1 \\ 1 \end{pmatrix} - \frac{c^+ \gamma}{(c^+ + c^-) q_1} \begin{pmatrix} 0 \\ 1 \end{pmatrix} \in \langle w^+ \rangle. \quad (38)$$

Now let

$$v^- = [v_1^-, v_2^-]$$
$$:= \text{a left eigenvector of } A_1 \text{ corresponding to } \lambda_-. \quad (39)$$

Since left and right eigenvectors corresponding to different eigenvalues are orthogonal, i.e., $v^- w^+ = 0$, (38) is true, if and only if,

$$v^- e^{A_1 z^*} \begin{pmatrix} 1 \\ 1 \end{pmatrix} - \frac{c^+}{c^+ + c^-} v^- \begin{pmatrix} 1 \\ 1 \end{pmatrix} - \frac{c^+ \gamma}{(c^+ + c^-) q_1} v^- \begin{pmatrix} 0 \\ 1 \end{pmatrix} = 0.$$

Since $v^- e^{A_1 z^*} = e^{\lambda_- z^*} v^-$, this reduces to

$$e^{\lambda_- z^*} = \frac{c^+}{c^+ + c^-} \left[1 + \frac{\gamma}{q_1} \frac{v_2^-}{v_2^- + v_1^-}\right].$$

Since $v^- A_1 = \lambda_- v^-$, by equating the second components of both sides,

$$v_1^- = -v_2^- \frac{(r-d)}{q_1} \left[\lambda_- + \frac{(\gamma + q_2)}{d}\right]$$

and now substituting for $v_2^-/(v_2^- + v_1^-)$, we get

$$e^{\lambda_- z^*} = \frac{c^+}{c^+ + c^-} \left[1 + \frac{\gamma d}{q_1 d - (\gamma + q_2 + \lambda_- d)(r-d)}\right]. \quad (40)$$

It is easy to check that

$$q_1 d - (\gamma + q_2 + \lambda_- d)(r-d) > 0 \quad (41)$$

and so the right-hand side of (40) is strictly positive. Taking logarithms, we therefore get

$$z^* = \frac{1}{\lambda_-} \log \left[\frac{c^+}{c^+ + c^-} \left(1 + \frac{\gamma d}{q_1 d - (\gamma + q_2 + \lambda_- d)(r-d)}\right)\right].$$

However, such a value of z^* may not be positive, and we already know from Section III, that if $z^* > 0$ is not optimal, then $z^* = 0$ is. Hence, we arrive at our conjecture

$$z^* = \max \left\{0, \frac{1}{\lambda_-} \log \left[\frac{c^+}{c^+ + c^-} \right.\right.$$
$$\left.\left. \cdot \left(1 + \frac{\gamma d}{q_1 d - (\gamma + q_2 + \lambda_- d)(r-d)}\right)\right]\right\}. \quad (42)$$

Note: Due to (41), it follows that $[1 + \gamma d/\{q_1 d - (\gamma + q_2 + \lambda_- d)(r-d)\}] > 1$. Hence, for *every* $c^- > 0$, there exists $c^* > 0$ such that if $c^+ \geq c^*$, then $z^* = 0$. Thus, the optimal inventory level may be zero even though $c^+ < +\infty$. This is somewhat counterintuitive and surprising.

VIII. Optimal Cost Function

In the previous section we have conjectured the optimal policy. In this section we shall conjecture the optimal cost function also.

Let us consider separately the two cases where z^* as given by (42) is zero, and where it is positive.

A. Case 1: $z^* = 0$

When $z^* = 0$, the optimal cost function is $V^0(\cdot)$ which satisfies (23) with $z = 0$, and has the boundary conditions (28) and (30) with $z = 0$. Defining w^+ as in (27), (28) says that

$$V(0) = kw^+ + A_1^{-2}b_1c^- \quad \text{for some constant } k.$$

Then, from (30), we get

$$0 = [1, \ 0]A_1 V(0) = k\lambda_+ + [1, \ 0]A_1^{-1}b_1c^- = k\lambda_+ - \frac{c^-}{\gamma}$$

and so $k = c^-/\gamma\lambda_+$. Hence,

$$V(0) = \frac{c^-}{\gamma\lambda_+} w^+ + A_1^{-2}b_1c^-. \tag{43}$$

Solving the differential equations (23) with the boundary condition (43), we get

$$V(x) = \frac{c^-}{\gamma\lambda_+} e^{\lambda_+ x} w^+ + A_1^{-1}b_1c^- x + A_1^{-2}b_1c^- \quad \text{for } x \le 0$$

$$= e^{A_2 x}\left[\frac{c^-}{\gamma\lambda_+} w^+ + A_1^{-2}b_1c^- + A_2^{-2}b_2c^+\right]$$

$$- [A_2^{-1}b_2c^+ x + A_2^{-2}b_2c^+] \quad \text{for } x \ge 0. \tag{44}$$

B. Case 2: $z^* > 0$

From (34) we have

$$V(z^*) = -A_1^{-1}b_1c^+ z^* - A_1^{-1}\begin{pmatrix}0\\1\end{pmatrix}\frac{c^+}{q_1} \tag{45}$$

and so, solving the piecewise linear differential equations (23) with boundary condition (45), we get

$$V(x) = e^{A_1 x}\left\{ e^{-A_1 z^*}\left[A_1^{-2}b_1c^+ - A_1^{-1}\begin{pmatrix}0\\1\end{pmatrix}\frac{c^+}{q_1}\right]\right.$$

$$\left. - A_1^{-2}b_1(c^+ + c^-)\right\} + A_1^{-1}b_1c^- x + A_1^{-2}b_1c^-$$

$$\text{for } x \le 0$$

$$= e^{A_1(x-z^*)}\left[A_1^{-2}b_1c^+ - A_1^{-1}\begin{pmatrix}0\\1\end{pmatrix}\frac{c^+}{q_1}\right]$$

$$- A_1^{-1}b_1c^+ x - A_1^{-2}b_1c^+ \quad \text{for } 0 \le x \le z^*$$

$$= e^{A_2(x-z^*)}\left\{A_2^{-1}b_2c^+ z^* - A_1^{-1}b_1c^+ z^* - A_1^{-1}\begin{pmatrix}0\\1\end{pmatrix}\frac{c^+}{q_1}\right.$$

$$\left. + A_2^{-2}b_2c^+\right\} - A_1^{-1}b_2c^+ x - A_2^{-2}b_2c^+ \quad \text{for } x \ge z^*. \tag{46}$$

By simplification, it can also be seen that

$$V(x) = e^{A_2(x-z^*)}\left\{ -A_2^{-1}\begin{pmatrix}0\\1\end{pmatrix}\frac{c^+}{q_1} + A_2^{-2}b_2c^+\right\}$$

$$- A_2^{-1}b_2c^+ x - A_2^{-2}b_2c^+ \quad \text{for } x \ge z^*. \tag{47}$$

Our conjectures for the optimal cost function in the two cases $z^* = 0$ and $z^* > 0$ are given by (44) and (46), respectively.

IX. SOLUTION OF HJB EQUATION

In the previous sections we have arrived at the conjecture that if z^* is as specified by (42), then $\pi^{z^*}(\cdot)$ defined as in (2) is the optimal policy and $V(\cdot)$ defined as in (44) and (46), for the two cases $z^* = 0$ and $z^* > 0$, is the optimal cost function.

Beginning with this section, we commence the rigorous proof of the validity of our conjectures.

In this section we will show that $V(\cdot)$ satisfies the Hamilton-Jacobi-Bellman dynamic programming equation (7).

Lemma 1: $V(\cdot)$ is continuously differentiable.

Proof: Consider first the case $z^* = 0$ where $V(\cdot)$ is specified by (44). We only need to check the continuous differentiability at $x = 0$. Denote by $\dot{V}(a^+)$, $\lim_{h\downarrow 0}(V(a+h) - V(a))/h$ and similarly for $\dot{V}(a^-)$. Now, from (44)

$$\dot{V}(0-) = \frac{c^-}{\gamma} w^+ + A_1^{-1}b_1c^- = \frac{c^-}{\gamma} w^+ - \frac{c^-}{\gamma}\begin{pmatrix}1\\1\end{pmatrix} = \frac{c^-}{\gamma}\begin{pmatrix}0\\w_2^+ - 1\end{pmatrix}$$

$$\dot{V}(0+) = \frac{c^-}{\gamma\lambda_+} A_2 w^+ A_2 A_1^{-2}b_1c^- = \frac{c^-}{\gamma}\begin{pmatrix}0\\w_2^+ - 1\end{pmatrix} \tag{48}$$

where we have used (27), (37), and

$$A_2 = \begin{pmatrix} -\frac{r-d}{d} & 0 \\ 0 & 1 \end{pmatrix} A_1. \tag{49}$$

Hence $V(\cdot)$ is continuously differentiable whenever $z^* = 0$. Now consider the case $z^* > 0$, where $V(\cdot)$ is specified by (46). We only need to check the continuous differentiability at $x = 0$ and $x = z^*$. Now, clearly, from (46),

$$\dot{V}(0-) = \dot{V}(0+) = e^{-A_1 z^*}\left[A_1^{-1}b_1c^+ - \frac{c^+}{q_1}\begin{pmatrix}0\\1\end{pmatrix}\right] - A_1^{-1}b_1c^+ \tag{50}$$

and so we proceed to consider $x = z^*$, for which,

$$\dot{V}(z^*-) = \dot{V}(z^*+) = -\begin{pmatrix}0\\1\end{pmatrix}\frac{c^+}{q_1} \tag{51}$$

as can be seen from (46) and (47). □

Lemma 2: $V(0) - A_1^{-2}b_1c^- \in \langle w^+\rangle$.

Proof: From (44), this is clearly true for the case $z^* = 0$. Considering $z^* > 0$, we see from (46) that

$$V(0) = e^{-A_1 z^*}\left[A_1^{-2}b_1c^+ - A_1^{-1}\begin{pmatrix}0\\1\end{pmatrix}\frac{c^+}{q_1}\right] - A_1^{-2}b_1c^+$$

and so noting that $\langle w^+\rangle$ is invariant under $e^{A_1 z^*}A_1$, we get

$$e^{A_1 z^*}A_1^{-1}b_1 + \frac{c^+}{(c^+ + c^-)q_1}\begin{pmatrix}0\\1\end{pmatrix} - \frac{c^+}{(c^+ + c^-)}A_1^{-1}b_1 \in \langle w^+\rangle.$$

Noting that left and right eigenvectors corresponding to different eigenvalues are orthogonal, i.e., $v^- w^+ = 0$, where v^- and w^+ are given by (39) and (27), we only need to verify that

$$v^-\left[e^{A_1 z^*}A_1^{-1}b_1 + \frac{c^+}{(c^+ + c^-)q_1}\begin{pmatrix}0\\1\end{pmatrix} - \frac{c^+}{(c^+ + c^-)}A_1^{-1}b_1\right] = 0.$$

Noting that $v^- e^{A_1 z^*} = e^{\lambda_- z^*} v^-$, by using (37) and simplifying, we see that we only have to show that

$$e^{\lambda_- z^*} v^-\begin{pmatrix}1\\1\end{pmatrix} - \frac{c^+}{(c^+ - c^-)q_1} v^-\begin{pmatrix}0\\1\end{pmatrix} - \frac{c^+}{(c^+ + c^-)} v^-\begin{pmatrix}1\\1\end{pmatrix} = 0.$$

But noting the equivalence of this and (40) and (42) when $z^* > 0$, the assertion follows. □

Now we are ready to consider the case $z^* = 0$ and show that $V(\cdot)$ satisfies the HJB dynamic programming equation (7).

Lemma 3: Suppose z^* given by (42) is equal to 0, and $V(\cdot)$ is defined by (44). Then

$$\begin{pmatrix} (\pi^{z^*}(x)-d)\dot{V}_1(x) \\ -d\dot{V}_2(x) \end{pmatrix} = \begin{bmatrix} \gamma+q_1 & -q_1 \\ -q_2 & \gamma+q_2 \end{bmatrix} \begin{pmatrix} V_1(x) \\ V_2(x) \end{pmatrix}$$

$$- \begin{pmatrix} 1 \\ 1 \end{pmatrix} (c^+ x^+ + c^- x^-) \quad \text{for all } x \quad (52.\text{i})$$

$$(\pi^{z^*}(x)-d)\dot{V}_1(x) = \min_{u \in [0, r]} (u-d)\dot{V}_1(x) \quad \text{for all } x. \quad (52.\text{ii})$$

Proof: It is easily checked that $V(\cdot)$ defined by (44) satisfies (23) for $x < 0$ and $x > 0$, and so (52.i) is valid for $x < 0$ and $x > 0$. Now considering $x = 0$ and using (48), we have

$$\begin{pmatrix} (\pi^{z^*}(0)-d)\dot{V}_1(0) \\ -d\dot{V}_2(0) \end{pmatrix} = -\frac{c^- d}{\gamma} \begin{pmatrix} 0 \\ w_2^+ - 1 \end{pmatrix}$$

while, by using (43), we have

$$\begin{pmatrix} \gamma+q_1 & -q_1 \\ -q_2 & \gamma+q_2 \end{pmatrix} V(0) = -dA_2 V(0)$$

$$= \frac{-dc^-}{\gamma\lambda_+} A_2 w^+ - dA_2 A_1^{-2} b_1 c^-$$

$$= \frac{-c^- d}{\gamma} \begin{pmatrix} 0 \\ w_2^+ - 1 \end{pmatrix}$$

where we have also used (49) and (37). Hence, (52.i) is also valid at $x = 0$. Now turning to (52.ii), since $\pi z^*(\cdot)$ satisfies (2), we only need to show that

$$\dot{V}_1(x) \leq 0 \quad \text{for } x < 0 \quad \text{and} \quad \dot{V}_1(x) \geq 0 \quad \text{for } x > 0. \quad (53)$$

Consider $x < 0$ first. Then, from (44) it follows that

$$\dot{V}_1(x) = \frac{c^- \lambda_+}{\gamma} e^{\lambda_+ x} \geq 0 \quad \text{for } x < 0$$

where, by $\dot{V}_1(0)$, we mean $\dot{V}_1(0-)$. Since (48) shows that $\dot{V}_1(0) = 0$, it follows that (53) holds for $x < 0$. Now turning to $x > 0$, from (43) and (44) we see that

$$V(x) = e^{A_2 x}[A_2 V(0) + A_2^{-1} b_2 c^+] - A_2^{-1} b_2 c^+ \quad \text{for } x \geq 0.$$

From (43), (49) and (37), we have

$$A_2 V(0) = \frac{c^-}{\lambda_+ \gamma} A_2 w^+ + c^- A_2 A_1^{-2} b_1 = \frac{c^-}{\gamma} \begin{pmatrix} 0 \\ w_2^+ - 1 \end{pmatrix}.$$

Recalling (22) we thus have

$$\dot{V}(x) = e^{A_2 x} \left\{ \frac{c^-}{\gamma(w_2^+ - 1)} \begin{pmatrix} 0 \\ 1 \end{pmatrix} + \frac{c^+}{d} A_2^{-1} \begin{pmatrix} 1 \\ 1 \end{pmatrix} \right\}$$

$$- \frac{c^+}{d} A_2^{-1} \begin{pmatrix} 1 \\ 1 \end{pmatrix} \quad \text{for } x \geq 0. \quad (54)$$

Now

$\begin{pmatrix} 1 \\ 1 \end{pmatrix}$ and $\begin{pmatrix} -q_1 \\ q_2 \end{pmatrix}$ are right eigenvectors of A_2

corresponding to the eigenvalues $\left(\frac{-\gamma}{d}\right)$

and $-\left(\frac{\gamma+q_1+q_2}{d}\right)$, respectively, (55)

and

$$\begin{pmatrix} 0 \\ 1 \end{pmatrix} = \frac{q_1}{q_1+q_2} \begin{pmatrix} 1 \\ 1 \end{pmatrix} + \frac{1}{q_1+q_2} \begin{pmatrix} -q_2 \\ q_2 \end{pmatrix}. \quad (56)$$

Moreover, equating the first components of both sides of the equation $A_1 w^+ = \lambda_+ w^+$ and noting (4), we have

$$w_2^+ = \frac{1}{q_1} [\gamma + q_1 - \lambda_+(r-d)]. \quad (57)$$

Using (55), (56), and (57) in (54), we get

$$\dot{V}_1(x) = \frac{c^+}{\gamma}(1 - e^{-(\gamma/d)x}) + \frac{c^-(\gamma-\lambda_+(r-d))}{\gamma(q_1+q_2)}$$

$$\cdot [e^{-(\gamma/d)x} - e^{-((\gamma+q_1+q_2)/d)x}] \quad \text{for } x \geq 0.$$

If $\gamma - \lambda_+(r - d) \geq 0$, then clearly $\dot{V}_1(x) > 0$ for $x > 0$ and (53) is valid. So suppose that $\gamma - \lambda_+(r - d) < 0$ and note, by differentiating, that

$$\ddot{V}_1(x) = \left[\frac{c^+}{d} - \frac{c^-(\gamma-\lambda_+(r-d))}{\gamma(q_1+q_2)} \cdot \frac{\gamma}{d}\right] e^{-(\gamma/d)x}$$

$$+ \frac{c^-(\gamma-\lambda_+(r-d))}{\gamma(q_1+q_2)} \frac{(\gamma+q_1+q_2)}{d} \quad \text{for } x \geq 0.$$

If $p \geq 0$ and $n \leq 0$ are constants, then $pe^{-(\gamma/d)x} + ne^{-[(\gamma+q_1+q_2)/d]x} \geq 0$ for all $x \geq 0$ if and only if $p + n \geq 0$. Hence, to show that

$$\dot{V}_1(x) \geq 0 \quad \text{for all } x \geq 0$$

we only need to verify that

$$\frac{c^+}{d} + \frac{c^-(\gamma-\lambda_+(r-d))}{\gamma(q_1+q_2)} \left[\frac{\gamma+q_1+q_2}{d} - \frac{\gamma}{d}\right] \geq 0 \quad \text{for all } x \geq 0$$

or equivalently, that

$$c^+ \gamma + c^-(\gamma-\lambda_+(r-d)) \geq 0.$$

It is easy to check that $(\lambda_+ + \lambda_-)(r - d) = (\gamma + q_1) - (\gamma + q_2)[(r - d)/d]$, and so, substituting for λ_+, we only need to verify that

$$c^- \left[\frac{\gamma r}{d} + q_2 \frac{(r-d)}{d} - (\gamma+q_1-\lambda_-(r-d))\right] + c^+ \gamma \geq 0.$$

But this is in turn equivalent to

$$\frac{c^+}{c^+ + c^-} \left[1 + \frac{\gamma d}{q_1 d - (\gamma+q_2+\lambda_-d)(r-d)}\right] \geq 1$$

which is in fact true, since z^* given by (42) satisfies $z^* = 0$. Hence, (53) is valid, proving (52.ii). □

Turning now to the case $z^* > 0$ we show a similar result.

Lemma 4: Suppose z^* given by (42) is strictly positive, and $V(\cdot)$ is defined by (46), then (52.i) and (52.ii) are valid.

Proof: It is easily checked that $V(\cdot)$ satisfies (23) for $x < 0$, $0 < x < z^*$ and $x > z^*$, and so in all three of these cases (52.i) is satisfied. At $x = 0$, (50) and (46) again show that $\dot{V}(0) = A_1 V(0)$ and so (52.i) is also valid at $x = 0$. Turning now to $x = z^*$, we note from (51) that

$$\begin{pmatrix} (\pi^{z^*}(z^*)-d)\dot{V}_1(z^*) \\ -d\dot{V}_2(z^*) \end{pmatrix} = \frac{c^+ d}{q_1} \begin{pmatrix} 0 \\ 1 \end{pmatrix} = -d\dot{V}(z^*).$$

Also,

$$\begin{bmatrix} \gamma+q_1 & -q_1 \\ -q_2 & \gamma+q_2 \end{bmatrix} V(z^*) - \begin{pmatrix} 1 \\ 1 \end{pmatrix} c^+ z^* = -d\{A_2 V(z^*) + b_2 c^+ z^*\}$$

$$= -d \lim_{x \to z^*} \{A_2 V(x) + b_2 c^+ x\} = -d \lim_{x \to z^*} \dot{V}(x) = -d\dot{V}(z^*)$$

where we have used the continuous differentiability of $V(\cdot)$. Hence, (52.i) is valid for all x. Now turning to (52.ii), since $\pi^{z*}(\cdot)$ is of the form shown in (2), we need to show that

$$\dot{V}_1(x) \leq 0 \quad \text{for } x < z^* \text{ and } \dot{V}_1(x) \geq 0 \quad \text{for } x > z^*. \quad (58)$$

Consider $0 \leq x < z^*$ first. From (46) we obtain

$$\ddot{V}(x) = e^{A_1(x-z^*)}\left[b_1 c^+ - A_1 \begin{pmatrix} 0 \\ 1 \end{pmatrix} \frac{c^+}{q_1}\right] \quad \text{for } 0 \leq x \leq z^*$$

where, by $\ddot{V}(0)$ and $\ddot{V}(z^*)$ we mean $\ddot{V}(0+)$ and $\ddot{V}(z^*-)$, respectively. Now

$$b_1 c^+ - A_1 \begin{pmatrix} 0 \\ 1 \end{pmatrix} \frac{c^+}{q_1} = \begin{pmatrix} 0 \\ \theta \end{pmatrix}$$

where $\theta := c^+/d + (c^+/d)(\gamma + q_2)/q_1 > 0$. Let $t := -(x - z^*) \geq 0$ and denoting by $\mathcal{L}^{-1}$ the inverse Laplace transform, we have

$$\ddot{V}_1(x) = [1, 0]e^{-A_1 t}\begin{pmatrix} 0 \\ \theta \end{pmatrix}$$

$$= \mathcal{L}^{-1}\left([1, 0](sI + A_1)^{-1}\begin{pmatrix} 0 \\ \theta \end{pmatrix}\right)$$

$$= \mathcal{L}^{-1}\left(\frac{\theta q_1}{(s - \lambda_+)(s - \lambda_-)(r - d)}\right) = \frac{q_1 \theta(e^{\lambda_+ t} - e^{\lambda_- t})}{(r - d)(\lambda_+ - \lambda_-)} \geq 0$$

with strict inequality for $x \neq z^*$. Moreover, from (51), $\dot{V}_1(z^*) = 0$, and so the validity of (58) for $0 < x < z^*$ is established. By continuity of $\dot{V}_1(x)$, we see now that $\dot{V}_1(0) < 0$, thereby establishing (58) for $x = 0$ also. Now we consider $x < 0$. Since (23) is satisfied for $x < 0$, we see that

$$V(x) = e^{A_1 x}[V(0) - A_1^{-2} b_1 c^-] + A_1^{-1} b_1 c^- x + A_1^{-2} b_1 c^- \quad \text{for } x \leq 0. \quad (59)$$

Hence,

$$\dot{V}(x) = A_1 e^{A_1 x}[V(0) - A_1^{-2} b_1 c^-] + A_1^{-1} b_1 c^- \quad \text{for } x \leq 0.$$

From Lemma 2, we know that $V(0) - A_1^{-2} b_1 c^- = kw^+$ for some constant k, and so

$$\dot{V}(x) = k\lambda_+ e^{\lambda_+ x} w^+ + A_1^{-1} b_1 c^- \quad \text{for } x \leq 0.$$

Noting (27) and (37), we have

$$\dot{V}_1(x) = k\lambda_+ e^{\lambda_+ x} - \frac{c^-}{\gamma} = e^{\lambda_+ x}\dot{V}_1(0) - \frac{c^-}{\gamma}(1 - e^{\lambda_+ x}) \quad \text{for } x \leq 0.$$

Since $\lambda_+ x \leq 0$ for $x \leq 0$ and since $\dot{V}_1(0) < 0$ as previously shown, it follows that $\dot{V}_1(x) < 0$ for $x \leq 0$ and so (58) is valid for $x \leq 0$ in addition to $0 \leq x < z^*$. Now we consider $x > z^*$. From (47) we have

$$\ddot{V}(x) = e^{A_2(x-z^*)}\left\{-A_2 \begin{pmatrix} 0 \\ 1 \end{pmatrix}\frac{c^+}{q_1} + b_2 c^+\right\} \quad \text{for } x \geq z^*$$

$$= \mu e^{A_2(x-z^*)}\begin{pmatrix} 0 \\ 1 \end{pmatrix} \quad \text{for } x \geq z^*$$

where $\mu := (c^+/d)(1 + (\gamma + q_2)/q_1) > 0$, and by $\ddot{V}(z^*)$ we mean $\ddot{V}(z^*+)$. Now using (55) and (56) we get

$$\dot{V}_1(x) = \frac{\theta q_1}{q_1 + q_2}\{e^{(-\gamma/d)(x-z^*)} - e^{(-(\gamma + q_1 + q_2)/d)(x-z^*)}\} \geq 0 \quad \text{for } x \geq z^*$$

with strict inequality except at $x = z^*$. Since $\dot{V}_1(z^*) = 0$, the validity of (58) is also established for $x > z^*$. □

Theorem 5: Let

z^* be defined as in (42) (60.i)

$\pi^{z*}(\cdot)$ be defined as in (2) (60.ii)

$V(\cdot)$ be defined as in (44) or (46)
depending on whether $z^* = 0$ or $z^* > 0$. (60.iii)

Then

$V(\cdot)$ is continuously differentiable (60.iv)

$\pi^{z*}(\cdot)$ and $V(\cdot)$ satisfy (52.i) and (52.ii) and hence the

Hamilton–Jacobi–Bellman dynamic programming equation
(60.v)

$V(x) = 0(|x|)$ as $x \to \pm\infty$. (60.vi)

Proof: We have already established (60.iv) and (60.v) in Lemmas 1, 3, and 4. To show (60.vi) note that (44) and (46), (26), (27) together with Lemma 2 show that $V(x) = 0(|x|)$ as $x \to -\infty$. Moreover, since both eigenvalues of A_2 are strictly negative, it follows from (44) and (46) that $V(x) = 0(x)$ as $x \to +\infty$ also. □

X. Admissible Policies

It is now time for us to address more general issues. We begin by defining the class of admissible policies.

Definition: A measurable function $\pi: R \to [0, r]$ will be called an *admissible* policy if, for every $(\tau, \xi) \in R^2$ with $\tau \geq 0$, there exists a function $y_\pi(t; \tau, \xi)$ which satisfies

$y_\pi(t; \tau, \xi)$ is absolutely continuous in t (61.i)

$y_\pi(t; \tau, \xi) = \xi + \int_\tau^t (\pi(y_\pi(s; \tau, \xi)) - d)\, ds$ for $t \geq \tau$ (61.ii)

$y_\pi(t; \tau, \xi)$ is continuous in (t, τ, ξ) (61.iii)

$y_\pi(\cdot)$ is the unique function satisfying (i and ii) above. (61.iv)

Given such an admissible π, we now describe the manner in which we interpret the differential equation (5.i). Let $\{s(t, \omega); t \geq 0\}$ be a realization of (5.ii) with, say, $s(0, \omega) = 1$ and suppose x_0 is the initial inventory level. Define $\tau_0(\omega) := \inf\{t > 0 : s(t, \omega) = 2\}$ and $\tau_{i+1}(\omega) = \inf\{t > \tau_i(\omega); s(t+, \omega) \neq s(t-, \omega)\}$. Then we construct the process $\{x_\pi(t, \omega)\}$ by

$$x_\pi(t, \omega) := y_\pi(t; 0, x_0) \quad \text{for } 0 \leq t \leq \tau_0(\omega)$$
$$:= x_\pi(\tau_i(\omega), \omega) - d(t - \tau_i(\omega)) \quad \text{for } \tau_i(\omega) \leq t \leq \tau_{i+1}(\omega)$$
$$\text{and } i = 0, 2, 4, \cdots$$
$$:= y_\pi(t; \tau_i(\omega), x_\pi(\tau_i(\omega), \omega)) \quad \text{for } \tau_i(\omega) \leq t \leq \tau_{i+1}(\omega)$$
$$\text{and } i = 1, 3, 5, \cdots.$$

Note that an immediate consequence is

$$x_\pi(t, \omega) = x_0 + \int_0^t (u_\pi(s, \omega) - d)\, ds \quad \text{for all } t \geq 0 \quad (62)$$

where

$$u_\pi(t, \omega) = 0 \quad \text{if } s(t, \omega) = 2$$
$$= \pi(x_\pi(t, \omega)) \quad \text{if } s(t, \omega) = 1.$$

Thus, the differential equation (5.i) is interpreted in integral form in (62).

One can use the theory of semigroups of nonlinear contractions in Banach spaces, see Barbu [6], to obtain sufficient conditions for a policy π to be admissible. We now use this to establish the admissibility of policies of the $\pi^z(\cdot)$ type.

Theorem 6: $\pi^z(\cdot)$ defined by (18) is admissible.

Proof: Let A, a multivalued operator, or equivalently a subset of R^2, be defined by

$$A(x) := \{r - d\} \quad \text{if } x < z$$
$$:= [-d, r - d] \quad \text{if } x = z$$
$$:= \{-d\} \quad \text{if } x > z.$$

Then $x_1 \leq x_2$ and $y_1 \in A(x_1)$, $y_2 \in A(x_2)$ implies that $(x_1 - x_2)(y_1 - y_2) \leq 0$ and so A is a dissipative operator; see [6, Definition 3.1, p. 71]. Moreover,

$$\bigcup_{x \in R} \bigcup_{y \in A(x)} \{x - y\} = R$$

and so A is m-dissipative; see [6, p. 71]. Also, for every $x \in R$

$(\pi^z(x) - d) \in A(x)$ and $|\pi^z(x) - d| \leq |y|$ for every $y \in A(x)$.

By Corollary 1.1 and [6, Theorem 1.6, p. 118] we see that (61.i), (61.ii), and (61.iv) are satisfied. Moreover, by [6, Proposition 1.2, p. 110], $y_{\pi z}(t; \tau, \xi)$ as a function of t, for each ξ, is a semigroup of nonlinear contractions, and so from [6, Definition 1.1, p. 98] we see that $|y_{\pi z}(t; \tau, \xi_1) - y_{\pi z}(t; \tau, \xi_2)| \leq |\xi_1 - \xi_2|$. Since by uniqueness $y_{\pi z}(t; \tau, \xi) = y_{\pi z}(t - \tau; 0, \xi)$ for all $t \geq \tau$, it follows from $|y_{\pi z}(t; \tau_1, \xi_1) - y_{\pi z}(s; \tau_2, \xi_2)| \leq |y_{\pi z}(t - \tau_1; 0, \xi_1) - y_{\pi z}(t - \tau_2; 0, \xi_1)| + |y_{\pi z}(t - \tau_2; 0, \xi_1) - y_{\pi z}(t - \tau_2; 0, \xi_2)|$ that $y_{\pi z}(\cdot)$ is a continuous function, and so (61.iii) is also satisfied. □

XI. INTEGRAL EQUATION FOR COST FUNCTION

In this section we will show that the cost function corresponding to a policy π satisfies a certain integral equation.

Let $\{\tau_i\}$ be the successive jump times of $\{s(t)\}$. If $\{x_\pi(t); t \geq 0\}$ is the trajectory resulting from a policy π, define

$$V_{i,\pi}^n(\xi) := E\left[\int_0^{\tau_n} c(x_\pi(t))e^{-\gamma t}\, dt \,\Big|\, s(0) = i, x_\pi(0) = \xi\right]$$

as the cost of using π up to the nth jump of $\{s(t)\}$. Here

$$c(x) := c^+ x^+ + c^- x^-.$$

Clearly

$$V_{i,\pi}(\xi) = \lim_{n \to \infty} V_{i,\pi}^n(\xi) \qquad (63)$$

is the corresponding expected cost of using π indefinitely.

Define

$$x_\pi^1(t, \xi) := y_\pi(t; 0, \xi)$$
$$x_\pi^2(t, \xi) := \xi - td.$$

Clearly $x_\pi^i(t, \xi)$ represents the inventory level at time t if initially the inventory level is ξ, $s(0) = i$, and there are no jumps of $\{s(t)\}$ in $[0, t)$. By a renewal argument it follows that

$$V_{i,\pi}^0(\xi) = 0$$

and

$$V_{i,\pi}^{n+1}(\xi) = \int_0^\infty q_i e^{-q_i \sigma} \left[\int_0^\sigma e^{-\gamma t} c(x_\pi^i(t, \xi))\, dt \right.$$
$$\left. + e^{-\gamma \sigma} V_{j(i),\pi}^n(x_\pi^i(\sigma, \xi))\right] d\sigma \qquad (64)$$

where

$$j(i) \in \{1, 2\}, \quad j(i) \neq i \quad \text{for } i = 1, 2.$$

For $\epsilon > 0$, let $\mathcal{F}_\epsilon$ be the Banach space of all measurable functions mapping R into R, with norm defined by $\|f\|_\epsilon := \sup_x |e^{-\epsilon|x|} f(x)|$, and let $\mathcal{F} := \cap_{\epsilon > 0} \mathcal{F}_\epsilon$. On $\mathcal{F}_\epsilon^2 = \mathcal{F}_\epsilon \times \mathcal{F}_\epsilon$, define $\|(f_1, f_2)\|_\epsilon := \max_i \|f_i\|_\epsilon$, and note that $\mathcal{F}^2 = \cap_{\epsilon > 0} \mathcal{F}_\epsilon^2$. For $(f_1, f_2) \in \mathcal{F}^2$, define $T_\pi(f_1, f_2) = (T_{1,\pi} f_2, T_{2,\pi} f_1)$ by

$$(T_{i,\pi} f_{j(i)})(\xi) := \int_0^\infty q_i e^{-q_i \sigma} \left[\int_0^\sigma e^{-\gamma t} c(x_\pi^i(t, \xi))\, dt \right.$$
$$\left. + e^{-\gamma \sigma} f_{j(i)}(x_\pi^i(\sigma, \xi))\right] d\sigma. \qquad (65)$$

Lemma 7:

i) If $(f_1, f_2) \in \mathcal{F}^2$, then $T_\pi(f_1; f_2) \in \mathcal{F}^2$.

ii) T_π is a contraction with respect to the norm $\|\cdot\|_\epsilon$ for every $\epsilon > 0$ sufficiently small.

Proof: It suffices to show that $T_{i,\pi} f_{j(i)} \in \mathcal{F}$ and that $T_{i,\pi}$ is a contraction for $i = 1, 2$. Note first that by (62)

$$|x_\pi^i(t, \xi)| \leq |\xi| + k_1 t.$$

In the above and what follows, all the k_i's are constants chosen appropriately. Since $c(x) \leq k_2 |x|$, it follows that

$$\int_0^\infty q_i e^{-q_i \sigma} \left[\int_0^\sigma e^{-\gamma t} c(x_\pi^i(t, \xi))\, dt \right] d\sigma \leq k_3 |\xi| + k_4.$$

Also, for $0 < \epsilon < (\gamma + q_1)/k_1$

$$\int_0^\infty q_i e^{-(q_i + \gamma)\sigma} |f_{j(i)}(x_\pi^i(\sigma, \xi))|\, d\sigma$$

$$\leq \int_0^\infty q_i e^{-(q_i + \gamma)\sigma} \|f_{j(i)}\|_\epsilon e^{\epsilon |x_\pi^i(\sigma, \xi)|}\, d\sigma$$

$$\leq \int_0^\infty q_i e^{-(q_i + \gamma)\sigma} \|f_{j(i)}\|_\epsilon e^{\epsilon(|\xi| + k_1 \sigma)}\, d\sigma$$

$$\leq k_5 e^{\epsilon |\xi|} + k_6.$$

Hence, $|T_{i,\pi} f_{j(i)}(\xi)| \leq k_5 e^{\epsilon|\xi|} + k_3|\xi| + k_7$ and so, $T_{i,\pi} f_{j(i)} \in \mathcal{F}_\epsilon$ for all $\epsilon > 0$ sufficiently small, i.e., $T_{i,\pi} f_{j(i)} \in \mathcal{F}$. To show that $T_{i,\pi}$ is a contraction, consider $0 < \epsilon < \gamma/k_1$, then

$$(T_{i,\pi} f - T_{i,\pi} g)(\xi)$$

$$\leq \int_0^\infty q_i e^{-(q_i + \gamma)\sigma} |f(x_\pi^i(\sigma, \xi)) - g(x_\pi^i(\sigma, \xi))|\, d\sigma$$

$$\leq \int_0^\infty q_i e^{-(q_i + \gamma)\sigma} e^{\epsilon|x_\pi^i(\sigma, \xi)|} \|f - g\|_\epsilon\, d\sigma$$

$$\leq \int_0^\infty q_i e^{-(q_i + \gamma)\sigma} e^{\epsilon(|\xi| + k_1 \sigma)} \|f - g\|_\epsilon\, d\sigma$$

$$= e^{\epsilon|\xi|} \|f - g\|_\epsilon \int_0^\infty q_i e^{-q_i \sigma} e^{(\epsilon k_1 - \gamma)\sigma}\, d\sigma$$

$$\leq \beta e^{\epsilon|\xi|} \|f - g\|_\epsilon$$

where $0 < \beta < 1$. Hence, $|T_{i,\pi} f - T_{i,\pi} g\|_\epsilon \leq \beta \|f - g\|_\epsilon$. □

From (64) it follows that if $V_\pi^n := (V_{1,\pi}^n, V_{2,\pi}^n)$, then $V_\pi^{n+1} = T_\pi V_\pi^n$ for $n = 0, 1, 2, \cdots$ with $V_\pi^0 := 0$.

Theorem 8: Let $V_{i,\pi}(\xi)$ denote the cost of using π starting in the state $(s(0) = 1, x(0) = \xi)$. Let $V_\pi(\xi) := (V_{1,\pi}(\xi), V_{2,\pi}(\xi))$. Then V_π is the unique solution in $\mathcal{F}^2$ of the integral equation. (66.i)

$$V_{i,\pi}(\xi) = \int_0^\infty q_i e^{-q_i \sigma} \left[\int_0^\infty e^{-\gamma t} c(x_\pi^i(t, \xi)) \, dt \right.$$
$$\left. + e^{-\gamma \sigma} V_{j(i),\pi}(x_\pi^i(\sigma, \xi)) \right] d\sigma \text{ for every } \xi \in R.$$

For every $f \in \mathcal{F}^2$, $\lim_{n \to \infty} T_\pi^{(n)} f = V$. (66.ii)

Proof: From (63) and (64), we see that $V_\pi = \lim_{n \to \infty} T_\pi^{(n)} 0$, where $T_\pi^{(n)}$ denotes the n-fold iterate of T_π, and 0 is the identically zero function. However, since T_π is a contraction, $\lim_{n \to \infty} T_\pi^{(n)} f$ is a unique fixed point of T_π, for every $f \in \mathcal{F}$. Hence, V_π is the unique solution, in, $\mathcal{F}$, of $V_\pi = T_\pi V_\pi$. □

It is important to note that the boundary condition for the integral equation is really the condition that the solution be in $\mathcal{F}_z$. This is a condition on the asymptotic growth rate, and serves to differentiate the infinite horizon problem treated here from the finite horizon problem in [3].

XII. Optimality of πz^*

We are now ready to prove the optimality of the suggested policy. We shall actually show that optimality is a consequence of πz^* and $V(\cdot)$ satisfying the HJB equation (60.iv), (60.v), (60.vi) for the *infinite* time problem and so our proof is quite general.

Theorem 9: Let z^* and $\pi z^*(\cdot)$ be as in (3) and (2). If $z^* = 0$, define $V(\cdot)$ by (44), while if $z^* > 0$, define $V(\cdot)$ by (46). Then

i) If $V_\pi(\cdot)$ represents the cost function corresponding to an admissible policy π, then

$$V_{i,\pi z^*}(\xi) \le V_{i,\pi}(\xi) \quad \text{for } i = 1, 2; \xi \in R \text{ and all admissible } \pi$$

ii) $V_{\pi z^*}(\xi) = V(\xi)$ for every $\xi \in R$.

Proof: We will show that

$$T_\pi V_{\pi z^*} \ge V_{\pi z^*} \quad \text{for every admissible } \pi, \quad (67)$$

i.e.,

$$T_{i,\pi} V_{j(i),\pi z^*}(\xi) \ge V_{i,\pi z^*}(\xi) \quad \text{for every } \xi \in R \text{ and } i = 1, 2. \quad (68)$$

Since T_π is monotone, (67) implies that $T_\pi^{(n)} V_{\pi z^*} \ge V_{\pi z^*}$. Taking the limit in n and using (66.ii), we obtain $V_\pi \ge V_{\pi z^*}$, that is i) for $i = 1, 2$. So our goal is to show (67), along with equality when $\pi = \pi z^*$. Considering (52.i), (52.ii) we have

$$\min_{u \in [0,r]} (u - d) \dot{V}_1(x) = (\gamma + q_1) V_1(x) - q_1 V_2(x) - c(x) \quad (69)$$

$$-d \dot{V}_2(x) = -q_2 V_1(x) + (\gamma + q_2) V_2(x) - c(x). \quad (70)$$

For any π, therefore, (69) implies that

$$c(x) \ge (\gamma + q_1) V_1(x) - q_1 V_2(x) - (\pi(x) - d) \dot{V}_1(x) \quad (71)$$

and so for any admissible π

$$c(x_\pi^1(t, \xi)) \ge (\gamma + q_1) V_1(x_\pi^1(t, \xi)) - q_1 V_2(x_\pi^1(t, \xi))$$
$$- (\pi(x_\pi^1(t, \xi)) - d) \dot{V}_1(x_\pi^1(t, \xi)). \quad (72)$$

Now noting that $x_\pi^1(t, \xi)$ is absolutely continuous in t, with derivative $(\pi(x_\pi^1(t, \xi)) - d)$, and $\dot{V}_1(\cdot)$ is continuous, we can apply [7, Corollary 7] showing that the chain rule is valid, and so obtain

$$\frac{d}{dt} V_1(x_\pi^1(t, \xi)) = \dot{V}_1(x_\pi^1(t, \xi))(\pi(x_\pi^1(t, \xi)) - d) \text{ a.e.} \quad (73)$$

Hence, from (72) and (73), we have

$$\int_0^\sigma e^{-\gamma t} c(x_\pi^1(t, \xi)) \, dt \ge \int_0^\sigma e^{-\gamma t} [(\gamma + q_1) V_1(x_\pi^1(t, \xi))$$
$$- q_1 V_2(x_\pi^1(t, \xi))] \, dt - \int_0^\sigma e^{-\gamma t} \frac{d}{dt} V_1(x_\pi^1(t, \xi)) \, dt \quad \text{for } \sigma \ge 0.$$
(74)

Integrating the last term in (74) by parts (see [9, p. 287]), we have

$$\int_0^\sigma e^{-\gamma t} c(x_\pi^1(t, \xi)) \, dt \ge \int_0^\sigma e^{-\gamma t} q_1 (V_1(x_\pi^1(t, \xi)) - V_2(x_\pi^1(t, \xi)) \, dt$$
$$+ V_1(\xi) - e^{-\gamma \sigma} V_1(x_\pi^1(\sigma, \xi)) \quad \text{for } \sigma \ge 0.$$

Hence,

$$\int_0^\infty q_1 e^{-q_1 \sigma} \left[\int_0^\sigma e^{-\gamma t} c(x_\pi^1(t, \xi)) \, dt \right] d\sigma$$
$$\ge V_1(\xi) - \int_0^\infty q_1 e^{-(\gamma + q_1)\sigma} V_1(x_\pi^1(\sigma, \xi)) \, d\sigma$$
$$+ \int_0^\infty q_1 e^{-q_1 \sigma} \left[\int_0^\sigma e^{-\gamma t} q_1 (V_1(x_\pi^1(t, \xi)) \right.$$
$$\left. - V_2(x_\pi^1(t, \xi))) \, dt \right] d\sigma$$
$$= V_1(\xi) - \int_0^\infty q_1 e^{-(\gamma + q_1)\sigma} V_1(x_\pi^1(\sigma, \xi)) \, d\sigma$$
$$+ \int_0^\infty q_1 e^{-\gamma t} (V_1(x_\pi^1(t, \xi))$$
$$- V_2(x_\pi^1(t, \xi)) \left[\int_t^\infty q_1 e^{-q_1 \sigma} \, d\sigma \right] dt$$
$$= V_1(\xi) - \int_0^\infty q_1 e^{-(\gamma + q_1)t} V_2(x_\pi^1(t, \xi)) \, dt.$$

Hence,

$$\int_0^\infty q_1 e^{-q_1 \sigma} \left[\int_0^\sigma e^{-\gamma t} c(x_\pi^1(t, \xi)) \, dt \right.$$
$$\left. + e^{-\gamma \sigma} V_2(x_\pi^1(\sigma, \xi)) \right] d\sigma \ge V_1(\xi),$$

i.e.,

$$(T_{1,\pi} V_2)(\xi) \ge V_1(\xi)$$

noting that from (60.iv), $V(x) = 0(x)$ and so $V \in \mathcal{F}^2 = $ domain (T_π). Using (70), similarly, we deduce that $(T_{2,\pi} V_1)(\xi) = V_2(\xi)$. Thus, we have shown $T_\pi V \ge V$. On the other hand, since $\pi z^*(\cdot)$ attains equality in (71), we have $T_{\pi z^*} V = V$. Thus, $V_{\pi z^*} = V \le T_\pi V = T_\pi V_{\pi z^*}$ with equality when $\pi = \pi z^*$. □

XIII. Concluding Remarks

There are two directions in which more work is needed. The first is to realize the full program for flexible manufacturing systems outlined in Kimemia and Gershwin [2]. Consider a flexible manufacturing system making p parts on m machines. Part j requires a_{ij} units of time on machine i. Thus, if a subset $s_k \subseteq \{1, 2, \cdots, m\}$ of machines is functioning, while the rest have failed, then a vector $u = (u_1, u_2, \cdots, u_p)^T$ of production rates is feasible if and only if $u \in U_k$ where $U_k := \{u : \Sigma_{j=1}^p a_{ij} u_j \le 1 \text{ if } i \in s_k, \text{ or } = 0 \text{ if } i \notin s_k\}$. Suppose now that each machine is subject to occasional failure and let $\{s(t); t \ge 0\}$ be a Markov chain with state-space $\{s_1, s_2, \cdots, s_{2^m}\}$. Given a demand rate vector $d = (d_1, d_2, \cdots, d_p)^T$, we have the problem

$$\dot{x} = u(t) - d$$

$\{s(t); t \ge 0\}$ is a Markov chain with state-space $\{s_1, \cdots, s_{2^m}\}$

$$u(t) \in U_k \text{ if } s(t) = s_k.$$

$$\text{minimize } E \int_0^\infty e^{-\gamma t} c(x(t)) \, dt$$

where $c(\cdot)$ is some convex cost function. Due to the multidimensional nature of $x(t)$, this problem is much more difficult than the one solved here. Reference [2] has proposed an approximation, but the optimal solution needs more study.

The other direction in which more research is needed is theoretical, and is the problem of optimal control of continuous-time systems with jump Markov disturbances. As in Rishel [3], we also have proved optimality only within the class of Markov policies. For discrete-time systems (see [10], for example) much more progress has been made on optimal control, and one usually considers a much more general class of policies within which optimality is proven. The question of existence of optimal controls also needs more study. Finally, more work needs to be done on the average cost problem for systems with jump Markov disturbances (see [8]).

Acknowledgment

The author is grateful to S. Gershwin for introducing him to the problem studied here, to S. Mitter for his friendly hospitality during his visit to M.I.T., and to T. Seidman for several useful discussions. The first author would like to acknowledge the support of C. Abraham, P. Summers, and the Manufacturing Research Center at IBM.

References

[1] J. G. Kimemia, "Hierarchial control of production in flexible manufacturing systems," Ph.D. dissertation, Mass. Inst. Technol., Cambridge, MA, Rep. L.I.D.S.-TH-1215, Sept. 1982.
[2] J. G. Kimemia and S. B. Gershwin, "An algorithm for the computer control of production in flexible manufacturing systems," *IEEE Trans. Automat. Contr.*, vol. AC-15, pp. 353–362, Dec. 1983.
[3] R. Rishel, "Dynamic programming and minimum principles for systems with jump Markov disturbances," *SIAM J. Contr.*, vol. 13, pp. 338–371, Feb. 1975.
[4] N. N. Krassovskii and E.A. Lidskii, "Analytical design of controllers in stochastic systems with velocity limited controlling action," *Appl. Math. Mechan.*, vol. 25, pp. 627–643, 1961.
[5] E. A. Lidskii, "Optimal control of systems with random properties," *Appl. Math. Mechan.*, vol. 27, pp. 33–45, 1963.
[6] V. Barbu, *Nonlinear Semigroups and Differential Equations in Banach Spaces*. Leyden, The Netherlands: Noordhoff International, 1976.
[7] J. Serrin and D. E. Varberg, "A general chain rule for derivatives and the change of variables formula for the Lebesgue integral," *Amer. Math. Monthly*, vol. 76, pp. 514–520, May 1969.
[8] J. N. Tsitsiklis, "Convexity and characterization of optimal policies in a dynamic routing problem," Mass. Inst. Technol., Cambridge, MA, Rep. L.I.D.S.-R-1178, Feb. 1982.
[9] E. Hewitt and K. Stromberg, *Real and Abstract Analysis*. New York: Springer-Verlag, 1965.
[10] D. Blackwell, "Discounted dynamic programming," *Ann. Math. Stat.*, vol. 36, pp. 226–235, 1965.

Ramakrishna Akella was born in Hospet, India, on March 31, 1955. He received the B.Tech degree in electronics from the Indian Institute of Technology, Madras, in 1976 and the Ph.D. degree in systems engineering from the Indian Institute of Science, Bangalore, in 1981.

In 1982 he was a CSIR Research Associate and Postdoctoral Visitor with the Decision and Control Group at Harvard University, Cambridge, MA. From 1983 to 1985 he was a Postdoctoral Associate at the Laboratory for Information and Decision Systems, Massachusetts Institute of Technology, Cambridge. Since September 1985, he has been an Associate Professor at Carnegie-Mellon University, Pittsburgh, PA.

P. R. Kumar (S'77–M'77) was born in India on April 21, 1952. He received the B.Tech degree in electrical engineering from the Indian Institute of Technology, Madras, in 1973 and the M.S. and D.Sc. degrees from Washington University, St. Louis, MO, in 1975 and 1977, respectively.

At present he is an Associate Professor in Electrical and Computer Engineering at the Coordinated Science Laboratory, University of Illinois, Urbana-Champaign. His current research interests include control systems and communications.

Dr. Kumar was an Associate Editor of the IEEE Transactions on Automatic Control from 1982 to 1983 and is currently an Associate Editor of *Systems and Control Letters*. In 1985 he received the Donald Eckman Award of the American Automatic Control Council.

Production Control of a Manufacturing System with Multiple Machine States

ALI SHARIFNIA

Abstract—The production control of a single-product manufacturing system with arbitrary number of machine states (failure modes) is discussed. The objective is to find a production policy that would meet the demand for the product with minimum average inventory or backlog cost.

The optimal production policy has a special structure and is called a hedging point policy. If the hedging points are known, the optimal production rate is readily specified.

Assuming a set of tentative hedging points, we utilize the simple structure of the optimal policy to find the steady-state probability distribution of the surplus (inventory or backlog). Once this function is determined the average surplus cost is easily calculated in terms of the values of the hedging points. This average cost is then minimized to find the optimum hedging points.

I. Introduction

THIS paper describes a method for solving the short-term production control problem of a failure prone manufacturing system. The system produces a single product and is characterized by a number of possible machine states. Each machine state corresponds to a failure mode of the system (some of the machines failed) and has a production possibility set. The objective is to determine an optimum production policy under each machine state, so that the demand is met with minimum surplus costs (surplus may be negative representing backlog).

In their pioneering work Kimemia and Gershwin [1] consider manufacturing systems with an arbitrary number of machine states and product types. Formulating this problem as a dynamic programming problem they show that solving the Hamilton–Jacobi–Bellman equation for the optimum production policy in each machine state can be reduced to the solution of linear programs, provided the minimum "cost-to-go" function is known. The linear program has the form

$$\text{minimize } \nabla J(x, \alpha) \cdot u \quad \text{subject to } u \in \Omega(\alpha). \quad (1)$$

Here x is the inventory level, α is the machine state, $\nabla J(x, \alpha)$ is the gradient of the minimum cost-to-go function, u is the production rate, and $\Omega(\alpha)$ is the set of production rate possibilities in machine state α. The approach provides production rates which are then translated into loading times of the parts into the system.

A very interesting and useful result of [1] is the concept of hedging points. It is shown that for each feasible machine state (in which enough can be produced to keep up with demand), the optimum production policy will move the inventory level toward a specific target (the hedging point) as quickly as possible, and when at this point, it will keep it there as long as the machine state remains unchanged.

Manuscript received March 24, 1987; revised December 3, 1987 and January 21, 1988. This work was supported by the National Science Foundation under Grant DMC-8615560. Paper recommended by Past Associate Editor, S. G. Gershwin.

The author is with the College of Engineering, Boston University, Boston, MA 02215.

IEEE Log Number 8821357.

Gershwin, Akella, and Choong [2] apply the method to scheduling of a flexible manufacturing system using approximate values of the hedging points and the cost-to-go function. However, one difficulty with the implementation is the determination of the minimum cost-to-go and the hedging points.

Akella and Kumar [3] find an exact solution for the case involving a single product and two machine states (up and down). Their analysis is quite complicated even for this simple case. In a recent work Bielecki and Kumar [4] use another approach for solving the single-product two-machine-state problem. Assuming that the hedging point (corresponding to the up state) is known, they calculate the steady-state probability density function of the surplus and the probability mass at the hedging point. The optimum hedging point is then found by minimizing the average cost per unit time over all possible values of the hedging point. This approach is simpler and more promising and is adopted in our work.

In this paper we solve the single product problem with an arbitrary number of machine states. We use sample path analysis and time averaging to derive the probability density function (PDF) of the surplus. The ergodic hypothesis holds since state transitions are generated by a finite irreducible Markov chain.

The single product case is of interest for its own sake, and also because it is a first step towards understanding the more general multiple product case in flexible manufacturing systems. A method has already been suggested by Kimemia and Gershwin [1] for decomposition of the multiple product case into single product problems.

In Section II we give some definitions and discuss the basic concepts used in the paper. The steady-state probability density and mass functions of the surplus are derived in Section III. Section IV describes how the average cost function is calculated and the optimum hedging points are determined. An example is given in Section V and we conclude in Section VI.

II. Definitions and Basic Concepts

We consider a manufacturing system that produces a single product. The machines can be in a number of different failure modes $(\alpha, \beta \cdots)$ and we assume that the machine state changes in continuous time according to an irreducible Markov chain with known transition rates $\lambda_{\alpha\beta}$. The surplus at time t is a random variable $X(t)$, and there is a constant demand d, per unit time, for the product. Let $g(x)$ denote the cost of maintaining a surplus x for one time unit. The objective is to find the optimum production rate u, for each machine state α and surplus level x, so as to minimize the average surplus cost per unit time

$$J^* = \lim_{T \to \infty} \frac{1}{T} E \int_{t_0}^{t_0 + T} g(X(t)) \, dt. \quad (2)$$

The machine states which are capable of producing the demand are called feasible machine states. We assume that the system is strictly feasible in the long run, in the sense that if we produce at maximum possible rate in each machine state, then the average production rate will exceed the demand rate d.

The solution of the problem defined above can be obtained from the linear program (1). Since we are using the average cost criterion (J^*), the functions $J(x, \alpha)$ in (1) will be the "value" functions (differential costs). Since X is scalar in our case, $\Omega(\alpha)$ is always the positive segment of the real line bounded by the maximum production rate in machine state α, denoted $\hat{u}_\alpha$

$$\Omega(\alpha) = [0, \hat{u}_\alpha].$$

For each machine state α the value function $J(x, \alpha)$ reaches its minimum at a "hedging point" $x = z_\alpha$. The partial derivative $\partial J(x, \alpha)/\partial x$ is positive for $x > z_\alpha$ and is negative for $x < z_\alpha$. It follows that the optimum production rate for machine state α is

$$u^*(x, \alpha) = \begin{cases} \hat{u}_\alpha & \text{if } x < z_\alpha, \\ d & \text{if } x = z_\alpha, \\ 0 & \text{if } x > z_\alpha. \end{cases} \quad (3)$$

From the preceding discussion it follows that if the optimum hedging points (HP's) z_α are known, the optimum policy is completely determined. To find the optimum HP's we follow the approach of Bielecki and Kumar [4]. First we find the distribution of X given tentative values of the HP's, then we calculate the average cost per period J^* as a function of these tentative values, and finally we find the optimum HP's by minimizing the function J^*.

Once the optimum production policy is known we can derive the probability distribution P of X in steady state. For the density of the continuous part of P we write $f(x) = \Sigma_\alpha f(x, \alpha) = \Sigma_\alpha \pi_\alpha f(x|\alpha)$. Here π_α is the stationary probability of machine state α, $f(x|\alpha)$ is the conditional density of X given that the machine state is α, and $f(x, \alpha)$ is the joint probability distribution of x and α. The distribution P of X also has a finite number of point masses $P(z_\alpha)$ located at the hedging points z_α of the feasible machine states.

To find the steady-state probability distribution P we use time averaging of the sample path $X(t)$ over a long period of time. This approach turns out to be much simpler than calculating the probabilities in the ensemble space. The analysis of the sample paths is also found very useful in obtaining the boundary conditions needed to determine the density $f(x)$ in regions between successive hedging points, and the probability masses at the hedging points which separate regions.

At the heart of our derivation of the distribution P are the functions $m(x, \alpha)$, which are defined as follows. Let us consider a time interval of length T and let $M(x, \alpha, T)$ denote the number of times $X(t)$ crosses the line $X = x$ while the machine state is α (Fig. 1). We define

$$m(x, \alpha) = \lim_{T \to \infty} \frac{M(x, \alpha, T)}{T}. \quad (4)$$

The time average on the right-hand side converges with probability one to a unique limit, namely the expected frequency with which the surplus $X(t)$ crosses the level $X = x$ and the machine state is α, provided that $X(t)$ is ergodic. The process $X(t)$ is ergodic once the machine states form an irreducible Markov chain and the production policy is such that $X(t)$ is stationary. For the hedging point policy $X(t)$ will be stationary if the system is strictly feasible in the long run, i.e.,

$$\sum_\alpha \pi_\alpha \hat{u}_\alpha > d.$$

This condition implies that if we produce at maximum rate in each machine state, then there will be a strictly positive long term trend in $X(t)$. It is then clear that the smallest hedging point will be recurrent. However, since the hedging points are finitely separated, it follows that all the hedging points will be recurrent, and hence $X(t)$ will be stationary.

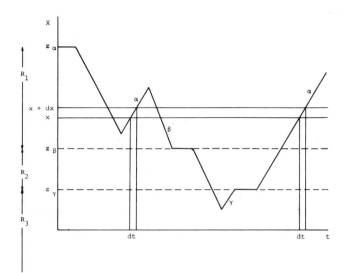

Fig. 1. A sample path of $X(t)$.

We can find $f(x, \alpha)$ in terms of $m(x, \alpha)$, again by time averaging. Indeed, suppose the inventory level $X(t)$ is observed over a long period of time T, and $\tau(x, \alpha)$ designates the accumulated sum of the short intervals during which X falls in $(x, x + dx)$ and the machine state is α. Then

$$f(x, \alpha)|dx| = \lim_{T \to \infty} \frac{\tau(x, \alpha)}{T}. \quad (5)$$

But $\tau(x, \alpha) = M(x, \alpha, T) \cdot |dt|$ where dt is the amount of time necessary for the surplus X to sweep through the interval $(x, x + dx)$ while the machine state is α (see Fig. 1). Substituting in (5) and using (4) we obtain

$$f(x, \alpha)|dx| = m(x, \alpha)|dt| \quad (6)$$

or

$$f(x, \alpha) = \frac{m(x, \alpha)}{\left|\dfrac{dx}{dt}\right|}. \quad (7)$$

Here dx/dt is the velocity with which the surplus $X(t)$ is changing when the machine state is α and the surplus is x. This velocity is denoted $v(x, \alpha)$ and is determined by the hedging point policy

$$\frac{dx}{dt} = v(x, \alpha) = u^*(x, \alpha) - d = \begin{cases} \hat{u}_\alpha - d & \text{if } x < z_\alpha, \\ 0 & \text{if } x = z_\alpha, \\ -d & \text{if } x > z_\alpha. \end{cases}$$

The optimum velocity is constant as long as the machine state α remains unchanged and the surplus X falls in a region between successive hedging points. We write R^i for the region in between ith and $(i + 1)$st hedging points, and v_α^i for the optimum velocity in region R^i. We find that

$$f(x, \alpha) = \frac{m(x, \alpha)}{|v_\alpha^i|}, \quad x \in R^i. \quad (8)$$

The density $f(x)$ is obtained by summing $f(x, \alpha)$ over all α

$$f(x) = \sum_\alpha f(x, \alpha) = \sum_\alpha \frac{m(x, \alpha)}{|v_\alpha^i|}, \quad x \in R^i. \quad (9)$$

In the next section we shall see that $m(x, \alpha)$ is more tangible and easier to work with than $f(x, \alpha)$. This is particularly true with regard to the boundary conditions required for solving the problem.

III. Distribution of the Surplus Level

In this section we derive an equation for $m(x, \alpha)$ in each region. We also derive boundary conditions for these functions at the hedging points which delimit regions. The PDF of X will then be determined by (9).

The machine failure states evolve according to a continuous-time Markov chain which is assumed irreducible. Let $\lambda_{\alpha\beta}$ be the rate of machine state transitions from α to $\beta \neq \alpha$, and let $\lambda_\alpha = -\lambda_{\alpha\alpha} = \Sigma_{\beta \neq \alpha}\lambda_{\alpha\beta}$. Thus, λ_α^{-1} is the mean holding time in state α.

We calculate the frequency with which the inventory X crosses a given level x in the interior of region R^i while the machine state is α, as follows. Let $\Delta x > 0$ be sufficiently small so that $x \pm \Delta x$ is contained in region R^i, and let us define σ_α^i and $\sigma_\beta^i (\beta \neq \alpha)$ as follows:

$$\sigma_\alpha^i = \operatorname{sgn} v_\alpha^i = \begin{cases} +1 & \text{if } v_\alpha^i > 0, \\ -1 & \text{if } v_\alpha^i < 0. \end{cases}$$

$$\sigma_\beta^i = \begin{cases} \sigma_\alpha^i & \text{if } v_\alpha^i v_\beta^i > 0, \\ 0 & \text{if } v_\alpha^i v_\beta^i < 0, \end{cases} \quad \beta \neq \alpha$$

One way to cross the level x while being in state α is to pass through level $x - \sigma_\alpha^i \Delta x$ while being in state α, and to make no state transition for a duration $\Delta t = \Delta x / |v_\alpha^i|$. The state transition fails to occur with probability

$$1 - \lambda_\alpha \Delta t + o(\Delta t) = 1 + \lambda_{\alpha\alpha} \frac{\Delta x}{|v_\alpha^i|} + o(\Delta x).$$

Another way to reach level x while being in state α is to start at level $x - \sigma_\beta^i \Delta x$ in state $\beta \neq \alpha$ and to make a transition from state β to state α during a time interval $\Delta t = \Delta x / |v_\beta^i| + o(\Delta x)$. Such a transition occurs with probability

$$\lambda_{\beta\alpha} \Delta t + o(\Delta t) = \lambda_{\beta\alpha} \frac{\Delta x}{|v_\beta^i|} + o(\Delta x).$$

We may neglect the possibility that level x is reached after several state transitions have taken place while X was in the range $x \pm \Delta x$, since the probability of such events is $o(\Delta x)$.

The following equation holds for $m(x, \alpha)$, which is the frequency with which $X(t)$ crosses level x while the machine state is α:

$$m(x, \alpha) = m(x - \sigma_\alpha^i \Delta x, \alpha)\left(1 + \lambda_{\alpha\alpha} \frac{\Delta x}{|v_\alpha^i|}\right)$$

$$+ \sum_{\beta \neq \alpha} m(x - \sigma_\beta^i \Delta x, \beta) \lambda_{\beta\alpha} \frac{\Delta x}{|v_\beta^i|} + o(\Delta x).$$

Simplification gives

$$\frac{m(x, \alpha) - m(x - \sigma_\alpha^i \Delta x, \alpha)}{\Delta x} = \sum_\beta m(x - \sigma_\beta^i \Delta x, \beta) \frac{\lambda_{\beta\alpha}}{|v_\beta^i|} + o(\Delta x).$$

Letting $\Delta x \to 0$ we obtain

$$\sigma_\alpha^i \frac{d}{dx} m(x, \alpha) = \sum_\beta m(x, \beta) \frac{\lambda_{\beta\alpha}}{|v_\beta^i|}.$$

Writing

$$g_{\beta\alpha}^i = \frac{\lambda_{\beta\alpha}}{|v_\beta^i|} \operatorname{sgn} v_\alpha^i,$$

we obtain the system of linear differential equations

$$\frac{d}{dx} m(x, \alpha) = \sum_\beta m(x, \beta) g_{\beta\alpha}^i, \quad x \in R^i. \tag{10}$$

In terms of the matrix $G^i = [g_{\alpha\beta}^i]$ and the row vector $\boldsymbol{m}(x) = [m(x, \alpha)]$, this system of differential equations can be written in matrix form as

$$\frac{d}{dx} \boldsymbol{m}(x) = \boldsymbol{m}(x) \cdot G^i, \quad x \in R^i. \tag{11}$$

The solution is $\boldsymbol{m}(x) = \boldsymbol{m}(x_0) e^{G^i(x - x_0)}$, $x \in R^i$. In particular, if the matrix G^i is diagonalizable with eigenvalues s_k^i and corresponding eigenvectors $\boldsymbol{a}_k^i$, then there exist constants C_k^i such that

$$\boldsymbol{m}(x) = \sum_k C_k^i e^{s_k^i x} \boldsymbol{a}_k^i, \quad x \in R^i. \tag{12}$$

Once the $m(x, \alpha)$ are known we can immediately compute the $f(x, \alpha)$. Indeed, the differential equation (10) for $m(x, \alpha)$ and relation (8) between $m(x, \alpha)$ and $f(x, \alpha)$ imply that

$$\frac{d}{dx} f(x, \alpha) = \sum_\beta f(x, \beta) \frac{\lambda_{\beta\alpha}}{v_\alpha^i}, \quad x \in R^i. \tag{13}$$

If we introduce the matrix

$$H^i = [h_{\alpha\beta}^i] \quad \text{where } h_{\alpha\beta}^i = \frac{\lambda_{\alpha\beta}}{v_\beta^i}, \tag{14}$$

then the row vector $\boldsymbol{f}(x) = [f(x, \alpha)]$ satisfies the linear system of differential equations

$$\frac{d}{dx} \boldsymbol{f}(x) = \boldsymbol{f}(x) \cdot H^i. \tag{15}$$

The solution is $\boldsymbol{f}(x) = \boldsymbol{f}(x_0) e^{H^i(x - x_0)}$, $x \in R^i$.

It is easy to show that G^i and H^i are similar matrices. Indeed,

$$h_{\alpha\beta}^i = g_{\alpha\beta}^i \frac{|v_\alpha^i|}{|v_\beta^i|},$$

and hence $H^i = (V_a^i) \cdot G^i \cdot (V_a^i)^{-1}$ where V_a^i is the diagonal matrix with diagonal entries $|v_\alpha^i|$. H^i and G^i are also similar to the matrix that appears in the Hamilton–Jacobi–Bellman equation for the system. This equation is (cf. [1])

$$\min_{u_\alpha} \left(\frac{\partial J(x, \alpha)}{\partial x}(u_\alpha - d) + \sum_\beta \lambda_{\alpha\beta} J(x, \beta) + g(x) \right) = 0, \quad \text{all } \alpha.$$

Since $v_\alpha^i = u_\alpha^* - d$ is the optimum control in region R^i, this equation reduces to

$$\frac{\partial J(x, \alpha)}{\partial x} v_\alpha^i + \sum_\beta \lambda_{\alpha\beta} J(x, \beta) + g(x) = 0, \quad x \in R^i, \text{ all } \alpha. \tag{16}$$

Now let $\boldsymbol{J}(x)$ denote the column vector $[J(x, \alpha), J(x, \beta), \cdots, J(x, \gamma)]^T$, $\boldsymbol{e}$ the column vector $[1, 1, \cdots, 1]^T$, V^i the diagonal matrix $[v_\alpha^i]$, and A^i the generator matrix $[\lambda_{\alpha\beta}/v_\alpha^i]$. Then we can express (16) in matrix form

$$\frac{d}{dx} \boldsymbol{J}(x) = -A^i \boldsymbol{J}(x) - g(x)(V^i)^{-1} \boldsymbol{e}, \quad x \in R^i. \tag{17}$$

The matrix H^i is in fact $V^i A^i (V^i)^{-1}$.

Notice that G^i and H^i depend on the $\lambda_{\alpha\beta}$ and v_α^i, but not on the location of the hedging points. Therefore, the functional form of the density $f(x) = \Sigma_\alpha f(x, \alpha)$ in each region R^i does not depend on the hedging points. This property facilitates solution of the problem if the hedging points are changed.

Boundary Conditions

To determine the constants C_k^i in (12) we write the boundary conditions at the hedging points. Consider any hedging point z_β, except the largest one, and a machine state $\alpha \neq \beta$ for which X will be increasing at z_β, i.e., $v(z_\beta, \alpha) > 0$ (Fig. 2).

Letting $z_\beta^+ = z_\beta + \epsilon$ and $z_\beta^- = z_\beta - \epsilon$ where $\epsilon > 0$ is infinitesimally small we may write

$$m(z_\beta^+, \alpha) = m(z_\beta^-, \alpha) + [m(z_\beta^+, \beta) + m(z_\beta^-, \beta)] \frac{\lambda_{\beta\alpha}}{\lambda_\beta},$$

$$\text{if } v(z_\beta, \alpha) > 0. \quad (18)$$

This equation states that $m(z_\beta^-, \alpha)$ and $m(z_\beta^+, \alpha)$ would be equal were it not for the transitions from state β to state α which will add to $m(z_\beta^+, \alpha)$. If X is decreasing at z_β in machine state α (i.e., if $v(z_\beta, \alpha) < 0$), then the boundary equation will be

$$m(z_\beta^+, \alpha) = m(z_\beta^-, \alpha) - [m(z_\beta^+, \beta) + m(z_\beta^-, \beta)] \frac{\lambda_{\beta\alpha}}{\lambda_\beta},$$

$$\text{if } v(z_\beta, \alpha) < 0. \quad (19)$$

For $\alpha = \beta$ the boundary condition is

$$[m(z_\beta^+, \beta) + m(z_\beta^-, \beta)] \frac{1}{\lambda_\beta} = P(z_\beta). \quad (20)$$

The expression inside the bracket on the left-hand side of (20) is the average number of times (per unit time) that the inventory reaches z_β while the machine state is β. Since the machine transition times have exponential distribution and have no memory, the expected time that X remains at z_β, given that it has reached z_β, is nothing but the expected holding time $1/\lambda_\beta$ of state β. Therefore, the left-hand side of (20) is the probability $P(z_\beta)$ of being at hedging point z_β.

For the largest hedging point z_γ the boundary conditions have a slightly different form, namely (see Fig. 3)

$$m(z_\gamma^-, \alpha) = m(z_\gamma^-, \gamma) \cdot \frac{\lambda_{\gamma\alpha}}{\lambda_\gamma} \quad \alpha \neq \lambda \quad (21)$$

$$m(z_\gamma^-, \gamma) \frac{1}{\lambda_\gamma} = P(z_\gamma). \quad (22)$$

Equation (21) is equivalent to (18) and (19), and (22) is equivalent to (20).

Suppose there are N machine states, out of which M are feasible. Then there will be M hedging points. The boundary conditions (18)–(22) give us N equations for each hedging point, or a total of $N \cdot M$ equations. We have $N \cdot M + M$ unknowns, namely C_k^i's and $P(z_\alpha)$'s. To find additional conditions we observe that

$$\sum_i \pi_\alpha^i + P(z_\alpha) = \pi_\alpha, \quad \alpha \text{ feasible} \quad (23)$$

where π_α is the steady-state probability of machine state α and π_α^i is the steady-state probability of being in region R^i and machine state α. But (8) implies that

$$\pi_\alpha^i = \int_{R^i} f(x, \alpha) \, dx = \frac{1}{|v_\alpha^i|} \int_{R^i} m(x, \alpha) \, dx.$$

Substituting π_α^i from this equation in (23) we find

$$\sum_i \frac{1}{|v_\alpha^i|} \int_{R^i} m(x, \alpha) \, dx + P(z_\alpha) = \pi_\alpha, \quad \alpha \text{ feasible.} \quad (24)$$

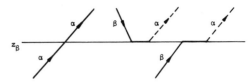

Fig. 2. Possible sample paths reaching or leaving a hedging point.

Fig. 3. Sample paths reaching or leaving largest hedging point.

This condition gives us M more equations. Since we have $NM + M$ equations in total we can find all the unknowns.

IV. OPTIMUM HEDGING POINTS

For tentative hedging points $z_\alpha, z_\beta, \cdots, z_\gamma$ we can calculate the steady-state PDF of the surplus X in each region and also the probability masses at the hedging points. The average cost per unit time is then given by

$$J^* = J^*(z_\alpha, z_\beta, \cdots, z_\gamma) = \int_{-\infty}^{\infty} g(x) f(x) \, dx + \sum_\delta g(z_\delta) \cdot P(z_\delta).$$

$$(25)$$

Here $g(x)$ is the cost (per unit time) of maintaining the surplus at level x. The function $J^*(z_\alpha, z_\beta, \cdots, z_\gamma)$ can be minimized over $z_\alpha, z_\beta, \cdots, z_\gamma$ to find the optimum values of the hedging points.

The minimization must be done subject to constraints of the form $z_\alpha \geq z_\beta$, to impose the order presumed for the hedging points. The function $J^*(z_\alpha, z_\beta, \cdots, z_\gamma)$ is in general nonlinear and the constraints are linear with a special structure. An appropriate algorithm for solving this problem would be the Frank–Wolfe algorithm. Clearly, the solution obtained depends on the order assumed for the tentative HP's $z_\alpha, z_\beta, \cdots, z_\gamma$, and may not be the true optimum. One may solve the problem for all plausible HP orders to find the true optimum. It may be possible, however, to use the solution obtained for a given order to choose the next HP order in such a way that the true optimum is found more efficiently. In particular, if it were assumed that $z_\alpha \geq z_\beta$ and in the optimum solution $z_\alpha = z_\beta$, this would be an indication that the right order is probably $z_\beta \geq z_\alpha$.

V. EXAMPLE

We apply the results derived in the previous sections to an example with three machine states (Fig. 4). The transition rate matrix is

$$\Lambda = \begin{pmatrix} -4 & 2 & 2 \\ 2 & -3 & 1 \\ 1 & 1 & -2 \end{pmatrix}.$$

The maximum production rate for the three states and the demand rate are:

$$\hat{u}_1 = 2, \quad \hat{u}_2 = 2, \quad \hat{u}_3 = 0, \quad d = 1.$$

Thus, states 1 and 2 are feasible and state 3 is infeasible. The steady-state probability of the states is found to be $\pi_1 = 5/19$, $\pi_2 = 6/19$, $\pi_3 = 8/19$. The overall system is strictly feasible, since $\sum_{i=1}^{3} \pi_i(\hat{u}_i - d) = 3/19$ is positive.

There are two hedging points, z_1 and z_2, corresponding to states 1 and 2, and we assume that $z_1 > z_2$. This assumption is plausible

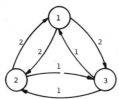

Fig. 4. The continuous-time Markov chain model of the production system in the example. Numbers on arcs are transition rates.

because state 1 is more susceptible to failure than state 2 ($\lambda_{13} > \lambda_{23}$), and the repair rate is the same for both states ($\lambda_{31} = \lambda_{32}$). There are two regions

$$R^1 = \{x | z_1 > x > z_2\},$$

$$R^2 = \{x | z_2 > x\}.$$

The optimum production strategy for each machine state is to produce at capacity if X is below the corresponding hedging point, and to produce nothing if it is above it. Therefore,

$$v_1^1 = \hat{u}_1 - d = 1, \quad v_1^2 = \hat{u}_1 - d = 1,$$
$$v_2^1 = 0 - d = -1, \quad v_2^2 = \hat{u}_2 - d = 1,$$
$$v_3^1 = 0 - d = -1, \quad v_3^2 = 0 - d = -1.$$

The matrices G^1 and G^2 corresponding to regions R^1 and R^2 are

$$G^1 = \left(\frac{\lambda_{\alpha\beta}}{|v_\alpha^1|} \text{sgn } v_\beta^1\right) = \begin{pmatrix} -4 & -2 & -2 \\ 2 & 3 & -1 \\ 1 & -1 & 2 \end{pmatrix},$$

$$G^2 = \left(\frac{\lambda_{\alpha\beta}}{|v_\alpha^2|} \text{sgn } v_\beta^2\right) = \begin{pmatrix} -4 & 2 & -2 \\ 2 & -3 & -1 \\ 1 & 1 & 2 \end{pmatrix}.$$

G^1 has eigenvalues 0, 3.54, and -2.54, and the eigenvalues of G^2 are 0, -5.54, and 0.54. After writing (12) for $m_1(x)$, $m_2(x)$, and $m_3(x)$ and solving the boundary equations (18)–(22) to calculate the constants, we find, with $\delta = z_1 - z_2$, that

$$m_1(x) = \begin{cases} \dfrac{1}{e^{2.54\delta} - 0.13}(0.0016 e^{3.54(x-z_1)} + 0.234 e^{-2.54(x-z_1)}) \\ \quad \text{if } x \in R^1, \\ \dfrac{e^{2.54\delta}}{e^{2.54\delta} - 0.13}(0.091 e^{0.54(x-z_2)}) \\ \quad \text{if } x \in R^2, \end{cases}$$

$$m_2(x) = \begin{cases} \dfrac{1}{e^{2.54\delta} - 0.13}(0.011 e^{3.54(x-z_1)} + 0.107 e^{-2.54(x-z_1)}) \\ \quad \text{if } x \in R^1, \\ \dfrac{e^{2.54\delta}}{e^{2.54\delta} - 0.13}(0.107 e^{0.54(x-z_2)}) \\ \quad \text{if } x \in R^2, \end{cases}$$

$$m_3(x) = \begin{cases} \dfrac{1}{e^{2.54\delta} - 0.13}(0.014 e^{3.54(x-z_1)} + 0.127 e^{-2.54(x-z_1)}) \\ \quad \text{if } x \in R^1, \\ \dfrac{e^{2.54\delta}}{e^{2.54\delta} - 0.13}(0.200 e^{0.54(x-z_2)}) \\ \quad \text{if } x \in R^2. \end{cases}$$

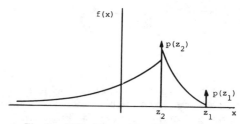

Fig. 5. PDF of the surplus X in the example.

The PDF of X is obtained by substituting these functions in (9). Including the probability masses at z_1 and z_2 as impulses $\delta(x - z_1)$ and $\delta(x - z_2)$ in $f(x)$ we have

$$f(x) = \begin{cases} \dfrac{0.468}{e^{2.54\delta} - 0.13}[e^{-2.54(x-z_1)} - 0.001 e^{3.54(x-z_1)} \\ \quad + 0.126\delta(x-z_1)] \quad \text{if } z_1 \geq x > z_2 \\ \dfrac{0.398 e^{2.54\delta}}{e^{2.54\delta} - 0.13}[e^{.54(x-z_2)} + 0.181\delta(x-z_2)] \\ \quad \text{if } x \leq z_2. \end{cases}$$

A discrete-event simulation was carried out for this example to confirm the validity of the results. In this simulation the probability of X being at the hedging points z_1 and z_2 as well as the probability of X being in regions R^1 and R^2 were measured by time averaging over 10 000 machine transitions. The same quantities were calculated from the equations obtained above and are compared to the simulation results (for 3 values of $\delta = z_1 - z_2$) in Table I. The results from the two methods are very close.

Assuming the cost function is linear with slopes $c^+ > 0$ and $c^- > 0$ for inventory and backlog

$$g(x) = \begin{cases} c^+ x & \text{if } x \geq 0, \\ -c^- x & \text{if } x \leq 0. \end{cases}$$

We calculate $J^*(z_1, z_2)$ as

$$J^*(z_1, z_2) = -c^- \int_{-\infty}^0 xf(x)\, dx + c^+ \int_0^\infty xf(x)\, dx$$

$$= -(c^+ + c^-) \int_{-\infty}^0 xf(x)\, dx + c^+ \int_{-\infty}^\infty xf(x)\, dx$$

$$= -(c^+ + c^-) \int_{-\infty}^0 xf(x)\, dx + c^+ E\{X\}$$

where $E\{X\}$ is the expectation of the surplus X. Evaluating the integrals and simplifying, we get

$$J^*(z_1, z_2) \cong \frac{c^+}{e^{2.54\delta} - 0.13}\left(\left(1.37\left(1 + \frac{c^-}{c^+}\right)e^{-0.54 z_2} + z_2 - 1.3\right)\right.$$
$$\left. \cdot e^{2.54\delta} - 0.126(z_2 + \delta) - 0.073\right).$$

We ignore the constraint $z_1 \geq z_2$ and minimize $J^*(z_1, z_2)$ with no constraints. We will see that $z_1 \geq z_2$ will be met automatically. We set $\partial J^*/\partial z_2$ and $\partial J^*/\partial \delta$ equal to zero to find optimum values for z_2 and δ. These equations yield

$$z_2 = 1.85 \log\left(\frac{1 + (c^-/c^+)}{1.35 - 0.17 e^{-2.54\delta}}\right)$$

$$e^{2.54\delta}(0.32\delta - 0.024 z_2 - 0.129) + 0.097 = 0.$$

Optimum values of z_2 and $z_1 = z_2 + \delta$ depend on the ratio (c^-/c^+). If $c^- \gg c^+$, z_1 and z_2 increase with $\log(1 + c^-/c^+)$ as shown in Fig. 6.

TABLE I
ANALYTICAL AND SIMULATION RESULTS FOR THE EXAMPLE IN SECTION V A = ANALYTIC, S = SIMULATION

$\delta = z_1 - z_2$	$P\{x \text{ in } R_1\}$		$P\{x \text{ in } R_2\}$		$P(z_1)$		$P(z_2)$	
	A	S	A	S	A	S	A	S
0.5	0.138	0.132	0.770	0.777	0.017	0.016	0.074	0.076
1.0	0.172	0.171	0.769	0.748	0.005	0.005	0.072	0.076
1.5	0.181	0.182	0.743	0.741	0.001	0.001	0.072	0.076

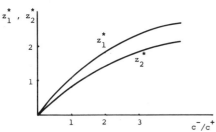

Fig. 6. Optimum HP's as functions of c^-/c^+.

VI. Conclusion

In this paper we have derived equations for the steady-state probability distribution of the surplus level in a single product manufacturing system with machine failures, when a hedging point control policy is used. The results make it possible to solve the optimum production scheduling of such systems with arbitrary number of failure states. The results are expected to be very valuable for solving the scheduling problem of the multiproduct case in flexible manufacturing systems.

Acknowledgment

The author is grateful to S. B. Gershwin for introducing him to the problem studied here and for his valuable comments on the manuscript. He is also indebted to his colleague P. H. Algoet for suggestions that led to substantial improvements in the presentation of Section III. Finally he thanks M. Caramanis for his valuable comments.

References

[1] J. G. Kimemia and S. B. Gershwin, "An algorithm for the computer control of production in flexible manufacturing systems," *IIE Trans.*, vol. AC-15, pp. 353–362, Dec. 1983.
[2] S. B. Gershwin, R. Akella, and Y. F. Choong, "Short-term production scheduling of an automated manufacturing facility," *IBM J. Research Development,* vol. 29, no. 4, July 1985.
[3] R. Akella and P. R. Kumar, "Optimal control of production rate in a failure prone manufacturing system," *IEEE Trans. Automat. Contr.*, vol. AC-31, no. 2, Feb. 1986.
[4] T. Bielecki and P. R. Kumar, "Optimality of zero-inventory policies for unreliable manufacturing systems," *Operat. Res.* (Special Issue in Manufacturing Systems), 1987.
[5] R. Akella, Y. Choong, and S. B. Gershwin, "Real-time production scheduling of an automated cardline," *Ann. Operat. Res.*, vol. 3, pp. 403–425, 1985.

Ali Sharifnia was born in Iran in 1948. He received the B.S. degree in electrical engineering from Tehran University of Technology, Tehran, Iran, in 1970, and the M.S. degree in industrial engineering, and the Ph.D. degree in engineering-economic systems, both from Stanford University, Stanford, CA.

He is currently an Assistant Professor of Manufacturing Engineering at Boston University, Boston, MA. His research interests are in capacity planning, scheduling, and control of manufacturing systems.

A Dynamic Optimization Model for Integrated Production Planning: Computational Aspects

WILLIAM A. GRUVER, MEMBER, IEEE, AND S. L. NARASIMHAN

Abstract—The formulation and algorithmic solution of a class of optimal control models is discussed for integrated production, inventory, and research and development (R&D) planning. The model is based on the constrained minimization of the cost of labor, capital, and R&D engineering subject to a production constraint utilizing a Cobb–Douglas type input–output function. The solution technique utilizes a modification of a gradient projection algorithm due to Demyanov and Rubinov. Numerical results include a sensitivity analysis of the cost structures.

I. Introduction

OPTIMAL control deals with the characterization of optimality and its computation for infinite dimensional minimization problems usually involving differential or difference equations for constraints, as described by Pontryagin et al. [1], and Canon et al. [2]. Although applications in traditional engineering disciplines are numerous, optimal control theory and particularly algorithmic methods of solution have received limited attention in production planning [3]–[10].

Manuscript received November 22, 1978; revised July 3, 1979.
W. A. Gruver was with North Carolina State University, Raleigh, NC. He is now with Logistics Technology International, Ltd., 2707 Toledo St., Suite 603, Torrance, CA 90503.
S. L. Narasimhan is with the Department of Management Science, University of Rhode Island, Kingston, RI 02881.

We consider deterministic models of production systems for manufacturing products of known demand. The products consume available resources known as "factors of production" which include labor, capital, and research and development (R&D) engineering. The input–output relationship is defined by an assumed or empirically verified production function. Decisions are to be made concerning the factors of production and the product mix for given objectives of the management.

A familiar example of this type of problem is the model due to Holt, Modigliani, Muth, and Simon (HMMS) [11]. The HMMS model determines the production rate and the workforce level for given demand while minimizing the cost of inputs and inventory. However, HMMS and other related models do not include the possibility of increasing the production rate by applied R&D engineering activities of the firm during the medium/long term planning period. Typically, engineering may include research in product and process engineering, industrial engineering, equipment, and labor training activities. Nelson and Winter [12] attribute the increased productivity toward improved inputs and capital equipment which often involve innovation.

According to Nelson and Winter [12] the conclusions of Mansfield were robust under different specifications; however, the magnitude displayed some sensitivity to the exact specification. The use of production functions for modeling the innovation process has been recently criticized by various researchers because the production function does not prescribe how to arrive at the goal. It also has been proposed that mechanization, when possible, provides an easy approach for innovation. Similarly, latent scale economics provide a route that is easy to follow [13]. This tendency to mechanize has been suggested by Piori [14] and documented by Setzer [15] in their work on the evolution of production processes at Western Electric. Nelson and Winter also point out that since models involving innovation vary greatly between sectors, uncertainty should be incorporated into the modeling process. At the present time, however, models which explicitly characterize uncertainty are not widely developed. The results described in this paper will be limited to deterministic models applicable to manufacturing processes, such as encountered in the chemical, drug, and electronic industries. In contrast to an earlier study based on a linear-quadratic formulation [10], these results incorporate a Cobb–Douglas type production function in the model.

Section II discusses the production function. Sections III and IV are concerned with a statement of the optimal control model and the characterization of optimality. In Section V we develop a gradient projection method which is effective for solving the model. Finally, Section VI presents a numerical example and sensitivity investigation of the model.

II. Production Functions

Production functions have been used extensively by economists in the theory of the firm. Applications are treated by Leontief [16] in planning and forecasting, Solow [17] in growth theory, Smith [18] in production planning, Griliches [19] in agricultural production, and Arrow and Kurz [20] in investment and returns.

Most empirical studies involving production functions can be classified according to the following two types: the Leontief input–output model

$$h_1(L, K) = a_1 L + a_2 K$$

or the Cobb–Douglas function

$$h_2(L, K) = L^\alpha K^\beta$$

where L and K are inputs of labor and capital, respectively, and the coefficients of the function are nonnegative numbers.

Mansfield et al. [21] proposed the following Cobb–Douglas type production function for planning R&D expenditures in the chemical and drug industries:

$$h_3(L, K, R) = Ae^{at}\left[\int_0^t e^{-\lambda \tau} R(\tau)\, d\tau\right]^c L(t)^a K(t)^{1-a} \quad (1)$$

where $R(\tau)$ represents engineering expenditures at time τ. Subsequently, through studies in the electronic manufacturing industries, Lele and O'Leary [22] approximated (1) by

$$h_4(L(t), K(t), E(t)) = L(t)^\alpha K(t)^\beta E(t)^\gamma \quad (2)$$

where $E(t)$ represents the cumulative engineering input at time t. However, use of this function in other industries may require extensive model verification and should be applied with caution. The production function (2) is based on the assumption that labor and capital are consumed as products are manufactured; however, engineering R&D contributes cumulatively towards production over the planning period.

III. Optimal Control Model

We consider a dynamical system model of a production system subject to the following assumptions: 1) the system is able to overcome all delays in processing of the product and, hence, there are no in-process goods; 2) no defective goods are manufactured; 3) the required amounts of inputs are delivered at the proper time and proper place; 4) inventory is not damaged or stolen, and does not become obsolete. The model described in this section is applicable to manufacturing processes such as occur in the chemical and electronic industries. Although the above assumptions may be restrictive in certain cases, appropriate modifications of the production-inventory constraint can be incorporated into the model and solution technique to accommodate specific planning situations.

The performance of the system is evaluated by a scalar valued objective which is composed of the cost of inputs, and quadratic penalties for backlog or inventory. It is required to minimize the objective

$$F(u) = \int_{t_0}^{t_1} \{c_1 u_1(t) + c_2 u_2(t) + c_3 u_3(t) + c_4 x_1(t)^2\}\, dt + c_5 x_1(t_1)^2 \quad (3)$$

while satisfying the system constraints

$$\dot{x}_1(t) = u_1(t)^\alpha u_2(t)^\beta x_2(t)^\gamma - d(t)$$
$$\dot{x}_2(t) = u_3(t) \quad (4)$$

where the dot over the function denotes derivative with respect to time, and subject to magnitude restrictions

$$\underline{u}_i \leq u_i(t) \leq \bar{u}_i, \quad t \in [t_0, t_1], \quad i = 1, 2, 3$$

where $x_1(t)$ and $x_2(t)$ represent the inventory and engineering R&D levels, respectively, for $t \in [t_0, t_1]$; $u_1(t), u_2(t)$, and $u_3(t)$ represent the rates of labor, capital, and engineering, respectively. The initial condition $x(t_0)$, the planning period $[t_0, t_1]$ and the demand rate $d(t)$ for $t \in [t_0, t_1]$ are specified. The coefficients c_1, c_2, and c_3 in the objective represent unit costs of labor, capital, and engineering for the normalized input variables u_1, u_2, and u_3, respectively. The costs c_4 and c_5 represent weighting factors for the quadratic penalty of the inventory or backlog. Parameters α, β, and γ are empirically determined technological coefficients of the production function.

IV. Characterization of an Optimal Control

The model described in the preceding section can be treated under the following general class of optimal control problems. Determine an optimal control function $\hat{u} \in U$ such that $F: U \to \mathbb{R}$ defined by

$$F(u) = \int_{t_0}^{t_1} L(x(t), u(t), t) \, dt + K(x(t_1)) \quad (5)$$

is minimized over U subject to

$$\dot{x}(t) = f(x(t), u(t), t), \quad x(t_0) = x_0 \in \mathbb{R}^n \quad (6)$$

where U is the space of admissible control functions $u(\cdot)$ having values in $\mathbb{R}^m$, and which are essentially bounded and measurable on $[t_0, t_1]$. Under suitable assumptions on the continuity of $L(x, u, t)$, $K(x)$, and $f(x, u, t)$ and its derivatives, the maximum principle of Pontryagin [1] characterizes an optimal control $\hat{u}$ and trajectory $\hat{x}$.

For the purposes of algorithmic implementation it is convenient to express the maximum principle in the following form.

Theorem 1

Given the optimal control problem described by (5) and (6), if $\hat{u} \in U$ is optimal then it is necessary that there exists a nonzero vector $\hat{p}(t) \in \mathbb{R}^n$, the solution of

$$\frac{d}{dt} \hat{p}(t) = -\frac{\partial H}{\partial x}(\hat{x}(t), \hat{p}(t), \hat{u}(t), t) \quad (7)$$

$$\hat{p}(t_1) = \frac{\partial K}{\partial x}(\hat{x}(t_1)) \quad (8)$$

such that the Frechet differential of F at $\hat{u}$ satisfies,

$$F'_{\hat{u}}(u - \hat{u}) = \int_{t_0}^{t_1} \frac{\partial H}{\partial u}(\hat{x}(t), \hat{p}(t), \hat{u}(t), t)$$

$$\cdot [u(t) - \hat{u}(t)] \, dt$$

$$\geq 0, \quad \text{for all } u \in U \quad (9)$$

where $H: \mathbb{R}^n \times \mathbb{R}^n \times \mathbb{R}^m \times [t_0, t_1] \to \mathbb{R}$ is defined by

$$H(x, p, u, t) = L(x, u, t) + p^T f(x, u, t). \quad (10)$$

Thus the vector $(\partial H/\partial u)(\hat{x}, \hat{p}, \hat{u}, t)$ represents the gradient ∇F in the normed space $L_\infty[t_0, t_1]$ on which is defined the standard L_2 inner product and derived norm $\|\cdot\|$.

In view of the particular constraint set of the model which can be expressed as

$$U = \{u \in L_2^m[t_0, t_1] | \underline{u} \leq u(t) \leq \bar{u},$$

$$t \in [t_0, t_1] \text{ almost everywhere}\}, \quad (11)$$

Theorem 1 implies that an optimal control satisfies for $t \in [t_0, t_1]$ almost everywhere,

$$\hat{u}(t) = \begin{cases} \underline{u}, & \frac{\partial H}{\partial u}(\hat{x}(t), \hat{p}(t), \hat{u}(t), t) > 0 \\ \underline{u} \leq u^*(t) \leq \bar{u}, & \frac{\partial H}{\partial u}(\hat{x}(t), \hat{p}(t), \hat{u}(t), t) = 0 \quad (12) \\ \bar{u}, & \frac{\partial H}{\partial u}(\hat{x}(t), \hat{p}(t), \hat{u}(t), t) < 0. \end{cases}$$

Thus an optimal control attains its extreme values except for the singular case in which the gradient vanishes on a subinterval of $[t_0, t_1]$ and the control bounds may be inactive.

V. Gradient Projection Algorithm

This section describes an algorithm for minimizing a differentiable functional F on a convex, weakly compact subset U of Hilbert space H; subsequently, the method will be applied to the production system model of Section III. The algorithm is originally due to Demyanov and Rubinov [23], who also prove convergence under the condition of an exact one-dimensional minimization for the step length. However, determination of an accurate step length in optimal control problems can be expensive because each evaluation of the objective functional and the gradient generally requires numerical integration of differential equations (6) and (7). Convergence and rate of convergence of the algorithm with an inexact step length, and its limiting case the conditional gradient method, are proved in Gruver and Sachs [24].

The method is based on projecting the unconstrained gradient descent point

$$u^i - \gamma_i \nabla F(u^i), \quad \text{for some } \gamma_i \in (0, \infty)$$

onto the feasible set. The projection is described by an operator $P: \mathcal{H} \to U$ defined by

$$\|Pv - v\| \leq \|u - v\|, \quad \text{for all } u \in U. \quad (13)$$

For each closed bounded convex set U, the vector Pv exists for all $v \in \mathcal{H}$, and satisfies

$$\langle Pv - v, u - Pv \rangle \geq 0, \quad \text{for all } u \in U. \quad (14)$$

We obtain, therefore, the following theorem.

Theorem 2

Let $F: \mathcal{H} \to U$ be Frechet differentiable. If $\hat{u} \in U$ is optimal, then for each $\alpha \in (0, \infty)$

$$P(\hat{u} - \alpha \nabla F(\hat{u})) = \hat{u}. \quad (15)$$

If, in addition, F is convex on U, the reverse implication also holds.

Define the normalized steepest descent direction

$$\tilde{u}^i = -\nabla F(u^i)/\|\nabla F(u^i)\|. \quad (16)$$

Selecting a fixed $\lambda \in (0, \infty)$, $u^i + \lambda \tilde{u}^i$ is projected onto U which yields the point

$$w^i = P(u^i + \lambda \tilde{u}^i). \quad (17)$$

Then, by (14) and using $u = u^i$, we obtain that

$$0 \leq \langle w^i - u^i - \lambda \tilde{u}^i, u^i - w^i \rangle$$

$$= -\|u^i - w^i\|^2 - \langle \lambda \tilde{u}^i, u^i - w^i \rangle$$

which can be written

$$\langle \nabla F(u^i), (w^i - u^i) \rangle \leq -\frac{1}{\lambda} \|\nabla F(u^i)\| \|w^i - u^i\|^2$$

$$< 0$$

TABLE I
INPUT DATA FOR NUMERICAL EXAMPLE

Initial States			
Inventory	$x_1(0) = 50$		
Engineering	$x_2(0) = 50$		
Control Bounds	lower, $\underline{u}_i$	upper, $\bar{u}_i$	
Labor, u_1	50.	65.	
Capital, u_2	45.	60.	
Engrg. rate, u_3	20.	25.	
Technological Coefficients			
$\alpha = .5$			
$\beta = .3$			
$\gamma = .2$			
Cost Coefficients			
Labor	$c_1 = (2., 2.5, 3., 3.5)$		
Capital	$c_2 = (5., 4.5, 4., 3.5)$		
Engrg. cost	$c_3 = (8., 15., 25.)$		
Inven. penalty	$c_4 = (1., 2., 3., 5., 10., 15., 25.)$		
Term. inven. penalty	$c_5 = 0$.		
Demand (per period)			
100,100,90,70,100,100,80,60,75,90,100,85,100,80,50,100,			
90,110,70,80,60,100,95,70,85,65,95,80,100,70,			
90,85,65,70,90,100,85,100,100,80			
Planning Horizon (40 subdivisions)			
$t_0 = 0$			
$t_1 = 2$			

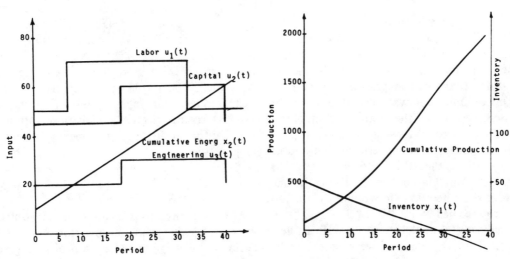

Fig. 1. Converged trajectories.

if u^i is not optimal and $u^i \neq w^i$. Under this assumption we obtain descent and, hence, feasibility of the iterate

$$u^i + \alpha(w^i - u^i) \in U, \quad \text{for all } \alpha \in [0, 1]$$

since U is convex.

Gradient Projection Algorithm

1) Select $u^0 \in U$, $\lambda > 0$, $\varepsilon > 0$ and set $i = 0$.
2) Determine the direction vector $\tilde{u}^i$, $\|\tilde{u}^i\| \leq 1$ such that $\langle \nabla F(u^i), \tilde{u}^i \rangle \leq \langle \nabla F(u^i), u \rangle$ for all $u \in U$, $\|u\| \leq 1$.
3) Determine $w^i \in U$ by $w^i = P(u^i + \lambda \tilde{u}^i)$ and set $v^i = w^i - u^i$.
4) If $\|w^i - u^i\| \leq \varepsilon$, stop.
5) Compute $\alpha_i \in [0, 1]$ such that $F(u^i + \alpha_i v^i) \leq F(u^i + \alpha v^i)$ for all $\alpha \in [0, 1]$.
6) Update u^i by $u^{i+1} = u^i + \alpha_i v^i$.
7) Increment i by one and go to step 2).

We remark that the constraint set (11) yields the projection defined by

$$(Pu)(t) = \text{sat } \{u(t)\} = \begin{cases} \underline{u}, & u(t) < \underline{u} \\ u(t), & \underline{u} \leq u(t) \leq \bar{u} \\ \bar{u}, & u(t) > \bar{u}. \end{cases}$$

Incorporating the gradient characterization in (9), we obtain the following implementation of the gradient projection algorithm for optimal control.

Optimal Control Algorithm

1) Select $u^0(t)$, $t \in [t_0, t_1]$; $\varepsilon > 0$; $\lambda > 0$; and set $i = 0$.
2a) Determine $x^i(t)$, $t \in [t_0, t_1]$ by integrating (6) forward from $x^i(t_0)$.
2b) Determine $p^i(t)$, $t \in [t_0, t_1]$ by integrating (7) backward from $p^i(t_1)$.
2c) Evaluate the negative gradient $\tilde{u}^i(t) = -g^i(t)/\|g^i\|$; $g^i(t) = (\partial H/\partial u)(x^i(t), p^i(t), u^i(t))$.
3) Determine the projection, $w^i(t) = \text{sat }\{u^i(t) + \lambda \tilde{u}^i(t)\}$, $v^i(t) = w^i(t) - u^i(t)$.
4) If $\|v^i\| \leq \varepsilon$, stop.
5) Compute $\alpha_i \in (0, 1]$ such that $F(u^i + \alpha_i v^i) \leq F(u^i + \alpha v^i)$, for all $\alpha \in [0, 1]$.
6) $u^{i+1}(t) = u^i(t) + \alpha_i v^i(t)$, $t \in [t_0, t_1]$.
7) Increment i by one and go to step 2).

The exact one-dimensional search in step 5) of the algorithms can be implemented by the step size rule due to Goldstein [25] which terminates within a finite number of iterations. Convergence and rate of convergence of the gradient projection algorithm with inexact search are treated in [24]. A modification of this algorithm is treated by Klein and Gruver [26] using a conjugate gradient modification. This latter approach has been especially useful for solving singular optimal control problems.

VI. Numerical Example

This section describes numerical experiments based on the model presented in Section III. A first order predictor-corrector scheme with a mesh size of 40 points was used to obtain a discrete-time optimal control problem with compatible adjoint. The optimization was carried out by the gradient projection version of the optimal control algorithm described in Section V, using an inexact step-length search. The computations were performed in double precision wordlength using the IBM370/165 facilities of the Triangle Universities Computation Center. Typically, one optimization required 34 iterations (direction updates) to obtain convergence specified by $\varepsilon = 10^{-5}$.

Data used in analyzing the model are displayed in Table I. The input data used for this example were chosen to be representative of expenses incurred by the electrical manufacturing industries during the period 1960–1970. Fig. 1 shows converged levels of labor, capital, engineering rate, and cumulative engineering during the planning period. The control is bang-bang with no singular arcs. Inventory $x_1(t)$ is rapidly driven to zero within the first half of the total interval, and then backlog accumulates. Although not studied in this example, the backlog could be decreased by appropriate selection of the penalty constant c_5.

The model was solved for three different sets of labor and capital costs: 1) $c_1 = 2.0$, $c_2 = 5.0$, 2) $c_1 = 2.5$, $c_2 = 4.5$, and 3) $c_1 = 3.5$, $c_2 = 3.5$. As the cost of capital equipment decreased more capital was used for the manufacturing of the products.

The cost of labor and capital equipment can be reasonably estimated in any manufacturing environment once the process has been established. However, considerable uncertainty exists in the cost of engineering and inventory. Therefore, the model was solved for different engineering and inventory costs. The terminal engineering, terminal inventory, and product cost

$$\sum_t \{c_1 u_1(t) + c_2 u_2(t) + c_3 u_3(t)\}$$

are displayed in Table II as a function of inventory costs for different costs of engineering input. Terminal backlog decreases initially and then increases slightly with inventory costs. Terminal inventory was relatively insensitive to changes in the cost of engineering. The backlog increases slightly as the cost of engineering is increased, implying that incurring additional backlog is cheaper than more production.

The objective function (3) is based on the assumption that the cost of inventory and backlog are equal. However, the cost of backlog may be lost sales which are more expensive than the cost of inventory. Such a modification in the objective can be accommodated by the gradient algorithm, assuming that the differentiability assumptions are satisfied, resulting in an inventory position instead of a backlog.

Table II indicates that terminal engineering initially increases as the cost of inventory increases and then slightly decreases as a function of the cost of inventory. Similarly the cost of product increases initially and then decreases slightly as a function of inventory (Fig. 2). This phenomenon can be attributed to the following consideration. As the inventory cost is increased, the backlog is decreased by more production, during which more inputs including engineering are consumed. The result of this change is reflected by the increased terminal engineering level and the increased total product cost, respectively. This observation is reasonable because the product cost also includes an increased cost of engineering.

VII. Conclusion

We have investigated an optimal control model for integrated production and inventory systems. Solutions for the model were obtained by a projected gradient descent algorithm with inexact step length search. Although other solution techniques such as dynamic programming can also be employed in this problem, the gradient projection algorithm described in this paper is easy to implement, requires relatively small core storage, and is specifically tailored to the structure of the dynamic constraints. Accuracy of the method can be increased by a smaller mesh size for the discretization and higher order integrators to obtain improved truncation error and stability [27].

No penalties were assigned for changes in the level of inputs; in practice, however, there is a cost associated with changes in hiring, training, and lay-offs. Such input derivative costs can be easily accommodated in the formulation without significant modification to the algorithm. Similarly, changes in the system constraints can be incorporated. For example, Mansfield [21] has proposed that productivity of engineering input decreases over the planning period, by

TABLE II
SENSITIVITY OF CONVERGED COSTS

c_1	c_2	c_3	c_4	Term. Inv. $x_1(t_1)$	Term. Engr. $x_2(t_1)$	Product Cost $\sum_t \{c_1 u_1(t) + c_2 u_2(t) + c_3 u_3(t)\}$	Inv. Cost $\sum_t c_4 x_2(t)^2$
3	4	8	1	−25.4	56.7	1065	1081
			2	−23.9	58.9	1097	2135
			3	−23.1	59.5	1107	3181
			5	−21.5	59.7	1119	5252
			10	−16.9	59.5	1144	10200
			15	−18.4	59.5	1135	25400
			25	−19.8	58.5	1113	25485
3	4	15	1	−25.7	54.0	1322	1085
			2	−24.3	55.6	1363	2145
			3	−23.2	58.0	1403	3187
			5	−21.6	59.1	1430	5257
			10	−16.9	59.0	1455	10205
			15	−18.4	59.5	1453	15405
			25	−19.8	58.5	1424	25484
3	4	25	1	25.5	54.0	1723	1084
			2	−25.8	54.0	1739	2147
			3	−23.7	54.0	1745	3203
			5	−21.8	57.7	1846	5267
			10	−16.9	58.1	1885	10204
			15	−18.4	59.0	1896	15411
			25	−19.9	58.1	1859	25493

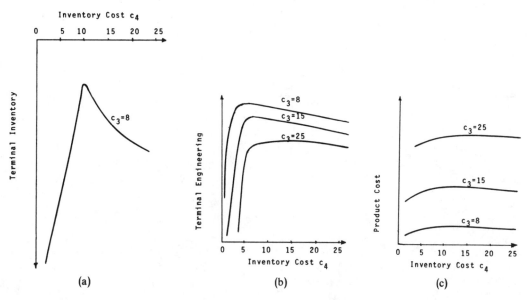

Fig. 2. Sensitivity of cost structures.

inclusion of a suitable discount factor in the production function.

This study has addressed optimal control in single product planning. The model may also be extended to encompass the multi-product situation [28]. However, measurement problems make interpretations of statistical results and empirical studies difficult. Therefore, parameters of production functions should be developed for particular applications and used with care in planning situations.

REFERENCES

[1] L. S. Pontryagin, V. G. Boltyanski, R. V. Gamkrelidze, and E. F. Mishenko, *The Mathematical Theory of Optimal Processes.* New York: Wiley Interscience, 1962.

[2] M. D. Canon, C. D. Cullum, and E. Polak, *Theory of Optimal Control and Mathematical Programming.* New York: McGraw-Hill, 1974.

[3] S. Axsäter, "Coordinating control of production-inventory systems," *Int'l. J. Prod. Res.*, vol. 14, no. 6, pp. 669–688, 1976.

[4] E. S. Lee and M. A. Shaikh, "Optimal production planning by a gradient technique, I. First variations," *Manag. Science*, vol. 16, no. 1, pp. 109–117, 1969.

[5] Z. Lieber, "An extension to Modigliani and Hohn's planning horizon results," *Manag. Sci.*, vol. 20, pp. 319–330, Nov. 1973.

[6] D. Pekelman, "Production smoothing with fluctuating price," *Manag. Science*, vol. 21, no. 5, pp. 576–590, 1975.

[7] P. R. Kleindorfer, C. H. Kriebel, G. L. Thompson, and G. B. Kleindorfer, "Discrete optimal control of production plans," *Manag. Sci.*, vol. 22, pp. 261–273, Nov. 1975.

[8] I. Adiri and A. Ben-Israel, "An extension and solution of Arrow-Karlin type production models by the Pontryagin Maximum Principle," *Cahiers du Centre d'Etudes de Recherche Operationelle*, vol. 8, no. 3, pp. 147–158, 1966.

[9] A. Bensoussan, E. G. Hurst, and B. Näslund, *Management Applications of Modern Control Theory.* New York: North-Holland/Elsevier, 1974.

[10] S. L. Narasimhan and W. A. Gruver, "Optimal control in integrated R&D, production and inventory planning systems," *AIEE Trans.*, vol. 11, no. 3, 1979.

[11] C. Holt, F. Modigliani, J. Muth, and H. Simon, *Planning Production,*

Investment and Workforce. Englewood Cliffs, NJ: Prentice-Hall, 1960.

[12] R. R. Nelson and S. G. Winter, "In search of useful theory of innovation," *Research Policy*, vol. 16, pp. 36–72, 1977.

[13] R. R. Nelson, M. J. Peck, and E. D. Kalacheck, *Technology, Economic Growth and Public Policy.* Washington, DC: Brookings, 1967.

[14] M. Piori, "The impact of the labor market upon the design and selection of production techniques within the manufacturing plant," *Quarterly J. of Economics*, vol. 82, p. 602, Nov. 1968.

[15] F. Setzer, "Technical change over the life of a product: Changes in skilled inputs and production processes," Unpublished Ph.D. dissertation, Yale Univ., New Haven, CT, 1974.

[16] W. Leontief, *Structure of the American Economy 1919–1939.* New York: Oxford Univ., 1951.

[17] R. M. Solow, *Growth Theory.* New York: Oxford Univ., 1970.

[18] V. Smith, *Theory of Investment and Production.* Cambridge, MA: Harvard Univ., 1961.

[19] Z. Griliches, "Research expenditure, education and the aggregate agricultural production functions," *Amer. Econ. Rev.*, vol. 54, no. 6, pp. 961–974, 1964.

[20] K. J. Arrow and M. Kurz, *Public Investment, the Rate of Return and Optimal Fiscal Policy.* Baltimore, MD: Johns Hopkins Univ., 1971.

[21] E. Mansfield *et al., Research and Innovation in the Modern Corporation*, New York: Norton, 1971.

[22] P. T. Lele and J. W. O'Leary, "Applications of production functions in management decisions," *AIIE Trans.*, vol. 4, pp. 36–42, Mar. 1972.

[23] V. F. Demyanov and A. M. Rubinov, *Approximate Methods in Optimization Problems.* New York: Elsevier, 1970.

[24] W. A. Gruver and E. Sachs, *Algorithmic Methods in Optimal Control, Research Notes in Mathematics.* London: Pitman, 1980.

[25] A. A. Goldstein, *Constructive Real Analysis.* New York: Harper and Row, 1967.

[26] C. F. Klein and W. A. Gruver, "Gradient projection methods for constrained optimal control," Operations Research Program, North Carolina State Univ., Raleigh, NC, OR Report No. 144, Feb. 1979.

[27] H. J. Kelley and W. F. Denham, "Modeling and adjoints for continuous systems," *J. Optimiz. Theory and Appl.*, vol. 3, pp. 174–183, Mar. 1969.

[28] S. L. Narasimhan, "Models for integrated research and development, production and inventory planning," Ph.D. dissertation, Ohio State Univ., 1973.

Chapter 4: Discrete Event Dynamic Systems

4.1: Introduction

One could easily claim that the whole problem of modeling and control of automated manufacturing systems has its roots in discrete-event dynamic systems. In a manufacturing system, machine failures and repairs, sudden changes in demand, and machine blockage or starvation are all events occurring at discrete instants in time. Furthermore, the occurrence of these events has an impact on the dynamic response of the manufacturing system. The initiation (or termination) of one of these events can propagate from machine to machine affecting numerous performance measures. It is the interaction of these events over time which is the difficult aspect of *discrete-event dynamic systems* (DEDS).

Perturbation analysis is one method for modeling and analyzing discrete-event dynamic systems. Another is the Petri net approach, which is described in the next chapter.

Perturbation analysis had its accidental beginning in 1977 [1] while researchers at Harvard University were studying the FIAT 131 engine production-line monitoring system [2]. This work resulted in several basic concepts that capture the flavor of perturbation analysis.

4.2: The Basic Idea of Perturbation Analysis

The FIAT production-line research lead to an efficient technique for calculating gradients (or sensitivities) in a serial-transfer line. This helps to find answers to a lot of "what if" questions.

First of all, what does the gradient mean for a discrete-event dynamic system? The answer to this question is related to the *perturbation of events*. For example, if a buffer size in a serial-transfer line is increased by one unit, then this perturbation will result in a gain in local production at machine M_i. This is turn may *propagate* through the line and result in a change in line production. This change may not always occur, which can be seen by considering a transfer line whose last machine has an extremely high failure rate. Thus, we would like to know when a gain in local production at machine M_i results in a gain in line production.

Intuitively, we can talk about a gradient

$$\frac{\Delta \tau}{\Delta b_i} = \text{expected gain in production } time \text{ due to a unit increase in the buffer size } b_i$$

where M_i feeds b_i. This gradient can be interpreted by noting that the addition of one buffer-storage unit at b_i allows M_i to produce the same number of parts in less time. This represents a local gain in production, which tends to propagate workpieces downstream. Similarly, since M_i can pull pieces from b_{i-1}, buffer vacancies can be propagated upstream. Depending upon the status of the other machines in the transfer line, these propagations may result in a line gain. It is the sensitivity of the line production gain to buffer size changes in which we are ultimately interested. Consequently, we wish to compute the sensitivity

$$\frac{\text{increase of transfer-line production}}{\text{buffer size increase at each buffer location}}$$

and then allocate the buffer size at each location to maximize some performance index.

Extensive simulation is one way to approach this problem. Suppose there are N machines and $N - 1$ buffers. Then one would need $N - 1$ simulations to compute the $N - 1$ elements of the gradient vector. A key contribution of perturbation analysis is the fact that it can determine the $N - 1$ gradients from *one* sample production-line history.

Perturbations are not just restricted to integer increases in buffer sizes.

4.3: Finite and Infinitesimal Perturbations

The previous discussion dealt with the sensitivity of the transfer line production to a buffer level change. This is an example of *finite perturbations* in which perturbation propagation rules (to be developed later) are used to determine the sensitivity of a performance measure with respect to parameters that take on discrete values.

On the other hand, there are many continuous parameter changes in a transfer line that could trigger a series of discrete events. For example, if the repair rate could be increased at machine M_i, then this would also result in a local production gain at M_i which may or may not propagate to the end of the line. Thus, *infinitesimal perturbations* are used to obtain the sensitivity of a performance measure with respect to continuous parameter changes.

4.4: "A Gradient Technique for General Buffer Storage Design in a Production Line" (Y. C. Ho, M.A. Eyler, and T.T. Chien)

This paper presents a method for computing the sensitivity $\Delta\tau/\Delta b_i$ discussed in Section 4.2. It also provides a set of rules that explain when a gain in production at machine M_i propagates to the end of the transfer line and results in an overall production gain for the system. The paper has been included here because it establishes the flavor of perturbation analysis.

Consider a transfer line with N machines, each with given failures and repair transition probabilities where

$1/q_i$ = mean time between failure of the i-th machine

$1/a_i$ = mean time to repair the i-th machine

Also, there are $N-1$ buffers between the machines. The problem is to choose the size of the buffer storages $b_1, b_2, \ldots b_{N-1}$ to maximize the production-line efficiency for given $q_i, a_i, i = 1, 2, \ldots N$ subject to a set of cost constraints.

4.4.1: Force Downs

We want to determine when a unit increase in buffer size at b_i will result in a production gain for the transfer line. This is related to the situation that can cause a machine to be forced down.

Inefficiency in a production line is caused by *force-down* (FDs) of an otherwise healthy machine. These machines are forced down because of a full output buffer or an empty input buffer (i.e., they are starved or blocked). There are two kinds of FDs in a production line.

Irreducible FDs are caused by differences in the production rates of the individual machines. If a fast machine is followed by a slow machine, then in the steady state the fast machine is forced down. No finite amount of buffer space can prevent this, and so these FDs are called irreducible.

Reducible FDs are caused by random-machine failures. A buffer placed between such machines can help to smooth out production. In fact, these FDs can be helped by making the appropriate increase of buffer capacity at the proper location in the line.

4.4.2: The Basic Bounce

The *basic bounce* refers to the trajectory (in time) of the buffer contents. Because of the random machine failures, the buffer contents bounces from full to empty.

Consider the following scenario in the context of Figure 4.1: A person, who is allowed to hold one extra workpiece, is sent to B_i to help manage it with the intention of increasing production. What strategy should this person take?

We make the following observations:

- If B_i is not full or empty, there's nothing that the person can do to help production.

- Suppose B_i is full. This will FD M_i and represents the first opportunity where the person can help. By holding one extra piece, the person allows M_i to produce one more unit. The person becomes useless again until he/she can get rid of the extra piece.

- The next opportunity to help productivity occurs when the machine is starved. Now the person gets rid of the extra piece by tossing it into the empty buffer, B_i.

Figure 1 in Ho, Eyler, and Chien illustrates this basic bounce. Here, $\Delta b_i(k) = 1$ represents the action of the person. The bounce occurs as the buffer contents swings from full to empty. Note that this is also the time between the first *full output* of M_i, (FO_i) and the first *no input* of M_{i+1}, (NI_{i+1}). In between these two times, anything can happen including other $FO_i's$. These other $FO_i's$ will cause more force downs, but these are irreducible force downs and we can't do anything more to help. This leads to Proposition 1, which states that the only opportunity for (local) improvement in production is between a full output (FO) force down and a no input (NI) force down for two adjacent machines. In other words, the only time you can contribute to production is by holding a piece from an otherwise full buffer and tossing it back in when the downstream machine is starved.

Referring to Figure 1 (in Ho, Eyler, and Chien), note that the new buffer-level trajectory, $B_i'(t)$, is ahead in time of the old trajectory $B_i'(t)$. Therefore, the same amount of production has occurred in less time. Propositions 1a through 1d formalize the shift in time for the buffer level, the machine state, and the FDs. These propositions summarize the gain in local production.

Propositions 2 through 4 comprise the *propagation rule*, which is the mechanism for determining when a local production gain results in a line-production gain. The idea behind the propagation rule is that any local-production gain by a machine is propagated downstream by NIs and upstream by FOs. It helps to think of this gain as a gain in production time (i.e., more time available to produce).

It is also possible that the local-production gain will not propagate to the end of the line. This is explained in Proposition 3, which states that the production gain will be canceled by an FO (NI) caused by a down state of a downstream (upstream) machine. Consideration is given to both the upstream and downstream machines, indicating that the propagation occurs in both directions. Let x_{ik} be the additional production piece that is trying to propagate to the end of the line. Similarly, let y_{ik} be the buffer vacancy that wants to propagate to the beginning of the line. Then according to Proposition 4, the local-production gain in-

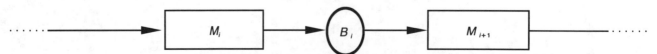

Figure 4.1: Buffer B_i Follows Machine M_i

duced by $\Delta b_i(k)$ becomes a line-production gain if both x_{ik} and y_{ik} are propagated to the end and beginning of the line, respectively.

4.4.3: Computing the Gradient

The gradient is computed by simulating the system and determining if x_{ik} and y_{ik} have propagated properly. Actual data from the system can be used in place of the simulation.

One simple way to compute the gradient relies only on Proposition 1. Let K_i equal the total number of bounces at B_i over the sample history. Then

$$\frac{\Delta \tau}{\Delta b_i} = \text{expected gain in production time with respect to a unit increase in } b_i$$

is taken as

$$\frac{\Delta \tau}{\Delta b_i} = K_i \; .$$

This assumes that the x_{ik}s and y_{ik}s have propagated properly.

A more accurate computation of the gradient can be obtained from

$$\frac{\Delta \tau}{\Delta b_i} = \sum_{j=1}^{K_i} \Delta \tau_i^s(j) = \hat{g}_i$$

where

$$\Delta \tau_i^s(j) = \text{actual gain in production time due to } \Delta b_i(j)$$
$$= \begin{cases} T & \text{if } x_{ij}, y_{ij} \text{ propagaged properly} \\ 0 & \text{if } x_{ij}, y_{ij} \text{ cancelled each other} \end{cases}$$

where T is the cycle time of the machines. Note that the calculation of $\hat{g}_i$, $i = 1,2...N - 1$ requires only one sample-production-line history. To determine if x_{ij} and y_{ij} have propagated to the end and to the beginning of the line, respectively, an algorithm, which is based on propositions 1 through 4, is provided in the paper.

4.4.4: Allocating Buffer Size

A gradient algorithm is used to obtain the buffer size. Specifically,

$$b_i \longleftarrow b_i + G(\hat{g}_i - \bar{g})$$

where

$$\bar{g} = \frac{1}{N-1} \sum_{i=1}^{N-1} \hat{g}_i$$

At the optimal value of the b_i^s, all gradients are equal indicating that no more improvement in production time is possible by re-allocating buffer sizes.

4.5: "Perturbation Analysis Explained" (Y.C. Ho)

The second reprint gives some intuitive explanations of what perturbation analysis is and why it works.

Basically, "perturbation analysis is a technique for the computation of the gradient of a performance measure (PM) with respect to system parameters by using only one sample path or Monte Carlo experiment of the system." Only one sample path is needed because information along a new trajectory can be inferred from the original sample path.

Consider the following finite state machine with states A,B,C, and system parameter θ such that

$$\theta = \begin{cases} 0 & \text{trajectory is periodic ABCABC}\ldots \\ 1 & \text{trajectory can go from one state to the other two states with equal probability.} \end{cases}$$

Let $\theta = 0$, and consider this *nominal path* (NP):

$$\text{A,B,C,A,B,C,A,B,C,}\ldots\ldots \quad \text{(NP)}$$

Now let $\theta = 1$, and consider this (one possibility) *perturbed path* (PP):

$$\text{A,B,A,B,C,A,B,C,B,C,B,C,}$$
$$\text{A,B,A,B,C,A,B,C,B}\ldots\ldots \quad \text{(PP)}$$

The claim of perturbation analysis is that we can construct the equivalent perturbed path from the data of the nominal path. This is actually a result of the memoryless property of Markov chains. Also, the fact that you are dealing with a finite number of states means that eventually you can visit all the states. Sub-chains form the perturbed path and the Markov property doesn't care how you got there. This is the cut-and-paste idea presented in the paper.

The key to perturbation analysis is the concept of *event chain invariance*. If a parameter θ is perturbed by $\Delta\theta$ and the event chain remains invariant under the perturbation, then only the nominal path (NP) is needed for the calculation and propagation of perturbations of any event in the system. The *propagation rules* in the buffer design paper are an example of event chain invariance.

In general, perturbation analysis is interested in the gradient of the expected value of a performance index. The cut-and-paste idea and the event-chain invariance concept allows the gradient and expectation operations to be interchanged, which leads to the desired gradient information from just one sample history. Ho derives this formally in the *third* reprint, "Parametric Sensitivity of a Statistical Experiment."

4.6: "A New Approach to the Analysis of Discrete Event Dynamic Systems" (Y.C. Ho and C. Cassandras)

One of the advantages of pertrubation analysis is that it allows us to work with real systems. This avoids many of the restrictive assumptions of queueing theory. Ho and Cassandras have studied serial transfer lines and an assembly line by using perturbation analysis.

The basic model consists of a set of machines, M_i, $i = 1,2,\ldots I$ and buffers, $Q_k, k = 1,2,\ldots K$ which are arbitrarily connected. Parts arrive at M_i, complete service there, and proceed to buffer Q_k. State space models are developed for these production networks where the state is given by

$$x^T(t) = [v_I(t) \ldots v_I(t) c_1(t) \ldots c_I(t)]$$

where

$v_i(t)$ = total amount of time that M_i has been operational up to time t

$c_i(t)$ = queue content at t including the customer at M_i

The perturbation analysis of these state equations is based on the concept of *similarity of event sequences*. This is the same idea as event-chain invariance presented in the second reprint by Ho. Let E_i be the event sequence of M_i in the nominal path and let E_i' be the event sequence in the perturbed path. Then, E_i and E_i' are similar, if all events in these sequences appear in the exact same sequential order. Note that the duration of these events is not necessarily equal.

The similarity of event sequences allows the derivation of an equation that determines how perturbations propagate through a system. This yields the perturbed trajectory from the nominal one. Thus, the perturbation is never actually realized.

The *perturbation-propagation equation* is different for each discrete-event dynamic system. Specific equations are derived by Ho and Cassandras for transfer lines and assembly lines. An example illustrating how perturbation analysis can be used on-line on a flexible manufacturing system to improve its performance has been reported by Suri and Dille [32].

The concept of similarity of event sequences and the perturbation-propagation equation are the two main contributions of this work.

4.7: A Brief History of Perturbation Analysis [1]

References [2-4] started the study of perturbation analysis of discrete-event dynamic systems. They illustrate the development of the idea. References [2-3] are included in this tutorial.

The papers [5-6] represent the first work of perturbation analysis on general queueing networks. Much research was stimulated by them.

This is the first paper [7] on the finite-perturbation analysis rules and experimental results. The importance of event order change was recognized then.

In [8], Cao and Ho present the first work that discovers the discontinuity of the sample-performance function and proposes the interchangeability problems for discrete-event systems. Perturbation-analysis algorithms are developed for sojourn-time-sensitivity estimation. Experimental results are presented. Eventually, a more rigorous version appeared as [27].

Cao [9] formalized mathematically the problem of interchangeability of the expectation and the differentiation operations for discrete-event systems. Conditions are found under which this interchangeability holds. It proved that under these conditions the sample derivative is the best estimate of the derivative of the expected value among the three kinds of estimates discussed in the paper.

The contribution of [10] is twofold: First, it gives an example that shows that the interchangeability does not hold for the throughput of multiclass systems. Second, it provides an algorithm that yields the exact estimate for the throughput sensitivity of a multiclass system by using first-order perturbation analysis.

Cassandras and Ho [11] provided a consistent formalism and proofs for the earlier results on infinitesimal PA. This is included as the *fifth reprint* in this tutorial.

Suri and Zazanis [12] did one of the earliest studies of consistency of infinitesimal perturbation analysis (IPA). It considers, for an M/G/1 queueing system, the sensitivity of mean system time of a customer to a parameter of the arrival of service distribution. It shows analytically that (1) the steady-state value of the perturbation-analysis estimate of this sensitivity is unbiased and (2) a perturbation-analysis algorithm implemented on a single sample path of the system gives asymptotically unbiased and strongly consistent estimates of this sensitivity.

In [13], Ho and Cao show that despite claims to the contrary [1], regular IPA rules can be applied to yield correct estimates of performance sensitivity to routing probabilities.

Later, Cao [14] proposed the concepts of sample performance functions and sample derivatives and proved that the interchangeability holds for the average time required to service one customer in any finite period as a function of the mean service time in a Jackson queueing network; and the perturbation analysis estimate of the sensitivity of throughput is a strongly consistent estimate.

Cassandras [15] placed IPA in the context of a family of PA estimation procedures, by showing the tradeoff between increased accuracy and state memory costs. The GI/G/1 model is analyzed to characterize the error properties of the simplest PA procedures, which under certain conditions, provide unbiased performance sensitivities. Extensions to tandem queueing networks and blocking effects are included.

In [16], Cao proved that the sample elasticity of throughput with respect to the mean service time obtained by perturbation analysis converges in mean to that of the steady-state mean throughput as the number of customers served goes to infinity.

Next, Cao [17] introduced the concept of *realization probability* for closed Jackson networks. This new concept provides an analytical solution to the sample elasticity of the system throughput and some other sensitivities. By using realization probability and the ergodicity of the system, it is proved that the sample elasticity of throughput with respect to the mean service time obtained by perturbation analysis also converges with probability one to that of the steady-state mean throughput as the number of customers served goes to infinity.

Ho and Yang [18] derived the PA algorithm for load-dependent queueing networks and showed that the idea of PA can be applied to aggregated systems.

In 1987, Cassandras [19] showed that a direct extension of PA, tracking queue lengths in addition to event times, can be used to estimate performance sensitivities in a simple state-dependent routing environment. This is done at the expense of state memory along the observed sample path. When a state-memory constraint is imposed, the estimates become biased but may still be sufficiently accurate.

Zazanis and Suri [20] examined the mean squared error (MSE) of PA estimates and compared it to that of estimates obtained by conventional methods. They considered two different experimental methods that are commonly used: independent replications and regenerative techniques. The analytic results obtained establish the asymptotic superiority of PA over conventional methods for both of these experimental approaches. Furthermore, it is shown that PA estimates have a MSE, which is of order $O(1/t)$ where t is the duration of the experiment in a regenerative system, whereas classical-finite-difference estimates have a MSE, that is at best $O(1/t^{1/2})$.

A PA algorithm is developed in Zazanis and Suri [21] for estimating second derivatives of the mean system time for a class of G/G/1 queueing systems, with respect to parameters of the interarrival and service distribution, from observations on a single sample path. The statistical properties of the algorithm are investigated analytically and it is proved that the estimates obtained are strongly consistent.

In [22], Cao shows that the perturbation-analysis estimate corresponds to the estimate of the difference of two random functions by using the same random variable: Thus, its variance is smaller than other one-sample-path-based-sensitivity estimates such as the likelihood-ratio estimate.

Cao [23] also derived equations for realization probability for systems with finite buffers and showed that the infinitesimal perturbation-analysis estimate is generally biased for these systems. However, examples indicate the bias is usually very small.

The concept of realization probability for multiclass closed networks is discussed in Cao [24].

An operational definition of realization probability is developed and is used to prove the sensitivity equations by using operational assumptions in Cao and Dallery [25].

Cao and Ho [26] proved that the perturbation-analysis estimate is strongly consistent for systems with finite-buffer capacities but with no simultaneous blocking. The perturbation-analysis estimate is used in the optimization of a production line. It is shown that perturbation analysis enables the use of the Robbins-Monro procedure instead of the conventional Kiefer-Wolfowitz procedure.

In Cao and Ho [27], convergence theorems for the perturbation-analysis estimate of sojourn times in closed Jackson networks is proved.

Simple PA algorithms are used to estimate performance sensitivities for communication-network models in Cassandras and Strickland [28]. Of particular interest is the application of IPA in estimating marginal delays in links modeled as GI/G/1 queues. These estimates are used in conjunction with a distributed minimum-delay algorithm to optimize routing in a quasi-static environment.

Reference [29] contains probably the most complete references on PA and related matters as of December 1986. It also contains references from other researchers and can be used to supplement this brief history.

In [30], Ho explains in simplest terms via an example the essence of PA and answers intuitively the question "How can one infer the performance of a discrete-event system operating under one parameter value from that of another with a different parameter value? Don't the two sample paths behave totally differently?" A final version of this paper is included in this tutorial as the *second reprint*.

Zazanis [31] uses classical Markovian analysis to establish the unbiasedness of IPA estimates for the M/M/1 system and refutes another public claim of the limitations of IPA. The restrictive Markovian assumption is the price paid for the simplicity of the arguments used.

Suri and Dille [32] have applied the PA approach to flexible-manufacturing systems (FMS). They give a simulation example illustrating how perturbation analysis could be used on-line on an FMS to improve its performance, including the reduction of its operating cost. Experimental results, validate the estimates obtained from this technique, are also presented.

A stochastic-optimization method to optimize a simulation model in a single simulation run is proposed in Suri and Leung [33]. Two algorithms are developed and evaluated empirically by using an M/M/1 queue problem. Experimental results show that an algorithm based on IPA provides extremely fast convergence as compared with a traditional Kiefer-Wolfowitz based method.

Cassandras and Strickland [34] present a new way of estimating performance sensitivities of Markov processes by direct observation. The parameters considered are discrete (integer-valued) for example, queue capacities, thresholds in routing policies, and number of customers of a specific class in a closed network model. The main idea is to construct an "augmented chain" whose state transitions are observable when the process itself is observed.

In Zazanis [35] an expression is given for the derivative of the mean-virtual-waiting time in a GI/G/1 queue with respect to the service rate.

The strong consistency of IPA estimates for the mean response time is shown [36] by using the regenerative structure of the GI/G/1 queue. The analysis throws some light on the conditions that are required for the consistency of IPA estimates in general systems with regenerative structure.

Suri [37] sets forth IPA in a general setting under deterministic similarity assumptions.

Gong et al. [38] show that by using the smoothing properties of conditional expectation, the problem of interchange between expectation and differentiation can be resolved to give consistent PA estimates for problems previously proclaimed to be unsolvable by PA (e.g., derivatives of throughput with respect to mean service time in multiclass queueing networks or mean number of customers served in a busy period with respect to mean service time).

Ho and Li [39] show another general approach to circumvent the difficulty of discontinuities in PM (θ,ω) with respect to q for Markov systems. This technique also puts in perspective earlier work on finite PA by showing it to be one member among a range of possible approximations from the crudest to the exact for handling the discontinuity problem. Robustness of these approximations is discussed and experimental supports are illustrated.

Glasserman [40] shows that the regular IPA rules can be applied to a birth-death process to yield correct sensitivity estimates despite written claims to the contrary.

4.8: "Performance Sensitivity to Routing Changes in Queueing Networks and Flexible Manufacturing Systems Using Perturbation Analysis" (Y.C. Ho and X.R. Cao)

This paper overcomes several of the shortcomings associated with queueing models. In particular, no closed form analytical solutions exist for queueing models with limited buffer size, nonexponential service-time distributions, or nonstandard queueing disciplines. The major contribution of this reprint is a method, based on perturbation analysis, which gives exact values of performance measure sensitivities for an arbitrary queueing network.

The problem under consideration is the sensitivity of the system throughput with respect to the routing probability of jobs in a network of machines. Direct application of perturbation analysis won't work here. Instead, the basic idea of the proposed method is that by using some existing formulas in queueing network theory, the estimation of the sensitivity of the throughput with respect to the routing probability can be converted to a problem of calculating the throughput sensitivity with respect to some parameters of service times.

By using perturbation analysis, Ho and Cao are able to handle systems with uniformly distributed service times and production networks where some machines have exponentially distributed processing times and others have uniformly distributed processing times.

The ideas developed in the paper rely heavily on the results obtained from the theory of networks of queues. The important definitions and results have been summarized in Section 2.5.

4.9: Perturbation Analysis: The State of the Art and Research Issues

A good summary and state-of-the-art review of perturbation analysis can be found in [53]. Within this special issue, Suri [54] presents a tutorial on perturbation analysis. He uses a simple queueing system example to motivate and explain the basic concepts of perturbation analysis. Several manufacturing examples are presented and an extensive up-to-date bibliography is included.

4.10: References

1. Y.C. Ho, "A Selected and Annotated Bibliography on Perturbation Analysis," *Proceedings of IIASA Conference on Discrete Event Systems*, Pergamon Press, London, England, 1988.
2. Y.C. Ho, A. Eyler and T.T. Chien, "A Gradient Technique for General Buffer Storage Design in a Serial Production Line," *International Journal on Production Research*, Vol. 17, No. 6, 1979, pp. 557–580.
3. Y.C. Ho, "Parametric Sensitivity of a Statistical Experiment," *IEEE Transactions on Automatic Control*, Vol. AC-24, No. 6, 1979, p. 982.
4. Y.C. Ho, A. Eyler and T.T. Chien, "A New Approach to Determine Parameter Sensitivities on Transfer Lines," *Management Science*, Vol. 29, No. 6, 1983, pp. 700–714.
5. Y.C. Ho and C.G. Cassandras, "A New Approach to the Analysis of Discrete Event Dynamic Systems," *Automatica*, Vol. 19, No. 2, 1983, pp. 149–167.
6. Y.C. Ho and X.R. Cao, "Perturbation Analysis and Optimization of Queueing Networks," *Journal of Optimization Theory and Applications*, Vol. 40, No. 4, 1983, pp. 559–582.
7. Y.C. Ho, X.R. Cao, and C.G. Cassandras, "Infinitesimal and Finite Perturbation Analysis for Queueing Networks," *Automatica*, Vol. 19, No. 4, 1983, pp. 439–445.
8. X.R. Cao and Y.C. Ho, "Estimating Sojourn Time Sensitivity in Queueing Networks Using Perturbation Analysis," *Technical Report*, Division of Applied Science, Harvard University, 1984; also *Journal of Optimization Theory and Applications*, Vol. 53, No. 3, 1987, 353–375.
9. X.R. Cao, "Convergence of Parameter Sensitivity Estimates in a Stochastic Experiment," *IEEE Transactions on Automatic Control*, Vol. AC-30, No. 9, 1985, pp. 834–843.
10. X.R. Cao, "First-Order Perturbation Analysis of a Single Multi-Class Finite Source Queue," *Performance Evaluation*, Vol. 7, 1987, pp. 31–41.

11. C.G. Cassandras and Y.C. Ho, "An Event Domain Formalism for Sample Path Perturbation Analysis of Discrete Event Dynamic Systems," *IEEE Transactions on Automatic Control*, Vol. AC-30, No. 12, 1985, pp. 1217–1221.

12. R. Suri and M.A. Zazanis, "Perturbation Analysis Gives Strongly Consistent Sensitivity Estimates for the M/G/1 Queue," *Management Science* Vol. 34, No. 1, Jan. 1988, pp. 39–64.

13. Y.C. Ho and X.R. Cao, "Performance Sensitivity to Routing Changes in Queueing Networks and Flexible Manufacturing Systems Using Perturbation Analysis," *IEEE Journal on Robotics and Automation*, Vol. 1, 1985, pp. 165–172.

14. X.R. Cao, "On Sample Performance Functions of Jackson Queueing Networks," *Operations Research*, Vol. 36, 1988, pp. 128–136.

15. C.G. Cassandras, "Error Properties of Perturbation Analysis for Queueing Systems," submitted to *Operations Research*.

16. X.R. Cao, "The Convergence Property of Sample Derivatives in Closed Jackson Queueing Networks," submitted to *Journal of Applied Probability*, 1987.

17. X.R. Cao, "Realization Probability in Closed Jackson Queueing Networks and Its Application," *Advances in Applied Probability*, Vol. 19, Sept. 1987, pp. 708–738.

18. Y.C. Ho and P.Q. Yang, "Equivalent Networks, Load Dependent Servers, and Perturbation Analysis—An Experimental Study," *Proceedings of the Conference on Teletraffic Analysis and Computer Performance Evaluation*, (eds. O.J. Boxma, J.W. Cohen, H.C. Tijms), North Holland, New York, 1986.

19. C.G. Cassandras, "On-Line Optimization for a Flow Control Strategy," *IEEE Trans. on Automatic Control*, Vol. AC-32, No. 11, Nov. 1987, pp. 1014–1017.

20. M.A. Zazanis and R. Suri, "Comparison of Perturbation Analysis with Conventional Sensitivity Estimate for Stochastic Systems," submitted to *Operations Research*, 1985.

21. M.A. Zazanis and R. Suri, "Estimating First and Second Derivatives of Response Time for GI/G/1 Queues from a Single Sample Path," submitted to *Queueing Systems: Theory and Applications*, 1985.

22. X.R. Cao, "Sensitivity Estimates Based on One Realization of a Stochastic System," *Journal of Statistical Computation and Simulation*, Vol. 27, 1987, pp. 211–232.

23. X.R. Cao, "Calculation of Sensitivities of Throughputs and Realization Probabilities in Closed Queueing Networks with Finite Buffers," manuscript submitted, 1987.

24. X.R. Cao, "Realization Probability in Multi-Class Closed Queueing Networks," *European Journal of Operations Research*, Vol. 36, 1988, pp. 393–401.

25. X.R. Cao and Y. Dallery, "An Operational Approach to Perturbation Analysis of Closed Queueing Networks," *Mathematics and Computers in Simulation*, Vol. 28, 1986, pp. 433–451.

26. X.R. Cao and Y.C. Ho, "Sensitivity Estimate and Optimization of Throughput in a Product Line with Blocking," *IEEE Transactions on Automatic Control*, Vol. AC-32, No. 11, 1987.

27. X.R. Cao and Y.C. Ho, "Perturbation Analysis of Sojourn Times in Queueing Networks," *Proceedings of the 22nd IEEE Conference on Decision and Control*, San Antonio, Tex., Dec. 1983, pp. 1025–1029.

28. C.G. Cassandras and S.G. Strickland, "Perturbation Analytic Methodologies for Design and Optimization of Communication Networks," *IEEE Journal of Selected Areas in Communications*, 1988, pp. 158–171.

29. Y.C. Ho, "Performance Evaluation and Perturbation Analysis of Discrete Event Dynamic Systems," *IEEE Transactions on Automatic Control*, Vol. AC-32, No. 6, July 1987, pp. 563–572.

30. Y.C. Ho, "Perturbation Analysis Explained," *Proceedings of the 26th IEEE Conference on Decision and Control*, Los Angeles, Calif., Dec. 1987, pp. 243–246.

31. M.A. Zazanis, "Unbiasedness of Infinitesimal Perturbation Analysis Estimates for Higher Moments of the Response Time of an M/M/1 Queue," *Technical Report 87–06*, Northwestern University, Evanston, Ill., 1987.

32. R. Suri and J. Dille, "A Technique for On-Line Sensitivity Analysis of Flexible Manufacturing Systems," *Annals of Operations Research*, Vol. 3, 1985, pp. 381–391.

33. R. Suri and Y.T. Leung, "Single Run Optimization of Discrete Event Simulations—An Empirical Study Using the M/M/1 Queue," *Technical Report #87-3*, Department of Industrial Engineering, University of Wisconsin, Madison, Wis., 1987.

34. C.G. Cassandras and S.G. Strickland, "An Augmented Chain Approach for On-Line Sensitivity Analysis of Markov Processes," *Proceedings of the 26th IEEE Conference on Decision and Control*, IEEE Press, New York, 1987.

35. M.A. Zazanis, "An Expression for the Derivative of the Mean Response Time of a GI/G/1 Queue," *Technical Report 87–08*, Northwestern University, Evanston, Ill., (see also "Extension to GI/G/1 Systems with a Scale Parameter," *Technical Report 87–07* Northwestern University, Evanston, Ill.,), 1987.

36. R. Suri and M.A. Zazanis, "Infinitesimal Perturbation Analysis and the Regenerative Structure of the GI/G/1 Queue," *Proceedings of the 1987 IEEE Decision and Control Conference*, Los Angeles, Calif., 1987, pp. 677–680.

37. R. Suri, "Infinitesimal Perturbation Analysis for General Discrete Event Dynamic Systems," *Journal of ACM*, Vol. 34, No. 3, July 1987, pp. 686–717.
38. W.B. Gong, and Y.C. Ho, "Smoothed (Conditional) Perturbation Analysis of Discrete Event Dynamic Systems," *IEEE Transactions on Automatic Control*, Vol. AC-32, No. 10., 1987, pp. 858–866.
39. Y.C. Ho and Shu Li, "Extensions of the Infinitesimal Perturbation Analysis Technique of Discrete Event Systems," *IEEE Transactions on Automatic Control*, Vol. AC-33, No. 5, 1988, pp. 427–438.
40. P. Glasserman, "IPA Analysis of a Birth-Death Process," *Operations Research Letters*, Vol. 7, No. 1, 1988.
41. Y.C. Ho and C. Cassandras, "Computing Co-State Variables for Discrete Event Systems," *Proceedings of the IEEE Conference on Decision and Control*, IEEE Press, New York, 1980, pp. 697–700.
42. R. Suri, "Infinitesimal Perturbation Analysis of Discrete Event Dynamic systems: A General Theory," *Proceedings of the IEEE Conference on Decision and Control*, IEEE Press, New York, 1983, pp. 1030–1038.
43. C.G. Cassandras, "Optimizing Recirculation in Flexible Manufacturing Systems," *Proceedings of the Second ORSA/TIMS Conference on Flexible Manufacturing Systems: Operations Research Models and Applications* (eds. K.E. Stecke and R. Suri), Elsevier Science Publishers, Amsterdam, The Netherlands, 1986, pp. 381–392.
44. R. Suri and M. Zazanis, "Perturbation Analysis Is Exact for the M/G/1 Queue," *Proceedings of the 23rd IEEE Conference on Decision and Control*, IEEE Press, New York, 1984, pp. 535–536.
45. C.G. Cassandras, "Sensitivity Analysis of a Simple Routing Strategy," *Proceedings of the 24th IEEE Conference on Decision and Control*, IEEE Press, New York, 1985, pp. 2022–2027.
46. R. Suri and M. Zazanis, "Robustness of Perturbation Analysis Estimates for Automated Manufacturing Systems," *Proceedings of the 1986 IEEE International Conference on Robotics and Automation*, IEEE Computer Society Press, Washington, D.C., p. 311.
47. M.J. Denham, "Control of Discrete Event Processes," *Proceedings of the 25th IEEE Conference on Decision and Control*, IEEE Press, New York, 1986, pp. 1717–1718.
48. M. Zazanis and R. Suri, "Comparison of Perturbation Analysis with Conventional Sensitivity Estimates for Regenerative Stochastic Systems," *Proceedings of the 24th IEEE Conference on Decision and Control*, IEEE Press, New York, 1985, pp. 2034–2035.
49. Q. Jiang, "A New Approach to Modelling Simulation and Perturbation Analysis for Manufacturing Scheduling Problems (MSP)," *Proceedings of the 25th IEEE Conference on Decision and Control*, IEEE Press, New York, 1986, pp. 1706–1707.
50. X.R. Cao, "First-Order Perturbation Analysis of Multi-Class Queueing Networks," *Proceedings of the 24th IEEE Conference on Decision and Control*, IEEE Press, New York, 1985, pp. 2028–2033.
51. P.J. Ramadge, "Observability of Discrete Event Systems," *Proceedings of the 25th IEEE Conference on Decision and Control*, IEEE Press, New York, 1986, pp. 1108–1112.
52. H. Mortazavian, "Modeling, Control and Verification of Flexible Manufacturing Systems," *Proceedings of the First ORSA/TIMS Conference on Flexible Manufacturing Systems*, Ann Arbor, Mich., 1984, pp. 359–371.
53. "Special Issue on Dynamics of Discrete Event Systems," *Proceedings of the IEEE*, Vol. 77, No. 1, Jan. 1989 (Y.-C. Ho, editor).
54. R. Suri, "Perturbation Analysis: The State of the Art and Research Issues Explained via the GI/G/1 Queue," *Proceedings of the IEEE*, Vol. 77, No. 1, Jan. 1989, pp. 114–137 (Y.-C. Ho, editor).

A gradient technique for general buffer storage design in a production line

Y. C. HO,† M. A. EYLER† and T. T. CHIEN‡

In this paper, we present a complete and novel solution to the well known buffer storage design problem in a serial production line. The key ingredient of the solution is the efficient calculation of the gradient vector of the throughput with respect to the various buffer sizes. Analytical and experimental results are presented.

Introduction

The problem of buffer storages in a production line has had a long history. Pertinent references date back to 1957. Briefly, machinery stations are assumed to have two states, namely UP (healthy and working) or DN (internal failure). The states of an isolated machine constitute a Markov chain with given failure and repair transition probabilities:

$$q \equiv Pr \text{ (machine DN next cycle/UP this cycle)}$$
$$a \equiv Pr \text{ (machine UP (repaired) next cycle/DN this cycle)}. \tag{1}$$

It is easily shown (Buzacott 1967), in this case via a Markov chain analysis, that:

$$\text{Mean Time Between Failure} = 1/q$$
$$\text{Mean Time To Repair} = 1/a \tag{2}$$

and the stand-alone efficiency of the machine is:

$$\tau = \frac{1}{1 + q/a}. \tag{3}$$

When the machines are connected in a serial production line, each performing a partial operation on the production piece sequentially, internal failures of one machine may be affecting the production of others (i.e., causing an otherwise healthy machine to be forced down (FD) due to full output buffer or empty input buffer). In the extreme case of direct connection of N machines, failure of any one will result in the 'force down' of the entire production line. A similar Markov chain analysis quickly shows (Buzacott 1967) that the efficiency of the zero-buffer line is:

$$\tau_0 = \frac{1}{1 + \sum_{i=1}^{N} \frac{q_i}{a_i}} < \frac{1}{1 + \frac{q_i}{a_i}} \equiv \tau_i \, \forall_i \tag{4}$$

where q_i, a_i are the parameters of the machine M_i. At the other extreme, if an infinite amount of buffer is provided between each two machines for intermediate storage of

---

Received 2 November 1978.
† Harvard University.
‡ The Charles Stark Draper Laboratory, Inc.

semi-finished pieces, then the production efficiency of such a line is bounded from above by (Okamura and Yamashina 1977):

$$\tau_\infty \leqslant \operatorname*{Min}_i [\tau_i] \tag{5}$$

for reasons of continuity of production material. For finite amount of buffer storages, the production line efficiency τ is bounded above and below by τ_∞ and τ_0. The buffer storage design problem can then be stated as:

Choose the size of buffer storages $b_1, \ldots, b_{N-1}$ to maximize τ for given $\{a_i, q_i; i = 1, \ldots, N\}$ subject to cost constraints.

It is clear that for finite size of all buffer storages $b_1, \ldots, b_{N-1}$, the 'states' of the production line constitute a Markov chain. Each machine can be healthy or under repair; each buffer can have content $0, 1, 2, \ldots, b_i$. The total number of states is:

$$N = 2^N \prod_{i=1}^{N-1} (b_i + 1). \tag{6}$$

The corresponding transition probability matrix P can be specified accordingly in terms of the various a_is and q_is. Once this is done, steady-state probability of various states π_i, $i \in 1, \ldots, N$ in the Markov chain can *in principle* be determined via the solution of the equation $\pi = P\pi$. Production efficiency of the line is then simply the sum of the π_i, $i \in P$. where P is the set of states that results in a piece being produced by the line. Thus, conceptually, line efficiency or throughput τ can be expressed as a function of all the b_is parameterized by the given a_is and q_is. This has actually been carried out for the two machines one buffer (Buzacott 1967, Artamonov 1976) and three machines two buffers case (Gershwin and Schick 1978). It is, however, equally clear that this approach is not feasible for any line with more than three machines or large storage size. The best effort currently is for $N = 3$, $b_1 + b_2 \leqslant 40$ (Gershwin and Schick 1978). For a general line with many machines, some other approach, approximate or otherwise, is necessary. Sevastyanov (1962) and Buzacott (1967) gave some approximate analysis for the general line by essentially introducing the idea of an equivalent machine to replace a segment of a production line and the repeated use of the two machines one buffer formula. However, no analysis or simulation of the range of validity of the approximation was presented. Other works in this area restrict themselves to two or three machine cases but introduce variations in machine characteristics and their specifications (Kraemer and Love 1970, Masso and Smith 1974, Rao 1975, Skeskin 1976, Ignall and Silver 1977) or direct brute force simulation of line production (Hanifin *et al.* 1975). It is fair to say that very little is known for the general production line buffer storage problems.

To illustrate the scope of the buffer design complexity in the real world, we present some relevant parameters of a typical transfer line:

Number of transfer machines: 27

total buffer capacity: 261 pieces

machine cycle time: from 27 to 81 seconds.

It might be noted that different cycle time is another complexity never discussed in the literature, although Hillier and Boling (1966) have studied the effect of random cycle time in the production line with three to four perfect machines.

With complexity of this nature, it seems doubtful whether a tractable closed form analytical solution of the design problem could ever be developed. An alternative approach is to design the buffer allocation by simulations. A simulation model has been developed by Hanifin et al. (1975). Their report studied the effect of buffer size and location on the production rate of a transfer line of 75 machine stations with identical cycle times. Among 75 stations, only three specified locations are assumed to be available for buffer allocation with a possible size increase up to 50 pieces. A brute force approach is applied in the simulation to evaluate the production rate gain on 18 combinations at three buffer locations, with size at each location ranging from 0 to 50 in increments of ten. The best buffer design in location and size is then selected on the basis of the economic analysis comparing capital investment with increasing production.

A more systematic approach for buffer sizing is to compute the sensitivity (gradient) of the increase of transfer line production per unit buffer size increase at each buffer location and then allocate the buffer size at each location by hill climbing procedure until all gradients become equal to an economic index defined by investment return. While this approach is conceptually straightforward, the method suffers from a heavy computation burden. For the above-mentioned transfer line of 27 machines, an array of 26 gradients must be evaluated, each requiring at least one long simulation. During the experiment on this approach, it was observed that the gradients required an extremely long computer run in order to show an appreciable change, not mentioning the requirement of statistical significance. A third way is to compute the gradients by a big increase in buffer size to shorten the run time, but this method suffers the inaccuracy due to the inherent nonlinearity characteristics of the buffer design problem as well as insufficient sample size.

The objective of this paper is to present a novel approach to the design problem derived from the insight of the fundamental role played by buffers in contributing to the production increase of the transfer lines.

Solution via 'successive improvement'
The basic 'gradient' approach

We begin by noting one fundamental fact: 'Inefficiency in production is caused by force-downs (FDs) of an otherwise healthy machine'. If we can keep each machine working except during internal failures, then surely average production cannot be further increased without changing the basic parameters of the machines. However, there are really two kinds of force-downs in a production line:

(i) Irreducible FDs. These FDs are caused by intrinsic differences in production rates of the machines. For example, suppose we have a perfect machine (no internal failures) connected to an imperfect machine. Then, no amount of buffer storage between the two machines can keep the perfect machine from being repeatedly forced down in the steady state.

(ii) Reducible FDs. These FDs are caused by random (usually alternate) failures of the machines. An extreme example is the situation where two machines with identical characteristics are connected in series. If they happen to fail alternately, then no production is possible if there is no buffer storage between them. Yet, with sufficient but finite buffer, the machines can, in fact, produce at a rate that approaches their stand-alone throughput rate.

The main difficulty here is how to distinguish between the two different kinds of FD when one looks at a sample history of a buffer storage content as a function of time. Because of the complex interplay among the different cycle times of the machines, the different failure and repair characteristics and the random nature of these events, it appears almost impossible to say whether or not a given force-down of a machine could be helped by an appropriate increase of buffer capacity somewhere in the line. It is here that we make a contribution.

In this section we shall treat the case of uniform cycle time only. The case of different cycle time will be discussed in a later section. Let us consider two consecutive machines, M_i and M_{i+1}, and their buffer storage B_i in a production line. For the moment, assume they are isolated from the rest of the line, i.e., infinite supply exists before M_i and infinite sink follows M_{i+1}. Now consider the following paradigm. A person is sent to B_i with the instruction that he is to help manage B_i where he can hold one extra piece. It is clear that so long as B_i is not full or empty, there is nothing he can do to help production. In fact, the first time he can be of use is when he encounters a FD of M_i due to full output (FO_i), i.e., B_i is full. At that time, the person can hold one extra piece of production on behalf of B_i, thus allowing one more unit of production by M_i. Thereafter, he is again useless when additional FO_is occur until he can get rid of this extra piece during a no-input phase of M_{i+1}, (NI_{i+1}), i.e., B_i becomes empty. After this, the cycle can repeat. Figure 1 graphically illustrates this phenomenon.

We denote the history of the buffer content during this cycle as a 'bounce' (the content bounced from full to empty), and the action of the person during this cycle as $\Delta b_i(k) = 1$, since this action is equivalent to adding unit capacity to the ith buffer size during the kth bounce. Note that the duration of a bounce is not constant. It is characterized only by the first full output of M_i, $FO_i(1)$, and the first no input of M_{i+1}, $NI_{i+1}(1)$. Between these two events, the buffer content history can be completely arbitrary. Yet the unit gain in production through the section M_i-B_i-M_{i+1} remains the same as represented by $FO'_i(1)$ and $NI'_{i+1}(1)$. All other FDs between these two FDs are irreducible. Similarly, all NI_{i+1}s between $NI_{i+1}(1)$ and the next FO_i are irreducible. Thus, we have:

Proposition 1. Only a pair of alternate FO and NI FDs of two adjacent machines represents an opportunity for improvement in production locally.

To formalize this, let us define some notations:

$B_i(t)$	contents of B_i at t
b_i	size of B_i (in pieces of production units)
$M_i(t)$	state of M_i at t: M_i can be working (denoted graphically by ·····), M_i can be forced down ($\times$——$\times$), or M_i can fail (———)
$NI_{i+1}[\xi, \tau]$	an interval of time $[\xi, \tau]$ during which $B_i(t) = 0$ *and M_{i+1} is forced down*
$FO_i[\xi, \tau]$	an interval of time $[\xi, \tau]$ during which $B_i(t) = b_i$ *and M_i is forced down*
$[Bounce]_i$	an interval $[t_b, t_f]$ of $B_i(t)$ defined by a FO_i where $\xi = t_b$ and the first following NI_{i+1} where $\tau = t_f$.
Condition (S)	a condition (S) is said to hold for $[Bounce]_i$ if in the corresponding $M_i(t)$ a FO_i state is followed immediately by a failed state.

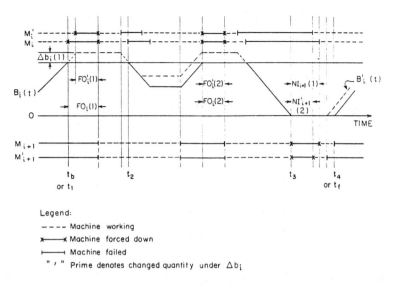

Figure 1. The basic bounce.

Finally, let the prime symbol denote a changed quantity, e.g., $B'_i(t)$ is the buffer content of B_i when $\Delta b_i(k)$ is added to the capacity; $M'_i(t)$ is the induced M_i state history, etc.

Proposition 1 (a). Let $\Delta b_i(k) = 1$ during a $FO_i[t_1, t_2]$, then:

$$M'_i(t) = M_i(t+T) \quad \forall t > t_2, \tag{7}$$

where T is the cycle time of the machines and provided condition(S) does not hold.

Proof: An analytical proof of the proposition hardly seems necessary. The exception for condition(S) is to rule out the situation where FO_i can completely disappear due to $\Delta b_i(k)$, resulting in a drastic change of $M_i(t)$. This phenomenon, known as 'shifting', will be discussed in the next section.

The main significance of proposition 1 (a) is the fact that $M'_i(t)$ is now translated forward in time (gain in production time) with respect to $M_{i+1}(t)$ and $M_{i-1}(t)$.

Proposition 1 (b) Let (7) hold and $NI_{i+1}[t_3, t_4]$ be the first no-input state of M_{i+1} after the action $\Delta b_i(k) = 1$, i.e., $b_i \geq B_i(t) > 0$, $t_1 < t < t_3$; then:

$$B'_i(t_3) = B_i(t_3). \tag{8}$$

Proof: Since $M'_i(t)$ is translated forward by T and M_{i+1} have not been affected by $\Delta b_i(k)$ for $t \leq t_3$, we must have:

$$B'_i(t) \geq B_i(t) \quad \forall t_1 < t \leq t_3. \tag{9a}$$

Now if (8) is not true, then $B'_i(t_3) > B_i(t_3)$. But this implies that M_{i+1} will enter the NI_{i+1} state later than t_3 or $M'_{i+1}(t) \neq M_{i+1}(t)$ for $t = t_3^+$. This is impossible since by definition M_{i+1} cannot be affected by $M'_i(t)$ as long as $B'_i(t) > 0$.

Proposition 1 (b) says that the initiation of the first NI state of M_{i+1} remains the same. But since the termination of NI_{i+1} is determined entirely by M_i, which is now translated forward, we have:

Proposition 1(c) (7) and $NI_{i+1}[t_3, t_4]$ implies:

$$NI'_{i+1}[t_3, t'_4] = NI_{i+1}[t_3, t_4] - T \tag{9b}$$

$$M'_{i+1}(t) = M_{i+1}(t+T). \tag{9c}$$

An exactly dual argument with respect to M_{i-1} and a $FO_{i-1}[t_3, t_4]$ yields:

Proposition 1(d) (7) and $FO_{i-1}[t_3, t_4]$ implies:

$$B'_{i-1}(t) \leq B_{i-1}(t) \quad t_1 < t \leq t_3 \tag{10a}$$

$$FO'_{i-1}[t_3, t'_4] = FO_{i-1}[t_3, t_4] - T \tag{10b}$$

$$M'_{i-1}(t) = M_{i-1}(t+T). \tag{10c}$$

Proposition 1 (a–d) says that the action of $\Delta b_i(k) = 1$ is first to create a forward translation (production time gain) of $M_i(t)$. This gain is propagated to M_{i+1} and M_{i-1} by the first occurrence of a NI_{i+1} and FO_{i-1}, respectively. Equivalently, to carry further the paradigm, we can imagine that, as a result of the human action $\Delta b_i(k)$ during $[Bounce]_i(k)$, we have created an additional production piece (say x_{ik}) in B_{i+1} and a vacancy or potential demand (say, y_{ik}) in B_{i-1} (see eqns. (9a) and (10a).

At this point, it is easy to see how these x_{ik}s and y_{ik}s (or equivalently, the gains of M_{i+1} and M_{i-1}) can be propagated further upstream or downstream in the production line. We can directly extend propositions 1 (c) and 1 (d) to:

Proposition 2 If M_{i+j} (M_{i-j}) is translated forward in time by T, then a NI_{i+j+1} (FO_{i-j-1}) will propagate the production time gain (T) to M_{i+j+1} (M_{i-j-1}).

Proof: Same as propositions 1 (c) and 1 (d).

On the other hand, a gain in M_{i+j} (M_{i-j}) can be cancelled by a FO_{i+j} (NI_{i-j}) as in:

Proposition 3(a) Suppose M_{i+j} is translated forward by T with respect to M_{i+j+1}, then the occurrence of a $FO_{i+j}[t_5, t_6]$ implies:

$$FO'_{i+j}[t'_5, t_6] = FO_{i+j}[t_5, t_6] + T \tag{11a}$$

$$M'_{i+j}(t) = M_{i+j}(t). \tag{11b}$$

Proof: Since M_{i+j} is ahead with respect to M_{i+j+1}, then $B'_{i+j} \geq B_{i+j}(t)$ and $B'_{i+j}(t_3 - T) = b_{i+j}$ by definition, i.e., FO_{i+j} must be initiated T seconds earlier. Since the termination of FO'_{i+j} is determined by M_{i+j+1}, which could not have been affected (it has been in DN state prior to t_3), eqns. (11a) and (11b) follow.

The dual of proposition 3(a) is

Proposition 3(b) Suppose M_{i-j} is translated forward by T with respect to M_{i-j-1}, then the occurrence of a $NI_{i-j}[t_5, t_6]$:

$$NI'_{i-j}[t'_5, t_6] = NI_{i-j}[t_5, t_6] + T \tag{12a}$$

$$M'_{i-j}(t) = M_{i-j}(t). \tag{12b}$$

Figures 2(a) and 2(b) illustrate graphically propositions 1 (a–d), 2, and 3.

Another way of visualizing propositions 1–3 is as follows: suppose machine M_i by some means gains one cycle of production with respect to the nominal sample production history. This local gain of M_i cannot affect either M_{i-1} (or M_{i+1}) so long as buffers B_{i-1} (B_i) are not empty or full. If $B_{i-1}(B_i)$ becomes fully (empty) owing to a DN state of M_i, then the machine M_{i-1} (M_{i+1}) will be tied to M_i and recover only when M_i recovers. In other words, M_{i-1} (M_{i+1}) will also recover earlier and gain in production. On the other hand if $B_{i-1}(B_i)$ becomes empty (full) owing to a DN state of M_{i-1} (M_{i+1}), then M_i will be forced down and have its history tied to that of M_{i-1} (M_{i+1}). Since by definition M_{i-1} (M_{i+1}) represents the nominal, M_i will lose its local production gain. Summarizing then, we have the

Propagation rule: Any local production gain with respect to the nominal by a machine is propagated downstream by NIs and upstream by FOs. The production gain will be cancelled by a FO(NI) caused by a DN state of a downstream (upstream) machine.

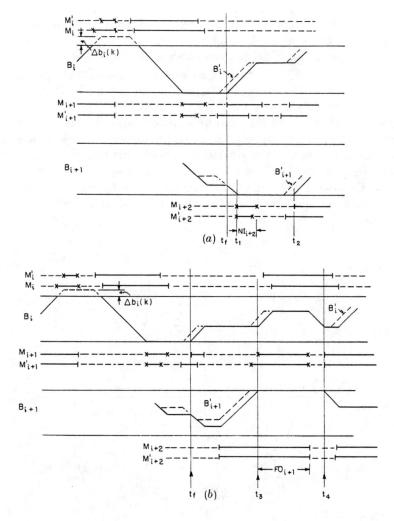

Figure 2. (a) Propagation due to NIFD; (b) propagation due to FOFD.

Conceptually, we can think of x_{ik} and y_{ik} created by the $[\text{Bounce}]_i(k)$ being continuously propagated back and forth among machines by FOs and NIs as the sample production line history unfolds (equivalently, the machine gains or loses production time). When finally both x_{ik} and y_{ik} are transmitted to the beginning and end of the line, one has then actually gained one unit of production (as compared to the sample production history). Now because of proposition 1 (a–d) we see that the action of $\Delta b_i(k)$ is to create a pair of opposite changes in B_{i+1} and B_{i-1}, respectively x_{ik} and y_{ik}. The x_{ik} and y_{ik} behave as +ve and −ve charges. Should they meet at any time, they cancel each other. For example, if an x_{ik} is propagated to the end of a production line, the production gain thus attained by x_{ik} may be cancelled if, through an unfortunate series of events, y_{ik} is also propagated to the end of the line.

Since we have agreed to always start and finish any production run with zero buffer content, a y_{ik} not propagated to the beginning of the line will eventually be transported to the end of the line, thus cancelling any temporary production gain. Thus, we have proved:

Proposition 4 The local production gain induced by $\Delta b_i(k)$ is realized as a line production gain if and only if both x_{ik} and y_{ik} are propagated to the end and beginning of the line, respectively, by the subsequent production line history.

A graphical illustration of proposition 4 is shown in Fig. 3, where ① denotes that the production gain x_{ik} or y_{ik} is sitting in B_j; and $\overrightarrow{\text{NI}_{j+1}}$ means that x_{ik} or y_{ik} will be propagated from B_j to the next buffer B_{j+1} if it encounters a NI FD event of M_{j+1} before a FO FD of M_j; similarly for other notations.

Propositions 1–4 spell out the conditions under which the effect of the human intervention $\Delta b_i(k)$ can actually result in a unit production increase. Since a sample production history is random, the expected gain in production time, $\Delta \tau$, due to $\Delta b_i(k)$ should be averaged over an ensemble of sample histories. However, if we further assume that the system is in steady state and the ergodic hypothesis holds, then we may perform the average over time rather than over ensembles. In other words, we assume:

$$\frac{\Delta\tau}{\Delta b_i(k)} = \frac{\Delta\tau}{\Delta b_i(j)} \quad \forall k, j \tag{13}$$

and

$$\frac{\Delta\tau}{\Delta b_i(k)} = \lim_{K_i \to \infty} \frac{1}{K_i} \sum_{j=1}^{K_i} \Delta\tau_i^s(j) \tag{14}$$

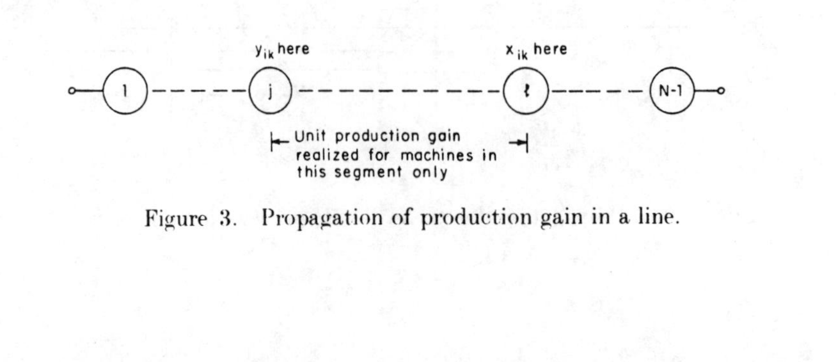

Figure 3. Propagation of production gain in a line.

where:

$\Delta \tau_i^s(j)$ = actual gain in production time due to $\Delta b_i(j)$ based on sample history (15)

$$= \begin{cases} T & \text{if } x_{ij} \text{ and } y_{ij} \text{ propagated to the end} \\ & \text{and to the beginning of the line, respectively.} \\ 0 & x_{ij} \text{ and } y_{ij} \text{ cancelled each other either} \\ & \text{in some buffer or at one end of the line.} \end{cases}$$

$K_i =$ total number of bounces at B_i over the sample history (assumed to be large enough for the ergodic hypothesis to hold).

If we introduce a permanent change Δb_i in the size of B_i through K_i bounces:

$$\frac{\Delta \tau}{\Delta b_i} = \text{expected gain in production time due to unit increase in } b_i \quad (16)$$

$$= K_i \frac{\Delta \tau}{\Delta b_i(k)}$$

$$= \sum_{j=1}^{K_i} \Delta \tau_i^s(j) \equiv \hat{g}_i$$

= estimated 'gradient' at B_i.

Note that the estimated gradient vector $\hat{g}_i, i = 1, \ldots N-1$ can be determined from *one* sample production line history under the ergodic hypothesis using the following conceptual algorithm.

Gradient algorithm

Step (1) For each B_i at the kth bounce, create x_{ik} and y_{ik}.

Step (2) $x_{ik} \rightarrow B_{i+1}$ at $t = t_f$
$y_{ik} \rightarrow B_{i-1}$ at $t = t_b$.

Step (3) For each FO_j, propagate all xs and ys currently residing at B_j to B_{j-1}. Similarly for each NI_j, propagate xs and ys from B_{j-1} to B_j. Repeat for all j.

Step (4) Count $\Delta \tau_i^s(k)$ according to proposition 3 and the propagation rule. Repeat for all i and k.

The above algorithm is to be contrasted with a brute-force gradient algorithm during which *one additional* sample production run must be simulated for *each* g_i.

One should point out that validity of (16) also depends on the Principle of Superposition. We are assuming that the expected line production increase for Δb_i (i.e., permanent increase in the size of B_i for all t) is equal to the sum of the expected increase for each $\Delta b_i(k)$. Finally, the $\hat{g}_i$ we determine is a kind of 'virtual' gradient, since everything is evaluated based on the nominal $B_i(t)$ and $M_i(t)$. If we actually introduce a $\Delta b_i(k)$, then the subsequent history will change slightly. What is robust about our procedure is that it only depends on the *number and occurrence* of FOs and NIs (indirectly, the probability of machine failures and repairs). They are not expected to change greatly even though their length and time of occurrence may change greatly. This has indeed been verified by experimental results in a later section.

Condition(S) and the 'shifting' correction

The above development all hinges on the assumption that condition(S), which is defined as the occurrence of a failed state of M_i immediately following an FO_i state, does not occur. Let this FO_i state be part of a $[\text{bounce}]_i(k)$. Then the following phenomenon, as illustrated in fig. 4 a, occurs.

What happens is that while M_i was originally forced down at t_1 (it did not have a chance to enter the failed state because of FO_i at t_1), now, because of $\Delta b_i(k) = 1$, M_i enters the failed state at t_1 instead of t_2. A 'shift' occurs, completely eliminating the FO_i interval (and the NI_{i+1} interval as well in this case). In other words, what originally occurs as separate alternate failed states of M_i and M_{i+1} now becomes overlapping and simultaneous. A large decrease in production time results from this saving not predicted by the 'bounce' theory in proposition 1 (a–d). To put it more suggestively, we can view the production gain predicted by proposition 1 (a) as the linear part of the gradient due to $\Delta b_i(k)=1$, while the 'shifting' gain represents the nonlinear gain of $\Delta b_i(k)=1$. Thus, we are led to:

Proposition 5 $\Delta b_i(k)=1$ plus condition(S) implies the creation of a shifting gain

$$\Delta_s(i) = \text{Min}[FO_i, DN_i],$$

where FO_i and DN_i are the length of the full output and down state of M_i involved in condition(S).

Far less obvious is the fact that a 'shifting' gain at M_i not only can occur owing to $\Delta b_i(k) = 1$, but also owing to $\Delta b_j(k) = 1$, for $j > i$ provided FO states are also present at $M_{i+1}, \ldots, M_l$ $l \geq j$, i.e.:

Proposition 6 Condition(S) at B_i plus $\Delta b_j(k) = 1$ for $j > i$ implies a shifting gain $\Delta_i(s)$ provided the machines $M_{i+1}, \ldots, M_l$ $l \geq j$ are in FO states.

Proof: The FO states of $M_i, \ldots, M_l$ must be caused by a long DN state of M_{l+1}. Now if $\Delta b_j(k) = 1$, then all machines M_j $j \leq l$ including M_i will have *one more* cycle of production. Consequently, a 'shift' will occur at M_i since condition(S) is assumed to be present.

Figure 4 (b) illustrates this situation, which we shall designate as a *secondary* shift to distinguish it from the *primary* one of proposition 5.

It is worthwhile giving a heuristic explanation of the 'shifting' phenomenon. We know that alternate stoppages of two adjacent machines are the causes of inefficiency. Thus, if machines must fail then it is to our advantage that they fail simultaneously or in overlap fashion. On the other hand, the best that can be hoped for is for the machines to fail independently and for them to be isolated from each other. The probability that both machines will be down simultaneously under isolation is the highest overlap that can be achieved. If the machines are connected by a finite buffer, then there will be occasions when failure of one machine will cause a FD of the other machine. This other machine is then deprived of the opportunity during the FD to fail concurrently with the first machine, which is DN. When we increase the buffer size, we decrease the FD time and hence increase the probability that both machines can fail together. Thus, the expected 'shifting' gain is the gain that resulted from increasing the probability of joint failure of two machines.

From an algorithmic point of view, there is no difficulty in detecting the occurrence of a shift, primary or secondary, during a simulation run. We merely have to allow the creation of production gain of more than unit value. In other words, the

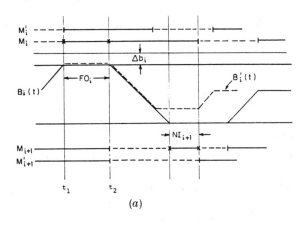

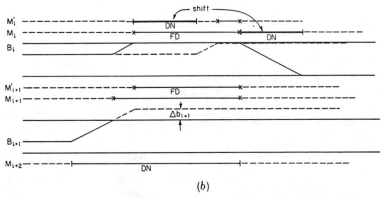

Figure 4. (b) illustrates the secondary shift at B_i due to $\Delta b_i + 1$.

x_{ik} and y_{ik} in proposition 4 can take on integer values other than 1. No basic difficulties arise in our algorithm in the previous section except for some modification in the propagation of the x_{ik} and y_{ik}s. These are detailed below.

Proposition 7 If $M_i(t)$ is translated forward by $\Delta_s(i)$ owing to shifting and another FO_i state occurs before an NI_{i+1} state, then all except one unit of the production gain is cancelled.

Proof: Since $M_i(t)$ is now ahead by $\Delta_s(i)$ with respect to $M_{i+1}(t)$, and the buffer B_i has increased in size by only one, $M_i(t)$ must now enter the FO_i state $\Delta_s(i) - 1$ cycles earlier since the FO_i state is governed entirely by $M_{i+1}(t)$, which has not been effected as yet.

Proposition 7 is in contrast but also consistent with the situation under proposition 1 (a) for the 'bounce' where a unit gain from $\Delta b_i(k) = 1$ is *not* cancelled by another FO_i state.

Proposition 8 If $M_j(t)$ is translated ahead by $\Delta_s > 1$ with respect to M_{j+1} (M_{j-1}), then the occurrence of $NI_{j+1}(FO_{j-1})$ implies that $M_{j+1}(t)$ ($M_{j-1}(t)$) will be translated ahead by $\text{Min}[\Delta_s, \text{duration of } NI_{j+1}(FO_{j-1})]$.

Proof: This is the direct extension of proposition 2 with essentially the same proof. The only difference is here the gain Δ_s may be larger than the duration of $NI_{j+1}(FO_{j-1})$.

On the other hand, because of the reasoning in proposition 7, we have:

Proposition 9 If $M_j(t)$ is translated ahead by $\Delta_s > 1$ with respect to M_{j+1} (M_{j-1}), then the occurrence of $FO_j(NI_j)$ implies that $M_j(t)$ will be moved back by Δ_s.

Propositions 7–9 describes how large gains are propagated. Operationally, this means that under normal circumstances we create a 'y' in B_{i-1} after FO_i and $\Delta b_i(k) = 1$, and similarly an 'x' in B_{i+1} when the following NI_{i+1} occurs, i.e. a bounce is completed. Under 'shifting' a 'y' with integer value is created in B_{i-1} but simultaneously an 'x' of the same size is created in B_i without waiting for NI_{i+1}. Furthermore, in forward propagation, we permit *fractional propagation* of xs and ys according to proposition 8. What is not propagated simply waits for the next occurrence of NI or FO state. Once again these requirements are easily implemented at practically no extra cost during a simulation run.

A more elaborate refinement concerning propagation as described by propositions 8 and 9 has to do with the fact that when large integer values of x_{ik} and y_{ik} are propagated, real *size and content* of the buffer may also become limiting factors. A FO(NI) when propagating a large x_{ik} and y_{ik} may induce additional FO(NI) states in succeeding buffers that may in turn produce additional propagations for the xs and ys. Thus, instead of one-stage propagation described in propositions 8 and 9, we may need to consider multi-stage propagation of the xs and ys. While this additional refinement is implementable, empirically we found that it is not necessary owing to apparent statistical averaging in a long production run. Hence it was not implemented in this study.

The above considerations can be summarized in the table below:

	Generation of local production gain	Propagation of local production gain
Linear	'Bounces' Proposition 1	Propositions 2, 3, and 4 Propositions 7, 8 and 9
Nonlinear	'Shifting' Propositions 5 and 6	Multistage propagation depending on size and content of all buffers

Table 1. Generation and propagation of production gain due to buffer size increase.

From a practical point of view, we can now visualize four different approaches to the approximate computation of the gradient $\Delta \tau / \Delta b_i$:

(i) Number of bounces in a simulation run, i.e., taking into account proposition 1 only.
(ii) $\hat{g}$ as described under the previous section (propositions 1–4 only).
(iii) $g^\#$ as described under this section accounting for both primarily and secondary shift (propositions 1–9 inclusive).
(iv) Brute-force calculation of g requiring one extra simulation run per component.

The important point is that both $\hat{g}$ and $g^\#$ can be determined in a single simulation run with little extra overhead. Whether or not $g^\#$ should be used instead of $\hat{g}$ depends on other considerations. Note that the occurrence of condition (S) or 'shifting' is directly proportional to and conditional on the number of 'bounces' at each buffer.

The product of this number with 'q', the failure probability, implies that condion(S) is infrequently occurring. Thus, if we were to compute the gradient by brute force comparison, an extremely long run is necessary before we can attain statistical regularity with respect to the *expected* contribution of 'shifting' to g_i. This has actually been observed experimentally where $\hat{g}_i$ approaches steady state long before g_i or $g_i^{\#}$. See the section on statistical significance for more details and explanations.

Algorithm for optimal allocation

The problem of maximizing the throughput of the production line without any constraints on buffers has a trivial solution: use of infinite buffers everywhere. An optimization is involved only when there is a cost on the buffer storage:

$$\max_{b_1,\ldots,b_{N-1}} \tau(b_1,\ldots,b_{N-1}) - C(b_1,\ldots,b_{N-1}), \tag{17}$$

where τ is the throughput of the system and C is the cost on buffers, as functions of buffer sizes.

A special case of (17) can be stated as follows:

$$\max_{b_1,\ldots,b_{N-1}} \tau(b_1,\ldots,b_{N-1}) \tag{18a}$$

subject to:

$$\sum_{i=1}^{N-1} b_i = B,$$

where B is the total buffer capacity. This form of the problem implicitly assumes a linear cost function with equal prices at each buffer, as (18a) can be written equivalently as:

$$\max_{b_1,\ldots,b_{N-1}} \tau(b_1,\ldots,b_{N-1}) - \lambda(b_1 + \ldots + b_{N-1} - B). \tag{18b}$$

A set of necessary conditions for (18b) follows immediately:

$$\frac{\Delta \tau}{\Delta b_i} = \lambda \qquad i=1,\ldots,N-1. \tag{19}$$

The section on the basic gradient approach gives a way of estimating $\Delta\tau/\Delta b_i$ as a number $\hat{g}_i$, at each buffer B_i, as a result of a real or simulated run of the system. The optimization, therefore, reduces to the necessary condition:

$$\hat{g}_i = \lambda \qquad i=1,\ldots,N-1, \tag{20}$$

which states that the estimated gradient at all buffer locations must be equal to each other at optimal allocation.

The following algorithm was used to solve equation (20):

(i) For initial $b_1,\ldots,b_{N-1}$, run the system to obtain $\tau_0, \hat{g}_1,\ldots,\hat{g}_{N-1}$.

(ii) Let $b_i \leftarrow b_i + G(\hat{g}_i - \bar{g})$ $i=1,\ldots,N-1$, where $\bar{g} = \frac{1}{N-1}\sum_{i=1}^{N-1}\hat{g}_i$ and G is the step size empirically determined.

(iii) Run the system to obtain $\tau, \hat{g}_i,\ldots,\hat{g}_{N-1}$.

(iv) If $\tau > \tau_0$, let $\tau_0 \leftarrow \tau$, go to (ii); otherwise, STOP. τ_0 is the maximum throughput.

Experimental results
Simulation details

All the experimental results presented in the next section were obtained by SIM, a digital simulator of transfer lines, written in FORTRAN, executed in PDP-11/70 under the Operating System RSX-11M.

SIM needs as input the characteristics of each machine, the buffer sizes and a set of random seeds to initialize the random number generator. Although not used in the systems described below, SIM can also simulate possible delays in input and output stations of the machines, as well as a production system with a network structure.

The simulation is based on a set of detailed but quite straightforward rules which determine the transitions from one state to another. Basically, each machine has four states: UP (working), DN (internal failure), NI (*no pieces in input buffer*) and FO (*output buffer full*); the last two stand for the forced-down states, the interactions between the components of the system. At the end of each UP cycle, which has a duration equal to the cycle time, the input and output buffers are checked for a possible force-down. If forced down, the machine remains so until the constraint in the buffer is released. When the force-down condition is absent, a biased coin is tossed to determine the next state. With probability q, a transition to DN occurs; otherwise, the next state is UP. If DN, another biased coin is tossed to determine the duration of DN cycle, which is taken from an exponential distribution with a given mean ($1/r$). The transition at the end of a DN cycle is always to a UP state. As for the buffer content, the input buffer is decremented by one at the beginning of a UP cycle; the output buffer is incremented by one at the end of a UP cycle.

The transient effects were also simulated using the following method. The simulation starts with all the buffers empty. There are p units (p is a fixed number, given as an input to the simulator) in the input of the first machine. The line operates until all these units pass through the line. The total production time PT, which marks the emergence of the last piece from the last buffer, is noted as the main result of the simulation. It is apparent that, by starting and ending with empty buffers, this method not only solves the problem of transient effects, but by picking p equal to the expected number of pieces produced between type changes in production, it is very realistic as well.

Other than PT, the simulator produces a run report showing K_i number of bounces at B_i, $\hat{g}_i$, expected gradient at B_i and other relevant run statistics, such as number of occurrences and time spent in each state for each machine.

Knowing p, number of pieces produced, and PT, production time, we can determine the throughput rate of the system as:

$\tau = p/(PT)$ in pieces per second, where 'second' denotes the time unit in simulations.

Since p is fixed through the different runs of a given system under different buffer distributions, small variations in production time are reflected into the variations in throughput rate through the equation:

$$\Delta\tau = -(p/(PT)^2)\Delta(PT).$$

This equation indicates that $\Delta(PT)$ can be used in place of $\Delta\tau$, for small variations in PT, introducing only a change in units with which throughput is measured. Thus, the symbol $\Delta\tau$ will be used to denote a change in either production time or throughput in the following section; the units will make the meaning clear.

Another feature of SIM is its ability to repeat a given number of runs, all with different but reproducible random seeds. This set of runs is called a batch. SIM produces a batch summary, similar in form to the run report, showing τ, K_i, $\hat{g}_i$ and other statistics accumulated through the batch.

Simulation results

In the experiments detailed below, four different systems are considered. The machine characteristics are shown separately in each case. The cycle time, T, of all the machines unless otherwise noted is 1 second.

(a) three machines, infrequent shifting

$a^{-1} = 200$ s
$q^{-1} = 1000$ s } for all the machines

Total buffer capacity $= 200$
Simulation length: 50 runs of 50 000 pieces each.

This is a case where the optimal solution is known beforehand. An equal division of the capacity among the two buffers will be optimal, since this is the only distribution satisfying symmetry. This distribution gave the following results:

i	b_i	K_i	$\hat{g}_i$	$g_i^{\#}$	g_i	occurrence of condition(S)
1	100	888	771	946	945	4
2	100	880	814	2362	2422	11

The numbers $K_i, \hat{g}_i, g_i^{\#}$ were all counted in one simulation, whereas g_is required one additional simulation per buffer by increasing the corresponding buffer size by one and calculating the difference in production time.

These results show that

(i) $g^{\#}$ predicts g very closely;
(ii) K and $\hat{g}$ show the expected symmetry, resulting from the symmetry of the system;
(iii) $g^{\#}$ and g are far from being symmetrical: thus, if an optimization is based on these values, b_2 would be favoured heavily, disturbing the equal distribution.

The apparent contradiction in (iii) is resolved by an examination of these numbers on a run-by-run basis. In 39 out of 50 runs, we see that $\hat{g}$ is an excellent predictor of g, while in the remaining 11 runs $\hat{g}$ underestimates g. Since $g^{\#}$ is a good predictor in these 11 runs while $\hat{g}$ is not, we conclude that these runs contain condition (S) and thus shifting.

The following is a summary of the results in the 39 runs without shifting:

i	b_i	K_i	$\hat{g}_i$	$g_i^\#$	g_i	occurrence of condition (S)
1	100	698	594	594	594	0
2	100	701	630	630	616	0

For these runs we can claim that

(i) both $\hat{g}$ and $g^\#$ predict g very closely;
(ii) the expected symmetry is present in all the columns;
(iii) this is the optimal distribution.

(b) three machines, frequent shifting

$$\left.\begin{array}{l} a^{-1} = 10\,\text{s} \\ q^{-1} = 20\,\text{s} \end{array}\right\} \text{for all the machines}$$

Total buffer capacity $= 10$
Simulation length: 10 runs of 25 000 pieces each.

In case (a) above, we claimed that the equal distribution was optimal, although both $g^\#$ and g indicated the opposite. This was due to the fact that the infrequent occurrence of condition(S) caused the gain due to shifting to be very unreliable for the purposes of optimization. Since the shifting was both rare and of a large magnitude, we chose to ignore it during the optimization.

Now, by increasing the value of q, we can make condition(S) occur more frequently, thus allowing the system to run long enough to average shifting effects. This gave the following results:

i	b_i	K_i	$\hat{g}_i$	$g_i^\#$	g_i	occurrence of condition (S)
1	5	3756	3333	5593	5350	805
2	5	3585	3190	5345	5147	728

We now have

(i) $g^\#$ predicts g (with some overestimation) within 5% accuracy;
(ii) all the columns show the expected symmetry (statistically);
(iii) this is the optimal distribution.

(c) five machines, infrequent shifting

$$\left.\begin{array}{l} a^{-1} = 200\,\text{s} \\ q^{-1} = 1000\,\text{s} \end{array}\right\} \text{for all the machines}$$

Total buffer capacity $= 400$
Simulation length: 50 runs of 50 000 pieces each.

In this case, we do not know much about the optimal distribution, except that it should be symmetrical. Starting from an equal distribution, the buffer capacity was redistributed, with the algorithm for optimal allocation, using

(i) K as the gradient, until no further improvement in τ;
(ii) $\hat{g}$ as the gradient, until no further improvement in τ.

$g^{\#}$ was not used in the optimization, because, as already shown in case (a) above, $g^{\#}$ is not reliable, shifting gains being so rare and of such a large magnitude in the system under consideration.

The results can be summarized as follows:

Iteration		b_1	b_2	b_3	b_4	τ (1000 s)
#0	Initial distribution	100	100	100	100	4218
#9	End of (i)	80	122	126	72	4203
#16	End of (ii)	60	146	142	52	4196

The history of hill climbing is shown in fig. 5, where the first two buffer sizes and $\Delta\tau$ (the improvement in production time as compared to the initial distribution) are plotted as a function of iteration number. Particularly interesting is the kink on $\Delta\tau$ curve at iteration #9, which marks a change in the definition of the gradient.

Sixteen iterations were sufficient, in this case, to come to a point where no more improvement in τ was possible in the direction of $\hat{g}$. Randomly selected changes in buffer sizes around this point showed only a slight improvement (at fifth significant digit) in τ, thus confirming the optimality.

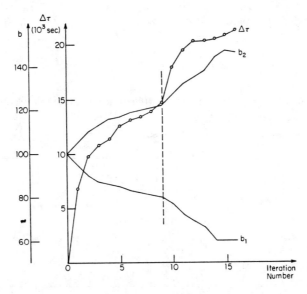

Figure 5. Hill climbing for the 5-machine case. K was used as gradient up to iteration 9, after which $\hat{g}$ replaced K.

The last distribution, claimed to be optimal, was examined to test the equality of $\hat{g}, g^{\#}$, and g. (g_is were determined by four additional simulations, one for each buffer.)

i	b_i	K_i	$\hat{g}_i$	$g_i^{\#}$	g_i	occurrence of condition (S)
1	60	1183	742	1214	1212	5
2	146	1003	796	958	957	5
3	142	1039	812	991	988	9
4	52	1225	783	1404	1404	6

It is seen that
(i) $g^{\#}$ predicts g very accurately;
(ii) $\hat{g}$ confirms optimality, whereas g favours the first and last buffers.

Since optimality was shown by random changes in buffer sizes, this supports our point that $g^{\#}$ (or g) is not reliable as the gradient in systems with infrequent shifting. Indeed, by eliminating the 10 runs where condition(S) occurred, we have the following results for 40 runs:

i	b_i	K_i	$\hat{g}_i$	$g_i^{\#}$	g_i	occurrence of condition (S)
1	60	952	610	610	609	0
2	146	809	644	644	644	0
3	142	834	655	655	654	0
4	52	960	606	606	598	0

(d) five machines, frequent shifting

This case will be presented as an example of non-uniform machine parameters, as well as quite frequent shifting, to show that $g^{\#}$ is still a good predictor of g.
Machine characteristics and buffer sizes:

i	a_i^{-1}(s)	q_i^{-1}(s)	b_i
1	11	20	5
2	19	167	11
3	12	22	8
4	7	22	7
5	7	26	—

Simulation length: 50 runs of 5000 pieces each.

Results:

i	b_i	K_i	$\hat{g}_i$	$g_i^{\#}$	g_i	occurrence of condition (S)
1	2821	1448	2592	2450		470
2	1746	1367	2479	2339		491
3	2310	1649	2702	2609		471
4	1780	870	1344	1304		264

The agreement between $g^{\#}$ and g is within 6%.

Statistical significance

In this section, three questions with respect to the significance of the results will be answered as an illustration of the application of statistical methods.

(a) In case (a) in the previous section, after cancelling the runs with 'shifting', it was claimed that $\hat{g}_1 = \hat{g}_2$. Does this hypothesis pass statistical tests?

There are 39 independent runs with individual $\hat{g}_i$s. Assuming 39 is large enough to apply central limit theorem, we can use normal distributions in each case. 95% interval estimates for them on per-run basis were calculated as follows:

i	$\hat{g}_i$
1	$15 \cdot 2 \pm 0 \cdot 8$
2	$16 \cdot 1 \pm 0 \cdot 7$

The hypothesis $\hat{g}_1 = \hat{g}_2$ is supported by these interval estimates. Moreover, it could not be rejected by applying an equality test on sample means at 95% level of confidence.

(b) Was the simulation interval long enough for the random number generator to generate the desired distributions?

This question will be answered for the estimation of a^{-1} in case (c) of the previous section. Let d_i be the random variable denoting the length of the ith DN cycle. d_i is taken from an exponential distribution:

$$Pr\ (d_i \geq t) = \exp(-at)$$

with:

$$E(d_i) = a^{-1},\ \text{var}(d_i) = a^{-2},\ \text{all } i.$$

Observing m DN cycles through the simulation, the average duration $\theta = 1/m\ (d_1 + \ldots + d_m)$ is an unbiased estimator of a^{-1}, mean time to repair. Since θ is a sum of m independent exponential random variables, it is distributed as Γ_m with $E(\theta) = a^{-1}$ and $\text{var}(\theta) = a^{-2}/m$.

In the case under study, since there are 50 runs with 50 000 pieces each, and since $q = 0 \cdot 001$, the expected number of DN cycles is 2500. (The actual number varied between 2447 and 2613 with the random seeds used.) With this many terms, the gamma distribution behaves like a normal distribution for all practical purposes.

Using $a^{-1} = 200$ and $m = 2500$, we have:

$$\theta \sim N(200, 4).$$

Therefore, at 95% level of confidence, we can claim:

$$\left| \frac{\theta - 200}{4} \right| \leqslant 1.96$$

or

$$192 \leqslant \theta \leqslant 208.$$

It was observed that for all five machines, θ was in this interval.

(c) The improvement in τ claimed in case (c) of the previous section was realized using one set of random seeds. Is the improvement statistically significant?

This question will be answered by a comparison of τ using two sets of random seeds, run at the initial and optimal distributions. The following 95% interval estimates on τ and $\Delta\tau$ are based on fifty runs, with the assumption of normal distribution:

	Seed no. 1	Seed no. 2
Initial distribution τ_0	4218 ± 47	4221 ± 47
Optimal distribution τ_1	4196 ± 47	4199 ± 47
Improvement $\Delta\tau = \tau_0 - \tau_1$	21.3 ± 5.5	21.5 ± 5.1

(Units in 1000s)

The analysis above shows that the variance on $\Delta\tau$ is much smaller than the variance on τ; therefore, the improvement $\Delta\tau$ is significant.

The effect of different cycle times

Let the cycle time of machine M_i be T_i. The main effects introduced by different cycle times in our theory are three. First in the generation of production gain in M due to $\Delta b_i(k)$, we have:

Proposition 10 Equation (7) of proposition 1 (a) remains valid, provided we replace T in (7) by:

$$T = \text{Max}(T_i, T_{i+1}) \tag{21}$$

and

$$FO_i[t_1, t_2] > \text{Min}(T_i, T_{i+1}).$$

Proof: If $T_i > T_{i+1}$, then it is clear that (7) remains true by definition. If $T_{i+1} > T_i$ then the termination of the FO_i (when both M_i and M_{i+1} start to produce) must be followed by at least another FO_i of duration $T_{i+1} - T_i$ which, in turn, is followed immediately by a DN state of M_i. Thus, M_i first gains T_i seconds due to $\Delta b_i(k) = 1$. But an additional $T_{i+1} - T_i$ seconds is gained when M_i goes into the DN state. The last FO_i before the DN state simply will not take place. This is illustrated by fig. 6.

The second effect of different cycle time is in the propagation of the production gains. A careful examination of the proofs of propositions 1 (b)–(d) shows that they remain valid even for different cycle times of M_i and M_{i+1}, provided the relevant NI and FO are of sufficiently long duration. However, in this case of different cycle ti...

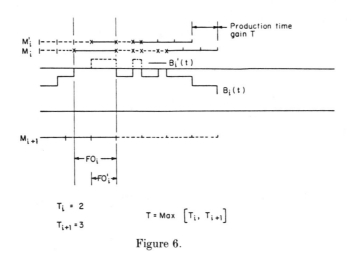

Figure 6.

it is possible to have NI and FO states of durations less than either T_i or T_{i+1}, e.g., both machines working with buffer empty or full. Consequently, we modify eqns. (9c) and (10c) by:

Proposition 11 Equations (9c) and (10c) and hence the conclusions of propositions 2, 3, and 4 remain valid if we replace T in equations (9c) and (10c) by:

$$M'_{i+1}(t) = M_{i+1}(t+T^*)$$
$$T^* = \text{Min }(NI_{i+1}, T) \qquad (22)$$

and

$$M'_{i-1}(t) = M_{i-1}(t+T^*)$$
$$T^* = \text{Min }(FO_{i-1}, T) \qquad (23)$$

respectively.

Proof: We only need to prove (22) since the argument for (23) is completely parallel. If $NI_{i+1} > T$, then no change in the arguments of propositions 1(b)–(d) is necessary. If $NI_{i+1} < T$, then $B_i(t_3) > 0$, since both M_i and M_{i+1} must be in UP state just prior to t_3 and since M_i is ahead by T with respect to M_{i+1}. Thus, $[NI'_{i+1}] = 0$, which implies (22)

The upshot of proposition 11 is that in the propagation of the production time gains (x_{ik} and y_{ik}), we must be prepared to subdive these gains and permit fractional propagations. Otherwise, all our results remain valid. This represents no difficulty in implementation since the requirement is already incorporated as a result of our earlier discussion on shifting. The only comment we should note is that because of the presence of these short NI or FO durations, we expect the nonlinear effect of propagation discussed earlier to be more pronounced. Hence we expect less accuracy in g^* in estimating the true g. Experimental data confirms this, although the accuracy is still quite respectable and completely adequate for hill climbing purposes.

Finally, we should note that FO of short duration $T_{i+1} - T_i$ cannot give rise to 'shifting' by definition since machine M_{i+1} is not in a DN state. Hence in the detection of condition(S) during a simulation, these short FO states are not considered.

These are the only three refinements that need to be implemented in an algorithm for the general case of different cycle time.

We now show some simulation results on a seven-machine line with different cycle times.

Machine characteristics and buffer sizes (where $T_i a_i^{-1}$ = mean time to repair and $T_i q_i^{-1}$ = mean time between failures)

i	T_i(s)	$T_i a_i^{-1}$(s)	$T_i q_i^{-1}$(s)	b_i
1	40	450	820	5
2	34	760	5700	11
3	39	460	870	8
4	38	270	830	7
5	37	270	970	19
6	40	650	1900	4
7	43	320	1100	—

Simulation length: 50 runs of 5000 pieces each.
Results are tabulated below.

i	b_i	K_i	$\hat{g}_i$	$g_i^\#$	g_i	occurrence of condition (S)
1		3207	1276	2294	2483	523
2		1837	1186	2137	2298	530
3		2665	1478	2539	2687	561
4		2497	1055	1756	1895	446
5		979	524	1006	1072	303
6		3326	817	1139	1140	259

As discussed earlier, we expect that $g_i^\#/T_i$ will be a less accurate predictor of g_i/T_i owing to the additional complexity introduced by the different cycle times. But even in this case, we observed that the agreement between $g_i^\#/T_i$ and g_i/T_i is still within 7·5%.

Conclusion and future work

In this paper, we presented a design algorithm for optimal buffer size allocation in transfer lines. Our algorithm has been shown to be both efficient and robust. In comparison with the brute-force gradient approach, our algorithm can generate the gradients at all buffer locations in a single simulation run. For real-world transfer lines, this savings in computation burden is usually significant, for example, a ratio of 1 : 27 for the quoted 27-machine transfer line. For a design problem of this scope, we have not yet known of any feasible analytical approach. The second significant feature of our approach is its robustness. Our algorithm, unlike the Markov-chain approach, can accommodate arbitrary distribution functions characterizing the machine failure and repair processes. There also exists strong experimental evidence from the simulation results and theoretical analysis that our algorithm is independent of the differences in the cycle times of the machines involved.

A natural extension for the application of the concept of estimated sensitivities is worth mentioning. An important aspect of the buffer sizing in shop practice is the on-line management of additional floor space for buffer storage, as well as the on-line assignment of the repair and the loading–unloading crews. To successfully resolve this 'dynamic' buffer management problem, one of the basic questions is the optimal choice of the on-line information. It is our belief that the on-line gradient is of fundamental information value. As a further demonstration of its flexibility, our algorithm can easily be implemented on the on-line monitoring system. There is, in fact, an ongoing tandem task of a comprehensive transfer-line monitoring and management system now under development, in collaboration with the production engineers of the Gruppo Auto of Fiat S.p.A.

Acknowledgments

This paper reports the results of a research programme sponsored by Fiat S.p.A. The authors wish to thank Mr. F. Guaschino, Director of Fiat Stabilimento Mirafiori Meccanica, and the engineers of Fiat/Gruppo Auto for their support and contributions to the development of this system. The authors are also indebted to many colleagues, in particular, Dr. James S. Rhodes, of The Charles Stark Draper Laboratory, Inc., for many profound and fruitful discussions which substantially stimulated our insight on the buffer size allocation problem.

Significant portions of the theoretical work were also supported by the Office of Naval Research under N00014-75-C-0648 and through the National Science Foundation Grant ENG 78-15231 administered through the Division of Applied Sciences, Harvard University, for basic research in decision and control.

Dans cet article, nous presentons une solution complète et nouvelle aux problèmes bien connus de la conception de dispositifs de stockage-tampons dans une ligne de fabrication en série. L'élément-clé de cette solution est un calcul efficace du vecteur de gradient des quantités traitées par rapport aux tailles de systemes-tampons. Les résultats analytiques et expérimentaux sont présentés.

In dieser Arbeit legen wir eine neuartige. umfassende Lösung des bekannten Problems der Pufferspeicherauslegung in eine Serienfertigungsanlage vor. Der Hauptbestandteil der Lösung ist die effiziente Berechnung des Neigungsvektors des Durchgangs in bezug auf die verschiedenen Puffergrößen. Es werden analytische und experimentelle Ergebnisse geliefert.

References

ARTAMONOV, G. T., 1976, Productivity of a two-instrument discrete processing line in the presence of failures, *Cybernetics*, **12**, 464. (Translation from Russian).
BUZACOTT, J. A., 1967, Automatic transfer lines with buffer stocks, *Int. J. Prod. Res.*, **5**, 183.
GERSHWIN, S. B., and SCHICK, I. C., 1978, Analytic methods for calculating performance measures of production lines with buffer storages, *Proceedings of the 1978 IEEE Conference on Decision and Control*.

HANIFIN, L. E., LIBERTY, S. G., and TARAMAN, K., 1975, *Improved transfer line efficiency utilizing systems simulation*, Society of Manufacturing Engineers, Technical Paper MR75-169.

HILLIER, F. S., and BOLING, R. W., 1966, The effect of some design factors on the efficiency of production lines with variable operation times, *J. Ind. Eng*, **17**, 651.

IGNALL, E., and SILVER, A., 1977, The output of a two-stage system with unreliable machines and limited storage, *AIIE Trans.*, **9**, 183.

KRAEMER, S. A., and LOVE, R. F., 1970, A model for optimizing the buffer inventory storage size in a sequential production system, *AIIE Trans.*, **2**, 64.

MASSO, J., and SMITH, M. L., 1974, Interstage storage for three-stage lines subject to stochastic failures, *AIIE Trans.*, **6**, 354.

OKAMURA, K., and YAMASHINA, H., 1977, Analysis of the effect of buffer storage capability in transfer line systems, *AIIE Trans.*, **9**, 127.

RAO, N. P., 1975, Two-stage productions systems with intermediate storage, *AIIE Tran*. **7**, 414.

SEVASTYANOV, B. A., 1962, Influence of storage bin capacity on the average standstill time of a production line, *Theory of Probability and Applications*, **7**, 429, (Translation from Russian).

SHESKIN, T. J., 1976, Allocation of interstage storage along an automatic production line, *AIIE Trans.*, **8**, 146.

Perturbation Analysis Explained

Y. C. HO

Abstract—This note provides a simple explanation of the ideas behind perturbation analysis.

Perturbation analysis (PA) is a technique for the computation of the gradient of performance measure (PM) of a discrete-event dynamic system with respect to its parameters (θ) using only one sample path or Monte Carlo experiment of the system. When first presented, one's immediate reaction has often been incredulity stemming from the belief that "one cannot get something for nothing." Later this disbelief may be developed into a more sophisticated and technical objection involving the legitimacy of interchanging differentiation and expectation operators or the probabilistic convergence of the PA estimate to its true value (more about these later). In nontechnical terms, these translate to "How can you squeeze out information about a trajectory/sample-path operating under one value of the system parameter from that of another operating under a different value? Don't the two trajectories behave entirely dissimilarly?" First we shall offer a conceptual explanation of the above query in terms of two basic ideas in random variable generation. Next a simplest example of PA will be given. This example covers the essence of PA in the simplest terms stripped of all of its computational trappings, efficiency tricks, etc. Once the basic ideas are made clear, we leave the question of "how efficient" to other papers and potential implementors. For similar tutorial reasons, the example used below is not at all the most general case for which PA can be applied. The purpose here is to convey the fundamental idea behind this approach in terms accessible to the largest audience. Interested readers are referred to other PA papers which at this point total over 40 (see [7], [8]).

First, the explanation in terms of random variable generation.

The transform method of generating samples of random variables with arbitrary distribution.

It is well known that given samples of uniform distribution between [0, 1], $u_1, u_2, \cdots$, etc., we can transform these into samples, $x_1, x_2, \cdots$, from arbitrary distribution $F(x)$ by using the inverse transform $F^{-1}(u)$ provided it exists. This is an elementary way of using samples from one distribution to get samples from, and hence information about, another distribution.

The so-called "likelihood ratio/score function/change of measure method of single run sample path analysis is based on the same principle [4]-[6]. Let $f(x; \theta)$ be the density function of the stochastic process under discussion and PM $(\theta) = \int L(x) f(x; \theta) \, dx$ be some performance measure. Let $g(x) = f(x; \theta + \Delta\theta)$, then we have

$$\text{PM}\,(\theta + \Delta\theta) = \int L(x) g(x) \, dx = \int L(x) \frac{f(x)}{f(x)} g(x) \, dx$$
$$= \int \left[L(x) \frac{g(x)}{f(x)} \right] f(x) \, dx.$$

In other words, we can compute PM ($\theta + \Delta\theta$) using the samples from $f(x)$ provided we appropriately *rescale* the sample performance measure $L(x)$ as above. This is conceptually exactly the same as rescaling the uniform random variables by the inverse transform to get $F(x)$-distributed random variables.

The rejection method of generating samples of random variable with arbitrary distribution.

The other well-known method of generating random variables of arbitrary distribution is the so-called rejection method which is best illustrated using Fig. 1.

The samples generated by the uniform random number generators in the shaded region are simply thrown away, and those in the unshaded region retained to yield samples from the density function $f(x)$. The "cut-and-paste" method of extended perturbation analysis (EPA) [2] is simply the application of the "rejection" idea to the case of stochastic processes from discrete-event simulations.

Now let us offer the simple example of PA. Picture a finite state machine consisting of three states only: A, B, C and a system parameter θ which can assume values of 0 and 1. Under $\theta = 0$, the trajectory of the machine is periodic, $A \to B \to C \to A \to B \to \cdots$. While under the condition $\theta = 1$, the trajectory from A has equal probability to go to the other two states, and similarly for the states B and C resulting in a completely symmetrical and stochastic system. Now consider a nominal path (NP) under $\theta = 0$

$$A, B, C, A, B, C, A, B, C, A, B, C, A, B, C,$$
$$A, B, C, A, B, C, A, B, C, A, B, C, \cdots \quad \text{(NP)}$$

and one realization of a perturbed path (PP) with $\theta = 1$ which may look like

$$A, B, A, B, C, A, B, C, B, C, B, C,$$
$$A, B, A, B, C, A, B, C, B, C, A, B, C, B, C, \cdots \quad \text{(PP)}$$

We claim we can construct an equivalent perturbed path (constructed perturbed path) from the data of (NP) as follows: At any state, say A, we flip a fair coin to see where the perturbed path could have gone. If the outcome of the flip is to B, then we continue to follow along the NP since no deviation from the NP took place. If on the other hand, the outcome is C, then we simply skip the next state along NP which is B and wait until the state C actually occurs on the NP. At that time we can follow the NP until the next occurrence of coin flip deviation and the process repeats $\cdots$. Exactly the same "coin toss–test–accept or wait" scheme can be followed for other states along the NP. Thus, by selectively (according to the outcome of the coin toss) discarding sections of the NP, we constructed an equivalent perturbed path, CPP, as below

$$A, B, A, B, C, A, B, A, B, C, B, C,$$
$$A, B, C, A, B, A, B, A, B, C, A, \cdots \quad \text{(CPP)}$$

which is obtained from NP by skipping segments of the state sequence

$$A, B, \mathcal{C}, A, B, C, A, B, \mathcal{C}, A, B, C, \mathcal{A}, B, C,$$
$$A, B, C, A, B, \mathcal{C}, A, B, C, A, B, C, A, B, C, A, \cdots \quad \text{(NP)}$$

Note that the CPP thus constructed need not be the same as the particular PP above. Nevertheless, we submit that CPP thus constructed is statistically indistinguishable from the PP since they were constructed out

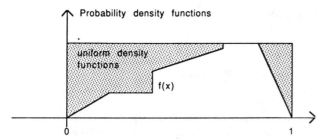

Fig. 1. The rejection method of random variable generation.

of the same probabilistic mechanism.[1] This idea, which we shall call "invariance of state sequence under cut-and-paste" depends on the following two things.

i) A long enough trajectory or a large enough ensemble of trajectory segments of an irreducible finite state machine eventually produces all conceivable behavior of the system under all parameter values. The only difference in trajectories under different parameter values is in the frequency of occurrence of different states and sequences.

ii) Using short term calculations (e.g., coin toss or next state transition function) which are given system knowledge, we can pick out selected segments of a trajectory to reconstruct a CPP under the operating condition of an arbitrary parameter value. Only state transition calculations are needed to enable us to construct the CPP. There is no need to extrapolate the path deviation indefinitely into the future which is what construction of a PP requires. The Markov property allows the construction of CPP to be repeatedly restarted.

Alternatively, we can squeeze out system behavior information under θ' from that of θ in the following way, which is the analog of the inverse transform method for random number generation. It is entirely possible for a path such as $ABCABC$ to be generated under $\theta = 1$ instead of under $\theta = 0$. Its probability of occurrence is different under the two different parameter values. Thus, if we appropriately rescale the contribution of any path to an ensemble average by weighting it with different probabilities (i.e., change of measure), then one can in fact compute statistical averages of system performance under one probability distribution θ' using samples from that of another distribution θ. This is the approach to single run gradient estimation by Rubinstein [4], Reiman and Weiss [5], and Glynn [6]. Instead of changing the measure, in PA under "cut-and-paste" we change the path directly. This can be thought of as a dual approach. The point is that there is nothing counterintuitive about a given trajectory under θ that can contain information about its behavior under θ' as long as one uses the given system knowledge to help analyze it.

Now for PA, most of the time we are interested in the case of $\Delta\theta = d\theta \rightarrow 0$. Let us consider a simple variation on the above example. Suppose now we can perturb the value of $\theta = 0$ by a small amount $\Delta\theta$. This perturbation does not change the behavior (order) of the state sequence $A \rightarrow B \rightarrow C \rightarrow A \cdots$ under $\theta = 0$. It merely changes the duration of the time interval between the occurrence of the states (events) A and B by an amount Δt_{AB} which is proportional to $\Delta\theta$, say $\Delta t_{AB} = a*\Delta\theta$.

Question: What is the change in the termination time of the 1 000 000th occurrence of the state (event) C with $\theta = 0 + \Delta\theta$?

---

[1] The alert reader may wish to point out that in this case the construction of CPP is computationally not that different from actually implementing PP itself. Two points should be made. An actual comparison can show that even in this case there is a small saving in terms of codes required to implement PP separately or to implement CPP as part of NP. This is because, for CPP one only has to mechanize the functions of "test" and "wait" which do not become any more complex with systems of increasing complexity. On the other hand, for PP one has to perform additional tasks such as "future event list generation," "process updating," and "trajectory evolution for the perturbed path" which can be complex for practical simulation experiments. More importantly, as the dimension of the parameter vector becomes larger than 1, say n, we can recreate equivalent CPP's for different parameter perturbations from a single NP without making the length of NP n times as long. This is because segments of NP skipped as a result of one parameter may very well be used in the construction of another CPP for a different parameter. Consequently, we preserve the inherent N:1 advantage of perturbation analysis. However, we emphasize that the purpose of this note is not to argue computational efficiency but to answer the intuitive disbelief about information contained in a stochastic sample path. For another view to this question see additional discussions below and more details about efficiency in [2].

Answer: Since the state sequence did not change, all Δt_{AB} generated will accumulate and propagate to other events intact. We have by inspection $\Delta T_C(1\,000\,000) = 1\,000\,000\,a*\Delta\theta$.

The crucial point here is the invariance of the state sequence between the NP and the PP. This makes the propagation of perturbation from one event to another using NP alone very easy. More generally, if the event chain (critical timing path) of a simulation/experiment remains invariant under perturbation, then the calculation and propagation of perturbations of any event in the system can be efficiently done using the NP alone.

The above two ideas: "state sequence invariance under cut-and-paste" and "perturbation propagation under event chain invariance" can now be combined to answer a nontrivial question about the system under discussion. Imagine now θ can be varied continuously from 0 to 1. As θ changes from 0 to $\Delta\theta$, not only the duration between event A and B change as described above, but the probability of state transition from A to B (similarly for $B \rightarrow C$ and $C \rightarrow A$) changes from 1 to $1 - 0.5\Delta\theta$ and that of A to C (similarly for $B \rightarrow A$ and $C \rightarrow B$) from 0 to $0.5\Delta\theta$ in complementary fashion.

Nontrivial Question: Using the NP trajectory with $\theta = 0$, what is the expected termination time of the 1 000 000th occurrence of the state C under $\theta = 0 + \Delta\theta =$, say, 0.01? (For simplicity, let us assume that the time durations from $A \rightarrow B$, $B \rightarrow C$, $C \rightarrow A$ are always the same and equal to 1 unit except for $A \rightarrow B$ when perturbed.)

Answer: Using the first idea we can identify the segments of the NP that should be skipped to construct a CPP. Using the second idea we only generate and propagate perturbations along the CPP. In other words, all calculations are frozen during the skipped segments. We leave as an exercise for the reader to prove that for any $\Delta\theta$ the expected change in the termination time is $1\,000\,000\,a(\Delta\theta - \Delta\theta^2/2)/3$.

Finally, for those readers who are more technically inclined, the seemingly intuitive objection mentioned in the opening paragraph can be translated into a mathematical question as follows. Let $\theta =$ system parameter, $\omega =$ sample path, and PM $(\theta, \omega) =$ sample performance measure, the basic question for PA can be posed as follows.

Question: Infinitesimal PA (i.e., $\Delta\theta = d\theta$) calculates the expected value of the derivative of PM (θ, ω). Yet one is really interested in the derivative of the expected value of PM (θ, ω) assuming it is well defined and exists. When are these two quantities equal, i.e., when is the interchange between E and $d/d\theta$ valid?

$$E\left\{\frac{d\,\text{PM}\,(\theta, \omega)}{d\theta}\right\} = ? = \frac{dE\{\text{PM}\,(\theta, \omega)\}}{d\theta}.$$

The reason this is the mathematical version of the same question is that for any trajectory or ensemble of paths event chain invariance between NP and PP will be violated with probability one if the trajectory is long enough or the ensemble is large enough regardless of the size of $d\theta$. Once violated, the NP and PP may look entirely different henceforth, i.e., arbitrarily small perturbation in θ will always lead to discontinuities in trajectory, and hence in PM (θ, ω). In fact, one can easily construct an example for which PM (θ, ω) is discontinuous in θ for every ω. Elementary analysis tells us that such discontinuities may rule out any possibility for the above-mentioned interchange to be valid. Hence, the mathematical objection that PA cannot provide a consistent estimate of the derivative of expected PM by computing the expectation of the derivatives.

There are several basic ways to answer this objection mathematically. First, one can show that for a large class of problems, such discontinuities do not occur frequently enough to amount to anything significant in the averaging process (see [7, sect. III]). Secondly, discontinuous path change (instead of $A \rightarrow B \rightarrow C$, we have $A \rightarrow B \rightarrow A$) can be avoided by "cut-and-paste." Lastly, they can be smoothed out by mathematical averaging. Note that even though PM (θ, ω) may be discontinuous, a sufficient amount of average can make $E[\text{PM}\,(\theta, \omega)]$ differentiable. In fact, the usual brute force way of computing sensitivity is via

$$\lim_{n \rightarrow \infty, \Delta\theta \rightarrow 0} \left\{\frac{1}{n}\sum_{k=1}^{n}\text{PM}\,(\theta + \Delta\theta, \omega_k) - \frac{1}{n}\sum_{j=1}^{n}\text{PM}\,(\theta, \omega_j)\right\}/\Delta\theta$$

which is a simple statement of the above fact. Now the trick with PA is to average just enough to avoid discontinuities but not to require the duplication of another experiment as in the brute force case. For this we consider

$$\frac{dE[PM(\theta, \omega)]}{d\theta} = \frac{dE_z E_{/z}[PM(\theta, \omega)]}{d\theta} \equiv \frac{dE_z[PM(\theta, z)]}{d\theta}$$

$$= ? = \frac{E_z d\, PM(\theta, z)}{d\theta}$$

where z is some conditioning variable computable along a sample path. We submit that because of the smoothing property of conditional expectation the interchange of E_z with $d/d\theta$ has much greater chance of succeeding.

Between the extremes of differentiating the expectation and taking expectation of the differentiation, a whole spectrum of possibilities exist. There should always exist an intermediate z to render PA possible.

For more analytical details, readers are referred to [2], [3], [7] and references therein.

REFERENCES

[1] X. R. Cao, "Convergence of parameter sensitivity estimates in a stochastic experiment," *IEEE Trans. Automat. Contr.*, vol. AC-30, pp. 845-853, Sept. 1985.
[2] Y. C. Ho and S. Li, "Extensions of infinitesimal perturbation analysis," *IEEE Trans. Automat. Contr.*, May 1988.
[3] W. B. Gong and Y. C. Ho, "Smoothed perturbation analysis of discrete event dynamic systems," *IEEE Trans. Automat. Contr.*, vol. AC-32, 1987.
[4] R. Y. Rubinstein, "Sensitivity analysis and performance extrapolation for computer simulation models," manuscript, Harvard Univ., Cambridge, MA, 1986.
[5] M. I. Reiman and A. Weiss, "Sensitivity analysis for simulations via likelihood ratios," preprint 1986.
[6] P. Glynn and J. L. Sanders, "Monte Carlo optimization in manufacturing systems: Two new approaches," in *Proc. ASME CIE Conf.*, Chicago, IL, 1986.
[7] Y. C. Ho, "Performance evaluation and perturbation analysis of discrete event dynamic systems," *IEEE Trans. Automat. Contr.*, July 1987.
[8] "A selected and annotated bibliography on perturbation analysis," in *Proc. IIASA Conf. Discrete Event Systems.* Springer-Verlag Lecture Notes on Control and Information Science #103, 1988.

Parametric Sensitivity of a Statistical Experiment

Y. C. HO

Abstract—We describe a general procedure for evaluating the parametric sensitivity of a Monte Carlo experiment and illustrate its effective applications.

Suppose we are interested in the expected value of a very complicated function $L(u,w)$,

$$J = E[L(u,w)] \qquad (1)$$

where u is a set of parameters, w is a given random vector with specified $p(w)$, and $L(\cdot,\cdot)$ represent the output of some complex system. Let us assume that the only feasible way to calculate (1) is via a Monte Carlo experiment on the *real* or *simulated* L. Now we wish to ask an important auxiliary question: what is $\Delta J/\Delta u$, the gradient of J?

A brute force approach would be to perform another statistical experiment with one of the parameter value, say u_i, slightly changed to $u_i + \Delta u_i$, then

$$\delta J_i \equiv E[L(u+\Delta u_i, w)] - E[L(u,w)]$$
$$\approx \sum_s L(u+\Delta u_i, w_s) - \sum_s L(u, w_s) \qquad (2)$$

where

$$\Delta u_i \equiv [0, \cdots, 0, \delta u_i, 0, \cdots, 0],$$

$\sum_s$ denotes sample averages[1] and w_s denotes samples of w.

We repeat (2) for each parameter u_i, $i=1,\cdots,n$ using a total of $n+1$ experiments to calculate $\Delta J/\Delta u$. This is often infeasible in terms of computing time requirement.

It is also prone to numerical difficulties since in (2) we are taking the difference between two sample averages which has, hopefully, small variances. A numerically less naive way to calculate (2) is

$$\delta J_i = \sum_s [L(u+\Delta u_i, w_s) - L(u, w_s)] \qquad (3)$$

where the difference is taken on a sample-by-sample basis. However, this nevertheless does not eliminate the requirement that we *repeat* the entire experiment n times (with identical samples) in order to calculate $\Delta J/\Delta u$.

On the other hand, suppose we can calculate the difference $L(u+\Delta u_i, w_s) - L(u, w_s)$ based on the sample $L(u,w_s)$ and Δu_i only *without* having to evaluate $L(u+\Delta u_i, w_s)$, then a considerable saving in computer time can be effected. Only the original statistical experiment plus some overhead need to be implemented. The purpose of this note is to point out some examples where this idea of parametric sensitivity calculation has been and can be effectively exploited.

Manuscript received April 16, 1979; revised August 1, 1979. This work was supported in part by the National Science Foundation under Grant ENG-8-15231 and in part by the U.S. Office of Naval Research under Contract N00014-75-C-0648.
The author is with the Division of Applied Sciences, Harvard University, Cambridge, MA 02138.

[1]For notational convenience, we use Σ_s to denote sample averages with the understanding that the sum is to be divided by appropriate constants.

The first proposed application concerns cases where we can write

$$L(u+\Delta u_i, w_s) - L(u, w_s) = \frac{\partial L(u, w_s)}{\partial u_i} \Delta u_i \qquad (4)$$

where L is differentiable and $\partial L/\partial u_i$ is easy to evaluate. Then along with the original Monte Carlo experiment we can calculate

$$E\left[\frac{\partial L(u,w)}{\partial u}\right] \approx \sum_s \frac{\partial L(u, w_s)}{\partial u} \triangleq g \qquad (5)$$

which is the gradient vector.

As an illustration consider a stochastic nonlinear dynamic system under feedback control. We model such a system by

$$\dot{x} = f(x, p, w) \qquad (6)$$

where

x is the state of the dynamic system
p is a set of parameters of the controller
w is some physically realizable noise or disturbance.

Such a model could easily be the result of applying an extended Kalman filter and deterministic controller to a stochastic nonlinear dynamic system or an adaptive control system.

Such a system is often designed by a combination of theory and heuristics. To test the actual performance of the system, a Monte Carlo simulation is performed, namely, we repeatedly simulate (6) from 0 to T with a different realization of $w(t)$ and fixed p. Let $x_s(t)$ and $w_s(t)$ be the actual sample trajectories, then we measure performance by

$$J = \sum_s \int_0^T L(x_s, p, w_s) dt + \phi(x_s(T)) \qquad (7)$$

where $\phi + \int L\, dt$ is the performance criterion.

Our point is that at small added cost we can in fact get as a byproduct of this Monte Carlo experiment the gradient

$$\frac{\partial J}{\partial p} \triangleq \frac{\partial}{\partial p} E\left\{\int_0^T L(x; p, w) dt + \phi(x(T))\right\}. \qquad (8)$$

The reason we can do this is because it is relatively easy to compute $\partial \mathcal{J}/\partial p$ where

$$\mathcal{J} = \int_0^T L(x_s, p, w_s) dt + \phi(x_s(T)) \qquad (9)$$

is the performance measured for a particular sample realization $x_s(t)$ and $w_s(t)$.

From control theory we have

$$\frac{\partial \mathcal{J}}{\partial p} = \int_0^T H_p\, dt \qquad (10)$$

where $H \triangleq \lambda^T f + L$

$$\dot{\lambda}^T = -\frac{\partial H}{\partial x} \qquad \lambda^T(T) = \frac{\partial \phi}{\partial x(T)} \qquad (11)$$

and all partials are evaluated with respect to (w.r.t.) $x_s(t)$ and $w_s(t)$. Thus, for each sample realization of $x_s(t)$, we need to solve (10) and (11) in addition. At the conclusion of the Monte Carlo experiment,

$$\frac{\partial J}{\partial p} = \sum_s \frac{\partial \mathcal{J}(x_s, w_s)}{\partial p}. \qquad (12)$$

Equation (12) can be used effectively to "tune" the design of such controllers.

A simple experiment was actually carried out using this idea. We took a simple first-order linear dynamic system under Gaussian white noise disturbance and measurement error. In the steady state the optimal LQG controller is determined by only two parameters, the Kalman gain K and the feedback gain C. Using (12) we in fact successfully converged onto the correct optimal value of K and C which can of course be determined analytically.[2]

A second and rather different application is associated with the calculation of average throughput of a serial production line whose component machines are subject to random failures and repairs. In order to smooth production, it is necessary to install intermediate buffer storage to prevent the occasional failure of one machine from forcing down another. The average throughput of a given production line with given buffer size [the parameters u in (1)] at various locations can be expressed as a function [i.e., the J in (1)] of the eigenvector of a certain Markov transition probability matrix [1]. For any realistic production line the size of this matrix easily reaches $10^n \times 10^n$ where n ranges from 4 to 30. Thus, the only reasonable approach to calculate $\Delta J / \Delta u$ is the one indicated here, i.e., we built a simulator to calculate $E[L(u, w)]$. It turns out for this general problem there in fact exist a simple but not at all obvious method to calculate $L(u + \Delta u_i, w_s) - L(u, w_s)$ based on $L(u, w_s)$ and Δu_i alone. A very successful design procedure for production line has been worked out based on this approach to the sensitivity $\Delta J / \Delta u$. The details are worked out elsewhere [1].

It is perhaps worth emphasizing another advantageous byproduct of this approach for *real* problems. Once we have decided to use simulation, we can dispense with many of the assumptions that are often necessary in an analytical stochastic model. The simulation can be made as realistic as we like, usually at no or little extra cost than to simulate the analytical model. In the case of experiments involving real system, data can be directly used without further reduction or processing. Thus, even for real problems where an *approximate* analytical model and solution exist, it is not at all clear that this proposed approach might not give better answers than calculations based on the analytical model.

Note added in proof: Chronologically, the idea of this note was derived in connection with [1]. After the writing of this note, it was brought to the author's attention that the approach to control system design was anticipated by Mayne [2]. In this sense, an earlier note [3] can also be viewed in this light.

References

[1] Y. C. Ho, A. Eyler, and T. T. Chien, "A gradient technique for general buffer storage design in a production line," presented at the IEEE 1979 CDC Conf., Jan. 1979; to appear in *Int. J. Prod. Res.*, 1980.
[2] D. Jacobson and D. Mayne, *Differential Dynamic Programming*. New York: Elsevier, 1970, pp. 168–178.
[3] Y. C. Ho, "Adaptive design of feedback controllers for stochastic systems," *IEEE Trans. Automat. Contr.*, vol. AC-10, pp. 367–368, July 1965.

[2]The author wishes to thank C. Cassandras for carrying out this experiment.

A New Approach to the Analysis of Discrete Event Dynamic Systems*

Y. C. HO† and C. CASSANDRAS†

On the basis of an appropriate state-space representation, perturbation equations around a nominal trajectory can be used to predict behavior by observing only single sample realizations for some classes of discrete event dynamic systems, typified by queueing networks and production systems.

Key Words—Discrete event system; sample path; simulation; queueing theory; state space; perturbation analysis.

Abstract—We present a new, time domain approach to the study of discrete event dynamical systems (DEDS), typified by queueing networks and production systems. A general state-space representation is developed and perturbation analysis is carried out. Observation of a *single sample realization* of such a system can be used to predict behavior over other sample realizations, when some parameter is perturbed, without having to make additional observations. Conditions under which this is always possible are investigated and explicit results for some special cases are included.

1. INTRODUCTION

A LARGE number of dynamical phenomena such as traffic flow in a road network, material (parts and fixtures) flow in a production system, messages in a communication network and generally non-standard queueing systems can be viewed as discrete event dynamical systems (DEDS). In contrast to the more familiar dynamical systems that are governed by ordinary differential or difference equations, discrete event systems evolve according to the occurrences of distinct events, such as the arrival of a message or customer, the unexpected or scheduled shutdown of a production machine and the completion of a task. Such events can be stochastic or deterministic in nature. They cause the system to move from one 'state' to another by initiating or terminating some activities. The interesting and difficult part of the analysis of such systems comes from the complex interaction of these events over time.

Traditionally, the analytical approach to DEDS is typified by queueing or Markov chain theory. A huge literature exists and a fair amount of success has been obtained. The major drawback, as is common with all theories, is the number of restrictive assumptions that must be satisfied for the theory to be valid. For example, almost all of queueing theory forbids the phenomena of 'blocking', or more generally the dependence of the service rate of one server on that of another or the state of the rest of the system. It is also customary to assume distributions of a known standard form over various random variables of interest. In some DEDS, discrete events may also initiate dynamics involving continuous variables (e.g. in command–control–communication problems), thus resulting in a mixed system that cannot be handled by queueing–theoretic tools alone. Finally, the state space of such systems can also become combinatorically large so as to preclude effective solution.

The other general purpose tool for attacking DEDS problems involves the use of simulation. Cost considerations aside, simulation models can be made as close to the real world as our understanding permits and as the needs require. Judging from the number of discrete event simulation languages, such as GPSS, GASP, SIMSCRIPT, SIMULA, that exist, it is fair to say that this approach represents one of the major tools for brute-force analysis of such systems. However, the main cost of this method is the computational burden, particularly in situations where prescriptive or design issues are involved. Since stochastic phenomena are often involved, a single simulation or experiment means a large-scale Monte Carlo run on the computer. Repeated simulation of a complex system over various parameter ranges can be very costly or even infeasible.

This paper proposes an emerging approach

---

* Received 6 October 1981; revised 27 April 1982; revised 14 June 1982. The original version of this paper was presented at the 8th IFAC Congress on Control Science and Technology for the Progress of Society which was held in Kyoto, Japan during August 1981. The published proceedings of this IFAC meeting may be ordered from Pergamon Press Ltd, Headington Hill Hall, Oxford OX3 0BW, U.K. This paper was recommended for publication in revised form by associate editor T. Başar under the direction of editor H. Kwakernaak.

† Division of Applied Sciences, Harvard University, Cambridge, MA 02138, U.S.A.

which promises a third alternative to the study of DEDS. The starting point of this approach is *experimental*. An experiment (Monte Carlo or otherwise) is performed on the real or simulated DEDS and the sample path of the dynamical system is observed. Analysis is then performed to derive relationships that must be satisfied by all the observed variables, such as machine busy time, throughput, average downtime, etc. These relationships together with information on other basic parameters of the system are used to *predict behavior of the DEDS along other sample paths without having to perform additional simulations or experiments*. This is the well-established scientific tradition of observation (experiment)–deduction–prediction. By simulation or experiment we work with real or almost real systems, thus avoiding much of the restrictive assumptions of analytic queueing theory. By analysis and deduction, we bypass the computational burden of brute-force simulation. The surprising thing is that this program can in fact be carried out in many cases. Furthermore, the analysis in some sense can be said to be much simpler and intuitive than the classical queueing theory approach, since the former requires no probabilistic considerations. In computer system performance evaluation studies, such an approach is known as 'operational analysis' and it has attained considerable success in the past few years (Buzen, 1973; Denning and Buzen, 1977, 1978, 1980). More recently, the techniques of operational analysis have been applied and extended (Suri, 1980) to automation systems. Independently, Ho, Eyler and Chien (1979, 1983) solved a long standing problem in serial production lines using a simulation–analysis approach motivated by control theoretic ideas. By hindsight, the solution technique can be viewed in the spirit of 'operational analysis', but where *blocking* and dynamics are the major concerns of the problem. Consequently, it appears to be applicable to nonclassical queueing problems. Some preliminary results can be found in Ho and Eyler (1980). Lastly, since simulation and operational analysis is essentially a time domain (vs probabilistic) methodology, one can draw upon the wealth of concepts built up from the study of dynamic systems governed by ordinary differential equations (Ho and Casandras, 1980).

There are many examples of nonstandard and nontrivial queueing systems: traffic flow, where traffic lights have mutually dependent service rates; production lines, where blocking due to machine failures is a major factor; communication networks, where customers of different types (messages, tokens and acknowledgements) travel together. For such systems, no known closed form solutions exist. To use a rough analogy, the product form solution for classical queueing networks may be likened to the closed form solution of ordinary linear dynamical systems. Once we step beyond its realm, we are in the domain of nonlinear dynamics. However, all is not lost. We must reduce our demands for a closed form solution. To push the analogy one step further, for ordinary nonlinear dynamical systems, we can always 'linearize' the solution about a nominal path and ask for perturbation solutions around the nominal path. The notions of co-state variables and multiplier equations familiar in control theory can be effectively employed to analyze and optimize performance of nonlinear systems. We shall draw on this analogy and develop in below sections a similar procedure for general DEDS.

2. DEFINITIONS AND INPUT TO A DEDS

2.1. *The basic model*

In its general form a DEDS is viewed as a set of SERVERS (MACHINES or MODULES) M_i, $i = 1, \ldots, I$ and QUEUES Q_k, $k = 1, \ldots, K$ interconnected in some arbitrary way. Through this network of servers and queues there are CUSTOMERS (ENTITIES) circulating, and each customer may belong to an ENTITY CLASS (TYPE) $E_j, j = 1, \ldots, J$.

As a first step in a long-range plan, we shall consider in this paper systems with a single customer type, i.e. $J = 1$. Furthermore, although the initial discussion (Sections 2 and 3) is applicable to general networks, our results focus on a special class of DEDS, i.e. production or assembly lines (Section 4).

Although the interconnection of servers and queues is arbitrary, it is simpler to adopt a model where every server M_i is preceded by a queue Q_i.* This is always possible by appropriately introducing fictitious queues of zero capacity and fictitious servers of zero service time so as to always pair up servers with queues.

In our model, customers arrive at servers, complete service there and normally proceed to some queue Q_k and hence server M_k. If, however, upon completion of service at M_i there is no space available at Q_k, the customer remains *blocked* at M_i; this situation is referred to as a FULL OUTPUT (FO) event. If a FO does not take place or has just come to an end, then the customer may proceed to Q_k. It is at this time that M_i may receive a new arriving

---

*This is the convention adopted by queueing theory. Note, however, that in production line terminology servers are always *followed* by queues (buffers). Also in queueing theory, the content of a queue by convention includes the customer currently under service, while a buffer may be empty even when the server (machine) is working on a customer (part) in production system terminology.

customer. If such a customer is not present, M_i remains *idle* until a new one arrives; this situation is referred to as a NO INPUT (NI) event. The occurrence of either a FO or a NI event is termed FORCE DOWN (FD) and constitutes an inefficiency in the proper functioning of the system.

In many DEDS, in particular production systems, a server occasionally 'fails' to provide service to an incoming customer, due to some kind of internal breakdown that needs to be repaired or to any other stoppage. The duration of this stoppage is by convention incorporated in the present customer's service time, which is therefore expected to be longer than normal. The occurrences of such failures are generally random events in time; we shall assume that all servers are *'operation dependent'*, which means that the random phenomena governing breakdowns and repairs (in general: service durations) are not affected by the FD events in their histories. The rationale behind this assumption is that servers (in particular, machines) do not deteriorate while they are idle (forced-down due to a FO or NI event). This is an important simplifying assumption, justified in practise [Buzacott (1967); Buzacott and Hanifin (1978). Also note later relevant remarks in Section 5.1.] It could be waived, by adopting more complex models of server deterioration with time, without affecting the essence of our subsequent analysis.

2.2. Input functions

In order to study a general DEDS as a dynamical system the following information pertaining to a single sample realization (trajectory) is required: (i) the characteristics of each server, i.e. the duration of service at each M_i for every arriving customer; (ii) the characteristics of each queue, in particular its size or capacity; and (iii) the various ways in which the routing of customers through the network is to take place.

We shall assume that for any given sample path being observed the following input information is made available:

(i) SERVICE TIME.* $S_i(n)$ indicates the time required for M_i to serve a customer upon its nth activation (i.e. this is the nth overall customer arriving at M_i). Since $i = 1, \ldots, I$, for a total of N possible activations of each server there are a total of $I \times N$ possible values of this function. As mentioned in the previous section, if a server fails to provide service following its nth activation until it

---

* The *ARRIVAL TIME* of a new customer to an input server of the DEDS can be viewed simply as the service time of a fictitious server preceding this input server. Thus, our model is compatible with queueing network terminology.

is repaired, the corresponding repair time is included in $S_i(n)$. In many cases of interest, server failures depend on the total operation time of the server, i.e. the total amount of time that this server spends being active as opposed to being forced down due to blocking (FO event) or lack of an incoming customer (NI event). To a first approximation, we may let the sample generation of $S_i(n)$ be dependent on n in some probabilistic manner by adding, for example, some random repair time to it every kth activation, on the average.

Thus, for a stochastic DEDS, $S_i(n)$, $n = 1, 2, \ldots$, $i = 1, \ldots, I$, simply become the sample realizations of random arrival and service times for different servers. These can be routinely generated in any simulation, given arbitrarily specified service time distributions or can be directly observed in a particular real experiment. They constitute the driving input to the DEDS.

Note that the generation of the service times $S_i(n)$, $n = 1, 2, \ldots$, $i = 1, \ldots, I$, may, in general, be dependent on the topology of the system or even the state history. It is only *after* a certain observation period is completed that the service time information can be regarded as given.

Given the $S_i(n)$ functions, we many define the ith ACTIVATION FUNCTION $\tau_i(K)$, $K = 1, 2, \ldots$ as follows:

$$\tau_i(K) \triangleq \sum_{k=1}^{K} S_i(k), \quad i = 1, \ldots, I; \quad \tau_i(0) = 0. \quad (1)$$

Note that $\tau_i(K)$ stands for the time of service completion of the Kth customer and hence of arrival of the $(K + 1)$th customer at M_i, if this server were never forced down.

(ii) QUEUE CAPACITY. b_i indicates the total number of customers Q_i can possibly accommodate, including a customer residing at M_i.

(iii) DESTINATION. $D_i^m(n)$ indicates the mth choice of the nth customer completing service at M_i as to which queue to go to next, with $m = 1, \ldots, M$. Normally, we are only interested in $D_i^1(n)$, the first choice of destination, but it may be that this particular queue is full, in which case $D_i^2(n)$ would have to be selected and so on. In what follows we shall drop m altogether and consider the destination dependent on n alone. It is emphasized, however, that this does not imply fixed routing. The $D_i(n)$ functions simply represent the sample realization of routing in a stochastic DEDS simulation; we may generate $D_i(n)$ according to any probabilistic law. (In the case of multiclass customer systems, i.e. $J > 1$, alternate views of this function can be considered.)

Since we are only dealing here with the case of a single customer type in the system, we shall not concern ourselves with issues such as 'queue discipline' and 'priority' in accessing a particular server; these are concepts whose importance emerges in multiclass customer systems.

Throughout this paper we shall work in discrete time, the time unit being sufficiently small such that no event of any consequence occurs at any time other than discrete instants 1, 2, ... Also, we maintain the symbols M_i, Q_i although they simply stand for the integer i, $i = 1, ..., I$; thus, when we set $D_i(n) = Q_j$, the actual value of the function $D_i(n)$ is the integer j.

Let $\mathbb{N}$ and $\mathbb{N}^+$ denote the sets of non-negative and positive integers, respectively. Let $\mathbb{N}_I^+$ denote a subset of $\mathbb{N}^+$ with precisely I elements. Then, we have

$$S: \mathbb{N}_I^+ \times \mathbb{N} \to \mathbb{N}$$

$$b: \mathbb{N}_I^+ \to \mathbb{N}$$

$$D: \mathbb{N}_I^+ \times \mathbb{N} \to \mathbb{N}_I^+.$$

3. STATE-SPACE REPRESENTATION OF A DEDS

3.1. *Definition of state variables*

To each server M_i, $i = 1, ..., I$ we associate an INTERNAL CLOCK $v_i(t)$ which indicates the total amount of time that M_i has been operational (i.e. not forced down due to FO or NI events) up to time t. In other words, this is the ACCUMULATED BUSY TIME at M_i.

Thus $v_i: \mathbb{N} \to \mathbb{N}$. Note that $v_i(t)$ is a monotonically nondecreasing function of t; it is constantly incremented, unless a FD event occurs at M_i. Clearly, if no FD event ever occurs at M_i, then $v_i(t) = t$, but, in general $v_i(t) \leq t$.

To each queue Q_i, $i = 1, ..., I$ we associate the corresponding QUEUE CONTENT at time t, denoted by $c_i(t)$, which includes a customer under service at M_i. Clearly, we have: $0 \leq c_i(t) \leq b_i$ for all t and $c_i: \mathbb{N} \to \{0, 1, ..., b_i\}$.

We assert that the variables defined here constitute the state of the system $\mathbf{x}(t)$

$$\mathbf{x}^T = [v_1 ... v_I \ c_1 ... c_I]$$

in the sense that knowing $\mathbf{x}(t)$ and the input to the system (Section 2.2) we can obtain $\mathbf{x}(t+1)$. Note that $\mathbf{x}(t)$ is a vector of dimension $2I$, where I is the total number of servers in the DEDS. For simplicity, we shall always assume that $\mathbf{x}(0) = 0$.

3.2. *Evolution of the state vector* $\mathbf{x}(t)$

To describe the evolution of a DEDS we shall first define several functions of the state $\mathbf{x}(t)$. All such functions are expressed in the form $F[\mathbf{x}(t)]$. As the state of the DEDS evolves, the values of $F(\mathbf{x})$ vary with time; we shall permit the abuse of notation to write $F(t)$, but it should be always understood that $F(t)$ is evaluated along a trajectory of $\mathbf{x}(t)$ and is dependent on it.

3.2.1. Activation index $k_i(t)$. This is a function of the internal clock of M_i alone and indicates the total number of customers that have been serviced, including the one currently under service, by time t

$$k_i[v_i(t)] \triangleq L \text{ such that } \tau_i(L-1) < v_i(t) \leq \tau_i(L)$$

$$L = 1, 2, ...; \quad k_i(0) = 0. \quad (2)$$

Note that the activation index is updated *after* the arrival of a new customer. It immediately follows that the condition: $v_i(t) < \tau_i[k_i(t)]$ implies M_i is currently serving the $k_i[v_i(t)]$th customer.

3.2.2. Ready to output indicator $h_i(t)$. This is a function that takes on values 0 or 1 to indicate whether M_i is currently ready to forward a customer to some queue Q_k

$$h_i(t) \triangleq \begin{cases} 1, & v_i(t) = \tau_i[k_i(t)], c_i(t) > 0 \\ 0, & \text{otherwise}, \quad i = 1, ..., I. \end{cases} \quad (3)$$

The condition $v_i(t) = \tau_i[k_i(t)], c_i(t) > 0$ indicates that service has been completed and a customer is currently present at M_i; hence M_i is ready to output this customer.

3.2.3. Full output indicator $f_i(t)$. This is a function that takes on values 0 and 1 to indicate whether M_i is forced down at t due to a blocked customer there

$$f_i(t) \triangleq \begin{cases} 1, & h_i(t) = 1; \\ & c_m(t) = b_m, m = D_i[k_i(t)]; \\ & h_m(t) = 0 \text{ or } [f_m(t) = 1, f_m(t) \text{ inde-} \\ & \text{pendent of } f_i(t)], m = D_i[k_i(t)] \\ 0, & \text{otherwise.} \end{cases} \quad (4)$$

The condition $h_i(t) = 1$ implies that M_i is currently ready to output a customer. However, $c_m(t) = b_m$ implies that this customer's destination is full. The third condition then indicates that either M_m is busy or it is in turn blocked due to a FO event; the requirement that $f_m(t)$ be independent of $f_i(t)$ is needed to exclude deadlocks in case where a loop leads back to M_i: if such a loop exists and it turns out that $f_i(t)$ depends on itself, no FO event is allowed to take place, since $h_i(t) = 1$ indicates that M_i is actually ready to output a customer.

Thus, if $f_i(t) = 1$, then a FO event is taking place at server M_i at time t.

3.2.4. *No input indicator* $n_i(t)$. Similar to $f_i(t)$, $n_i(t)$ is used to indicate whether M_i is forced down at t due to lack of incoming customers

$$n_i(t) \triangleq \begin{cases} 1, & c_i(t) = 0 \text{ or } [c_i(t) = 1, h_i(t) = 1, f_i(t) = 0]; \\ & h_l(t) = 0 \text{ for all } l \text{ such that } Q_i \text{ follows } M_l \\ 0, & \text{otherwise;} \quad i = 1, \ldots, I. \end{cases} \quad (5)$$

If $n_i(t) = 1$, a NI event is taking place at server M_i at time t.

3.2.5. *State equations*. With the state variable functions defined in Sections 3.2.1–3.2.4 we are in a position to describe the evolution of a DEDS in terms of the following state equations

$$v_i(t+1) = \begin{cases} v_i(t) + 1, & f_i(t) = n_i(t) = 0 \\ v_i(t), & \text{otherwise;} \quad i = 1, \ldots, I \end{cases} \quad (6)$$

$$c_i(t+1) = \begin{cases} c_i(t) + 1, & c_i(t) < b_i, [h_i(t) = 0 \text{ or } f_i(t) = 1]; \\ & h_l(t) = 1, i = D_l[k_l(t)]; \\ c_i(t) - 1, & c_i(t) > 0, h_i(t) = 1, f_i(t) = 0; \\ & h_l(t) = 0 \text{ for all } l \text{ such that } Q_i \text{ follows } M_l \\ c_i(t), & \text{otherwise;} \quad i = 1, \ldots, I. \end{cases} \quad (7)$$

These state equations are of the general form

$$\mathbf{x}(t+1) = F[\mathbf{x}(t), \mathbf{u}(t), t]$$

where $\mathbf{u}(t)$ stands for the input function specified by $S_i(n), D_i(n), b_i, n = 1, 2, \ldots, I$, as described in Section 2.2. It is again emphasized that this state evolution refers to a particular sample realization of the DEDS under consideration. Thus, strictly speaking, one should write

$$\mathbf{x}(t+1) = F[\mathbf{x}(t), \mathbf{u}(t), t; \omega]$$

with ω indexing over all possible sample paths of the DEDS.

Conceptually, the state evolution obeys the flow-chart in Fig. 1.

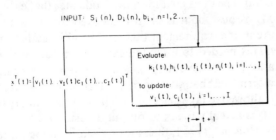

Fig. 1. State evolution.

4. APPLICATION OF STATE-SPACE REPRESENTATION TO SOME SIMPLE DEDS

We shall consider two examples of DEDS, which may be viewed as special types of queueing networks, but are usually encountered as manufacturing systems: the *serial transfer line* and the *assembly line*.

4.1. Serial transfer line

This simple kind of DEDS (see Fig. 2) is obtained from the general DEDS model presented in Sections 2 and 3 by introducing the additional constraint on the destination function

$$D_i(n) = Q_{i+1}, \quad \forall n = 1, 2, \ldots \text{ and } i = 1, \ldots, I-1 \quad (8)$$

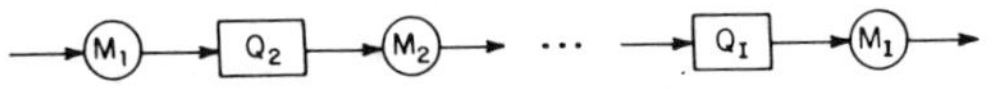

Fig. 2. A serial transfer line.

This is the simplest possible nonstandard queueing problem and the object of much research in production control problems (Buzacott, 1967; Buzacott and Hanifin, 1978; Sevastyanov, 1962; Okamura and Yamashina, 1977; Soyster, Schmidt and Rohrer, 1979). Usually, service times $S_i(n), n = 1, 2, \ldots$ for all servers are constant. However, during every service event, M_i has probability p_i of failing, in which case it has probability q_i of being repaired during the next cycle. Thus, we can model each server as having a service time distribution as shown in Fig. 3.

In other words, instead of distinguishing between a working or failed server (machine), we simply consider M_i as a server with probability $(1 - p_i)$ of having a known constant value (machine cycle time) CT_i and probability p_i of having service time $CT_i + RT_i$, where RT_i is geometrically distributed with parameter q_i.

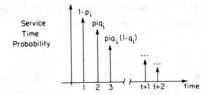

Fig. 3. Service time distribution for a production line server.

A state-space representation for the serial transfer line can be obtained from the general expressions of Section 3.2 using the constraint (8) above. Equation (3) remains unchanged, while (4) and (5) reduce to

$$f_i(t) = \begin{cases} 1, & h_i(t) = 1; c_{i+1}(t) = b_{i+1}; h_{i+1}(t) = 0 \\ & \text{or } f_{i+1}(t) = 1 \\ 0, & \text{otherwise} \end{cases} \quad (9)$$

$$n_i(t) = \begin{cases} 1, & c_i(t) = 0 \text{ or } [c_i(t) = 1, h_i(t) = 1, \\ & f_i(t) = 0]; h_{i-1}(t) = 0. \\ 0, & \text{otherwise} \end{cases} \quad (10)$$

where Q_1 is taken to be continuously nonempty (hence $h_0(t) = 1$, for all t by definition) and $b_{I+1} = \infty$ (hence an infinite sink exists at the end of the line).

The state equations (6) and (7) reduce to

$$v_i(t+1) = \begin{cases} v_i(t) + 1, & f_i(t) = n_i(t) = 0 \\ v_i(t), & \text{otherwise}; \quad i = 1, \ldots, I \end{cases} \quad (11)$$

with $f_i(t)$, $n_i(t)$ now defined as in (9) and (10) above

$$c_i(t+1) = \begin{cases} c_i(t) + 1, & c_i(t) < b_i, [h_i(t) = 0 \text{ or } \\ & f_i(t) = 1]; h_{i-1}(t) = 1. \\ c_i(t) - 1, & c_i(t) > 0, h_i(t) = 1, \\ & f_i(t) = 0; h_{i-1}(t) = 0. \\ c_i(t), & \text{otherwise}; \quad i = 1, \ldots, I. \end{cases} \quad (12)$$

4.2. Assembly line

It is sometimes the case that a server provides service to two or more arriving customers simultaneously; if any one of these customers is not present at some time, then the server cannot be activated at all. We call this an 'assembly server', since it effectively combines two or more customers to output an assembled customer. Similarly, there are 'disassembly servers', which must break up an outcoming customer to two or more parts; if any one of all of these parts cannot be accommodated by the next destination server, the disassembly server is blocked.

A serial transfer line that contains assembly and/or disassembly servers is called an assembly line. A typical example of such a system is shown in Fig. 4.

In this example, M_4 is as assembly server, while M_5 is both an assembly and disassembly server. Note that there is a loop in the system consisting of M_5 and M_6. Also note that two separate queues provide customers to assembly servers M_4, M_5.

For an assembly server we shall assume that there are $m = 1, \ldots, M$ servers feeding into it; for a disassembly one there are $l = 1, \ldots, L$ servers to which it feeds. Again, to obtain a state-space representation for an assembly line, (3)–(5) become

$$h_i(t) = \begin{cases} 1, & v_i(t) = \tau_i[k_i(t)], c_{im}(t) > 0, \\ & \forall m = 1, \ldots, M. \\ 0, & \text{otherwise} \end{cases} \quad (13)$$

$$f_i(t) = \begin{cases} 1, & h_i(t) = 1; \\ & c_l(t) = b_l; h_l(t) = 0 \text{ or } [f_l(t) = 1, \\ & f_l(t) \text{ independent of } f_i(t)] \text{ for any} \\ & l = 1, \ldots, L \\ 0, & \text{otherwise} \end{cases} \quad (14)$$

$$n_i(t) = \begin{cases} 1, & c_{im}(t) = 0 \text{ or } [c_{im}(t) = 1, h_i(t) = 1, \\ & f_i(t) = 0]; h_m(t) = 0 \text{ for any} \\ & m = 1, \ldots, M. \\ 0, & \text{otherwise}. \end{cases} \quad (15)$$

The state equations (6) and (7) become

$$v_i(t+1) = \begin{cases} v_i(t) + 1, & f_i(t) = n_i(t) = 0 \\ v_i(t), & \text{otherwise}; \quad i = 1, \ldots, I \end{cases} \quad (16)$$

with $f_i(t)$, $n_i(t)$ as defined in (14) and (15)

$$c_{im}(t+1) = \begin{cases} c_{im}(t) + 1, & c_{im}(t) < b_{im}, [h_i(t) = 0 \text{ or } \\ & f_i(t) = 1]; h_m(t) = 1 \text{ for} \\ & \text{some } m = 1, \ldots, M. \\ c_{im}(t) - 1, & c_{im}(t) > 0, h_i(t) = 1, \\ & f_i(t) = 0; h_m(t) = 0 \text{ for} \\ & \text{some } m = 1, \ldots, M. \\ c_{im}(t), & \text{otherwise}; \quad i = 1, \ldots, I, \\ & m = 1, \ldots, M. \end{cases} \quad (17)$$

The additional subscript m is included in the above to cover cases where M_i is an assembly server.

It is worthwhile to note the highly complex, nonlinear and discontinuous nature of these state-space equations (11), (12), (16), and (17).

5. PERTURBATION ANALYSIS OF DEDS

Having developed the state-space equations for a DEDS and given examples of such equations for some production lines, in this section we shall develop a linearization procedure for these equations. However, it is obvious that the conventional method of taking partial derivatives of the right-hand side of the state-space equations will not be feasible or productive. Instead, we shall first introduce some new concepts concerning the representation of the state space trajectory of a DEDS. Then a perturbation analysis can be carried out based on this representation.

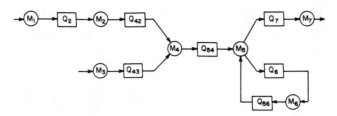

Fig. 4. Example of an assembly line.

5.1. Event sequences of a DEDS

In Section 2.2 we introduced a sequence of service times $S_i(1)$, $S_i(2)$, ..., $S_i(n)$, ... which characterizes server M_i for a particular sample realization ω. If the system consisted of this server alone, this sequence would be sufficient to provide the entire sample history of M_i. When two or more servers are present in a network, however, their interaction creates FO and NI events which occasionally interrupt the sequence of service times. Thus, a server M_i can experience at any time one of three types of events: a Service event, whose duration is denoted by $S_i(n)$, a full output (FO) event (i.e. blocking) or a NO input (NI) event (i.e. idling). Let $FO_i(n)$, $NI_i(n)$ denote the duration of the nth FO and nth NI event, respectively. In what follows, for the purpose of notational simplicity, an event and its duration shall be left indistinguishable, e.g. $FO_i(n)$ must be understood to stand for an ordered pair denoting the type of event, a FO, and the duration of the nth such event taking place. For a particular sample path ω of a given DEDS, the history of server M_i is contained in its *EVENT SEQUENCE* $\mathscr{E}_i^\omega$; this is defined to be the *sequence of service times and of durations of any FO or NI event taking place at M_i in their exact order of occurrence*. A typical $\mathscr{E}_i^\omega$ would look as follows:

$$\mathscr{E}_i^\omega = \{S_i(1), S_i(2), NI_i(1), S_i(3), S_i(4), FO_i, \ldots\}.$$

Here we are making use of the 'operation dependence' of servers assumption, discussed in Section 2.1, in that the service times characterizing M_i remain unaffected by the introduction of FO and NI events in constructing $\mathscr{E}_i^\omega$, regardless of the actual durations of these events.

When all event sequences of a DEDS are collected together, we may construct a 'tableau' which actually summarizes the operating history of the entire system for a given ω. A typical tableau would appear as follows:

$$\begin{Bmatrix} S_1(1), S_1(2), FO_1(1), \ldots, S_1(n), \ldots \\ NI_2(1), S_2(1), S_2(2), \ldots \\ \vdots \\ NI_i(1), S_i(1), S_i(2), NI_i(2), \ldots \\ \vdots \\ NI_I(1), S_I(1), NI_I(2), S_I(2), \ldots \end{Bmatrix}$$

Henceforth, we shall omit ω in our notation and, unless otherwise specified, understand that we always refer to a given sample realization of a DEDS.

Now suppose we have obtained a tableau of event sequences $\mathscr{E}_i$, $i = 1, \ldots, I$; this tableau specifies a *nominal* path (or represents a sample trajectory) of our DEDS. Next, suppose that a specific, known perturbation is introduced in the system tableau (e.g. some service time $S_i(n)$ is changed) but everything else remains unchanged. A perturbed tableau of event sequences will result, specifying a *perturbed* path of the system, relative to the same ω. We are interested in the relationship between the nominal and the perturbed tableaus. In other words, we are interested in answers to questions of the type 'what if we perturb a particular sample trajectory of the DEDS?' or 'what if we replicate an experiment of the DEDS exactly, except for the perturbation...?'

5.2. Similarity of event sequences

For a given sample realization of a DEDS we shall denote by $\mathscr{E}_i$ the event sequence of M_i in the nominal path and by $\mathscr{E}_i'$ the corresponding one in the perturbed path; the perturbation is fixed and known, as explained in the previous section.

We then define two event sequences $\mathscr{E}_i$ and $\mathscr{E}_i'$ of M_i (pertaining to the same ω) to be *SIMILAR if all events in these sequences appear in the exact same sequential order*.

Note that the actual values (durations) of those events are not necessarily equal. Thus, as an example, if the Kth event in $\mathscr{E}_i$ is $FO_i(m)$, then the Kth event in $\mathscr{E}_i'$ must be the mth overall FO event taking place at M_i, if $\mathscr{E}_i$ and $\mathscr{E}_i'$ are indeed similar.

Intuitively, similarity between a nominal, $\mathscr{E}_i$, and a perturbed, $\mathscr{E}_i'$, event sequence for server M_i implies the following: in the perturbed path no FO or NI event will be completely eliminated and no new FO or NI event will be introduced between two successive service events in $\mathscr{E}_i$. In others words, if $S_i(n)$ if followed by $S_i(n+1)$ in $\mathscr{E}_i$ then $S_i'(n)$ must be followed by $S_i'(n+1)$; if it is followed by $FO_i(m)$ in $\mathscr{E}_i$, then $S_i'(n)$ must be followed by $FO_i'(m)$; if it is followed by $NI_i(l)$ in $\mathscr{E}_i$, then $S_i'(n)$ must be followed by $NI_i'(l)$. This must be true for any n. All primed quantities indicate that the actual duration of the event may be different in $\mathscr{E}'_i$.

An alternative, more quantitative, view of similarity is obtained by introducing: $I_i(n) \triangleq$ time of *initiation* of service for the nth customer at M_i and $C_i(n) \triangleq I_i(n) + S_i(n) \triangleq$ time of *completion* of service for the nth customer at M_i.

Then, $I_i(n+1)$ indicates the initiation of service for the $(n+1)$th customer at M_i and, according to the construction of our DEDS model, this event depends on:

(a) the availability of the server, which occurs at $C_i(n)$;
(b) the availability of queueing space for the nth customer after time $C_i(n)$. Let: $q_i(n) \geq C_i(n)$ denote this *time of queueing space availability*;
(c) the availability of a new, the $(n+1)$th, customer after $C_i(n)$. Let: $r_i(n) \geq C_i(n)$ denote this *time of customer availability*.

It follows immediately that we can write a fundamental update equation for $I_i(n)$

$$I_i(n+1) = \max\{C_i(n), q_i(n), r_i(n)\}^* \quad (18a)$$

or

$$I_i(n+1) = \begin{cases} C_i(n), & \text{if no FD event follows } n\text{th} \\ & \text{customer service completion} \\ q_i(n), & \text{if a FO event follows } n\text{th} \\ & \text{customer service completion} \\ r_i(n), & \text{if a NI event follows } n\text{th} \\ & \text{customer service completion} \end{cases} \quad (18b)$$

or

$$I_i(n+1) = I_i(n) + S_i(n) + y_i(n) \quad (18c)$$

where, by definition $I_i(n) + S_i(n) = C_i(n)$, and we set

$$y_i(n) \triangleq \begin{cases} 0, & \text{if no FD event follows } n\text{th customer} \\ & \text{service completion} \\ q_i(n) - C_i(n), & \text{if a FO event follows } n\text{th} \\ & \text{customer service completion} \\ r_i(n) - C_i(n), & \text{if a NI event follows } n\text{th} \\ & \text{customer service completion.} \end{cases}$$

The notation introduced here is linked directly to the state-space representation of Section 3 and is only used for convenience. From (18c) above, we get

$$I_i(n) = I_i(1) + \tau_i(n-1) + \sum_{k=1}^{n-1} y_i(k)$$

---

*Because we have defined: $q_i(n) \geq C_i(n)$, $r_i(n) \geq C_i(n)$, it is true that: $\max\{C_i(n), q_i(n), r_i(n)\} = \max\{q_i(n), r_i(n)\}$. $C_i(n)$ is included in (18a) to emphasize the fact that the $(n+1)$th activation of M_i follows one of *three* types of events at M_i: service, FO or NI.

where we have used the definition of the activation function $\tau_i(\cdot)$ in (1). Observe that

$$I_i(1) + \sum_{k=1}^{n-1} y_i(k)$$

stands for the total amount of time server M_i spends in FD events after $(n-1)$ activations; that is exactly the same as: $t - v_i(t)$, the difference between real time and the internal clock of M_i which measures operational time only. Hence, setting: $n = k_i(t)$, it can be easily seen that $I_i[k_i(t)]$ is a function of t, $v_i(t)$ and $\tau_i[k_i(t) - 1]$.

Since $\tau_i[\cdot]$ is simply an input function to the DEDS, $I_i[k_i(t)]$, given any time instant t, is directly obtainable from the internal clock state variable $v_i(t)$ and $k_i[v_i(t)]$. Thus, knowing the state variables $v_i(t)$ we can reconstruct $I_i[k_i(t)]$. In this sense, (18c) can be viewed as part of a state-space equation for the DEDS in question. $S_i(n)$ is an input term and $y_i(n)$ is a nonlinear term of the state variable. The complexity of the equation lies in the severe non-linear nature of $y_i(n)$. Specific examples of $y_i(n)$ for the serial and assembly line will be given below and in Sections 5.4.1 and 5.4.2.

Returning to the definition of similarity using the notation just developed, it is easy to see that similarity may be interpreted as follows:

$$\left. \begin{array}{l} I_i(n+1) = C_i(n) \Rightarrow I'_i(n+1) = C'_i(n) \\ I_i(n+1) = q_i(n) \Rightarrow I'_i(n+1) = q'_i(n) \\ I_i(n+1) = r_i(n) \Rightarrow I'_i(n+1) = r'_i(n) \end{array} \right\} \begin{array}{l} \text{for all} \\ n = 1, 2, \ldots \end{array} \quad (19)$$

Here, $I_i(n+1)$ determines the type of event to follow the nth service event at M_i. If, for all $n = 1, 2, \ldots$, this event type remains unchanged in the perturbed path (indicated by a prime), then the order and type of all events in $\mathscr{E}_i$ are preserved in $\mathscr{E}'_i$.

Condition (19) above implies the following two conditions:

(C1) [*No elimination of FD*]:

$$I_i(n+1) \neq C_i(n) \Rightarrow I'_i(n+1) \neq C'_i(n). \quad (20)$$

(C2) [*No creation of FD*]:

$$\left. \begin{array}{l} I_i(n+1) = C_i(n) \Rightarrow I'_i(n+1) = C'_i(n) \\ I_i(n+1) = q_i(n) \Rightarrow I'_i(n+1) \neq r'_i(n) \\ I_i(n+1) = r_i(n) \Rightarrow I'_i(n+1) \neq q'_i(n). \end{array} \right\} \quad (21)$$

It is also easy to see that (C1) and (C2) together imply similarity as expressed in condition (19).

The following two results provide specific expressions for $q_i(n)$ and $r_i(n)$ in the cases of a serial transfer line and an assembly line, which were discussed in Section 4.

Lemma 1. In a serial transfer line

$$q_i(n) = \begin{cases} C_i(n), & c_{i+1}[C_i(n)] < b_{i+1} \\ I_{i+1}(n+1-b_{i+1}), & \text{otherwise} \end{cases} \quad (22)$$

$$r_i(n) = \begin{cases} C_i(n), & c_i[C_i(n)] > 1 \\ C_{i-1}(n+1), & \text{otherwise}; \end{cases} \quad i = 1, \ldots, I. \quad (23)$$

Proof: Consider (22) first. It is obvious that if $c_{i+1}[C_i(n)] < b_{i+1}$, queueing space is immediately available to the nth outcoming customer, hence $q_i(n) = C_i(n)$. On the other hand, if $c_{i+1}[C_i(n)] = b_{i+1}$, 'conservation of customers' implies that

$$k_i[C_i(n)] = b_{i+1} + k_{i+1}[C_i(n)]$$

and since at time $C_i(n)$ the activation index of M_i is, by the definition of $C_i(\cdot)$, n, we have

$$k_{i+1}[C_i(n)] = n - b_{i+1}.$$

Now, queueing space is next available when M_{i+1} is ready to receive a customer from Q_{i+1}, i.e. at time $q_i(n) = I_{i+1}(x)$, where $x = k_{i+1}[C_i(n)] + 1$, since the activation index of M_{i+1} is incremented by 1 at that time. It follows that: $x = n + 1 - b_{i+1}$ and

$$q_i(n) = I_{i+1}(n+1-b_{i+1}).$$

Next, consider (23). It is obvious again that if $c_i[C_i(n)] > 1$, a new customer is immediately available to M_i, hence $r_i(n) = C_i(n)$. If, however, $c_i[C_i(n)] = 1$, then a new customer is made available to M_i as soon as the next service completion at M_{i-1} takes place, i.e. at time $r_i(n) = C_{i-1}(y)$, for some y. 'Conservation of customers' then implies

$$k_{i-1}[C_{i-1}(y)] = k_i[C_{i-1}(y)] + 1$$

Since, at time $C_{i-1}(y)$, M_{i-1} contains the yth customer. At that time also, the activation index of M_{i-1} is y and that of M_i is still n [since $k_i(t)$ is incremented *after* the arrival of a new customer]. Thus, the above equation gives

$$y = n + 1$$

and therefore

$$r_i(n) = C_{i-1}(n+1). \quad \blacksquare$$

Lemma 2. In an assembly line

$$q_i(n) = \begin{cases} C_i(n), & c_l[C_i(n)] < b_l, \ \forall l = 1, \ldots, L \\ \max_{l=1,\ldots,L} \{I_l(n+1-b_l)\}, & \text{otherwise} \end{cases} \quad (24)$$

$$r_i(n) = \begin{cases} C_i(n), & c_{im}[C_m(n)] > 1, \ \forall m = 1, \ldots, M \\ \max_{m=1,\ldots,M} \{C_m(n+1)\} & \text{otherwise}; \end{cases} \quad i = 1, \ldots, I \quad (25)$$

where there are M servers feeding into assembly server M_i and disassembly server M_i feeds into L servers.

Proof: The proof is exactly the same as that of Lemma 1 for a serial transfer line, except that when $c_l[C_i(n)] = b_l$ we are interested in the last of the L servers following M_i that receives its $(n+1-b_l)$th customer; similarly, when $c_{im}[C_i(n)] = 1$ for one or more Q_{im}, we are interested in the last of the M servers preceding M_i that completes service of the $(n+1)$th customer. $\blacksquare$

5.3. A descriptive view of perturbation propagation in DEDS

Our purpose here is to examine whether, given a nominal sample path and a specific perturbation, we can immediately derive conclusions about the perturbed path without actually having to realize the perturbation. Once again, the reader is reminded that the sample path ω is fixed; thus, we attempt to answer the following type of question: 'what if the experiment we have just performed were repeated under identical conditions, but with a given perturbation put to effect?'

Perturbation analysis can be carried out for a number of performance measures of interest. We shall restrict ourselves here to THROUGHPUT, i.e. the number of customers completing service at some specified server per unit time; this specified server we shall call OUTPUT SERVER and always choose to denote by M_I. Thus, throughput, T, can be viewed as a function of the activation index, $k_I(t)$, of server M_I

$$T[k_I(t)] \triangleq \frac{k_I(t)}{C_I[k_I(t)]}. \quad (26)$$

If throughput alone is of interest, perturbation analysis need only be carried out for $v_i(t)$, $i = 1, \ldots, I$, i.e. half the state variables, as will be seen below. Nevertheless, throughput analysis is only one application of our approach, which we expect to be valid for other performance measures as well (e.g. mean queue length, utilization).

We shall also choose to distinguish between 'generation of a perturbation' and 'propagation of a perturbation': the former refers to a specific change in some parameter of a DEDS along a nominal path that is ultimately reflected as a change in service time initiation $I_i(n)$ [e.g. by perturbing $S_i(n-1)$], for M_i and some activation index n; the latter refers to the way in which this change in $I_i(n)$ can propagate through the DEDS, affect service, FO or NI events occurring at various servers and eventually cause a change in the throughput $T[k_I(t)]$. We shall only be concerned with the propagation of a perturbation. Furthermore, we shall restrict ourselves, for the time being, to a single perturbation created at some server for a given activation index and whose size δ is infinitesmal.

The next few sections will attempt to formalize the rules governing perturbation analysis in a DEDS. Here, we shall motivate formalization by providing a descriptive view of these rules.

Suppose that a perturbation is introduced to $I_i(n)$, due to a change in $S_i(n-1)$. As will be shown (in Theorems 1-3), this perturbation—which will be termed GAIN—can propagate to other servers: (i) at times when a new customer arrives at this other server from M_i; and (ii) provided an FO or NI event has preceded this arrival.

This is clear since M_i controls the termination of the FO or NI at this other server. Any perturbation in the event sequence of M_i will be PROPAGATED to the event sequence of this other server. In particular, for the case of a serial transfer line, we have the following 'propagation rule':

Gain at M_i is propagated to M_{i+1} when a NI event occurs at M_{i+1}; it is propagated to M_{i-1} when a FO event occurs at M_{i-1}.

Clearly, if gain at M_i is propagated to M_I, the output server of the system, then the throughput will be affected. We refer to this phenomenon as REALIZATION of gain for the entire system (as opposed to realization of gain for a single server M_i).

Referring to the tableau of event sequences $\mathscr{E}_i^\omega$, which represents the realization of sample path ω (Section 5.1), the 'propagation rule' above can be visualized as follows: since only FO and NI events may cause a change in the gain of server M_i, for the purpose of propagation, a tableau can be reduced to the representation of $FO_i(n)$, $NI_i(n)$ alone, $i = 1, \ldots, I, n = 1, 2, \ldots$ An example is shown in Fig. 5.

Looking at Fig. 5, a perturbation is introduced at $S_i(j)$. This perturbation can be thought of as a marble which starts rolling along the ith row of the tableau, since all subsequent service events will be affected by it. Then:

(1) The marble keeps on rolling horizontally from left to right until a NI or FO event is encountered in the ith row.

(2) If this marble detects a FO in the row right above it or a NI in the row below it, a new identical marble is started at that point in that row and rolls horizontally from left to right as well.

(3) A rolling marble is stopped when a FO or NI is encountered in its own row. (Note this does not mean there can be no more perturbation in any $\mathscr{E}_i^\omega$ after a NI or FO event, since such events can bring in perturbations from adjacent sequences $\mathscr{E}_{i-1}^\omega$ or $\mathscr{E}_{i+1}^\omega$.)

(4) If a marble eventually reaches the end of the Ith row, then the perturbation is realized for the system as a whole, after $k_I(t) = N$ service events in the output server M_I.

It is also worthwhile emphasizing the generality of the tableau representation: (i) the only important events are FO and NI ones, which are directly observable from a nominal path; and (ii) the tableau may represent any collection of servers, arbitrarily interconnected. The way in which the marbles will roll will be dependent on the topology of the given DEDS, which also determines how FO and NI events are introduced in the service time sequences of servers. Fig. 5 represents the simplest case of a serial transfer line.

Thus, the idea of rolling marbles in a given tableau can readily be extended to systems other than serial transfer lines, such as an assembly line (Fig. 4). The topology of the DEDS will then determine the way in which marbles can roll from row to row. The 'propagation rule' above will be modified to:

Gain at M_i is propagated to M_j when a NI event is terminated at M_j by the arrival of a customer from M_i; it is propagated to M_k when a FO event is terminated at M_k by an activation of M_i creating queueing space at Q_i.

Naturally, this descriptive view needs to be formalized; this issue is addressed in later sections and is the subject of continuing research.

5.4. *A formal view of perturbation propagation in DEDS*

We now attempt to formalize the discussion of the previous section. Note that all quantities pertaining to the nominal path are nonprimed and all those pertaining to the perturbed path are primed.

We define a function of the activation index of a server M_i, which we shall call *GAIN FUNCTION* (or just *GAIN*) of M_i, as follows:

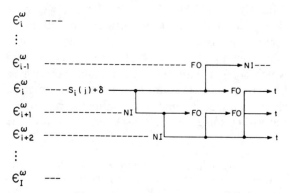

FIG. 5. Example of propagation in a tableau.

$$G_i[k_i(t)] \triangleq I_i[k_i(t)] - I'_i[k_i(t)]. \quad (27)$$

This function measures the difference in service initiation at M_i between the nominal and a perturbed path for a given value of the activation index. If we know the evolution of $G_i[k_i(t)]$, then we can completely determine the perturbed trajectory from the nominal one. Theorems 1 and 2 below characterize this. Also, note that $G_i[k_i(t)]$, in measuring the difference in service initiation between a nominal and a perturbed path, really reflects the difference between nominal and perturbed internal clocks, $v_i(t)$ and $v'_i(t)$, i.e. the effective perturbation on a state variable, as discussed in Section 5.2 above. We shall once again make abuse of notation (see introduction to Section 3.2) and express $G_i[k_i(t)]$ as $G_i(t)$.

An immediate consequence of the above definition is the following.

Theorem 1. $G_i(t+1) =$

$$G_i(t) \text{ for all } t \in (I_i[k_i(t)], I_i[k_i(t)+1]. \quad (28)$$

Proof: Note in (27) that the value of $G_i(\cdot)$ does not change as long as $k_i(t)$ does not change. The activation index is only updated after a new customer arrives, i.e. at $t = I_i[k_i(t)] + 1$ and remains unaffected until the next customer arrives, i.e. until $t = I_i[k_i(t) + 1]$. Thus, the equality above holds for all

$$t = I_i[k_i(t)] + 1, \ldots, I_i[k_i(t) + 1]. \quad \blacksquare$$

The idea reflected by this theorem is that the gain of M_i can *only* be affected at points in time when a new customer arrives at M_i.

Next, exploiting the concept of 'similarity', introduced in the previous section, the following general result is obtained.

Theorem 2. If $\mathscr{E}_i$ and $\mathscr{E}'_i$ are similar event sequences of M_i, then

$$G_i(n+1) = \begin{cases} q_i(n) - q'_i(n), & \text{if a FO event} \\ & \text{follows } n\text{th} \\ & \text{customer service} \\ & \text{completion at } M \\ r_i(n) - r'_i(n) & \text{If a NI event} \\ & \text{follows } n\text{th} \\ & \text{customer service} \\ & \text{completion at } M_i \\ G_i(n) + \Delta S_i(n), & \text{otherwise} \end{cases} \quad (29)$$

where

$$\Delta S_i(n) \triangleq S_i(n) - S'_i(n), \; n = 1, 2, \ldots.$$

Proof: Similarity, as expressed in equations (19) and the definition of gain in (27) imply that

$$G_i(n+1) = \begin{cases} q_i(n) - q'_i(n), & \text{if a FO event follows} \\ & n\text{th customer service} \\ & \text{completion at } M_i \\ r_i(n) - r'_i(n), & \text{if a NI event follows} \\ & n\text{th customer service} \\ & \text{completion at } M_i \\ C_i(n) - C'_i(n), & \text{otherwise} \end{cases}$$

and since

$$C_i(n) - C'_i(n) = I_i(n) - I'_i(n) + S_i(n) - S'_i(n)$$
$$= G_i(n) + \Delta S_i(n)$$

we immediately get (29). ∎

From a control theoretic point of view, the development of equation (29) indicates that for a given DEDS whose dynamics are described in general by a state equation (see Section 3.2.5) of the form

$$\mathbf{x}(t+1) = F[\mathbf{x}(t), \mathbf{u}(t), t; \omega]$$

it is possible (as will be explained below) to find a function $\phi(\cdot)$ such that

$$\delta\mathbf{x}(t+1) = \phi[\delta\mathbf{x}(t), \delta\mathbf{u}(t), t]^* \quad (30)$$

where $\phi(\cdot)$ is dependent on $\mathbf{x}(t)$, $\mathbf{u}(t)$ and hence ω. This equation describes the evolution of a state perturbation, $\delta\mathbf{x}(t)$, when the input is changed by $\delta\mathbf{u}(t)$. The analysis holds for a fixed sample path ω. This is the exact analog of linearizing a nonlinear dynamical system, as mentioned in the end of the Introduction. More on this will be discussed in Section 5.5.

Note that equation (30) is perfectly general, for any DEDS, provided similarity holds. Its applicability can be seen when the expressions for $q_i(n)$, $r_i(n)$ are known and if $q'_i(n)$, $r'_i(n)$ can be expressed in terms of nonprimed quantities alone. (See Sections 5.4.1 and 5.4.2 below and Lemmas 1 and 2.) Still, a simple fact that is immediately revealed is the following: $G_i(n)$ can only be affected by a given perturbation $\Delta S_i(n)$, $n = 1, 2, \ldots$ and by FO and NI events. Suppose that $G_i(1) = \delta \neq 0$ and $\Delta S_i(n) = 0$ for all $n \geq 1$. Then $G_i(n)$ for all $n \geq 1$ will only change from this value when a FO or NI event occurs; every other service event, regardless of duration, is completely irrelevant. The precise way in which $G_i(\cdot)$ is initialized to δ will not concern us here. In general, it is not a state variable (internal clock or queue content) that is directly perturbed, but rather some system parameter, **p** (e.g. a queue

---

*In particular, we are concerned here only with part of the perturbed state vector: $\delta v_1(t), \ldots, \delta v_I(t)$, the perturbed internal clocks of the servers in the system, which is what the $G_i(t)$ functions reflect, as explained before.

capacity or the destination function of some server), giving rise to $\delta\mathbf{p}$. This, in turn, must be translated to $\delta\mathbf{x}(\tau)$, which will initialize gain. Thus, there is an intermediate step which entails evaluating $\Delta x(\tau)/\Delta p$, so as to obtain

$$\delta\mathbf{x}(\tau) = \frac{\Delta x(\tau)}{\Delta p}\delta\mathbf{p} \quad (31)$$

for a given $\delta\mathbf{p}$. This 'gain generation'—as opposed to 'gain propagation' once it is generated—process depends entirely on the nature of the specific perturbation under consideration and is not dealt with in this paper. [See, however, Eyler (1979); Ho, Eyler and Chien (1979, 1983).]

In view of Lemma 1, where we saw that in a serial transfer line: $q_i(n) = I_{i+1}(n+1-b_{i+1})$, when a FO event follows the nth customer service completion at M_i and: $r_i(n) = C_{i-1}(n+1)$, when a NI event follows the nth customer service completion at M_i, the interpretation of Theorem 2 in this case reveals the following simple rule: *the gain of M_i remains unaffected as long as no FD event occurs; if a FO event occurs, then the gain of M_{i+1} is propagated to M_i; if a NI event occurs, then the gain of M_{i-1} is propagated to M_i*. This will be shown explicitly in the next section. Note that this rule is equivalent to obtaining a transition function for (30) above, $\Phi(t,\tau)$ so that

$$\delta\mathbf{x}(t) = \Phi(t,\tau)\delta\mathbf{x}(\tau), \quad t \geq \tau$$

where $\Phi(t,\tau)$ only depends on the nominal path history.

One may question the validity of the similarity assumption in Theorem 2. In the spirit of 'linearization' assumptions in continuous systems, similarity is dependent on the magnitude of the perturbation; if δ above is small, then it will not have any effect on the occurrence of FO and NI events at M_i (i.e. it will not eliminate such events from $\mathscr{E}_i$ neither will it create new ones).

However, it must be pointed out that an update equation for $G_i(n)$ can still be obtained even if similarity is violated. The actual value of $G_i(n+1)$ is dependent on $I_i(n+1)$, $I'_i(n+1)$ and each one of these two quantities can only take three values: $C_i(n)(C'_i(n))$, $q_i(n)(q'_i(n))$ and $r_i(n)(r'_i(n))$.

Therefore, $G_i(n+1)$ can acquire any of nine possible values, of which similarity simply covers three and yields a simple equation. This issue will be addressed again in the next section.

Before proceeding to special cases of Theorem 2, it is appropriate to outline our overall general approach in evaluating the gain function of a server, independent of similarity assumptions. We shall make use of:

(1) $I_i(n)$ and $C_i(n)$, as defined in Section 5.2.
(2) The definition of gain: $G_i(n) = I_i(n) - I'_i(n)$, in (27), where n stands for the activation index of M_i. (Note that the evolution of gain as a function of time is immediately obtainable by setting: $n = k_i(t)$ and invoking Theorem 1.)

(3) The fundamental update equation for $I_i(n)$, (18)

$$I_i(n+1) = \max\{C_i(n), q_i(n), r_i(n)\}.$$

Our methodology is then summarized in the following three steps:

(1) Look at the nominal path of a DEDS. For each n, the value of $I_i(n+1)$ is known, since it is directly observable.

(2) Next, look at the perturbed path, for a given perturbation. In general

$$I'_i(n+1) = \max\{C'_i(n), q'_i(n), r'_i(n)\}$$

while, under similarity, condition (19) holds.

(3) Finally, all primed quantities are replaced by their nonprimed (nominal) equivalents and the appropriate value of gain, e.g. $I'_i(n+1) = I_i(n+1) - G_i(n+1)$. This will allow us to derive the evolution equations for various $G_j(\cdot)$, $j = 1, \ldots, I$ using quantities obtainable *from the nominal path alone*.

Explicit equations resulting from this approach require explicit expressions for $q_i(n)$, $r_i(n)$. this is easily accomplished in the cases of a serial transfer line and an assembly line, as shown in Lemmas 1 and 2, and the exact results are the subject of the next two sections.

5.4.1. Propagation of perturbations in a serial transfer line. In the case of a serial transfer line, expressions for $q_i(n)$ and $r_i(n)$ were obtained in Lemma 1 of Section 5.2. Thus, an explicit equation for the propagation of $G_i(t)$ can be obtained. We shall make use of the following definitions

$$\mathscr{F}_i(t+1) \triangleq \begin{cases} \mathscr{F}_i(t) + 1, & \text{if } f_i(t) = 1 \\ 0, & \text{otherwise}; \end{cases} \quad i = 1, \ldots, I \quad (32)$$

with $\mathscr{F}_i(0) = 0$

$$\mathscr{N}_i(t+1) \triangleq \begin{cases} \mathscr{N}_i(t) + 1, & \text{if } n_i(t) = 1 \\ 0, & \text{otherwise}; \end{cases} \quad i = 1, \ldots, I \quad (33)$$

with $\mathscr{N}_i(0) = 0$.

Thus, $\mathscr{F}_i(t)(\mathscr{N}_i(t))$ simply keeps track of the duration of a FO(NI) event taking place at M_i at time t. When this event is terminated $\mathscr{F}_i(t)(\mathscr{N}_i(t))$ is reset to zero.

The next theorem is the specialization of Theorem 2 to the case of a serial transfer line. We are, however, only interested in how a single specific perturbation propagates in time and for this reason we shall set $G_j(\bar{n}) = \delta \neq 0$ for some $n = \bar{n}$ and $G_i(\bar{n}) = 0$, $i \neq j$ and $\Delta S_i(n) = 0$ for all $n \geq \bar{n}$ and $i = 1, \ldots, I$. Also note that we do not discuss the exact nature of this perturbation, but only its ultimate effect on the $\bar{n}$th service initiation time of M_i.

Theorem 3. If $\mathscr{E}_i$ and $\mathscr{E}'_i$ are similar event sequences of M_i in a serial transfer line and some perturbation is introduced to the nominal path so that: $G_j(\bar{n}) = \delta \neq 0$ for some j, then, for all t such that $k_j(t) > \bar{n}$

$$G_i(t+1) = \begin{cases} G_{i+1}(t+1), & \mathscr{F}_i(t) > 0; f_i(t) = 0 \\ G_{i-1}(t), & \mathscr{N}_i(t) > 0; n_i(t) = 0 \\ G_i(t), & \text{otherwise}; \quad i = 1, \ldots, I. \end{cases} \quad (34)$$

Proof: Using the results of Lemma 1 in Section 5.2, (22) and (23), and the assumption of similarity, as expressed in (19), we have

$$\begin{aligned} q_i(n) - q'_i(n) &= I_{i+1}(n+1-b_{i+1}) \\ &\quad - I'_{i+1}(n+1-b_{i+1}) \\ &= G_{i+1}(n+1-b_{i+1}) \end{aligned}$$

if a FO event follows the nth customer service completion at M_i; and

$$\begin{aligned} r_i(n) - r'_i(n) &= C_{i-1}(n+1) - C'_{i-1}(n+1) \\ &= I_{i-1}(n+1) - I'_{i-1}(n+1) \\ &\quad + S_{i-1}(n+1) - S'_{i-1}(n+1) \\ &= G_{i-1}(n+1) + \Delta S_{i-1}(n+1) \end{aligned}$$

if a NI event follows the nth customer service completion at M_i.

Thus, recalling that $\Delta S_i(n) = 0$ for all $i = 1, \ldots, I$ and all $n \geq \bar{n}$, Theorem 2 in this case gives

$$G_i(n+1) = \begin{cases} G_{i+1}(n+1-b_{i+1}), & \text{if a FO event follows the } n\text{th customer service completion at } M_i \\ G_{i-1}(n+1), & \text{if a NI event follows the } n\text{th customer service completion at } M_i \\ G_i(n), & \text{otherwise} \end{cases}$$

for all $n > \bar{n}$.

Next, recall Theorem 1, which says that the gain of M_i can only change at points in time $t = I_i(n)$, $n = 1, 2, \ldots$ Thus, we shall focus on such a time instant $t = I_i(n+1)$, $n+1 > \bar{n}$.

If a FO event precedes the $(n+1)$th customer arrival at M_i, we know from Lemma 1 that: $I_i(n+1) = I_{i+1}(n+1-b_{i+1})$. Since the activation index is updated after the arrival of a new customer, we have

$$k_{i+1}[I_i(n+1) + 1] = n + 1 - b_{i+1}$$

Similarly, if a NI event precedes the $(n+1)$th customer arrival at M_i, we know that: $I_i(n+1) = C_{i-1}(n+1)$. Hence

$$k_{i-1}[I_i(n+1)] = n + 1.$$

If none of the above events occur at M_i, then

$$k_i[I_i(n+1) + 1] = n + 1.$$

Therefore, our previous gain update equation gives

$$G_i[I_i(n+1) + 1] = \begin{cases} G_{i+1}[I_i(n+1) + 1], & \text{if a FO event follows } n\text{th customer service completion at } M_i \\ G_{i-1}[I_i(n+1)], & \text{if a NI event follows } n\text{th customer service completion at } M_i \\ G_i[I_i(n+1)], & \text{otherwise} \end{cases}$$

for all $n + 1 > \bar{n}$.

Furthermore, using the definitions in (32) and (33), note that the condition: 'a FO (NI) event follows the nth customer service completion at M_i' can be replaced by '$\mathscr{F}_i[I_i(n+1)] > 0 \, (\mathscr{N}_i[I_i(n+1)] > 0)$ and $f_i[I_i(n+1)] = 0 \, (n_i[I_i(n+1)] = 0)$' which indicates that a FO (NI) event of duration $\mathscr{F}_i[I_i(n+1)] \, (\mathscr{N}_i[I_i(n+1)])$ has just ended at time $I_i(n+1)$.

Finally, we already mentioned that $G_i(t+1) = G_i(t)$ for any $t \neq I_i(n)$, $n > \bar{n}$, as shown in Theorem 1. Thus

$$G_i(t+1) = \begin{cases} G_{i+1}(t+1), & \mathscr{F}_i(t) > ; f_i(t) = 0 \\ G_{i-1}(t), & \mathscr{N}_i(t) > 0, n_i(t) = 0 \\ G_i(t), & \text{otherwise} \end{cases} \quad (35)$$

for all t such that $k_j(t) > \bar{n}$, $G_j(\bar{n}) = \delta \neq 0$ and $i = 1, \ldots, I$. ∎

5.4.2. Propagation of perturbations in an assembly line. The results of the previous section can be extended to the case of an assembly line by using Lemma 2, instead of Lemma 1, in proving the equivalent of Theorem 2. For simplicity, we further assume that the amount of perturbation δ to be realized is sufficiently small so that, referring to (24), if

$$c_l[C_l(n)] = b_l \text{ and } l^* \text{ is such that } I_{l^*}(n+1-b_l) = \max_{l=1,\ldots,L}\{I_l(n+1-b_l)\}$$

then

$$c_i'(C_i'[n]) = b_l \text{ and } l^* \text{ is such that } I_{l_*}'(n+1-b_l) = \max_{l=1,\ldots,L} \{I_l'(n+1-b_l)\}.$$

In other words, the server, M_{l_*}, that causes the FO event at M_i remains the same in the perturbed path. A similar assumption can be made pertaining to M_{m_*}, the server causing the NI event at M_i in (25).

Under these assumptions, the result of Theorem 3 can be easily extended to the case of an assembly line. We state the result without proof

$$G_i(t+1) = \begin{cases} G_{l_*}(t+1), & \mathscr{F}_i(t) > 0; f_i(t) = 0 \\ G_{m_*}(t), & \mathscr{N}_i(t) > 0; n_i(t) = 0 \\ G_i(t), & \text{otherwise}; \quad i = 1, \ldots, I \end{cases}$$
(36)

where l^* is such that $I_{l_*}[k_i(t)+1-b_l] = \max_{l=1,\ldots,L} \{I_l[k_i(t)+1-b_l]\}$ and m^* is such that $I_{m_*}[k_i(t)+1] = \max_{m=1,\ldots,M} \{I_m[k_i(t)+1]\}$.

The discussion of the previous section, in reference to extensions of Theorem 3, is relevant to possible extensions of (36) as well.

5.5. Extensions of perturbation analysis

Several questions can be raised:

(1) How large can the perturbation, δ, become before similarity is violated? As long as similarity is not violated, it is evident, from (35), that $G_I(N)$, the gain at M_I after N customer arrivals, is either δ—if the perturbation propagates to M_I—or 0—if it does not. Thus, $G_I(N; \delta)$ is a *linear* function of δ in the sense that, as long as similarity holds

$$G_I(N; K\delta) = \begin{cases} K\delta, & \text{if the perturbation is realized} \\ 0, & \text{otherwise} \end{cases}$$
$$= K \cdot G_I(N; \delta)$$

for arbitrary K.

The point here is that similarity is an assumption in the same spirit as *linearization assumptions in standard continuous functions*: the assumption is exact as long as the size of the perturbation remains small. While, however, in nonlinear continuous systems perturbation analysis is exact only for infinitesimal perturbations, in DEDS perturbation analysis may be exact for finite perturbations from the nominal. This is because similarity may hold for a considerable range of perturbations, e.g. service times $S_i(n)$ may be perturbed for finite amounts without causing the elimination of any FD event of the creation of any new one in $\mathscr{E}_i^\omega$, $i = 1, \ldots, I$.

(2) If $S_i(n)$ is now perturbed several times, for more than a single activation index n, does superposition hold? In other words, if two separate perturbations are realized for the system independently, will their combined effect also be realized?

(3) Probably the most interesting question is one that addresses the validity of results on one sample path for other sample paths: can we expect that $G_I(N; \delta)$ obtained over some sample realization is the same over other ones, for the same DEDS? We have found the answer to this question to be affirmative for a great variety of experiments performed via simulations. This is extremely important since the implication is: *given a single sample path of a DEDS, we can study the effect of different kinds of perturbations and predict the system's performance for any sample path, without having to realize these perturbations or observe other sample trajectories.*

We shall attempt to address questions (2) and (3) in the next section.

6. STATISTICAL PERTURBATION ANALYSIS

In Section 5 we dealt with perturbations generated on a given sample trajectory. A nominal path (tableau) was observed and from its history alone we were able to draw conclusions about some perturbed path, given a specific perturbation. In this section, we would like to adapt this analysis to some realistic cases to yield useful results. Typically, in a DEDS, we are interested in the sensitivity of some performance criterion with respect to a parameter, e.g. throughput with respect to the mean service time of a server. Let the mean service time of server i be S_i. If we introduce a perturbation ΔS_i in S_i, then this amounts to changing the service time duration of the jth activation of server M_i by the amount $S_i(j)(\Delta S_i/S_i)$ for all i along the sample path (Suri, 1983).

Throughput on the other hand can be measured by the time it takes the output server to serve a specified number of customers. By following the propagation rules developed in Section 5, we can determine the proportion of perturbations thus generated that propagate to the output server. From this proportion we can directly determine the desired sensitivity. The validity of the sensitivity thus determined rests on the following two assumptions:

(1) *Stochastic similarity*. In terms of the descriptive analogy used in Section 5.3, we visualized the series of perturbations in service time generated by ΔS_i as an ensemble of marbles (of possible different sizes) released at each service activation of the sample tableau. These marbles will be propagated or cancelled by the random distribution of the FOs and NIs in the tableau. Eventually a certain proportion of them will be realized at the output server event sequence. This proportion constitutes the sensitivity under discussion. Now stochastic simi-

larity has to do with the question 'what happens if we released the same ensemble of marbles in a different sample tableau of the same DEDS?'. Intuitively, if the sample tableaus are of sufficiently long duration (or equivalently they are statistically 'representative'), then we expect the same proportion of perturbations will be realized. This is because we expect the ensemble to encounter the same distribution of NIs and FOs. In this sense, we can use the results derived from one sample tableau to make predictions for another. Note this issue of stochastic similarity is fundamental to every Monte Carlo experiment. Our approach cannot avoid this requirement either, if we were to use our analysis in prediction.

Our contribution in this paper is an efficient means for calculating these sensitivities. The validity of these sensitivities to other sample trajectories is a separate issue. There are many more or less standard techniques for the analysis of this issue. We shall not enter into a long discussion of it here.*

(2) *Statistical linearity.* Our propagation analysis of perturbations treats each perturbation separately. When we generate an ensemble of perturbations and use the proportions realized as a measure of sensitivity we are implicitly assuming that superposition or linearity holds for the system under study. This is certainly true if deterministic (as opposed to stochastic) similarity as defined in Section 5 holds. Deterministic similarity will hold so long as all perturbations considered are infinitesimal. The set of sample tableaus on which a series of infinitesimal perturbations could cause violation of deterministic similarity must have measure zero. However, this is not so if we consider finite perturbations no matter how small they are. A sufficient number of small perturbations can eventually cause:

(a) creation of new NIs and FOs or the elimination of nominal ones;
(b) the server which terminates a NI at another server may change as a result of finite perturbations;
(c) the customer which enters a server at the end of a FO may change.† These possibilities can change permanently the tableau from the nominal one.

Any perturbations that come after the change will be propagated by a different set of FO and NI occurrences. However, this does not necessarily mean that the sensitivities derived using the nominal tableau are totally useless in such cases.

Let θ be the parameter whose sensitivity is sought. Furthermore, let TB(θ) denote the nominal tableau from which we calculate the sensitivity. What is meant by 'statistical linearity' is that the tableaus represented by TB($\theta + \Delta\theta$) for $0 \leq \Delta\theta \leq \Delta$ all have essentially the same distribution of FOs and NIs. In other words, the tableaus will propagate the same proportion of perturbations, i.e. they have the same property as far as propagation of perturbations is concerned despite the difference in details of trajectory. The satisfaction of this requirement, of course, depends on the systems involved. However, it is in fact quite reasonable as later experiments demonstrate. At this point we have no analytical means to determine such 'statistical linearity' for a given system; this must form the subject of a separate study. [See, however, Ho, Cao and Cassandras (1982).]

Note also that the assumption of linearity is again fundamental in any use of these sensitivities for optimization purposes. Our assumption of 'statistical linearity' is no stronger than other similar applications of sensitivity (or gradient) information. In fact, because of the inherent averaging process in the sensitivity calculation, it can be argued that we stand a far better chance of encountering 'statistically linear' DEDS than in purely deterministic systems.

7. EXPERIMENTAL VERIFICATION

The results of Sections 5.4.1 and 5.4.2 have been experimentally and extensively verified for a number of simple fictitious serial transfer lines and assembly lines, via computer simulations, as well as on real world production lines (Ho, Eyler and Chien, 1979a). The way this is done is as follows: first, a nominal sample path (one sufficiently long simulation run) is generated. During this run, for a fixed parameter perturbation, Gain is updated according to (35) or (36) as the case may be. The simulation run is stopped for a given value, N, of the activation index $k_I(t)$ of M_I, the output server of the system; we shall refer to N as the *output target*. Thus, at the end of this run, we have at our disposal $G_I(N)$, which stands for an estimate of the difference to achieve the same output target if the specified perturbation were to be realized. We also have the *output target time*, $C_I(N)$, i.e. the time of completion of the Nth customer at M_I. Next, a perturbed path is obtained by actually putting the specified perturbation to effect, while all other parameters and initial conditions (random seeds included) are left unchanged. The product of this second run is $C'_I(N)$, the new real output target time. We can then compare $[C_I(N)] - C'_I(N)$. [For simplicity, assume $\Delta S_I(N) = 0$ always.]

We present here some representative results of

---

*An often used simple solution involves comparing results obtained from one simulation with the same but with experimental duration half as long.

†Note possibilities (b) and (c) cannot arise in a serial transfer line but only in lines with merging nodes or feedback loops.

our experiments. All service times, S_i, are constant. Servers fail with probability p_i, in which case their repair time is exponentially distributed with mean μ_i. All queue capacities are denoted by b_i.

Example 1. Consider an assembly line with parameters specified in Fig. 6. We reduced the mean repair time of M_2 by one time unit, so that $\Delta \mu_2 = -1$. We obtained

$$C_I(N) = 8184, \qquad G_I(N) = 63, \qquad C'_I(N) = 8121.$$

Note that $G_I(N) = C_I(N) - C'_I(N)$.

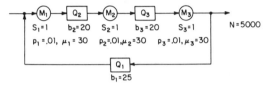

Fig. 6. System of Example 1.

Example 2. For the same system as that of Fig. 6, but with $b_1 = 10$, and for $\Delta \mu_2 = -1$, we obtained

$$C_I(N) = 8953, \qquad G_I(N) = 67, \qquad C'_I(N) = 8886.$$

Again, note that: $G_I(N) = C_I(N) - C'_I(N)$ exactly.

Example 3. For the same system as that of Fig. 6, but with an output target $N = 20\,000$, we have perturbed the queue capacity of Q_3 by increasing it by one unit, i.e. $\Delta b_3 = 1$. We obtained

$$C_I(N) = 33\,123, \qquad G_I(N) = 56,$$
$$C'_I(N) = 33\,067.$$

Again, $G_I(N) = C_I(N) - C'_I(N)$.

Example 4. Consider an assembly line as shown in Fig. 7. Note that service times are now different among the three servers. We have created a perturbation: $\Delta \mu_2 = -1$ and obtained

$$C_I(N) = 20\,670, \qquad G_I(N) = 22,$$
$$C'_I(N) = 20\,648.$$

The gain prediction is once again exact.

What makes the above examples particularly simple, so that our gain predictions are exact, is the fact that deterministic similarity is never violated, since the cumulative effects of the perturbations considered cannot eliminate nominal FD events nor create new ones. Some similar experimental results, verifying the correctness of gain prediction under similarity, can be found in Ho and Eyler (1980).

The nonlinear effects of the violation of similarity can be seen in more complex systems with service time perturbations, whose cumulative effect is more pronounced. All subsequent experimental results were obtained by using a simulator developed at Charles Stark Draper Laboratory, Cambridge, Massachusetts. We shall now be comparing throughputs, $T(N)$ and $T'(N)$, corresponding to a given output target N. Gain predictions from a nominal path are now expressed as SENSITIVITY COEFFICIENTS, directly indicating the change in throughput due to a given perturbation. We shall denote the predicted throughput by $T_{\text{pred}}(N)$.

Example 5. Consider a five-server serial transfer line, as shown in Fig. 8. Note that M_3 and M_4 have normally distributed service times. The perturbation of interest here is: $\Delta S_2 = -1$ (or 25%) over all activations, which yielded the following results

$$T(N) = 5.095, \qquad T_{\text{pred}}(N) = 5.274,$$
$$T'(N) = 5.238.$$

Most importantly, the essence of our approach was used in solving the long standing problem of general buffer storage design in a real serial transfer line (Eyler, 1979; Ho, Eyler and Chien, 1979), i.e. based on actual data obtained from a FIAT plant in Turin, Italy. The assumptions made in Section 6 pertaining to stochastic similarity and superposition were justified in the final result of the solution to the buffer problem.

The following examples further illustrate the validity of our assumptions. We consider various types of systems under a continuous perturbation: a 5% reduction in the service times of all servers. We then record the throughput and respective sensitivity coefficients, as well as the fractions of NI and FO events observed in a given sample path. (Note that our approach allows the computation of many sensitivity coefficients at a time from a single nominal path history at minimal extra cost.)

Our claim is that if the given sample path is sufficiently long, and hence representative, then the sensitivity coefficients thus derived should be applicable to other sample paths. One way to validate this is to compare the result from one long simulation with the average value of the results from an ensemble of shorter simulations. If our assumptions in Section 6 are correct, then these two methods of calculating sensitivity coefficients should be equal.

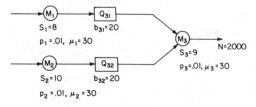

Fig. 7. System of Example 4.

Example 6. Consider the serial transfer line of

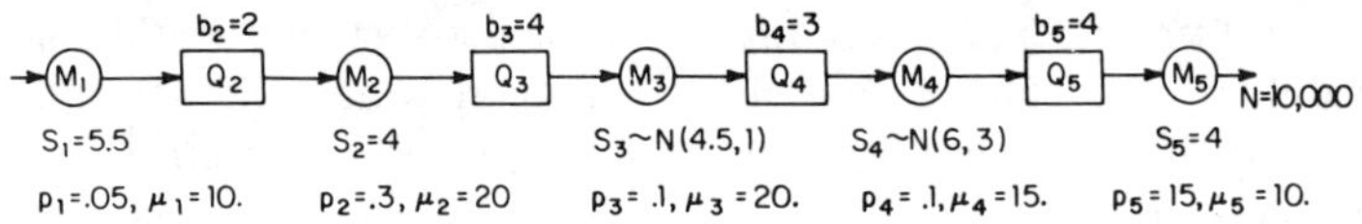

FIG. 8. System of Example 5.

TABLE 1. TEN RUNS, $N = 10\,000$

	NI (%)		FO (%)		Sensitivity coefficients	
M_1		0	[47.6, 49.4]	48.5	[0.0154, 0.0199]	0.0182
M_2	[3.1, 3.7]	3.4	[9.7, 11.4]	10.4	[0.0333, 0.0373]	0.0356
M_3	[33.7, 38.1]	36.2	[7.3, 8.5]	8.0	[0.0090, 0.0110]	0.0100
M_4	[33.6, 35.9]	34.6	[0.8, 1.2]	1.0	[0.0180, 0.0220]	0.0194
M_5	[51.5, 53.5]	52.8		0	[0.0003, 0.0011]	0.0007

$T(N)$: [5.089, 5.253] 5.152.

TABLE 2. ONE RUN, $N = 20\,000$

	NI (%)	FO (%)	Sensitivity coefficients
M_1	0	48.6	0.0163
M_2	3.2	10.1	0.0374
M_3	35.9	8.2	0.0100
M_4	34.0	1.1	0.0191
M_5	52.3	0	0.0007

Fig. 8 once again. One experiment consists of running ten different sample paths with $N = 10\,000$. Lower and upper bounds are recorded in brackets, accompanied by the mean value of each variable of interest (see Table 1). In another experiment, a single path is obtained with $N = 20\,000$, i.e. the observation interval is doubled (Table 2).

Typical standard deviations in the variables recorded in Table 1 are under 5%. Note that the sensitivity coefficients obtained from a single long run (Table 2) differ very little from the ensemble averages shown in Table 1 and could be taken as reliable representatives of any sample path.

Example 7. Consider a network as shown in Fig. 9. We have recorded results analogous to the ones of Example 6 in Tables 3 and 4 below. Note that routing takes place in this system: a customer from M_2 is directed to M_3, unless M_3 is busy, in which case it is directed to M_4. Similarly, a customer from M_3 is routed to Q_{23} with probability 0.25, otherwise it is routed to Q_{53}; the same applies for M_4. M_2 is activated by a customer from Q_{21}, Q_{23} or Q_{24} and M_5 by a customer from either Q_{53} or Q_{54}.*

In this example, customers follow random routing. Yet sensitivity coefficients obtained from one long run (Table 4) agree well with the mean value of the coefficients in Table 3. Once again, one has experimentally validated the discussion in Section 6.

---

* This is another example of the difference in terminology between production lines and queueing systems. See also the footnote on p. 150.

Example 8. We set $p_i = 0.$ in the network of Fig. 9 and record the fraction of customers routed from M_2 to M_3 over a range of different output targets (Table 5).

Note in Table 5 that the routing frequency at the branching node of this system remains essentially unchanged as the length of the observation interval increases.

8. CONCLUSIONS AND FUTURE RESEARCH

We have attempted to transcend traditional approaches to discrete event dynamical systems by creating a state space model and thus place such systems in the context of control theory. As seen in Section 3, the state equations obtained are highly nonlinear. In the spirit of linearization of ordinary dynamical systems, we have introduced the concept of *similarity* (a direct analog of linearity) in Section 5.2 and have carried out perturbation analysis for DEDS: a sample trajectory of the system is observed and its history recorded; based on this information alone, the performance of the system under a given perturbation can be predicted with accuracy, if the sample path were to remain the same. This is reflected by the gain evolution equations developed in Section 5.4 (general case) and 5.4.1 and 5.4.2 for the cases of a serial transfer line and an assembly line, respectively. Furthermore, under certain reasonable assumptions (Section 6), predictions made for one sample realization are applicable to other (stochastically similar) realizations as well. Our assumptions and analysis are supported and

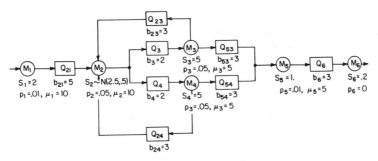

Fig. 9. System of Example 7.

Table 3. Ten runs, $N = 5000$

	NI (%)		FO (%)		Sensitivity coefficients	
M_1	0		[47.3, 49.4]	48.5	0	
M_2	[5.9, 6.2]	6.1	0		[0.1099, 0.1397]	0.1238
M_3	[22.2, 24.1]	22.9	0		[0.0506, 0.0726]	0.0607
M_4	[95.0, 95.1]	95.1	0		0	
M_5	[0.0, 0.7]	0.5	[1.5, 2.3]	1.9	[0.2777, 0.3440]	0.3132
M_6	[74.1, 74.7]	74.6	0			

$T(N)$: [14.536, 14.924] 14.662.

Table 4. One run, $N = 10\,000$

	NI (%)	FO (%)	Sensitivity coefficients
M_1	0.0	48.3	0.0
M_2	5.8	0	0.1250
M_3	22.2	0	0.0588
M_4	95.1	0	0.0
M_5	0.3	1.8	0.3173
M_6	74.5	0	0.0

Table 5. Routing frequency as a function of observation period

N	Routed from M_2 to M_3
5 000	50.34
15 000	50.45
30 000	50.53
50 000	50.54

verified by a variety of experiments and simulations on real and fictitious systems of different kinds (Section 7).

Thus, our approach can be summarized in the following steps:

(1) Record the history (event sequence tableau) of a (nominal) sample path.

(2) Translate a given parameter or input perturbation (e.g. service time, queue capacity, destination) to a state perturbation [how this is done depends on the nature of the perturbation alone; see discussion accompanying (31) in Section 5.4]. This is what initializes the gain function $G_i(t)$, $i = 1, \ldots, I$.

(3) Track the propagation of this perturbation in the system via the gain update equations. This can be done while the nominal sample path is being observed.

(4) Performance of the DEDS (in our case: throughput) for the perturbation specified is predicted through $G_I[k_I(t)]$ for any given value of the activation index $k_I(t)$.

Note that in predicting performance: (i) the perturbation does not have to be actually realized, and (ii) predictions apply to all stochastically similar sample path of the DEDS.

It was already mentioned in Section 2.1 that we are only considering systems with a single customer type being served by all servers. In the multiclass customer case we have found that a different—more complex—state-space characterization is required. This is currently the subject of further research.

Furthermore, it was pointed out in the discussion following Theorem 2 that the gain update equation (31) is of interest when explicit expressions for $q_i(n)$, $r_i(n)$, the times of queueing space and customer availability at M_i respectively, are known. Such is the case for a serial transfer line (tandem queueing network) and an assembly line. Our aim is to extend concrete results to more complex systems, such as a nonstandard queueing network model.

Despite the limiting assumption of similarity, it was mentioned that gain update equations dependent on the nominal path alone can still be developed even if this assumption is not made. The only problem is in having to store more information from the nominal sample path history, which may result in a costly process. Furthermore, the discussion in Section 6 suggests that deterministic similarity may not necessarily be required for superposition to hold. Statistical linearity may yet save the day.

The extent to which stochastic similarity is valid seems to require further investigation both in an experimental and a theoretical framework. The exact reasons for which stochastic similarity has been found to hold have yet to be completely analyzed and understood. It would appear that the nature of DEDS is such that special properties exist, which need to be revealed. We have seen that 'linearization' can take place for *finite* perturbations and that the evolution of a state perturbation depends on the nominal state history alone in a remarkably simple way (see 'propagation rule' of Section 5.3). More similar properties may be seen as the precise nature of DEDS in a system theoretic setup is studied.

Finally, throughput may not necessarily be the system performance criterion of main interest; we believe that our approach can be used to carry out perturbation analysis (and hence design optimization) with respect to other criteria, such as server utilization, work-in-process inventory, etc. In general, perturbation analysis can be a powerful tool to carry out optimization of various performance criteria with respect to different parameters. This was successfully accomplished for the buffer design problem (Eyler, 1979; Ho, Eyler and Chien, 1979). The reader is also referred to Ho and Cassandras (1980).

Acknowledgements—The results reported in this paper represent the culmination of effort, both experimental and analytical, over the past four years on the subject. Much of the motivation, support, and experimentation were furnished by the C. S. Draper Laboratory through their work with the FIAT Motor Company, Italy, the Army FMS Program DAAK 30-80-C-0016 and the Tailored Clothing Technology Corporation. Analytical research support for the authors was furnished by the National Science Foundation ENG 78-15231, the Office of Naval Research under the Joint Services Electronics Program N00014-75-C-0648 and the Office of Naval Research N00014-79-C-0776.

REFERENCES

Buzacott, J. A. (1967). Automatic transfer lines with buffer stocks. *Int. J. Prod. Res.* **5**, 183.

Buzacott, J. A. and L. E. Hanifin (1978). Models of automatic transfer lines with inventory banks: a review and comparison. *AIEE Trans* **10**, 197.

Buzen, J. (1973). Computational algorithms for closed queueing networks with exponential servers. *Commun. ACM* **16**, 527.

Denning, P. and J. Buzen (1977). Operational analysis of queueing networks. In H. Beilner and E. Belenbe (Eds), *Proceedings of Third International Symposium in Computer Performance, Modelling, Measurement and Evaluation*. North Holland.

Denning, P. and J. Buzen (1978). The operational analysis of queueing networks. *ACM Comp. Surveys* **10**, 225.

Denning, P. and J. Buzen (1980). Operational treatment of queue distributions and mean value analysis. *Computer Performance*, **1**, 6.

Eyler, M. A. (1979). A new approach to the productivity study of transfer lines. Ph.D. thesis, Harvard University, Division of Applied Sciences.

Ho, Y. C. (1979). Parametric sensitivity of a statistical experiment. *IEEE Trans Aut. Control* **AC-24**, 982.

Ho, Y. C. (1981). Discrete event dynamical system and its application to manufacturing. *IFAC Congress Proceedings*.

Ho, Y. C. and C. Cassandras (1980). Computing costate variables for discrete event systems. *Proc. of IEEE Conf. on Decision and Control*.

Ho, Y. C., X. Cao and C. Cassandras (1982). Zeroth and first order perturbation analysis of queueing networks. *1982 IEEE Conference on Decision and Control*.

Ho, Y. C. and M. A. Eyler (1980). Analysis of large scale discrete event dynamical systems. *Proc. of 1980 IEEE Conf. on Circuits and Systems*.

Ho, Y. C., M. A. Eyler and T. T. Chien (1979). A gradient technique for general buffer storage design in a production line. *Int. J. Prod. Res.* **17**, 557.

Ho, Y. C., M. A. Eyler and T. T. Chien (1983). A new approach to determine parameter sensitivities on transfer lines. *Management Sci.*, to be published.

Okamura, K. and H. Yamashina (1977). Analysis of the effect of buffer storage capacity in transfer line systems. *AIIE Trans* **9**, 127.

Sevastyanov, B. A. (1962). Influence of storage bin capacity on the average standstill time of a production line. *Theory Probab. Applic.* **7**, 429.

Soyster, A. L., J. W. Schmidt and M. W. Rohrer (1979). Allocation of buffer capacities for a class of fixed cycle production lines. *AIIE Trans* **11**, 140.

Suri, R. (1980). Robustness of analytical formulae for performance prediction in certain nonclassical queueing networks. Harvard University, Division of Applied Sciences, technical report.

Suri, R. (1983). Implementation of sensitivity calculations on a Monte Carlo experiment. *J. Optimiz. Theory & Appl.*, to be published.

An Event Domain Formalism for Sample Path Perturbation Analysis of Discrete Event Dynamic Systems

CHRISTOS G. CASSANDRAS AND Y. C. HO

Abstract—We present a state-space representation of a general discrete event dynamic system (DEDS) model on the basis of a natural *event* domain, which replaces time. Perturbation analysis of a sample path of a DEDS is then formalized using some fundamental concepts introduced to establish conditions for its validity. A set of simple perturbation equations is then derived under these conditions, so that performance gradients with respect to parameters are estimated from observation of real or simulated data on a *single* sample path.

Manuscript received August 18, 1983; revised March 22, 1984 and February 7, 1985. Paper recommended by Associate Editor, J. Walrand. This work was supported in part by the U.S. Office of Naval Research under Contracts N00014-75-C-0648 and N00014-79-C-0776 and in part by the National Science Foundation under Grant ECS 82-13680.

C. G. Cassandras was with ITP Boston, Inc., Cambridge, MA 02138. He is now with the Department of Electrical and Computer Engineering, University of Massachusetts, Amherst, MA 01003.

Y. C. Ho is with the Division of Applied Sciences, Harvard University, Cambridge, MA 02138.

I. Introduction

Discrete event dynamic systems (DEDS) are characterized by state transitions occurring at discrete points in time, when various activities are initiated and terminated. Traditional approaches for analyzing such systems involve classical queueing theory and simulation techniques, the former being limited by a number of restrictive assumptions and the latter by significant computational costs. Recently, we introduced a new approach [1]–[4] termed *sample path perturbation analysis,* which combines experiment (simulation or direct observation) with analysis. The main idea is to study specific stochastic realizations (sample paths) of DEDS, from which we can derive equations (algorithms) for the perturbed behavior of the system about the nominal trajectory observed. Thus, we are able to *predict performance of a perturbed DEDS along the same sample path without having to implement given parameter changes*; and by standard techniques of statistical inference, we can extend our predictions to other sample paths as well. Extensions of this new approach can be found in [5], [6]. Successful applications are presented in [7]–[12].

Our purpose here is to provide a new formalism for sample path perturbation analysis. By replacing the conventional time domain by a more natural "event" domain, it is shown that the resulting state-space representation is simpler than that of earlier work [3]. Furthermore, the fundamental result of sample path perturbation analysis (9) is formally derived under well-defined conditions.

II. DEDS State-Space in the Event Domain

The model we use for a DEDS is that of a queueing network (open or closed). A detailed description may be found in [3]. It is important to recall that a server (or module) M_i is always in one of three possible states: 1) *busy* (BY); 2) *blocked,* also termed *full output* (FO) (it is assumed that queues Q_i have finite capacity B_i, $i = 1, \cdots, N$); 3) *idle,* also termed *no input* (NI). FO or NI constitute the "interesting" phenomena in the system history, since they are the means by which one server becomes coupled to another.

Our motivation for selecting an event domain becomes clear when one considers how discrete events generate state transitions: such transitions take place instantaneously at particular points in time, while in between such points the state remains unaffected. From this viewpoint, one may also regard the development that follows as an attempt to formalize the descriptions of event-driven simulations [13]–[15].

For simplicity, we shall henceforth refer to the term "event" as a "customer service completion at some server." All other occurrences of importance shall be termed "induced events." We introduce

k = event order, $k = 1, 2, \cdots$ for all events in the DEDS. It is assumed that no two events occur at the same time instant.
$e_i(k)$ = time of next event occurrence at ith server.
$c_i(k)$ = ith queue content

which are shown to provide a natural state-space X for a DEDS (they are the information needed at the time of the kth event in order for a simulation to proceed). Note that $e_i(k)$ and $c_i(k)$ are evaluated at any finite time after the $(k - 1)$th event occurrence and *before* the kth event occurrence.

The input space U consists of two kinds of information, as it unfolds along an element ω of the sample space Ω:

$S_i(k)$ = service time at ith server.
$D_i(k)$ = destination of customer completing service at ith server.

$S_i(k)$ and $D_i(k)$ are updated *after* the kth event, if this event occurs at the ith server; they remain unchanged otherwise. It is emphasized that $S_i(k)$, $D_i(k)$ are generally stochastic variables whose realizations are observed along a sample path. $S_i(k)$ is defined over the nonnegative real numbers and $D_i(k)$ over the set $\{1, \cdots, N\}$ for a closed network and $\{0, \cdots, N\}$ for an open one (where 0 represents the "outside world").

State equations can be derived to represent the state transition logic outlined in Fig. 1. Definitions 2.1–2.3 are simply convenient functions of state and input information that facilitate the formulation of (3) and (4).

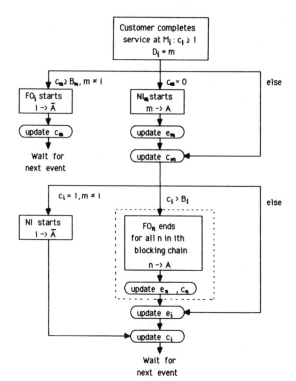

Fig. 1. DEDS model state transition logic.

Our aim here is only to present the equations, a detailed development of which may be found in [18].

Definition 2.1a: The ith server is said to belong to the *set of inactive servers* $\bar{A}(k)$ if: $c_i(k) = 0$ or $c_m(k) > B_m$, $m = D_i(k)$ (i.e., the ith server is experiencing an NI or FO).

Definition 2.1b: The ith server is said to belong to the *set of active servers* $A(k)$ if it does not belong to $\bar{A}(k)$.

Remark: Note that $c_i(k)$, $i = 1, \cdots, N$, is allowed to take on values greater than the capacity B_i. It is convenient to do so in order to be able to easily detect an FO: if $c_i(k) = B_i$, then the queue is full but no FO has yet occurred; only when $c_i(k) > B_i$ does an FO actually take place.

Definition 2.2: The next event *generating function* is a mapping: $X \times U \times \Omega \to \{1, \cdots, N\}$:

$$s[\underline{e}(k), \underline{c}(k), \underline{D}(k); \omega] = \arg\min_{i \in A(k)} \{e_i(k)\}. \tag{1}$$

Definition 2.3: The ith *blocking function* is a mapping: $X \times U \times \Omega \to \{1, \cdots, N\}$:

$$d_i[\underline{e}(k), \underline{c}(k), \underline{D}(k); \omega] = \begin{cases} d_m(k) & c_m(k) > B_m, \ m = D_i(k) \\ i & \text{otherwise.} \end{cases} \tag{2}$$

For simplicity, we denote the last two functions by $s(k)$, $d_i(k)$ with the understanding that they are functions of state and input information over a given sample path $\omega \in \Omega$. Note that $s(k)$ is the server where the next event occurs, evaluated after the occurrence of the $(k - 1)$th event. $d_i(k)$ is recursively defined to indicate the queue-server pair blocking the ith server, with the possibility of a chain of blockings taking place.

Remark: The blocking function as defined in (2) does not specify which server is unblocked first in case two or more are blocked by one queue. This would require the introduction of a "prioritization" scheme, which does not concern us here. For simplicity, one can assume that the first server blocked is always the first one to be freed.

We can now formulate the following state equations for all $i = 1, \cdots, N$:

$$c_i(k+1) = \begin{cases} c_i(k) + 1 & s(k) = n, \ D_n(k) = i, \ d_i(k) \neq n \\ c_i(k) - 1 & s(k) = d_i(k) = b, \ D_b(k) \neq i \\ c_i(k) & \text{otherwise} \end{cases} \tag{3}$$

$$e_i(k+1) = \begin{cases} e_n(k) + S_i(k) & s(k) = n, \ D_n(k) = i; \\ & c_i(k) = 0 \\ e_b(k) + S_i(k) & s(k) = d_i(k) = b; \\ & [c_i(k) \neq 1, \ c_m(k) \neq B_m, \ D_i(k) = m] \\ & \quad \text{or } D_b(k) = i \\ e_i(k) & \text{otherwise.} \end{cases} \tag{4}$$

Details on the interpretation of (3), (4) and their advantages over the formulation in [3] may be found in [18]. Briefly, they provide significant economy over the time domain representation and bypass the problem of events and induced events occurring at the same time.

A more succinct representation is obtained by defining transition functions

$$\sigma_i(k) \triangleq c_i(k+1) - c_i(k) \tag{5a}$$

$$\tau_i(k) \triangleq e_i(k+1) - e_i(k) \tag{5b}$$

which from (3), (4) implies that

$$\sigma_i(k) = \begin{cases} 1 & s(k) = n, \ D_n(k) = i, \ d_i(k) \neq n \\ -1 & s(k) = d_i(k) = b, \ D_b(k) \neq i \\ 0 & \text{otherwise} \end{cases} \tag{6}$$

$$\tau_i(k) = \begin{cases} e_n(k) + S_i(k) - e_i(k) & s(k) = n, \ D_n(k) = i; \\ & c_i(k) = 0 \\ e_b(k) + S_i(k) - e_i(k) & s(k) = d_i(k) = b; \\ & [c_i(k) \neq 1, \ c_m(k) \neq B_m, \ D_i(k) = m] \\ & \quad \text{or } D_b(k) = i \\ 0 & \text{otherwise.} \end{cases} \tag{7}$$

III. Perturbation Analysis

In developing state perturbation equations in the event domain, the following difficulty arises: the kth event in the nominal path does not necessarily correspond to the kth event in the perturbed path. Imbedded in the idea of a sample path for a DEDS is the order in which events take place. The concept of "event ordering" is captured by Definitions 3.1–3.3, which attempt to provide an alternative definition for an "event."

Definition 3.1: The sequence $G(\omega; K) = \{s(k), k = 1, \cdots, K\}$, with $s(k)$ defined by (1), is called the *event generating sequence* (EGS) corresponding to a specific sample realization of a DEDS limited to K events.

Definition 3.2: The number of occurrences of i, $i \in \{1, \cdots, N\}$, within the first k elements of $G(\omega; K)$ is called the *activation index* of the ith server after k events and denoted by $n_i(k)$.

Note that the activation index may also be defined through the iterative relationship

$$n_i(k+1) = \begin{cases} n_i(k) + 1 & s(k) = i \\ n_i(k) & \text{otherwise} \end{cases}$$

with $n_i(0) = 0$. Thus, $n_i(k)$ is a monotonically nondecreasing function of k.

Definition 3.3: The kth *event* within a sample path ω is defined as the ordered pair

$$E_\omega(k) = (s(k), n_i(k)) \quad \text{with } i = s(k).$$

Thus, the kth event is characterized by the server where it occurs, given by $s(k)$, and by its order of occurrence at that server, given by $n_i(k)$. A direct consequence of Definition 3.3 is that for two sample realizations of a DEDS, ω and ω', one can always find a unique $k' = r(k)$ such that $E_\omega(k) = E_{\omega'}(r(k))$. Thus, $r(k)$ is the order of the perturbed path event uniquely corresponding to the kth event in the nominal path. In general, $r(k) \neq k$.

We can now provide a precise definition for state perturbations in DEDS, given a nominal sample path ω and a perturbed one ω'. Of particular interest is the event occurrence time part of the state and we initially limit ourselves to studying perturbations in those variables (sometimes referred to as server *gains*)

$$\delta e_i(k) \triangleq e_i(k) - e_i'(r(k)) \quad \text{with } E_\omega(k) = E_{\omega'}(r(k)), \ k = 1, 2, \cdots. \tag{8}$$

A. Infinitesimal Perturbations

What we are effectively pursuing is a linearization of a DEDS about a nominal trajectory. We must, therefore, be careful to define the range of perturbations allowed so that the resulting predictions are exact. In DEDS the most stringent constraint we may impose is that perturbations never change the order of event occurrences in the nominal path.

Definition 3.4: Perturbations introduced to a sample path ω of a DEDS are said to be *infinitesimal* if: $s(k) = s'(k)$, $k = 1, \cdots, K$ for all resulting perturbed paths ω'.

This definition implies that infinitesimal perturbations are such that the EGS of the nominal sample path always remains unchanged: $G(\omega; K) = G(\omega'; K)$ for all perturbed paths ω'. Thus, the order of all events is unaffected. However, the magnitude of the perturbation giving rise to some ω' is not necessarily infinitesimal.

Assuming initial values $\delta e_i(0)$, $i = 1, \cdots, N$ are given (i.e., we do not concern ourselves with the "generation" aspect of perturbation analysis), the next result is easily derived.

Theorem 3.1: For infinitesimal perturbations introduced to the nominal sample path of a DEDS

$$\delta e_i(k+1) = \begin{cases} \delta e_n(k) & s(k) = n, \ D_n(k) = i; \\ & c_i(k) = 0 \\ \delta e_b(k) & s(k) = d_i(k) = b; \\ & [c_i(k) \neq 1, \ c_m(k) \neq B_m, \ D_i(k) = m] \text{ or } D_b(k) = i \\ \delta e_i(k) & \text{otherwise} \end{cases} \tag{9}$$

Proof: Definition 3.4 applied to (6), (7) implies that $\tau_i'(r(k)) = \tau_i'(k) = \tau_i(k)$ for all k. Also, $r(k+1) - 1 = r(k)$, for otherwise there must exist some $r(q) = r(k+1) - 1$ with $r(k) < r(q) < r(k+1)$, and hence some q with $k < q < k+1$ which is impossible. Then, combining (8) and (5b) with the three cases in (7), we get

1) $\delta e_i(k+1) = e_n(k) + S_i(k) - e_n'(r(k)) - S_i(r(k)) = \delta e_n(k)$

2) $\delta e_i(k+1) = e_b(k) + S_i(k) - e_b'(r(k)) - S_i(r(k)) = \delta e_b(k)$

3) $\delta e_i(k+1) = e_i(k) - e_i(r(k)) = \delta e_i(k)$. Q.E.D.

Corollary 3.1.1: The result of Theorem 3.1 can also be expressed as

$$\delta e_i(k+1) = \begin{cases} \delta e_n(k) & s(k) = n, \ D_n(k) = i; \\ & c_i(k) = 0 \\ \delta e_b(k) & s(k) = d_i(k) = b; \\ & [c_i(k) \neq 1, \ c_m(k) > B_m, \ D_i(k) = m] \text{ or } D_b(k) = i \\ \delta e_i(k) & \text{otherwise} \end{cases} \tag{9a}$$

Proof: The second case in (9) can be separated into two subcases: $c_m > B_m$ and $c_m < B_m$. The first subcase corresponds to an FO and the second to a normal service completion. The latter implies that $d_i(k) = b = i$, and hence yields $\delta e_b(k) = \delta e_i(k)$, which is included in the third case of (9). Q.E.D.

Intuitively, since infinitesimal perturbations do not affect the global event order and (3) and (4) are piecewise linear between events, (9) follows directly. This result is usually interpreted as the "perturbation propagation rule" [3] stating that "*a perturbation at server i is propagated to a server j through an NI at server j terminated by i; it is propagated to a server k through an FO at server k terminated by i.*" It is worthwhile emphasizing the simplicity of (9a) from an implementation point of view: FO and NI states alone need to be detected along the nominal sample path, in order to predict the change in event occurrence times throughout the system (see [3], [4] for pictorial representations of perturbation propagation and experimental results). Also note that $\delta e_i(k)$ propagates independently of any queue content perturbations $\delta c_i(k)$; this is because server gains can propagate only at boundary values of $c_i(k)$, i.e., when $c_i(k) = 0$ or $c_i(k) = B_i$.

B. Deterministic Similarity of DEDS Sample Paths

Infinitesimal perturbations impose a "total order" restriction on the EGS. It turns out that this is far too restrictive an assumption to make in order to derive (9). Weaker conditions may be imposed by requiring a "local" rather than "global" event order property to hold. Thus, we introduce a "local" EGS for each server:

Definition 3.5: The *event generating sequence for the ith server* (*i*th EGS), $G_i(\omega; K) = \{s_i(j)\}$, is a subsequence of $G(\omega; K) = \{s(k), k = 1, \cdots, K\}$ defined by the algorithm:

1) $k = 1$, $j = 1$.

2) If: $\tau_i(k) > 0$, then: $s_i(j) = s(k)$, $j \leftarrow j+1$. Else: go to step 3.

3) $k \leftarrow k+1$.

4) If: $k > K$, then: stop. Else: go to step 2.

We are thus creating N separate sequences, with the elements of the *i*th EGS imposing an order on events affecting the *i*th server alone; this is seen in step 2 above, where only servers causing a positive transition in $e_i(k)$ are entered in $G_i(\omega; K)$. It should therefore be clear from (7) that $G_i(\omega; K)$ is not restricted to contain all those $s(k)$ with $s(k) = i$ (see the example at the end of this section).

In order to formally define deterministic similarity, we will first establish the fact that the number of elements in the sequence $\{s_i(j)\}$ corresponding to k events in $G(\omega; K)$ is given by the activation index of server i, $n_i(k)$. The proof of Theorem 3.2 is based on Lemmas 3.1a and 3.1b and may be found in [18]. The two lemmas state that within a sample path defined by K events, to every termination of NI (FO) there corresponds an event that initiates that NI (FO).

Lemma 3.1a: If a NI state at the *i*th server is terminated by the kth event, then there exists some $k' < k$ such that the k'th event initiated the same NI at the *i*th server.

Proof: Let $c_i(k) = 0$. Since $n_i(k) > 1$ there is at least one event that has occurred at the *i*th server prior to k. Equation (3) implies that the only transition to $c_i(k) = 0$ can take place when $s(k') = d_i(k') = b$, $D_b(k') = i$ and provided $c_i(k') = 1$ for some $k' < k$. Q.E.D.

Lemma 3.1b: If an FO state at the *i*th server is terminated by the kth event, then there exists some $k' < k$ such that the k'th event initiated the same FO at the *i*th server.

Proof: Similar to Lemma 3.1a.

Theorem 3.2: Let $s_i(j) = s(\bar{k})$ for some server i and some $\bar{k}, j$. Then $j = n_i(k)$.

As a result of this theorem, we can henceforth replace j in $s_i(j)$ by $n_i(k)$ and present the following definition.

Definition 3.6: Two sample paths ω and ω' are said to be *deterministically similar* if $s_i(j) = s_i'(j)$ for all $j = n_i(k) = 1, 2, \cdots$ and $i = 1, \cdots, N$.

Thus, the property that held for all members of $G(\omega; K)$ now holds only for members of the subsequences $G_i(\omega; K)$. The interpretation of deterministic similarity is that event order invariance is restricted to events directly influencing $G_i(\omega; K)$. An extensive discussion can be found in [1], [3].

Our aim now is to derive (9) under the weaker condition of deterministic similarity. Lemma 3.2 states that if server states are never eliminated or added, then the event causing a transition at server i must be the same in both ω and ω'. Lemma 3.3 further asserts that events causing a transition at some server preserve their global relative order in deterministically similar paths (proofs may be found in [18]).

Lemma 3.2: If ω and ω' are deterministically similar sample paths, then for any event k such that $s(k) \in G_i(\omega; K)$: $s(k) = s'(r(k))$.

Lemma 3.3: Let k_1, k_2 be such that $k_1 < k_2$, $s(k_1) \in G_i(\omega; K)$ and $s(k_2) \in G_i(\omega; K)$. Then, for any deterministically similar path ω', $r(k_1) < r(k_2)$.

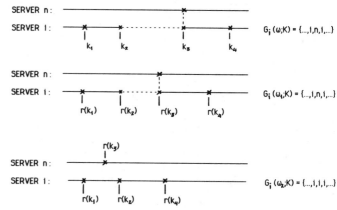

Fig. 2. Example of deterministically similar path (ω_1) and nondeterministically similar path (ω_2).

Theorem 3.3: If perturbations introduced to the nominal sample path of a DEDS result in a deterministically similar perturbed path, then the evolution of $\delta e_i(k)$ is still given by (9).

Proof: The proof is similar to that of Theorem 3.1 once we show that $\tau_i'(r(q)) = 0$ for all $r(k) < r(q) < r(k + 1)$. By Lemma 3.2 and Theorem 3.2: $s'(r(k)) = s(k) = s_i(n_i(k))$. If $k_1 < k_2$ with $\tau_i(k_1) > 0$, $\tau_i(k_2) > 0$ and $\tau_i(k) = 0$ for all $k_1 < k < k_2$, then $n_i(k_2) = n_i(k_1) + 1$. By Lemma 3.3, $r(k_1) < r(k_2)$ and there cannot exist any $r(q)$ with $r(k_1) < r(q) < r(k_2)$ and $\tau'(r(q)) > 0$, for otherwise: $s_i'(n_i(q)) = s'(r(q))$ with $n_i(k_1) < n_i(q) < n_i(k_1) + 1$ which is impossible. Hence, the statement above holds with $k_1 = k$, $k_2 = k + 1$. The remainder of the proof follows by combining (8) and (5b) with the three cases in (7). Q.E.D.

Corollary 3.3.1: The result of Theorem 3.3 can be rewritten as (9a).

Proof: Same as the proof of Corollary 3.1.1.

Deterministic similarity and Theorem 3.3 are illustrated in the example of Fig. 2 (solid lines indicate a server is busy, dotted lines represent FO and NI states, and X indicates an event occurrence). Here, ω is the nominal path with an NI at server i terminated by server n. In ω_1, deterministic similarity holds as the *i*th EGS remains unaffected. In ω_2, deterministic similarity is violated as the order of the $r(k_2)$th and $r(k_3)$th events is interchanged and the NI is eliminated.

IV. CONCLUSIONS

The event domain state-space representation presented provides a natural environment for the study of DEDS and in particular sample path perturbation analysis. Thus, we have attempted to formalize some earlier results [3] and establish rigorous foundations for simple cases that will hopefully enhance our understanding of more complex situations. These results may be further extended to include higher order perturbation analysis [6]. For large perturbations, one must resort to stochastic arguments, which in many cases (although not all) support the superiority of sample path perturbation analysis over conventional simulation techniques [16]–[17].

REFERENCES

[1] C. G. Cassandras, "Sample path perturbation analysis of discrete event dynamic systems," Ph.D. dissertation, Division of Appl. Sci., Harvard Univ., Cambridge, MA, Oct. 1982.

[2] Y. C. Ho and C. G. Cassandras, "Computing costate variables for discrete event systems," in *Proc. 19th IEEE Conf. Decision Contr.*, Dec. 1980.

[3] ———, "A new approach to the analysis of discrete event dynamic systems," *Automatica*, vol. 19, pp. 149–167, Mar. 1983.

[4] "SPEEDS: A new technique for the analysis and optimization of queueing networks," Y. C. Ho, Ed., Division of Appl. Sci., Harvard Univ., Cambridge, MA, Tech. Rep. 675, Feb. 1983.

[5] Y. C. Ho and X. Cao, "Perturbation analysis and optimization of queueing networks," *J. Optimiz. Theory Appl.*, vol. 40, Aug. 1983.

[6] Y. C. Ho, X. Cao, and C. Cassandras, "Infinitesimal and finite perturbation analysis for queueing networks," *Automatica*, vol. 19, July 1983.

[7] M. A. Eyler, "A new approach to the productivity study of transfer lines," Ph.D. dissertation, Division of Appl. Sci., Harvard Univ., Cambridge, MA, 1979.

[8] Y. C. Ho, M. A. Eyler, and T. T. Chien, "A gradient technique for general buffer

storage design in a serial production line," *Int. J. Prod. Res.*, vol. 17, pp. 557–580, 1979.

[9] ——, "A new approach to determine parameter sensitivities on transfer lines," *Management Sci.*, vol. 29, pp. 700–714, June 1983.

[10] R. Suri, "Implementation of sensitivity calculation on a Monte Carlo experiment," *J. Optimiz. Theory Appl.*, vol. 40, pp. 625–630, Aug. 1983.

[11] R. Suri and X. Cao, "Optimization of flexible manufacturing systems using new techniques in discrete event systems," in *Proc. 20th Allerton Conf. Commun., Contr., Comput.*, Oct. 1982.

[12] ——, "The phantom customer and marked customer methods for optimization of closed queueing networks with blocking and general service times," *ACM Performance Evaluation Rev.*, pp. 243–256, Aug. 1983.

[13] G. S. Fishman, *Principles of Discrete Event Simulation*. New York: Wiley, 1978.

[14] D. P. Gaver, "Simulation theory," in *Handbook of Operations Research*. New York: Van Nostrand, 1978.

[15] G. Gordon, "Simulation-computation," *Handbook of Operations Research*. New York: Van Nostrand, 1978.

[16] X. Cao, "Convergence of parameter sensitivity estimates in a stochastic experiment," *IEEE Trans. Automat. Contr.*, submitted for publication.

[17] P. R. Kumar, "On estimating derivatives of cost functions from simulated or real data," preprint.

[18] C. G. Cassandras, "Perturbation analysis of discrete event systems in the event domain," Dep. Elect. Comp. Eng., Univ. Mass., Amherst, MA, Tech. Rep., 1984.

Performance Sensitivity to Routing Changes in Queueing Networks and Flexible Manufacturing Systems Using Perturbation Analysis

YU-CHI HO AND XI-REN CAO, MEMBER, IEEE

Abstract—A new approach for estimating the throughput sensitivity with respect to routing probabilities in queueing networks and flexible manufacturing systems is proposed. The approach converts the problem to a problem of calculating the sensitivity of throughput with respect to service distributions using perturbation analysis.

I. Introduction

FLEXIBLE manufacturing systems (FMS) are a class of automated systems in which parts are automatically transported under computer control from one machine to another for processing. Flexible manufacturing systems can be used to improve the productivity in discrete parts manufacturing [1]–[3]. It represents a relatively new development in the field of automated manufacturing. Queueing network models and simulation methods are widely used in analysing and optimizing FMS's [5], [6]. Queueing network models give reasonable estimates of performance of a FMS under certain stochastic assumptions. However, these assumptions are usually not satisfied in a real system. Hence for other than "ball park figure" design, queueing network models do not always work well in many cases. On the other hand, simulation models can be made very accurate. But they impose relatively large computational load requirements especially in optimization procedures in which repeated simulations are needed in order to obtain the performance measures for systems with slightly different parameters.

Another technique of optimizing FMS's, based on a new theory called perturbation analysis of discrete event dynamic systems, has been developed recently [7]. The basic idea of this new technique is to observe one sample path of a system, either through simulation or for the real system, for one set of parameters. By analyzing this sample path, this technique gives estimates of the gradients of system performance measures with respect to these parameters. The important point is that this method obtains all these gradients at the same time as one estimates the performance measures from one sample path. Since the method is based on a real sample path of a system, it does not suffer from the constraining assumptions required by the analytical queueing models. On the other hand, it does not require re-running the system (or simulation) to obtain the performance for slightly different parameters. Thus it offers tremendous computational savings over the "brute-force" simulation approach of estimating gradients. Moreover, the method offers a real-time applications to the sensitivity analysis of FMS's. This feature makes the method unique among other approaches to FMS analysis.

Much research has been done on the theoretical analysis as well as practical applications of perturbation analysis. Two kinds of perturbations, infinitesimal and finite perturbations, are studied [8]. Propagation rules of these perturbations are developed. The infinitesimal perturbation propagation rules are used to obtain the sensitivity of system throughput with respect to continuous parameter changes, e.g., changes in service time parameters [9], [10]. The finite perturbation propagation rules are used to determine the throughput sensitivity with respect to parameters that take discrete values (e.g., the numbers of customers in a multiclass closed network [11]) or where discontinuous perturbations result from continuous parameter changes (e.g., waiting times for different customer classes can change discontinuously due to changes in arrival rates.) A recent work [12] deals with the sensitivity of another performance measure, the average sojourn time of a customer in a server or a system, with respect to mean service times. A list of reference about this subject can be found in [7].

In this paper we consider the sensitivity of system throughput (TP) with respect to another parameter: the routing probability of customers in a system. For a M-server single-class closed queueing network, the routing probabilities form a $M*M$ matrix $P = [p_{i,j}]'$ (with the prime symbol denoting the transposition of a matrix). Each entry of the matrix $p_{i,j}$, $i, j = 1, 2, \cdots, M$, represents the probability that a customer goes to server j after the completion of its service at server i. The estimation of the sensitivity of TP with respect to $p_{i,j}$ is of importance in practice. For example, in a FMS, parts may go to different machines to continue their processing. The proportions of parts going to different machines are represented by probabilities in queueing models. In this case the sensitivity estimates can be used to optimize the proportions of the routes of these parts. The above model can be extended to systems with more than one type of parts. These different types of parts usually have different routes. If such systems are modelled by single-class queueing networks, then routing probabilities of customers in the queueing networks reflect the effect of these different routes. For example, consider a FMS consisting of two types of parts. Suppose that upon the completions of their services at machine 1, type 1 parts go to machine 2, and type 2 parts go to machine 3. Assume that the numbers of these two types of parts served by machine 1 have the ratio 1:2. Then in its queueing network model we take $p_{1,2} = 1/3$ and $p_{1,3} = 2/3$. In this case the sensitivity of TP with respect to $p_{i,j}$ is actually the sensitivity of TP with respect to the ratio of numbers of different types of parts. This information is certainly important to a FMS designer or an operation manager for obtaining the best performance by choosing an optimum value of the mixing proportions of different types of parts.

The work in this paper is different from previous works of perturbation analysis. The conventional method in perturbation analysis mainly consists of two aspects: 1) introducing perturbations along a given path, either infinitesimal or finite, according to some rules corresponding to the perturbations of parameters involved and 2) propagating these perturbations along the sample path. However, this procedure does not work for analyzing the sensitivity of TP with respect to routing probabilities $p_{i,j}$, assuming infinitesimal changes in $p_{i,j}$ will result in obtaining zero as an estimate of the sensitivity of the mean TP with respect to $p_{i,j}$. This does not give us any useful information. This fact will be further explained in Section II. On the other hand, creating finite perturbations due to routing changes of customers makes the problem too difficult to handle even for the finite propagation rules.

We propose a new method for applying the perturbation analysis. The basic idea of this method is by using some existing formulas in queueing network theory, the estimation of the sensitivity of TP with respect to $p_{i,j}$ can be converted to a problem of calculating the sensitivity of TP with respect to some parameters of service times. The latter can be obtained by applying perturbation analysis results we already have developed. Section II gives some preliminaries. Section III illustrates this idea for the case of product form networks. The result of this section will confirm the basic validity of our approach since for this case, the sensitivity result can be independently derived to provide a check. Numerical examples demonstrating the high accuracy of this approach are included there. In Section IV, an approximate method is proposed for general queueing systems. Examples are given for systems with finite buffers and/or uniformly distributed service times. The method provides a way to estimate the sensitivity of TP with respect to $p_{i,j}$ by observing only one sample path of a system. The results are quite accurate in most cases. Section V concludes with some discussions.

II. PRELIMINARIES

The existing analytical models for FMS's are basically queueing network models. FMS's are modelled as open or closed queueing systems in which the customers are parts to be processed by the system and the servers are machines [1]. The queueing network models can be used to obtain both qualitative and quantitative information about FMS's. It provides a considerable variety of useful performance measures with good accuracy for preliminary design purposes.

However, for more realistic modelling of FMS's we often are forced to use nonstandard queueing network models for which no closed form analytical expressions exist for performance measures. For example, limited queue size, nonexponential service time distribution, and nonstandard queueing discipline are features easily found on realistic FMS's. On the other hand, under very broad assumptions perturbation analysis gives an exact value of the sensitivity of performance measure of an arbitrary queueing network for a sample path with finite observation period. Specifically, let $L(\theta, \xi)$ be the performance measure of a sample path with finite length, where θ is a parameter and ξ is a random vector that represents the random effect involved in the system during the observation period. In particular, we shall let ξ represent all the random service times, routing decisions, and initial conditions of a queueing network. More technically, $\xi = (\eta_1, \eta_2, \cdots, \eta_M; \zeta_1, \zeta_2, \cdots, \zeta_M; \varphi)$, where $\eta_i = (\eta_{i,1}, \eta_{i,2}, \cdots)$, and $\eta_{i,l}$ is a random variable uniformly distributed over $[0, 1]$ determining the service time of the lth customer served at server i, by the following equation

$$s_{i,l} = F^{-1}(\eta_{i,l}), \qquad i = 1, 2, \cdots, M; \; l = 1, 2, \cdots.$$

In the equation $F(x)$ is the distribution function of the ith server's service time; $\zeta_i = (\zeta_{i,1}, \zeta_{i,2}, \cdots)$; and $\zeta_{i,l}$ is a random variable that is also uniformly distributed over $[0, 1]$. The destination of the customers are decided as follows. If $\sum_{k=0}^{j-1} p_{i,k} < \zeta_{i,l} \leq \sum_{k=0}^{j} p_{i,k}$, $p_{i,0} = 0$, then the lth customer at server i goes to server j after its service completion at server i. φ is a random vector representing the initial condition, i.e., the number of customers in each server at time $t = 0$. All these random variables $\eta_{i,l}$, $\zeta_{i,l}$, and ϕ are independent. Perturbation analysis calculates the following "sample" gradient:

$$\frac{\partial}{\partial \theta}[L(\theta, \xi)] = \lim_{\Delta\theta \to 0} \frac{L(\theta + \Delta\theta, \xi) - L(\theta, \xi)}{\Delta\theta} \qquad (1)$$

where the same random vector ξ is used for both parameters $\theta + \Delta\theta$ and θ. This reduces the variance of $\{L(\theta + \Delta\theta, \xi) - L(\theta, \xi)\}/\Delta\theta$ ([13]). However, our purpose is to get an estimate of the sensitivity of the mean value of the performance measure. Let

$$J(\theta) = E[L(\theta, \xi)]$$

where E denotes the expect value. Then we have

$$\frac{\partial J}{\partial \theta} = \frac{\partial}{\partial \theta}\{E[L(\theta, \xi)]\}$$

$$= \lim_{\Delta\theta \to 0} E\left\{\frac{L(\theta + \Delta\theta, \xi) - L(\theta, \xi)}{\Delta\theta}\right\}. \qquad (2)$$

Thus we encounter the following problem: is the value (1) obtained by perturbation analysis gives an unbiased estimate of (2)? This is equivalent to the following question: does the

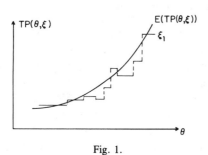

Fig. 1.

following equation hold for $L(\theta, \xi)$?

$$E\left\{\frac{\partial}{\partial \theta}[L(\theta, \xi)]\right\} = \frac{\partial}{\partial \theta}\{E[L(\theta, \xi)]\}. \quad (3)$$

Equation (3) is important. If it holds, then by the strong law of large numbers, (1) gives an unbiased and strongly consistent estimate of the left-hand side of (3), and hence of $\partial J/\partial \theta$, provided $E\{|\partial/\partial\theta\, L(\theta, \xi)|\} < \infty$.

For many functions $L(\theta, \xi)$, this equation does hold. It is proved in [14] that (3) hold for the system throughput (as function L) and mean service times (as parameter θ) in a system with exponentially distributed service times. However, if $L(\theta, \xi)$ is discontinuous as a function of θ, (3) may not hold. The problem is rigorously studied in [15].

Unfortunately, (3) does not hold for the present problem. If we take the system throughput TP as L, and routing probability $p_{i,j}$ as θ, then for any sample path with a finite length we have[1]

$$\frac{\partial}{\partial \theta}[\text{TP}(\theta, \xi)] = 0, \text{ w.p. } 1. \quad (4)$$

Equation (4) can be easily proved. In fact, for any sample path with a finite length, every server in the system serves a finite number of customers with probability one. Therefore, we can always choose $\Delta\theta$ small enough such that no values of $\zeta_{i,l}$ range in between θ and $\theta + \Delta\theta$. This means that the routes of customers in both systems with θ and $\theta + \Delta\theta$ are the same. Thus two sample paths for both systems are identical. Of course they give the same values for both TP $(\theta + \Delta\theta, \xi)$ and TP (θ, ξ). Hence we get (4). On the other hand, if we makes $\Delta\theta$ large, $\theta + \Delta\theta$ will meet some $\zeta_{i,l}$. At that point one customer will change its route, and the value of TP $(\theta + \Delta\theta, \xi)$ changes. From this discussion, we know that the sample function TP (θ, ξ) is a piecewise constant function of θ. This is shown in Fig. 1. However, the derivative of the mean throughput (2) is obviously not zero. Thus (3) does not hold in this case. The derivative of TP (θ, ξ) with respect to θ does not give any information about $\partial J/\partial \theta$. We can not use the conventional procedure of perturbation analysis to obtain an estimate for $\partial J/\partial \theta$.

In this paper we propose a new approach to estimate $\partial J/\partial \theta$. The approach is based on an observation about Jackson networks, and it is extended to general queueing networks. The discussions are in the following sections.

[1] Because of the restriction $\sum_{j=1}^{M} p_{i,j} = 1$, we must have $\sum_{j=1}^{M} dp_{i,j} = 0$. In the following derivative we understand that this restriction holds.

III. PRODUCT-FORM NETWORKS

A. Jackson Networks

We start the discussion with the simplest networks. Consider a single-class closed Jackson network consisting of M servers and N customers. Customers are circulating among servers with routing probabilities $p_{i,j}$. The service times of these servers are exponentially distributed with mean $\bar{s}_i = 1/\mu_i$, $i = 1, 2, \cdots, M$. The visiting ratio, v_i, $i = 1, 2, \cdots, M$, is decided by

$$v_i = \sum_{j=1}^{M} p_{j,i} v_j. \quad (5)$$

From the convolution algorithm (Buzen [16]), the utilization of server i, $u_i = p(n_i \geq 1)$, where n_i is the number of customers in server i (including the customer being served), is

$$u_i = Y_i \frac{G(N-1)}{G(N)} \quad (6)$$

where

$$Y_i = v_i \bar{s}_i$$

$$G(N) = \sum_{n_1 + n_2 + \cdots + n_M = N} \prod_{i=1}^{M} Y_i^{n_i}$$

$$G(N-1) = \sum_{n_1 + n_2 + \cdots + n_M = N-1} \prod_{i=1}^{M} Y_i^{n_i}. \quad (7)$$

One simple but quite important observation from (6) and (7) is that u_i depends on $p_{i,j}$, and $\bar{s}_i$, $i, j = 1, 2, \cdots, M$, only through the product $v_i \bar{s}_i$. To denote this point, we write

$$u_i = f(Y_1, Y_2, \cdots, Y_M) = f(v_1 \bar{s}_1, v_2 \bar{s}_2, \cdots, v_M \bar{s}_M). \quad (8)$$

Differentiating both sides of (8) yields

$$\Delta u_i = \sum_{j=1}^{M} \frac{\partial f}{\partial Y_j} \Delta Y_j \quad (9)$$

in which

$$\Delta Y_j = \Delta(v_j \bar{s}_j) = \bar{s}_j \Delta v_j + v_j \Delta \bar{s}_j \quad (10)$$

where the change Δv_j is due to the change in the routing probability matrix $P = [p_{i,j}]'$. Suppose that our purpose is to calculate the change in utilization of a server in a system in which v_i change by Δv_i, $i = 1, 2, \cdots, M$, and $\bar{s}_i$ remains the same. Then by (9) and (10), so far as the utilization is concerned, the problem is equivalent to calculating the change of utilization for a system in which v_j, $j = 1, 2, \cdots, M$, do

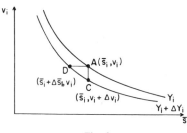

Fig. 2.

not change but $\bar{s}_j$, $j = 1, 2, \cdots, M$, change by

$$\Delta \bar{s}_j = \bar{s}_j \frac{\Delta v_j}{v_j}, \quad j = 1, 2, \cdots, M. \quad (11)$$

As was discussed in the previous section, the sample gradient of performance measure with respect to $p_{i,j}$ (hence v_i) is always zero. It gives no useful information about the gradient of the expected value of the performance measure. However, equation (11) converts the problem to another one of estimating the gradient for mean service times. These estimates can be easily obtained by perturbation analysis as demonstrated in [9].

The above proposed approach is illustrated in Fig. 2. The two curves corresponding to Y_i and $Y_i + \Delta Y_i$ are two contours of function $f(Y_i)$. Point H in the figure represents the system with parameters $\bar{s}_i$ and v_i. We intend to estimate the value of $f(Y_i)$ at point C with parameters $(\bar{s}_i, v_i + \Delta v_i)$. Equation (11) suggests that if $f(C)$ is hard to obtain, one can estimate $f(D)$ instead. D is the point with parameters $(\bar{s}_i + \Delta \bar{s}_i, v_i)$, since C and D are on the same contour.

Let $\Delta P = [\Delta p_{i,j}]'$ be the matrix of the changes in routing probabilities which satisfies

$$\sum_{j=1}^{M} \Delta p_{i,j} = 0, \quad i = 1, 2, \cdots, M. \quad (12)$$

Let Δu_i, $i = 1, 2, \cdots, M$, be the change in u_i due to this ΔP. Then we can calculate this Δu_i via its sensitivity to Δv_j, and equivalently $\Delta \bar{s}_j$. Thus

$$\Delta u_i = \sum_{j=1}^{M} \frac{\Delta u_i}{\Delta v_j} \Delta v_j$$

$$= \sum_{j=1}^{M} \frac{\Delta u_i}{\Delta \bar{s}_j} \Delta \bar{s}_j$$

$$= \sum_{j=1}^{M} \frac{\Delta u_i}{\Delta \bar{s}_j} \frac{\bar{s}_j}{v_j} \Delta v_j.$$

Or in differential form

$$du_i = \sum_{j=1}^{M} \frac{du_i}{d\bar{s}_j} \frac{\bar{s}_j}{v_j} dv_j. \quad (13)$$

Let TP_i be the throughput of server i, i.e., the average number of customers served by server i per unit time in the observation period. Then $TP_i = u_i/\bar{s}_i$.

From this

$$dTP_i = \frac{1}{\bar{s}_i} du_i \quad (14)$$

and

$$\frac{du_i}{d\bar{s}_j} = \begin{cases} \bar{s}_j \dfrac{d\,TP_i}{d\bar{s}_j}, & \text{if } i \neq j \\ \bar{s}_i \dfrac{dTP_i}{d\bar{s}_i} + TP_i, & \text{if } i = j. \end{cases} \quad (15)$$

$dTP_i/d\bar{s}_j$ can be obtained by perturbation analysis. The only problem remained is to decide dv_i in (13). Equation (5) can be written in the vector form

$$V = P*V$$

where $V = (v_1, v_2, \cdots, v_M)'$ is a $M*1$ vector. Differentiating both sides of this equation yields

$$dV = (dP)*V + P*dV$$

or

$$(I - P)*dV = (dP)*V \quad (17)$$

where I is the $M*M$ identical matrix. Note that the matrix $(I - P)$ is not of full rank. Thus it is not invertible. The solution to (17) is not unique. We arbitrarily choose

$$dv_M = 0. \quad (18)$$

Equations (17) and (18) have a solution since (12) holds for P.

The new approach for estimating dTP_i due to dP can be summarized in the following algorithm.

Algorithm:

1) Solve equation (17) for dV with condition (18).

2) Apply perturbation analysis to a sample path (either from simulation or a real system) to obtain an estimate of $dTP_i/d\bar{s}_j$, $i, j = 1, 2, \cdots, M$

3) dTP_i, $i = 1, 2, \cdots, M$, are then obtained from (13), (14), and (15)

Remark: The nonuniqueness of the solution to equation (17) corresponds to the non-uniqueness of Y_i in (6). In fact, if we use $(V + \Delta V) = (P + \Delta P)(V + \Delta V)$ to obtain many sets of Δv_i and let $Y_i = (v_i + \Delta v_i)\bar{s}_i$, then (6) still gives the correct values of $u_i + \Delta u_i$ for all these sets of Δv_i. We choose $dv_M = 0$ for the following reasons: 1) (17) holds only for infinitesimal dV and 2) $d\bar{s}_i$ should be infinitesimal for implementing perturbation analysis rules. Otherwise if dv_M is finite, then by (17) all dv_i, $i = 1, 2, \cdots, M$, would be finite. By (11) the equivalent values of $d\bar{s}_i$ would also be finite, for which the infinitesimal analysis may not give accurate enough results.

Example 1): This is a typical example of calculating the sensitivity of throughput with respect to routing probability in a closed Jackson network using the above approach. The system consists of three servers and nine customers. The service times at these servers are exponentially distributed with mean service times of ten, eight, and six, respectively.

The routing probability matrix and its change are

$$P = \begin{pmatrix} 0 & 0.5 & 0.5 \\ 0.3 & 0 & 0.7 \\ 0.6 & 0.4 & 0 \end{pmatrix}$$

and

$$\Delta P = \begin{pmatrix} 0 & 0.01 & -0.01 \\ 0.01 & 0 & -0.01 \\ 0.01 & -0.01 & 0 \end{pmatrix}.$$

We run one computer simulation for the system with routing probability matrix P and applied the new approach to predict the change in throughput ΔTP_i, $i = 1, 2, 3$, due to this ΔP. In the simulation, the servers in the system served 100 000 customers altogether. The results obtained follow.

svr(i)	1	2	3
TP	0.09429	0.09215	0.1120
$\frac{\bar{s}_i}{TP}\frac{\partial TP}{\partial s_i}$	0.6951	0.1839	0.1210
ΔTP	$0.522*10^{-3}$	$-0.766*10^{-3}$	$-2.15*10^{-3}$

In order to verify these results, we calculated these values using Jackson formulas. They are

$$(\Delta TP)_f \quad 0.513*10^{-3} \quad -0.771*10^{-3} \quad -2.14*10^{-3}.$$

We also did another simulation for the same system except with routing probability matrix $P + \Delta P$, and obtained the "brute force" value of these ΔTP as

$$(\Delta TP)_b \quad 0.55*10^{-3} \quad -0.70*10^{-3} \quad -2.0*10^{-3}.$$

B. Networks with Product-Form Distributions

There exists a large class of networks in which the system throughput depends on P and service time distributions only through $v_i \bar{s}_i$, $i = 1, 2, \cdots, M$. For example, it is proved that if a network only consists of servers with either FCFS (first come, first served) and equal exponential service times for all classes, LCFSPR (last come, first served, preemptive resume), PS (processor sharing), or IS (infinite servers), then the marginal distribution of customer numbers in every server depends only on $v_i \bar{s}_i$ [17], [18]. This property also holds for other networks with produc-form equilibrium distributions.

Consider a multiclass closed queueing network. Let there be K different classes of customers in the system, and number them by $k = 1, 2, \cdots, K$. Chandy et al. [18] proved that if each queue of the network satisfies station balance, then the equilibrium state probability density function takes the following product form

$$p(\bar{X}) = \frac{1}{C} \prod_{i=1}^{M} p_i(X_i) \tag{19}$$

and

$$p_i(X_i) = cq_i(S_i) \prod_{j=1}^{L_j} \lambda_{i,k_i(j)} [1 - F_{i,k_i(j)}(X_{i,j})] \tag{20}$$

where G is a normalizing constant; $\bar{X} = (X_1, X_2, \cdots, X_M)$ is the state of the system; $X_i = (S_i, X_{i,1}, \cdots, X_{i,L_i})$ is the state of queue i; and L_i is the length (the number of customers) of this queue. The $k_i(j)$, $j = 1, 2, \cdots, L_i$, denote the classes of the jth customer in queue i. $S_i = (k_i(1), k_i(2), \cdots, k_i(L_i))$, X_{ij} is a positive real supplementary variable giving the remaining service requirement of the jth customer of queue i. $F_{i,k}(X)$ is the distribution function of class k customers' service requirements at server i. $\lambda_{i,k}$ is class k customers' visiting ratio at server i, which satisfies

$$\lambda_{i,k} = \sum_{j=1}^{M} \sum_{l=1}^{K} \lambda_{j,l} b_{j,l;i,k},$$

$$i = 1, 2, \cdots, M, \quad k = 1, 2, \cdots, K$$

where $b_{j,l;i,k}$ is the probability that a customer of class l departing from queue j will change to class k and enter queue i. Let $\bar{S} = (S_1, S_2, \cdots, S_M)$ be the discrete part of the system state. From (19) and (20) one can obtain the marginal distribution of $\bar{S}$

$$p(\bar{S}) = \int_0^{\infty} \cdots \int_0^{\infty} p(\bar{X}) \{\{dX_{i,j}\}_{j=1}^{L_i}\}_{i=1}^{M}$$

$$= \frac{1}{G'} \prod_{i=1}^{M} \left\{ q_i(S_i) \prod_{j=1}^{L_i} [\lambda_{i,k_i(j)} \bar{s}_{i,k_i(j)}] \right\} \tag{21}$$

where

$$\overline{s_{i,k}} = \int_0^{\infty} [1 - F_{i,k}(x)] \, dx$$

is the mean service requirement of class k customers at server i. By theorem 1 and (5') in [18], $q_i(S_i)$ depends only on the queueing discipline at queue i. It does not depend on either routing probabilities $b_{j,l;i,k}$ or the service requirement distribution $F_{i,k}$. Equation (21) clearly shows that the system throughput depends on $b_{j,k;i,k}$ and $F_{i,k}$ only through the product $\lambda_{i,k} \bar{s}_{i,k}$. Therefore, if each queue of a network satisfies station balance, the sensitivity of TP with respect to $b_{j,l;i,k}$ can be calculated by the sensitivity of TP with respect to $\bar{s}_{i,k}$. The latter can be obtained by perturbation analysis.

Note that in such kinds of systems the throughput depends only on the mean service times. It is insensitive to the service time distributions. When applying perturbation analysis to calculate the sensitivity of TP with respect to $\bar{s}_{i,k}$, one can essentially perturb the distribution function $F_{i,k}$ in an arbitrary manner as long as $\bar{s}_{i,k}$ is perturbed by $\Delta \bar{s}_{i,k}$. However, the simplest way is using the method described in the next section.

Remark: In a multiclass queueing network, customers of different classes may change their order of entering a server due to the changes in mean service times of some other servers. This leads to discontinuous changes in throughputs. Therefore, finite perturbation propagation rules are needed to obtain the sensitivities of throughputs of different classes of customers with respect to mean service times [8], [15].

IV. General Networks

Note that the insensitivity of TP to the service time distribution as mentioned above does not hold for more general networks. Thus the statement about that the system utilization depends on routing probability matrix P and service requirement distributions only through $v_i \bar{s}_i$ has to be modified. Consider single-class networks. Let $F_i(s)$ be the service time distribution function at server i. $F_i(s)$ is characterized by the moment generating function

$$E(e^{ts_i}) = \int_0^\infty e^{ts_i} \, dF_i(s_i)$$

$$= \sum_{k=0}^\infty \frac{t^k}{k!} E(s_i^k)$$

$$= \sum_{k=0}^\infty m_{i,k} \frac{t^k}{k!}$$

where $m_{i,k} = E(s_i^k)$ is the kth moment of the service time at server $i(s_i)$. The first moment $m_{i,1} = E(s_i) = \bar{s}_i$ is the mean service time at server i.

If we consider $m_{i,1} = \bar{s}_i$ as a variable having the time dimension $[t]$, then $m_{i,2} = E(s_i^2)$ is a variable having dimension $[t]^2$, and $m_{i,k} = E(s_i^k)$ has dimension $[t]^k$. Inspired by the dimensional theory in physics, we conjecture that for networks with general service time distributions the system utilization depends on P and $F_i(x)$ only through $v_i m_{i,1}$, $v_i^2 m_{i,2}$, $v_i^3 m_{i,3}$, $\cdots$, $v_i^k m_{i,k}$, $\cdots$, $i = 1, 2, \cdots, M$, where $m_{i,k}$ is the kth moment of the service time at server i. To write formally, we have the following equation, which is the analog of (8).

Conjecture:

$$u_i = f(v_1 m_{1,1}, v_1^2 m_{1,2}, \cdots, v_1^k m_{1,k}, \cdots; \cdots;$$
$$v_M m_{M,1}, v_M^2 m_{M,2}, \cdots, v_M^k m_{M,k}, \cdots),$$
$$i = 1, 2, \cdots, M. \quad (22)$$

From (22), if the routing probability matrix P changes such that v_i changes by Δv_i, $i = 1, 2, \cdots, M$, then the change in utilization is just the same as if P does not change but all moments $m_{i,k}$, $i = 1, 2, \cdots, M$; $k = 1, 2, \cdots$ change according to the following equations.

$$(v_i + \Delta v_i) m_{i,1} = v_i (m_{i,1} + \Delta m_{i,1})$$
$$(v_i + \Delta v_i)^2 m_{i,2} = v_i^2 (m_{i,2} + \Delta m_{i,2})$$
$$\vdots$$
$$(v_i + \Delta v_i)^k m_{i,k} = v_i^k (m_{i,k} + \Delta m_{i,k})$$
$$\vdots$$

where $i = 1, 2, \cdots, M$. Neglecting the high order terms, we get

$$\Delta m_{i,1} = \frac{\Delta v_i}{v_i} m_{i,1}$$
$$\Delta m_{i,2} = 2 \frac{\Delta v_i}{v_i} m_{i,2}$$
$$\vdots$$
$$\Delta_{m,k} = k \frac{\Delta v_i}{v_i} m_{i,k}$$
$$\vdots \quad (23)$$

where $i = 1, 2, \cdots, M$. Thus if (22) holds then the sensitivity

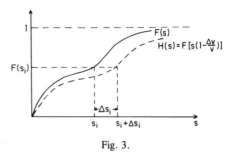

Fig. 3.

of TP with respect to P can be calculated through the sensitivity of TP with respect to all these changes $\Delta m_{i,k}$. Fortunately, this sensitivity can be obtained by applying perturbation analysis on one sample path of the system. This is based on the following lemma.

Lemma 1): the kth moment of the distribution function $H(s) = F[s(1 - \Delta v/v)]$ is $m_k + \Delta m_k$, where m_k is the kth moment of $F(s)$, and $\Delta m_k = k (\Delta v/v) m_k$.

Proof:

$$E(e^k) = \int_0^\infty s^k \, dH(s)$$

$$= \int_0^\infty s^k \, dF\left[s\left(1 - \frac{\Delta v}{v}\right)\right]$$

$$= \int_0^\infty \left(\frac{s}{1 - \frac{\Delta v}{v}}\right)^k dF(s)$$

$$= \int_0^\infty s^k \left(1 + k \frac{\Delta v}{v}\right) dF(s)$$

$$= \left(1 + k \frac{\Delta v}{v}\right) m_k$$

$$= m_k + \Delta m_k.$$

This lemma provides a way to implement perturbation analysis for obtaining the sensitivity of TP with respect to all changes of $\Delta m_{i,k}$. In fact, when you analyses a given (nominal) sample path of a system using perturbation analysis, you must answer the following question: what would be the perturbed path if the distribution functions were $H_i(s) = F_i[s(1 - \Delta v/v)]$? This is a standard problem in perturbation analysis. The perturbation generation rule for this problem is shown in Fig. 3. Whenever server i, $i = 1, 2 \cdots, M$, completes its service to a customer, a perturbation Δs_i is generated for server i. According to the perturbation generation rule ([10]) $\Delta s_i = H_i^{-1}[F_i(s_i)] - s_i$. Let

$$s_i' = \frac{s_i}{1 - \frac{\Delta v_i}{v_i}}.$$

Then

$$\Delta s_i = H_i^{-1}\left\{F_i\left[s_i'\left(1 - \frac{\Delta v_i}{v_i}\right)\right]\right\} - s_i$$

$$= s_i' - s_i = \frac{\Delta v_i}{v_i} s_i. \qquad (24)$$

This equation shows that using $H_i(s)$ instead of $F_i(s)$ is just changing the time scale of server i by multiplying a factor $(1 + \Delta v_i/v_i)$. The perturbation generated at the completion of each service is proportional to the customer's service time. The rules for systems with exponential service times are special cases of this general approach.

In summary, (22) can be viewed as a natural power series generalization of (8). If (22) holds, then the algorithm in Section III-A is exact. If not, then (22) can be regarded as a power series approximation to the calculation of u_i. The interesting point to be emphasized is that for the purpose of calculating $\Delta u_i/\Delta p_{i,j}$ via $\Delta u_i/\Delta \bar{s}_j$ using perturbation analysis, we do not need to know the explicit form of the function f. The only modification of the algorithm in Section III-A is using (24) as the perturbation generation rule.

To decide for what kinds of systems the conjecture holds is a difficult topic for further research. However, experimentally we applied the above proposed approach to many non-product-form systems and obtained quite accurate results. These systems include systems with finite buffers, for which we know that conjecture (22) does not hold exactly. Our experiments show that for a wide range of systems the approach provides a quite accurate approximate method of estimating the sensitivity of TP with respect time routing probabilities by analysing one sample path of a system. The following are some typical examples.

Example 2): The system considered here is the same as the one in example 1), except that the queues are limited by queue sizes $Q_1 = Q_2 = Q_3 = 5$. We run two computer experiments with different seeds. A total of 300,000 customers are served by all servers in each experiment. The average results follow.

svr	1	2	3
TP	0.08794	0.08548	0.1040
ΔTP	0.658*10⁻³	−0.543*10⁻³	−1.79*10⁻³
(ΔTP)$_b$	0.61*10⁻³	−0.54*10⁻³	−1.8*10⁻³

ΔTP are values obtained using our approach. (ΔTP)$_b$ are values from the brute force method, i.e., the difference of throughputs in two simulations with P and $P + \Delta P$ respectively.

Example 3): This example applies the approach to a system with uniformly distributed service times and finite buffers. The system consists of three servers and nine customers. The matrices P and ΔP are the same as that in example 1) of Section III. All buffers are limited by a size of five. The service times at servers are uniformly distributed with the following bounds.

svr	1	3	3
Upper bound	16	20	12
Lower bound	0	0	0

The servers in the system serve a total of 300,000 customers in the experiment. The experimental results follow.

TP	0.09775	0.09449	0.1150
ΔTP	0.113*10⁻²	−0.021*10⁻²	−0.15*10⁻²
(ΔTP)$_b$	0.109*10⁻²	−0.021*10⁻²	−0.15*10⁻²

Example 4): This is an example combining many features. The system consists of five servers and 15 customers. The buffer sizes are five for all servers. The service times at servers 1–3 are uniformly distributed, and at servers 4–5 are exponentially distributed. The parameters of these distributions follow.

svr	1	2	3	4	5
Upper bound	16	12	14	—	—
Lower bound	0	0	0	—	—
Mean	8	6	7	4	7

The matrices P and ΔP are

$$P = \begin{bmatrix} 0 & 0.25 & 0.25 & 0.25 & 0.25 \\ 0.2 & 0 & 0.2 & 0.3 & 0.3 \\ 0.4 & 0.1 & 0 & 0.4 & 0.1 \\ 0 & 0.5 & 0.2 & 0 & 0.3 \\ 0.15 & 0 & 0.25 & 0.4 & 0.2 \end{bmatrix}$$

and

$$\Delta P = \begin{bmatrix} 0 & 0.01 & -0.01 & 0 & 0 \\ 0.015 & 0 & -0.005 & -0.005 & -0.005 \\ 0.01 & 0 & 0 & -0.01 & 0 \\ 0.01 & -0.015 & 0.005 & 0 & 0 \\ 0 & 0 & 0.005 & -0.015 & 0.01 \end{bmatrix}$$

Note that the probability $p_{4,1}$ changes from 0 to 0.01, which is an infinite percentage change. Note also that the customers at server 5 have a probability 0.2 to go back to the same server, and this feedback rate changes by 0.01.

A total of 500,000 customers are served by all servers in the experiment. The experimental results follow.

svr	1	2	3	4	5
TP	0.07626	0.09639	0.09632	0.1357	0.1226
ΔTP	0.337*10⁻²	−0.195*10⁻²	−0.008*10⁻²	−0.302*10⁻²	0.011*10⁻²
(ΔTP)$_b$	0.351*10⁻²	−0.173*10⁻²	0.005*10⁻²	−0.29*10⁻²	0.03*10⁻²

The values of ΔTP and (ΔTP)$_b$ for servers 1, 2, and 4 are quite close. The relative differences are less than ten percent. The values for servers 3 and 5 are too small in absolute value to be significant. One can hardly say which one (either ΔTP or $(\Delta TP)_b$) is correct.

V. Conclusion

In optimizing the throughput of a flexible manufacturing system, it is important to estimate the sensitivity of throughput with respect to routing probabilities of different parts. A new approach for estimating this sensitivity is proposed. The approach converts the problem to a problem of calculating the sensitivities of TP with respect to service distributions and uses perturbation analysis to obtain these sensitivities. The approach possesses all the advantages of perturbation analysis. In particular it saves many simulation runs and applies to real systems on line. The approach gives unbiased, strongly consistent estimates of the sensitivity of TP with respect to routing probabilities for systems with product-form equilibrium distributions. Numerical examples show that it is also quite accurate for more general networks.

The implication of this result for perturbation analysis is also significant. It shows that while perturbation analysis is sample path based, its predictive power is far from limited to the particular sample path under question. It can be appropriately combined with other purely probabilistic results (i.e., queueing theoretic formulas) to enhance its appliability to predict average behavior about other sample paths.

References

[1] J. A. Buzacott and D. D. W. Yao, "Flexible manufacturing systems: A review of models," University of Toronto, Toronto, Canada, 1982.
[2] C. Dupont and Gatelmand, "A survey of flexible manufacturing systems," *J. Manufacturing Syst.*, vol. 1, no. 1, pp. 1-16, 1982.
[3] W. Eversheim and P. Hermann, "Recent trends in flexible automated manufacturing," *J. Manufacturing Syst.*, vol. 1, no. 2, pp. 139-148, 1982.
[4] R. Suri and R. R. Hildebrant, "Modelling flexible manufacturing systems using mean-value analysis," *SME J. Manufacturing Syst.*, vol. 3, no. 1, pp. 27-38, 1984.
[5] R. Suri, "An overview of evaluative models for flexible manufacturing systems," in *Proc. First ORSA/TIMS Conf. FMS*, 1984. (See also *Annals of Operations Research*, 1986.)
[6] J. J. Solberg, "A mathematical model of computerized manufacturing systems," in *Proc. 4th Int. Conf. Production Research*, pp. 22-30, 1977.
[7] Y. C. Ho et al., "Optimization of large multiclass (nonproduct-form) queueing networks using perturbation analysis," *Large Scale Systems*, 7, pp. 165-180, 1984.
[8] Y. C. Ho, X. R. Cao, and C. Cassandras, "Infinitesimal and finite perturbation analysis for queueing networks," *Automatica*, vol. 19, no. 4, pp. 439-445, 1983.
[9] Y. C. Ho and X. R. Cao, "Perturbation analysis and optimization of queueing networks," *J. Optim. Theory Applic.* vol. 40, no. 4, pp. 559-582, Aug. 1983.
[10] R. Suri, "Implementation of sensitivity calculation on a Monte Carlo experiment," *J. Optim. Theory and Applic.* vol. 40, no. 4, pp. 625-630, 1983.
[11] R. Suri and X. R. Cao, "Optimization of flexible manufacturing systems using new techniques in discrete event systems," *Proc. 20th Allerton Conf. Communic. Control and Computing*, pp. 434-443, 1982.
[12] X. R. Cao and Y. C. Ho, "Perturbation analysis of sojourn times in queueing networks," in *Proc. 22nd IEEE Conf. Dec. and Control*, pp. 1025-1029, Dec. 1983. See also *Operations Research*.
[13] W. Whitt, "Bivariate distributions with given marginals," *The Annals of Statistics*, vol. 4, pp. 1280-1289, 1976.
[14] X. R. Cao, "On the sample functions of Jackson networks," submitted to *Operations Research*.
[15] ——, "Convergence of parameter sensitivity estimates in a stochastic experiment," *IEEE Trans. Automat. Contr.*, 30, pp. 845-853, 1985.
[16] J. P. Buzen, "Computational algorithms for closed queueing networks with exponential servers," *Commun. ACM*, vol. 16, no. 9, pp. 527-531, 1973.
[17] F. Baskett, K. M. Chandy, R. R. Muntz, and F. G. Palacios, "Open, closed, and mixed networks of queues with different classes of customers," *J. ACM*, vol. 22, no. 22, pp. 248-260, 1975.
[18] K. M. Chandy, J. H. Howard, Jr., and D. F. Towsley, "Product form and local balance in queueing networks," *J. ACM*, vol. 24, no. 2, pp. 250-263, 1977.

Yu-Chi Ho (S'54-M'55-SM'62-F'73) was born in China. He received the S.B. and S.M. degrees in electrical engineering from M.I.T. and the Ph.D. degree in applied mathematics from Harvard University.

Except for three years in industry, he has been with Harvard University, where he is currently the Gordon McKay Professor of Engineering and Applied Mathematics and chairman of the Business Studies Program, a joint Ph.D. program in decision sciences between the Harvard Business School and the Faculty of Arts and Sciences. He is on the editorial board of seven international journals in various aspects of control, decision, and optimization, and for the past twenty years in various administrative capacities for the IEEE Control System Society. Professor Ho served in a consulting capacity for several government boards and panels, and he is currently the board member of the Massachusetts Foundation for Humanities and Public Policy. He has been elected to a Guggenheim fellowship (1970), a Churchill College Fellowship (1970), a U.K. Science Research Fellowship (1973, 1977), and a USA/USSR Exchange Fellowship (1973).

His current research interest is in discrete event dynamical systems and manufacturing automation. He holds four patents and has published two books and over 80 papers. His book on optimal control has been translated into Russian and Chinese. He is active in government and professional and community services.

Xi-Ren Cao (S'82-M'84) received the diploma from the Department of Physics, China University of Science and Technology in 1967, the M.S. degree in engineering in 1981, and the Ph.D. degree in applied mathematics in 1984, both from Harvard University, Cambridge, MA.

From 1968 to 1978 he served as an electrical engineer in China. He was a research assistant and a teaching assistant and is presently a research fellow at the Division of Applied Sciences, Harvard University. His current research interests are in the fields of queueing network theory, perturbation analysis of discrete-event dynamic systems and its applications, and stochastic approximation and optimization.

Chapter 5: Modeling and Control Using Petri Nets

5.1: Introduction

Petri nets are useful tools for the modeling and analysis of discrete-event dynamic systems. They are particularly valuable when state and control information are distributed throughout the system. Petri nets have been widely used to model and evaluate the performance of computer systems, especially operating systems. In this chapter, the theory and application of Petri nets are presented with emphasis on their application to the modeling, analysis, and control of automated manufacturing systems.

5.2: Why Petri Nets?

Carl A. Petri developed a net-theoretic approach to model and analyze communication systems [1]. More recently, Petri nets have proven to be useful for the modeling, performance evaluation, and control of manufacturing systems [2,3]. Specifically, they are useful for modeling systems with the following characteristics:

- *Concurrency or parallelism:* In a manufacturing system, many operations take place simultaneously.
- *Asynchronous operations:* Machines complete their operations in variable amounts of time and so the model must maintain the ordering of the occurrence of events.
- *Deadlock:* In this case, a state can be reached where none of the processes can continue. This can happen when two processes share two resources. The order by which these resources are used and released could produce a deadlock.
- *Conflict:* This may occur when two or more processes require a common resource at the same time. For example, two workstations might share a common transport system or might want access to the same database.
- *Event driven:* The manufacturing system can be viewed as a sequence of discrete events. Since operations occur concurrently, the order of occurrence of events is not necessarily unique; it is one of many allowed by the system structure.

These types of systems have been difficult to accurately model with differential equations and queueing theory. Petri nets can provide accurate models for the following reasons:

- Petri nets capture the precedence relations and structural interactions of concurrent and asynchronous events.
- They are logical models derived from the knowledge of how the system works. As a result, they are easy to understand and their graphical nature is a good visual aid.
- Deadlocks, conflicts, and buffer sizes can be modeled easily and concisely.
- Petri net models have a well developed mathematical foundation that allows a qualitative and quantitative analysis of the system.
- Finally, Petri net models can also be used to implement real-time control systems for an automated manufacturing system. They can sequence and coordinate the subsystems like a programmable logic controller does.

5.3: Ordinary Petri Nets

A Petri net graph uses circles to represent *places* and bars to represent *transitions*. Input-output relationships are represented by directed *arcs* between places and transitions. Figure 5.1 shows a simple Petri net. Here, *tokens* reside in places, travel along arcs, and their flow through the net is regulated by the transitions.

Tokens flow through the net depending on the present *marking* of the net. The marking of a Petri net is contained in an $n \times 1$ vector, $m(p)$, which represents the number of tokens in the i-th place, $p_i, i = 1,2...n$. When there is a token in each of the input places of a transition, that transition is enabled and fires. The transition fires by removing a token from each of its input places and by placing a token in each of its output places. This is referred to as the execution of a Petri net and it causes the marking to change and, therefore, tokens flow through the net. Figure 5.1 also illustrates the execution of a simple Petri net.

5.4: Modeling Manufacturing Systems Using Petri Nets

The tokens, places, and transitions must be assigned a meaning for proper interpretation of the model. In a manufacturing system, places usually represent resources, (e.g., machines, parts, data, and communication lines). Whereas a token in a place means that the resource is available, no token indicates that it is unavailable. A place can also be used to represent a condition. A token in such a place implies that a particular condition is true, and no token means that it is false.

Transitions are generally used to represent the initiation or termination of an event (e.g., a machine has been repaired or the part program has been downloaded).

5.4.1: Example

Consider the assembly cell shown in Figure 5.2. Any conveyor and two neighboring robots are needed to carry out an assembly task. Each conveyor requests the left robot first and, after acquiring it, requests the right one. Then the

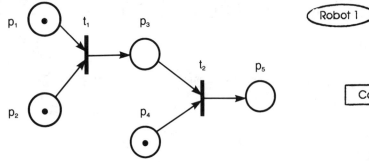

Figure 5.1a: Initial Marking

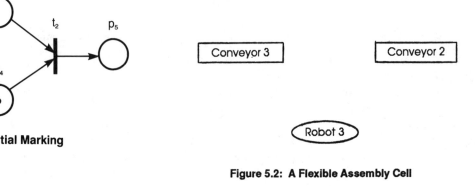

Figure 5.2: A Flexible Assembly Cell

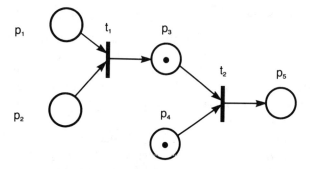

Figure 5.1b: Marking after t_1 Fires

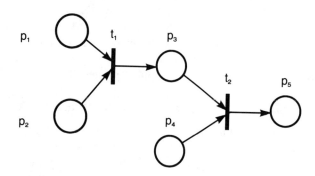

Figure 5.1c: Marking after t_2 Fires

Figure 5.1: A Simple Petri Net

For the assembly cell of Figure 5.2, the following places can be assigned:

p_1	C_1	requesting its left robot $R_1(p_{10})$
p_4	C_2	requesting its left robot $R_2(p_{11})$
p_7	C_3	requesting its left robot $R_3(p_{12})$
p_2	C_1	requesting its right robot $R_2(p_{11})$
p_5	C_2	requesting its right robot $R_3(p_{12})$
p_8	C_3	requesting its right robot $R_1(p_{10})$
p_3	C_1	and its two robots R_1 and R_2 are in use
p_6	C_2	and its two robots R_2 and R_3 are in use
p_9	C_3	and its two robots R_3 and R_1 are in use

and transitions

t_1	C_1	acquires its left robot R_1
t_4	C_2	acquires its left robot R_2
t_7	C_3	acquires its left robot R_3
t_2	C_1	acquires its right robot R_2
t_5	C_2	acquires its right robot R_3
t_8	C_3	acquires its right robot R_1
t_3	C_1	releases R_1 and R_2
t_6	C_2	releases R_2 and R_3
t_9	C_3	releases R_3 and R_1

Note: left and right are determined by standing in the center of Figure 5.2 while facing each conveyor.

At the start of the assembly task, all conveyors and robots are free, and the three conveyors are concurrently requesting their left robots. This produces the initial marking

$$m_0 = (1\ 0\ 0\ 1\ 0\ 0\ 1\ 0\ 0\ 1\ 1\ 1)^T$$

The initial marking along with the logical requirements of the assembly task leads to the Petri net model shown in Figure 5.3.

The initial marking has enabled transitions t_1, t_4, and t_7. If these transitions are allowed to fire concurrently, the new marking will be

assembly operation starts. When the task is completed, the conveyor releases both robots.

Requesting or releasing a robot are two events that can occur concurrently and asynchronously. For example, the three conveyors could simultaneously request their left robots (concurrency). Also, the release of two allocated robots occurs when the assembly task is completed but the time of this event cannot be accurately predicted because of possible delays or errors (asynchronous). The Petri net model must capture the concurrent and asynchronous behavior of the system.

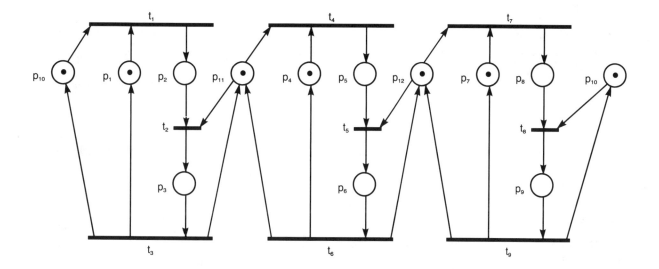

Figure 5.3: Petri Net Model for the Flexible Assembly Cell

$$m' = (0\ 1\ 0\ 0\ 1\ 0\ 0\ 1\ 0\ 0\ 0\ 0)^T$$

This means that the three conveyors have acquired their left robots and are now waiting for their right one. However, all three robots are already committed and so the process is deadlocked; it cannot proceed because m' enables no transitions. As a result, Petri nets can detect the presence of deadlock, which can lead to the re-design of systems that will be deadlock-free.

5.5: Petri Nets vs. Finite State Machines

Suppose that each marking of a Petri Net represents a state in a finite state machine. Then starting with the initial marking, we can enumerate all possible future markings and construct an equivalent finite state machine.

5.5.1: Example

Figure 5.4 shows a simple Petri net (note that concurrency is modeled here). For the initial marking $m_0 = (1\ 0\ 0)^T$, we can compute all of the possible future markings (i.e., the *reachability space*). This space will consist of m_0 and $m_1 = (0\ 1\ 1)^T$. The equivalent state machine is shown in Figure 5.4b with labeled arcs indicating the fired transition. Changing the initial marking to $m_0 = (2\ 0\ 0)^T$ results in a completely different state machine, even though the logical structure of the net has not changed. Starting at m_0, t_1 is enabled and results in $m_1 = (1\ 1\ 1)^T$. This enables t_1 and t_2. If t_1 fires first, then $m_2 = (0\ 2\ 2)^T$. If t_2 fires first, then $m'_2 = (2\ 0\ 0)^T$, which is the same as m_0. For the marking m_2, t_2 fires, resulting in $m_3 = (1\ 1\ 1)^T = m_1$. The equivalent state machine is shown in Figure 5.4c.

Suppose that p_1 represents the number of machines of a specific type that are available. Then our change in the initial marking from $(1\ 0\ 0)^T$ to $(2\ 0\ 0)^T$ represents the addition of a new machine to the system. Thus, the Petri net can represent many different states with the same structure while a state machine represents a fixed number of states and must be completely modified every time the state informa-

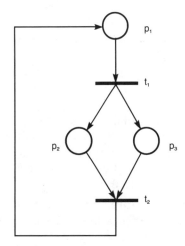

Figure 5.4a: Petri Net

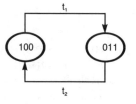

5.4b: State Machine with Initial Marking $(1,0,0)^T$

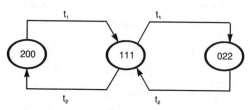

Figure 5.4c: State Machine with Initial Marking $(2,0,0)^T$

Figure 5.4: Petri Nets vs. Finite State Machines

tion changes. Also, the addition of new places and transitions is done easily, while such a change in a finite state machine requires altering the entire machine. Therefore, Petri nets are a simple and more powerful modeling tool than state machines.

5.6: Analysis of Petri Nets

Once a system has been modeled by using a Petri net, we want to analyze the net to determine which properties the net possesses.

5.6.1: Petri Net Properties

The concept of *reachability* is essential to understanding the definitions of the Petri net properties. A marking, m_r, is reachable from m_0 if there exists a firing sequence (a sequence of transition firings) that will yield m_r. The *reachability set*, $R(m_0)$, is the set of all possible markings that are reachable from m_0. This set can also be described by a reachability graph or a reachability tree.

Liveness: A Petri net is live with respect to a marking, m_0, if for any marking in $R(m_0)$, it is possible to fire any transition in the net. Liveness guarantees the absence of deadlock. The model in the flexible assembly cell example is not live.

Boundedness: A Petri net is k-bounded with respect to m_0 if each place has at most k tokens for all of the markings in $R(m_0)$. In a manufacturing system, a bounded net implies that resource constraints have been met (e.g., buffer levels are finite. If $k = 1$, then the net is said to be *safe*).

Conservativeness: A Petri net is conservative if, for any initial marking m_0 and a reachable marking $m \in R(m_0)$, there exists an $n \times 1$ nonnegative integer vector x such that

$$x^T m = x^T m_0$$

This says that the sum of the tokens weighted by x is constant.

Reversibility: A Petri net is reversible if for every $m \in R(m_0)$ then $m_0 \in R(m)$. Reversibility means that the initial marking is reachable from all reachable markings. This is important in a manufacturing system where failures occur and the system has to be reinitialized.

5.6.2: The Incidence Matrix

The first method of Petri net analysis is based on the *incidence matrix*. For a Petri net with n places and m transitions, the incidence matrix A is an $n \times m$ matrix of integers defined as

$$A = [a_{ij}] \quad \begin{array}{l} i = 1,2...n \\ j = 1,2...m \end{array}$$

where

$$a_{ij} = a_{ij}^+ - a_{ij}^-$$

and

a_{ij}^+ = number of arcs from transition j to output place i

a_{ij}^- = number of arcs from input place i to transition j

Example: The incidence matrix for the Petri net in Figure 5.3 is

$$A = \begin{bmatrix} -1 & 0 & 1 & 0 & 0 & 0 & 0 & 0 & 0 \\ 1 & -1 & 0 & 0 & 0 & 0 & 0 & 0 & 0 \\ 0 & 1 & -1 & 0 & 0 & 0 & 0 & 0 & 0 \\ 0 & 0 & 0 & -1 & 0 & 1 & 0 & 0 & 0 \\ 0 & 0 & 0 & 1 & -1 & 0 & 0 & 0 & 0 \\ 0 & 0 & 0 & 0 & 1 & -1 & 0 & 0 & 0 \\ 0 & 0 & 0 & 0 & 0 & 0 & -1 & 0 & 1 \\ 0 & 0 & 0 & 0 & 0 & 0 & 1 & -1 & 0 \\ 0 & 0 & 0 & 0 & 0 & 0 & 0 & 1 & -1 \\ -1 & 0 & 1 & 0 & 0 & 0 & 0 & -1 & 1 \\ 0 & -1 & 1 & -1 & 0 & 1 & 0 & 0 & 0 \\ 0 & 0 & 0 & 0 & -1 & 1 & -1 & 0 & 1 \end{bmatrix}$$

5.6.3: The Reachability Graph

This represents all of the possible reachable markings. Starting with the initial marking m_0, all of the enabled transitions are fired, which leads to a new marking that might enable other transitions. Taking each of these new markings as a new root, one can recursively generate all of the reachable markings. This tree structure preserves the firing order of the transitions.

5.6.4: "Putting Petri Nets to Work" (T. Agerwala)

A good summary of Petri-net-analysis techniques is presented in the first reprint of this chapter. Properties of the net are then verified. The reader is advised to study this paper before continuing with the other tutorial information.

Read the first reprint: "Putting Petri Nets to Work" by T. Agerwala.

5.7: "Applications of Petri Net Based Models in the Modeling and Analysis of Flexible Manufacturing Systems" (M. Kamath and N. Viswanadham)

The reprint by Kamath and Viswanadham provides a good survey of Petri nets for flexible manufacturing systems. It emphasizes modeling and analysis techniques and also contains an extensive list of references.

Dubois and Stecke [10] were the first to use Petri nets to model and analyze a flexible manufacturing system. Al-Jaar and Desrochers [2] present an introductory tutorial on the use of Petri nets in automation and manufacturing and a shorter survey [3] of Petri nets in automated manufacturing systems.

5.7.1: Ordinary Petri Nets

Narahari and Viswanadham [9] provide a good treatment of Petri nets in the modeling and analysis of FMSs. The various operations (machines, jobs, and part-types) are modeled by using Petri nets and their invariants are computed. Then a Petri net model for the whole system is obtained and analysed for deadlocks and buffer overflows. Narahari and Viswanadham demonstrate this approach by using a transfer line with three machines and two buffers, and an FMS with three machines and two part-types.

Martinez et al. [11] give a general and intuitive overview about the modeling of FMSs by using Petri nets and colored Petri nets which point out their respective advantages and disadvantages. The applicability of these models to the design and simulation of an FMS and the implementation of its control system is demonstrated by using two examples: a flowshop production system and a transport system with automatic guided vehicles.

Valette et al. [12] used the car production facility at Renault as an example of the control of FMSs by using Petri nets. Based on structuring, refinement, and reduction of the net (top-down approach), they were able to specify and validate a model of interconnected controllers for this facility.

Beck and Krogh [13] introduced modified Petri nets (MPNs) and presented a general methodology for modeling a manufacturing process by decomposing it into operations and resources. These nets serve as a basis for the modeling, simulation, and design of the discrete-control logic. The construction of MPNs is demonstrated by using a two-robot arm-assembly process. Krogh and Sreenivas [14] extended the previous work and suggested the use of essentially decision-free Petri nets for real-time resource allocation and the synthesis of the required control logic.

5.7.2: Colored Petri Nets

This is an extension to ordinary Petri nets in which colors are associated with the tokens. Now transitions fire according to a set of rules that match the appropriate colors. A colored token is analogous to a subscripted variable.

The advantage of colored Petri nets is that they provide compact models of large systems. More details are given in the reprint, included in this chapter, by Kamath and Viswanadham.

Viswanadham and Narahari [15] give two detailed examples on the use of colored Petri nets (CPNs) in automated manufacturing. The first is concerned with the modeling and analysis of two machines that process two part-types. The second describes a manufacturing cell that consists of three machines and three robots that process two part-types by using a limited number of shared tools.

Alla and Ladet [18] illustrated the use of CPNs as a modeling, validation, and simulation tool by using a flexible manufacturing line with first-in first-out queues.

Martinez et al. [16] turned their attention to the level above the local control: the co-ordination (cell) level. Monitoring and real-time scheduling was the third level. By using the Renault FMS layout, a CPN model for the co-ordinator was derived and analyzed. To resolve any conflicts in the local control-level model, the use of an expert system was suggested, especially since production rules can be modeled by using CPNs. Also, fault/error detection and recovery can be similarly modeled, either at the same or higher levels.

Gentina and Corbeel [17] proposed the use of structured adaptive and structured-colored adaptive Petri nets to model FMSs and their control systems, at the two lowest levels. The third level was modeled as a rule-based (declarative) expert system. The method of analysis and validation is illustrated by using an FMS.

5.8: Performance Evaluation of Manufacturing Systems

The Petri net model of the logical and causal dependencies in a manufacturing system is not sufficient to answer time-related questions. The addition of time allows for such a temporal performance evaluation. Timed Petri nets (TPNs) and stochastic (timed) Petri nets (SPNs) are used in this context. The main application of these nets has been in the performance evaluation of computer systems and more recently in manufacturing systems.

5.8.1: Timed Petri Nets

Time can be included in a Petri net model by associating time with the transitions, to form a timed transition Petri net (TTPN), or with the places, resulting in a timed place Petri net (TPPN). Both representations are equivalent [48].

In a TTPN, the firing of a transition takes a certain amount of time. Note that this time is fixed, which makes TTPNs deterministic.

In a TPPN, a token enters a place and is unavailable for time $= d_i$, after which it becomes available. In this net, only available tokens in a marking can enable a transition. During the unavailable period, another token may arrive in the place. A TPPN with all delays set to zero reduces to an ordinary Petri net.

The properties of Petri nets analyzed in the previous section can be applied to timed Petri nets by using the incidence matrix. An alternate approach attempts to cast these nets in a system-theoretic framework [5,49]. A mini-max algebra is applied to the Petri net models and concepts analogous to transfer functions, input-output models, feedback, etc., are developed. The potential for this approach lies in its ability to build on analogies with traditional control theory concepts. It is still to be proven that useful analogies exist and can be extended to systems that include shared resources.

5.8.2: "Performance Analysis Using Stochastic Petri Nets" (M.K. Molloy)

Stochastic Petri nets (SPNs) are timed Petri nets in which the transition times have exponentially distributed firing rates. In SPNs, the time delays are associated with the transitions.

The value of SPNs becomes apparent when one examines the following theorem due to Molloy.

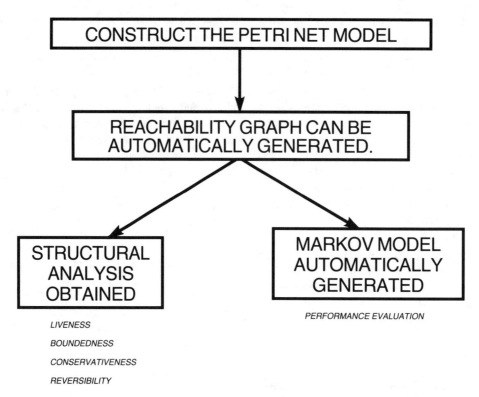

Figure 5.5: Petri Net Theory

THEOREM: Any finite place, finite transition, marked stochastic Petri Net is isomorphic to a one-dimensional discrete space Markov process. Note: A one-dimensional Markov process is one that has only one random variable, (e.g., time is the only random variable).

This opens the way to performance analysis of automated manufacturing systems. Specifically, all one has to do is model the system with a Petri net. Then, based on the initial marking, the reachability tree can be generated and the equivalent Markov chain can be obtained and analyzed. Figure 5.5 summarizes the benefits of Petri net theory.

The main advantage of the Petri net is that it simplifies the generation of the Markov chain. For example, in Figure 5.7 the Petri net for this production network can generate the equivalent 1820 state Markov chain. The large dimensionality of the problem is still there, but the Petri net provides a more compact and logical representation of the system.

Solving the equivalent Markov chain involves the solution of a set of linear algebraic equations (i.e., the steady-state probabilities (see Section 2.2.4). In many cases, this is the only solution available since analytic solutions are difficult, if not impossible, to find.

The steps involved in going from the Petri net model to the reachability tree and then to the Markov chain have all been automated and can be found in several software packages.

5.8.3: Software for Petri Net Analysis

Chiola [37] has developed GreatSPN for the construction and analysis of complex generalized SPN models. This software accepts deterministic delays or exponentially distributed firing rates. It also computes the transient and steady-state solutions to the Markov chains.

Dugan et al. [38] have developed the Duke extended SPN evaluation package (DEEP) for the performance analysis of SPN models. This has led to a more recent version [39]. Holiday and Vernon [40] have developed the GTPN analyzer for the performance evaluation of generalized timed Petri net models.

It should be emphasized, that these software packages are *not* simulations of Petri net models. They are solving the equivalent Markov chain. The value of Petri nets lies in the fact that they are conceptually easy to use as a modeling tool, yet they can represent very large (and equivalent) Markov chains. The examples in the next section will illustrate this powerful advantage.

5.8.4: Analysis of Transfer Lines and Production Networks Using Generalized Stochastic Petri Nets

Generalized stochastic Petri nets (GSPNs) [50] incorporate both timed transitions (drawn as white boxes) and immediate transitions (drawn as thin black bars). The timed transitions have an exponentially distributed firing rate λ and fire $1/\lambda$ time units after being enabled. The immediate transitions fire in zero time.

GSPNs also permit the use of inhibitor arcs, priority functions, and random switches. Inhibitor arcs are used to prevent transitions from firing when certain conditions are true. A priority function is defined for the marking in which both timed and immediate transitions are enabled. Usually, immediate transitions are given the higher priority. The

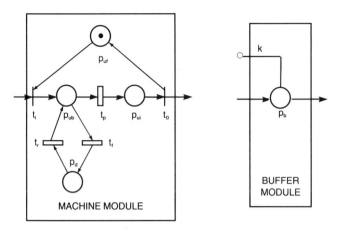

Figure 5.6: Machine and Buffer Modules

been defined [42]. These modules are used as building blocks to model and analyze any arbitrary connection of machines and buffers.

The machine and buffer modules are shown in Figure 5.6. Note that these modules use tokens to model the discrete parts in the system. When tokens accumulate beyond the buffer capacity, k, the inhibitor arc is activated. This arc will be connected to t_i and will prevent that transition from firing (which prevents parts from entering the machine) when the buffer has k tokens (parts) in it. Tokens are also used in the machine module to represent random failures and repairs of the machine.

The places in the net have the following interpretation:

p_{uf} = machine is up and free
p_{ub} = machine is up and busy
p_{ui} = machine is up and idle
p_d = machine is down
p_b = machine buffer

and for the transitions

t_i = input to the machine (immediate)
t_p = part processing rate
t_o = output from the machine (immediate)
t_r = machine repair rate
t_f = machine failure rate

random switch is used to resolve conflicts between two or more immediate transitions. The random switch is basically a discrete probability distribution. These additional modeling capabilities do not destroy the equivalence with Markov chains.

The steady-state probabilities obtained from the Markov chain are used to compute the expected number of tokens in a place, the probability that a place is not empty, and the probability that a transition is enabled. Performance measures such as average production rate, average in-process inventory, and average resource utilization can also be computed from the steady-state probabilities.

To analyze the performance of transfer lines and production networks, basic modules of machines and buffers have

Example: A production network (Figure 5.7) that consists of four machines and three buffers was analyzed by using the Petri net approach. Machine three is an assembly station

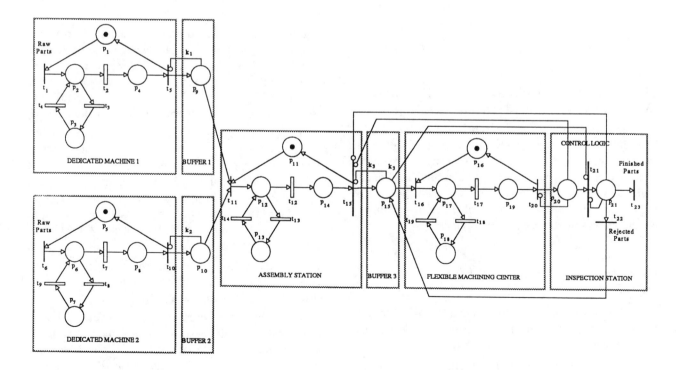

Figure 5.7: A Four Machine Three Buffer Production Network

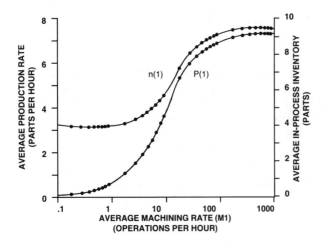

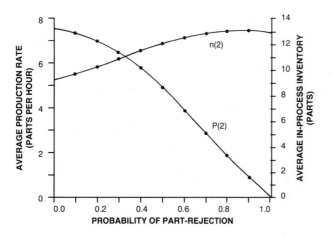

Figure 5.8: Performance Results

and machine four is a flexible machine capable of inspection and of reworking the rejected parts (see t_{21} and p_{15}).

Figure 5.8 shows the results obtained from SPNP [39]. The equivalent Markov chain has 1820 states. Modeling this system from the beginning as a finite state machine would be an absurd exercise prone to many errors. Similarly, finding a closed form solution to this problem is likely to prove impossible. The Petri net approach offers a valuable tool for obtaining the solution to the Markov chain.

5.8.5: "Performance Evaluation of Asynchronous Concurrent Systems Using Petri Nets" (C.V. Ramamoorthy and G.S. Ho)

In the next reprint, Ramamoorthy and Ho use deterministic timed Petri nets to find the maximum cycle time for processing a task.

In a deterministic timed Petri net, an execution time r is associated with each transition. A transition takes exactly r units of time to complete its execution. For these nets, time is associated with the transitions.

The paper deals with a special case of Petri nets called decision free Petri nets (or sometimes called marked graphs).

Definition: a Petri Net is decision free if for each place in the net there is one input arc and one output arc. For this class of systems the minimum cycle time can be computed easily.

Ramamoorthy and Ho present examples for computer systems but applications to manufacturing can be found in [2,5,10].

5.9: Petri Nets as Real Time Controllers

Petri nets can be used to sequence the operations in a manufacturing system. As transitions fire, events are made to occur and so the flow of tokens through the net can be used as a real-time controller. The controller can also be analyzed for deadlocks, boundedness, etc. and it can be evaluated against various performance measures.

A Petri-net controller for a machining workstation is reported in Crockett et al. [26]. This approach directly implements the Petri-net-based controller on a general-purpose computer. No programmable logic controllers are used.

5.9.1: Example

Some graphical modifications and alterations to Petri nets have been devised to facilitate the understanding and implementation of the controller. Figure 5.9 shows these additional place types and Figure 5.10 presents the Petri-net controller by using the proposed graphical representation.

The workstation is based on a Cincinnati Milacron 5VC Machining Center and is controlled by the Petri net that runs on a VAX 11/780. The VAX is used because it can handle the concurrency in Figure 5.10. The actual machine controller is an IBM PC/XT. The VAX communicates to the IBM PC and the operator interface by using mailboxes.

A machining cycle is initiated by issuing a part request (ID and description). Next, two concurrent commands are issued: one for the parts handler (fixture) and another for downloading and initializing the NC program. After both of these commands have been carried out, the machining operation starts. Upon completion of the machining the part is either unloaded or remains in the machine for additional machining operations.

A workcell, consisting of this machining center and a plastic injection molding machine, has also been controlled by using the Petri net approach [34].

Grafcet: Valette [53] presents an excellent review on the use of Petri nets in the design and implementation of controllers for manufacturing systems. The Grafcet formalism is introduced first. This is a European Petri net based graphic-tool standard intended to apply to all software con-

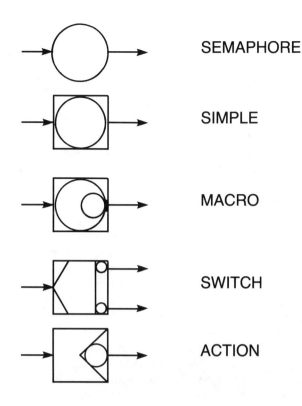

Figure 5.9: Controller Place Types

trol systems for industrial automation. In 1975, this represented the first attempt to use Petri nets in manufacturing.

In the United States, Grafcet has been used to help design generic manufacturing system controllers for the Automated Manufacturing Research Facility (Section 1.3.1) at the National Institute of Standards and Technology (NIST) [35]. It has proven to be a powerful graphical language for expressing control flow. Grafcet helps the designer determine modularization of the code, functions that can be performed in parallel, communication between parallel processes, and problems in control flow. Grafcet shows how the controller operates instead of how it is implemented [36].

The NIST researchers [35] have reported on their experience with using Grafcet. They site the following specific benefits that can be gained from Grafcet:

- The graphical representation of control allows easy understanding, documenting, and prototyping
- Programs are encoded symbolically (i.e., by using pictures)
- The language has parallel constraints that allows the user to represent distributed systems
- Both parallel and sequential control logic can be expressed
- Both synchronous and asynchronous processes can be simulated on one machine

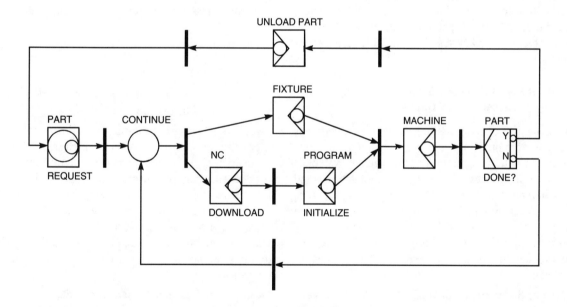

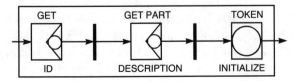

Figure 5.10: Petri Net Controller

- A design can be represented as a dynamic entity that can also be used as a functional model of the working system
- It can determine which functions in a controller can be executed in parallel
- It has a MAP network interface
- A mechanism is provided for communicating between parallel processes linked over a network
- A comparison between shared data and peer-to-peer communication can be made

Basically, Grafcet represents the software-engineering approach to the production-control problem. It maintains and monitors the flow of data in a way that is similar to that of an operating system. Several Grafcet models are included in Thomas and McLean [35] who conclude that it is an excellent tool for documenting and demonstrating the control flow in a manufacturing system.

Other Petri net based controllers: Silva and Velilla [29] compare various implementation strategies for programming PLCs (programmable logic controllers) based on Petri net models. The net provides a structured model of the control logic. This logic can then be analyzed and verified by using Petri net theory before the actual implementation, which helps to reduce the number of software and hardware errors.

Vallete et al. [32] have used Petri nets in a three-level hierarchical real-time control structure that includes a local machine-control level, coordination level, and real-time shop scheduling. Merabet [7] has also implemented a three-level hierarchical control system for a flexible manufacturing system.

5.10: "Generalized Petri Net Reduction Method" (H. Lee Kwang, J. Favrel, and P. Baptiste)

In some practical cases, modeling a system with Petri nets can lead to a large number of places, transitions, and arcs. To gain insight with the operation of the original system, it is desirable to find a net with fewer places, transitions, and arcs that still retains the liveness and boundedness properties of the original net. By studying and analyzing the reduced net, it is possible to make conclusions about the token flow and structural properties of the original net. The authors propose a reduction method to accomplish this for generalized Petri nets.

The goal of the reduction method is to establish a set of rules for combining places, transitions, and arcs that preserve the number and direction of flow of tokens in the original net. This must be the goal since creating or destroying tokens will alter the liveness and boundedness properties of the net. In summary, the flow of tokens into and out of a reduced net, must be the same as in the portion of the original net (or subnet) that has been replaced.

Definition: A macronode is a node that represents a subnet. Macrotransitions and macroplaces are two possible macronodes.

5.10.1: Example

Consider the two nets in Figure 2c of the Lee-Kwang et al. reprint. The basic idea is that the transition-place-transition sequence can be combined, and therefore reduced to a single macrotransition. To see this, consider the left net of Figure 2c. When transition t_i fires, a token is placed in the left place and in P_i causing t_0 to fire (eventually). These two transitions can be replaced by a macrotransition, t^*, since the firing of t^* corresponds to the firing of t_0 and also puts a token in the left place. Thus, from an input-output point of view, the flow of tokens is indistinguishable and the nets are equivalent.

Next, consider doing the same thing with the nets of Figure 3c in the reprint. Combining the transition-place-transition sequence is not possible here because of the path from t_i to the left place. This path is lost when replacing $t_i - P_i - t_0$ with a macrotransition. To see this, suppose the left net in Figure 3c has a token in the leftmost place. This enables the output transition. In the right net of Figure 3c, a token in the leftmost place will fire the macrotransition (which includes t_i), which is incorrect. This difficulty arises because there is more than one path from t_i to t_0. Any attempt to combine transitions will not be able to distinguish between these flows.

The authors present numerous situations under which places, transitions, and arcs can be combined to form a simpler net that retains the properties of liveness and boundedness. An example is also included for reducing a Petri net model of a factory producing corrugated fiberboard boxes.

5.11: Summary

The reprints are a variety of tutorial papers, in which applications and theory are discussed.

Agerwala's paper is a tutorial on Petri nets. It is a concise introduction to Petri net analysis, which includes topics on net invariants, the incidence matrix, boundedness, conservativeness, mutual exclusion, buffer overflow, and deadlock. There is no discussion of performance analysis since this did not appear until three years later. This paper also shows the modeling power of Petri nets by considering models of operating systems, chemical reactions, computer programs, computer systems, databases, and communication protocols.

Kamath and Viswanadham also provide a good review of Petri nets while focusing their presentation on applications to flexible manufacturing systems. Colored Petri nets are also introduced in his paper.

Molloy in "Performance Analysis Using Stochastic Petri Nets," establishes the connection between stochastic Petri nets and discrete-space Markov processes. This paper forms the basis for the performance analysis of systems that can be modeled with SPNs. It also provides a good explanation of the limitations of queueing and other models while emphasizing the advantages of Petri nets.

Ramamoorthy and Ho, in "Performance Evaluation of Asynchronous Concurrent Systems Using Petri Nets," use deterministic timed Petri nets for performance evaluation. Obtaining performance information from Petri net models is potentially one of the most significant contributions to automated manufacturing systems.

In the next paper, "Generalized Petri Net Reduction Method," Hyung et al. recognize that for real applications the Petri net will become very large. Consequently, model reduction methods are needed to help avoid the reachable state explosion problem. The authors also present an application to a complex flexible-manufacturing system that was originally modeled with 92 places, 59 transitions, and 174 arcs. Their procedure was used to find a reduced net (12 transitions and 20 places), which allowed them to study the original system in the reduced reachable state space. The results of the reduction suggested five subsystems for the flexible manufacturing system.

The reprint by Komoda et al., "An Autonomous, Decentralized Control System for Factory Automation," is a good example of how Petri nets can be used to control an automated factory. The paper is mainly an overview of the controller concepts used at Hitachi.

The final reprint, "A Petri Net-Based Controller for Flexible and Maintainable Sequence Control and Its Applications in Factory Automation," by Murata et al., is a detailed description of the actual application of Hitachi's Petri-net-based controller. They modify Petri nets into control nets to facilitate the supervision and the interaction with the physical components as well as the development and maintainability of the control programs. The control system consists of station controllers for control sequencing and station coordination for monitoring and diagnosis. Whereas the station controllers are programmed by using Petri nets, the station coordinators use IF-THEN rules. This work has resulted in a commercial system for Hitachi.

5.12: References

1. C.A. Petri, "Kommunikation mit Automaten," Ph.D. Dissertation, University of Bonn, Bonn, West Germany, 1962.
2. R.Y. Al-Jaar and A.A. Desrochers, "Petri Nets in Automation and Manufacturing," in *Advances in Automation and Robotics* (ed. G.N. Saridis), JAI Press, Greenwich, Conn., Vol. 2, 1989.
3. R.Y. Al-Jaar and A.A. Desrochers, "A Survey of Petri Nets in Flexible Manufacturing Systems," *Proceedings of the 1988 IMACS Conference*, Paris, France, July 1988.
4. R. Ravichandran and A.K. Chakravarty, "Decision Support in Flexible Manufacturing Systems Using Timed Petri Nets," *Journal of Manufacturing Systems*, Vol. 5, No. 2, 1986, pp. 89–101.
5. G. Cohen, D. Dubois, J.P. Quadrat, and M. Viot, "Linear System Theory for Discrete Event Systems," *Proceedings of the 23rd IEEE Conference on Decision and Control*, Las Vegas, Nev., Dec. 1984, pp. 539–544.
6. T. Murata, "State Equation, Controllability, and Maximal Matchings of Petri Nets," *IEEE Transactions on Automatic Control*, Vol. AC-22, June 1977, pp. 412–416.
7. A.A. Merabet, "Synchronization of Operations in a Flexible Manufacturing Cell: The Petri Net Approach," *Journal of Manufacturing Systems*, Vol. 5, No. 3, 1986, pp. 161–169.
8. T. Murata, "Circuit Theoretic Analysis and Synthesis of Marked Graphs," *IEEE Transactions on Circuits and Systems*, Vol. CAS-24, June 1977, pp. 400–405.

5.12.1: Petri Net Models for Manufacturing Systems

9. Y. Narahari and N. Viswanadham, "A Petri Net Approach to the Modelling and Analysis of Flexible Manufacturing Systems," *Annals of Operations Research*, Vol. 3, 1985, pp. 449–472.
10. D. Dubois and K.E. Stecke, "Using Petri Nets to Represent Production Processes," *Proceedings of the 22nd IEEE Conference on Decision and Control*, San Antonio, Tex., Dec. 1982, pp. 1062–1067.
11. J. Martinez, H. Alla, and M. Silva, "Petri Nets for the Specification of FMSs," *Modelling and Design of Flexible Manufacturing Systems* (ed. A. Kusiak), Elsevier, New York, 1986, pp. 389–406.
12. R. Valette, M. Courvoisier, and D. Mayeux, "Control of Flexible Production Systems and Petri Nets," *Proceedings of the 3rd European Workshop on the Application and Theory of Petri Nets*, Varenna, Italy, Sept. 1982, pp. 264–277.
13. C.L. Beck and B.H. Krogh, "Models for Simulation and Discrete Control of Manufacturing Systems," *Proceedings of the 1986 IEEE International Conference on Robotics and Automation*, San Francisco, Calif., April 1986, pp. 305–310.
14. B.H. Krogh and R.J. Sreenivas, "Essentially Decision Free Petri Nets for Real-Time Resource Allocation," *Proceedings of the 1987 IEEE International Conference on Robotics and Automation*, Raleigh, N.C., April 1987, pp. 1005–1011.
15. N. Viswanadham and Y. Narahari, "Colored Petri Net Models for Automated Manufacturing Systems," *Proceedings of the 1987 IEEE International Conference on Robotics and Automation*, Raleigh, N.C., April 1987, pp. 1985–1990.

16. J. Martinez, P. Muro, and M. Silva, "Modeling, Validation and Software Implementation of Production Systems Using High Level Petri Nets," *Proceedings of the 1987 IEEE International Conference on Robotics and Automation*, Raleigh, N.C., April 1987, pp. 1180–1185.

17. J.C. Gentina and D. Corbeel, "Colored Adaptive Structured Petri Net: A Tool for the Automatic Synthesis of Hierarchical Control of Flexible Manufacturing Systems," *Proceedings of the 1987 IEEE International Conference on Robotics and Automation*, Raleigh, N.C., April 1987, pp. 1166–1173.

18. H. Alla and P. Ladet, "Colored Petri Nets: A Tool for Modeling, Validation, and Simulation of FMS," in *Flexible Manufacturing Systems: Methods and Studies* (ed. A. Kusiak), Elsevier, New York, 1986, pp. 271–281.

5.12.2: Reduction of Petri Nets

19. T. Murata and J.Y. Koh, "Reduction and Expansion of Live and Safe Marked-Graphs," *IEEE Transactions on Circuits and Systems*, Vol. CAS-27, No. 1, Jan. 1980, pp. 68–70.

20. R. Johnsonbaugh and T. Murata, "Additional Method for Reduction and Expansion of Marked Graphs," *IEEE Transactions on Circuits and Systems*, Vol. CAS-28, No. 10, Oct. 1981, pp. 1009–1014.

21. I. Suzuki and T. Murata, "A Method for Stepwise Refinement and Abstraction of Petri Nets," *Journal of Computer and System Sciences*, Vol. 27, 1983, pp. 51–76.

22. K.H. Lee et al., "Hierarchical Reduction Method for Analysis and Decomposition of Petri Nets," *IEEE Transactions on Systems, Man, and Cybernetics* Vol. SMC-15, No. 2, 1985, pp. 272–280.

5.12.3: Petri Net Based Controllers

23. J. Ayache, J. Courtiat, and M. Diaz, "REBUS, a Fault Tolerant Distribution System for Industrial Real Time Control," *IEEE Transactions on Compters*, Vol. C.-31, No. 7, July 1982, pp. 637–647.

24. B.H. Krogh and C.L. Beck, "Synthesis of Place/Transition Nets for Simulation and Control of Manufacturing Systems," *Proceedings of the IFAC/IFIP Symposium on Large Scale Systems*, Zurich, Switzerland, Aug. 1986.

25. A.L. Hopkins, Jr. and G.R. Walker, "The State Transition Diagram as a Sequential Control Language," *Proceedings of the 25th IEEE Conference on Decision and Control*, Dec. 1986, pp. 1096–1101.

26. D. Crockett, A. Desrochers, F. DiCesare, and T. Ward, "Implementation of a Petri Net Controller for a Machining Workstation," *Proceedings of the 1987 IEEE International Conference on Robotics and Automation*, Raleigh, N.C., 1987, pp. 1861–1867.

27. M. Courvoisier, R. Valette, J.M. Bigou, and P. Esteban, "A Programmable Logic Controller Based on a High Level Specification Tool," *Proceedings of the IEEE IECON Conference on Industrial Electronics*, IEEE Press, New York, 1983, pp. 174–179.

28. D. Chocron and E. Cerny, "A Petri-Net-Based Industrial Sequencer," *Proceedings of the IEEE International Conference and Exhibition on Industrial Control and Instrumentation*, IEEE Press, New York, 1980, pp. 18–22.

29. M. Silva and S. Velilla, "Programmable Logic Controllers and Petri Nets: A Comparative Study," *Proceedings of the IFAC Software for Computer Control*, Pergamon Press, 1982.

30. R. Masuda and K. Hasegawa, "Mark Flow Graph and Its Application to Complex Sequential Control System," *Proceedings of the 13th Hawaii International Conference on System Science*, Honolulu, Haw., 1980, pp. 194–203.

31. T. Murata, N. Komoda, and K. Matsumoto, "A Petri-Net Based FA (Factory Automation) Controller for Flexible and Maintainable Control Specifications," *Proceedings of the IEEE IECON Conference on Industrial Electronics*, IEEE Press, New York, 1984, pp. 362–366.

32. R. Valette, M. Courvoisier, J.M. Bigou, and J. Albukerque, "A Petri Net Based Programmable Logic Controler," *Proceedings of the IFIP Conference on Computer Applications in Production and Engineering*, North-Holland, New York, 1983, pp. 103–116.

33. J.C. Gentina, J.P. Bourey, and M. Kapusta, "Colored Adaptive Structural Petri Net," *Computer Integrated Manufacturing Systems*, Vol. 1, No. 1, Feb. 1988, pp. 39–47.

34. E. Kasturia, F. DiCesare, and A.A. Desrochers, "Real Time Control of Multilevel Manufacturing Systems Using Colored Petri Nets," *Proceedings of the 1988 IEEE International Conference on Robotics and Automation*, Philadelphia, Penn., April 1988, pp. 1114–1119.

35. B.H. Thomas and C. McLean, "Using Grafcet to Design Generic Controllers," *Proceedings of Rensselaer's First International Conference on Computer Integrated Manufacturing*, IEEE Computer Society Press, Washington, D.C., Troy, New York, May 1988, pp. 110–119.

36. A.D. Baker, T.J. Johnson, D.I. Kerpelman, and H.A. Sutherland, "Grafcet and SFC as Factory Automation Standards, Advantages and Limitations," *Proceedings of the 1987 American Control Conference*, Minneapolis, Minn., June 1987, pp. 1725–1730.

5.12.4: Software for Petri Net Analysis

37. G. Chiola, "A Graphical Petri Net Tool for Performance Analysis," *Proceedings of the 3rd International Workshop on Modeling Techniques and Performance Evaluation,* AFCET, Paris, France, March 1987.

38. J.B. Dugan, A. Bobbio, G. Ciardo, and K. Triverdi, "The Design of a Unified Package for the Solution of Stochastic Petri Net Models," *Proceedings of the IEEE International Workshop on Timed Petri Nets,* Torino, Italy, July 1985, pp. 6–13.

39. G. Ciardo, "Manual for the SPNP Package," Duke University, Durham, N.C., July 1988.

40. M.A. Holliday and M.K. Vernon, "A Generalized Timed Petri Net Model for Performance Analysis," *Proceedings of the IEEE International Workshop on Timed Petri Nets,* Torino, Italy, July 1985, pp. 181–190.

5.12.5: Performance Analysis

41. G. Balbo, G. Chiola, G. Franceschinis, and G. Molinar Roet, "Generalized Stochastic Petri Nets for the Performance Evaluation of FMS," *Proceedings of the 1987 IEEE International Conference on Robotics and Automation,* Raleigh, N.C., 1987, pp. 1013–1018.

42. R.Y. Al-Jaar and A.A. Desrochers, "Modeling and Analysis of Transfer Lines and Production Networks Using Generalized Stochastic Petri Nets," *Proceedings of the Sixth National Conference on University Programs in Computer-Aided Engineering, Design and Manufacturing,* Georgia Institute of Technology, Atlanta, Ga., June 1988, pp. 16–21.

43. R.Y. Al-Jaar and A.A. Desrochers, "Evaluation of Part-Type Mix for a Machining Workstation Using Generalized Stochastic Petri Nets," *Proceedings of the 1988 IEEE Conference on Decision and Control,* Austin, Tex., Dec. 1988, pp. 2307–2313.

44. A. Seidmann, P.J. Schweitzer, and S.Y. Nof, "Performance Evaluation of a Flexible Manufacturing Cell with Random Multiproduct Feedback Flow," *International Journal of Production Research,* Vol. 23, No. 6, 1985, pp. 1171–1184.

45. B. Maione, Q. Semeraro, and B. Turchiano, "Closed Analytical Formulae for Evaluating Flexible Manufacturing System Performance Measures," *International Journal of Production Research,* Vol. 24, No. 3, 1986, pp. 583–592.

46. A. Arbel and A. Seidmann, "Performance Evaluation of Flexible Manufacturing Systems," *IEEE Transactions on Systems, Man, and Cybernetics,* Vol. SMC-14, No. 4, July/Aug. 1984, pp. 606–617.

47. Y. Dallery and R. David, "A New Approach Based on Operational Analysis for Flexible Manufacturing Systems Performance Evaluation," *Proceedings of the 22nd IEEE Conference on Decision and Control,* San Antonio, Tex., Dec. 1983, pp. 1056–1061.

48. J. Sifakis, "Use of Petri Nets for Performance Evaluation," in *Measuring, Modelling, and Evaluating Computer Systems,* (eds. H. Beilner and E. Gelenbe), North Holland, New York, 1977, pp. 75–93.

49. G. Cohen, D. Dubois, J.P. Quadrat, and M. Viot, "A Linear-System—Theoretic View of Discrete Event Processes and Its Use for Performance Evaluation in Manufacturing," *IEEE Transactions on Automatic Control,* Vol. AC-30, No. 3, March 1985, pp. 210–220.

50. M. Ajmone Marsan, G. Balbo, and G. Conte, "A Class of Generalized Stochastic Petri Nets for the Performance Evaluation of Multiprocessor Systems," *ACM Transactions on Computer Systems,* Vol. 2, No. 2, May 1984, pp. 93–122.

Today's modeling tools, appropriate for conventional sequential systems, will be inadequate for the complex concurrent systems of the 80's. Petri nets may offer a solution.

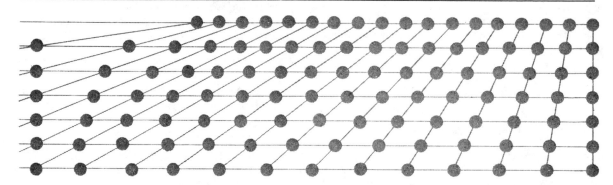

Special Feature:
Putting Petri Nets to Work

Tilak Agerwala*
IBM

Petri nets have developed over the last decade into a suitable model for representing and studying concurrent systems. Petri nets can be viewed as a special structure to be studied as an intellectual exercise, as another automaton capable of accepting or generating formal languages, or as a representation scheme for describing, analyzing, and synthesizing different kinds of "real" systems. This paper brings together a large body of work on useful applications of Petri nets.

Modeling a system using (interpreted) Petri nets has three potential advantages: First, the overall system is often easier to understand due to the graphical and precise nature of the representation scheme. Secondly, the behavior of the system can be analyzed using Petri net theory,[1] which includes tools for analysis such as marking trees and invariants, and established relationships between certain net structures and dynamic behavior. Techniques developed for the verification of parallel programs[2-4] can also be applied. Finally, since Petri nets can be synthesized using bottom-up and top-down approaches, it is possible to systematically design systems whose behavior is either known or easily verifiable.

It is expected that many future computer-based systems will incorporate multiple, communicating units. Such systems can exhibit very complex interactions and behaviors. Tools for modeling, representing, and analyzing conventional sequential systems will be totally inadequate. Future computer system designers and users will require new conceptual mechanisms and theories to deal with their systems. Petri nets incorporate the fundamental concepts which can be used as a basis for these models and theories.

What are Petri nets?

A Petri net may be identified as a bipartite, directed graph $N=(T,P,A)$ where

$T=\{2_1 t_2, \ldots, t_n\}$ is a set of *transitions*
$P=\{p_1 p_2, \ldots, p_m\}$ is a set of *places*
   ($T \cup P$ form the nodes of N)
$A \subseteq \{T \times P\} \cup \{P \times T\}$ is a set of directed arcs.

A *marking* M of a Petri net is a mapping:

$M: P \rightarrow I$

where $I = \{0, 1, 2, \ldots \}$. M assigns *tokens* to each place in the net. Where convenient M can also be viewed as a vector whose ith component $(M)_i$ represents the number of tokens M assigns to p_i. A Petri net $N = (T,P,A)$ with marking M is a *Marked Petri Net* $C = (T,P,A,M)$.

Pictorially, places are represented by circles, transitions by bars, and tokens by small black dots. Figure 1 presents an example of a Petri net.

The above definition of a marked Petri net may be viewed as the *syntax* of a language for system representation. The *semantics* of the language (which give the behavior of the system) are specified by defining certain *simulation rules:* The set of *input places* of a transition t is given by $I(t) = \{p \,|\, (p,t) \varepsilon A\}$. The set of *output places* of a transition t is given by $O(t) = \{p \,|\,$

---

*This work was performed while the author was at the University of Texas at Austin.

$(t,p) \varepsilon A$ }. A transition t is said to be *enabled* in a Petri net $N = (T,P,A)$ with marking M if $M(p) > 0$ for all $p \varepsilon I(t)$. An enabled transition can *fire* by removing a token from each input place and putting a token in each output place. This results in a new marking M' where

$$M'(p) = \begin{cases} M(p) + 1 & \text{if } p \varepsilon O(t), p \notin I(t) \\ M(p) - 1 & \text{if } p \varepsilon I(t), p \notin O(t) \\ M(p) & \text{otherwise} \end{cases}$$

Tokens are indivisible—i.e., a token can be removed from a place by only one transition. Except for the above restrictions, firing of transitions proceeds in an asynchronous manner.

In Figure 1a, t_1 is the only transition that can fire. On completion, p_1 is empty and p_2 and p_3 each contain a token. At this stage t_2 and t_3 are both enabled and can fire *concurrently* since they do not share any input places. When these two firings are completed, p_4 and p_5 are the only places containing tokens. (See Figure 1b.) This situation represents a *conflict:* Both t_4 and t_5 are enabled, but firing of either disables the other. In such a case, the decision as to which one fires is completely arbitrary. The ability to represent both concurrency and conflict makes Petri nets very powerful.

Transitions in a Petri net could represent events in a real system. A marked net then represents the coordination or synchronization of these events. The movement of tokens clearly shows which conditions cause a transition to fire and which conditions come into being on the completion of firing. Moreover, the nets are not based on any concept of a central system state. The nets provide a natural representation of systems where control and state information is distributed. The use of finite state machines in such cases often leads to unmanageably large single states.

Additional concepts. A marking M' is *immediately reachable* from M if the firing of some t in M yields M'. M' is *reachable* from M if it is immediately reachable from M or is reachable from any marking which is immediately reachable from M or is M itself. The *reachability set* $R(M)$ of a marked Petri net (T,P,A,M) is the set of all markings reachable from M. A place in a marked Petri net (T, P, A, M) is k-*bounded* if and only if there exists a fixed k such that $M'(p) \leq k$ for all $M' \varepsilon R(M)$. A marked Petri net is k-bounded if for some fixed k each place is k-bounded. A place is safe if it is 1-bounded and a marked Petri net is safe if each place in it is safe. A transition t in a marked Petri net (T,P,A,M) is *live* if for each $M' \varepsilon R(M)$ there exists a marking reachable from M' in which t is enabled. A marked Petri net is live if each transition is live. Boundedness and liveness are important net properties.

Petri nets find their basis in a few simple rules yet can exhibit very complex behavior. This paper will introduce some analysis techniques later. Analysis in general requires some knowledge about the reachability set and can often be quite complex. As a result, restricted classes of Petri nets have been introduced and their properties studied. Two important subclasses are marked graphs and free choice nets. A *marked graph* is a Petri net in which each place is an input place of at most one transition and the output place of at most one transition. Marked graphs can represent concurrency but not conflict. A *free choice* Petri net is a Petri net where every place p is either the only input place of a transition or there is at most one transition which has p as an input place. Analysis of such substructures have provided relationships between structure and marking on the one hand and dynamic behavior (liveness, safeness, etc.) on the other.

For a more comprehensive study of Petri nets the reader is referred to Peterson.[1]

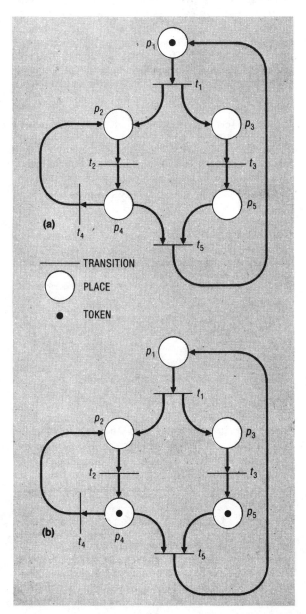

Figure 1. An example of a Petri net showing the ability to represent both concurrency (a) and conflict (b).

Interpretation

Petri nets, as defined in the previous section, are an abstract model. A net represents a system when a meaning or interpretation is assigned to various entities in the net—namely, the places, transitions, and tokens. Petri nets can be used in many different environments by using appropriate interpretations. Figure 2 represents a computer system where a processor is devoted to servicing two devices that are gathering data from the outside world. The cycle on the left represents device I_1 and the cycle on the right I_2. Device I_1 obtains new data (firing of t_1) only when the previous data has been transmitted (token in p_1). Completion of this activity is signaled by a token in p_2. Under these conditions, if a processor is available, it executes the service routine for I_1 and signals that the transmission is complete by placing a token in p_1. The whole cycle for I_1 can then repeat. The cycle for I_2 is quite similar. Notice that the net represents both concurrency and conflict.

Interpretations do not have to be computer related. A net could represent a chemical process where input places represent reacting chemicals, transitions represent reactions, output places the results of a reaction, and tokens the number of molecules of a given type. Figure 3 represents a particular reaction. Other novel interpretations will be discussed later.

Within the same field of application, different degrees of interpretation can be used. Thus, for a net representing parallel computation the interpretation may be minimal, such as simply assigning an operation name to each transition. A complete interpretation can also be used where for each transition, the exact transformation, the input and output memory locations, and the initial memory contents are specified. The degree of interpretation depends on the type of information to be obtained. Complete interpretation is required, for example, to establish that data integrity is maintained in a given parallel computation. For the derivation of general properties such as liveness and safeness, a partial interpretation should be used.

Based on the interpretations provided, Petri nets can be used to represent systems in a top-down fashion at various levels of abstraction and detail. For example, a single transition at a higher level such as "floating point add" may be expanded at a lower level into a series of transitions, "extract exponents," "compare exponents," "shift mantissa," etc. Petri nets have been used to model the CDC 6400 operating systems using this approach.[5] The approach is extremely useful in systems analysis where high-level descriptions which provide good perspective can be used, for example, to identify potential bottlenecks which may then be analyzed in detail using lower-level descriptions.

Operating systems and compilers

Petri nets can represent, in a straightforward manner, the flow of control in programs containing constructs such as IF-THEN-ELSE, DO-WHILE, GO-TO, and PARBEGIN-PAREND (see Figure 4). The net is uninterpreted: A token in place p indicates initiation of the IF-THEN-ELSE statement; it is not specified whether S_1 will execute or S_2. The representation is straightforward and will not be elaborated on further. Petri nets can clearly and explicitly represent the interaction between concurrent processes coordinated using P and V operations on semaphores.[6]* Such processes are frequently found in operating systems. As an example consider a system of two processes: (1) a producer (input process or device) that obtains data and places it in a bounded buffer (of size B); (2) a consumer (computing process or device) that removes data from the buffer and operates on it. The processes are asynchronous but must be prevented from accessing the buffer simultaneously; buffer overflow and underflow must also be prevented. Figure 5 gives the programs of the two processes coordinated using P and V primitives. (E represents the number of empty buffer positions, F the number of full buffer positions, and M is used for mutual exclusion.) Figure 6 gives a compact net for the system. As an example of system verifications, this net will be analyzed in two different ways.

---

*P and V operations are indivisible, operate only on special variables called semaphores, and can be logically defined as follows:

$P(S) \equiv \underline{\text{if}} \ S > 0 \ \underline{\text{then}} \ S := S-1 \ \underline{\text{else}} \ \text{wait}$
$V(S) \equiv S := S+1$

A waiting process can be scheduled at some later time when $S > 0$. At this time S is decremented and the process continues.

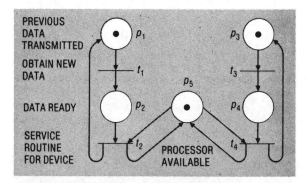

Figure 2. The processor in this system services two devices that are gathering data from the outside world.

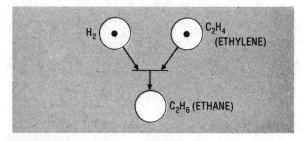

Figure 3. Petri nets can also be used to model chemical reactions.

Net invariants are used in the first approach. An invariant is a set of places, I, such that

$$\sum_{p \,\varepsilon\, I} M(p)$$

is a constant for each reachable marking M, and I does not have any proper subsets that are invariants. Let $N = (T,P,A)$ be a Petri net. The incidence matrix C of N is defined as $C = (C(t,p))$ where $t \,\varepsilon\, T$ and $p \,\varepsilon\, P$ such that

$$C(t,p) = \begin{cases} -1 \text{ if } (p,t) \,\varepsilon\, A,\ (t,p) \not\varepsilon\, A \\ +1 \text{ if } (t,p) \,\varepsilon\, A,\ (p,t) \not\varepsilon\, A \\ 0 \quad \text{otherwise} \end{cases}$$

Let y be a solution of the system of equations

$$C \cdot y = 0$$

where each element of y is either 0 or 1 and y cannot be obtained additively from other solutions. The set of places corresponding to the non-zero elements of y is an invariant.[7] For the net in Figure 6, the system of equations (above) is equivalent to:

$$\begin{array}{c} \\ t_1 \\ t_2 \\ t_3 \\ t_4 \end{array} \begin{array}{c} \begin{array}{ccccccc} p_1 & p_2 & p_3 & p_4 & p_5 & p_6 & p_7 \end{array} \\ \begin{bmatrix} -1 & 1 & 0 & 0 & 0 & 0 & 0 \\ 1 & -1 & 0 & 0 & -1 & 0 & 1 \\ 0 & 0 & -1 & 1 & 1 & 0 & -1 \\ 0 & 0 & 1 & -1 & 0 & 0 & 0 \end{bmatrix} \end{array} \begin{bmatrix} y_1 \\ y_2 \\ y_3 \\ y_4 \\ y_5 \\ y_6 \\ y_7 \end{bmatrix} = \begin{bmatrix} 0 \\ 0 \\ 0 \\ 0 \\ 0 \\ 0 \\ 0 \end{bmatrix}$$

The corresponding equations are:

$$\begin{aligned} -y_1 + y_2 &= 0 \\ y_1 - y_2 - y_5 + y_7 &= 0 \\ -y_3 + y_4 + y_5 - y_7 &= 0 \\ y_3 - y_4 &= 0 \end{aligned}$$

The solutions which cannot be additively obtained from other solutions are:

[1 1 0 0 0 0 0]
[0 0 0 0 1 0 1]
[0 0 1 1 0 0 0]
[0 0 0 0 0 1 0]

The invariants are:

$$\{p_1,p_2\}, \{p_3,p_4\}, \{p_5,p_7\}, \{p_6\}$$

Given the set of invariants, some properties about the dynamic behavior of the net can be deduced.[7,8] Assume M_0, the initial marking, is as shown in Figure 6 with $B > 0$. Let NP_i represent the total number of tokens in p_i.

Boundedness. Since each place is in some invariant and the net starts with a bounded marking, the net is bounded.

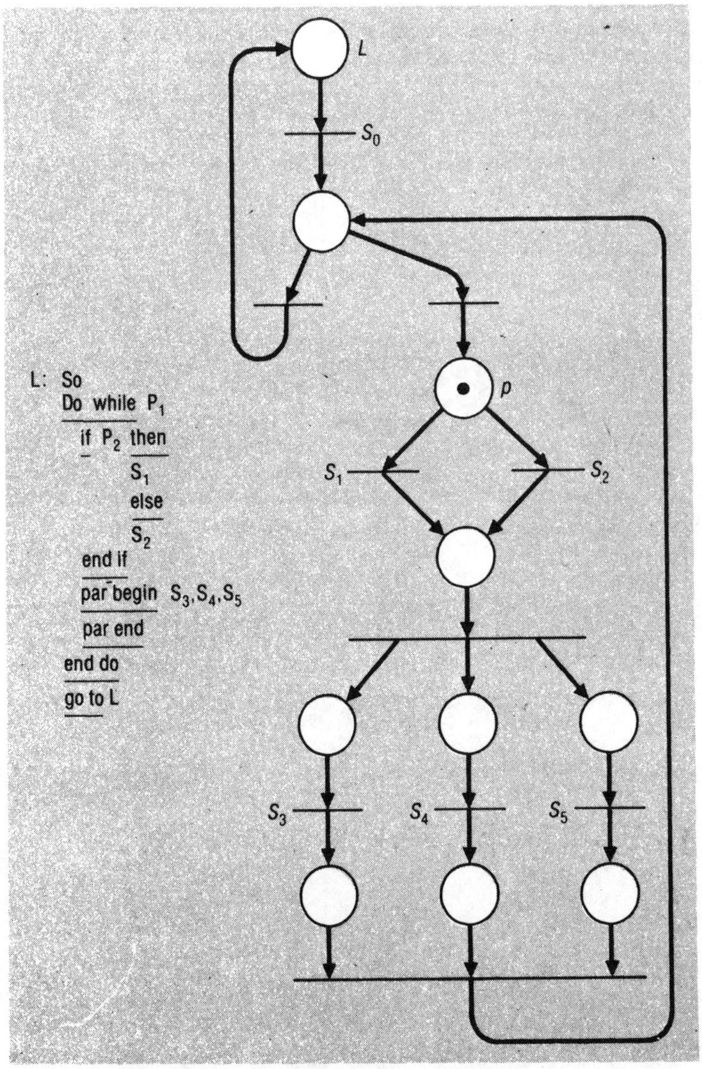

Figure 4. Example of a Petri net used to represent the flow of control in programs containing certain kinds of constructs.

```
Semaphore E ;   initial (E)  =  B
Semaphore F ;   initial (F)  =  0
Semaphore M ;   initial (M)  =  1
Producer:                        Consumer:
  PROD: "produce                 CONS: P(F)
        P(E)                           P(M)
        P(M)                           "remove from buffer"
        "place in buffer"              V(M)
        V(M)                           V(E)
        V(F)                           "consume"
        Go to PROD                     Go to CONS
```

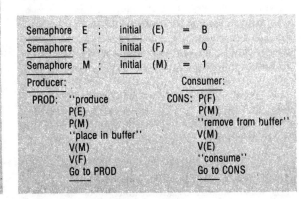

Figure 5. Programs of two concurrent processes coordinated using P and V primitives.

Conservativeness. Since the set of places can be partitioned into disjoint subsets each of which is an invariant, the net is conservative and the total number of tokens in the net remains constant.

Mutual exclusion. If an input or output place of a transition t is contained in an invariant I, t is said to be a transition of I. If two transitions are transitions of the same invariant and the initial marking is such that the sum of the tokens in the places of the invariant is 1, then the transitions are mutually exclusive and cannot fire simultaneously. Thus, the initial marking and invariant $\{p_6\}$ guarantee that t_2 and t_3 are mutually exclusive.

No buffer underflow. Buffer underflow is impossible since t_3 cannot fire if the buffer is empty ($NP_7 = 0$).

No buffer overflow. Since $\{p_5, p_7\}$ is an invariant, the initial marking guarantees that $NP_5 + NP_7$ is always B. Therefore, $NP_7 \leq B$ and buffer overflow cannot occur.

No deadlock. A deadlock will be said to have occurred if the net reaches a marking where no transition can fire. If the net is deadlocked, t_2 cannot fire. This implies that $NP_5 = 0$ or $NP_2 = 0$. In the former case, from the initial marking M_0 and invariant $\{p_5, p_7\}$ it can be concluded that $NP_7 > 0$; if $NP_3 = 0$ then $NP_4 = 1$ (from invariant $\{p_3, p_4\}$) and t_4 can fire, else $NP_3 = 1$ and t_3 can fire. In the latter case, $NP_2 = 0$, from invariant $\{p_1, p_2\}$ and M_0 it can be concluded that $NP_1 = 1$ and t_1 can fire. Thus, if t_2 cannot fire, then either t_4 or t_3 or t_1 can and the net can never be deadlocked.

Invariants are a useful aid to verifying the behavior of Petri nets. Rather than solving a system of equations, all the invariants can be systematically obtained by following certain rules during the construction of a net.[9]

A basic approach to analyzing Petri nets is to use the *reachability tree*. The nodes of a reachability tree of a marked Petri net represent reachable markings of the net. Let $N = (T, P, A, M_0)$ be a marked Petri net. Let ω be a special quantity such that $\omega \pm x = \omega, x < \omega$ and $\omega \leq \omega$ for every integer x. M is considered to be a vector below. The reachability tree for N is constructed as follows:

Let the initial marking be the root node and tag it "new."

While new markings exist *do*

Select a new marking M.

If M is identical to another node in the tree which is not new, then tag M to be "old" and stop processing M.

If no transition is enabled in M, tag M to be "terminal."

For every transition t enabled in M

(1) Obtain the marking M' which results from firing t in M.

(2) If there exists a path from the root to M containing a marking M'' such that $M' > M''$, then replace $(M')_i$ by ω wherever $(M')_i > (M'')_i$.

(3) Introduce M' as a node, draw an arc from M to M' labeled t, and tag M' to be "new."

End

It can be shown that the above procedure always terminates in a finite number of steps resulting in a finite tree. Also, a place p_i is unbounded if and only if the tree contains a marking with $(M)_i = \omega$. For bounded nets, each node is a reachable marking and the tree contains all reachable markings. The reachability tree for Figure 6 and $B = 1$ is given in Figure 7. Analysis of this tree yields some useful in-

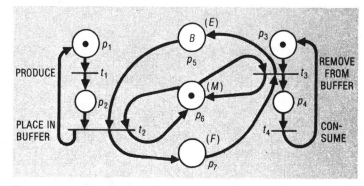

Figure 6. A compact net for the two-process system.

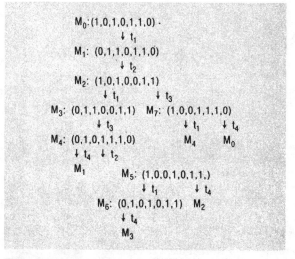

Figure 7. Reachability tree for Figure 6 and B = 1.

formation. It indicates exactly the set of reachable markings. Since no node contains an ω, the net is bounded. In fact, the net is safe since no reachable marking assigns more than one token to any place. The net is conservative: the sum of tokens in each marking is 4. There is no buffer overflow since no marking assigns more than 1 token to p_7. An analysis of the tree indicates that from any marking in the tree any transition can be enabled by an appropriate firing sequence. The net is thus live. With $B = 1$, t_2 and t_3 are never enabled simultaneously and they are therefore mutually exclusive. Detailed information about a system's behavior can be obtained from an analysis of its reachability tree. The major disadvantage is that the reachability tree can easily become complex, large, and unmanageable. This would be true for our example if B was chosen to be greater than 1.

Resource allocation is a major activity of operating systems. Consider a system of N processes each of which requires exclusive access to a subset of m resources $R_1, R_2, ...R_m$.[10] Processes are granted access without any consideration of priorities. If two processes use disjoint subsets of resources they may execute simultaneously. On completion, each process voluntarily releases its acquired resources. The reader is invited to construct a Petri net for a particular system where there are three processes X, Y, and Z competing for a card reader, printer, and tape. Process X requires the reader and printer, process Y the reader and tape, and process Z the printer and tape. This problem has also been referred to as the "Cigarette Smoker's Problem."[11]

Petri nets have been applied to compiler modeling to determine whether existing compilation algorithms are suitable for parallel processing.[12,13] An XPL/S compiler was first directly modeled (using an extension of Petri nets) resulting in a net that had little concurrency; the application of known methods for the automatic detection of parallelism yielded poor results. However, it was noticed that the compilation process could be modeled as a three-stage software pipeline. The net representation facilitated the subsequent restructuring of the compiler and its associated data structures. The three-stage pipeline was simulated assuming a three-processor system, and the results indicated that a speedup of 2:1 could be obtained over the sequential case. Different degrees of interpretation were used at different stages: initial analysis required relatively little interpretation; for detailed simulation, complete interpretation was used. In general, the Petri net-based model was found to be a useful tool.

Distributed data bases and communication protocols

The applicability of Petri nets to distributed data base systems is demonstrated below by modeling a duplicate file update protocol. A copy of the data base exists at each site in the distributed system. Requests to update the data base can originate at any site. The update protocol must guarantee that all copies are identical except for transient update times (mutual consistency).

Centralized and decentralized control schemes have been described by Ellis[14] and by Noe and Nutt using E-nets.[15] A data base controller is associated with each site, and it is the only entity allowed to update the local data base. An update request by a user is channeled through the local controller to a central supervisor. If some other update could be in progress the supervisor rejects the request; the user is informed of this and may request the update at a later time. Otherwise, permission is granted to the local controller and the update is performed by all controllers in the system.

A Petri net representing the local controller is given in Figure 8. A user request for update is signaled by a token in IR. The local controller is notified of acceptance (rejection) by the supervisor by a token in ACCEPT (DENY). "Broadcase update" causes a token to appear in place UPD in each controller. Each "transmit ACK" causes a token to appear in place ACK in the originating controller. A slight generalization has been introduced in this net: "Transmit 'done' to supervisor and user" occurs only when the controller is waiting for an acknowledgement and place ACK contains n-1 tokens (n-1

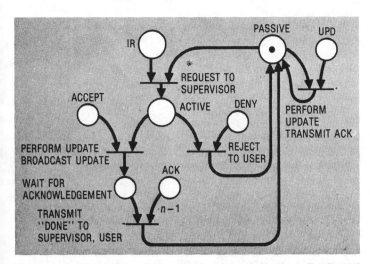

Figure 8. Petri net representing the local controller in a distributed system.

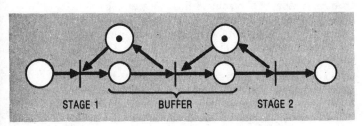

Figure 9. Simple two-stage pipeline with a buffer of size 2 between stages.

acknowledgements received from the n-1 remote controllers).

Petri net analysis tools can be used to verify the protocol. Invariants can establish that the sum of tokens in "ACTIVE," "PASSIVE," "Wait for acknowledgement" is always 1. Thus, even though there could be more than one token in IR, only one request at a time is serviced. Detailed analysis of the entire system can be used to establish that all data bases perform the same sequence of updates; mutual consistency is guaranteed. In the absence of failures the net can be shown to be live, thus ruling out the presence of "hang-up states." A supervisor crash could prevent tokens from ever being placed in ACCEPT or DENY, and this could soon halt all activity in the system. By assigning further interpretation (such as transition execution times) the net can be simulated to obtain throughput and response times. The applicability of other related models such as E-nets and RESQ for distributed data base modeling has been informally discussed by Chandy.[16]

Petri nets can be used to model and analyze communication protocols.[17,18] Analysis can be performed using a token machine (TM) which is a directed graph whose nodes represent unique markings. An arc (m_i, m_j) labeled t indicates that m_j is reachable from m_i by firing transition t. A well-behaved protocol (WBP) can be defined as one whose TM satisfies certain conditions: (1) the number of states is finite; (2) from any state there is a directed path to the initial state; (3) there is no directed loop containing only exceptional states; (4) etc. The following validation methodology is then applicable: Model the protocol as a Petri net. Obtain the TM and analyze it for WBP properties. If the protocol is not well-behaved, determine whether the model is incorrect or the protocol is inherently not well-behaved. If the former condition holds, adjust the Petri net and continue iteratively. This methodology was proposed by Merlin and applied to a telephone signaling protocol.[17] A similar approach was taken by Postel and Farber[18] to analyze computer communication protocols using the UCLA graph model.[19]

Computer hardware

Though computer hardware modeling has not been discussed explicitly, the application of Petri nets in this area should be evident. Figure 9 represents a simple two-stage pipeline with a buffer of size two between stages. Pipelines have frequently been used in high-performance computer systems such as the Texas Instruments Advanced Scientific Computer, the IBM 360/91, and the CRAY-1. A multiple functional unit computer resembling the CDC 6600 has also been modeled.[20] Shapiro and Saint[21] used Petri nets to generate efficient CDC 6600 programs: An algorithm originally expressed in a conventional high-level language is represented as a net to remove incidental sequencing constraints imposed by the language. The sequencing constraints required by the target hardware are then introduced into the net. All sequences of which this net is capable are realizable on the target hardware and perform the desired functional mapping.

Speed-independent circuit design

A circuit in which the presence of arbitrary delays in elements and connections has no effect upon circuit operation is called a speed-independent circuit. Such designs have been used where speed is critical since the circuit is not constrained by a clock but operates "as fast as possible." An example is the design of a processor for the synthesis of music.[22] A conventional processor cannot approach the necessary speed in this application, which requires a very large number of computations.

One of the major problems with speed-independent circuit design in the past was the lack of a formal model to represent, analyze, and synthesize such circuits. A suitable model should have the following characteristics: (1) the model should provide a clear and understandable representation of the circuit; (2) there should be a direct correlation between elements of the model and circuit realizations; and (3) the model should not serve only as a description scheme but should be accompanied by mathematical tools which allow analysis and synthesis. Clearly, Petri nets have a potential for being useful in this environment: places, tokens, and transitions could represent wires, signals, and actions.

For direct implementation as a speed-independent circuit, a Petri net should be live, safe, and persistent (in any reachable marking an enabled transition can be disabled only by being fired). A non-safe net would be difficult to implement since the circuit would have to keep track of the number of tokens present. Also, non-safeness and non-persistence can lead to critical races and will generally not result in speed-independence.[22] A live net guarantees that all parts of the corresponding circuit realization are utilized.

Speed-independent circuits were modeled by Patil and Dennis using Petri nets.[23] The hardware is represented as two structures: the control structure which is a Petri net and the data flow structure which consists of registers, operation units, decision elements, and data links. Firing of a transition in the control structure corresponds to the execution of an operation in the data flow structure. The sequence of steps is (1) remove tokens from input places; (2) send a ready signal to the operator over a control link; (3) execute the operation; (4) receive acknowledge signal from operator on the control link; (5) put tokens in output places.

In addition to the work on modeling, the implementation of Petri nets has also been studied.[22-25] The approach where circuit modules are directly substituted for places and transitions[24] usually leads to fairly complex circuit realizations. Misunas[22] uses an alternative approach to reduce complexity: a basic collection of Petri net functions and their corresponding speed-independent hardware implementation have ben developed. Complete circuits can be obtained either directly from the circuit modules or a Petri net can first be synthesized using the net modules and directly translated into the circuit realization. The latter has the advantage that the circuit can be analyzed and verified using the Petri net description. A transistor level implementation which has desirable fault properties has also been studied.[25]

Speed-independent circuits have some useful properties: they operate at maximum speed, are insensitive to delays in circuit elements and connections (these delays could vary based on environmental conditions such as humidity and temperature or with aging), and do not exhibit races and hazards. In addition, the circuits have advantages with respect to design verification, simulation, and fault detection.[25] Design verification of ordinary circuits which may exhibit races and hazards requires five-valued simulation: 0(true logical zero), 1(true logical one), X(unknown), U(rising signal), and D (falling signal). Such simulators use complex timing analysis and involved data structures. A simulator that uses only the first three values is much simpler but useful only for logic verification in conventional circuits. For speed-independent circuits, a simple three-valued simulator, which performs the simulation assuming a unit delay in each element and zero delay in the interconnections, is adequate for both logic and design verification. Also, speed-independent implementations of live, safe, and persistent Petri nets are inherently fail-secure[25]: most failures cause circuit operation to cease and fault propagation is prevented. This property is useful in highly secure computer systems and circuits that interface with expensive peripherals or sophisticated weapons systems.

Some major problems with speed-independent circuits are that testing is not fully understood and fault detection and isolation require the use of timing information. More significant, however, is the fact that the development of Petri net theory has provided impetus to research in the important area of speed-independent circuits.

Petri nets as a uniform design language

It should be clear from the preceding discussion that in addition to being a very suitable model for concurrent systems, Petri nets exhibit two other useful properties. First, the nets are equally suited for the representation of hardware and software systems. This is particularly useful given today's microprocessor technology wherein a large number of conventional hardware systems now contain an intimate mix of hardware and software. Secondly, Petri nets can be used at all levels including network, PMS, register-transfer, functional, and gate. The interpretation can be varied to suit the particular requirements of each level and the nets can be analyzed or simulated. The nets thus have an advantage over existing design languages which are generally not applicable across the entire spectrum. In conventional approaches different languages, simulators, and analytical tools have to be used at different levels.

The best example of the utility and feasibility of the Petri net approach as a design language is provided by the powerful LOGOS system,[26] which is Petri net-based. Another example of the use of Petri nets for design verification is described by Azema et al.[27]

In this context, it is important to note that Petri nets can be designed in a top-down manner so that certain properties are preserved.[28] Consider a net, N, which is live and safe with respect to an initial marking M_0. Let there be a place p (called an idle place) which has exactly one input transition and one output transition, which is the only place marked in M_0, and which is not marked in any other marking. Let M_0 be reachable from every marking in its reachability set. Under these conditions, N is said to be a well-behaved net. (Necessary and sufficient conditions for a net to be well-behaved can easily be given with respect to the corresponding token machine.) Let N_1 and N_2 be well-behaved nets. Let t be a transition in N_1, p the idle place in N_2, and t_1 and t_2 the corresponding input and output transitions of p. Substitute N_2 for t as follows: delete place p and arcs (t_1, p) and (p, t_2) from N_2. Delete t and its input and output arcs from N_1. Cause the input (output) places of t to become input (output) places of t_2 (t_1). The resulting net N can be shown to be well-behaved. The substitution preserves precedence and independence relationships between activities represented by transitions at the previous level of abstraction. Thus, important properties can be established during synthesis without the necessity of complex a posteriori verification.

A design methodology (for speed-independent systems) has been developed.[25] This methodology incorporates a common description language, a top-down synthesis procedure which preserves important properties, and a procedure for direct translation into hardware implementations. The work establishes the potential of Petri nets in this area.

Novel interpretations of nets

Since a live system is free from deadlock, the property of liveness has received much attention in net literature. On the other hand, a dead transition (not enabled in any reachable marking) can be viewed as an invariant assertion about the modeled system's behavior.[29] Using this interpretation, a net calculus, isomorphic to propositional calculus, has been derived by Thieler-Mevissen.[30] In this approach a transition is enabled only if each of its input places contains a token and all its output places are empty. A single transition with input and output places (an elementary net) can then represent an elementary disjunction of literals (a clause), as shown in Figure 10. An assignment of truth values to literals is called a case, and a correspondence between markings and cases can be established as follows: $M(p) = 1$ if and only if p is True. The transition t in Figure 10 is enabled when $M(a) = M(b) = M(c) = 1$ and $M(d) = M)(e) = 0$. The corresponding assignment of truth values is $a = b = c =$ True and $d = e =$ False, and the clause C is False. Any marking other than M causes transition t to be dead; the corresponding case makes clause C True.

The union of two elementary net $N_1 = (T_1, P_1, A_1)$ and $N_2 = (T_2, P_2, A_2)$ is defined as $(T_1 \cup T_2, P_1 \cup P_2, A_1 \cup A_2)$. If N_1 and N_2 represent clauses C_1 and C_2 respectively, then $N_1 \cup N_2$ represents $C_1 \wedge C_2$. The net in Figure 11 represents the proposition $(\bar{a} \vee \bar{b} \vee c) \wedge (\bar{c} \vee \bar{d} \vee e)$. Since every proposition can be expressed in conjunctive normal form, every proposition can be represented as a Petri net and every Petri net represents a proposition. When interpreted as a proposition a Petri net is called a fact net. If the proposition holds in a set of cases S, the net is a fact concerning S. To complete the calculus of fact nets and establish its isomorphism to propositional calculus, two net transformation rules can be shown to be consistent and complete[30]—i.e.,

(1) If a net is a fact concerning a set of cases S, then all derived nets N^* are also facts concerning S.
(2) If a proposition B follows (in propositional logic) from proposition A, then the net for B is derivable from the net for A by the rules.

Petri nets have also been used to represent mathematical knowledge.[31] Each token in a place has a definite structure and represents a well-defined construct of mathematics: a set, a group, etc. The transitions represent specifications which generate new constructs. Used in this manner the nets can be a valuable teaching and working tool for mathematics.

A subclass of Petri nets, free choice nets, have been found useful for modeling industrial production environments.[32] An "assembly line," for example, can be represented by a net similar to that in Figure 9. Necessary and sufficient conditions for a free choice net to be live and safe can be used to analyze production systems which are modeled using these nets. The nets have even been used in the representation of legal systems.[33]

The modeling power of Petri nets

It has been established that Petri nets cannot model certain priority situations.[34] Many extensions to Petri nets have been proposed, some of which yield no increase in power but do facilitate modeling.[35] A fundamental extension is to allow a special arc from a place p to a transition t such that t can fire only if p is empty.[8] This extension not only allows a more natural representation in many situations but also leads to Turing completeness.[35,36] Other extensions such as switches,[12] constraint sets,[24] and external firing priorities[37] are equivalent to zero-testing.[36] If nets are restricted to be bounded then all extensions are equivalent to finite state machines, and different extensions are merely better suited for different applications.

Conclusions

Petri nets can be used in many disciplines as a tool for representation, analysis, and synthesis. However, Petri nets are difficult to analyze. Bounded systems can be completely analyzed using marking trees, but these can rapidly become unmanageable. For general Petri nets the reachability problem, though decidable, has been shown to be exponentially time- and space-hard. However, if analysis problems are encountered, this reflects on the complexity of the system being modeled and should not be considered to be a disadvantage of the Petri nets. ∎

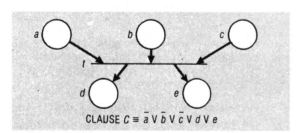

Figure 10. A Petri net representation of an elementary disjunction.

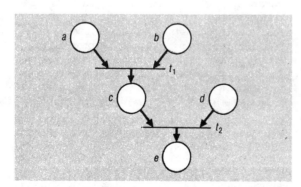

Figure 11. A Petri net representation of the proposition $(\bar{a} \vee \bar{b} \vee c) \wedge (\bar{c} \vee \bar{d} \vee e)$.

References

1. J. Peterson, "Petri Nets," *Computing Surveys*, Vol. 9, No. 3, Sept. 1977, pp. 223-252.
2. C. A. R. Hoare, "Parallel Programming, an Axiomatic Approach," Tech. Report C5-73-394, Stanford University, Oct. 1973.
3. R. M. Keller, "Formal Verification of Parallel Programs," *Comm. ACM*, Vol. 19, No. 7, July 1976, pp. 371-384.
4. S. Owicki and D. Gries, "Verifying Properties of Parallel Programs: An Axiomatic Approach," *Comm. ACM*, Vol. 19, No. 5, May 1976, pp. 279-285.
5. J. D. Noe, "A Petri Net Model of the CDC 6400," Report 71-04-03, Computer Science Dept., University of Washington, 1971; also in *Proc. ACM SIGOPS Workshop on System Performance Evaluation*, N. Y., 1971, pp. 362-378.

6. E. W. Dijkstra, "Cooperating Sequential Processes," *Programming Languages,* F. Genuys (ed.), Academic Press, N. Y., 1968, pp. 43-112.

7. K. Lautenbach and H. A. Schmid, "Use of Petri Nets for Proving Correctness of Concurrent Process Systems," *Proc. IFIP Congress 74,* North-Holland Pub. Co., Amsterdam, The Netherlands, 1974, pp. 184-191.

8. T. Agerwala and M. Flynn, "Comments on Capabilities, Limitations and 'Correctness' of Petri Nets," *Proc. First Ann. Symp. Computer Architecture,* ACM, N. Y., 1973, pp. 81-86.

9. T. Agerwala and Y. Choed-Amphai, "A Synthesis Rule for Concurrent Systems," *Proc. 15th Design Automation Conf.,* Las Vegas, June 1978, pp. 305-311.

10. T. Agerwala, "Some Extended Semaphore Primitives," *Acta Informatica,* Aug. 1977, pp. 201-220.

11. S. S. Patil, "Limitations and Capabilities of Dijskstra's Semaphore Primitives for Coordination Among Processes," Computation Structures Group Memo 57, Project MAI, MIT, Cambridge, Mass., Feb. 1971.

12. J. L. Baer, "Modeling for Parallel Computation: a Case Study," *Proc. 1973 Sagamore Computer Conf. Parallel Processing,* Springer-Verlag, N. Y., 1973, pp. 13-22.

13. J. L. Baer and C. S. Ellis, "Model, Design, and Evaluation of a Compiler for a Parallel Processing Environment," *IEEE Trans. Software Eng.,* Vol. SE-3, No. 6, Nov. 1977, pp. 394-405.

14. C. A. Ellis, "A Robust Algorithm for Updating Duplicate Data Bases," *Second Berkeley Workshop on Distributed Data Management,* 1977.

15. J. D. Noe and G. J. Nutt, "Macro E-Nets for Representation of Parallel Systems," *IEEE Trans. Computers,* Vol. C-22, No. 8, Aug. 1973, pp. 718-727.

16. K. M. Chandy, "Models of Distributed Systems," *Proc. 1977 Int'l Conf. on Very Large Data Bases,* pp. 105-120.

17. P. M. Merlin, "A Methodology for the Design and Implementation of Communication Protocols," *IEEE Trans. Communications,* Vol. COM-24, No. 6, June 1976, pp. 614-621.

18. J. B. Postel and D. Farber, "Graph Modeling of Computer Communication Protocols," *Proc. Fifth Texas Conf. on Computing Systems,* Univ. of Texas, Austin, Oct. 1976, pp. 66-77.

19. K. P. Gostelow, "Flow of Control, Resource Allocation and the Proper Termination of Programs," PhD diss., Computer Science Dept., UCLA, Dec. 1971.

20. J. B. Dennis, "Modular Asynchronous Control Structures for a High Performance Processor," *Record of the Prohect MAC Conf. Concurrent Systems and Parallel Computation,* ACM, N. Y., 1970, pp.55-80.

21. R. M. Shapiro and H. Saint, "A New Approach to Optimization of Sequencing Decisions," *Ann. Review of Automatic Programming,* Vol. 6, No. 5, 1970, pp. 257-288.

22. D. Misunas, "Petri Nets and Speed Independent Design," *Comm. ACM,* Vol. 16, No. 8, Aug. 1973, pp. 474-481.

23. S. S. Patil and J. B. Dennis, "The Description and Realization of Digital Systems," *Digest of Papers, COMPCON 72,* IEEE Computer Society, San Francisco, 1972, pp. 223-226.

24. S. S. Patil, "Coordination of Asynchronous Events," MAC TR-72, Project MAC, MIT, June 1970.

25. E. Pacas-Skewes, "A Design Methodology for Digital Systems Using Petri Nets," PhD diss., University of Texas at Austin, 1979.

26. C. W. Rose and M. Albarran, "Modeling and Design Description of Hierarchical Hardware/Software Systems," *Proc. 12th Design Automation Conf.,* Boston, Mass., June 1975, pp. 421-430.

27. P. Azema et al., "Petri Nets as a Common Tool for Design Verification and Hardware Simulation," *Proc. 13th Design Automation Conf.,* San Francisco, June 1976, pp. 109-116.

28. R. Valette and R. Prajoux, "A Model for Parallel Control Systems and Communication Systems," *Proc. 1976 Conf. on Information Sciences and Systems,* The Johns Hopkins University, Baltimore, Md., pp. 313-318.

29. C. A. Petri, "Interpretations of Net Theory," Interner Bericht 75-07, Gesellschaft fur Mathematik und Datenverabeitung, Bonn, W. Germany, July 1975.

30. G. Thieler-Mevissen, "The Petri Net Calculus of Predicate Logic," Interner Bericht ISF-76-09, Institut fur Informationssystemforschung, Gesellschaft fur Mathematic und Datenverabeitung, Birlinghoven, W. Germany, Dec. 1976.

31. H. J. Genrich, "The Petri Net Representation of Mathematical Knowledge," GMD-ISF Internal Report 75-06, Institut for Informationssystemforschung, Gesellschaft fur Mathematik und Datenverabeitung, Birlinghoven, W. Germany, 1975.

32. M. Hack, "Analysis of Production Schemata by Petri Nets," MAC TR-94, Project MAC, MIT, Feb. 1972.

33. J. A. Meldman and A. W. Holt, "Petri Nets and Legal Systems," *Jurimetrics J.,* Vol. 12, No. 2, Dec. 1971, pp. 65-75.

34. S. R. Kosaraju, "Limitations of Dijkstra's Semaphore Primitives and Petri Nets," Tech. Report 25, The Johns Hopkins University, May 1973; also in *Operating Systems Review,* Vol. 7, No. 4, Oct. 1973, pp. 122-126.

35. T. Agerwala, "Towards a Theory for the Analysis and Synthesis of Systems Exhibiting Concurrency," PhD diss., The Johns Hopkins University, 1975.

36. T. Agerwala and M. Flynn, "On the Completeness of Representation Schemes for Concurrent Systems," *Conf. on Petri Nets and Related Methods,* MIT, 1976.

37. M. Hack, "Petri Net Languages," Computation Structures Group Memo 124, Project MAC, MIT, Cambridge, Mass., June 1975.

Tilak Agerwala is a research staff member at the IBM T. J. Watson Research Center in Yorktown Heights, New York. From April 1975 to January 1979 he was an assistant professor in the Departments of Electrical Engineering and Computer Sciences and a faculty affiliate of the Electronics Research Center at the University of Texas at Austin. During the summer of 1977 he was a member of the technical staff at Bell Telephone Laboratories, Murray Hill, New Jersey. His research interests are in the areas of computer architecture, distributed and parallel computer systems, operating systems, microprocessors, and Petri nets.

A member of ACM, IEEE, and Sigma Xi, Agerwala received a B. Tech. degree in 1971 from the Indian Institute of Technology at Kanpur, and his PhD in 1975 from The Johns Hopkins University.

APPLICATIONS OF PETRI NET BASED MODELS IN THE MODELLING AND ANALYSIS OF FLEXIBLE MANUFACTURING SYSTEMS

Manjunath Kamath
Department of Industrial Engineering
University of Wisconsin - Madison
Madison, WI 53706, USA

N. Viswanadham
School of Automation
Indian Institute of Science
Bangalore 560012, India

Abstract

Petri nets have evolved into a powerful tool for analyzing asynchronous concurrent systems. In this paper we review the applications of Petri nets to FMSs. We present applications of Petri net models in the design of simulators for flexible manufacturing systems (FMSs) and in the verification (by investigating the presence/absence of deadlock) of the logic used to design the hardware/software for controllers used in FMSs.

1. Introduction

Petri nets [2,5,8] have evolved into an elegant and powerful graphical modelling tool for asynchronous concurrent systems. The wide ranging application areas of Petri nets [5,18] include multiprocessing, distributed processing, a host of hardware and software systems, and recently [15,19,23,24,25,26,27,29,30] modern manufacturing systems. There are many advantages in modelling a system using Petri nets. Examples of these advantages are: (1) They describe the modelled system graphically and hence enable an easy visualization of complex systems, (2) Petri nets can model a system hierarchically; systems can be represented in a top-down fashion at various levels of abstraction and detail, (3) A systematic and complete *qualitative* analysis of the system is possible by well-developed Petri net analysis techniques [8], (4) The existence of well-formulated schemes for Petri net synthesis facilitates system design and synthesis [4,26], and (5) Performance evaluation of systems is possible using timed Petri nets [1,3,6,7,10,11,15].

The use of *flexible manufacturing systems* (FMSs) has led to a marked increase in productivity and reduction in inventory costs of industries that manufacture a set of related parts at low to medium volumes. An FMS is basically a computer-controlled configuration of semi-independent workstations coupled by a palletized material handling system. Modelling the complex interactions in an FMS is a formidable task. Queueing network models and simulation techniques have been quite useful in the planning and control of FMS operations.[32,34] The perturbation analysis technique (a hybrid analytical-simulation methodology) has been used for the optimization of FMSs.[33] The recent applications of Petri net theory in the study of FMSs[15,19,23,24,25,26,27,29,30] look quite promising.

In this paper we present two important applications of Petri nets to FMSs. First, we explain the use of *colored Petri net* [9] models in developing simulation programs for FMSs. Secondly, we use recent results[24,26] in Petri nets to verify the logic used in the construction of controllers used in FMSs by checking for the presence/absence of deadlock. We also review the applications of Petri net models in the study of FMSs. A brief discussion of the possible impact of the recent advances in timed Petri nets[10,11,12] on the performance evaluation of FMSs is then presented.

The rest of the paper is organized as follows. In § 2, we present a brief introduction to Petri nets. §3 contains a review of the applications of Petri nets to FMSs. In the next two sections, we investigate the use of Petri net models in FMS simulation and in deadlock detection. Finally, in § 6, we present a brief discussion on the performance evaluation of FMSs using timed Petri nets.

2. An Introduction to Petri nets

Petri nets, also known as *place-transition nets (PTNs)*, are a formal graph model for the description and analysis of systems that exhibit both asynchronous and concurrent properties.[2,5,8] PTNs serve as a means for a natural representation of the flow of information and control in such systems. An expository introduction to PTNs may be found in literature.[2,5]

2.1 Place-transition nets (PTNs)

PTNs are special bipartite directed graphs. The standard PTN model is defined by a set of *places*, a set of *transitions*, and a set of *directed arcs* which connect places to transitions or vice vesa. Pictorially, places are represented by circles and transitions by bars. Places may contain *tokens* (drawn as dots). A PTN with tokens is a *marked* PTN. A *marking* of a marked PTN is a vector, the elements of which give the distribution of tokens in the places of the net. A marking represents a state of the system being modelled. Generally, places represent conditions and transitions represent events. If a place represents a condition then the presence (absence) of tokens connotes the truth (falsity) of the condition. A place is an *input* (*output*) place of a transition if an arc exists from the place (transition) to the transition (place).

The dynamic behavior of a system is modelled as follows. The occurrence of an event is represented by the *firing* of the corresponding transition. The movement of tokens in the net resulting from the firing of one or more transitions represents a change in the system state. The following are the firing rules for marked PTNs.

1) A transition is *enabled* when each of its input places contains at least one token.
2) A transition can fire only if it is enabled.
3) When a transition fires:
   - a token is removed from each of its input places, and
   - a token is deposited into each of its output places.

A marking M' is *immediately reachable* from a marking M if we can fire some enabled transition in marking M resulting in the marking M''. A marking M' is *reachable* from M if it is immediately reachable from M or is reachable from any marking which is immediately reachable from M. The *reachability tree (set)* of a marked PTN is the set of all markings that are reachable from a given *initial marking*. The initial marking and the reachability tree represent the initial state and the state space respectively, of the system being modelled. Fig. 1 models a single server queueing system with a perpetual source of jobs. The reachability set of the PTN model consists of just two markings, M and M'.

2.1.1 Extended PTNs:
One of the first extensions is to permit the use of *multiple* arcs[2] so that a place may contribute or receive more than one token from the firing of a transition. PTNs that allow multiple arcs have been called *generalized Petri nets*.[2] Pictorially, k arcs between a place and a transition or vice versa are represented by an arc labelled with the *multiplicity factor* k (if $k=1$ the arc is unlabelled). Although the inclusion of multiple arcs may increase the convenience of use, it does not increase the fundamental modelling power of PTNs. Hence, the distinction between generalized Petri nets and PTNs is often ignored.

To facilitate the modelling of priority systems with PTNs, *inhibitor* arcs[2,8] are used. An inhibitor arc from a place to a transition in a marked PTN is drawn with a small circle rather than an arrowhead at the transition. The firing rule for the transition is changed such that the transition is *disabled* if there is at least one token present in the corresponding inhibiting input place. There is no movement of tokens along inhibitor arcs. The addition of inhibitor arcs represents a fundamental extension of the concept of PTNs.

2.1.2 Structural properties of PTNs: A PTN is said to be *pure* or *self-loop free* if and only if no place is an input place and an output place for the same transition. A pure PTN can be completely defined by its *incidence matrix*. For a PTN with n places, $\{p_1, p_2, ..., p_n\}$ and m transitions, $\{t_1, t_2, ..., t_m\}$, the incidence matrix is defined by an $n \times m$ matrix, C, whose ij th element, c_{ij} is equal to 0 if no arc exists between place p_i and transition t_j; is equal to $-k$ if an arc with multiplicity factor k exists from p_i to t_j, and is equal to $+k$ if an arc with multiplicity factor k exists from t_j to p_i.

A place in a marked PTN is said to be *k-bounded* if and only if there exists a positive integer k such that there are no more than k tokens in the place at the same time for all markings contained in the reachability set of the PTN. If $k=1$ then we say that the place is *safe*. If all places in a marked PTN are k-bounded (safe), the PTN itself is said to be k-bounded (safe). The PTN model given in Fig. 1 is safe. *Boundedness* is an important property of PTNs. If we model an FMS by a PTN and if, for example, some places in the net represent buffers, then since physically realizable buffers can hold only a finite number of jobs, the PTN model must be bounded.

A transition in a marked PTN is said to be *live* if and only if for all markings contained in the reachability set of the PTN, there exists a sequence of firings that results in a marking in which the transition is fireable. A marked PTN is said to be live if all transitions in it are live. The PTN model given in Fig. 1 is live. *Liveness* is tied to the concepts of *deadlock*. If a transition in a PTN model is not live, then the occurrence of the event corresponding to the transition could lead to a possible deadlock in the system. If a PTN model of an FMS is live, we can conclude that the operation of the FMS is free of deadlock.

Consider a pure PTN and let C be its incidence matrix. A ($1 \times n$) row-vector X, where n is the number of places in the PTN, is said to be a *place invariant* (*P-invariant*) of the PTN if and only if $XC = 0$. It can be easily shown that for all reachable markings of the PTN, a weighted sum of tokens is constant, the weights being given by any P-invariant. A knowledge of the P-invariants of a PTN is useful in investigating certain properties of the PTN such as boundedness and liveness.[5,8]

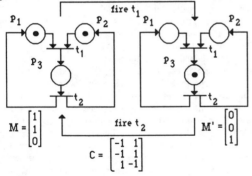

p_1: a job is ready; p_2: the server is idle; p_3: a job is being processed
t_1: job processing is started; t_2: job processing is completed

FIGURE 1. Modelling of a single server queueing system

2.1.3 Timed PTNs: To facilitate performance evaluation of a system, the PTN model is extended to include the notion of time.[1,3,6,7,10,11] Two schools of thought exist as regards timed PTNs: most researchers associate time with transitions[1,6,7,10,11], whereas a few others associate time with places[3,12] in order to preserve the classical PTN notion of transitions as instantaneous events, with the presence of a token at a place indicating the activity of a process. A good review of timed nets can be found in literature.[12] Early work in the area of timed PTNs[1,3,6] was confined to deterministic times and a restricted class of PTNs. Recently, timed PTNs have assumed a stochastic nature and also include generalized PTNs.[10,11,12]

2.2 Colored Petri Nets (CPNs)

PTNs elegantly capture the details of a concurrent system. However, if the modelled system is large and fairly complex, the number of places and transitions in the PTN model becomes very large and the representation will be unwieldy. Also, if the system under consideration has several identical processes, it will be necessary to have several identical subnets in the corresponding PTN model. This necessitated the development of more powerful net types to describe complex systems in a compact and manageable way. A *Colored Petri Net (CPN)* [9] is in this respect, a significant improvement.

A CPN is a generalization of a PTN in which a color is associated with each token, indicating its identity. Also, a set of colors is associated with each place and each transition. The set of colors associated with a place indicates the colors of tokens that can be present at the place. A transition can fire with respect to each of its colors. When a transition is fired, tokens are removed and added at its input and output places respectively, as in a marked PTN. In addition, a functional dependency is specified between the color of the transition firing and the colors of the involved tokens. The color attached to a token involved in a transition firing may be changed and it often represents a complex information unit.

Graphically, places are represented by ellipses and transitions by rectangular boxes and tokens are not explicitly represented. The color set associated with a place or a transition is indicated on one of its sides. The input and output arcs of a transition are labelled with functions that decide the colors of the tokens removed from its input places and the colors of the tokens deposited in its output places when the transition fires with respect to a particular color. The structure of a CPN is described by its incidence matrix whose elements are functions.

A formal definition of CPNs can be found in literature[9] along with examples where CPN models are quite useful. CPNs have the same modelling power as generalized PTNs. The main advantage of CPNs over PTNs is the possibility of getting a compact representation of a large and complex system. The definitions of liveness, boundedness, P-invariants and other properties in CPNs are similar to that in PTNs. However, the computation of invariants in CPNs can be quite involved since the elements of the incidence matrix are functions and not integers.

We conclude this section with an example (Fig. 2) which illustrates certain important aspects, such as, definitions of color sets and color functions, in the construction of a CPN model of a system. Fig. 2 gives the CPN model of the FCFS (first come, first serve) queueing discipline with multiple job classes. This example is similar to the one discussed in literature.[29] The queueing discipline in the input and output buffers of a workstation in an FMS is generally FCFS.

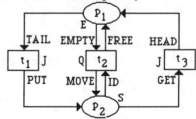

FIGURE 2. CPN model of the FCFS Queueing Discipline with multiple job classes.

Consider a buffer consisting of n consecutive places. There are p job classes, $\{j_1, j_2, ..., j_p\}$. Jobs belonging to different job classes can be simultaneously present in the buffer. The arrival and the departure processes of jobs are not modelled. Interpretation of the places and transitions and the definitions of the color sets and the color functions of the CPN model in Fig. 2 are given below.

<u>Places:</u> p_1: buffer places free; p_2: buffer places full;
<u>Transitions:</u> t_1: add a job to the buffer ; t_2: move a job one place ahead in the queue and free the place which was occupied by it
t_3: remove a job from the buffer

Color sets:
E : { e_k | k = 1, 2, ..., n}; a token of color e_k indicates that the kth place in the buffer is empty.
J : {j_i | i = 1, 2, ..., p}; a set of job classes.
Q : { <j_i, e_k> | i = 1, 2, ..., p; k = 2, ..., n}; a token of color <j_i,e_k> represents a situation in which a job belonging to class j_i is occupying the kth place in the buffer.
S : Q U {<j_i, e_1> | i = 1, 2, ..., p}; similar to the set Q; in addition to the elements of the set Q, the set S also contains the details of the job occupying the first place in the buffer.

Color functions:
∀ j_i ∈ J,   HEAD(j_i) = e_1; TAIL(j_i) = e_n ;
         PUT(j_i) = <j_i, e_n>; GET(j_i) = <j_i, e_1>
∀ <j_i, e_k> ∈ Q, ID(<j_i, e_k>) = <j_i, e_k>; FREE(<j_i, e_k>) = e_k ;
         EMPTY(<j_i, e_k>) = e_{k-1}; MOVE(<j_i, e_k>) = <j_i,e_{k-1}>

The initial marking of the CPN corresponding to an empty buffer is, no tokens in the place p_2 and n tokens, $e_1, e_2, ..., e_n$, in the place p_1. Transition t_1 is enabled with respect to color j_i, i = 1, 2, ..., p, if the place p_1 contains a token with the color TAIL(j_i) = e_n (last place in the buffer empty). When t_1 fires, a token with the color e_n is removed from the place p_1 and a token with the color PUT(j_i) = <j_i, e_n>, is deposited in the place p_2, to indicate that the last place (nth) now contains a job of class j_i. The movement of jobs in the buffer is captured by the transition t_2. Assuming that the place p_2 has a token with the color <j_i, e_k> ∈ Q, transition t_2 will be enabled with respect to the color <j_i, e_k>, if the place p_1 contains a token with the color EMPTY(<j_i, e_k>) = e_{k-1} (empty place in front). When t_2 fires, a token with the color FREE(<j_i, e_k>) = e_k is deposited in the place p_1, to indicate that the kth place in the buffer is empty and a token with the color MOVE(<j_i, e_k>) = <j_i, e_{k-1}> is deposited in the place p_2 to indicate that the (k-1)th place is full. Firing of transition t_3 removes a job from the first place in the buffer.

3. Applications of Petri Nets to FMSs

In this section, we present a brief review of the literature available on the use of Petri nets in modelling, analyzing, evaluating, and controlling FMSs.

Suitability of Petri nets for FMS modelling is investigated in literature.[21,23] Petri nets are described as an ideal tool to model the complex interactions among different processes in FMSs. A preliminary investigation of the use of timed PTNs in the study of real-time control and performance evaluation of FMSs is carried out by Dubois et al.[23] Also, manufacturing systems that can be modelled by a restricted class of PTNs (safe, decision-free, and strongly connected PTNs)[23] are analyzed and a simple performance evaluation algorithm, based on a special algebraic structure, is presented for the restricted class of PTN models.

Petri nets have been used in the design, verification, and construction of control systems for FMSs.[19,27,30] Martinez et al[27] discuss the design of a coordinator of a hierarchical control system for a flexible workshop. A CPN model of the coordinator is first developed. The CPN model is then described using a special language which facilitates validation of the CPN model. Validation is through the invariant analysis technique.[9] Alla et al [30] develop a technique for the calculation of the P-invariants of a subclass of CPNs. A CPN model of the flexible workshop[27] is presented and validation of the model is carried out using the invariant analysis technique developed. The design of programmable logic controllers (PLCs) based on Petri net models of control flow is presented in Valette et al.[20] The control of FMSs requires a complex interconnection of sophisticated PLCs. Vallete et al [20] show Petri nets to be useful for a formal specification and validation of the control procedures. Petri net models allow a direct implementation of the specifications, so that introduction of bugs in the preliminary design phase of PLCs can be avoided; the controller is loaded with the net description and the controller functions by emulating the net execution. On similar lines, Thuriot et al [20], consider the design of the bottom level (machine level control performed by PLCs) of a hierarchical, real-time control system of FMSs. In particular, CPNs are presented as a tool for expressing the inter-task cooperation and a method is described for direct coding of the Petri net based specification. The possibility of validation prior to implementation is again stressed.

The power of Petri nets lies in their ability to prove certain *qualitative* properties of systems (§ 2.1.2). Invariant analysis is often used to prove the properties of Petri net models. It is in this respect that the recent results[24,26] for fast and easy computation of the invariants assume much significance. The main result developed makes it possible to determine the invariants of the union of two Petri nets when the invariants of the individual Petri nets are known. A PTN model of a given FMS is synthesized by constructing the union of simpler PTN models that correspond to certain primitive operations into which the given FMS structure can be decomposed. At each stage of the synthesis, the properties of the PTN model are evaluated; analysis and synthesis proceed in parallel. More on this methodology will be discussed in §5. Recently[31], the result has been extended to include the calculation of invariants of CPNs.

Development of simulation programs for complex real world systems such as FMSs can be a formidable task. The difficulty arises primarily due to the concurrent nature of FMS operations and the complex interactions among various processes in an FMS. The utility of timed Petri net models of FMSs in designing FMS simulators is studied.[25,29] We present this application of Petri net models in §4.

Performance evaluation of FMSs using Petri nets models has not gained much ground primarily due to the fact that performance analysis tecniques [3,7], were available only for a very restricted class of PTNs. Recently[10,11,12], performance evaluation methods have been developed based on the theory of stochastic processes (mainly Markov processes) for generalized PTNs. Bruno et al[15] analyze the problem of tool handling in an FMS, where the tools are transported between machines and a common tool storage area by means of a limited number of conveyers, using *generalized stochastic Petri nets* [11] (GSPNs). §6 contains a brief discussion on the performance analysis of FMSs using the new techniques[12] developed for timed nets.

4. Use of Petri net models in designing FMS simulators

This section is concerned with the applications of timed Petri nets in the modelling and simulation of FMSs.[25,29] Timed Petri net models have been used to design simulators for dataflow computers.[14] The translation of Petri net models into simulation programs is a straightforward procedure.[14,23,29] To illustrate this we first develop a CPN model of a special class of FMSs.

We consider a fixed-route FMS with p job classes (part types), $J_1, J_2, ..., J_p$. The system has m machines, $M_1, M_2, ..., M_m$. We denote the load/unload station by M. A part belonging to a job class J_c, c = 1, 2, ..., p, goes through O_c operations excluding the load and unload operations. There is a limited central storage space and a limited number of pallets of each type. In the CPN model, the number of buffer spaces in the central storage and the number of pallets of each type available are specified by the initial markings of the places, p_1 and p_5 (Fig. 3), respectively. We make the following assumptions regarding the operation of the FMS to keep the CPN model simple, as the main idea is to illustrate the construction of the model and its use in developing simulation programs for the FMS. A part, after being processed by a machine is transported to the common storage, provided a place is free (transition t_2). From the common storage it is either transported to another machine for further processing (transition t_4) or to the load/unload station for unloading (transition t_3), depending on whether or not all the required operations are complete. We assume that transporters are always available. We also assume that there is a perpetual source of new parts and a new part is let into the system as soon as a pallet of that type becomes available. Also, transport of parts, loading and unloading operations, and tool and workpiece setup operations are not modelled explicitly. They can be easily modelled by replacing certain transitions in the CPN model (Fig. 3) by *sub*-CPN models.

The following notation is adopted to facilitate the definition of the color sets and color functions. R(i,k), k = 1, 2, ..., O_i, denotes the machine for the kth operation of a part belonging to the job class J_i. This information is got from the routing table. R(i,O_i+1) is always M.

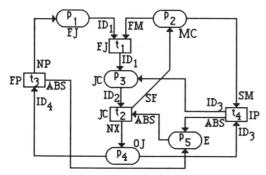

p_1: new parts waiting on pallets; p_2: machines free;

p_3: parts being machined;

p_4: parts waiting in the central storage;

p_5: space available in the central storage area;

t_1: machining of new parts is started;

t_2: machining completed and parts transported to the central storage;

t_3: finished parts unloaded and new parts loaded on pallets;

t_4: parts sent to machines for their next operations; machining is started;

FIGURE 3. CPN model of a fixed-route FMS.

Color sets:

MC : $\{M_j \mid j = 1, 2, ..., m\}$; set of machines;

FJ : $\{<J_i,1,R(i,1)> \mid i = 1, 2, ..., p\}$; a set of 3-tuples giving complete information about fresh jobs waiting to enter the system. The three components of these tuples respectively stand for job class (part type),

JC : $\{<J_i,1,R(i,1)>, <J_i,2,R(i,2)>, ..., <J_i,O_i,R(i,O_i)> \mid i = 1, 2, ..., p\}$; a set of 3-tuples that give details of parts undergoing machining. The three components are part type, current operation number, and the identity of the machine currently processing the part.

IP : JC - FJ; a set of tuples describing in-process parts.

FP: $\{<J_i,O_i+1,M> \mid i = 1, 2, ..., p\}$; a set describing finished products.

OJ : IP U FP; a set of tuples describing the status (details of the next operations) of parts waiting in the central storage.

E : a set containing only one element e; if a place has E as its associated color set, then tokens in that place are treated as ordinary tokens.

Color functions :

ID_1, ID_2, ID_3, and ID_4 are *identity functions* (color of token removed/deposited is the color of the transition firing) on the sets FJ, JC, IP, and FP respectively.

∀ $<J_i,1,R(i,1)>$ ∈ FJ,   FM($<J_i,1,R(i,1)>$) = R(i,1);

∀ $<J_i,k,R(i,k)>$ ∈ JC,   NX($<J_i,k,R(i,k)>$) = $<J_i,k+1,R(i,k+1)>$

SF($<J_i,k,R(i,k)>$) = R(i,k)

ABS($<J_i,k,R(i,k)>$) = e

∀ $<J_i,k,R(i,k)>$ ∈ IP,   SM($<J_i,k,R(i,k)>$) = R(i,k) and

ABS($<J_i,k,R(i,k)>$) = e

∀ $<J_i,O_i+1,M>$ ∈ FP,   NP($<J_i,O_i+1,M>$) = $<J_i,1,R(i,1)>$

ABS($<J_i,O_i+1,M>$) = e

Places in the CPN model either represent the availability of resources (e.g. place p_5) or model activities (e.g. place p_3). Times are associated with places that model activities. For example, processing (machining) times are associated with the place p_3. Uncertainty in activity times is handled by giving the time durations associated with places a stochastic characterization. Tokens arriving at a place with a non-zero place time are delayed by certain time units before they become available. Transitions can be enabled only by available tokens.

To facilitate the development of a simulation program, dummy places and transitions are introduced into the CPN model, if necessary, so that the termination of every activity in the system is explicitly modelled. In other words, every place associated with a non-zero place time should be the only input place of the transition that corresponds to the event denoting the termination of the activity associated with the place. For example, the transition, t_2, in the CPN model in fig. 3, can be replaced by a sub-CPN consisting of transition t_2', place p_x, and transition t_2'', as shown in Fig. 4. The resulting CPN model is a more detailed version of the original CPN model.

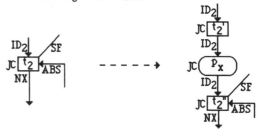

FIGURE 4. Refining transition t_2

The CPN model is used as a framework from which the simulation program is constructed. In literature[14,25,29] Pascal is the language chosen. The color sets aid in the construction of the data structures. Each transition in the CPN model gives rise to a Pascal procedure. A procedure simulates the firing of the corresponding transition by updating the values of certain variables, which can be imagined to correspond to the movement of tokens. The color functions indicate the actions to be performed by the various procedures. A few methods for resolving conflicts in transition firing are, 1) assigning priorities to transitions; for example, transition t_4 is fired before transition t_2 (Fig. 3), so that the in-process parts get preference over the new parts and 2) randomly choosing from a set of enabled transitions that are in conflict. It is worth noting that the firing of a transition, in a CPN model, with respect to two or more of its colors could be in conflict, as the transition represents two or more transitions in the equivalent PTN model. Again, conflicts are resolved by methods similar to those mentioned above. Henceforth, the firing (enabling) of a transition always stands for the firing (enabling) of a transition with respect to one of its colors. The main steps involved in the construction of a simulation program are as follows:

i) An *event list* containing the future occurrences of transitions, that model terminations of activities, is maintained. Procedures needed for inserting and deleting elements (events) of this event list are similar to those used in any event driven simulation program. The simulation clock is updated when a transition taken from the event list is fired.

ii) Simulation principles:

a) Initialize the values of the variables to reflect the initial marking of the CPN model. Initialize clock.

b) Establish a *transition list* containing transitions enabled by the current marking. All conflicts are resolved while creating this list. Firing of transitions taken from the transition list would have no effect on the simulation clock. For each place with zero place time, a list of transitions for which this place is an input place is maintained, in order to simplify the task of creating the transition list. Only the output transitions of the marked places (with zero place time) need be checked.

c) If the transition list is empty (no fireable transitions left at the current value of the clock) go to step c else get the next transition from the transition list and fire it by invoking the corresponding procedure and go to step b. Whenever tokens are deposited in places with non-zero place times, the output transitions (events representing the termination of the activities modelled by the places) are inserted into the event list.

c) Get the next transition to be fired from the event list, update the simulation clock, and go to step b.

iii) Procedures to collect statistics such as the following ones:

- mean number of firings (number of procedure calls) of a CPN transition, with respect to each of its colors,
- maximum and mean value of the number of tokens of a particular color simultaneously present in a place, and
- maximum and mean value of the time spent by tokens of a particular color, in a place with a non-zero place time.

Work of a similar nature has been reported.[20,29]

The entire procedure, starting from the construction of the CPN

model to simulating the net model can be automated. The task of developing software for automating the process of modelling and simulation of FMSs using CPNs is simplified due to the following:
1) Neither the color sets nor the color functions depend on any specific details (number of machines or part types or the routing table) of FMSs.
2) The color sets can be constructed easily; most of the sets contain tuples that can be derived from the routing table.
3) In CPN models of FMSs and FMS like networks, there exists a set of very frequently used functions. For example, identity functions (ID_1, ID_2, ID_3, and ID_4 in our model), counting functions (ABS), and projection functions (FM, SM, SF). Functions such as NX (Fig. 3), can be realized by using the routing table.
4) Sub-CPN models of the various activities in FMSs, like loading and unloading operations, transportation from machines to storage and vice versa, and processing parts, can be created and stored in a database. A CPN model can be obtained as a union of such sub-CPN models based on the specific structure and operating rules of the FMS being modelled. This step will involve the design of a good user interface.
Similar software tools for modelling and simulation, based on *function nets* (Petri net based structures), have been reported in literature.[16]

5. Deadlock detection using Petri nets

In FMSs there are several situations where processes compete for shared resources. Examples of these situations are, machines in a flexible manufacturing cell (FMC) sharing a finite capacity storage and a finite capacity material handling system being shared by parts in an FMS. In real-time systems where resources are shared by several processes a major concern is deadlock. Petri net models have been widely used[9,15,16,25,28,30] for the investigation of deadlock in real-time applications. If a Petri net model of the system being investigated is live (§2.1.2) then the system is free of deadlock. A non-live Petri net model indicates that deadlock could potentially exist in the system. P-invariants (§2.1.2) of a Petri net together with the initial marking of the net facilitate investigation of liveness in most cases by applying elementary intuition.[4,9,25] Linear algebraic methods[5,8] are generally used to compute P-invariants. Recently, a more efficient method[24,26] based on union of Petri nets has been developed to compute the invariants. The basic idea is the synthesis of a PTN model for the system, starting from primitive PTN models. Invariants of the PTN model can be evaluated at each stage of the synthesis procedure. This alleviates the need for constructing large PTN models since, often times our main aim is to evaluate certain system properties. The results proved by Narahari et al [24,26] enable us to derive invariants of the union of PTNs starting from the invariants of the individual PTNs.

The system we consider is a small FMC with two machines, M_1 and M_2. The cell also consists of a buffer with finite capacity to hold in-process parts. Parts belonging to two job classes visit the cell.

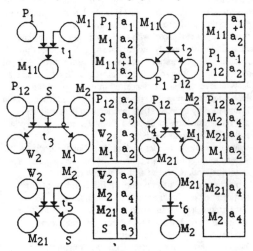

FIGURE 5. PTN models and P-invariants for type 1 part operations

Parts of type 1 have their first operation on M_1 and the second operation on M_2. Type 2 parts first visit M_2 and then M_1. We assume that there is a perpetual supply of new parts and that parts waiting in the storage can be accessed in a random manner. To build a controller for this cell we have to define certain operating rules. A possible set of rules is :
i) If a machine is free then a part is loaded (in-process or new) and processing is started; all parts, in-process or new, have the same priority.
ii) If a part finds the machine for the second operation unavailable then it is transported to the storage space, provided, a place is available.

The interpretation of places in the PTN model is given below. It may be noted that i,j can take values 1 and 2.
M_i : machine M_i is free
P_i : a part of type i is waiting to enter the cell
M_{ij} : machine M_i is processing a part of type j
P_{ij} : a part on machine M_i is waiting for machine M_j
W_i : a part in the storage is waiting for machine M_i
S : space is available in the central storage

Fig. 5 gives the PTN models (and thier P-invariants) that make up the overall processing of a part of type 1 only. The PTN models and the invariants corresponding to part type 2 can be similarly obtained. In fact , the PTN models for part type 2 can be obtained from fig. 5, by relabeling the places and transitions. The transitions in the PTN models for part type 2, corresponding to $t_1, t_2, ..., t_6$, are $t_1', t_2', ..., t_6'$ respectively. The P-invariants of the overall PTN model are obtained using the union of Petri nets theorem in Narahari et al. [26] The P-invariants are listed below. $a_1, a_2, ..., a_5$, are all non-negative.

M_1	M_2	P_1	P_2	M_{11}	M_{12}	M_{21}	M_{22}	P_{12}	P_{21}	W_1	W_2	S
a_2	a_4	a_1	a_5	a_1+a_2	a_2	a_4	a_4+a_5	a_2	a_4	a_3	a_3	a_3

An initial marking of the PTN model is one token in each of the following places M_1, M_2, P_1, and , P_2, to indicate that both machines are free and that a part of each type is waiting to enter the cell. Also, n (equal to buffer capacity) tokens in the place S and no tokens present at the other places in the initial marking. Let M denote any reachable marking. Consider some of the invariants. In the following all a_i's except those indicated are zero.

$a_2 \neq 0; M(M_1) + M(M_{11}) + M(M_{12}) + M(P_{12}) = 1$ (I1)
$a_3 \neq 0; M(W_1) + M(W_2) + M(S) = n$ (I2)
$a_4 \neq 0; M(M_2) + M(M_{21}) + M(M_{22}) + M(P_{21}) = 1$ (I3)

We consider a reachable marking in which there are no buffer spaces available (M(S) = 0), a part is waiting on machine M_1 to go to machine M_2 ($M(P_{12}) = 1$) and vice versa ($M(P_{21}) = 1$). The non-availability of storage space disables the following transitions : t_3 and t_3' (in PTN models of part type 2). As $M(P_{12}) = 1$, we get from invariant I1, $M(M_1) = 0$ which disables t_1, t_4', and t_5'; $M(M_{11}) = 0$ which disables t_2, and $M(M_{12}) = 0$ which disables t_6'. $M(p_{21}) = 1$ and invariant I3 together imply that transitions t_1', t_4, t_5, t_2', and t_6 are disabled. In the marking considered, no transition is fireable and hence, the PTN model is not live. We conclude that system can reach a deadlocked state. The technique we have demonstrated provides a systematic way of investigating systems for the presence/absence of deadlock.

6. Performance evaluation using timed Petri nets

Performance evaluation (PE) of FMSs using Petri nets has not received much attention primarily due to the fact that PE techniques were available till recently only for a restricted class of PTNs.[3,6,7] Recently, stochastic Petri nets (SPNs) have received much attention from researchers in the field of computer sciences.[10,11,13] The SPNs studied can handle generalized PTNs; the firing times have either an exponential or a phase-type distribution. SPN models with exponentially distributed transition rates are isomorphic to Markov models.[10] PE techniques for SPNs are based on Markovian analysis methods. SPN models have basically two advantages over Markovian models: i) simplicity in specification and 2) enable automatic generation of state space which is not a trivial task in the case of large distributed systems. Software packages have been developed for the solution of SPNs.[13] The only application of SPNs to FMSs has been the study of tool handling by Bruno et al[15]. An analysis tool combining queueing networks and generalized SPNs (GSPNs) has been reported in literature.[17] It will be interesting to see whether or not the combined power of GSPNs and queueing networks can handle the problems encountered while modelling FMSs by queueing networks, such as blocking due to the presence of finite buffers, simultaneous resource possession (a part needs a pallet, tools, and a machine simultaneously), and shared resources (tools and material handling systems).

7. Conclusions

Petri nets, a tool widely used to analyze asynchronous cocurrent systems, have been shown to be useful in modelling FMSs. The various applications of Petri net models in the study of FMSs have been reviewed. Two important application areas, design of FMS simulators and verification of the control logic through investigation of deadlock, have been discussed. The compactness of the CPNs makes them an ideal tool to model FMSs. We have shown how CPN models of FMSs can be directly transformed into simulation software. Petri nets are a major tool for proving qualitative properties of systems. A recent technique for fast and efficient computation of invariants based on union of Petri nets has been used to calculate the invariants of a PTN model of the operations in a small manufacturing cell. The invariants have been used to prove the presence of a deadlocked state in the operation of the cell. The possibility of developing powerful performance analysis techniques for FMSs based on the recent advances in SPNs indicate that Petri nets could become a complete tool for qualitatively and quantitatively analyzing FMSs.

8. Acknowledgements

We wish to thank Y. Narahari for his active involvement at various stages of this work. We also wish to acknowledge the useful suggestions offered by J. L. Sanders during the preparation of this paper.

References

[1] C. Ramchandani, "Analysis of asynchronous concurrent systems by timed Petri nets," Ph.D. dissertation, MIT, Cambridge, Project Mac Rep. MAC-TR-120, 1974.

[2] J. L. Peterson, "Petri nets," *ACM Computing Surveys*, vol. 9, no. 3, pp. 223-252, September 1977.

[3] J. Sifakis, "Petri nets for performance evaluation," in *Measuring, Modelling, and Evaluating Computer Systems, Proc. 3rd Int. Symp. IFIP Working Group 7.3*, H. Beilner and E. Gelenbe, Eds., North-Holland Publishing Company, pp. 75-93, 1977.

[4] T. Agerwala and Y. Choed-Amphai, "A synthesis rule for concurrent systems," *Proc. 15th Design Auto. Conf.*, pp. 305-311, June 1978.

[5] T. Agerwala, "Putting Petri nets to work,"*IEEE Computer*, vol. 12, no. 12, pp. 85-94, December 1979.

[6] C. V. Ramamoorthy and G.S. Ho, "Performance evaluation of asynchronous concurrent systems by Petri nets," *IEEE Trans. on Software Engg*, vol. SE-6, no. 5, pp. 440-449, September 1980.

[7] W. M. Zuberek, "Timed Petri nets and preliminary performance evaluation," in *Proc. 7th Annu. Symp. Comput. Architecture*, pp. 88-96, 1980.

[8] J. L. Peterson, *Petri Net Theory and the Modelling of Systems*, Prentice-Hall Inc., Englewood Cliffs, NJ, 1981.

[9] K. Jensen, "Colored Petri nets and the invariant method," *Theoretical Computer Science*, vol 14., pp. 317-336, 1981.

[10] M. K. Molloy, "Performance analysis using stochastic Petri nets," *IEEE Trans. on Comp.*, vol. C-31, no. 9, pp. 913-917, Sept. 1982.

[11] M. Ajmone-Marsan, G. Balbo, and G. Conte, "A Class of generalized stochastic Petri nets for the performance evaluation of multiprocessor systems," *ACM Trans. on Comp. Syst.*, vol. 2, no. 2, pp. 93-122, May 1984.

[12] *Proceedings of the International Workshop on Timed Petri Nets*, Torino, Italy, IEEE Computer Society Press, 1985.

References 13 through 17 are from the above mentioned proceedings.

[13] J. B. Dugan, A. Bobbio, G. Ciardo, and K. Trivedi, "The design of a unified package for the solution of stochastic Petri net models," pp. 6-13.

[14] T. Smigelski, T. Murata, and M. Sowa, "A timed Petri net model and simulation of a dataflow computer," pp. 56-63.

[15] G. Bruno and P. Biglia, "Performance evaluation and validation of tool handling in flexible manufacturing systems using Petri nets," pp. 64-71.

[16] M. Leszak and H. P. Godbersen, "DEAMON: A tool for performance-availability evaluation of distributed systems based on function nets," pp. 152-161.

[17] G. Balbo, S. C. Bruell, and S. Ghanta, "Combining queueing network and generalized stochastic Petri net models for the analysis of a software blocking phenomenon," pp. 208-225.

[18] M. Diaz, "Modelling and analysis of communication and cooperation protocols using Petri net based models," *Computer Networks*, vol. 6, no. 6, pp. 419-441, December 1982.

[19] R. Valette, M. Courvoisier, and D. Mayeux, "Control of flexible production systems and Petri nets," *3rd European Workshop on Appl. and Theory of Petri Nets*, Varenna, Italy, September 1982.

[20] R. Valette, M. Courvoisier, JM. Bigou, J. Albukerque, "A Petri net based programmable logic controller," in *Computer Applications in Production and Engineering (CAPE' 83)*, E. A. Warman (editor), North-Holland Publishing Company, pp. 103-116, 1983.

[21] Y. Narahari and M. Kamath, "Modelling and analysis of flexible manufacturing systems by Petri net based models," Award winning paper, All India annual student paper contest for 1983 conducted by IEEE India Council.

[22] E. Thuriot, R. Valette, and M. Courvoisier, "Implementation of a centralized synchronozation concept for production systems," *Proc. Real-Time Syst. Symp.*, Arlington, VA, pp. 163-171, Dec. 1983.

[23] D. Dubois and K. E. Stecke, "Using Petri nets to represent production processes," *Proc. 22nd IEEE Conf. Decision and Control*, pp. 1062-1067, December 1983.

[24] Y. Narahari, "Modelling and analysis of flexible manufacturing systems using Petri nets," Masters thesis (unpublished), School of Automation, I. I. Sc., Bangalore, India, June 1984.

[25] M. Kamath, "Simulation and analysis of flexible manufacturing systems using Petri net models," Masters thesis (unpublished), School of Automation, I. I. Sc., Bangalore, India, July 1984.

[26] Y. Narahari and N. Viswanadham, "Analysis and synthesis of flexible manufacturing systems using Petri nets," *Proc. First ORSA/TIMS Conf. on Flexible Manufacturing Systems*, University of Michigan, Ann Arbor, pp. 346-358, August 1984.

[27] J. Martinez and M. Silva, "A language for the description of concurrent systems modelled by colored Petri nets: application to the control of flexible manufacturing systems," *1984 IEEE Workshop on Lang. for Automation*, New Orleans, pp. 72-77, Nov. 1984.

[28] A. Datta, D. Harms, and S. Ghosh, "Deadlock avoidance in real-time resource sharing distributed systems : An approach using Petri nets," *Proc. Real-Time Syst. Symp.*, Austin, TX, pp. 49-61, Dec. 1984.

[29] P. Alanche, K. Benakour, F. Dolle, P. Gillet, P. Rodrigues, and R. Valette, "PSI: A Petri net based simulator for flexible manufacturing systems," in *Advances in Petri Nets 1984*, G.Goos and J. Hartmanis (editors), LNCS, vol. 188, Springer Verlag, pp. 1-14, 1985.

[30] H. Alla, P. Ladet, J. Martinez, and M. Silva-Suarez, "Modelling and validation of complex systems by colored Petri nets: application to a flexible manufacturing system," in *Advances in Petri Nets 1984*, G.Goos and J. Hartmanis (editors), LNCS, vol. 188, Springer Verlag, pp. 15-31, 1985.

[31] Y. Narahari and N. Viswanadham, "On the invariants of coloured Petri nets," *6th European workshop on the applications and theory of Petri Nets*, Helsinki, Norway, June 1985.

[32] J. A. Buzacott and J. G. Shantikumar, "Models for understanding flexible manufacturing systems," *AIIE Transactions*, vol. 12, no. 4, pp. 339-350, December 1980.

[33] R. Suri, "New techniques for modelling and control of flexible manufacturing systems," *Proc. IFAC 8th Triennial World Congress*, Kyoto, Japan, vol. 14, pp. 175-181, 1981.

[34] J. A. Buzacott and D. D. W. Yao, "Flexible manufacturing systems: A review of models," Working paper no. 82-007, Dept. of Industrial Engineering, University of Toronto, Canada, March 1982.

Reprinted from *IEEE Transactions on Computers*, Volume C-31, Number 9, September 1982, pages 913-917. Copyright © 1982 by The Institute of Electrical and Electronics Engineers, Inc. All rights reserved.

Performance Analysis Using Stochastic Petri Nets

MICHAEL K. MOLLOY

Abstract—An isomorphism between the behavior of Petri nets with exponentially distributed transition rates and Markov processes is presented. In particular, k-bounded Petri nets are isomorphic to finite Markov processes and can be solved by standard techniques if k is not too large. As a practical example, we solve for the steady state average message delay and throughput on a communication link when the alternating bit protocol is used for error recovery.

Index Terms—Alternating bit protocol, Markovian systems, performance analysis, Petri nets, Petri nets with time.

Introduction

In the past, the modeling of systems has taken several different forms depending upon the viewpoint of the system analyst. The basic differences between each form are found in the mathematics in which each class of models is constructed. First, the models used by the software analysts are based on formal logic such as the predicate calculus. These models form a foundation for proving sequential programs correct. Second, the models now being presented for hardware systems involve precedence concepts typically found in

Manuscript received April 9, 1981; revised December 4, 1981 and March 3, 1982. This work was supported by the Office of Naval Research under Grant N00014-79-C-0866. and was performed while the author was with the Department of Computer Science, University of California, Los Angeles, CA 90032.

The author is with the Department of Computer Science, School of Natural Sciences, University of Texas, Austin, TX 78712.

graph models. These models present a formal specification of the design of asynchronous operations. Third, the models of computer systems used to determine average performance are stochastic in nature and are founded on the theories of probability and statistics. These models provide a framework for the study of time-varying systems.

Each type of model concentrates on different aspects of a system. However, the models still have many common features and constraints. An ideal model for such systems would take the problem definitions and create a structure to which each of the analyses could be applied separately. This would facilitate consistent analysis of a single system by several analysts with various viewpoints and requirements.

Attempts to merge some aspects of these different modeling techniques have been made. In performance modeling, load dependent servers which are blocked during certain conditions can incorporate some precedence relations such as the need to acquire multiple resources. In hardware systems, by adding a time value to the steps in precedence graphs, some performance characteristics can be modeled [21], [19].

While both of these attempts extend the available information in the models, the limits to their application are quickly reached. The queueing model requires such a complex analysis that only a two resource blocking model has been attempted. The fixed time graph model cannot properly model processes at a higher level of abstraction where time becomes a random variable and the system behaves stochastically. Models at any level of abstraction are possible if we extend graph models and deal with time as a random variable. There are many graph models in the literature [16] but here we select Petri nets as the graph model of interest.

STOCHASTIC PETRI NETS

Petri net models have been studied extensively over the last decade [18], [5], [6], [16]. These models have been applied to many types of systems [13], [15], [17], [8], [1], [19], [4], [2]. Most of the analysis has been made on the set of possible states a system may occupy, called the reachability set. These analyses have omitted any study of timing considerations. Recently some attempts have been made to include timing as a specification [19], [21]. The work of Merlin and Farber [14] discussed timed Petri nets where a time threshold and maximum delay were assigned to a transition. This was done to allow the incorporation of timeouts into a protocol model. Zuberek [21] used a fixed time for each transition in his work to model the performance of a computer system at the register level. In another case, probability was introduced to allow a random switching of flow through the graph [20]. The work of Shapiro limits the model to discrete time and a maximum of one token in each place.

In this paper, Petri nets are extended by assigning an exponentially distributed firing rate to each transition for continuous time systems or a geometrically distributed firing rate to each transition for discrete time systems. These new stochastic Petri nets (SPN's) are isomorphic to homogeneous Markov processes [12]. The proof leads to a simple, albeit practically limited, procedure to generate the reachability set of the underlying Petri net. This construction technique was programmed in Fortran in order to examine its practicality. The author found that by numbering the places in the SPN in decreasing order of boundedness, moderately sized (approximately 50 places) SPN's can easily be dealt with.

The state of a given Petri net is the ordered n-tuple of the number of tokens in the n places of the Petri net, called the marking. For any Petri net with a finite number of places, the markings make up a countable, possibly infinite set. Since there is some probability that a Markov model changes state in a period of time, there is more structure to the Markov model than the Petri net model. Therefore, the extension of Petri nets to stochastic Petri nets allows the extraction of additional information on behavior. The countability of the markings and the memoryless property of exponential distributions are the key factors in allowing an isomorphism to be constructed between the stochastic Petri net model and the Markov model.

The Petri net itself allowed the modeling of sequential and concurrent actions including phenomena such as contention and synchronization. These models could then be analyzed for such properties as deadlocks, boundedness and self regulation to solve problems involving state verification. The SPN model also allows the calculation of the steady state probabilities of marking occurrences. This opens up an area of analysis for such performance measures as average delay and average throughput. All of this analysis is performed using the equivalent Markov model.

The Markov model is complex (multidimensional or from a one-dimensional view it is not a nearest system) and may be much larger than the SPN model. This increase in size and complexity would hamper the direct Markov analysis of such systems by making the generation of the Markov model difficult for the analyst. By modeling the system in a manner which retains much of the character of the system, the analyst is more apt to generate a proper and complete model of the system. Modeling with SPN's does retain the machine like character of a system.

Since this modeling still requires the generation of the reachability set, the same analysis performed on normal Petri nets can be done before calculating performance measures.

THE MODEL

Recall the formal description of Petri nets [1] where the model PN has places P, transitions T, input and output arcs A and an initial marking M.

$$PN \triangleq (P, T, A, M)$$
$$P = \{p_1, p_2, \cdots, p_n\}$$
$$T = \{t_1, t_2, \cdots, t_m\}$$
$$A \subset \{P \times T\} \cup \{T \times P\}$$
$$M = \{\mu_1, \mu_2, \cdots, \mu_n\}. \quad (1)$$

The marking may be viewed as a mapping from the set of places P to the natural numbers N:

$$M:P \rightarrow N \quad \text{where } M(p_i) = \mu_i \text{ for } i = 1, 2, \cdots, n.$$

Define for a Petri net PN the set function I of input places for a transition t:

$$I(t) \triangleq \{p \mid (p, t) \subset A\}. \quad (2)$$

Define for a Petri net PN the set function O of output places for a transition t:

$$O(t) \triangleq \{p \mid (t, p) \subset A\}. \quad (3)$$

As is common in practice, a Petri net can be drawn using circles to represent places and bars to represent transitions. Tokens are represented as dots inside the circles (places). A five place, five transition Petri net could look like the one shown in Fig. 1.

The continuous-time stochastic Petri net SPN $\triangleq (P, T, A, M, \lambda)$ is extended from the Petri net PN $\triangleq (P, T, A, M)$ by adding the set of average, possibly marking dependent, transition rates $\lambda = \{\lambda_1, \lambda_2, \cdots, \lambda_m\}$ for the exponentially distributed transition firing times.

Definition: Two stochastic transition systems are isomorphic iff the following hold:
1) $\exists$ a 1-1 and onto mapping F between the state spaces of the two systems.
2) $\exists$ a transition in one system $S_u \rightarrow S_v \leftrightarrow \exists$ a transition in the other system $F(S_u) \rightarrow F(S_v)$.
3) The probability $P[S_u \rightarrow S_v, \tau] = P[F(S_u) \rightarrow F(S_v), \tau]$ for $\forall$ state.

Note: This definition of an isomorphism does not consider the transition sequence in the SPN, only the marking sequence. This implies that if $\exists$ more than one transition which maps a marking M_u into a marking M_v, these transitions are indistinguishable and act as a single transition.

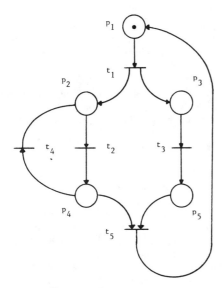

Fig. 1. A simple Petri net.

Theorem: Any finite place, finite transition, marked stochastic Petri net is isomorphic to a one-dimensional discrete space Markov process [12].

EXAMPLE—A SMALL SPN

Consider the five place, five transition continuous-time stochastic Petri net shown in Fig. 1. That SPN displays sequential operation (t_5, t_1), parallel operation (t_2, t_3), forking (t_1), joining (t_5), and contention (t_4, t_5). Include the expected firing rates of $\lambda_1 = 2$, $\lambda_3 = 1$, $\lambda_4 = 3$, and $\lambda_5 = 2$. Assuming an initial marking of one token in place p_1 and no tokens in the remaining places, then solving for the reachability set, we find five states:

```
        p1  p2  p3  p4  p5
M1       1   0   0   0   0
M2       0   1   1   0   0
M3       0   0   1   1   0
M4       0   1   0   0   1
M5       0   0   0   1   1
```

Solving the ergodic Markov chain shown in Fig. 2 we obtain the following steady-state marking probabilities:

$P[M_1] = 0.1163$

$P[M_2] = 0.1860$

$P[M_3] = 0.0465$

$P[M_4] = 0.5349$

$P[M_5] = 0.1163$.

Using the marking probabilities and the number of tokens in each place in a particular marking we can deduce the steady state probability of there being μ_i tokens in each place for any marking. This is precisely the token probability density function.

$P[\mu_1 = 1] = 0.8837 \qquad P[\mu_1 = 1] = 0.1163$

$P[\mu_2 = 0] = 0.2791 \qquad P[\mu_2 = 1] = 0.7209$

$P[\mu_3 = 0] = 0.7675 \qquad P[\mu_3 = 1] = 0.2325$

$P[\mu_4 = 0] = 0.8372 \qquad P[\mu_4 = 1] = 0.1628$

$P[\mu_5 = 0] = 0.3488 \qquad P[\mu_5 = 1] = 0.6512$.

If we assume a different initial marking then we obtain a different Markov chain. The more tokens we have in the system the larger the Markov chain. If we start with two tokens in place p_1 we would find 14 states in the reachability set. Similarly, if we started with 3 tokens in place p_1 we would find 30 states in the reachability set.

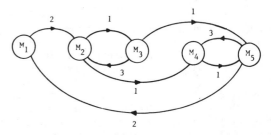

Fig. 2. The Markov equivalent.

The analysis of the delay in an arbitrary system yields to one of two techniques. First, if the equivalent Markov chain has trapping states, one can calculate the expected number of steps until one of the trapping states is reached [7], [11], [21]. Second, if the chain is ergodic then the delay for components of the net may be calculated using Little's result and flow balance.

Consider the example of the small SPN in Fig. 1. Let us apply Little's result to the subsystem made up of places p_2, p_3, p_4, and p_5 and transitions t_2, t_3, t_4, and t_5. Since t_1 is only enabled when p_1 contains a token the utility of transition t_1 is 11.63 percent. Using the average service time of 0.5 units for t_1, the average rate at which tokens flow through p_1 is 0.2326 tokens per unit time.

By the conservation of flow, we know that the number of tokens entering the subsystem per unit time λ is 0.4652 which is double the flow through t_1 since t_1 is a fork transition. Since t_5 is a join transition the flow in and flow out will be balanced. Since the subsystem conserves tokens (it neither destroys nor creates tokens) we may apply Little's result $\overline{N} = \lambda T$.

The average number of tokens in the subsystem is the sum of the average number of tokens in each place in the subsystem:

$$\overline{N} = \overline{\mu}_2 + \overline{\mu}_3 + \overline{\mu}_4 + \overline{\mu}_5 = 1.7674$$

Therefore, on the average, the time until a token returns to p_1 after leaving is 3.8 units of time.

For the special case above, where the object is to determine the mean recurrence time of a marking M in a continuous time SPN, another method can be applied which does not need token conservation. By making the observation that the time until an enabled transition fires is the time spent in the marking, the mean time $\overline{\tau}_M$ is simply $(\sum_i \lambda_i)^{-1}$ for each enabled transition i, in that marking. After solving for the steady state probabilities, $P[M_i]$, the mean time to return to the marking M_i is simply $\overline{\tau}_{M_i}/P[M_i] - \overline{\tau}_{M_i}$. Using the same values for the SPN as before, we find the same results $\overline{\tau}_{M_i}/P[M_i] - \overline{\tau}_{M_i} = 3.8$.

EXAMPLE—ALTERNATING BIT PROTOCOL

A practical application of Petri nets was made by Merlin in various papers [14], [15]. He used Petri nets to study a communication protocol, the alternating bit protocol [3]. This protocol establishes a means by which duplicate messages (due to acknowledgment loss) and their acknowledgements may be distinguished from the original messages. To use the protocol, each time a new message is sent, a check bit in the header is complemented. Therefore, a sequence of messages with no duplicates would have alternating values of this check bit. By viewing a sequence of check bits from six successive messages, such as 101101, we can determine that the fourth message received was a duplicate of the third message.

The stochastic Petri net which completely models the alternating bit protocol is shown in Fig. 3. This differs from the timed Petri net described by Merlin [14] in several ways. First, there is a logical difference between the handling of states when a timeout occurs. Merlin had the timeout create a new message at the destination. The stochastic Petri net in Fig. 3 returns to the send message state so that the delay incurred by the message due to retransmission is properly modeled. Second, the timed Petri net described by Merlin omitted

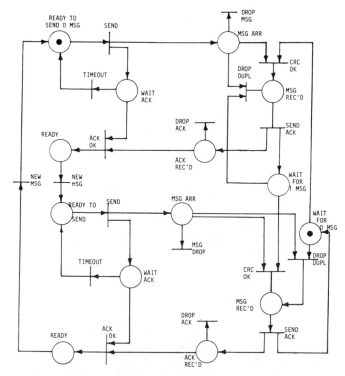

Fig. 3. SPN for the alternating bit protocol.

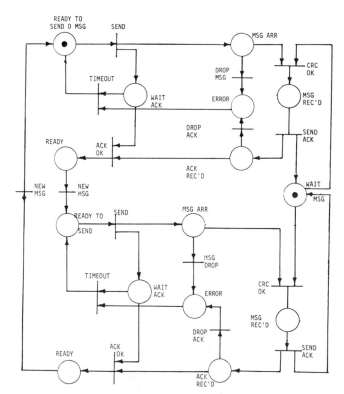

Fig. 4. Restricted SPN for the alternating bit protocol.

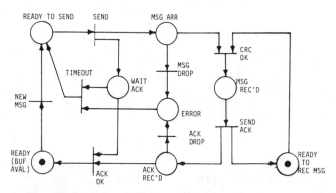

Fig. 5. Reduced SPN for a communication protocol.

the possibility of acknowledgment loss. Acknowledgment loss is included as a possibility in the stochastic Petri net in Fig. 3.

The reachability set of the SPN in Fig. 3, unfortunately, is infinite. This is correct for the protocol since acknowledgment of a message could take an arbitrarily long time. However, it does complicate the analysis of the SPN. Several approaches to the problem are possible. First, we may obtain approximate results if we truncate the infinite Markov chain [9]. (This is equivalent to arbitrarily bounding the message arrival place.) Second, by changing the model to make certain states impossible, the token creation cycle can be broken. This would be accurate if the protocol was being applied to a single channel where reception within a bounded time was guaranteed.

The latter approach was taken by Merlin. He assigned ranges of time to the firing of each transition. Under his rules when two or more transitions are enabled in a marking, a transition cannot fire if its minimum firing time is greater than any another transition's maximum firing time. In the SPN such restrictions are not possible. Such conditioning destroys the Markovian property in the SPN, so the change must be made in the structure of the model. It seems best to make the logical restrictions on the state space appear in the structure of the graph, rather than in some secondary constraint. The SPN shown in Fig. 4 models the alternating bit protocol where timeouts happen only when messages are really dropped.

Now that the model is limited to a finite reachability set the analysis of its performance is easier. We assume that the transmission, CRC calculation and probability of error are the same for both 0 and 1 type messages. This assumption allows the SPN to be "folded" onto itself to further simplify the calculations. The reduced SPN is shown in Fig. 5.

The normal assumptions of exponential message length and interarrival times make the distributions of the NEW MSG, SEND and SEND ACK transitions exponential. The CRC OK and MSG DROP transitions occur when the message is received and the error detection routine either okays or drops the message. If the CRC check is not embedded in the line driver routine, this calculation would take a constant time for a particular message length and therefore would also have an exponentially distributed execution time. If the CRC check is embedded in the line driver, the net can be further reduced by collapsing the CRC OK and MSG DROP transitions onto the SEND transition. Similarly, the ACK OK and ACK DROP transitions have exponentially distributed execution times.

The TIMEOUT transition in the real system is not exponential if a fixed timeout value is used. Notice that the TIMEOUT transition is enabled only after the message was transmitted and the CRC was checked. That means that the probability density function of the random variable representing the time until the TIMEOUT transition fires would be Erlangian instead of an impulse at the fixed timeout value. In this analysis the time until the transition TIMEOUT fires is assumed to be exponentially distributed. That implies that the timeout value is actually random.

Solving for the reachability set we find 6 states. Notice that the probability ρ of no token being in the READY place P_{READY} is the probability that the subsystem is busy and cannot accept new messages. Solving for the steady state probabilites of markings, we can then find the probability of the subsystem being busy. In this model messages arrive to the system at a Poisson rate λ and are dropped if they arrive when no buffers are available. Therefore, the actual throughput of the system is $\lambda(1 - \rho)$.

Consider a system using this protocol which has a 9600 baud line with a 5 percent error probability and 1024 bit packets. Then we can analyze the performance of this protocol by using the values for the transition rates in Table I.

TABLE I

Transition	Rate (firings/s)
SEND, SEND ACK	9.375
MSG DROP, ACK DROP	3.91
CRC OK, ACK OK	74.22
TIMEOUT	1.000

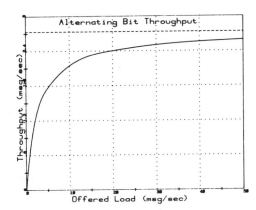

Fig. 6. Throughput versus offered load for the protocol.

Using these values and varying the rate of the NEW MSG transition λ we can plot the throughput S versus offered load. The plot is shown in Fig. 6.

The average delay may be calculated when the observation is made that when a single buffer system is busy, it has exactly one message in it. Therefore, the average delay can be calculated from Little's result $N = ST$. This delay is the delay seen by the messages that actually enter the system and gives no weight to the blocked messages. For the values used above, the average delay in the protocol would be 0.3662 s.

Conclusions

A model has been presented which gives a formal method of generating Markov models omitting the states which are considered as blocked. The actual Markov analysis is not unusual and has been carried out in similar models by other authors. However, the use of the stochastic Petri nets does have two basic advantages. The major feature of the model lies in the simplicity of its specification. In addition, the automatic generation of the state space creates a verification step not previously available in Markovian analysis.

References

[1] T. Agerwala, "Putting Petri nets to work," *IEEE Comput.*, pp. 85–94, Dec. 79.
[2] J. L. Baer, G. Gardarin, C. Girault, and G. Roncairol, "The two-step commitment protocol: Modeling, specification and proof methodology," in *Proc. 5th Conf. Software Eng.*, May 1980.
[3] K. A. Bartlett, R. A. Scantlebury, and P. T. Wilkinson, "Complex transmission over half-duplex links," *Commun. Ass. Comput. Mach.*, vol. 12, p. 260, May 1969.
[4] J. Y. Cotronis and P. E. Lauer, "Verification of concurrent systems of processes," Univ. Newcastle-upon-Tyne, Tech. Rep. 97, 1977.
[5] M. Hack, "The equality problem for vector addition systems is undecidable," *Theoretical Comput. Sci.*, vol. 2, June 1976.
[6] A. W. Holt *et al.*, "Information system theory project report," Rome Air Development Center, Griffiss Air Force Base, NY, Tech. Rep. RADC-TR-68-305, 1968.
[7] D. L. Isaacson and R. W. Madsen, *Markov Chains: Theory and Applications*. New York: Wiley, 1976.
[8] R. M. Keller, "Formal verification of parallel programs," *Commun. Ass. Comput. Mach.*, vol. 19, pp. 371–384, July 1976.
[9] P. J. B. King and I. Mitrani, "Numerical method for infinite Markov processes," in *Proc. 1980 Performance ACM Sigmetrics IFIPS Conf.*, May 28–30, 1980, pp. 277–282.
[10] L. Kleinrock, *Queueing Systems, Vol. 1: Theory*. New York: Wiley, 1975.
[11] D. Martin and G. Estrin, "Models of computations and systems—Evaluation of vertex probabilities in graph models of computations," *J. Ass. Comput. Mach.*, vol 14, pp. 281–299, Apr. 1967.
[12] M. Molloy, "On the integration of delay and throughput measures in distributed processing models," Ph.D. dissertation, Univ. California, Los Angeles, 1981.
[13] J. A. Meldman and A. W. Holt, "Petri nets and legal systems," *Jurimetrics J.*, vol. 12, Dec. 1971.
[14] P. M. Merlin and D. J. Farber, "Recoverability of communication protocols—Implications of a theoretical study," *IEEE Trans. Commun.*, pp. 1036–1043, Sept. 1976.
[15] ——, "Specification and validation of protocols," *IEEE Trans. Commun.*, pp. 1671–1680, Nov. 1979.
[16]. R. F. Miller, "A comparison of some theoretical models of parallel computation," *IEEE Trans. Comput.*, vol. C-22, pp. 710–717, Aug. 1973.
[17] J. D. Noe, "A Petri net model of the CDC 6400," in *Proc. ACM SIGOPS Workshop System Performance Evaluation*, 1971, pp. 362–378.
[18] J. L. Peterson, "Petri nets," *ACM Comput. Surveys*, vol. 9, pp. 223–252, Sept. 1977.
[19] C. V. Ramamoorthy and G. S. Ho, "Performance evaluation of asynchronous concurrent systems using Petri nets," *IEEE Trans. Software Eng.*, vol. SE-6, pp. 440–449, Sept. 1980.
[20] S. D. Shapiro, "A stochastic Petri net with applications to modeling occupancy times for concurrent task systems," *Networks*, vol. 9, pp. 375–379, 1979.
[21] W. M. Zuberek, "Timed Petri nets and preliminary performance evaluation," in *Proc. IEEE 7th Ann. Symp. Comput. Arch.*, 1980, pp. 89–96.

Performance Evaluation of Asynchronous Concurrent Systems Using Petri Nets

C. V. RAMAMOORTHY, FELLOW, IEEE, AND GARY S. HO, MEMBER, IEEE

Abstract—Some analysis techniques for real-time asynchronous concurrent systems are presented. In order to model clearly the synchronization involved in these systems, an extended timed Petri net model is used. The system to be studied is first modeled by a Petri net. Based on the Petri net model, a system is classified into either: 1) a consistent system; or 2) an inconsistent system. Most real-world systems fall into the first class which is further subclassified into i) decision-free systems; ii) safe persistent systems; and iii) general systems. Procedures for predicting and verifying the system performance of all three types are presented. It is found that the computational complexity involved increases in the same order as they are listed above.

Index Terms—Asynchronous, concurrent, performance, Petri net, real time.

I. Introduction

THE RECENT advances in solid-state technology have provided computer designers with powerful functional capabilities at low cost. This enables computer designers to isolate the functions of a computing element and add intelligence to functional units that can optimize their circuits locally. Such an approach eliminates the cost involved in time and bandwidth to submit these local events for a central judgment in real time. This trend has led to the emergence of distributed systems. Furthermore, the recent growth in microprocessor architectures, their capabilities, and their low cost, have motivated system designers to design computer systems as distributed networks of microprocessors. By using multiple processing elements, system throughput can be improved and processing requirements and capabilities unobtainable by uniprocessors can be satisfied. However, the success of multiple processor systems greatly depends on the effectiveness of the synchronization among the processing elements.

In this paper, we concentrate our discussion on techniques for the prediction and verification of performance of distributed systems. We consider a distributed system as a set of loosely or tightly coupled processing elements working cooperatively and concurrently on a set of related tasks. In general, there are two approaches for performance evaluation [6], [2]: 1) deterministic models and 2) probabilistic models. In deterministic models, it is usually assumed that the task arrival times, the task execution times, and the synchronization involved are known in advance to the analysis. With this information, a very precise prediction of the system performance can be obtained. This approach is very useful for performance evaluation of real-time control systems with hard deadline requirements.

In probabilistic models, the task arrival rates and the task service times are usually specified by probabilistic distribution functions. The synchronization among tasks is usually not

---

Manuscript received February 00, 1979; revised December 00, 1979. This work was supported by the Ballistic Missile Defense Advanced Technology Center under Contract DASG60-77-C-0138, the U.S. Air Force under Grant F49620-79-C-0173, and the National Science Foundation under Grant MCS77-27293. The work of G. S. Ho was done while he was at the University of California, Berkeley, CA.

C. V. Ramamoorthy is with the Department of Electrical Engineering and Computer Sciences, University of California, Berkeley, CA 94720.

G. S. Ho is with Bell Laboratories, Naperville, IL 60540.

modeled, because otherwise the number of system states becomes so large that it would be impossible to perform any analyses. Probabilistic models usually give a gross prediction on the performance of a system and are good for early stages of system design when the system characteristics are not well understood. In this paper, we focus on performance analysis of real-time systems and therefore we have chosen the deterministic approach. In particular, in order to model clearly the synchronization involved in concurrent systems, the Petri net model [16] is chosen.

In this approach, the system to be studied is first modeled by a Petri net. Based on the Petri net model, a given system is classified as either (Fig. 1): 1) a consistent system; or 2) an inconsistent system (the definitions are given in later sections of the paper). Most real-world systems fall into the first class and so we focus our discussion on consistent systems. Due to the difference in complexity involved in the performance analyses of different types of consistent systems, they are further subclassified into: i) decision-free systems; ii) safe persistent systems; and iii) general systems. Procedures for predicting and verifying the system performance of all three types are presented. It is found that the computational complexity involved increases in the same order as they are listed above.

The paper is divided into four sections. In Section II, a brief introduction to Petri nets is given. In Section III, consistent systems, decision-free systems, safe persistent systems, and general systems are defined. Analysis techniques for each type of systems are discussed. Examples are used extensively to illustrate the realism and the applicability of the approaches. Finally, in Section IV, the results are summarized and some areas of future research are discussed.

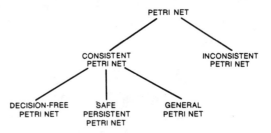

Fig. 1. System classification.

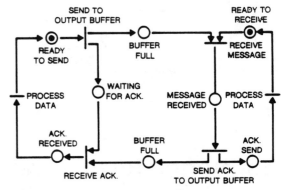

Fig. 2. Petri net model of a communication protocol between two processes.

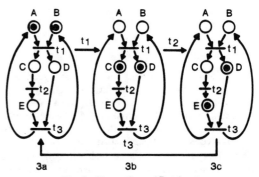

Fig. 3. Execution of Petri net.

II. An Introduction to Petri Nets

Petri nets [16], [1] are a formal graph model for modeling the flow of information and control in systems, especially those which exhibit asynchronous and concurrent properties. A Petri net contains two types of nodes (Fig. 2): the circles (called *places*) represent conditions and the bars (called *transitions*) represent events. A black dot (called a *token*) at a place indicates the holding of the condition of the place. A pattern of tokens in a Petri net (called a *marking*) represents the state of the system. For example, Fig. 2 models a communication protocol between two processes. The process on the left is ready to send and the process on the right is ready to receive.

To model the dynamic behavior of a system, the execution of a process is represented by the firing of the corresponding transition. The changes in system state are represented by the movements of tokens in the net. The firing rules of Petri nets are as follows.

1) A transition is enabled if and only if each of its input places has at least one token.
2) A transition can fire only if it is enabled.
3) When a transition fires:
   i) a token is removed from each of its input places; and
   ii) a token is deposited into each of its output places.

Fig. 3 shows the execution of a Petri net. At the beginning [Fig. 3(a)], transition t_1 is enabled because both of its input places, A and B, have a token in them. Firing transition t_1 removes one token from places A and B, and deposits a token into each of its output places, namely, C and D [Fig. 3(b)]. At this point, transition t_3 remains disabled because one of its input place, E, still has no token in it. However, transition t_2 is enabled. Firing t_2 removes a token from place C and deposits a token into place E [Fig. 3(c)]. Now transition t_3 is enabled. Firing transition t_3 removes a token from places D and E and deposits a token into places A and B returning the system to its initial configuration.

A. Control Flow Analyses

Petri nets have been used extensively to study the control flow of computer systems. By analyzing the liveness, boundedness, and proper termination properties of the Petri net model of a computer system, many desirable properties of the system can be unveiled.

A Petri net is *live* [9], [11] if there always exists a firing sequence to fire each transition in the net. By proving that the Petri net is live, the system is guaranteed to be deadlock free.

A Petri net is *bounded* [12], [14] if for each place in the next, there exists an upper bound to the number of tokens that can be there simultaneously. If tokens are used to represent intermediate results generated in a system, by proving that the Petri net model of the system is bounded, the amount of buffer space required between asynchronous processes can be determined and therefore information loss due to buffer overflow can be avoided. If the upper bound on the number of tokens at each place is one, then the Petri net is *safe*. Programming constructs like critical regions [3] and monitors [4], [10] can be modeled by safe Petri nets.

A Petri net is *properly terminating* [8], [17] if the Petri net always terminates in a well-defined manner such that no tokens are left in the net. By verifying that the Petri net is properly terminated, the system is guaranteed to function in a well behaved manner without any side effects on the next initiation.

B. Extended Timed Petri Nets

In order to study the performance of a system, the Petri net model is extended to include the notion of time [18]. In such extended nets, an execution time r is associated with each transition. When a transition initiates its execution, it takes r units of time to complete its execution. With the extended Petri net model, the performance of a computer system can be studied.

For example, Fig. 4 shows a simple computer system with a processor and two tape units. The input queue contains a set of similar tasks to be processed by the system. Each task first requires some computation by the processor together with a tape unit for r_1 units of time. Then the system does some input and output on the tape unit for r_2 time units while the processor continues its computation for r_3 time units. After that, the processor works on the task with the tape unit for r_4 time units. The processor is then ready to start the next task with the other tape unit while the current tape unit is being rewound. Strictly speaking, Fig. 4 is not a Petri net as queues are not defined in the Petri net model. However, when the system performance is evaluated, only the control portion of a system (i.e., the portion inside the square in Fig. 4) is used and therefore is a Petri net. For the rest of this paper, queues will be used to represent the system structure but the readers must keep in mind that only the control portion of a system is used in the evaluation of the system performance.

The problem studied in this paper is to find the maximum performance of a system, i.e., to find the minimum cycle time (for processing a task) of a system. As pointed out in the introduction, different computational complexities are involved in the analyses of systems of different types. The approaches for analyzing each type of system are studied separately in detail in the following section. Before we come to the analyses, some definitions are in order.

Definition: In a Petri net, a sequence of places and transitions, $P_1 t_1 P_2 t_2 \cdots P_n$, is a *directed path* from P_1 to P_n if transition t_i is both an output transition of place P_i and an input transition of place $P_{(i+1)}$ for $1 \leq i \leq n-1$.

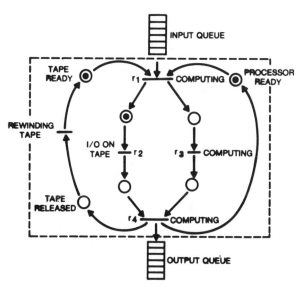

Fig. 4. Petri net model of a computer configuration.

Definition: In a Petri net, a sequence of places and transitions, $P_1 t_1 P_2 t_2 \cdots P_n$, is a *directed circuit* if $P_1 t_1 P_2 t_2 \cdots P_n$ is a directed path from P_1 to P_n and P_1 equals P_n.

Definition: A Petri net is *strongly connected* if every pair of places is contained in a directed circuit.

In the next section, we discuss performance analysis techniques for strongly connected nonterminating [14] Petri nets. Extensions to analyze weakly connected Petri nets are quite straightforward and are left to the readers.

III. PERFORMANCE EVALUATION

A. Consistent and Inconsistent Systems

The first step involved in our approach to analyze the performance of a system is to model it by a Petri net. A system is a consistent (inconsistent) system if its Petri net model is consistent (inconsistent). A Petri net is *consistent* (condition A) if and only if there exists a nonzero integer assignment to its transitions such that at every place, the sum of integers assigned to its input transitions equals the sum of integers assigned to its output transitions; otherwise, the system is inconsistent. If a transition has n input arcs to a place, it is counted as n input transitions to that place.

Fig. 5(a) is an inconsistent system and Fig. 5(b) is a consistent system. In Fig. 5(a), there does not exist an integer assignment to its transitions to satisfy condition A. This can be verified by assigning an integer variable to each transition and getting a contradiction in trying to solve the simultaneous equations provided by condition A:

for place A: $\quad x + y = z \quad$ (1)
for place B: $\quad x = z \quad$ (2)
for place C: $\quad y = z \quad$ (3)
(2) + (3) $\quad x + y = 2z \quad$ (4)

and therefore (4) contradicts (1).

Fig. 5(b) is a consistent Petri net. If each transition is assigned an integer of value 1, condition A is satisfied.

The practical implication behind this system classification is that the integer assigned to a transition is the relative num-

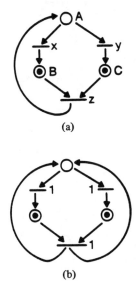

Fig. 5. (a) An inconsistent system. (b) A consistent system.

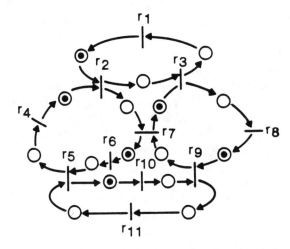

Fig. 6. Petri net model of a train configuration.

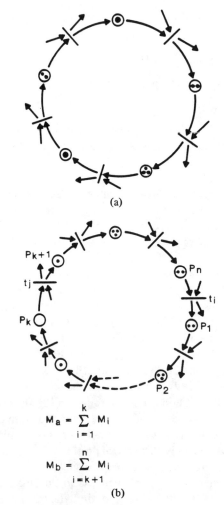

Fig. 7. (a) A cycle in a decision-free Petri net.

ber of executions of that transition in a cycle. If a system is live and consistent, the system goes back to its initial configuration (state) after each cycle and then repeats itself. If a system is inconsistent, either it produces an infinite number of tokens (i.e., it needs infinite resources) or consumes tokens and eventually comes to a stop. Most real-world systems which function continuously with finite amount of resources fall into the class of consistent systems. For the rest of this paper, we focus our discussion on consistent systems and further subclassify them into decision-free systems, persistent systems, and general systems. Performance analysis techniques for each subclass are discussed in the following subsections.

B. Decision-Free Systems

A system is a *decision-free* system if its Petri net model is a decision-free Petri net (also known as marked graph [5], [15].) A Petri net is decision-free if and only if for each place in the net, there is one input arc and one output arc. This means that tokens at a given place are generated by a predefined transition (its only input transition) and consumed by a predefined transition (its only output transition).

The computer configuration shown in Fig. 4 is a decision-free system. The train system shown in Fig. 6 is another decision-free system. The tokens in the net are used to represent trains. For the convenience of the passengers, trains wait at stations for the next train to arrive so as to allow passengers to transfer between trains before leaving stations. Similarly, the chaining operations in the CRAY-1 computer [20] can be modeled by a decision-free Petri net as shown in Fig. 6. The results issued from one functional unit are immediately fed into another functional unit and so on. For a decision-free system, the maximum performance can be computed quite easily. However, before we come to that result, we need the following two theorems.

Theorem 1: For a decision-free Petri net, the number of tokens in a circuit remains the same after any firing sequence.

Proof: Without loss in generality, a circuit containing five places and five transitions in a decision-free Petri net is shown in Fig. 7(a). Tokens in the circuit can only be produced or consumed by transitions in the circuit. When a transition consumes a token, it produces one back into the circuit; therefore, the number of tokens in a circuit remains the same after any firing sequence. Q.E.D.

This result has been proven by many researchers [5], [15]. The proof is included here just for the completeness of this paper.

Definition: Let $S_{i(n_i)}$ be the time at which transition t_i

initiates its n_ith execution. The *cycle time* C_i of transition t_i is defined as

$$\lim_{n_i \to \infty} \frac{S_i(n_i)}{n_i}.$$

Theorem 2: All transitions in a decision-free Petri net have the same cycle time.

Proof: For any two transitions t_i and t_j in a decision-free Petri net, choose a circuit that contains both transitions. Such a circuit must exist because the net is strongly connected. Without loss in generality, assume that there are M_i tokens in place i in the initial marking, Fig. 7(b). Let

$$M_a = \sum_{i=1}^{k} M_i \quad \text{and} \quad M_b = \sum_{i=k+1}^{n} M_i.$$

At time $S_{i(n_i)}$,

$$n_i - M_b \leqslant \text{number of initiations of transition } t_j \leqslant n_i - M_a$$

$$\lim_{n_i \to \infty} \frac{S_i(n_i)}{n_i - M_b} \geqslant C_j \geqslant \lim_{n_i \to \infty} \frac{S_i(n_i)}{n_i - M_a}.$$

Since M_a and M_b are finite, as $n_i \to \infty$, the left- and right-hand side expressions approach C_i, i.e.,

$$C_i \geqslant C_j \geqslant C_i$$

$$C_i = C_j.$$

Therefore, all transitions in a decision-free Petri net have the same cycle time C. Q.E.D.

Theorem 3: For a decision-free Petri net, the minmum cycle time (maximum performance) C is given by

$$C = \max \left\{ \frac{T_k}{N_k} : k = 1, 2, \cdots, q \right\}$$

such that $S_{i(n_i)} = a_i + C n_i$ where

$T_k = \sum_{t_i \in L_k} r_i$ = sum of the execution times of the transition in circuit k

$N_k = \sum_{P_i \in L_k} M_i$ = total number of tokens in the places in circuit k

q = number of circuits in the net

a_i = constant associated with transition t_i

L_k = loop (circuit) k

M_i = number of tokens in place P_i.

Similar results have been obtained in [19], [15]. The proof given here uses a graph theoretical approach and is different from the previous approaches. Based on this approach, we develop a very fast procedure to verify the performance of a system.

Before we prove Theorem 3, we would like to give an example of the usage of Theorem 3 to clarify our notations. The computer configuration with the execution time of each transition is shown in Fig. 8. According to Theorem 3, for

circuit $At_1 C t_2 E t_4 G t_5$: $\quad \dfrac{T_k}{N_k} = \dfrac{5 + 20 + 3 + 2}{2} = 15$

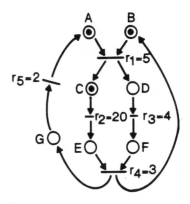

Fig. 8. Computer configuration with execution times.

circuit $At_1 D t_3 F t_4$: $\quad \dfrac{T_k}{N_k} = \dfrac{5 + 4 + 3 + 2}{1} = 14$

circuit $Bt_1 C t_2 E t_4$: $\quad \dfrac{T_k}{N_k} = \dfrac{5 + 20 + 3}{2} = 14$

circuit $Bt_1 D t_3 F t_4$: $\quad \dfrac{T_k}{N_k} = \dfrac{5 + 4 + 3}{1} = 12.$

By enumerating all circuits in the net, the minimum cycle time is 15.

Proof of Theorem 3: The proof is in two parts.

a) Minimum cycle time, $C \geqslant \max \{T_k/N_k, k = 1, 2, \cdots, q\}$.

b) For $C = \max \{T_k/N_k, k = 1, 2, \cdots, q\}$ there exists a_i, such that $S_i(n_i) = a_i + C n_i$ and the firing rules are not violated.

Proof of a):

number of transitions that are enabled simultaneously $\leqslant$ number of tokens in circuit
$= N_k \quad$ (Theorem 1)

processing power required by circuit per cycle $= T_k = \sum_{t_i \in L_k} r_i$

$\leqslant$ maximum processing power of the circuit per cycle time
$= C N_k$.

Therefore,

$$T_k \leqslant C N_k \quad \text{for every circuit}$$

and

$$C \geqslant \max \left\{ \frac{T_k}{N_k}, k = 1, 2, \cdots, q \right\}.$$

Lemma: For $C = \max \{T_k/N_k, k = 1, 2, \cdots, q\}$

$$0 \geqslant T_k - C N_k \quad \text{for all circuit } k.$$

Proof:

$$C \geqslant \frac{T_k}{N_k} \quad k = 1, 2, \cdots, q$$

$$0 \geqslant T_k - C N_k.$$

Proof of b): Let

$$C = \max\left\{\frac{T_k}{N_k}, k = 1, 2, \cdots, q\right\}$$

$$\begin{array}{cccccc}
t_i & M_{ij} & t_j & M_{jk} & t_k \\
| \longrightarrow & \odot & \longrightarrow | \longrightarrow & \odot & \longrightarrow | \\
S_i(n_i) & & S_j(n_j) & & S_k(n_k) \\
= a_i + Cn_i & & = a_j + Cn_j & & = a_k + Cn_k.
\end{array}$$

In order not to violate the firing rules:

finish time of the n_ith   initiation time of the $n_i + M_{ij}$
execution of transition t_i ⩽ execution of transition t_j

$$S_i(n_i) + r_i \leqslant S_j(n_i + M_{ij})$$
$$a_i + Cn_i + r_i \leqslant a_j + C(n_i + M_{ij})$$
$$a_i - CM_{ij} + r_i \leqslant a_j. \quad (1)$$

Similarly,

$$a_j - CM_{jk} + r_j \leqslant a_k \quad (2)$$

(1) + (2) $a_i - C(M_{ij} + M_{jk}) + r_i + r_j \leqslant a_k.$

In general,

$$a_i - C \sum_{(u,v)\in R} M_{uv} + \sum_{w\in R} r_w \leqslant a_s \quad (3)$$

where R is a path from transition i to transition s. In order not to violate the firing rules, we have to find a_i's such that (3) is satisfied.

Procedure for assigning a_i's such that

$$a_i - C \sum_{(u,v)\in R} M_{uv} + \sum_{w\in R} r_w \leqslant a_s. \quad (3)$$

1) Define the distance from transition t_i to transition t_j (t_i adjacent to t_j) to be $r_i - CM_{ij}$:

$$\begin{array}{ccc}
t_i & & t_j \\
| \longrightarrow & \odot & \longrightarrow | \\
r_i & M_{ij} & r_j
\end{array}$$

2) Find a transition t_s, which is enabled initially and assign 0 to a_s.

3) Assign a_u to each transition t_u such that a_u is the greatest distance from t_s to t_u,

i.e., $a_u = \max\left\{\sum_{w\in R} r_w - C \sum_{(u,v)\in R} M_{uv}\right\}$

where R is a path from t_s to t_u.

Such an assignment of a_i's exists because by the lemma, $T_k - CN_k \leqslant 0$, the greatest distance between any two nodes is finite and the corresponding path would never contain a loop.   Q.E.D.

A drawback of the above approach is that all circuits in the net must be enumerated; this can be very tedious. In the design of computer systems, the required performance is usually given. With this information, the performance of a system can be verified very efficiently. By the lemma, the performance requirement (expressed in cycle time C) can be satisfied if and only if $CN_k - T_k \geqslant 0$ for all circuits. This can be verified by the following procedure.

A Procedure for Verifying System Performance:

1) Express the token loading in an $n \times n$ matrix P, where n is the number of places in the Petri net model of the system. Entry (A, B) in the matrix equals x if there are x tokens in place A, and place A is connected directly to place B by a transition; otherwise (A, B) equals 0. Matrix P of the example system in Fig. 8 is shown below:

	A	B	C	D	E	F	G
A	0	0	1	1	0	0	0
B	0	0	1	1	0	0	0
C	0	0	0	0	1	0	0
D	0	0	0	0	0	0	0
E	0	0	0	0	0	0	0
F	0	0	0	0	0	0	0
G	0	0	0	0	0	0	0

Matrix P

2) Express transition time in an $n \times n$ matrix Q. Entry (A, B) in the matrix equals to r_i (execution time of transition i) if A is an input place of transition i and B is one of its output places. Entry (A, B) contains the symbol "w" if A and B are not connected directly as described above. Matrix Q for the example system is

	A	B	C	D	E	F	G
A	w	w	5	5	w	w	w
B	w	w	5	5	w	w	w
C	w	w	w	w	20	w	w
D	w	w	w	w	w	4	w
E	w	3	w	w	w	w	3
F	w	3	w	w	w	w	3
G	2	w	w	w	w	w	w

Matrix Q

3) Compute matrix $CP - Q$ (with $n - w = \infty$ for $n \in N$), then use Floyd's algorithm [7] to compute the shortest distance between every pair of nodes using matrix $CP - Q$ as the distance matrix. The result is stored in matrix S. There are three cases.

a) All diagonal entries of matrix S are positive (i.e.,

$CN_k - T_k > 0$ for all circuits)—the system performance is higher than the given requirement.

b) Some diagonal entries of matrix S are zero's and the rest are positive (i.e., $CN_k - T_k = 0$ for some circuits and $CN_k - T_k > 0$ for the other circuits)—the system performance just meets the given requirement.

c) Some diagonal entries of matrix S are negative (i.e., $CN_k - T_k < 0$ for some circuits)—the system performance is lower than the given requirement.

In the example, for $C = 15$, $CP - Q$ is

	A	B	C	D	E	F	G
A	∞	∞	10	10	∞	∞	∞
B	∞	∞	10	10	∞	∞	∞
C	∞	∞	∞	∞	-5	∞	∞
D	∞	∞	∞	∞	∞	-4	∞
E	∞	-3	∞	∞	∞	∞	-3
F	∞	-3	∞	∞	∞	∞	-3
G	-2	∞	∞	∞	∞	∞	∞

After applying Floyd's algorithm to find the shortest distance between every pair of places, we have

	A	B	C	D	E	F	G
A	0	2	10	10	5	6	2
B	0	2	10	10	5	6	2
C	-10	-8	0	0	-5	-4	-8
D	-9	-7	1	1	-4	-4	-7
E	-5	-3	5	5	0	1	-3
F	-5	-3	5	5	0	1	-3
G	-2	0	8	8	3	4	0

Matrix S

Since the diagonal entries are nonnegative, the performance requirement of $C = 15$ is satisfied. Moreover, since entries (A,A), (C,C), (E,E), and (G,G) are zero's, $C = 15$ is optimal (i.e., it is the minimum cycle time). In addition, when a decision-free system runs at its highest speed, CN_k equals to T_k for the bottleneck circuit. This implies that the places that are in the bottleneck circuit will have zero diagonal entries in matrix S. In the example, the bottleneck circuit is $At_1Ct_2Et_4Gt_5$. With this information, the system performance can be improved by either reducing the execution times of some transitions in the circuit (by using faster facilities) or by introducing more concurrency in the circuit (by introducing more tokens in the circuit). Which approach should be taken is application dependent and beyond the scope of this paper.

The above procedure can be executed quite fast. The formulation of matrices P and Q takes $O(n^2)$ steps. The Floyd algorithm takes $O(n^3)$ steps. As a whole, the procedure can be executed in $O(n^3)$ steps. Therefore, the performance requirement of a decision-free system can be verified quite efficiently.

C. Safe Persistent Systems

A system is a safe persistent system if its Petri net model is a safe persistent Petri net. A Petri net is a safe persistent Petri net if and only if it is a safe Petri net and for all reachable markings, a transition is disabled only by firing the transition. It differs from a decision-free Petri net in that it may have more than one input (output) arcs to (from) a place. However, like a decision-free Petri net, it models a deterministic system. In a persistent Petri net, if a token enables a transition, it will be consumed by that transition only, i.e., a token will never enable two or more transitions simultaneously. As a result, a safe persistent Petri net can always be transformed into a decision-free Petri net.

Fig. 9(a) shows a persistent Petri net. It models the operations of a double buffer input port. Transitions t_1 and t_2 represent fetching the contents of buffer 1 and buffer 2, respectively. Transition t_3 represents storing the input into the memory. To compute the performance of the system, we first transform it into a decision-free system and then use the algorithm discussed in the previous subsections to compute the system performance.

A persistent Petri net can be transformed into a decision-free Petri net by tracing the execution of the system for one cycle. For example, Fig. 9(b) is the decision-free system corresponding to the persistent Petri net shown in Fig. 9(a). Places A_1 and A_2 in Fig. 9(b) represent two different occurrences of place A in Fig. 9(a) in a cycle. Condition A_1 holds when transition t_1 is enabled and condition A_2 holds when transition t_2 is enabled. Similarly, place D is duplicated into D_1 and D_2. Condition D_1 holds when transition t_3 is enabled and transition t_2 will be enabled after firing t_3. Condition D_2 holds when transition t_3 is enabled and transition t_1 will be enabled after firing t_3.

Initially, there is a token in places A and B and transition t_1 is enabled in Fig. 9(a), and therefore there is a token in places A_1 and B in Fig. 9(b). After firing t_1, a token is deposited into places C and D in Fig. 9(a). This is represented by depositing a token in places D_1 and C in Fig. 9(c). By following the execution of the system for a cycle, the corresponding decision-free system can be generated. The system performance can then be computed by the procedure discussed in Section III-B.

D. General Systems

A system is a general system if its Petri net model is a general Petri net. A Petri net is a *general Petri net* if it is a consistent Petri net and there exists a reachable marking such that the firing of a transition disables some other transitions. Figs. 10 and 11 show two general Petri nets. Fig. 10 models the communication protocol from process P_1 to processes P_2 and P_3, such that the difference in the number of messages sent from P_1 to P_2 and P_3 is always less than three. It is not a decision-free Petri net because place A has more than one

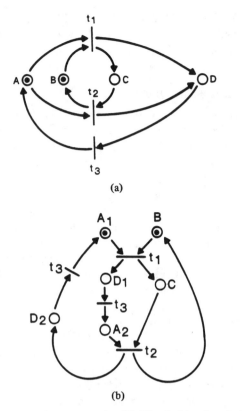

Fig. 9. (a) A persistent system. (b) The decision-free system corresponding to (a).

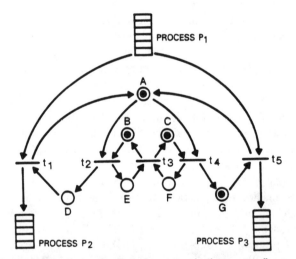

Fig. 10. A general system (a communication protocol).

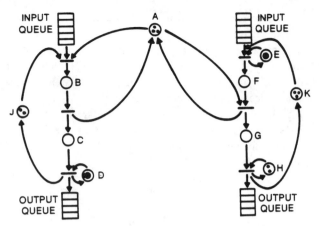

Fig. 11. A general system (shared resource pipelines).

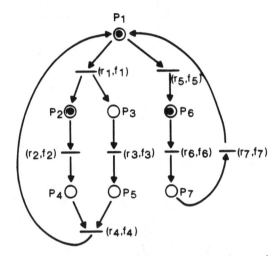

Fig. 12. A general system with specified execution times and relative firing frequencies.

input and output arcs. It is not a persistent Petri net because, in the configuration shown, the execution of either transition t_2 or transition t_4 disables the other transition. This introduces the nondeterministic characteristic of a system.

Fig. 11 models two shared resource pipelines. Place A represents the condition that resource A is free. The system has three units of resource A and they are used by the first and second stages of the pipeline on the left, and by the second stage of the pipeline on the right. Places J and K represent the conditions that buffers for the left and right hand side pipelines are free, respectively. By the same reasons given for Fig. 10, it is a general system.

Fig. 12 shows another general Petri net. Tokens at place P_1 can be consumed by either transition t_1 or t_5. In addition to this, the execution frequencies of the two transitions are independent of each other. This introduces another degree of freedom into the system. In order to fully specify the dynamic characteristics of the system, the number of executions of each transition per cycle has to be defined. In the example, it is specified by the ordered pair, (r_i, f_i), where r_i is the execution time and f_i is the execution frequency per cycle of transition t_i. General systems are very difficult to analyze. In the next theorem, we show that it is unlikely that a fast algorithm exists to verify the performance of a general system. A method of computing the upper and lower bounds of the performance of a conservative general system [14] is proposed. For a nonconservative general system, no good heuristics are known to the authors and further research is needed.

Theorem 4: Verifying the performance of a conservative general Petri net is an *NP*-complete problem [12].

Proof:

i) It is in *NP* because we can guess the optimal schedule (execution sequence). The nondeterminisms of the general Petri net are resolved and therefore the general Petri net can be transformed into a decision-free Petri net according to the guessed schedule. The performance can then be verified by the procedure discussed in Section III-B.

ii) It is *NP*-complete because the set partitioning problem (an *NP*-complete problem) can be reduced to the above problem.

The Set Partition Problem [13]: Given a set of integers, $S = \{x_1, x_2, \cdots, x_n\}$, partition it into two subsets, S_1 and S_2 such that (condition *B*)

$$S = S_1 \cup S_2 \quad \text{and} \quad S_1 \cap S_2 = \phi$$

$$\text{and} \quad \sum_{i \in S_1} x_i = \sum_{i \in S_2} x_i = \frac{\sum_{i=1}^{n} x_i}{2}.$$

Reduction: Given $S = \{x_1, x_2, \cdots, x_n\}$, generate the general Petri net shown in Fig. 13. It is easy to see that the system has minimum cycle time, $C = (\Sigma_{i=1}^{n} x_i)/2$, if and only if the set S satisfies condition B.

A Method to Compute Upper and Lower Bounds of the Performance of a Conservative General System:

1) Upper Bound: We choose a schedule which satisfies the execution frequency requirement and then use the algorithm discussed in Section III-B to find the cycle time of the system.

2) Lower Bound (Fig. 12):

a) Find a nonzero integer assignment to the places such that the sum of integers assigned to the input places of a transition equals the sum of integers assigned to the output places. Such an integer assignment must exist because the system is conservative [14]. Intuitively, the integer assigned to a place represents the relative processing capability of a token at that place. The weighted execution time of a transition is equal to the product of the transition execution time and the sum of the integers assigned to its input places. For example,

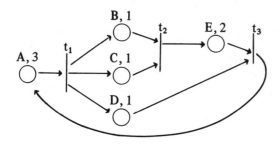

A token at place E has twice the processing capability of a token at either place $B, C,$ or D and transition t_2 has weighted execution time $r_2(1 + 1)$.

b) Assume that all tokens in the net are busy all the time. Then

$C \times$ weighted processing capability $\geq$ sum of weighted execution time

$$C \times \Sigma s_i M_i \geq \Sigma s_i f_i r_i$$

$$C \geq \frac{\Sigma s_i f_i r_i}{\Sigma s_i M_i}$$

where s_i = sum of integers assigned to the input places of transition i.

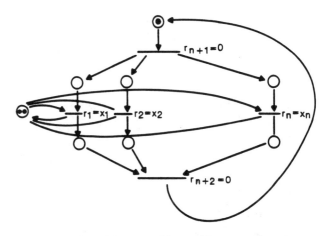

Fig. 13. Reduction of the set partition problem to a general Petri net.

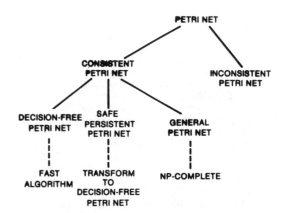

Fig. 14. Summary of results.

IV. CONCLUSIONS

In this paper, we have discussed a systematic method to evaluate and verify the performance of concurrent systems. The system to be studied is first modeled by a Petri net. Based on the Petri net model, the system is classified into either 1) a consistent system or 2) an inconsistent system. A consistent system is further subclassified into: i) a decision-free system; ii) a safe persistent system; and iii) a general system. The system classification and the results are summarized in Fig. 14. The performance of decision-free systems and safe persistent systems can be computed quite efficiently. In the case of general systems, we have proven that the verification of system performance is *NP*-complete. An approach for computing the upper and lower bounds of the performance of a conservative general system is proposed. However, the bounds produced may be loose. For a nonconservative general system, no good heuristics are known. Further research is needed.

Another difficulty that may arise in the proposed approach is the inaccuracy in estimating the execution times of the processes in a system. However, in real-time systems such as the air traffic control systems, chemical plant control systems, nuclear power plant control systems, etc., the execution times of the processes may be predicted quite accurately. In the case that the execution times cannot be estimated accurately, the worst case execution times of the processes can be used. The performance prediction obtained will be the worst case system performance.

Acknowledgment

We would also like to thank Dr. C. R. Vick and J. E. Scalf for many helpful discussions related to this work.

References

[1] T. Agerwala and M. J. Flynn, "On the completeness of representation schemes for concurrent systems," presented at the Conf. Petri Nets and Related Methods, M.I.T., Cambridge, MA, July 1975.
[2] F. Baskett, K. M. Chandy, R. R. Muntz, and F. Palacios-Gomez, "Open, closed and mixed networks of queues with different classes of customers," *J. Ass. Comput. Mach.*, vol. 22, Apr. 1975.
[3] P. Brinch Hansen, "Structured multiprogramming," *Commun. Ass. Comput. Mach.*, vol. 15, July 1972.
[4] P. Brinch Hansen, "Concurrent programming concepts," *Computing Surveys*, vol. 5, Dec. 1973.
[5] F. Commoner *et al.*, "Marked directed graphs," *J. Comput. Syst. Sci.*, vol. 5, 1971.
[6] D. Ferrari, *Computer Systems Performance Evaluation*. Englewood Cliffs, NJ: Prentice-Hall, 1978.
[7] R. W. Floyd, "Algorithm 97, shortest path," *Commun. Ass. Comput. Mach.*, vol. 5, 1962.
[8] K. P. Gostelow, "Flow of control, resource allocation, and the proper termination of programs," Ph.D. dissertation, School Eng. Appl. Sci., Univ. California, Los Angeles, Dec. 1971.
[9] M. Hack, "Decidability questions for Petri nets," Ph.D. dissertation, Dep. Elec. Eng., M.I.T., Cambridge, MA, Dec. 1975.
[10] C. A. R. Hoare, "Monitors: An operating system structuring concept," *Commun. Ass. Comput. Mach.*, vol. 17, Oct. 1974.
[11] R. C. Holt, "On deadlock in computer systems," Ph.D. dissertation, Dep. Comput. Sci., Cornell Univ., Ithaca, NY, Jan. 1971.
[12] R. M. Karp and R. E. Miller, "Properties of a model for parallel computation: Determinacy, termination, queuing," *SIAM J. Appl. Math.*, vol. 14, Nov. 1966.
[13] R. M. Karp, "Reducibility among combinatorial problems," in *Complexity of Computer Computations*. New York: Plenum, 1972.
[14] Y. E. Lien, "Termination properties of generalized Petri nets," *SIAM J. Comput.*, vol. 5, no. 2, June 1976.
[15] T. Murata, "Petri nets, marked graphs, and circuit-system theory," *Circuits Syst.*, vol. 11, June 1977.
[16] J. L. Peterson, "Petri nets," *Comput. Surveys*, vol. 9, Sept. 1977.
[17] J. B. Postel, "A graph model analysis of computer communication protocols," Ph.D. dissertation, Dep. Comput. Sci., Univ. California, Los Angeles, Jan. 1974.
[18] C. Ramchandani, "Analysis of asynchronous concurrent systems by Petri nets," Project MAC, TR-120, M.I.T., Cambridge, MA, 1974.
[19] R. Reiter, "Scheduling parallel computations," *J. Ass. Comput. Mach.*, vol. 15, Oct. 1968.
[20] R. M. Russell, "The CRAY-1 computer system," *Commun. Ass. Comput. Mach.*, vol. 21, Jan. 1978.

C. V. Ramamoorthy (M'57–SM'76–F'78), for a photograph and biography, see page 117 of the March 1980 issue of this TRANSACTIONS.

Gary S. Ho (S'77–M'79) was born in Hong Kong on March 2, 1953. He received the B.S., M.S., and Ph.D. degrees in computer science from the University of California, Berkeley, in 1975, 1977, and 1979 respectively.

Currently he is working for Bell Laboratories, Indian Hill, IL. His research interests include distributed computer system, design methodology, performance evaluation, and computational complexity.

Dr. Ho is a member of the Association for Computing Machinery.

Generalized Petri Net Reduction Method

HYUNG LEE-KWANG, MEMBER IEEE, JOEL FAVREL, AND
PIERRE BAPTISTE

Abstract — A reduction method of generalized Petri nets is proposed. This method is a generalization of the reduction method which was previously given by Lee-Kwang and Favrel. The proposed method is defined not on the basis of the dynamic behavior but of the structure of the net, and thus the test of reducible subnet can be done by a deterministic approach. The reduction preserves the properties such as liveness, boundedness, and proper termination, and allows easy analysis of generalized Petri nets.

I. INTRODUCTION

Petri nets [1] have been used to model and analyze a large class of systems, especially parallel/concurrent computation systems [6]–[8]. A Petri net which allows multiple arcs is called a generalized Petri net [3]–[7], and this correspondence is concerned with generalized Petri nets. Petri nets have some interesting properties such as liveness, boundedness, and proper termination. After modeling a system with a Petri net, many desirable properties of the system can be unveiled by analyzing the Petri net [9]. However, the complexity of the model is drastically increased with the number of reachable states and events in the net [22]. Our fundamental principle in managing the complexity is to reduce the size of reachable state space by Petri net reduction.

Petri net reduction is a procedure that homomorphically transforms Petri nets to their reduced nets while preserving some desirable properties of the original nets. This procedure reduces the reachable state space; analysis of the reduced net can provide useful information for understanding of the original net. We have surveyed several existing reduction methods of Petri nets [13]–[23] and proposed a reduction method of ordinary Petri nets in [9]–[11]. In this correspondence, as an extension of the method, we develop a reduction method of generalized Petri nets.

We have pointed out in [9] that the existing methods are defined on the basis of the dynamic behavior of the net such as liveness and well-behaved module. Therefore, the test of a reducible subnet is often complex and has not been automated. The reduction method which will be developed in this correspondence is based upon graph theory, that is, the structure of the net. Therefore, the test of a reducible subnet is independent from the dynamic state of net, and thus it can be programmed easily. In Section II, we present basic notations of Petri nets and some terminology. Section III introduces reduction rules of generalized Petri nets. In Section IV, properties of the proposed method are studied, and an example of its application is discussed.

II. PRELIMINARIES

A. Petri Nets

In this section, we present some notations of Petri nets following the approaches of [7]. A generalized Petri net N is a 5-tuple defined by

$$N = (P, T, B, F, M)$$

where

$P = \{p_1, \cdots, p_r\}$, finite set of places,
$T = \{t_1, \cdots, t_s\}$, finite set of transitions,
$B: P \times T \to \mathbb{N}$ backward incidence function,
$F: P \times T \to \mathbb{N}$ forward incidence function,
$M \in \mathbb{N}^r$ marking.

As usual, $\mathbb{N}$ denotes the set of nonnegative integers.

A Petri net is represented by a bipartite graph containing two types of nodes: places and transitions. Graphically, we use circles for places and bars for transitions (Fig. 1). The relationships from places to transitions and from transitions to places are represented by directed arcs. If p is a place and t a transition, the graph has n arcs directed from p to t when $B(p_i, t_j) = n$ or $B(i, j) = n$ (the weight of the arc in n). Similarly, there are m arcs from t to p when $F(p_i, t_j) = F(i, j) = m$. (The node t_j is called as an input of p_i, p_i an output of t_j. The arc (t_j, p_i) is called an output arc of t_j and input arc of p_i.)

A marking is a mapping $M: P \to \mathbb{N}$, represented by a vector $M \in \mathbb{N}^r$. $M(p)$ is the number of tokens in p for the marking M. A transition t can fire in a marking M if $M \geq B(\cdot, t)$.

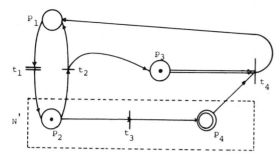

Fig. 1. Petri net.

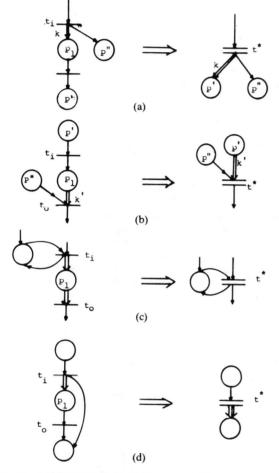

Fig. 2. GRSN-1T. (a) $k = 2$. (b) $k' = 2$. (c) $k = k' = 1$. (d) $k = 2$.

B. Terminology

In this section, we introduce some terminology with examples which were given in [9]. In the Petri net of Fig. 1 we consider a subnet N' whose nodes are $\{p_2, t_3, p_4\}$.

A *macro node* is node that represents a subnet. A *macrotransition* is denoted by a double bar, as t_1 in Fig. 1, and a *macroplace* by a double circle, as p_4.

An *input door* (ID) is formed by nodes related by incoming arcs from the exterior of the subnet. The input door is denoted by a set of the nodes. For example, in Fig. 1, the input door of the subnet N' is ID = $\{p_2\}$.

An *output door* (OD) is formed by nodes related by outgoing arcs to the exterior of the subnet.

The OD is denoted by a set of the nodes. In Fig. 1, the output door of the subnet N' is OD = $\{p_2, p_4\}$.

A *between door* (BD) is formed by nodes in the subnet which are not involved in the input door nor the output door. In Fig. 1, the between door of the subnet N' is BD = $\{t_3\}$.

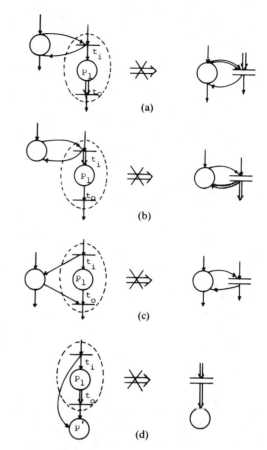

Fig. 3. Subsets which do not satisfy conditions of GRSN-1T. (a) $k' = 2$ when circuit containing t_i exists. (b) $k = 2$ when circuit containing t_i exists. (c) Path $t_i - p_1 - t_o$ is not only one path from t_i to t_o. (d) $F(p', t_i) \neq 0$ when $k' > 1$.

A *subnet of transition* (SNT) is a subnet taking only transitions as its input and output doors.

A *subnet of place* (SNP) is a subnet taking only places as its input and output doors. The subnet N' in Fig. 1 is an SNP because $\not\exists t \in$ ID $\cup$ OD.

A *subfiring sequence* is a firing sequence that can fire all of the transitions in the subnet consecutively.

III. REDUCTION RULES

In this section, generalized reducible subnets (GRSN's) are defined, and their reduction rules are given. A generalized Petri net $N = (P, T, B, F, M)$ is transformed into a reduced net $N' = (P', T', B', F', M')$.

A. GRSN-1

1) GRSN-1T (cf. Figs. 2 and 3):

Definition: A GRSN-1T is an SNT satisfying the following conditions:

a) ID $\cup$ OD = $\{t_i, t_o\}$, BD = $\{p_1\}$, $M(p_1) = 0$;
b) the path $t_i - p_1 - t_o$ is only one path from t_i to t_o;
c) $\exists k \in \mathbb{N}^+$, $F(p_1, t_i) = k \cdot B(p_1, t_o)$, or $\exists k' \in \mathbb{N}^+$, $k' \cdot F(p_1, t_i) = B(p_1, t_o)$, where $\mathbb{N}^+$ is the set of positive integers (the firing of t_i and t_o produces tokens k times more or k' times less);
d) $k = k' = 1$, if $\exists$ circuit containing t_i;
e) $\forall p \neq p_1$, $F(p, t_i) = 0$ when $k' > 1$.

Reduction: A GRSN-1T is replaced by a macrotransition t^* as follows (cf. Fig. 2):

a) $P' = P - \{p_1\}$;
b) $T' = T - \{t_i, t_o\} + \{t^*\}$;

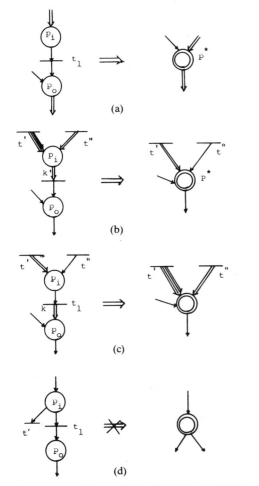

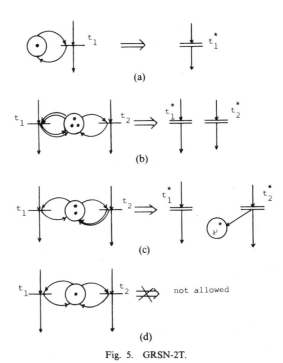

Fig. 5. GRSN-2T.

Fig. 4. GRSN-1P. (a) $k = k' = 1$. (b) $k' = 2$. (c) $k = 2$. (d) Not allowed because $B(p_i, t') \neq 0$.

c) $B'(p, t) = k' \cdot B(p, t_i) + B(p, t_o)$, if $t = t^*$,
   $\quad = B(p, t)$, otherwise
(The weight of input arcs of t_i is multiplied by k');

d) $F'(p, t) = k \cdot F(p, t_o) + F(p, t_i)$, if $t = t^*$
   $\quad = F(p, t)$, otherwise
(the weight of output arcs of t_o is multiplied by k);

e) $M'(p) = M(p)$.

This reduction rule reduces a sequence of transition-place-transition into a macrotransition.

2) GRSN-1P:

Definition: A GRSN-1P is an SNP satisfying the following conditions (cf. Fig 4):

a) $ID \cup OD = \{p_i, p_o\}$, $BD = \{t_1\}$, $p_i \in ID$, $p_o \in OD$, $M(p_i) = 0$;

b) $\forall t \neq t_1, B(p_i, t) = 0$ (t_1 is the only one output of p_i);

c) $\exists k \in \mathbb{N}^+, k \cdot B(p_i, t_1) = F(p_o, t_1)$ (the transition t_1 increases tokens by k times) or $\exists k' \in \mathbb{N}^+, B(p_i, t_1) = k' \cdot F(p_o, t_1)$, and $\exists l_a \in N, \forall t_a \in T, F(p_i, t_a) = k' \cdot l_a$ (t_1 decreases tokens by k' times, and the weight of input arcs can be divided by k').

Reduction: A GRSN-1P is replaced by a macroplace p^* as shown in Fig. 4:

a) $P' = P - \{p_i, p_o\} + \{p^*\}$;
b) $T' = T - \{t_1\}$;
c) $B'(p, t) = B(p_o, t)$, if $p = p^*$,
   $\quad = B(p, t)$, otherwise;
d) $F'(p, t) = (k/k') \cdot F(p_i, t) + F(p_o, t)$, if $p = p^*$,
   $\quad = F(p, t)$, otherwise;
e) $M'(p) = M(p_o)$, if $p = p^*$,
   $\quad = M(p)$, otherwise.

This reduction rule reduces a sequence of place-transition-place into a macroplace.

B. GRSN-2

1) GRSN-2T:

Definition: A GRSN-2T is an SNT satisfying the following conditions (Fig. 5):

a) $ID \cup OD = \{t_1, t_2, \cdots, t_a, \cdots, t_q\}$, $q = 1, 2, 3, \cdots$, and $BD = \{p_2\}$;

b) $\forall t_a \in ID, \exists p \in P - \{p_2\}, B(p, t_a) > 0$ (p_2 is not only one input of t_a);

c) $\forall t_a \in ID, F(p_2, t_a) - B(p_2, t_a) = l_a \geq 0$;

d) $M(p_2) \geq \sum_{a=1}^{q} B(p_2, t_a)$.

Reduction: A GRSN-2T is reduced as follows (Fig. 5):

a) $P' = P - \{p_2\} + \{p^*\}$, if $\exists l_a > 0, a = 1, 2, 3, \cdots, q$,
   $\quad = P - \{p_2\}$, otherwise;
b) $T' = T - \{t_1, \cdots, t_q\} + \{t_1^*, t_2^*, \cdots, t_q^*\}$;
c) $B'(p, t) = 0$, if $p = p^*$,
   $\quad = B(p, t)$, otherwise;
d) $F'(p, t) = l_a$, if $t = t_a, p = p^*$,
   $\quad = F(p, t)$, otherwise;
e) $M'(p) = 0$, if $p = p^*$,
   $\quad = M(p)$, otherwise.

This reduction rule eliminates a redundant representation of resource sharing.

2) GRSN-2P:

Definition: A GRSN-2P is an SNP satisfying the following conditions (Fig. 6):

a) it forms a circuit $p_1 - t_1 - \cdots - p_a - t_a - \cdots - p_q - t_q$;
b) $\forall p \in ID \cup OD, \forall t \in BD, B(p, t) = F(p, t) = w$ (the weights of arcs in the circuit are same);
c) $\forall t, F(p_a, t) = w$ if $F(p_a, t) > 0$ or $B(p_a, t) = w$ if $B(p_a, t) > 0$.

Reduction: A GRSN-2P is replaced by a macroplace p^* (Fig. 6):

a) $P' = P - \{p_1, \cdots, p_q\} + \{p^*\}$;
b) $T' = T - \{t_1, \cdots, t_q\}$;
c) $B'(p, t) = \sum_{a=1}^{q} B(p_a, t)$, if $p = p^*$,
   $\quad = B(p, t)$, otherwise;

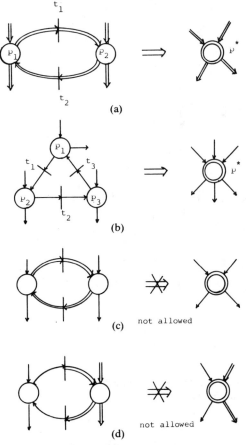

Fig. 6. GRSN-2P.

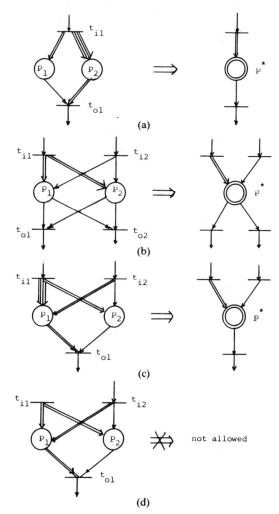

Fig. 7. GRSN-3T.

d) $F'(p, t) = \sum_{a=1}^{q} F(p_a, t)$, if $p = p^*$,
    $= F(p, t)$, otherwise;
e) $M'(p) = \sum_{a=1}^{q} M(p_a)$, if $p = p^*$,
    $= M(p)$, otherwise.

This reduction rule reduces a circuit having places at its input or output doors into a macroplace.

C. GRSN-3

1) GRSN-3T:

Definition: A GRSN-3T is an SNT satisfying the following conditions (Fig. 7):

a) $\{t_{i1}, t_{i2}, \cdots, t_{ia}, \cdots, t_{im}\} \subseteq \text{ID}$, $m = 1, 2, \cdots$,
    $\{t_{o1}, t_{o2}, \cdots, t_{ob}, \cdots t_{on}\} \subseteq \text{OD}$, $n = 1, 2, \cdots$,
    $\text{BD} = \{p_1, p_2, \cdots, p_l, \cdots, p_q\}$, $q = 1, 2, \cdots$,
    $M(p_l) = 0$;
b) $F(p_l, t_{ia}) > 0$ for $l = 1, 2, \cdots, q, a = 1, 2, \cdots, m$,
    $B(p_l, t_{ob}) > 0$ for $l = 1, 2, \cdots, q, b = 1, 2, \cdots, n$;
c) $\forall t_{ia}, t_{ob} \; \exists k_l \in \mathbb{N}^+$ such that
    $$k_l = (1/k_{xa}) \cdot F(p_l, t_{ia}) = (1/k'_{xb}) \cdot B(p_l, t_{ob})$$
    for $l = 1, 2, \cdots, q$

(the least common multiple exists) where
$$k_{xa} = \min_{l=1,\cdots,q} [F(p_l, t_{ia})] = F(p_x, t_{ia})$$

(the minimum weight of arcs between t_{ia} and p_l is given by p_x) and
$$k'_{xb} = \min_{l=1,\cdots,q} [B(p_l, t_{ob})] = F(p_x, t_{ob})$$

(the minimum weight of arcs between p_l and t_{ob} is given by p_x).

Reduction (cf. Fig. 7):

a) $P' = P - \{p_1, \cdots, p_q\} + \{p^*\}$;
b) $T' = T$;
c) $B'(p, t) = k'_{xb}$, if $p = p^*, t = t_{ob}$,
    $= B(p, t)$, otherwise;
d) $F'(p, t) = k_{xa}$, if $p = p^*, t = t_{ia}$,
    $= F(p, t)$, otherwise;
e) $M'(p) = 0$, if $p = p^*$,
    $= M(p)$, otherwise.

2) GRSN-3P:

Definition: A GRSN-3P is an SNP satisfying the following conditions (Fig. 8):

a) $\{p_{i1}, p_{i2}, \cdots, p_{ia}, \cdots, p_{im}\} \subseteq \text{ID}$,
    $\{p_{o1}, p_{o2}, \cdots, p_{ob}, \cdots, p_{on}\} \subseteq \text{OD}$,
    $\{t_1, t_2, \cdots, t_l, \cdots, t_q\} = \text{BD}$;
b) $B(p_{ia}, t_l) > 0$ for $a = 1, 2, \cdots, m, l = 1, 2, \cdots, q$,
    $F(p_{ob}, t_l) > 0$ for $b = 1, 2, \cdots, n, l = 1, 2, \cdots, q$;
c) $\forall p_{ia}, p_{ob} \; \exists k_a, k'_b \in \mathbb{N}^+$ such that
    $$B(p_{ia}, t_l) = k_a$$
    $$F(p_{ob}, t_l) = k'_b, \quad l = 1, 2, \cdots, q.$$

Reduction: (Fig. 8):

a) $P' = P$;
b) $T' = T - \{t_1, \cdots, t_q\} + \{t^*\}$;
c) $B'(p, t) = k_a$, if $t = t^*, p = p_{ia}$,
    $= B(p, t)$, otherwise;

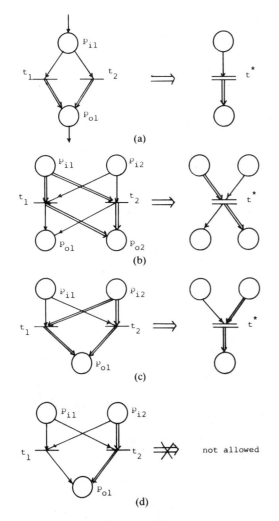

Fig. 8. GRSN-3P.

d) $F'(p, t) = k'_b,$ if $t = t^*, p = p_{ob},$
   $\quad\quad\quad = F(p, t),$ otherwise;
e) $M'(p) = M(p).$

Note that in all the foregoing reduction rules, the value of k and k' is 1 when the net is an ordinary Petri net (that is, $F(p, t)$ and $B(p, t)$ are 0 or 1).

IV. ANALYSIS

A. Properties of Reduction Method

Some important properties are defined upon Petri nets, such as liveness, boundedness and proper termination. By analyzing such properties, modeling errors are often detected, and many desirable properties of the system modeled by a Petri net can be unveiled. However, the analysis of a large Petri net is often complex. The basic reason of Petri net reduction is to reduce the complexity of analysis. Therefore, a reduction has to preserve the properties of the net. To verify the homogeneity between the original net and its reduced net, we develop the following theorems.

Theorem 1: A Petri net is live if and only if its reduced net is live.

Proof: The liveness of Petri nets is determined by the firing sequence of transitions and the number of tokens received and produced by transitions. If a reduction does not change any firing sequence and the number of tokens, the liveness is preserved. In other words, the proposed reduction method must retain the number or direction of flow of tokens in the original net. Therefore, we prove this theorem by checking whether a reduction changes the firing sequence and the number of tokens for each reduction rule.

GRSN-1T: The transitions in a GRSN-1T can be fired consecutively, that is, a GRSN-1T can be fired by a subfiring sequence. In the firing sequence of transitions, a reduction of GRSN-1T by a macrotransition means a replacement of the subfiring sequence by a macrotransition. Therefore, the reduction does not change the firing sequence. From the viewpoint of token number, the macrotransition produces tokens by k times if the output arc weight of ID is k times greater than the input arc weight of OD (i.e., $F(p_1, t_1) = k \cdot B(p_1, t_o)$); the macrotransition demands tokens by k times if the input arc weight of OD is k' times greater than output arc weight of ID (i.e., $k \cdot F(p_1, t_i) = B(p_1, t_o)$). Therefore, the number of tokens received and produced by the macrotransition is equal to that number of the GRSN-1T.

GRSN-1P: A GRSN-1P can be fired by a subfiring sequence. A reduction of GRSN-1P by a macroplace means a deletion of the subfiring sequence. Therefore, the firing sequence is not changed. If the BD of GRSN-1P has output arcs k times more than its input (i.e., $k \cdot B(p_i, t_1) = F(p_o, t_1)$), it has the function of increasing tokens by k times. After reduction, the function of increasing is done by the input transitions of the macroplace ($F(p, t) = k \cdot F(p_i, t) + F(p_o, t)$, if $p = p^*$). If the BD has input arcs k' times more than output (i.e., $B(p_i, t_1) = k' \cdot F(p_o, t_1)$), it decreases the token flow by k' times. After reduction, the input transitions of the macroplace decreases tokens by k' times ($F'(p, t) = (1/k') \cdot F(p_i, t) + F(p_o, t)$, if $p = p^*$). Therefore, the number of tokens entering and leaving the macroplace is the same as the number of the GRSN-1P.

GRSN-2T: The reduction rule of GRSN-2T does not delete any transition or replace any transition by a macrotransition. Therefore, the firing sequence is not changed through the reduction. In a GRSN-2T, the number of tokens in the BD is not less than the sum of output arc weight of BD ($M(p) \geq \sum_{a=1}^{q} B(p_2, t_a)$). The transitions can be fired without considering the state of BD. The BD is redundant and thus eliminated. A place without output arc is added only if $F(p_2, t_a) - B(p_2, t_a) > \theta$. Therefore, the reduction does not change the number of tokens entering and leaving the GRSN-2T.

GRSN-2P: A GRSN-2P is a circuit. The weights of arcs are all the same ($\forall p \in \text{ID} \cup \text{OD}, \forall t \in \text{BD}, B(p, t) = F(p, t)$). Tokens in a place can always go to any place in the GRSN if they enable a transition. The transitions in the GRSN can thus be fired consecutively by a subfiring sequence. The reduction of GRSN-2P means a deletion of the subfiring sequence in the firing sequence. Therefore, the reduction does not change the firing sequence. The weights of input and output arcs of GRSN-2P are all same, and they are conserved through the reduction. The number of tokens entering and leaving the GRSN-2P is equal to the number of macroplaces.

GRSN-3T: A reduction of the GRSN-3T is merging places which present parallel states. It does not delete any transition or replace any transition by a macrotransition. Therefore, the reduction does not change the firing sequence. The firing of t_{ia} gives tokens to all the places in BD simultaneously ($F(p_l, t_{ia}) > \theta$ for $l = 1, \cdots, q, a = 1, \cdots, m$), and the firing of t_{ob} removes tokens from all the places ($B(p_l, t_{ob}) > \theta$ for $l = 1, \cdots, q, b = 1, \cdots, n$). Therefore, multiple places in BD are redundant. The reduction eliminates all places in BD except for a place having the least common multiple weight of input and output arcs. Therefore, the number of tokens received by ID and tokens produced by OD is not changed after reduction.

GRSN-3P: A reduction of the GRSN-3P eliminates nondeterministic situations. Transitions having the same input and output place through same weight of arcs are merged in a macrotransition. The macrotransition has the same input and output place; and the weight of arcs is not changed. Therefore,

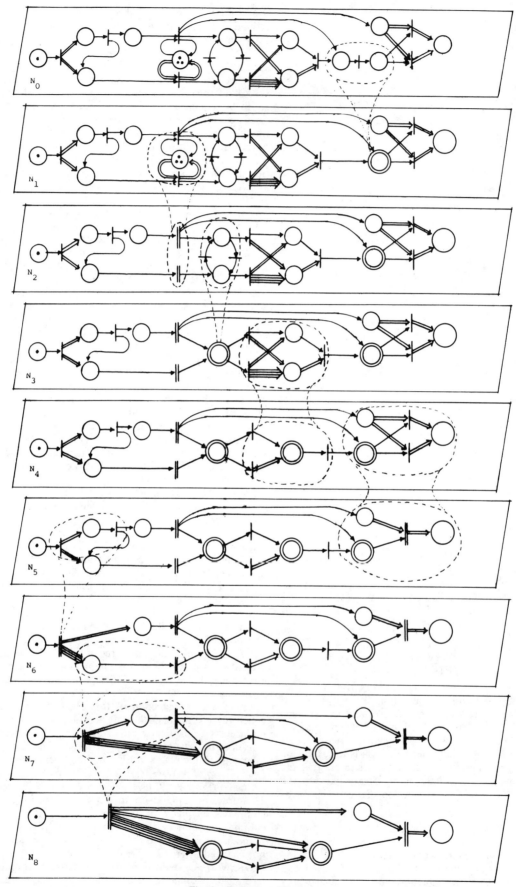

Fig. 9. Reduction of Petri net.

the reduction does not change the firing sequence and the number of tokens entering ID and leaving OD.

Theorem 2: A Petri net is bounded if and only if its reduced net is bounded.

Proof: The boundedness of a Petri net is determined by the number of tokens which flow in the net. If the number of tokens is not changed, the property of boundedness is preserved. In the proof of Theorem 1, we have seen that all of the reduction rules do not change the number of tokens which flow in the net. Therefore, the property of boundedness is preserved.

Theorem 3: A Petri net is properly terminating if and only if its reduced net is properly terminating.

Proof: In the same way as the earlier theorems, the property of proper termination is preserved because the reduction does not change the number or direction of flow of tokens in the net.

B. Example and Application

We have programmed the proposed reduction procedure in Prolog on an IBM-PC (MS-DOS). In this automated reduction procedure, a generalized reducible subnet is reduced into a macroplace or macrotransition [9] in each iteration. A GRSN to reduce is selected randomly. An example illustrating this reduction procedure is given in Fig. 9. The original Petri net N_0 in Fig. 9 is reduced into the irreducible net N_8 after eight iterations.

We have also applied the reduction procedure to reduce a Petri net modeling a factory producing corrugated fiberboard boxes in Korea. In this model, a place represents a condition, a transition an event, and a token a signal or physical material. The Petri net has 151 nodes (92 places, 59 transitions) and 174 arcs. We have obtained a reduced net containing 20 places, 12 transitions, and 35 arcs. We have analyzed the system with this reduced net, and the net was live, bounded, and properly terminating.

V. Conclusion

A reduction method of generalized Petri nets is proposed. The properties of the proposed reduction method are also discussed.

The reduced nets are homogeneous with the original net from the viewpoint of the properties of liveness, boundedness, and proper termination. A complex Petri net can be analyzed with its reduced net. Depending upon the properties of the system, the Petri net can be reduced by macrotransition, macroplace, or both. The reduction method is defined on the basis of the structure of nets. Therefore, the test of reducible subnet is easy, and the reduction procedure can be automated.

References

[1] C. A. Petri, "Communication with automata," Ph.D. dissertation, Bonn Univ., 1964. (Translated: Grifiss Air Force Base. New York, RADC-TR-65-377, 1966).
[2] F. Commoner, A. W. Holt, S. Even, and A. Pnueli, "Marked directed graphs," *J. Comp. Syst. Sci.*, vol. 5, pp. 511–523, 1971.
[3] J. L. Peterson, "Petri nets," *Ass. Comput. Mach. Comput. Surveys*, vol. 9, pp. 223–252, Sept. 1977.
[4] ——, *Petri Net Theory and the Modeling of Systems*. Englewood Cliffs, NJ: Prentice-Hall, 1981.
[5] G. W. Brams, *Reseaux de Petri Theorie et Pratique*. Paris, France: Masson, 1983.
[6] W. Brauer, Ed., *Net Theory and Applications*, Lecture Notes in Computer Science 84. Berlin, Germany: Springer-Verlag, 1980.
[7] M. H. Hack, "Analysis of production schemata by Petri nets," Project MAC, Mass. Inst. Technol. Cambridge, Tech. Rep. 94, 1972.
[8] G. Rozenberg, Ed., *Advances in Petri Nets 1984*, Lecture Notes in Computer Science, 188. Berlin, Germany: Springer-Verlag, 1985.
[9] H. Lee-Kwang and J. Favrel, "Hierarchical reduction method for analysis and decomposition of Petri nets," *IEEE Trans. Syst., Man, Cybern.*, vol. SMC-15, pp. 272–280, Mar. 1985.
[10] H. Lee-Kwang, "Analyse et modelisation des systeme production par les reseaux de Petri," Dr-Ing, INSA de Lyon, 1985.
[11] H. Lee-Kwang and J. Favrel, "Analysis of Petri nets by hierarchical reduction and partition," in *IASTED Modelling and Simulation*. Zurich, Switzerland: Acta Press, 1982, pp. 363–366.
[12] ——, "Hierarchical reduction and decomposition of graphs for system analysis," in *Proc. IEEE*, pp. 94–101, Oct. 1984.
[13] ——, "Hierarchical decomposition of Petri net languages," *IEEE Trans. Syst., Man, Cyber.*, to appear.
[14] C. Andre, F. Boeri, and J. Marin, "Synthese et realisation des systemes logiques a evolution simultanees," *RAIRO*, vol. 10, pp. 67–86, Apr. 1976.
[15] M. Silva, "Sur le concept de macro place et son utilisation pour l'analyse des reseaux de Petri," *RAIRO Automat.*, vol. 15, pp. 335–345, 1981.
[16] Y. S. Kwong, "On reduction of asynchronous systems," *Theorit. Comput. Sci.*, vol. 5, pp. 25–50, 1977.
[17] G. Berthelot, "Une methode de verification de reseaux de Petri," in *Proc. Afcet Conf. Reseaux de Petri*, pp. 33–54, 1977.
[18] ——, "Preuve de non blocage de programmes paralleles par reduction de reseaux de Petri," in *Proc. First Conf. Parallel and Distributed Processing*, Toulouse, France, 1979, pp. 251–259.
[19] ——, "Transformation et analyse de reseaux de Petri: Application aux protocoles," These d'Etat, Univ. Paris, VI, 1983.
[20] T. Murata and J. Y. Koh, "Reduction and expansion of live and safe marked-graphs," *IEEE Trans. Circuits Syst.*, vol. CAS-27, pp. 68–70, Jan. 1980.
[21] R. Johnsonbaugh and T. Murata, "Additional method for reduction and expansion of marked graphs," *IEEE Trans. Circuits Syst.*, vol. CAS, pp. 1009–1014, Oct. 1981.
[22] S. T. Dong, "The modelling, analysis and synthesis of communication protocols," Ph.D. dissertation Univ. California, Berkeley, 1983.
[23] E. W. Mayr and A. R. Meyer, "The complexity of the finite containment problem for Petri nets," *J. Ass. Comput. Mach.*, no. 3, pp. 561–567, 1981.

An Autonomous, Decentralized Control System for Factory Automation

Norihisa Komoda, System Development Laboratory, Hitachi, Ltd.
Kazuo Kera and Takeaki Kubo, Omika Works, Hitachi, Ltd.

Reprinted from *Computer*, December 1984, pages 73-83. Copyright © 1984 by The Institute of Electrical and Electronics Engineers, Inc. All rights reserved.

Hitachi's total factory automation systems are based on distributed workstations adaptable enough for use in a variety of industries. Key components are the flexible manufacturing cell controller and a process local area network.

Computer systems are a must for managing modern complex production lines. However, the drastic progress in software engineering and semiconductor technologies—especially microcomputer technologies—has meant that automated managerial and control systems must alter configurations almost every year.

The concept of total factory automation (Figure 1) includes computer-aided design, or CAD, and production control, such as flexible manufacturing systems and corresponding production scheduling systems. These systems, which are provided as one loosely coupled system, are gradually integrated and tightened into a total FA system.

Modern FA systems consist of CAD and computer-aided manufacturing, which includes a flexible manufacturing cell controller and laboratory automation. Some system builders are approaching this total FA system market from CAD technology, some from FMS, and others from robot application technology.[1,2] In some industries, computerized systems have been categorized as laboratory automation systems because of plant characteristics. Semiconductor production processes are a good example.

Hitachi has been producing process computer systems since the early 1960's for both domestic and overseas markets. One of its major applications fields is production control, which was the inspiration for Hitachi's factory management system, or FMS, products.[3] More than 600 production-control systems are supplied for applications such as automobile assembly, iron- and steel-making, assembly of machinery and equipment, and warehouse control. The company has started producing total FA systems, adopting, as FMS, its existing production-control systems and integrating Hitachi-designed CAE workstations.

The autonomous, decentralized FA system

Overview. A distributed total FA system requires (1) unified management and control from overall production scheduling to on-line control at the smallest unit in production lines; (2) a high degree of expandability and maintainability in system hardware and software; (3) a high level of system reliability and fault tolerance; (4) minimum engineering manpower and

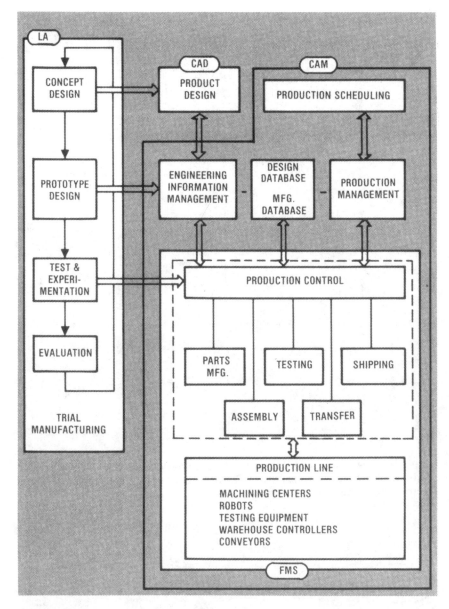

Figure 1. A total factory automation system; LA = laboratory automation and FMS = factory management system.

short-term manpower for software system development and enhancement; and (5) minimum system cost made possible by the adoption of suitable microcomputer systems for use as subsystems.

To meet these requirements, Hitachi developed a hierarchical system configuration, referred to as an autonomous decentralized control system for factory automation (Figure 2). Figure 2 shows the various stages and corresponding functions for such a system. For each stage, subsystems are provided, which can be anything from business computers for high-ranking managerial procedures to PLCs (programmable logic controllers) for robots and/or warehouse controllers.

The autonomous, decentralized FA system has two key components. The first is composed of the Hitachi FMC, or flexible manufacturing cell controller, and a process local area network called the Micro-Sigma-Network. FMC, a controller based on a 16-bit microprocessor, has been developed as the work center and station levels controller. Even during scheduled or nonscheduled outages of the upper level business/process computer, processes at the FMC level must continue. Should the FMC fail, the surviving FMC(s) needs to continue cooperative control and monitor functions. This fault tolerance also allows system expansion and modification without total system shutdown.

Computers at lower levels and FMCs are connected through the Micro-Sigma-Network LAN. To ensure a high degree of expandability and maintainability in the network structure, function-coded broadcasting is incorporated in data transmission.

The second key component of the autonomous, decentralized FA system is problem-oriented languages for FA applications. Control software developed through conventional methods, such as the use of general languages or a ladder diagram, is incomprehensible and hard to maintain. To resolve such difficulties and to provide a high level of system flexibility and maintainability, a new concept adapted from the knowledge engineering field is becoming more popular. Here, control logics are picked up as data statements from program flows, and procedures are automatically selected in accordance with the system situation to be controlled. From this concept, the SCR (Station Controller) for sequence control and SCD (Station Coordinator) for constraint combinational control were developed. In these control software description tools, control logics are embedded in control software as data, not as procedures, and take such problem-oriented forms as IF-THEN rules. Control logics are thus easily comprehended and can be modified independently of other logics, facilitating control software enhancement and upgrades. As processes or system requirements change, SCR and SCD descriptions can be easily updated in the interactive mode or in the higher ranked system.

Hardware specifications. The FMC, designed to monitor and control work centers and workstations, is linked to the Hitachi V-90 process computer family and other FMCs through the Micro-Sigma-Network process LAN. The FMC executes its direct-control data acquisition as monitoring and human-machine communication over FA control equipment by means of the process input-output interface (PI/O) and/or PLCs, both of which are linked to the FMC through the S-bus. FMC peripherals, such as CRTs and printers, can be controlled directly or through the Micro-Sigma-Network.

Although FMCs are designed as subsystems of total, integrated FA systems, they can be installed as stand-alone controllers permanently or temporarily at the earliest stage of system construction. A typical one-FMC system configuration is given in Figure 3; multiple FMCs are shown in Figure 4. Table 1 summarizes major FMC hardware specifications.

Token-passing is the access method used in the Micro-Sigma-Network. Transmission speed is 1M bps, with a variable message length facility (up to 512 bytes in one message) that can raise data transmission speed among stations to 20K bytes per second. Data transmission incorporates function-coded broadcasting to allow the sending station to inform many stations of its message.

Another network feature is loop-back, which ensures data transmission security.

Table 2 lists major hardware specifications for the Micro-Sigma-Network.

Table 1. FMC processor specifications.

FMC COMPONENT	SPECIFICATION
Processor HD68000	
Instruction set	56 types
Data length	8/16/32 bits
Instruction speed	ADD 0.8 μs
	MUL 9.0 μs
Main memory	
Capacity	512K-1024K bytes
Cycle time	0.63 μs
Error check	with ECC
Interrupt levels	7 levels
Bus	
Interface	based on IEEE 796
Data speed	1.5M bytes per second

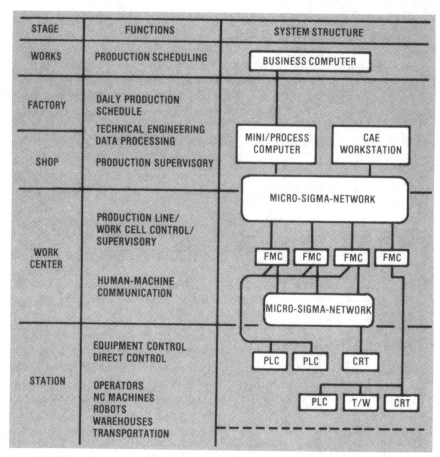

Figure 2. A decentralized, total factory automation system; FMC = flexible manufacturing cell controller, PLC = programmable logic controller, and Micro-Sigma-Network is a process LAN.

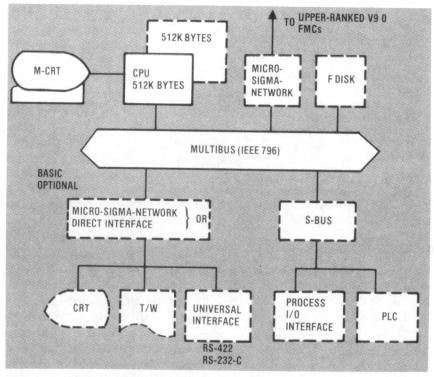

Figure 3. Configuration of a stand-alone flexible manufacturing cell controller.

Another data transmission scheme provided in FMC systems is the S-bus interface, which controls process I/O interface equipment and PLCs. Data transmission is 19.2K bps. As shown in Figure 4, one PLC can be controlled by multiple FMCs with party-line control via the S-bus.

Software specifications. In FMC software, the Process Monitor System, or PMS, is implemented as an on-line, real-time operation system. (FMC software specifications are given in Table 3.) Its design concept has been adopted for Hitachi process computer families within the last two decades. As mentioned earlier, function-coded broadcasting is provided to support distributed FMC systems. Also, for centralized software maintenance and development, both downloading and uploading are supported.

For programming development, AT&T Bell's Unix or Digital Research's CP/M can be implemented as optional off-line operating systems.

Fortran 77 and C are available as languages, as are two problem-oriented languages, SCR and SCD, which we will talk about later. Programs can be developed and maintained on FMCs and on the higher ranked Hitachi V-90 process computer (which has faster computing speed and more memory address space) as centralized software maintenance.

Standardized software packages are available in general-purpose languages or SCR/SCD for human-machine interface, database handling, and application-oriented processing for transfer control, data acquisition, and production control.

Rule-based control languages for factory automation. The two types of control logic in FA systems are *sequence control* with synchronization and exclusion, such as control on tools in a workstation, and *constraint combinational control*, in which several conditions are checked and the optimum action is selected.

Traditionally, control software is written in general-purpose languages, such as Fortran, or in a ladder diagram. However, understanding and modifying such software is difficult and time-consuming, even for trained persons. Thus, methods for improving the productivity, flexibility, and maintainability of control software are needed.

Hitachi introduced a new concept from the knowledge engineering field when it developed Station Controller, or SCR, and Station Coordinator, or SCD, for the two types of control logic just described. To express sequential logic plainly, an enhanced Petri-net[4] and high-performance interpretation method for control-net have been developed. This type of controller is the SCR.[5] SCD, on the other hand, is a high-speed inference mechanism that handles constraint logic for deciding actions represented by IF-THEN rules.

Station controller. We define a safe Petri net as a subset of a Petri net PN in which each place can have at most one token; that is, the tuple $PN = (P, T, I, O, M)$, where P is the set of places $(p_1, p_2, \ldots, p_n)$, T is the set of transitions $(t_1, t_2, \ldots t_n)$, I represents "input," O represents "output," and M is a marking. A place such that $p \in I(t)$ (or $p \in O(t)$) is called an input place (or an output place) of t. M is called a marking, where $M(p_i) = 1$ (or $M(p_i) = 0$) means that the place p_i has a token (or has no token). A transition is *enabled* at marking M_1, if and only if $M_1(p_i) = 1$ for all $p_i \in I(t)$, and if $M_1(p_j) = 0$ for all $p_j \in O(t)$. Regarding an enabled transition and its input/output places, a new marking $M_2 : M_2(p_i) = 0$ and $M_2(p_j) = 1$ for all $p_i \in I(t)$ and $p_j \in O(t)$ is reachable by firing t.

In a Petri-net model, if each place is assumed to represent an action of machines to be controlled, sequence-control specifications—including mutual synchronization and exclusion control of many machines' actions—can be described explicitly as a directed graph model. The execution sequence of the machines' actions can be defined basically as a Petri-net structure, and a synchronous and concurrent progress of the sequence is realized as the movement of tokens on the Petri net. To apply the Petri-net model to describe sequence-control specifications, Hitachi has developed the control-net model.

Control-net is defined by the tuple $CN = (P, T, I, O, \delta, \varphi, \eta, U, V, M)$, where $P, T, I, O,$ and M are the same as those of the Petri-net model. $\delta, \varphi,$ and η are process input/output functions for establishing a correspondence between the places of the con-

Table 2. FA network specifications.

FEATURE	SPECIFICATION
Name	Micro-Sigma-Network
Maximum system scale	32 Workstations/loop
Station-to-station distance	Max. 1000 meters (optical fiber) Max. 100 meters (metal cable)
Transmission speed	1M bps
Transmission processing	Based on IEEE 802
Message length	Max. 512 bytes/message (variable)
Access method	Token-passing
System RAS	Automatic retrial Loopback Tracing

Table 3. FMC software specifications.

FEATURE	SPECIFICATION
Real-time OS	
PMS	Multitask, on-line, real-time monitor system
Micro DPCS	Function-coded broadcast transmission Network support (Micro-Sigma-Network) Uploading/downloading
Programming language	
General	Fortran 77, C
SCR	Sequence language on Control-net (C-net)
SCD	Rule-based language
Application support packages (not limited to)	
NSS	Network support
FMS-R	Real-time file management
IOS	I/O execution support
PDCS	PI/O execution support
CMS	Graphic CRT support

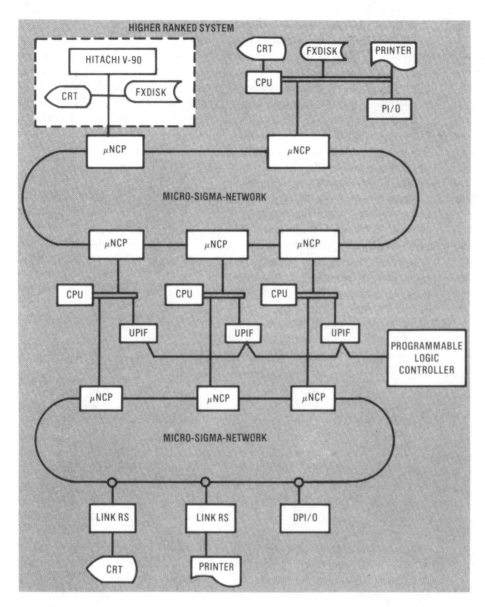

Figure 4. Configuration of a network of flexible manufacturing cell controllers; μNCP = micro-node control processor, PI/O = process input-output interface, and UPIF = universal peripheral interface.

trol net and the control and observable signals of the system to be controlled. δ is a command function, φ is a response function, and η is a gate function. U and V are system status functions to supervise the execution status of actions at a place and to manage transition open/close statuses. These five new functions are introduced for describing process interfaces and supervising system statuses.

When a token moves to place p_i, a control signal x_i defined by $\delta(p_i)$ is generated, which triggers an action. After a control signal is given, the token remains in the box until one of the input signals y_{i1}, y_{i2}, y_{in} defined by $\varphi(p_i)$ is detected as the completion of the action concerned. Input signal y_{ik} defined by $\eta(t_i)$ is used for open/close operations on transitions. Since tokens cannot advance until input signal y_{ik} is detected, token movement can be easily coded with a self-explanatory, transition control rule. System status functions U and V are defined as follows:

$U(p_i) =$
   O: actions associated with p_i under execution.
   i_n: actions associated with p_i that are completed with return code y_{in}.

$V(t_i) =$
   0: t_i is closed.
   1: t_i is open.

Places or transitions defined as process input/output functions and systems status functions are called *boxes* or *gates*.

A gate $t_i \epsilon T$ is enabled at marking M_1, if and only if $V(t_i) = 1$, $M_1(p_i) = 0$, $U(p_i) = 0$, and $M_1(p_j) = 0$ for all $p_i \epsilon I(t_i)$ and $p_j \epsilon O(t_i)$. When t_i is fired, all output boxes are provided with a token, and all tokens in input boxes are deleted. After this firing, the output commands specified by the output boxes are sent out, and $U(p_i)$ becomes 0. A box in which process input/output functions are defined is called an *act box*.

For more complicated sequence-control descriptions, control-net has a number of other functions. The first is the *receive box,* which detects a request to start a job and generates a token in itself. The *timer box* holds a token for a specified time interval. An associated act box counts the execution time of an action and signals an alarm for time-out error.

A third function describes repetition of some sequence blocks, introducing *source boxes* and *count gates*. A source box generates a specified number of subtokens, while a count gate sums up the number of tokens that pass over it. The count gate absorbs the tokens it receives until the count reaches a preassigned number. When it has counted up the specified number of tokens, the count gate will pass tokens to all appropriate output boxes.

The last function involves numerical calculation routines that can be implemented in the form of user subroutines linked to program boxes. When a token drops into a program box, the associated subroutine will be executed.

Figure 5 is an example of an assembly station. At this station, robot A fetches two parts from a feeder and assembles them on the table. Robot B picks up four screws and turns them. Control specifications for two robots are described with control-net as shown in Figure 6. Boxes and gates are given unique box and gate names. The boxes GRAS1, PUT, and RETN represent actions of robot A. GRAS2 and SCREW are for robot B. When a job-start request signal is detected at receive box RCEVE, a token is generated in it. This token moves through control-net, as shown in Figure 6, according to control-net firing rules, and the robots' actions are started and stopped in accordance with the token's movement.

For practical system response, an efficient control-net interpretation schema called the selective scanning method (Figure 7) has been developed. This method has four execution steps, S1, S2, S3, and S4, with four stacks, H1, H2, H3, and H4. At first, all boxes p_i with $M(p_i) = 1$ and $U(p_i) = 0$ are picked up in H1. The number of boxes and gates in Hi is indicated as h_i. The steps of the scanning method are

S1: Complete judgment.
   Select p_i with $U(p_i) = 1$ from H1 and store them in H2.
S2: Choose candidate gates for enable test.
   For all $p_i \epsilon$ H2, check whether T_j satisfies $p_i \epsilon I(T_j)$ and $M(p_i) = 1$, or $p_i \epsilon O(T_j)$ and $M(p_i) = 0$.
   Pick up all T_j that satisfy above conditions into H3 and delete all p_i that bring T_j from H2.
S3: Enable test.
   For all $T_j \epsilon$ H3, select all T_k that satisfy the enable condition from H3 and save them in H4

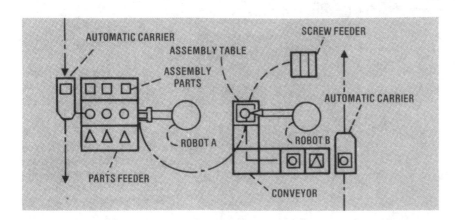

Figure 5. A sample assembly station.

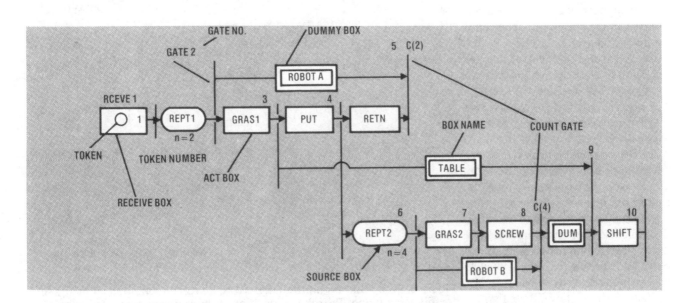

Figure 6. Control-net representation of sequence control specification for the assembly station in Figure 5.

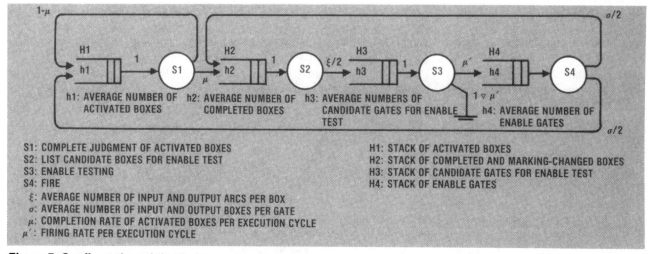

Figure 7. Configuration of the limited (reduced) scanning method.

S4: Fire.

For all $T_k \in H4$, fire T_k, select $p_k \in O(T_k)$ onto H1, select $p_j \in I(T_k)$ onto H2, and delete T_k from H4.

With these execution steps, the test is executed only for gates that have possibility for firing, thus reducing total execution time and the response time of control-net interpretation.

We obtain the total execution time of S1 through S4 by

$$Z = h_1 \cdot t_j + h_2 \cdot t_i + h_3 \cdot t_e + h_4 \cdot t_f$$
$$= (t_j + (2t_i + \epsilon t_e + 2t_f/\sigma)\mu)h_1$$

where $t_j, t_i, t_e,$ and t_f are the average runtimes for S1 through S4; ϵ is the average number of arrows per box; σ is the average number of input and output boxes per gate; and μ is the ratio τ/u, where τ is the cycle time of the control-net interpreter and u is an average time of machine acting cycle. For example, if $\tau = 50$ ms and $u = 500$ ms, then μ becomes 0.1, the number of arcs 1100, and the number of boxes 550. The response time of Z is expected to be approximately 12 ms for 10 machines in parallel operation. It should be sufficient to control ordinary assembly station sequences. Even if the control specifications become much more complicated with more boxes and gates, system response will not get proportionally worse.

Figure 8 shows the SCR configuration. Programmers, who are usually

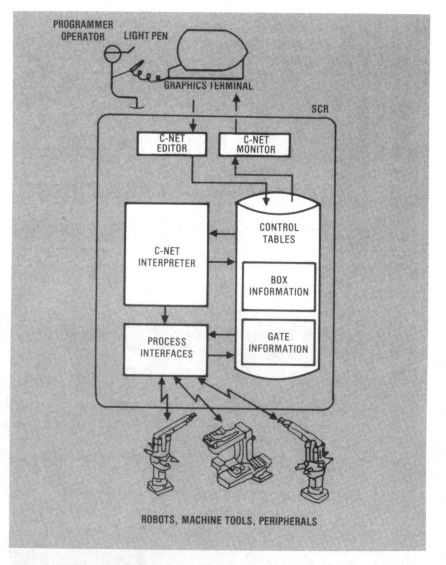

Figure 8. Using Station Controller, or SCR, a problem-oriented language for factory automation. The robots, machine tools, and peripherals are elements of a typical workstation.

also plant operators, can easily draw a control-net on the graphics terminal with a light pen and enter it into the control-net control tables after syntax checking.

SCR has been successfully applied in several situations. For example, a parts assembly station consisting of one robot and six machine tools is under control of SCR; SCR reduces the man-months of software development by about half those expended when the conventional relay ladder diagram methods are used; and, finally, an SCR monitor on the graphics terminal provides useful information for troubleshooting and resuming normal operation.

Station coordinator. In systems that control discrete events, the state of a component is identified separately. A detected system component state is called a *fact* and denoted by $f_i \in F$. Here, the subscript i is used as an identifier to distinguish facts, as are subscripts j and k.

Accordingly, a system situation S is defined as a set of facts that are detected by the controller all at one time. A command to the system to be controlled is called a *control action* and is denoted by $a_j \in A$. Control actions are determined according to the current system situation and are sent to the system to be controlled.

Control strategies are given to the controller by describing D, and control actions are decided by calculating $D(S)$ for each S. A control strategy D is specified with a set of rules as follows:

$$r_k : \{f_{i1}, f_{i2}, \ldots, f_{in}\} \rightarrow a_j$$

The part enclosed with { } is called the *condition* and the right-hand side of the arrow is called the *action*.

In this rule expression, all rules have to be written explicitly for all conditions. Consequently, as the number of rules grows, a condition may become long, incomprehensible, and difficult to maintain. To solve this problem, the concept of deduced facts is introduced. Deduced facts are defined as action and used in conditions in combination with other facts. The rule expression is enhanced as follows:

$$r_k : \{f_{i1}, f_{i2}, \ldots, f_{in}, m_{l1}, m_{l2}, \ldots\} \rightarrow m_l \text{ or } a_j$$

Furthermore, we can introduce variables such as x, y, and z to reduce the number of rules. Expressions like "Reclaimer x is free" or "Leading train y on lane z" can be used both in the condition and in the action parts of the rules.

With these enhancements, the control strategy can be obtained from the reduced rules. Also, rules are more easily understood and modified, and control strategies can be modified with little difficulty by the personnel in charge of machine operations.

The extended rules can be expressed by any string of characters. The rules used in SCD are called IF-THEN rules (see examples in Figure 9). The character string in parentheses corresponds to a fact, a deduced fact, or a control action and is divided into two parts: parameter (enclosed in brackets) and fixed. The parameter is used to define variables and to describe constants. It can also be used to specify objects, by

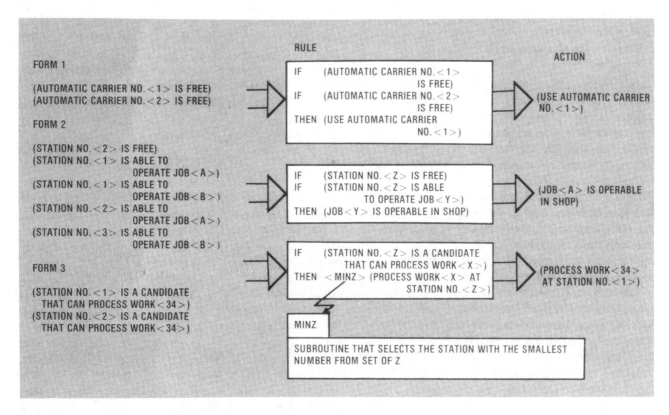

Figure 9. An example of IF-THEN rules for use in controlling workstations.

a machine or job name or a machine number, for example. The fixed part is used to identify a fact, a deduced fact, or a control action.

IF-THEN rules are classified into three patterns:

(1) In the first pattern, each parameter contains only constants, such as an existing machine name, or the parameter parts do not exist (Form 1).

(2) In the second pattern, parameter parts have variables, such as x, y, and z (Form 2).

(3) The last pattern, the open type, integrates user subroutines for, say, numerical calculations or optimization processing (Form 3).

The procedure name is written in the THEN part as "THEN procedure name (a character string)." The calculation method of these rules is basically the same as in the two other forms except that the user subroutines described in the THEN part are called up prior to the processing to add character strings. At this time, the variables needed to satisfy the combinational condition in the IF portion are given to the subroutines. The subroutine modifies these values for the interpreter. Character strings in the THEN part, which replace the variables with the modified values, are then added to the database.

A production system automatically organizes the extended rules according to the current system situation.[6,7] The production system consists of production memory, a database, and an interpreter. If a condition part of a rule exists in the database, where the system's current situation is stored, the action part is added by the interpreter. After iterations of this process, control actions are finally inferred.

To carry out this function in realtime applications, a high-speed interpreter called a control action generator has been developed. The software architecture of the control action generator is shown in Figure 10. The database of the production system consists of a control situation table and a control action table. The predecided control actions are stored in the control table, and the production

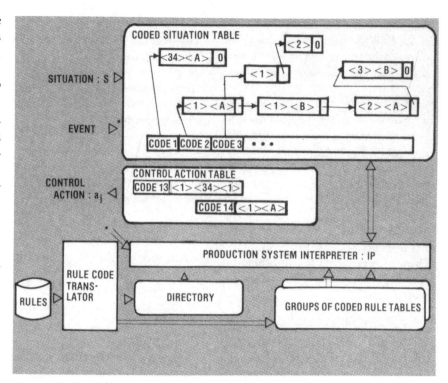

Figure 10. Control action generator architecture.

memory is constructed from directory and coded rule tables. The rules are divided into multiple rule sets and stored in the coded rule tables. The directory is provided for rule selection. In the control action generator, the fixed parts of each character string are coded into one number. The parameters that have the same coded number are linked together by pointers and stored in the coded situation table. When a character string is added to the database, as mentioned earlier, only the parameter parts of the character string are kept in the coded situation table and linked with other parameter parts of the same coded number. The facts and deduced facts, which have been unified, are extracted from the coded situation table with tracing pointers. Time-consuming pattern-matching processing is thus eliminated. The average calculation time of the action generator is approximately 5 ms per rule for typical examples.

The left-hand side of Figure 11 shows the configuration of SCD. Events in the controlled system are detected, and through the sensor input module, the system situation is reflected onto the tracking table by the tracking module. The control action generator generates the control action a_j on the basis of rules for each event and corresponding to input system situation S. Predecided control actions are sent to the system to be controlled by the actuator output module. The control rules are stored in the auxiliary file and taken into the control action generator when the control system is initialized. The rules in the auxiliary file can be easily modified according to the changes in the control system or by means of a rule editor in managerial situations.

SCD has been successfully applied in warehouse control, rolling mill (billet mill) control, and automatic carrier control for assembly shops. Some IF-THEN rules adopted to control a stacker crane are given in the right-hand side of Figure 11. (Materials arriving at the stations are stacked in storage areas by stacker cranes.)

As Figure 12 shows, SCD reduces the number of man-months needed for ordinary applications by two thirds, as compared with the number expended in a general-purpose programming environment. □

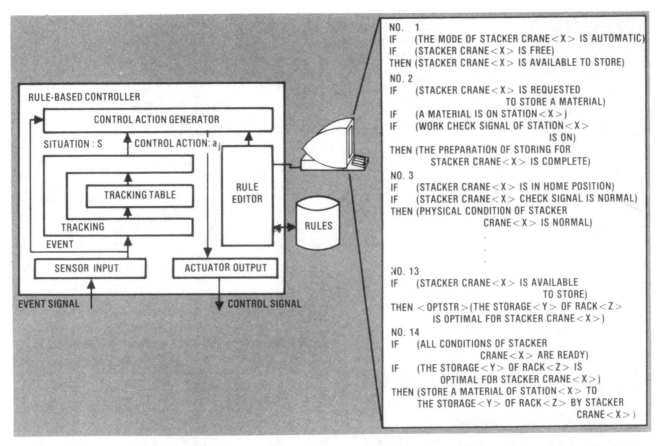

Figure 11. The use of station coordinator, a problem-oriented language for factory automation, and sample rules for controlling a crane that stacks material in storage areas as it arrives at stations.

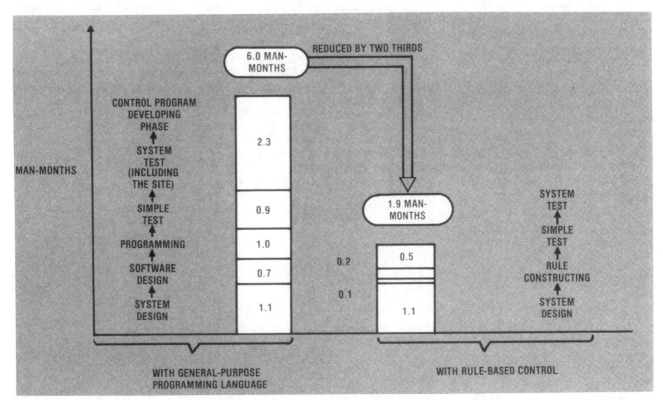

Figure 12. Productivity of the Rule-Based Control Method in system development. Use of the rule-based method instead of a general-purpose language or ladder diagram reduces man-months by two thirds.

Acknowledgments

We express our gratitude to A. Ichikawa of the Tokyo Institute of Technology for his valuable discussions, and J. Kawasaki of the Systems Development Laboratory, Hitachi, Ltd. We also thank H. Kuwahara of Omika Works, Hitachi, Ltd., for giving us a chance to work on this project.

References

1. *Proc. IEEE,* special issue on Robotics and the Factory of the Future, R. Reddy, ed., Vol. 71, No. 7, July 1983, pp. 787-900.
2. C. L. Keller, "Computerizing the Factory Floor," *Electronics,* Oct. 20, 1983, pp. 120-132.
3. H. Nakanishi et al., "System Architecture of Hitachi Control Computers," *Hitachi Review,* Vol. 32, No. 6, 1983, pp. 275-280.
4. J. L. Peterson, *Petri-Net Theory and the Modeling of Systems,* Prentice-Hall, Inc., Englewood Cliffs, N.J., 1983.
5. T. Murata et al., "A Petri-Net Based FA (Factory Automation) Controller for Flexible and Maintainable Control Specifications," *Proc. Iecon 84,* Oct. 1984, pp. 362-366.
6. N. J. Nilsson, *Principles of Artificial Intelligence,* Tioga Publishing, Los Altos, Calif., 1980.
7. A. Barr and E. A. Figenbaum, *The Handbook of Artificial Intelligence,* William Kaufman, Los Altos, Calif., 1981.

Norihisa Komoda has been with Systems Development Laboratory, Hitachi, Ltd., since 1974. His current interests include end-user programmable software development in factory automation, discrete-event system control, and artificial intelligence techniques. His previous research interests were requirement analysis and structuring modeling. Komoda received his BS, ME, and PhD in electrical engineering from Osaka University, Japan, in 1972, 1974, and 1982. He spent a year as a postdoctoral scholar at the Engineering Systems Department, UCLA. He is a member of the IEEE, IEE of Japan, and Society of Instrument and Control Engineers of Japan.

Kazuo Kera is a senior engineer at Omika Works, Hitachi, Ltd., where he has helped develop computer control systems for factory automation in iron and steel-making plants and has worked on a transportation system. Since 1982, he has been responsible for developing the decentralized computer system for factory automation. Kera is a member of the Information Processing Society of Japan and the IEE of Japan. He received the BS degree from Yamagata University in 1970.

Takeaki Kubo is a chief engineer at Omika Works, Hitachi, Ltd., where he has helped develop computer control systems for a thermal power plant, a nuclear power plant, and an iron and steel plant. His current interests include developing a computer system for factory automation. Kubo received his BS and MS from Tokyo University in 1963 and 1965, respectively. He is a member of the Information Processing Society of Japan.

Questions about this article can be directed to Takeaki Kubo, Omika Works, Hitachi, Ltd., 2-1, Omika-cho 5-chome, Hitachi-shi, Ibaraki, 319-12 Japan.

A Petri Net-Based Controller for Flexible and Maintainable Sequence Control and its Applications in Factory Automation

TOMOHIRO MURATA, NORIHISA KOMODA, MEMBER, IEEE, KUNIAKI MATSUMOTO, AND KOICHI HARUNA, MEMBER, IEEE

Abstract—A new type of software system for an industrial sequence controller is proposed. In this system, a control program is described with the Petri net-like language named Control-net (C-net). This language improves control software maintainability and flexibility. An efficient C-net interpretation schema for real-time control is presented and an overhead time evaluation model of the proposed schema is developed. Through the model analysis and measurement of the response time of an interpreter on a microcomputer, it is proved that the interpretation schema satisfies the required response time. Finally, a microcomputer-based controller named Station Controller (SCR) in which the presented C-net interpreter was installed is described and several applications of this controller to real systems are illustrated.

I. INTRODUCTION

RECENTLY, in sequence control systems for automatic assembly processes in factory automation, high flexibility and maintainability of control software have become of greater necessity. This is because product life cycles have become shorter and control specifications must be changed more frequently. Usually, programmable logic controllers and microcomputer-based controllers have been used for sequence control of assembly process. Their control programs have been developed with relay ladder diagrams or procedural languages (like assembler). However, control software written by these programming languages do not have sufficient flexibility and maintainability because comprehending control specifications from such a program is very difficult even for trained engineers.

Sequence control is mainly asynchronous, with concurrent control to maintain synchronization and exclude conflicting machine actions. With conventional languages, sequence control specifications are implemented as programs that include many control flags implicitly representing the synchronization and exclusion conditions of sequential process. These control flags make it difficult to understand the control program's behavior. In addition, it is not easy to change their functions flexibly corresponding to changes in control specifications.

To improve understandability of concurrent programs, concurrent programming languages like concurrent Pascal have been proposed. These languages have syntax for description of concurrent processes such as POST, WAIT, FORK, JOIN, etc.. However, these languages are fundamentally procedural languages and it is still difficult to understand the program's behavior when program size becomes large.

To resolve this difficulty, the Petri net model and language [1] have recently been experimentally applied as tools for describing sequence control specifications. For example, in France, a Petri net-like representation method named GRAFCET was proposed as a standard representation method for sequence control specifications. However, GRAFCET is simply a model for sequence control specifications and cannot be executed directly. Another Petri net-like representation model called Mark Flow Graph (MFG) [2], [3] and Petri net-based sequence description languages [4]–[6] have also been proposed and can be executed directly on microcomputer-based controllers. However, it has not been proven that they can be applied to real process control, and evaluation of their response time for real-time control has not been examined sufficiently.

In this paper, a new industrial sequence control system based on an enhanced Petri net interpretation is proposed [7]. In Section II Control-net (C-net), an enhancement of Petri net, is defined as a model for describing sequence control specifications. In Section III, an efficient C-net interpretation schema called the selective scanning method is described, and its overhead time evaluation model built to forecast its response time is analyzed. Finally, in Section IV, a 16-bit microcomputer-based controller called the Station Controller (SCR) on which the C-net interpreter presented here was realized, is described. Several examples of the SCR's successful applications to real systems are illustrated.

II. DESCRIPTION OF CONTROL SPECIFICATIONS BY C-NET

A. Safe Petri Net Model

A Petri net [1] is a simple and powerful graph model for describing and analyzing asynchronous and concurrent systems behavior. Fig. 1 shows a simple Petri net. The graph contains two types of nodes; circles (called places) and bars (called transitions). These nodes are connected by directed arcs. The execution of Petri net is controlled by the position and movement of markers (called tokens). A Petri net in which each place can contain at most one token is called a safe Petri net.

A safe Petri net is defined formally as the tuple $PN = (P,$

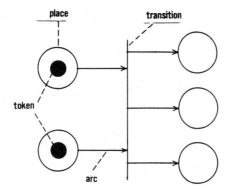

Fig. 1. Safe Petri net model.

T, I, O, M), where P is the set of places ($p_1, p_2, \cdots, p_n$), T is the set of transitions ($t_1, t_2, \cdots, t_n$), $I:T \to P$, $O:T \to P$ and $M:P \to N(N = 0, 1)$ are functions. A place $p \in P$ such that $p \in I(t)$ (or $p \in O(t)$) is called an input place (or an output place) of $t \in T$. A transition $t \in T$ such as $p \in I(t)$ (or $p \in O(t)$) is called an input transition (or output transition) of $p \in P$. M is called marking, where $M(p_i) = 1$ (or $M(p_i) = 0$) means place p_i has a token (or has no token). A transition $t \in T$ can be enabled at a marking M_1, if and only if $M_1(p_i) = 1$ and $M_1(p_j) = 0$ for all $p_i \in I(t)$ and $p_j \in O(t)$. Regarding an enable transition and its input/output places, a new marking $M_2(p_i) = 0$ and $M_2(p_j) = 1$ for all $p_i \in I(t)$ and $p_j \in O(t)$ is reachable by firing t.

In a safe Petri net model, if each place can represent a machine's action, sequence control specifications including mutual synchronization and exclusion control of several machine actions, can be described explicitly as a directed graph model. Namely, the execution sequence of the machines' actions can be defined as a Petri net structure, and asynchronous and concurrent progress of the sequence can then be realized as a movement of tokens. However, this description needs a lot of places to describe more complicated control specifications which include many machine actions. This is because a place in a safe Petri net can represent only two statuses corresponding to the token's existence or nonexistence, while machine actions usually have plural statuses depending on the results of its execution. To avoid this problem and to apply the Petri net model for describing sequence control specifications, a C-net model is introduced as follows.

B. C-net Model

A C-net is defined by the tuple $CN = (P, T, I, O, \delta, \varphi, \eta, U, V, M)$, where P, T, I, O, and M are the same as those of the Petri net model. δ, φ, and η are called process i/o functions, and U and V are called process status functions. These functions are introduced to define process interfaces and process statuses.

1) Process i/o Functions: To define a correspondence between the places in a C-net and the controllable and observable process signals in a controlled system, process i/o functions are introduced. Let A be a set of control signals (x_i) and E be a set of observable signals (y_{ij}). These signals are usually defined as 0/1 bit patterns or code data. Process i/o functions $\delta: P \to A$, $\varphi: P \to E$, and $\eta: T \to E$ are defined as follows:

$$\delta(p_i) = x_i, \quad (x_i \in A, p_i \in P) \quad (1)$$

$$\varphi(p_i) = (y_{i1}, y_{i2}, \cdots, y_{in}), \quad (y_{ij} \in E, p_i \in P) \quad (2)$$

$$\eta(t_i) = y_{ik}, \quad (y_{ik} \in E, t_i \in T). \quad (3)$$

When a token enters into a place p_i, a control signal x_i defined by $\delta(p_i)$ is put out and a machine action is triggered. Next, the token stays in a box until one of the input signals $y_{i1}, \cdots, y_{in}$ defined by $\varphi(p_i)$ is detected as a response signal of a completed action. Because the answer-back signal takes various values depending on the execution result, $\varphi(p_i)$ corresponds to plural input signals in each place. With these functions, the contents of the machine action and its completion judgement conditions can be defined and supervised at places in the C-net. Input signal y_{ik} defined by $\eta(t_i)$ is used for open/close operations on transitions. Since tokens cannot pass through a transition until input signal y_{ik} is detected, an operator can directly control the movement of tokens using transitions.

2) Process Status Functions: To define the action's execution status at a place and to manage transition open/close statuses, process status functions $U:P \to L(L = 0, 1, \cdots, m)$, $V:P \to N(N = 0, 1)$ are introduced as follows:

$$U(p_i) = \begin{cases} 0 & \text{(action associated with } p_i \text{ is executing now)} \\ in & \text{(action associated with } p_i \text{ is completed} \\ & \text{with return code } y_{in}) \end{cases} \quad (4)$$

$$V(t_i) = \begin{cases} 0 & (t_i \text{ is closed}) \\ 1 & (t_i \text{ is opened}) \end{cases}. \quad (5)$$

When an output signal x_i defined by $\delta(p_i)$ has been put out, $U(p_i)$ is set at 0. When one of the input signals y_{in} defined by $\varphi(p_i)$ is detected, $U(p_i)$ is set at the value of in. If an input signal y_{ik} defined by $\eta(t_i)$ has not yet been detected, the value of $V(t_i)$ is set at 0; otherwise, $V(t_i)$ is set at 1.

By introducing these functions, action execution statuses or transition operation modes can be supervised at a place or transition. Hereafter in this paper, places and transitions will be called boxes and gates since process i/o functions and process status functions can be defined at places and transitions.

3) Gate Firing Rule: A gate $t_i \in T$ can be enabled at marking M_1, if and only if,

$$V(t_i) = 1, \ M_1(p_i) = 1, \ U(p_i) \neq 0,$$
$$\text{and } M_1(p_j) = 0 \text{ for all } p_i \in I(ti) \text{ and } p_j \in O(t_i). \quad (6)$$

This condition is called an enable condition of t_i. By firing t_i, tokens are generated in all output boxes and are deleted from all input boxes. When the tokens are generated, the output signals defined at the output boxes of t_i are put out and $U(p_i)$ is set at 0. A box in which process i/o functions are defined is called an act box.

4) Other Functions for Sequence Control Description: The following functions have been introduced to describe more complicated sequence control specifications, such as conditional branches based on job request type or the result of an executed action and timing control for job starting.

a) Detection of job start requests and discrimination of job type: When a job start request signal is detected, a numbered token corresponding to the job kind is generated in a box. This number (called the token number) is used as an identifier to define what type of action should be started at each place. A box which detects the request signal to start a job and generates a numbered token within itself is called a receive box.

b) Conditional branch of tokens: A box which has plural output gates is called a conflict box. At a conflict box, the output gate to be fired is selected according to the result of an action's execution represented by the value of $U(p_i)$. With this function, conditional branching of tokens is easily realized.

c) Timer: A box called a timer box holds a token for a specified length of time. This function can be used as a delay timer. Furthermore, an act box can measure the execution time and detect time-out errors. This function can be used as a watchdog timer.

d) Similar job sequence repetition block: In sequence control, there are sometimes repetition blocks in which the execution order of actions is the same but the content of each action is different in each execution iteration. This kind of job sequence is called a similar job sequence. To describe this kind of job sequence, source boxes and count gates have been used. A source box is a box which generates a specified number of sub-tokens having sequential numbers. When these sub-tokens pass through the same box in the same sequence block, they trigger different actions associated with the process i/o functions defined for each sub-token number. A count gate is a gate which defines the end of the sequence repetition block and counts the number of tokens passing through it. The count gate only deletes tokens in its input boxes until the count number reaches the specified number. When the count number reaches the specified number it makes the tokens pass through to its output boxes. Using source boxes and count gates, similar job sequence repetition blocks are easily defined.

e) Subroutine call: Subroutines for numerical calculation or information processing can be defined and executed in a box called a program box. Object modules of user subroutines are linked to program boxes. When a token enters this box, the user subroutine associated with it is executed. The results of subroutine executions are passed back to the program box as a return code, and can be refered to using the process status function $U(p_i)$.

C. An Example of C-net Description

Fig. 2(a) is an example of an assembly station. At this station, robot A gets two assembly parts from a parts feeder and assembles them on the assembly table. Robot B gets four screws from the screw feeder and inserts them at the specified points of the assembled parts. The control specifications for these two robots can be described with C-net as shown in Figure 2(b). Each box and gate has a unique box name and gate number. The boxes GRAS1, PUT, and RETN represent

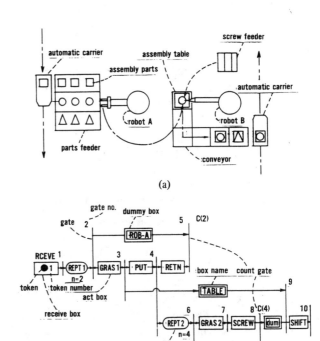

Fig. 2. (a) A station. (b) C-net representation of sequence control specification for the station.

robot A's actions and the boxes GRAS2 and SCREW represent robot B's actions. When a job start request signal is detected at receive box RCEVE, the numbered token is generated in it according to job request type. This token moves through the boxes in a C-net according to the C-net firing rules and the robot's actions are triggered according to the movement of the tokens.

III. An Efficient Interpretation Schema of C-net

To apply the C-net as a real-time process control language, an efficient C-net interpretation schema which realizes quick response time is essential. To move tokens, the C-net interpreter chooses a set of enabled gates and fires them. The simplest method for this operation is to test all gates to determine if they satisfy the enable condition explained in Section II-B. However, this method requires a large overhead time if the number of gates in C-net becomes large.

This enable testing overhead can be considerably reduced by using information of C-net structure and marking. According to the C-net firing rule, the gates which have possibility of firing can be limited to such gate that (1) a token has entered into one of its input boxes, (2) a token has gone out from one of its output boxes, or (3) one of input signals defined by $\varphi(p_i)$ has been detected at one of its input boxes p_i. In the following section, configuration of an efficient C-net interpretation method based on this idea called the selective scanning method is proposed.

A. Selective Scanning Method

The selective scanning method configuration is shown in Fig. 3. This method is constructed from four execution steps $S1$, $S2$, $S3$, and $S4$, and uses four stacks, H_1, H_2, H_3, and

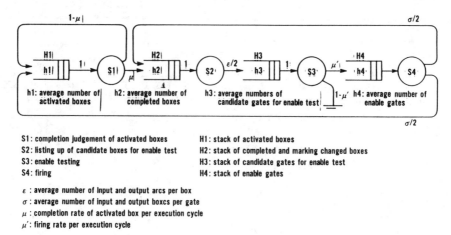

Fig. 3. Selective scanning method configuration.

H_4. At first, box p_i in which $M(p_i) = 1$ and $U(p_i) = 0$ is pushed down in H_1. The content of each execution step is explained as follows, where the number of boxes or gates in H_i is indicated as h_i:

S1: Completion judgement.
Pop p_i such that $U(p_i) \neq 0$ from H_1 and push it down on H_2 for all $p_i \in H_1$.

S2: Choice of candidate gates for enable test.
For all $p_i \in H_2$, search T_j such that $p_i \in I(T_j)$ and $M(p_i) = 1$ or $p_i \in O(T_j)$ and $M(p_i) = 0$.
Push down T_j on H_3 and delete p_i from H_2.

S3: Enable test.
For all $T_j \in H_3$, pop T_k such that it satisfies the enable condition from H_3, and push it down on H_4.

S4: Fire.
For all $T_k \in H_4$, fire T_k, push down $p_j \in I(T_k)$ on H_2, push down $p_k \in O(T_k)$ on H_1, and delete T_k from H_4.

In these execution steps, gate enable test is executed only for T_j such as at least one of T_j's output box tokens has vanished or at least one of T_j's input box status has changed to completion in the previous execution cycle. Namely, the enable test is executed only for gates that have a possibility of firing. With this method, the execution time for enable test can be reduced and, as a result, the overhead time of C-net interpretation can be reduced considerably.

B. Overhead Time Analysis of the Selective Scanning Method

Using Fig. 3 the performance of the proposed method can be evaluated as follows, with ϵ and σ defined as follows:

$$\epsilon = (\text{numbers of all arcs})/(\text{numbers of all boxes}) = ma/mb \quad (7)$$

$$\sigma = 2*(\text{number of all boxes})/(\text{number of all gates})$$
$$= 2*mb/mg \quad (8)$$

where ϵ, σ are characteristic parameters of the C-net structure. ϵ represents the average number of arrows per box, and σ represents the average number of input and output boxes per gate. Now, h_1 can be regarded as the number of acting machines in a controlled process. Let μ (completion rate) be the rate of completed boxes per h_1 in an execution cycle; h_2 is the average number of boxes in which tokens have been deleted or generated in an execution cycle; h_3 is the average number of candidate gates for enable testing; and h_4 is the average number of enable gates. $\mu'(=h_3/h_4)$ is called the firing rate. Between h_2 and h_3, there is a relation such that $h_3 = \epsilon*h_2/2$ regarding the C-net structure. In S4 the average value of $\sigma*h_4/2$ tokens are generated and that of $\sigma*h_4/2$ tokens are vanished in an execution cycle. Since the runtimes of S1 $\cdots$ S4 are much smaller than machine action execution time, $h_1, \cdots, h_4$ can be regarded as being balanced. Thus the following equation must be completed:

$$h_1 = h_1(1-\mu) + (\sigma/2)h_4 \quad (9)$$

$$h_2 = h_1\mu + (\sigma/2)h_4 \quad (10)$$

$$h_3 = (\epsilon/2)h_2 \quad (11)$$

$$h_4 = \mu' h_3. \quad (12)$$

If these simultaneous equations have answers other than $h_1 = h_2 = h_3 = h_4 = 0$, the following matrix equation must be completed:

$$\begin{vmatrix} -\mu & 0 & 0 & \sigma/2 \\ \mu & -1 & 0 & \sigma/2 \\ 0 & \epsilon/2 & -1 & 0 \\ 0 & 0 & \mu' & -1 \end{vmatrix} = 0. \quad (13)$$

From (13), $\mu(1 - \mu'\epsilon\sigma/2) = 0$ is obtained. μ must not be 0, so that $\mu' = 2/\epsilon\sigma$ can be obtained. This equation is called a balance condition. Using this condition, h_2, h_3, and h_4 are obtained as follows:

$$h_2 = 2\mu h_1, \quad h_3 = \epsilon\mu h_1, \quad h_4 = 2\mu h_1/\sigma. \quad (14)$$

The total execution time of S1, $\cdots$, S4 is calculated as follows:

$$Z = h_1 \cdot t_j + h_2 \cdot t_i + h_3 \cdot t_e + h_4 \cdot t_f$$
$$= (t_j + (2t_i + \epsilon t_e + 2t_f/\sigma)\mu)h_1 \quad (15)$$

where t_j, t_i, t_e, and t_f are the average runtimes for $S1$, $S2$, $S3$, and $S4$. Hereafter, h_1 is represented as P. A C-net interpreter using this method was realized on a 16-bit microcomputer system, and $t_j, \cdots, t_f$ was measured. Using these values, Z can be calculated as

$$Z = (0.91 + 3.12\mu)P \text{ ms} \qquad (16)$$

where $ma = 1100$, $mb = 550$ (i.e., $\epsilon = 2.0$, $\sigma = 2.2$). Z has also been measured on the developed interpreter with $= 1$ and 0.5 for $P = 2, \cdots, 12$. The result is shown in Fig. 4. The measured value of Z shows good coincidence with the calculated value of Z by (16).

In a real system, μ is given by the ratio v/u. Here v is the interval time of an interpreter execution cycle and u is an average time of the machine's acting cycle time. For example, in a case where $v = 50$ ms and $u = 500$ ms, $\mu = 50/500 = 0.1$; if 10 machines are acting concurrently (i.e., $P = 10$), response time Z is calculated as 12.2 ms. This is enough speed for assembly station sequence control. Furthermore, if the control specification becomes more complicated, the response time of this interpretation method is not influenced by an increase in the number of boxes and gates.

IV. C-NET-BASED INDUSTRIAL SEQUENCE CONTROLLER: SCR

A. Configuration of the SCR

The C-net interpreter and its peripheral software were installed on a 16-bit microcomputer system. This system was named Station Controller (SCR). The SCR consists of an editor/monitor, an interpreter, control tables, and process interfaces. Operators can easily draw a C-net on the graphic display with a light pen and input it to the control tables after simple syntax checking. Fig. 5 shows the editing screen. On this editor, several editing commands for box or gate definition can be provided. The available commands are displayed on the editing screen and operator can choose suitable command by pointing with a light pen. Symbols of box and gate are automatically drawn at the pointed position by an operator. Operation guidances are also indicated interactively. By these functions operator can easily input C-net even if he or she has no skill in computer programming. C-net can be also defined hierarchically, so a large-scale C-net can be easily defined. After making the C-net, operator can generate initial marking and simulate the movemeant of tokens for testing.

The control tables are the executable expression of the C-net model. These control tables consist of a gate information table (Table I) and a box information table (Table II). The interpreter executes C-net interpretation with the selective scanning method based on these tables. The monitor displays the controlled system's status by displaying the C-net and its marking in real time. Machine actions executing simultaneously in a station are indicated by tokens in the boxes. Thus operators can easily understand the controller's behavior and status. In the offline mode, the operator can change marking for setting system restart condition on the monitor display screen.

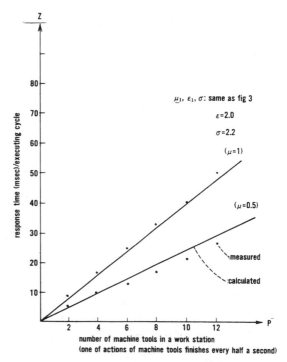

Fig. 4. Response time of C-net interpreter.

B. Applications of SCR

The SCR was successfully applied to several applications. Three applications using different types of SCR installations are illustrated below.

1) Control of Parts Assemble Station: A parts assembly station consisting of one industrial robot and six machine tools was controlled by the SCR. Fig. 6 shows the configuration of this assembly station and control system. An SCR program was written in S-PL/H [8]. The SCR controls the robot and machine tools through a digital i/o interface which contains 200 points for process input and output, and the execution cycle time is about 10 ms. In this application, the scale of the C-net is about 1000 boxes and 1000 gates. Each box and gate is assigned a list of process output and input signals.

Initial software development man-hours have been reduced to about 50 percent compared with the relay ladder diagram method. Furthermore, several changes in control specifications were easily carried out. Since the SCR's monitor provides suitable information about the station, trouble shooting in abnormal situations, and returning to normal operation status were performed rapidly.

2) Industrial Robot Controller: SCR was also installed as a microcomputer-based industrial robot controller which has both a robot control function for robot axises control and a station control function for coordinating robots and peripheral equipments in a work station. The configuration of this controller is shown in Fig. 7. C-net has been adopted as the language for describing the station control logic. C-net programs can be inputted through a list form editor on an operator console or a graphic editor on a personal computer connected via the communication line. SCR program was written in language C. Each box is assigned a robot number and a robot program number which should be executed when a token

Fig. 5. C-net editor display format.

TABLE I
GATE INFORMATION

input box list	
output box list	
pointer of associated input events (gate condition)	
maximum count number for count gate	counter for count gate

TABLE II
BOX INFORMATION

input gate list	
output gate list	
pointer of associated output events (control outputs)	
pointer of associated input events (box condition)	
maximum execution time	timer counter

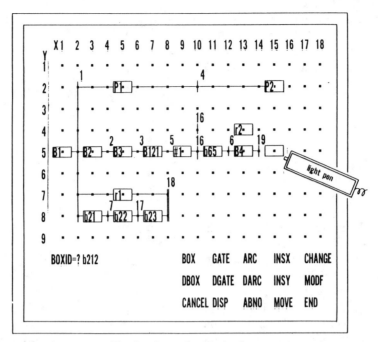

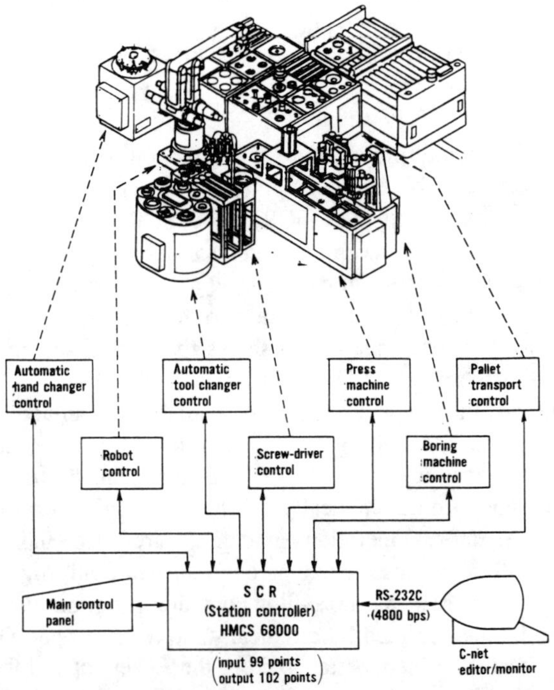

Fig. 6. Application of SCR to parts assembly station.

comes into the box. Robot programs are put in using a teaching box. Since robot and peripheral control functions are executed by different tasks from SCR, activation of the robot programs and information reporting of program completion are performed by using task communication function. Ten C-net models can be interpreted simultaneously. Each C-net execution can be separately started by the console operation.

This controller was applied to palletizing system control, and heavy material transfer system control. For instance, in palletizing system control, one robot, a marshalling system, and a conveyor were controlled by this controller.

3) General Purpose FA Controller [9]: Flexible Manufacturing Cell Controller (FMC) is a microcomputer-based general purpose controller designed to supervise and control various types of work stations. In FMC, two problem-oriented languages, C-net for sequence control and a rule-based language for constraint combinational control, are available as well as Fortran and C. All SCR programs (editor, loader, interpreter, and monitor) are written in C. In the boxes and gates of C-net several transactions for control operation can be defined. For instance, a job trigger transaction, a normal job completion detect transaction, and several abnormal job

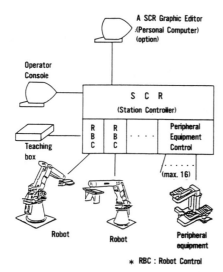

Fig. 7. An industrial robot controller.

completion detect transactions are defined in an act box. These transactions have names assigned to standard predefined subroutines or user-developed subroutines which performed process *i/o* or information processing. Assignment of subroutines is supported by the SCR editor.

To describe the conditional branch, a decision box function can be used. In a decision box several "IF (condition transaction name) THEN (gate identifier)" form logics are assigned. If the conditional transaction returns "true" as a result of its execution, a token in the decision box passes the gate defined by a gate identifier. Furthermore, to manage plural job request queues, a dummy box which can hold plural tokens is supported. In this box, the maximum number of tokens held in, and the method of storage and removal of tokens (first-in–first-out or first-in–last-out) can be specified.

By assigning transactions which perform various functions to boxes and gates, SCR on FMC can be applied to a wide variety of factory automation control functions, including information processing as well as sequential control of machines. As an example of FMC's applications, control of an automatic warehouse load/unload system, whose layout is shown in Fig. 8(a), is briefly illustrated below. Part of the control logics of a loading conveyor described by C-net is shown in Fig. 8(b). In this application, by applying SCR, control programs could be developed graphically and interactively on the SCR editor, and transactions corresponding to each box and gate could be defined as independent modules. Therefore, initial software development man-hours were reduced to about 80 percent and system maintenance man-hours were also reduced to about 60 percent.

V. Concluding Remarks

An enhanced model of the safe Petri net called C-net was proposed for describing industrial sequence control specifications. Process *i/o* functions and process status functions were introduced for describing process interface specifications and supervising machines' execution status. An efficient C-net interpretation schema called the selective scanning method was presented. This method realizes quick response time in

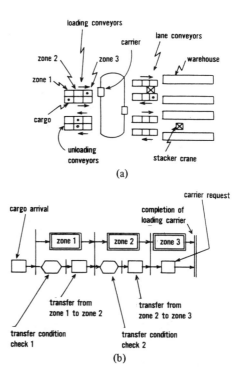

Fig. 8. (a) Layout of automatic warehouse system. (b) A part of control logic on loading conveyor.

real-time control. Through overhead time analysis of this method, its overhead time calculation equation was introduced. Using this equation, it was proved that the overhead time of the presented interpretation method was within 15 ms/execution cycle on a 16-bit microcomputer system when controlling an assembly station which contains 10 machine tools.

A microcomputer-based controller called the SCR based on the presented C-net interpretation method was successfully applied to several applications. Through these applications, it was proved that man-hours for initial software development and maintenance was reduced considerably.

VI. Acknowledgment

The authors want to express their gratitude to Prof. A. Ichikawa of the Tokyo Institute of Technology for his valuable discussions, and Dr. J. Kawasaki, General Manager of Systems Development Laboratory (SDL), Hitachi Ltd., for providing the opportunity to do this research. They also thank the relevant members of SDL, Omika Works, Narashino Works, and Tuchiura Works of Hitachi Ltd., for their valuable discussion and cooperation.

References

[1] J. L. Peterson, *Petri-Net Theory and the Modeling of Systems*. Englewood Cliffs, NJ: Prentice-Hall Inc., 1981.
[2] R. Masuda and K. Hasegawa, "Mark flow graph and its application to complex sequential control system," in *Proc. of 13th Hawaii Int. Conf. on System Science*, pp. 194–203, Jan. 1980.
[3] K. Hasegawa *et al.*, "On programming of conventional programmable controller by using mark flow graph," in *Proc. of ISCAS '85*, pp. 933–936, June 1985.
[4] D. Chocron and E. Cerny, "A Petri net based industrial sequencer," in *Proc. of IECI (IEEE Int. Conf. and Exhibition on Industrial Control and Instrumentation)*, pp. 18–22, Mar. 1980.

[5] M. Silva and S. Velilla, "Programmable logic controller and Petri nets: A comparative study," *IFAC Software for Computer Control*, pp. 83–88, 1982.
[6] R. Valette *et al.*, "Putting Petri nets to work for controlling flexible manufacturing systems," in *Proc. of ISCAS '85*, pp. 929–932, June 1985.
[7] T. Murata *et al.*, "A Petri net based FA (factory automation) controller for flexible and maintainabnle control specifications," in *Proc. IECON '84*, pp. 362–366, Oct. 1984.
[8] T. Kohno *et al.*, "Development of high level languages S-PL/H for 16-bit microcomputers and its application to 68000 system software," in *Proc. 3rd Int. Microcomputer Applications Conf.*, pp. 193–202, 1982.
[9] N. Komoda *et al.*, "An autonomous and decentralized control system for factory automation," *IEEE Trans. Comput.*, vol. C-17, pp. 73–83, Dec. 1984.

Chapter 6: Mathematical Methods for Improving Performance

6.1: Introduction

This collection of papers addresses several scheduling, planning, and control problems that have not been treated elsewhere in this tutorial. These include parts routing, satisfying delivery time, balancing work loads in a transfer line, and coordinating subsystems within the entire manufacturing facility. In addition, measures of flexibility are presented along with a reprint on concurrent routing and job sequencing. Various optimization and mathematical programming techniques are used to solve these problems. The reader is likely to be familiar with these techniques but is probably not aware of their applicability to manufacturing systems.

One major goal of this chapter is to point out there is no widespread agreement which optimization technique to use with what model. As a result, a diverse set of approaches emerges depending on what aspect of the production hierarchy is being studied. These papers are representative of these approaches and, in most cases, provide new and interesting directions for fruitful areas of research.

In each of the following sections, the purpose and contribution of each paper is explained. This is immediately followed by the important mathematical background and the appropriate references. Reading the paper first and then checking the background section is the suggested approach to understanding these papers.

6.2: "Optimal Control of a Queueing System with Two Heterogeneous Servers" (W. Lin and P.R. Kumar)

This paper address a problem that is very basic to manufacturing: the routing of individual parts to specific machines. This activity takes place at the lowest level of the production-control hierarchy shown in Figure 1.8.

Specifically, Lin and Kumar address the routing of parts through two machines that share a common buffer. Their results are obtained by using methods from queueing theory and Markov chains.

Their formulation is applicable to a broad range of systems including computers, communications, and manufacturing production networks. In their terminology, servers are machines, customers are parts, and service times are machining times.

Although, the one buffer that serves two-machines problem has limited practical application, the authors conjecture about the multiserver case (and others) in their concluding remarks.

6.2.1: Mathematical Background

The authors make extensive use of the Markov theory presented in Section 2.2 of this tutorial. In addition, their proofs are based on the policy-iteration method and the value-iteration method. The development of these techniques is a bit lengthy for this tutorial but a good treatment can be found in Howard [8,9].

6.3: "Cooperation among Flexible Manufacturing Systems" (O. Berman and O. Maimon)

A complete manufacturing system often consists of several identifiable subsystems. If they operate independently, the entire system is likely to operate inefficiently. Therefore, the problem is to integrate these subsystems so their activities are coordinated. A method for achieving this coordination (cooperation) for a system under complete computer control is presented by Berman and Maimon.

Most of the papers in this tutorial have dealt with systems consisting of parts, buffers, and machines. Here, a system is studied that also uses automatically guided vehicles (AGVs) and individual tools. Thus, a further dimension of reality is added.

Note that the subsystems are considered to be flexible manufacturing systems while their union forms the complete automated manufacturing system. Therefore, each subsystem is very versatile and the coordination of these numerous degrees of freedom is essential for achieving good overall system performance.

The integration of the subsystems occurs at the dispatching level of the production-control hierarchy shown in Figure 1.8. In particular, resources are dispatched to the FMSs by the AGVs. When a resource is summoned, and all suitable AGVs are busy, then cooperation should take place. This results in a vector of decision alternatives defined as a strategy. The goal is to find the strategy that minimizes the cost of responding to a random call for service subject to the performance constraints of the FMSs. This should minimize machine idle time and, hence, increase machine utilization.

6.3.1: Mathematical Background

Linear programming: The problem formulation leads to the investigation of a linear-cost-function subject to a set of inequality and equality constraints. This is a linear programming problem and techniques for solving it can be found in [5–7].

6.4: "Measuring Decision Flexibility in Production Planning" (J.B. Lasserre and F. Roubellat)

This paper directly addresses the issues that were described earlier in Chapter 1, namely, how to carry out *production planning* when there is

- Uncertainty in demand
- Uncertainty in manufacturing capacity because of random machine failures and repairs

As a result of these uncertainties, decisions must be made that keep options open. Thus, a "good" decision is one that has high flexibility.

6.4.1: Decision Flexibility

The *flexibility* of a decision is related to the size of the set of all feasible plans. It is the volume of the set that provides an analytical measure of decision flexibility. Such a measure is a step toward clarifying the meaning and definition of a flexible manufacturing system. The main purpose for including this paper is to motivate the need for quantifying the concept of flexibility. There are several other approaches [20–24] and it is not clear which one is best.

It is important to note that the authors are working at the medium-term level of the production-control hierarchy (Figure 1.8). Consequently, their techniques for measuring decision flexibility are based on aggregate production models (Section 3.2).

6.4.2: Measure of Flexibility: Basic Concepts

The development of their measures centers around the (familiar) production planning model,

$$x_t = x_{t-1} + u_t - d_t$$

where x_t is the inventory level at time t, u_t is the control or decision vector, and d_t is the demand. The control is constrained by $0 \leq u_t \leq \bar{u}_t$ and the state must always be nonnegative (i.e., $x_t \geq 0$ for all $t = 1, 2, \ldots T$).

Next, the state variables are eliminated which results in a set of equations that keeps only the decision variables. This can be done by first considering,

$$x_1 = x_0 + u_1 - d_1 \quad (6.5.1)$$

and

$$\begin{aligned} x_2 &= x_1 + u_2 - d_2 \\ &= x_0 + u_1 + u_2 - d_1 - d_2 \end{aligned} \quad (6.5.2)$$

or in general

$$x_t = \sum_{i=1}^{t} u_i - \sum_{i=1}^{t} d_i + x_0$$

Since $x_t \geq 0$ in their formulation, this requires

$$\sum_{i=1}^{t} u_i \geq \sum_{i=1}^{t} d_i - x_0$$

Defining the appropriate vectors results in a set of inequalities that can be written (in general form) as

$$Au \leq b \quad (6.5.3)$$

where $u = (u_1 u_2 \ldots u_T)$ and u_i is the decision vector at period i.

The basic idea is to make a first decision u_1 such that all subsequent decisions satisfy the constraints (6.5.3). The measure of flexibility of decision u_1 is done by calculating the volume of the convex polyhedron [6] defined by the set of inequalities resulting from decision u_1.

The paper uses a method, previously developed by the authors, for calculating the volume of a convex polyhedron in R^n. As a by-product of their method, they obtain the sensitivity of this volume to b_i, which represents the sensitivity of the flexibility measure to small changes in machine capacity and product demand.

6.4.3: Flexibility and Economic Criteria

The goal here is to find u^* to optimize flexibility and also to minimize some cost criterion such as

$$x_1 + x_2 + \ldots x_T \quad (6.5.4)$$

This linear criteria can also be written in terms of the decision variables. Adding (6.5.1) and (6.5.2) yields

$$x_1 + x_2 = 2x_0 + 2u_1 + u_2 - 2d_1 - d_2$$

which can be extended to represent (6.5.4) as

$$x_1 + x_2 + \ldots x_T = Tx_0 + Tu_1 + (T-1)u_2 + \ldots u_T \\ - Td_1 - (T-1)d_2 - \ldots d_T \quad (6.5.5)$$

Now suppose that the plans (or decisions) that result in costs higher than some threshold are eliminated. Specifically, costs that are higher than $C^* + \delta$ ($\delta > 0$) where C^* is the cost corresponding to the optimal plan are eliminated. This leads to the final problem formulation.

First, Tx_0 is the same for all possible plans and so it can be eliminated from (6.5.5). Then, those decisions that meet the constraints must satisfy,

$$x_1 + x_2 + \ldots x_T \leq C^* + \delta$$

which implies

$$\begin{aligned} Tu_1 + (T-1)u_2 + \ldots u_T \\ \leq C^* + \delta + Td_1 + (T-1)d_2 + \ldots d_T \\ \leq C^* + \delta + \sum_{i=1}^{T} (T - i + 1)d_i \end{aligned}$$

The flexibility of the decision is the volume (in R^{T-1}) of

$$u_2 + \ldots u_T \geq d_1 + d_2 + d_T - x_0 - u_1$$

subject to

$$(T-1)u_2 + (T-2)u_3 + \ldots u_T \leq C^* + \delta - Tu_1 + \sum_{i=1}^{T}(T-i+1)d_i$$

Examples of this measure are presented in the paper. In addition, the authors consider some alternate economic criteria as well as one other measure of flexibility.

6.4.4: Related Research

Other researchers have also looked at the problem of measuring flexibility [20–24]. An information theoretic point of view [22] is one of the approaches.

6.5: "Optimal Scheduling for Load Balance of Two-Machine Production Lines" (S. Mitsumori)

In this paper, machines are scheduled in a two-machine production line so that the overall production rate of the system is maximized. Multiple products, which means that the line must be able to switch from product to product, are made. For this type of problem, scheduling can be defined as

$$\text{scheduling} = \text{lot sizing} + \text{sequencing}$$

where lot sizing refers to the number of type-i products that are to be produced, and sequencing determines when the i-th product type is released into the system. This is usually determined by a criteria such as the due date of part type i or its priority. The scheduling problem should also take into account the production loss when product changeover occurs.

6.5.1: Objectives

The overall goal is to maximize the production rate of a two-machine production line by the proper scheduling of each machine. Since multiple products are being manufactured, this introduces another objective: minimize the changeover loss subject to

- Meet the due date
- Queue length of in-process parts must be shorter than the maximum space between the machine
- No idle time for either machine

To meet these objectives requires the determination of the lot size for each commodity and the production sequence. Mitsumori presents a branch and bound algorithm to solve this problem.

The ideas presented in this paper are based on results obtained from the author's previous work [2,3].

The basic concept here is to determine production lots and their flow sequences at the same time. Other methods do the lot sizing first and then try to determine the sequence that will produce these lots. The latter approach neglects the balancing of the work load between the machines.

6.5.2: Mathematical Background: Branch and Bound Optimization Algorithms [5]

Branch-and-bound methods are particularly useful for solving large classes of mathematical programming problems. Consider the following problem,

$$\text{maximize } f(x) = \sum_{j=1}^{N} c_j x_j$$

subject to

$$\sum_{j=1}^{N} a_{ij}x_j \leq b_i \qquad i = 1,2,\ldots M$$

$$\begin{aligned} U_j &\geq x_j \geq L_j & j &= 1,2,\ldots N \\ x_k &\geq 0 & k &= 1,2,\ldots \bar{N} \\ x_l \text{ integer} & & l &= \bar{N}+1, \bar{N}+2, \ldots N \end{aligned}$$

Since some of the x_j are required to be integers, this is called a *mixed-integer programming* problem. Note that if $\bar{N}$ is zero, then we have an all *integer-linear program*. Thus, the formulation is very general.

The basic idea behind branch-and-bound strategies is the simple fact that for any value of x_j

$$[x_j] + 1 \geq x_j \geq [x_j]$$

where $[x_j]$ is the largest integer less than or equal to x_j.

Now suppose that we are seeking an integer solution for x_j and by the normal-simplex procedure [6] we obtain $x_j^* = 4.3$. Then we branch to two new problems: one that adds $U_j \geq x_j \geq 5.0$ to the original-problem constraints and one that adds $4.0 \geq x_j \geq L_j$. Note that the original solution, $f(x_j^*)$, provides an upper bound on the optimal solution. This is the case because the additional constraints can only decrease the value of $f(\cdot)$. This process of creating new problems creates a tree of possible alternatives.

At each branch of the tree, one of two events will occur. First, a solution might be obtained that does not satisfy the integer requirement and has an objective-function value equal to or less than another branch whose solution is all integer. At this point, we would abandon this branch. The second possibility is that we may generate an integer solution with an objective function value that is equal to or less than that of another solution, which also satisfies the integer constraints. This branch would also be abandoned.

The process is complete when only one solution dominates every other branch of the tree. This corresponds to the optimal solution. An example will serve to illustrate these ideas.

Example [5]: Maximize $f(x) = 9x_1 + 6x_2 + 5x_3$ subject to

$$2x_1 + 3x_2 + 7x_3 \leq 35/2$$
$$4x_1 + 9x_3 \leq 15$$
$$x_1 \quad \text{nonnegative integer}$$

Step 1: Solve the problem as an ordinary linear program and obtain

$$x_1 = 15/4 \quad x_2 = 10/3, \quad x_3 = 0, \quad f(x) = 215/4$$

Step 2: The solution to the linear program does not satisfy the integer constraint. We now branch to two new problems.

(a) Maximize $f(x) = 9x_1 + 6x_2 + 5x_3$

subject to

$$2x_1 + 3x_2 + 7x_3 \leq 35/2$$
$$4x_1 + 9x_3 \leq 15$$
$$x_1 \geq 4 \text{(integer)}$$
$$x_2, x_3 \geq 0$$

This problem has no feasible solution since x_1 can't satisfy the second and third constraints simultaneously.

(b) Maximize $f(x) = 9x_1 + 6x_2 + 5x_3$ subject to

$$2x_1 + 3x_2 + 7x_3 \leq 35/2$$
$$4x_1 + 9x_3 \leq 15$$
$$0 \leq x_1 \leq 3 \text{(integer)}$$
$$x_2, x_3 \geq 0$$

The solution to this problem is $x_1 = 3$, $x_2 = 23/6$, $x_3 = 0$, and $f(x) = 50$. All constraints are satisfied, so this is a feasible and optimal solution.

6.5.3: Application to a Home Appliance Production Line

The examples in this paper are more realistic than in most other papers. The author points out that "computational experience with the proposed method indicates it is capable of solving problems involving 20 commodities over a planning horizon of 300 time periods." This is probably due to the use of branch and bound techniques.

6.6: "Concurrent Routing, Sequencing, and Setups for a Two-Machine Flexible Manufacturing Cell" (E.J. Lee and P.B. Mirchandani)

This is one of the few papers in this tutorial that deals specifically with the problem of scheduling, which can be defined as

$$\text{scheduling} = \text{routing} + \text{sequencing}$$

where routing refers to *which* machines will do *what* operation on a partpiece and sequencing indicates the *order* in which the partpieces will be dispatched to the machines. This problem focuses on the lower level of the production-control hierarchy, while the papers in Chapter 3 were concerned with the upper levels.

In this paper, a specific two-machine manufacturing system is studied. The problem resembles the conventional two-machine flowshop case [38] except here the machines can perform more than one operation (versatile) as opposed to being dedicated to a specific one. Thus, the contribution of this paper is its generalization of the classic two-machine flowshop problem. The final scheduling method for this two-versatile-machine flowshop scheduling problem is based on a "concurrent strategy."

6.6.1: The "Concurrent" Scheduling Strategy

Concurrency in routing and sequencing exists because of the versatility of the machine. If the machine can only process one particular part-type then there is no routing problem. The only thing left to do is to sequence the parts through the machines in an order that satisfies some criteria (e.g., due date of the finished part, priority of the finished part).

With versatile machines, the routes and sequences should be obtained simultaneously (i.e., concurrently).

6.6.2: Setups

The versatile machines, as defined here, have more than one tool that resides in a tool magazine or carousel. When a workpiece arrives at a machine and the required tool is already in the tool magazine, then the tool setup time is negligible. On the other hand, if the tool is not there, then a new tool magazine is needed, resulting in a magazine setup involving significant time. The consideration of these magazine setups has only been considered by a few authors. This fact, along with the extension to the conventional two-machine flowshop problem, represent the main contribution of this paper.

Finally, the concurrent schedule must make decisions about routes and sequences depending on when and if in-process magazine setups take place.

6.6.3: Optimality and Johnson's Algorithm

The two-machine flowshop problem has been solved optimally by the polynomial-time algorithm of Johnson [4]. In the present paper, it is shown that the two-versatile-machine flowshop scheduling problem is an NP-complete problem [1]. This class of problems defies attempts to find exact solutions. Thus, one must resort to clever heuristic methods to obtain solutions.

The final scheduling method presented by Lee and Mirchandani, including the heuristics, relies heavily on Johnson's algorithm. It is repeated here for convenience.

Johnson's algorithm: In the two-machine flowshop problem, n jobs must be processed by machine M_1 followed by M_2. Each job may require a different processing time, which is assumed to be known and deterministic. The problem is to find the optimal job sequence that minimizes the time to complete all n jobs (i.e. minimize the makespan). If the jobs are not released into the system in the proper sequence, the idle time of the machines will be unnecessarily high. Johnson's algorithm [4] finds this optimal sequence. The steps of his algorithm are summarized next:

1. List all processing times of all jobs to be scheduled on machine M_1 and M_2.
2. Locate the job with the minimum processing time. This job can be on either M_1 or M_2.
3. If the minimum processing time is on $M_1(M_2)$, place that job first (last) in the sequence.
4. Eliminate the jobs assigned by step 3 from the list. Repeat steps 2 and 3 for the remaining jobs on the list.
5. If two processing times are equal, then the two can be resolved arbitrarily or according to some priority, since the machines won't be affected by the sequencing.

Step 3 is the key to the algorithm and can be explained by two intuitive rules [4]:

a. The lower limit on the total processing time may be obtained as

$$L_1 = \sum_{i=1}^{n} t_{i1} + t_{n2} \quad (6.5.1)$$

where

t_{i1} = processing time of job i on machine 1
t_{n2} = processing time of job n on machine 2

This lower bound represents the extreme case where the processing of the $(n - 1)$ job at M_2 is completed in less or equal time than it takes to process all n jobs at M_1. This is illustrated in Figure 6.1 and can also be written as

$$\sum_{i=1}^{n-1} t_{i2} \leq \sum_{i=1}^{n} t_{i1} \quad (6.5.2)$$

Therefore, in this case, the only additional processing time is due to job n at M_2. Consequently, if the job with the shortest processing time is on M_2, then it should be placed last in the sequence in order to minimize (6.5.2).

b. A second lower limit can occur by considering Figure 6.2. Here, the shortest processing time is job 1 on M_1. The total processing time can be minimized by processing job 1 first resulting in a makespan of

$$L_2 = t_{11} + \sum_{i=1}^{n} t_{i2} \quad (6.5.3)$$

As a result, if the job with the shortest processing time is on M_1, place it first in the sequence (since $\sum_{i=1}^{n} t_{i2}$ is a constant).

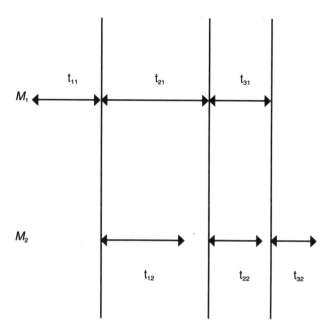

Figure 6.1: The L_1 Rule for n = 3

Example: The operation times for four jobs are given next

Job	J_1	J_2	J_3	J_4
M_1	6	7	4	6
M_2	8	5	9	3

Solution: By checking the processing times for all jobs on the list, we see that J_4 has the shortest processing time and occurs on M_2. Since it occurs on M_2, it is sequenced as late as possible resulting in

			J_4
—	—	—	

and a new listing

	J_1	J_2	J_3
M_1	6	7	4
M_2	8	5	9

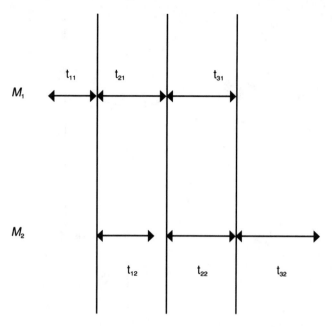

Figure 6.2: The L_2 Rule for $n = 3$

Next, we repeat step 2 and find that J_3 has the shortest processing time and it occurs on M_1, which puts it as early as possible in the sequence:

$$\underline{J_3} \quad \underline{} \quad \underline{} \quad \underline{J_4}$$

J_3 is deleted from the job list.

	J_1	J_2
M_1	6	7
M_2	8	5

J_2 is now the shortest and is included as late as possible in the sequence since it occurs at M_2

$$\underline{J_3} \quad \underline{} \quad \underline{J_2} \quad \underline{J_4}$$

and the final sequence is

$$\underline{J_3} \quad \underline{J_1} \quad \underline{J_2} \quad \underline{J_4}$$

6.7: "Selection of Process Plans in Automated Manufacturing Systems" (A. Kusiak and G. Finke)

This is the first paper in this tutorial that deals with the *process planning* problem. The process plan specifies *how* a part is to be made by answering

- What are the tool requirements?
- What operations are needed (e.g., milling, boring, and pocketing for metal removal problems)?
- What is the order in which the operations should be performed?

Usually, there is more than one way to make the finished part, and so the problem is to select one of the process plans that meets some optimization criteria, such as minimum total cost. This is the *process planning selection* problem. Kusiak and Finke present two methods for making the selection.

6.7.1: The Process Planning Problem

Before we can devise a way to select a process plan, the process planning problem must be understood and then formulated.

In the context of machining systems, a process plan requires three things:

- A set of material volumes to be removed, $V_l, l = l_1 \ldots l_n$
- A set of tools to do the removing, $T_l, l = l_1 \ldots l_k$
- A fixture, f_i, for holding the part

A simple process plan, P_i, can than be defined as

$$P_i = \{\underbrace{(V_1; T_1; f_1)}_{\substack{\text{a fixture and a set of} \\ \text{tools to remove volume l}}} \ldots \ldots, \underbrace{(V_m; T_m; f_m)}_{\substack{\text{a fixture and a set of} \\ \text{tools to remove volume m}}}\}$$

Example 1 in Kusiak and Finke indicates two ways to achieve the removal of a set of volumes $v_1, v_2, \ldots v_8$. In general, each process plan has a different cost, c_1 and c_2, corresponding to plans P_1 and P_2. These costs are the result of a measure that is assigned to P_1 and P_2 that quantifies their dissimilarity. In this paper, the weighted Hamming distance (d_{ij}) is used (equation 1).

Note that the process plan, the schedule, and the tool selection problem are all intertwined.

6.7.3: Problem Formulation and Mathematical Background

This paper formulates the process plan selection problem by using (1) a graph-theoretic formulation and (2) an integer programming formulation.

Graph theory: A *graph*, G, is defined as

$$G = (V, A)$$

where V is a set of vertices (nodes) and A is a set of arcs connecting the vertices. Suppose the nodes of the graph can be partitioned into m sets. Furthermore, suppose that each arc is directed from an element of one set to an element of the other set. Then we have an *m—partite directed* graph.

In the process plan selection problem, the authors define their graph as

$$G = (N,A)$$

where N is the set of all process plans and A is the set of directed arcs connecting the process plans. Each arc has a weight associated with it, d_{ij}, defined earlier. In addition, each process plan has a cost c_i associated with it. The process plan selection problem has now become the problem of finding the minimum cost $(\sum d_{ij} + \sum c_i)$ path, in the m-partite graph $G = (N,A)$, that will produce the m parts.

Integer programming: The integer programming problem has been formulated and discussed in Section 6.4.2. The mixed integer programming problem also appeared there. Additional details, including algorithms for solving these types of problems can be found in many optimization textbooks, such as [5].

6.7.4: Heuristics

The authors point out that these integer and mixed-integer programming problems are too complex to solve and so they introduce some heuristic methods that obtain quick and efficient solutions to the process planning selection problem.

6.8: Summary

This chapter has examined several different mathematical approaches for planning, scheduling and controlling an automated manufacturing system.

The first reprint, "Optimal Control of a Queueing System with Two Heterogeneous Servers" by Lin and Kumar, addressed a problem that is very basic to manufacturing: The routing of parts through two machines that share a common buffer. It is representative of other queueing applications in manufacturing [10-19]. In addition, it deals with the problem of allocating buffer sizes [53–56].

The second reprint, "Cooperation among Flexible Manufacturing Systems" by Berman and Maimon, dealt with the issue of integrating subsystems to form the complete manufacturing system. This is also related to the issue of flexibility, discussed in the third paper, "Measuring Decision Flexibility in Production Planning" by Lasserre and Roubellat. Several other approaches have been taken [20–24] in an effort to quantify flexibility. A good measure would help in the design of flexible manufacturing systems as well as their analysis.

Mitsumori has investigated the use of branch and bound algorithms for optimal scheduling [2]. In the fourth reprint, "Optimal Scheduling for Load Balance of Two-Machine Production Lines," he presented a scheduling method for balancing work loads in a two machine transfer line. Variations of this problem have been treated in other papers in this tutorial and the reader is encouraged to compare the methodologies.

Scheduling is treated in the next reprint, "Concurrent Routing, Sequencing, and Setups for a Two-machine Flexible Manufacturing Cell" by Lee and Mirchandani. This is an important problem in manufacturing, yet it is not the main theme of this tutorial. This is due to the fact that there is already a large body of literature in his area [25–52].

Kusiak and Finke in "Selection of Process Plans in Automated Manufacturing Systems," presented the process-planning selection problem and used several techniques from optimization theory to obtain their algorithms. Optimization theory has been used to solve numerous problems in production control [57–65].

6.9: References

6.9.1: General References

1. M.R. Garey and D.S. Johnson, *Computers and Intractability: A Guide to the Theory of NP-Completeness*, Freeman, San Francisco, Calif., 1979.
2. S. Mitsumori, "Optimum Production Scheduling of Multicommodity in Flow Line," *IEEE Transactions on Systems, Man, and Cybernetics*, Vol. SMC-2, No. 4, Sept. 1972, pp. 486–493.
3. S. Mitsumori, "Optimum Schedule Control of Conveyor Line," *IEEE Transactions on Automatic Control*, Vol. AC-14, Dec. 1969, pp. 633–639.

6.9.2: Books

4. E.A. Elsayed and T.O. Boucher, *Analysis and Control of Production Systems*, Prentice Hall, Englewood Cliffs, N.J., 1985.
5. C.S. Beightler, D.T. Phillips, and D.J. Wilde, *Foundations of Optimization*, Prentice Hall, Englewood Cliffs, N.J., 1979.
6. D.G. Luenberger, *Introduction to Linear and Nonlinear Programming*, Addison Wesley, Reading, Mass., 1973.
7. D.G. Luenberger, *Optimization by Vector Space Methods*, John Wiley and Sons Inc., New York, 1969.
8. R.A. Howard, *Dynamic Programming and Markov Processes*, M.I.T. Press, Cambridge, Mass., 1960.
9. R.A. Howard, *Dynamic Probabilistic Systems*, Vol. 1 (Markov Models) and Vol. II (Semi-Markov and Decision Processes), John Wiley and Sons Inc., New York, 1971.

6.9.3: Queueing Systems and Networks

10. S. Stidham, Jr., "Optimal Control of Admission to a Queueing System," *IEEE Transactions on Automatic Control*, Vol. AC-30, No. 8, Aug. 1985, pp. 705–713.

11. B. Hajek, "Optimal Control of Two Interacting Service Stations," *IEEE Transactions on Automatic Control*, Vol. AC-29, No. 2, June 1989, pp. 491–499.

12. F.H. Moss and A. Segall, "Optimal Control Approach to Dynamic Routing in Networks," *IEEE Transactions on Automatic Control*, Vol. AC-27, No. 2, April 1982, pp. 329–339.

13. A Lazar, "Optimal Flow Control of a Class of Queueing Networks in Equilibrium," *IEEE Transactions on Automatic Control*, Vol. AC-28, No. 11, Nov. 1983, pp. 1001–1007.

14. Y.C. Tay and R. Suri, "Error Bounds for Performance Prediction in Queueing Networks," *ACM Transactions on Computer Systems*, Vol. 3, No. 3, Aug. 1985, pp. 227–254.

15. R. Suri "Robustness of Queueing Network Formulas," *Journal of the ACM*, Vol. 30, No. 3, July 1983, pp. 564–594.

16. D.D. Yao, "An FMS Network Model with State-Dependent Routing," *Proceedings of the First ORSA/TIMS Conference on Flexible Manufacturing Systems*, Ann Arbor, Mich., 1984, pp. 129–143.

17. M.H. Ammar and S.B. Gershwin, "Equivalence Relations in Queueing Models of Assembly/Disassembly Networks," *Technical Report GIT-ICS-87/45*, Georgia Institute of Technology, Atlanta, Ga., Dec. 1987.

18. W.G. Marchal, "The Performance of Approximate Queueing Formulas: Some Numerical Remarks," *Proceedings of the First ORSA/TIMS Conference on Flexible Manufacturing Systems*, Ann Arbor, Mich., 1984, pp. 123–128.

19. K.R. Pattipati and M.P. Kastner, "Iterative Queueing Network Techniques for the Analysis of Large Maintenance Facilities," *Proceedings of the 22nd IEEE Conference on Decision and Control*, IEEE Press, New York, Dec. 1983, pp. 1045–1055.

6.9.4: Flexibility

20. N. Slack, "Manufacturing System Flexibility—An Assessment Procedure," *Computer—Integrated Manufacturing Systems*, Vol. 1, No. 1, Feb. 1988, pp. 25–31.

21. A. Chatterjee, M.A. Cohen, W.L. Maxwell, and L.W. Miller, "Manufacturing Flexibility: Models and Measurements," *Proceedings of the First ORSA/TIMS Conference on Flexible Manufacturing Systems*, Ann Arbor, Mich., 1984, pp. 49–64.

22. V. Kumar, "On Measurement of Flexibility in Flexible Manufacturing Systems: An Information Theoretic Approach," *Proceedings of the Second ORSA/TIMS Conference on Flexible Manufacturing Systems*, (eds. K.E. Stecke and R. Suri), Elsevier, New York, 1986, pp. 131–143.

23. C.H. Falkner, "Flexibility in Manufacturing Plants," *Proceedings of the Second ORSA/TIMS Conference on Flexible Manufacturing Systems*, (eds. K.E. Stecke and R. Suri), Elsevier, New York, 1986, pp. 95–106.

24. M.F. Carter, "Designing Flexibility into Automated Manufacturing Systems," *Proceedings of the Second ORSA/TIMS Conference on Flexible Manufacturing Systems,* (eds. K.E. Stecke and R. Suri), Elsevier, New York, 1986, pp. 107–118.

6.9.5: Scheduling

25. F. Villarreal and R.L. Bulfin, "Scheduling a Single Machine to Minimize the Weighted Number of Tardy Jobs," *IIE Transactions*, Vol. 15, No. 4, Dec. 1983, pp. 337–343.

26. T.T. Sen, L.M. Austin, and P. Ghandforoush, "An Algorithm for the Single-Machine Sequencing Problem to Minimize Total Tardiness," *IIE Transactions*, Vol. 15, No.4, Dec. 1983, pp. 363–366.

27. K. Phillips, "Aspects of Job Scheduling," *ASME Journal of Engineering for Industry,* Vol. 101, Feb. 1979, pp. 17–22.

28. R.S. Russell and B.W. Taylor, "An Evaluation of Scheduling Policies in a Dual Resource Constrained Assembly Shop," *IIE Transactions*, Vol. 17, No. 3, Sept. 1985, pp.219–232.

29. J.S. Panwalkar and W. Iskander, "A Survey of Scheduling Rules," *Operations Research*, Vol. 25, No. 1, 1977, pp. 45–61.

30. T. Yoshida and K. Hitomi, "Optimal Two-Stage Production Scheduling with Setup Times Separated," *AIIE Transactions*, Vol. 11, No. 3, Sept. 1979, pp. 261–263.

31. S. Axsater, "On Scheduling in a Semi-Ordered Flow Shop without Intermediate Queues," *IIE Transactions*, Vol. 14, No. 2, June 1982, pp.128–130.

32. N. Dridi, J.L. Menaldi, and J.M. Proth, "A Real Time Scheduling Algorithm," *Proceedings of the 23rd Conference on Decision and Control*, Las Vegas, Nev., Dec. 1984, pp. 855–858.

33. R. Hildebrant, "Generating and Implementing Schedules for Time-Critical Manufacturing Processes," *Proceedings of the 23rd Conference on Decision and Control*, Las Vegas, Nev., Dec. 1984, pp. 236–240.

34. C.S. Tang, "A Job Scheduling Model for a Flexible Manufacturing Machine," *Proceedings of the 1986 International Conference on Robotics and Automation*, San Francisco, Cal., April 1986, pp. 152–155.

35. K.R. Baker, *Introduction to Sequencing and Scheduling*, John Wiley and Sons, Inc., New York, 1974.

36. M. Berrada and K.E. Stecke, "A Branch-and-Bound Approach for Machine Loading," *Proceedings of the First ORSA/TIMS Special Interest Conference on Flexible Manufacturing Systems*, Ann Arbor, Mich., 1984, pp. 256–271.

37. A.K. Chakravarty and A. Shtub, "Selecting Parts and Loading Flexible Manufacturing Systems," *Proceedings of the First ORSA/TIMS Special Interest Conference on Flexible Manufacturing Systems*, Ann Arbor, Mich., 1984, pp. 284–289.

38. R.W. Conway, W.L. Maxwell, and L.W. Miller, *Theory of Scheduling*, Addison-Wesley, Reading, Mass., 1967.

39. S.C. Graves, "A Review of Production Scheduling," *Operations Research*, Vol. 29, 1984, pp. 646–675.

40. T.J. Greene and R.P. Sadowski, "A Mixed Integer Program for Loading and Scheduling Multiple Flexible Manufacturing Cells," *European Journal of Operations Research*, Vol. 24, 1986, pp. 379–386.

41. K.E. Stecke and J.J. Solberg, "Loading and Control Policies for a Flexible Manufacturing System," *International Journal of Production Research*, Vol. 19, 1981, pp. 481–490.

42. C.L. Haines, "An Algorithm for Carrier Routing in a Flexible Material-Handling System," *IBM Journal of Research and Development*, Vol. 29, No. 4, July 1985, pp. 356–362.

43. S.B. Gershwin, "Stochastic Scheduling and Set-Ups in Flexible Manufacturing Systems," *Proceedings of the Second ORSA/TIMS Conference on Flexible Manufacturing Systems*, Ann Arbor, Mich., Aug. 1986, pp. 431–442.

44. O.Z. Maimon and Y.F. Choong, "Dynamic Routing in Reentrant FMS," *Journal of Robotics and Computer Aided Manufacturing*, Vol. 4, No. 1, 1987.

45. E.G. Coffman, Jr. (Ed.), *Computer and Job Shop Scheduling Theory*, John Wiley and Sons, Inc., New York, 1976.

46. A.C. Hax and H.C. Meal, "Hierarchical Integration of Production Planning and Scheduling," *Studies in Management Science, I: Logistics* (ed. M.A. Geisler), North-Holland Elsevier, New York, 1978.

47. W.L. Maxwell, J.A. Muckstadt, L.J. Thomas, and J. Vander Ecken, "A Modeling Framework for Planning and Control of Production in Discrete Parts Manufacturing and Assembly Systems," *Interface 13*, 1983, pp. 92–104.

48. E.L. Lawler, J.K. Lenstra, and A.H.G. Rinnooy Kan, "Recent Developments in Deterministic Sequencing and Scheduling: A Survey," *Deterministic and Stochastic Scheduling* (eds. M.A.H. Dempster, J.K. Lenstra, and A.H.G. Rinnooy Kan), Reidel/Kluwer, Dordrecht, The Netherlands, 1982.

49. M. Pinedo, and L. Schrage, "Stochastic Shop Scheduling: A Survey," *Deterministic and Stochastic Scheduling* (eds. M.A.H. Dempster, J.K. Lenstra, and A.H.G. Rinnooy Kan), Reidel/Kluwer, Dordrecht, The Netherlands, 1982.

50. M. Berrada and K.E. Stecke, "A Branch and Bound Approach for FMS Machine Loading," *Proceedings of the First ORSA/TIMS Conference on Flexible Manufacturing Systems*, Ann Arbor, Mich., 1984, pp. 256–271.

51. Y.F. Choong and O.Z. Maimon, "On Dynamic Routing in FMS," *Proceedings of the 1986 IEEE International Conference on Robotics and Automation*, San Francisco, Calif., April 1986, pp. 1476–1481.

52. L.K. Platzmen and J.J. Bartholdi, III, "Real-Time Scheduling of Material Storage and Retrieval Operations," *Proceedings of the 24th IEEE Conference on Decision and Control*, Ft. Lauderdale, Fla., Dec. 1985, pp. 1807–1809.

6.9.6: Buffer Allocation Strategies

53. J. Wijngaard, "The Effect of Interstage Buffer Storage on the Output of Two Unreliable Production Units in Series, with Different Production Rates," *AIIE Transactions*, Vol. 11, No. 1, March 1979, pp. 42–47.

54. A. Altiok and S. Stidham, Jr., "The Allocation of Interstage Buffer Capacities in Production Lines," *IIE Transactions*, Vol. 15, No. 4, Dec. 1983, pp. 292–299.

55. J. Malathronas, J. Perkins, and R.L. Smith, "The Availability of a System of Two Unreliable Machines Connected by an Intermediate Storage Tank," *IIE Transactions*, Vol. 15, No. 3, Sept. 1983, pp. 195–201.

56. A.L. Soyster, J.W. Schmidt, and M.W. Rohrer, "Allocation of Buffer Capacities for a Class of Fixed Cycle Production Lines," *AIIE Transactions*, Vol. 11, No. 2, June 1979, pp. 140–146.

6.9.7: Optimal Production Strategies

57. J. Kimemia and S.B. Gershwin, "Flow Optimization in Flexible Manufacturing Systems," *International Journal of Production Research*, Vol. 24, No. 1, 1986, pp. 81–96.

58. B. Vinod and J.J. Solberg, "The Optimal Design of Flexible Manufacturing Systems," *International Journal of Production Research*, Vol. 23, No. 6, 1985, pp. 1141–1151.

59. D.R. Hansen and S.G. Taylor "Optimal Production Strategies for Plants with Multiple Shutdown Levels and Multiple Production Lines," *IIE Transactions*, Vol. 15, No. 1, March 1983, pp.46–53.

60. D.R. Hansen and S.G. Taylor, "Optimal Production Strategies for Identical Production Lines with Mini-

mum Operable Production Rates," *IIE Transactions*, Vol. 14, No. 4, Dec. 1982, pp. 288–295.

61. P. Afentakis, "Maximum Throughput in Flexible Manufacturing Systems," *Proceedings of the Second ORSA/TIMS Conference on Flexible Manufacturing Systems* (eds. K.E. Stecke and R. Suri), Elsevier, New York, 1986, pp. 509–520.

62. M.H. Han and L.F. McGinnis, "Throughput Maximization in Short Cycle Automated Manufacturing," *Proceedings of the 1986 International Conference on Robotics and Automation*, San Francisco, Calif., April 1986, pp. 147–151.

63. K. Hitomi and M. Yokoyamo, "Optimization Analysis of Automated Assembly Systems," *Transactions of the ASME*, Vol. 103, May 1981, pp. 224–232.

64. P.J. O'Grady and U. Menon, "A Multiple Criteria Approach for Production Planning of Automated Manufacturing," *Engineering Optimization*, Vol. 8, No. 3, 1985, pp. 161–175.

65. R. Suri and Y.C. Ho, "Resource Management for Large Systems: Concepts, Algorithms, and an Application," *IEEE Transactions an Automatic Control*, Vol. AC-25, No. 3, Aug. 1980, pp. 651–662.

Optimal Control of a Queueing System with Two Heterogeneous Servers

WOEI LIN, MEMBER, IEEE, AND P. R. KUMAR, MEMBER, IEEE

Abstract — The problem considered is that of optimally controlling a queueing system which consists of a common buffer or queue served by two servers. The arrivals to the buffer are Poisson and the servers are both exponential, but with *different* mean service times. It is shown that the optimal policy which minimizes the mean sojourn time of customers in the system is of threshold type. The faster server should be fed a customer from the buffer whenever it becomes available for service, but the slower server should be utilized if and only if the queue length exceeds a readily computed threshold value.

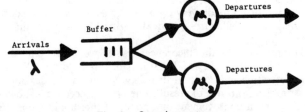

Fig. 1. Queueing system.

I. Introduction

THE queueing system shown in Fig. 1 is considered. Arrivals to the buffer form a Poisson process of rate λ. The buffer is served by two servers with *different* mean service times. The service time of a customer at server i is exponentially distributed with rate parameter μ_i ($i = 1, 2$). Without loss of generality we assume $\mu_1 > \mu_2$. To ensure stability we shall also assume that $\lambda < \mu_1 + \mu_2$. We wish to minimize the mean sojourn time of customers in the queueing system. Note that the sojourn time = waiting time in buffer + service time. By Little's theorem [1], this is equivalent to minimizing the mean number of customers in the system.

Manuscript received January 3, 1983; revised June 15, 1983, July 29, 1983, and August 31, 1983. Paper recommended by A. S. Willsky, Associate Editor for Estimation. This work was supported by the U.S. Army Research Office under Contract DAAG29-80-K0038.

W. Lin was with the Department of Mathematics and Computer Science, University of Maryland Baltimore County, Catonsville, MD 21228. He is now with the Department of Mathematics and Computer Science, Morgan State University, Baltimore, MD.

P. R. Kumar is with the Department of Mathematics and Computer Science, University of Maryland Baltimore County, Catonsville, MD 21228.

If server i is available (i.e., idle) and the buffer is nonempty (i.e., there is a customer waiting for service) should a customer from the buffer be provided to server i? We show that the optimal policy governing the dispatching of customers from the buffer to an available server is of *threshold type*, i.e., the faster server, whenever it is available and whenever the buffer is nonempty, should be dispatched a customer, but the slower server should be dispatched a customer only when, at the instant of dispatching, the number of customers in the buffers exceeds a certain readily computed *threshold* value.

This problem, which is a generalization of the $M/M/2$ queue incorporating different service rates at the two servers, was first posed by Larsen [2], who also conjectured that the optimal policy is of threshold type, and proceeded to do a detailed performance analysis of policies of threshold type. The motivation for the queueing system considered here lies in its application to the dynamic routing problem in computer systems or communication networks. For example, what is here called a "server," could be a communication line over which messages can be sent. The "service time" alluded to in this paper is then just the time taken for the message to traverse the line. Messages arriving at the buffer then have to be routed over one of several communication lines, each with a different mean transmission delay, and the goal now is to choose among the several alternatives in such a way as to minimize the average overall delay of a message.

An extensive motivation in the context of computer systems design is provided by Larsen [2] who surveys several applications. Recently, Sarachik [3] has also conjectured the optimality of the threshold policy and in addition to analyzing the performance of such policies, has exhibited the role they could play in a decentralized routing scheme for large traffic networks. Problems of this sort which feature multiple servers sharing common buffers also arise in manufacturing networks, where a part typically has to be processed at several work stations before it is a finished product, and the goal is again to find efficient routing algorithms; see Hahne [4] and Tsitsiklis [5].

Recently, there has been much interest in problems relating to optimal control of queueing systems. Hajek [6] considers the situation where there are two interacting service stations and in the process has unified and extended several earlier results. In the notation of [6], if we assume that only λ, μ_1, μ are nonzero and the only control is d, then we have a special situation where a server of rate μ_1 is always serving the buffer (called station 1 in [6]) and the question is whether an additional server of rate μ should also be used. However, in the framework of [6] customers need neither complete all of their service at one server nor be attended to by only one server at any given time. It is then clear that the optimal policy is to use the maximum service rate $\mu + \mu_1$ always and the problem becomes trivial. In this paper, however, the restriction that a customer has to complete all its service at just one server is imposed. This nontrivial problem is motivated by the fact that a message, once transmitted over one line, cannot thereafter be switched to another transmission line. Hajek also considers yet another problem where there is a single station and the cost is a convex, but not necessarily monotone, function of the state of the system and shows that the optimal policy governing whether to admit a customer or not and use a server or not is of threshold type. As above, this again is different from our problem. For a broad overview of the general area, the reader is referred to Crabill, Gross, and Magazine [7] and Stidham and Prabhu [8].

Turning now to the theoretical issues, we prove that there exists an optimal policy and that it is of threshold type. We use policy-iteration (or iteration in policy space) on the discounted cost problem to prove the result. The proof proceeds by showing first that each improvement of a policy of threshold type is *also of threshold type*, and furthermore the improvement *has a threshold at most one unit more than the original threshold*. Starting with the zero-threshold policy, we are thus monotically led either to a *finite* optimal threshold, or the policy iteration scheme never terminates. Taking the average cost problem as the limit of the discounted cost problems and by showing that the policy iteration scheme does terminate, the result is proved. Certain intermediate results needed in the above are however established by using value iteration. We also obtain a closed-form expression for the cost of a threshold policy.

III. The Discrete-Time Problem Formulation

We first show that the original continuous-time problem can be converted into an equivalent discrete-time problem by sampling the system at certain random instants of time. If a server is idle assume that it is serving a "dummy" (or "imaginary") customer. Now consider the original continuous-time system sampled at the sequence of random times when either 1) an arrival occurs, 2) a "real" customer departs, or 3) a "dummy" customer departs. This sampled system is a discrete-time system which is a faithful replica of the original continuous-time problem. This procedure of converting a continuous-time problem into a discrete-time problem is well recorded in the literature and rigorous validations are readily available in Lippman [9] or Rosberg, Varaiya, and Walrand [10] and Serfozo [12]. We shall not repeat the technical details here, but instead to supplement the already existing arguments, we will provide a heuristic and easily understood argument for why this procedure is valid.

Suppose, for instance, that in contrast to the sampling instants given above, we only sample the system when "real" transitions take place, i.e., when 1) an arrival occurs or 2) when a "real" customer departs. Let t be a time instant when both servers are idle. Then the probability that a real event takes place in the time interval $(t, t + dt)$ or equivalently, because of choice of sampling instants, the probability that the system will be sampled in the time interval $(t, t + dt)$ is $\lambda\, dt + o(dt)$. This is because the only event that can take place when both servers are idle is the arrival of a customer. On the other hand if t is a time instant at which server i is busy and the other server is idle, then the probability of sampling the system in the time interval $(t, t + dt)$ is $(\lambda + \mu_i)\, dt + o(dt)$. Finally, if both servers are busy at the time instant t, then the probability of sampling the system in the time interval $(t, t + dt)$ is $(\lambda + \mu_1 + \mu_2)\, dt + o(dt)$. Thus, when the system is in some states, sampling will be done less frequently than when it is in some other states, and so the sampled discrete-time system is not a faithful replica of the original continuous-time system. On the other hand if we assume, as in the previous paragraph, that an idle server is serving a "dummy" customer and we sample at the random times when either an arrival or a "real" or "dummy" customer departs, then the probability of sampling in the interval $(t, t + dt)$ is $(\lambda + \mu_1 + \mu_2)\, dt + o(dt)$ irrespective of what state the system is in at time t. Thus, the discrete-time system is a faithful replica of the original continuous-time system. In [9], [10], [12] this argument is made precise to show that the discrete- and continuous-time problems are equivalent in that for infinite-horizon cost criteria the optimal policies for the two formulations coincide.

Let us now normalize λ, μ_1, and μ_2 so that while they are still in the same proportion to one another, they now sum to unity. We rename the normalized values again as $\lambda, \mu_1,$ and μ_2.

We now proceed to provide a full description of the discrete-time problem. This is therefore the starting point of our rigorous formulation.

A. The Discrete-Time Problem

Let $x = (x_0, x_1, x_2)$ be the state of the queueing system. Here

x_0 = number of customers in the buffer (or queue)

x_i = 1 or 0 depending on whether server i is busy serving a real customer or idle.

Let $X = Z^+ \times \{0,1\} \times \{0,1\}$ be the state space of the system.

Define operations A, D_1, and D_2 mapping X into X as follows:

$$A(x_0, x_1, x_2) = (x_0 + 1, x_1, x_2)$$
$$D_1(x_0, x_1, x_2) = (x_0, (x_1 - 1)^+, x_2)$$
$$D_2(x_0, x_1, x_2) = (x_0, x_1, (x_2 - 1)^+)$$

where $n^+ := \max(n, 0)$. Here A is the arrival operator while D_i (for $i = 1, 2$) is the departure from server i operator. Note that the departure of a dummy customer does not change the state of the system.

Now define the following operators to denote assignment of the customer(s) from the buffer to the server(s):

$$P_h(x_0, x_1, x_2) = (x_0, x_1, x_2)$$
$$P_1(x_0, x_1, x_2) = (x_0 - 1, x_1 + 1, x_2)$$

defined on $\text{Dom}(P_1) = \{x_0 \geq 1, x_1 = 0\}$

$$P_2(x_0, x_1, x_2) = (x_0 - 1, x_1, x_2 + 1)$$
defined on $Dom(P_2) = \{x_0 \geq 1, x_2 = 0\}$
$$P_b(x_0, x_1, x_2) = (x_0 - 2, x_1 + 1, x_2 + 1)$$
defined on $Dom(P_b) = \{x_0 \geq 2, x_1 = x_2 = 0\}$.

Note that (mnemonically), P_h is the decision which holds the system state as is (no assignment of customers from buffer to any servers), P_i for $i = 1, 2$ is the decision to feed a customer from the buffer to server i, while P_b is the decision to feed both the servers simultaneously.

Let $U = \{u = (u_0, u_1, u_2): u_i \in \{h, 1, 2, b\}\}$ be the set of available controls. Let $U(x) = \{u \in U: Ax \in Dom(P_{u_0}), D_i x \in Dom(P_{u_i})$ for $i = 1$ and $2\}$ be the set of admissible controls when the system state is x.

Finally, the specification of the Markov decision process is completed by the definition of the transition probability function

$$\text{Prob}(x(t+1)|x(t) = x, u(t) = u) = \lambda \quad \text{if } x(t+1) = P_{u_0}Ax$$
$$= \mu_i \quad \text{if } x(t+1) = P_{u_i}D_i x.$$

An interpretation of our model is as follows. Suppose at a certain discrete-time instant t, the system state $x(t) = x$. We then choose an admissible control $u(t) = u = (u_0, u_1, u_2) \in U(x)$ with the understanding that if an arrival occurs we will reassign customers according to the operator P_{u_0}, while if a departure from server i occurs (real or dummy) we will reassign customers according to P_{u_i}. At every time instant $t = 0, 1, 2, \cdots$ exactly one of the three events, arrival (A), real or dummy departure from server 1 (D_1), or real or dummy departure from server 2 (D_2) occurs, each with a probability λ, μ_1, and μ_2, respectively. (Note $\lambda + \mu_1 + \mu_2 = 1$.) Thus, the new state $x(t+1)$ at time $t+1$ will either be $P_{u_0}Ax$, $P_{u_1}D_1 x$, or $P_{u_2}D_2 x$ each with a probability λ, μ_1, and μ_2, respectively.

Our goal is to choose the control actions nonanticipatively so as to minimize:

$$\limsup_{t \to \infty} \frac{1}{t} \sum_{n=1}^{t} x(n) e \qquad (1)$$

where $x(n)$ is the state of the system at time n, $e^T := (1, 1, 1)$, and thus $x(n)e$ is the total number of customers in the system at time n.

III. THE DISCOUNTED COST PROBLEM

Instead of considering directly the minimization of the average cost criterion (1) we first consider the discounted cost criterion

$$E \sum_{t=0}^{\infty} x(t) e \beta^t \qquad (2)$$

where $0 < \beta < 1$ is a discount factor. (In Section IV we obtain the result for (1) by letting $\beta \uparrow 1$.)

By a stationary policy π we shall mean a function $\pi: X \to U$ with $\pi(x) \in U(x)$ for every $x \in X$. When a stationary policy π is adopted, the control $u = \pi(x)$ is applied whenever the system is in state x.

Define the Banach space $\mathcal{F}$ of all functions $f: X \to R$ with norm $\|\cdot\|$ defined by

$$\|f\| = \sup_{x \in X} \left| \frac{f(x)}{\max(xe, 1)} \right|.$$

For any stationary policy π define $T_\pi: \mathcal{F} \to \mathcal{F}$ (the dynamic programming operator) by

$$(T_\pi f)(x) = xe + \beta \lambda f(P_{u_0}Ax) + \beta \mu_1 f(P_{u_1}D_1 x) + \beta \mu_2 f(P_{u_2}D_2 x)$$

where $\pi(x) = (u_0, u_1, u_2)$. For each x define $(Tf)(x)$ (and thus, the operator T) by

$$(Tf)(x) = \min_\pi (T_\pi f)(x).$$

The following results are well-known (see Lippman [11]).

For some n, $T^{(n)}$ is a contraction. (3.i)

If J^β is the optimal cost function (it is a function because it depends on the initial starting state), then $J^\beta = TJ^\beta$. (3.ii)

For any $f \in \mathcal{F}$, $\lim_{n \to \infty} T^{(n)} f = J^\beta$. (3.iii)

There always exists an optimal policy which is stationary.
A stationary policy π is optimal iff $J^\beta = T_\pi J^\beta$. (3.iv)

Note for future reference that (3.ii) says that

$$J^\beta(x) = \min_{u \in U(x)} \{xe + \beta \lambda J^\beta(P_{u_0}Ax) + \beta \mu_1 J^\beta(P_{u_1}D_1 x) + \beta \mu_2 J^\beta(P_{u_2}D_2 x)\}. \qquad (4)$$

A. Keep the Faster Server Busy if Possible

We prove in this section that if the buffer is nonempty and the faster server is available (idle), then it should be dispatched a customer from the buffer. This proof, in contrast to what follows, is based on value iteration.

Lemma 1:

i) $J^\beta(P_1 x) \leq J^\beta(P_h x)$ if $x \in Dom(P_1)$

ii) $J^\beta(P_1 x) \leq J^\beta(P_2 x)$ if $x \in Dom(P_1) \cap Dom(P_2)$.

Proof: Consider an $f \in \mathcal{F}$ which satisfies i) and ii) above and also:

iii) $f(x) \geq f(y)$ whenever $x \geq y$ (i.e., monotone).

(Here, and in the sequel, inequalities are meant component by component.) Note that the identically zero function ("0") satisfies all three properties and further $\lim_{n \to \infty} T^{(n)} 0 = J^\beta$ by (3.iii). Hence, if we show that Tf also satisfies all three properties we will be through. Suppose $x \in Dom(P_1)$, i.e., $x = (x_0, 0, x_2)$ where $x_0 \geq 1$. Then

$$f(P_h A P_1 x) = f(x_0, 1, x_2) = f(P_1(x_0 + 1, 0, x_2))$$
$$= \min\{f(x_0 + 1, 0, x_2), f(P_1(x_0 + 1, 0, x_2))\}$$
$$= \min\{f(P_h Ax), f(P_1 Ax)\}.$$

If in addition $x_2 = 0$, then

$$f(P_2 A P_1 x) = f(x_0 - 1, 1, 1) = f(P_1(x_0, 0, 1))$$
$$\leq \min\{f(x_0, 0, 1), f(x_0 - 1, 1, 1)\}$$
$$= \min\{f(P_2 Ax), f(P_b Ax)\}.$$

Hence, if $x \in Dom(P_1)$ then whether $x_2 = 0$ or 1,

$$\min_{u_0} \{f(P_{u_0} A P_1 x)\} \leq \min_{u_0} \{f(P_{u_0} Ax)\}. \qquad (5)$$

Similarly, we can prove by invoking the monotonicity property iii) that

$$\min_{u_i}\{f(P_{u_i}D_iP_1x)\} \leq \min_{u_i}\{f(P_{u_i}D_ix)\} \quad \text{for } i=1,2. \quad (6)$$

Since

$$Tf(P_1x) = \min_u\{P_1xe + \beta\lambda f(P_{u_0}AP_1x) + \beta\mu_1 f(P_{u_1}D_1P_1x)$$
$$+ \beta\mu_2 f(P_{u_2}D_2P_1x)\}$$
$$Tf(P_hx) = \min_u\{xe + \beta\lambda f(P_{u_0}Ax) + \beta\mu_1 f(P_{u_1}D_1x)$$
$$+ \beta\mu_2 f(P_{u_2}D_2x)\}$$

and since the above minimizations over u can be performed by doing separate minimizations over u_0, u_1, and u_2, from (5) and (6) we obtain

$$Tf(P_1x) \leq Tf(P_hx) \quad \text{for all } x \in \text{Dom}(P_1). \quad (7)$$

Now suppose $x \in \text{Dom}(P_1) \cap \text{Dom}(P_2)$, i.e., $x = (x_0,0,0)$ where $x_0 \geq 1$. Then, similar to the above, we can show

$$\min_{u_0}\{f(P_{u_0}AP_1x)\} \leq \min_{u_0}\{f(P_{u_0}AP_2x)\}$$
$$\min_{u_1}\{f(P_{u_1}D_1P_1x)\} = \min_{u_2}\{f(P_{u_2}D_2P_2x)\}$$
$$\min_{u_2}\{f(P_{u_2}D_2P_1x)\} \leq \min_{u_1}\{f(P_{u_1}D_1P_2x)\}$$

and

$$\min_{u_2}\{f(P_{u_2}D_2P_2x)\} \leq \min_{u_1}\{f(P_{u_1}D_1P_2x)\}.$$

Noting that $\mu_1 > \mu_2$, we have

$$Tf(P_1x) = xe + \beta\lambda \min_{u_0} f(P_{u_0}AP_1x) + \beta\mu_1 \min_{u_1} f(P_{u_1}D_1P_1x)$$
$$+ \beta\mu_2 \min_{u_2} f(P_{u_2}D_2P_1x)$$
$$\leq xe + \beta\lambda \min_{u_0} f(P_{u_0}AP_2x) + \beta\mu_1 \min_{u_2} f(P_{u_2}D_2P_2x)$$
$$+ \beta\mu_2 \min_{u_1} f(P_{u_1}D_1P_2x)$$
$$\leq xe + \beta\lambda \min_{u_0} f(P_{u_0}AP_2x) + \beta\mu_1 \min_{u_1} f(P_{u_1}D_1P_2x)$$
$$+ \beta\mu_2 \min_{u_2} f(P_{u_2}D_2P_2x)$$
$$= Tf(P_2x).$$

Hence, properties i) and ii) of the lemma hold. Turning now to monotonicity, i.e., iii), we have

$$Tf(x_0+1, x_1, x_2)$$
$$= \min_u\{(x_0+x_1+x_2+1) + \beta\lambda f(P_{u_0}A(x_0+1,x_1,x_2))$$
$$+ \beta\mu_1 f(P_{u_1}D_1(x_0+1,x_1,x_2))$$
$$+ \beta\mu_2 f(P_{u_2}D_2(x_0+1,x_1,x_2))\}$$
$$\geq \min_u\{(x_0+x_1+x_2) + \beta\lambda f(P_{u_0}A(x_0,x_1,x_2))$$
$$+ \beta\mu_2 f(P_{u_2}D_2(x_0,x_1,x_2))\}$$
$$= Tf(x_0,x_1,x_2).$$

Note here that the domain of admissible u's over which minimization is performed may differ from one inequality to another, but the comparison argument is straightforward. Similarly, we can show

$$Tf(x_0,1,x_2) \geq Tf(x_0,0,x_2)$$

and

$$Tf(x_0,x_1,1) \geq Tf(x_0,x_1,0).$$

Since monotonicity in each of x_0, x_1, and x_2 separately implies monotonicity in x, we have shown that Tf possesses properties similar to f. This completes our induction and the proof. □

Theorem 2: Whenever the faster server is idle it is optimal to supply it a customer if one is waiting for service.

Proof: From the form of the right-hand side of (4) it follows that u is minimizing if and only if u_0, u_1, and u_2 each minimize $J^\beta(P_{u_0}Ax)$, $J^\beta(P_{u_1}D_1x)$, and $J^\beta(P_{u_2}D_2x)$, respectively. Lemma 1 now shows that if $y \in \text{Dom}(P_1)$, then either P_1 or P_b minimizes $J^\beta(Py)$. This shows that whenever it is possible to feed a customer to server 1, it is optimal to do so. This completes the proof. □

From now on we shall restrict attention to policies which satisfy Theorem 2, i.e., keep the faster server busy whenever possible. This is done by restricting the set of admissible controls to

$$\overline{U}(x) = \{u \in U(x) : (P_{u_0}Ax)_1 = 1, (P_{u_1}D_1x)_1 = 1 \text{ if } x_0 \geq 1,$$
$$(P_{u_2}D_2x)_1 = 1 \text{ if } x_0 \geq 1\}.$$

Note that by Theorem 2, when the original set of admissible controls is changed from $U(x)$ to $\overline{U}(x)$, the optimal value function remains unchanged, and any policy optimal with this restriction is therefore also optimal without the restriction. The advantage of imposing this restriction is that the only decision which now needs to be taken is whether to feed server 2 or not. That is, we need only consider which of the two states $(k,1,0)$ or $(k-1,1,1)$ is preferable. It is proved in the next section that this decision should be made according to a threshold policy.

B. The Optimal Policy is of Threshold Type

In this section we show that for the discounted cost problem, the optimal policy is of threshold type. The threshold may be $+\infty$ (which just means that server 2 should never be used). When we turn to the average cost problem in Section IV we show, as a byproduct, that for all β sufficiently close to 1 the optimal threshold is finite. However, we have been unable to show that the optimal threshold is finite for all $\beta \in [0,1)$.

In Lemma 3 we show that any limit point of policies generated by policy iteration is an optimal policy. In Lemma 4 we show two facts: 1) each iteration of the policy improvement algorithm applied to a policy of threshold type again results in a policy of threshold type (or equivalently, the set of all policies of threshold type is closed under the policy improvement operation) and 2) the new policy obtained by policy improvement has threshold *at most* one more than the threshold of the original policy. These facts are used in Theorem 5 to show that the optimal policy is of threshold type. We have been unable to prove this result by other means such as value iteration.

For any given policy π it can be shown (as in [11]) that the resulting cost function J_π^β satisfies $J_\pi^\beta = T_\pi J_\pi^\beta$. Consider now the following standard policy iteration algorithm. Let π_0 be any policy. Define π_{n+1} recursively by $T_{\pi_{n+1}} J_{\pi_n}^\beta = T J_{\pi_n}^\beta$. (In the sequel all limits are taken in the pointwise sense.)

Lemma 3: Let π^* be a limit point of $\{\pi_n\}$, i.e., $\lim_{k \to \infty} \pi_{n_k}(x) = \pi^*(x)$ for every $x \in X$, then π^* is optimal.

Proof: Since $T_{\pi_{n+1}} J_{\pi_n}^\beta \leq T_{\pi_n} J_{\pi_n}^\beta$, it follows by the monotonicity of $T_{\pi_{n+1}}$ that $T_{\pi_{n+1}}^{(i)} J_{\pi_n}^\beta \leq J_{\pi_n}^\beta$ for all i. But since $J_{\pi_{n+1}}^\beta =$

$\lim_{i \to \infty} T_{\pi_{n+1}}^{(i)} f$ using a variation of (3.iii), for any $f \in \mathcal{F}$, it follows that $0 \leq J_{\pi_{n+1}}^{\beta} \leq J_{\pi_n}^{\beta}$. Let $J = \lim_{n \to \infty} J_{\pi_n}^{\beta}$. Since

$$J_{\pi_{n+1}}^{\beta} = T_{\pi_{n+1}} J_{\pi_{n+1}}^{\beta} \leq T_{\pi_{n+1}} J_{\pi_n}^{\beta} = T J_{\pi_n}^{\beta} \leq T_{\pi_n} J_{\pi_n}^{\beta}$$

by taking the limit in n for each fixed $x \in X$ we obtain $TJ = J$ showing that $J = J^{\beta}$. Since $J_{\pi_n}^{\beta} = T_{\pi_n} J_{\pi_n}^{\beta}$, taking the limit along the subsequence $\{n_k\}$ for each fixed x gives $J = T_{\pi^*} J$ showing that π^* is optimal. □

Now we define a threshold policy with threshold m, denoted t_m, as a policy which dispatches a customer to server 2 (when it is idle) if and only if the *total* number of customers in the system (= number in buffer + number, if any, in service) is strictly larger than m. Server 1, the faster server, should be kept busy whenever possible. More precisely, define

$$F_m(x) = P_h(x) \quad \text{if } \{x_0 = 0\} \text{ or } \{xe \leq m, x_1 = 1\}$$
$$\text{or } \{x_1 = 1, x_2 = 1\}$$
$$= P_1(x) \quad \text{if } \{xe \leq m+1, x_0 \geq 1, x_1 = 0\}$$
$$\text{or } \{x_0 \geq 1, x_1 = 0, x_2 = 1\}$$
$$= P_2(x) \quad \text{if } \{xe \geq m+1, x_0 \geq 1, x_1 = 1, x_2 = 0\}$$
$$= P_b(x) \quad \text{if } \{xe \geq m+2, x_0 \geq 2, x_1 = 0, x_2 = 0\}$$

and t_m, the threshold policy with threshold m, as

$$P_{(t_m(x))_i}(\cdot) = F_m(\cdot) \quad \text{for } i = 0, 1, 2.$$

Remark: A threshold policy, instead of being thought of as a 3-triple of contingency plans, can be viewed as a policy which depends only upon the intermediate "state" just after A, D_1, or D_2 happens, but just before a control is applied.

Note that

$$J_{t_m}^{\beta}(x) = xe + \beta \lambda J_{t_m}^{\beta}(F_m A x) + \beta \mu_1 J_{t_m}^{\beta}(F_m D_1 x) + \beta \mu_2 J_{t_m}^{\beta}(F_m D_2 x). \quad (8)$$

The following lemma is *crucial* and shows that policy iteration on a policy with threshold j produces a new policy which is 1) also of threshold type and 2) the new threshold is less than or equal to $j+1$.

Lemma 4: For any finite $i \geq 1$, there exists a j, $1 \leq j \leq i+1$, such that $T_{t_j} J_{t_i}^{\beta} = T J_{t_i}^{\beta}$.

Proof: For a given threshold policy t_i, we want to show that the policy iteration yields, as the improvement, t_j for some $1 \leq j \leq i+1$. To do this, define

$$h_0 = J_{t_i}^{\beta}(0, 1, 0) - J_{t_i}^{\beta}(0, 0, 1)$$
$$h_k = J_{t_i}^{\beta}(k, 1, 0) - J_{t_i}^{\beta}(k-1, 1, 1) \quad \text{for } k \geq 1.$$

Note that, $h_l < 0$ for $l \leq j-1$ and $h_l \geq 0$ for $l \geq j$, then the new policy obtained by one step of the policy iteration scheme is t_j. This is because $h_l < 0$ implies that the new policy will not feed a customer to server 2 when there are l in the buffer, while $h_l \geq 0$ implies that it will feed a customer to server 2. We intend to show that there is a j, $1 \leq j \leq i+1$, such that $h_l < 0$ for $l \leq j-1$ and $h_l \geq 0$ for $l \geq j$.

We consider the case $i \geq 4$ first. From (8), subtracting the equation for $J_{t_i}^{\beta}(k-1, 1, 1)$ from that for $J_{t_i}^{\beta}(k, 1, 0)$ for $k \geq 1$ and after some algebraic manipulation, we obtain

$$-(1-\beta)h_k + \beta \lambda (h_{k+1} - h_k)$$
$$= \beta \mu_1 (h_k - h_{k-1}) - \beta \mu_2 \left[J_{t_i}^{\beta}(k-1, 1, 1) - J_{t_i}^{\beta}(k-1, 1, 0) \right]$$
$$1 \leq k \leq i-2 \quad (9)$$

$$h_k = \beta \mu_1 h_{k-1} + \beta \mu_2 \left[J_{t_i}^{\beta}(k-1, 1, 1) - J_{t_i}^{\beta}(k-1, 1, 0) \right]$$
$$i-1 \leq k \leq i$$

$$h_k = \beta \mu_2 \left[J_{t_i}^{\beta}(k-1, 1, 1) - J_{t_i}^{\beta}(k-2, 1, 1) \right] \quad k \geq i+1.$$

Proceeding as in the proof of Lemma 1, it can be shown that if f is monotone, then $T_{t_i}^{\beta} f$ is also. Thus, it is monotone increasing in each argument. Hence, we have

$$-(1-\beta)h_k + \beta \lambda (h_{k+1} - h_k) \leq \beta \mu_1 (h_k - h_{k-1})$$
$$i \leq k \leq i-2 \quad (10.\text{i})$$

$$h_k \geq \beta \mu_1 h_{k-1} \quad i-1 \leq k \leq i \quad (10.\text{ii})$$

$$h_k \geq 0 \quad k \geq i+1 \quad (10.\text{iii})$$

Case 1: Suppose $h_i < 0$, then $h_{i-1} < 0$ by (10.ii). In fact, then the stronger statement $h_{i-1} < h_i < 0$ holds. Similarly, by (10.ii), we can show $h_{i-2} < h_{i-1} < h_i < 0$. By (10.i) with $k = i-2$, it follows that $h_{i-3} < h_{i-2} < 0$. Repeating (10.i) with $k = i-3, \cdots, 1$, we can obtain

$$h_0 < h_1 < \cdots < h_i < 0.$$

However by (10.iii), $h_k \geq 0$ for $k \geq i+1$. Thus, policy iteration on t_i yields the new policy t_{i+1}.

Case 2: Suppose $h_i \geq 0$, $h_{i-1} < 0$ then $h_{i-2} < h_{i-1} < 0$ by (10.ii). Again by (10.i), we can show

$$h_0 < h_1 < \cdots < h_{i-1} < 0 \leq h_i.$$

This together with (10.iii) implies that the new policy obtained by policy iteration is t_i.

Case 3: Suppose there exists some integer j, $1 \leq j \leq i-2$ such that $h_i \geq 0$, $h_{i-1} \geq 0, \cdots, h_{i-j} \geq 0$ but $h_{i-j-1} < 0$ then by using (10.i) for $k = i-j-1, i-j-2, \cdots, 1$ it follows that

$$h_0 < h_1 < \cdots < h_{i-j-1} < 0 \leq h_{i-j}$$

which implies that the new policy is t_{i-j}.

In all cases, we have shown that the new policy is t_j for some integer j, $1 \leq j \leq i+1$. A similar argument also holds for the cases $i = 1$, $i = 2$, and $i = 3$. Hence, the proof is complete. □

Theorem 5:
 i) There exists an optimal stationary policy which is of threshold type with a threshold $m^* \leq \infty$.
 ii) If $J_{t_i}^{\beta}(x) < J_{t_{i+1}}^{\beta}(x)$ for some $x \in X$, then $m^* \leq i$.

Proof: Start policy iteration with policy t_0. If $T_{t_0} J_{t_0}^{\beta} = T J_{t_0}^{\beta}$, then t_0 is optimal. If not, then from Lemma 4 and the proof of Lemma 3 it follows that $J_{t_1}^{\beta} \leq J_{t_0}^{\beta}$ with strict inequality for some $x \in X$. Again by policy iteration either t_1 is optimal or we will obtain a new policy t_2 which is better than t_1 and so on. The results now follow directly from Lemmas 4 and 3. □

Note that we also see, as a byproduct of our analysis, the behavior of the policy iteration algorithm. If policy iteration is started with t_0, then each iteration will increase the threshold by exactly one unit until the optimal threshold is reached. Thus, if t_j is the optimal policy, then it will be reached in exactly j steps of the policy iteration procedure.

IV. THE AVERAGE COST PROBLEM

Let π be a stationary policy which results in a Markov chain with a single positive recurrence class. Then if J_π is the average cost of this policy, i.e., the value of (1) when π is used, it is well known that J_π does not depend on the starting state $x(0)$. We now have the following.

Lemma 6: For $m \geq 4$

$$J_{t_m} = \begin{cases} \dfrac{\sum_{i=1}^{2} c_i \eta_i^m \left\{ \left[\dfrac{\rho}{1-\rho} - \dfrac{\eta_i}{1-b}\right] m + \dfrac{\eta_i}{(1-b)^2} b^{-m} + \left[\dfrac{1}{(1-\rho)^2} - \dfrac{\eta_i}{(1-b)^2} - \dfrac{1}{1-\eta_i}\right] \right\}}{\sum_{i=1}^{2} c_i \eta_i^m \left\{ \dfrac{\eta_i}{b(1-b)} b^{-m} + \left[\dfrac{\rho}{1-\rho} - \dfrac{\eta_i}{1-b}\right] \right\}} & \text{if } \lambda \neq \mu_1 \\[2em] \dfrac{\sum_{i=1}^{2} c_i \eta_i^m \left\{ \dfrac{\eta_i}{2} m^2 + \left[\dfrac{\rho}{1-\rho} + \dfrac{\eta_i}{2}\right] m + \left[\dfrac{1}{(1-\rho)^2} - \dfrac{1}{1-\eta_i}\right] \right\}}{\sum_{i=1}^{2} c_i \eta_i^m \left\{ \dfrac{\eta_i}{2} m + \left[\dfrac{\rho}{1-\rho} + \dfrac{\eta_i}{2}\right] \right\}} & \text{if } \lambda = \mu_1 \end{cases} \quad (11)$$

where

$$c_1 = \frac{1-\eta_1}{\eta_2 - \eta_1} \quad c_2 = \frac{\eta_2 - 1}{\eta_2 - \eta_1}$$

$$\eta_1 = \frac{1 - \sqrt{1 - 4\mu_1 \lambda}}{2\mu_1} \quad \eta_2 = \frac{1 + \sqrt{1 - 4\mu_1 \lambda}}{2\mu_1}$$

$$\rho = \frac{\lambda}{\mu_1 + \mu_2} \quad b = \frac{\lambda}{\mu_1}.$$

Proof: In the Appendix.

Lemma 7: There exists an m such that $J_{t_m} < J_{t_{m+1}}$.

Proof: If $\lambda \geq \mu_1$, it follows from (11) that $\lim_{m \to \infty} J_{t_m} = \infty$. Hence, the result follows for $\lambda \geq \mu_1$. If $\lambda < \mu_1$, it follows from (11) that

$$\lim_{m \to \infty} J_{t_m} = \frac{b}{1-b} = \frac{\lambda}{\mu_1 - \lambda}$$

(which is the average number in an $M/M/1$ queue, as is to be expected). Forming the difference $J_{t_{m+1}} - J_{t_m}$, we find

$$J_{t_{m+1}} - J_{t_m} = \frac{g \left[m\eta_2^{2m} b^{-m} + o(m\eta_2^{2m} b^{-m}) \right]}{\{\text{denominator of } J_{t_m} \text{ in (11)}\}\{\text{denominator of } J_{t_{m+1}} \text{ in (11)}\}}$$

where

$$g = b^{-1} c_2^2 \eta_2 \left(\eta_2^{-1} - \lambda^{-1}\right) \left[\rho(1-\rho)^{-1} - b(1-b)^{-1}\left(\lambda^{-1} - \eta_2^{-1}\right)\right]$$

and

$$\lim_{m \to \infty} \left[o(m\eta_2^{2m} b^{-m}) / m\eta_2^{2m} b^{-m} \right] = 0.$$

Since the denominator of the above expression is strictly positive and since $g > 0$ (which involves some algebra) it follows that for all m large enough, $J_{t_{m+1}} > J_{t_m}$. □

We are ready for the main result.

Theorem 8: For the average cost problem there exists an optimal policy which is of threshold type with a finite threshold.

Proof: For any given policy π which results in a Markov chain with a single positive recurrent class, Lippman [11] can be used to show that $\lim_{\beta \to 1}(1-\beta) J_\pi^\beta(x) = J_\pi$ for any $x \in X$. From Lemma 7, it follows that for some $x \in X$, $J_{t_m}^\beta(x) < J_{t_{m+1}}^\beta(x)$ for all $\beta > 1 - \delta$ where $0 < \delta < 1$. By Theorem 6 it follows that for each discounted cost problem with discount factor $\beta > 1 - \delta$, there is an optimal policy which is of threshold type with threshold less than or equal to m. It thus follows that some policy in $\{t_1, t_2, \cdots, t_m\}$ is optimal for every $\beta > 1 - \delta$. Since the sufficient conditions of [11] Theorem 3 are met, i.e., that there is a t_j which is optimal for a sequence $\{\beta_n\}$ of discount factors with $\beta_n \uparrow 1$, the results in [11] show that the average cost problem has an optimal policy which is also a member of the set $\{t_1, t_2, \cdots, t_m\}$. The proof is complete. □

Remark: Note that the previous result provides a simple algorithm to determine an optimal policy, i.e., an optimal threshold. Starting with t_0, use (11) to determine an n such that

$$J_{t_0} \geq J_{t_1} \geq J_{t_2} \geq \cdots \geq J_{t_n} \quad \text{and} \quad J_{t_n} < J_{t_{n+1}}.$$

Then t_n is the optimal policy.

V. CONCLUDING REMARKS

We have shown here that the optimal policy is of threshold type and that the threshold is readily computable. Many interesting questions have arisen.

i) Can these results be generalized to the multiserver case (i.e., more than 2 servers)? We suspect not. The threshold for each server may be dependent on which other servers are busy. Our conjectures for the k-server case (with $\mu_1 > \mu_2 > \mu_3 > \cdots > \mu_k$) are the following.

a) Server j should not be used if some server i for $i < j$ is idle.

b) If servers $B \subset \{1, 2, \cdots, k\}$ are busy, then there is a threshold $t_j(B)$ at which server j should be used. We strongly suspect both that $t_j(B)$ does in fact depend on B, but that the variation of $t_j(B)$ with B will not be so much as to be significant in practice.

ii) Note that if we cascade systems of the type considered in this paper, then the output of one stage is the input of the next stage. Thus, it is useful to know what the output process of these systems is. The next question is whether we can find an input process (i.e., arrival process) which reproduces itself at the output (i.e., departure process).

iii) Can we generalize the service time distributions?

iv) Does the optimal threshold decrease as the arrival rate increases? We conjecture that it is so.

v) Is the optimal threshold bounded above by μ_1/μ_2? Consider $\lambda = 0$, i.e., we just want to clear a system which already contains n customers, with the cost being the discounted total number of customers in the system. Let t_∞ be the policy which never uses server 2. Then

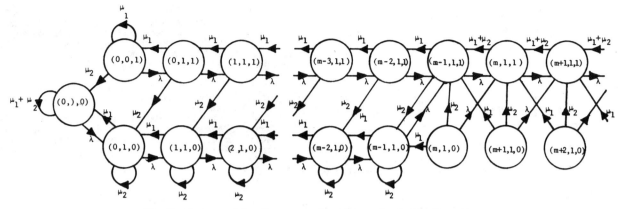

Fig. 2. State transition diagram resulting from a threshold policy with threshold m.

$$J_{t_\infty}^\beta(n) = n + \beta\mu_1 J_{t_\infty}^\beta(n-1) + \beta(1-\mu_1) J_{t_\infty}^\beta(n).$$

This can be recursively solved since

$$J_{t_\infty}^\beta(1) = \frac{1}{1-\beta(1-\mu_1)} =: a$$

to yield

$$J_{t_\infty}^\beta(n) = an + \frac{abn}{1-b} - \frac{ab}{(1-b)^2}(1-b^n)$$

where

$$b := \frac{\beta\mu_1}{1-\beta(1-\mu_1)}.$$

Let t_{n-1} be the policy which uses server 2 when the total number in the system is $\geq n$. Then

$$J_{t_{n-1}}^\beta(n) = c(\beta) + J_{t_\infty}^\beta(n-1)$$

where

$$c(\beta) := \frac{1}{1-\beta(1-\mu_2)}.$$

It then follows that if $n > \mu_1/\mu_2$, then $J_{t_{n-1}}^\beta(n) \leq J_{t_\infty}^\beta(n)$ for all $\beta > 1 - \delta$ where $\delta > 0$. Thus, if iv) can be established, then the bound on the threshold follows for all λ.

vi) Is the optimal threshold bounded for all $0 < \beta < 1$? Does the optimal threshold increase as β increases? We suspect that the answer to both questions is in the affirmative.

Appendix
Proof of Lemma 6

Let $p(x)$ be the steady-state probability of the state $x \in X$ when policy t_m is used. (A steady-state distribution exists because $\lambda < \mu_1 + \mu_2$ and it is unique since the state $(0,0,0)$ can be reached from all states.)

$$p(x) = \lambda \sum_{y \in (F_m A)^{-1}(x)} p(y) + \mu_1 \sum_{y \in (F_m D_1)^{-1}(x)} p(y) + \mu_2 \sum_{y \in (F_m D_2)^{-1}(x)} p(y). \quad (12)$$

The transition structure is given in Fig. 2.

Step 1: From (12), we obtain

$$p(0,0,1) = \mu_1 p(0,0,1) + \mu_1 p(0,1,1)$$
$$p(0,1,1) = \lambda p(0,0,1) + \mu_1 p(1,1,1)$$
$$p(k,1,1) = \lambda p(k-1,1,1) + \mu_1 p(k+1,1,1) \quad 1 \leq k \leq m-2.$$

Let $p(0,0,1) =: p_1$ and solving the equations, we obtain

$$p(k,1,1) = p_1 \left[c_1 \eta_1^{k+1} + c_2 \eta_2^{k+1} \right] \quad 0 \leq k \leq m-1$$

where

$$c_1 = \frac{1-\eta_1}{\eta_2 - \eta_1} \quad c_2 = \frac{\eta_2 - 1}{\eta_2 - \eta_1}$$

$$\eta_1 = \frac{1 - \sqrt{1-4\mu_1\lambda}}{2\mu_1} \quad \eta_2 = \frac{1 + \sqrt{1-4\mu_1\lambda}}{2\mu_1}.$$

Step 2: From (12), we can obtain (with $\mu := \mu_1 + \mu_2$)

$$p(k,1,1) = \lambda p(k-1,1,1) + \mu p(k+1,1,1) \quad k \geq m$$

solving these equations and taking the stable solution

$$p(k,1,1) = \rho^{k-m+1} p(m-1,1,1) \quad \text{for } k \geq m$$

where

$$\rho = \frac{\lambda}{\mu_1 + \mu_2}.$$

Step 3: From (12), we can obtain

$$p(m-1,1,1) = \lambda p(m-2,1,1) + (\mu_1+\mu_2) p(m,1,1) + \lambda p(m-1,1,0)$$
$$p(m-1,1,0) = \mu_2 p(m-1,1,0) + \lambda p(m-2,1,0) + \mu_2 p(m-1,1,1)$$
$$p(k,1,0) = \mu_2 p(k,1,0) + \lambda p(k-1,1,0) + \mu_1 p(k+1,1,0) + \mu_2 p(k,1,1) \quad 1 \leq k \leq m-2$$
$$p(0,1,0) = \mu_2 p(0,1,0) + \lambda p(0,0,0) + \mu_1 p(1,1,0) + \mu_2 p(0,1,1).$$

Solving these equations we obtain

$$p(k,1,0) = p_1 \left[db^k - c_1 \eta_1^{k+1} - c_2 \eta_2^{k+1} \right] \qquad 0 \leq k \leq m-1$$

$$p(0,0,0) = p_1 \left[\frac{d}{b} - 1 \right]$$

where

$$b = \frac{\lambda}{\mu_1} \qquad d = \frac{c_1 \eta_1^{m+1} + c_2 \eta_2^{m+1}}{b^m}.$$

Step 4: Noting that $p(k,1,0) = 0$ for $k \geq m$ we can obtain p_1 from the condition that the sum of the probabilities is unity. Thus

$$p_1^{-1} = \begin{cases} \sum_{i=1}^{2} c_i \eta_i^m \left\{ \frac{\eta_i}{b(1-b)} b^{-m} + \left[\frac{\rho}{1-\rho} - \frac{\eta_i}{1-b} \right] \right\} & \lambda \neq \mu_1 \\ \sum_{i=1}^{2} c_i \eta_i^m \left\{ \frac{\eta_i}{b} b^{-m} + \left[\frac{\rho}{1-\rho} + \frac{\eta_i}{b} \right] \right\} & \lambda = \mu_1. \end{cases}$$

Step 5: Since $J_{t_m} = \Sigma_x(xe)p(x)$, after some algebraic manipulation and simplification, we obtain (11).

Acknowledgment

The authors are grateful to Prof. A. Agrawala and Prof. S. Tripathi for posing this problem and for several useful discussions.

References

[1] L. Kleinrock, *Queueing Systems, Volume I: Theory*. New York: Wiley, 1975.
[2] R. L. Larsen, "Control of multiple exponential servers with application to computer systems," Ph.D. dissertation, Dep. Comput. Sci., Univ. Maryland, College Park, Tech. Rep. TR-1041, Apr. 1981.
[3] P. Sarachik, "A dynamic alternate route strategy for traffic networks," in *Proc. 21st IEEE Conf. on Decision and Contr.*, vol. 1, Dec. 1982, pp. 120–124.
[4] E. L. Hahne, "Dynamic routing in an unreliable manufacturing network with limited storage," Mass. Inst. Technol., Cambridge, Rep. LIDS-TH-1063, 1981.
[5] J. N. Tsitsiklis, "Convexity and characterization of optimal policies in a dynamic routing problem," Mass. Inst. Technol., Cambridge, Rep. LIDS-R-1178, 1982.
[6] B. Hajek, "Optimal control of two interacting service stations," in *Proc. 21st Conf. on Decision and Contr.*, vol. 2, Dec. 1982, pp. 840–845.
[7] T. Crabill, D. Gross, and M. J. Magazine, "A classified bibliography of research on optimal design and control of queues," *Operat. Res.*, vol. 25, pp. 219–232, 1977.
[8] S. Stidham and N. Prabhu, "Optimal control of queueing systems," in *Mathematical Methods in Queueing Theory* (Springer Verlag Lecture Notes in Economics and Mathematical Systems, vol. 98), A. B. Clarke, Ed. New York: Springer, 1974, pp. 263–294.
[9] S. Lippman, "Applying a new device in the optimization of exponential queueing systems," *Operat. Res.*, vol. 23, pp. 687–710, 1975.
[10] Z. Rosberg, P. Varaiya, and J. Walrand, "Optimal control of service in tandem queues," *IEEE Trans. Automat. Contr.*, vol. AC-27, pp. 600–610, June 1982.
[11] S. A. Lippman, "Semi-Markov decision processes with unbounded rewards," *Management Sci.*, vol. 19, pp. 717–731, Mar. 1973.
[12] R. F. Serfozo, "An equivalence betwen continuous and discrete time Markov decision processes," *Operat. Res.*, vol. 27, pp. 616–620, 1979.

Woei Lin (S'81–M'83) was born in Taipei, Taiwan, Republic of China, in 1954. He received the B.S. degree in 1976 from the National Taiwan University, Taiwan, and the M.S. and Ph.D. degrees, in 1981 and 1983, respectively, from the University of Maryland Baltimore County, all in mathematics.

From 1979 to 1983 he served as a Teaching/Research Assistant in the Department of Mathematics and Computer Science, University of Maryland Baltimore County. He is currently an Assistant Professor in the Department of Mathematics and Computer Science at Morgan State University, Baltimore, MD. His current research interests are in adaptive control and the stochastic control of queueing systems.

Dr. Lin is a member of SIAM. He was the recipient of the 1983 Michael J. Pelczar Award from the University of Maryland System.

P. R. Kumar (S'77–M'77) was born in India, on April 21, 1952. He received the B. Tech. degree in electrical engineering (electronics) in 1973 from the Indian Institute of Technology, Madras, and the M.S. and D.Sc. degrees in systems science and mathematics from Washington University, St. Louis, MO, in 1975 and 1977.

Since 1977 he has been on the faculty of the Department of Mathematics and Computer Science at the University of Maryland Baltimore County, where he is currently an Associate Professor. His current research interests are in the area of stochastic systems.

Cooperation Among Flexible Manufacturing Systems

ODED BERMAN AND ODED MAIMON

Abstract—Integration of an automated manufacturing system is now recognized as one of the key issues for improving productivity. The separation of a large complex system into smaller subsystems that operate individually contributes to an efficient integration and control of the entire system. However, the overall performance of the system may be reduced if each subsystem operates independently and if no cooperation between the subsystems' resources exists. How the subsystems should cooperate to improve system performance is investigated.

I. INTRODUCTION

A FLEXIBLE manufacturing system (FMS) typically consists of a set of cells; a material handling system (MHS) that connect those cells; and service centers (e.g., material warehouse, tool room, repair equipment). The cell is an autonomous unit that performs certain manufacturing functions (e.g., a machining center, inspection machine, and a load–unload robot). The MHS is used for distributing the appropriate input to the cells, so that the cell can performs its tasks and remove from the cell its output, e.g., ready products, worn tools.

A FMS is characterized by the versatility of its components (see [3] for a general survey of FMS). The work stations (e.g., robots and NC machines) are capable of performing many different operations, given the proper tooling, fixtures, and materials. Another feature is small switching time between different operations assigned to a work station (among a family of operations), for example, downloading of an NC program. The MHS can be directed via different routes to various destinations, carrying a variety of loads.

Hardware and software to carry out particular tasks automatically are rather well developed and in existence in manufacturing. However, even with adequate automation equipment, the manufacturing, as a system, usually does not live up to its performance expectation. One main reason of this phenomenon is a lack of appropriate integration and control between the various cells.

Hatvany [4] refers to unforeseen situations that the system has to cope with and stresses the importance of cooperation among manufacturing systems. He refers to the cooperation feature as part of the overall system intelligence. With FMS integration and control, the system operations are completely controlled by a computer (for example, determining which part will move to which cell, when, how, and to what process) so that all manufacturing tasks are performed and an overall system performance criteria is achieved. This integration and control has several levels, from system decisionmaking (such as cooperation), to determining the specific movement of parts, and to the actual sending of commands to the machine controllers to execute those movements.

Operational control of a FMS is very complicated. It involves large static and dynamic data sets (representing machine configuration, system status, and process plans) and complex control algorithms (determining the system operations) [4], [9], [11], and [12]. In [12] it was shown that the complexity of a small part of the operational control of FMS, namely, the calculation of the off-line decision tables including a linear program, where the number of constraints increases exponentially with the size of the FMS (e.g., number of machines). Thus, for control efficiency and feasibility, smaller FMS's are desired.

Therefore, a large system should be divided into several subsystems that are controlled independently. This action, however, may reduce the overall system performance since shared resources are preassigned to the subsystems. Thus the system may evolve such that, at certain times, a particular subsystem needs more resources, while another subsystem has too many. The reason for such an evolution is the versatility of the FMS, and inherent system uncertainties such as machine or tool failure that cause random demand on resources. To compensate for this partitioning, some cooperation among the subsystems is required. Such cooperation should ensure that the overall system performance is improved, while at the same time the subsystems' performance is not worsened. Henceforth we refer to the subsystems as the FMS's and to their union as the system.

The purpose of the cooperation among the various FMS's, is to increase the utilization of each FMS by reducing the amount of time an element of an FMS (e.g., an NC machine or a robot) has to wait for its required resources (e.g., material). This is accomplished by reducing the expected response time to get the resources. The MHS is viewed as consisting of several types of units including automated guided vehicles (AGV).

An AGV is assumed to have the capability of carrying a

Manuscript received April 29, 1985; revised December 30, 1985.

O. Berman is with The Faculty of Management, The University of Calgary, Calgary, Alberta T2N 1N4, Canada.

O. Maimon is with the Massachusetts Institute of Technology, Laboratory for Information and Decision Systems, Cambridge, MA 02139, USA.

IEEE Log Number 8607694.

certain type of resources (e.g., raw materials, tools) and is allocated primarily to one FMS, usually at a service center. The cooperation strategy may determine that, at a particular state of the system (defined later), a certain type of AGV should serve another FMS. The objective here is to find an optimal cooperation policy.

We refer to a methodology [1] that was developed in the context of urban planning. Independent urban entities (e.g., cities and districts) quite often use cooperation between their service units (e.g., police, ambulance, emergency repair).

In order to rigorously express the cooperation policy, the next section presents the notation and definitions. Section III describes the problem and its solution. Section IV illustrates the content of Sections II and III by an example. Finally, Section V brings conclusions and suggestions for future research.

II. The Model

The system is described as a graph $G(V, A)$, with a set of nodes V and a set of links A connecting them.

Typically, $v \in V$ is a cell in FMS G, but it can also denote a location of an AGV; $a = (u, v) \in A$ is the transportation link between two cells u and v.

The system is divided into n subsystems (FMS's), each represented by

$$G_i(V_i, A_i), \quad \text{where } G = \bigcup_{i=1}^{n} G_i$$

i.e.

$$V = \bigcup_{i=i}^{n} V_i, \quad A = \left(\bigcup_{i=i}^{n} A_i\right) \cup \left(\bigcup_{i,j} A_{ij}\right)$$

where $A_{ij} = \{a = (u, v): v \in V_i, u \in V_j\}$ are the inter FMS's links. Also assume the graphs are disjoint, i.e., $G_i \cap G_j = \emptyset$, in the sense that $V_i \cap V_j = \emptyset$. We refer to G_i as FMS i whenever it contributes to clarify the text.

When the FMS is in operation, resources are dispatched to the cells by the AGV's. Requests for the resources are modelled as a Poisson process with the rate λ^i per units of time from G_i. The fraction of calls for resource of type j, from $k \in V_i$ is denoted by h^i_{kj}, where

$$\sum_{\substack{k \in V_i \\ j \in J}} h^i_{kj} = 1$$

and J is the set of resources types. There may be several types of AGV's in the system. Let L be the set of all the AGV's types (i.e., there are $|L|$ different AGV types). For example, we can distinguish between an AGV that can carry only materials, an AGV that can carry only tools, an AGV that can carry a repair kit, etc. The service time of an AGV is assumed to behave according to a negative exponential distribution with parameter μ^i_j (which depends on the type j requests and FMS i).

The service time includes a travel time component (to the cell) in addition to the actual service. However, we assume (for the time being) that the travel component is very small relative to the actual service and therefore can be neglected. Later in the paper we will discuss the case when this assumption does not hold. The negative exponential distribution reflects a service which is usually very fast but may be slow on occasions due to interuptions, such as temporary block of the path, or a need to choose an alternative path. Another reason for the variable service time is a recovery operation that may take place, if the automatic loading/unloading of the vehicle encounters some difficulties. Support for the above probabilistic assumption can be found in [2], [8], [13]–[15].

For simplicity we also assume a zero-capacity queueing situation; that is, if a request arrives when all appropriate servers are busy, then there is a penalty P for not satisfying the request—namely, there is a special server that comes to help in such cases (e.g., a supervisor, as is often the case in manufacturing), but the cost of such a dispatch is very high.

Let us assume that there are Q AGV's in the whole system. Define by $S = \{b_Q, b_{Q-1}, \cdots, b_1\}$, the state of the system, where $b_j, j = 1, \cdots, Q$ is a binary variable that describes the status of AGV j (busy when $b_j = 1$ and free when $b_j = 0$). Note that there are 2^Q states. Let us denote by $S_1 = (0, 0, \cdots, 0)$, the state where all the servers are free, and by $S_{2^Q} = (1, 1, \cdots, 1)$, the state where all the servers are busy.

Whenever a call for resource j in V_i occurs while all suitable AGV's in G_i are busy, cooperation may take place. We define a *decision alternative* for the call to be either a decision not to dispatch any server from other FMS's (no cooperation decision), or a specification of which AGV from a different FMS is to be dispatched.

We define a *strategy* for state S_k to be a vector of decision alternatives. The components of this vector correspond to each possible call for service in each FMS.

For each state S_k there are several possible strategies. We denote by D_k the set of all strategies that can be used when the system is in state S_k. The goal of this work is to find the best strategy to apply to all states of the system.

Finally, X_{kj} denotes the steady state probability that the system is in state k while strategy j is being carried out ($k = 1, \cdots, 2^Q; j = 1, \cdots, d_k, |D_k| = d_k$).

III. The Problem

The cooperation policy among FMS's is a set of strategies that define the system cooperation behavior. The objective function is to minimize the cost of responding to a random call for service subject to particular FMS's performance constraints. By minimizing the cost of responding, we can achieve minimization of machine idle time due to waiting time for required resources, which means increasing machine utilization and productivity.

It is important to emphasize that there are many other factors which affect the utilization and productivity of the systems. This paper deals with only one of them, namely, the possibility of a certain type of cooperation.

The objective function can be expressed as

$$\min_{X_{kj}} \left\{ \sum_{k=1}^{2^Q-1} \sum_{j=1}^{d_k} x_{kj} \sum_{i=1}^{n} \cdot \left(\lambda^i \bigg/ \sum_{m=1}^{n} \lambda^m \right) \sum_{v \in V_i} \sum_{f \in J} h_{vf}^i d_j(S_k, v, f) + X_{2^Q} P \right\} \quad (1)$$

where $d_j(S_k, v, f)$ is the time it takes to reach node v to serve a call type f by strategy j given S_k. If there are several AGV's that can be dispatched, then a shortest time/distance can determine which AGV will be dispatched. Expression (1) takes into consideration that any state S_k can occur with any strategy j in D_k for S_k. The fraction of time strategy j in S_k that will be used is X_{kj}. Then at any FMS i a call for any service type f might occur in any cell v and a cost (in time units) of the dispatch to node v (to be determined by strategy j for state S_k) will be incurred. We also include in (1) the possibility to be in state 2^Q and to incur a penalty P.

It should be emphasized that (1) (and the rest of the analysis) can be easily modified when there is a queueing capacity greater than zero. For example, when there is an infinite capacity, instead of $X_{2^Q}P$ in (1) we have the following term:

$$\sum_{k=0}^{\infty} Y_k \frac{(k+1)}{\sum_{i,j} \mu_j^i}$$

where Y_k is a state where all the servers are busy and k requests are waiting in the queue. Given state Y_k, the average service time of each call waiting in line is $1/\Sigma_{i,j} \mu_j^i$ where $\Sigma_{i,j} \mu_j^i$ is the total service rate. Note that for simplicity an implicit assumption we have here is that we forbid a queue to occur unless all AGV's are busy. If this is not the case, (1) can be modified easily. In the example of Section IV we relax this assumption since we allow a no cooperation decision when all AGV's in either FMS 1 or FMS 2 are busy).

Other performance measures can appear as constraints (or replace the objective function). The constraints that are required in order to maintain a certain level of service for a particular FMS are of the form $DX \leq r$, where r is a vector of prespecified level of service. For example, the following constraint:

$$\sum_{v \in V_i} \sum_{f \in J} \sum_{k=1}^{2^Q-1} \sum_{j=1}^{d_k} X_{kj} h_{vf}^i d_j(S_k, v, f) + X_{2^Q} P \leq r_i \quad (2)$$

ensures that the expected response time to FMS i must not exceed a given threshold r_i. Note that this type of constraint is extremely important, if the system wants to prefer a certain FMS, or part of it, and tune the cooperation, such that a particular area will receive faster response than the other. For example, a "hot" demand for a particular product arrives, which causes a certain FMS to have priority upon others for a period of time. We see that in this case, the model will be modified to include this requirement. Solving the model will result in a strategy that reflects this priority. Upon satisfying the demand, the priorities return to normal, and the old strategy should then be used. Also, priorities for service could be set according to priorities of the products they serve. That is, if a tool or raw material is needed by a machine that services several products, the service would be given according to priorities that are determined according to, for example, due date or cost of the various products.

In some types of manufacturing certain areas have some restriction with respect to a certain AGV. This is dealt with another type of constraints which is to limit the fraction of service of a certain AGV to particular areas. Then the constraint is of the type

$$\sum_{v \in W} \sum_{f \in J} h_{vf}^i \sum_{S_k \in E_{l,v}} \sum_{j=1}^{d_k} X_{kj} \leq \alpha_l, \quad 0 \leq \alpha_l \leq 1 \quad (3)$$

where W is the restricted area in a certain FMS i, and E_{lv} is the set of all states in which AGV type l is dispatched to node v.

Certain constraints can be imposed on the various AGV types. For example a constraint that states that the utilization of an AGV type l_1 must not exceed the utilization of AGV type l_2 (or any other preference). Such a constraint can be stated as follows:

$$\sum_{k \in E_1} \sum_{j=1}^{d_k} X_{kj} \leq \sum_{k \in E_2} \sum_{j=1}^{d_k} X_{kj} \quad (4)$$

where E_i ($i = 1, 2$) is the set of all states where AGV type l_i is busy, $i = 1, 2$. Such a requirement occurs, for example, if newer equipment is preferred over older equipment because of performance reasons. Note, however, that the older equipment is still functioning and can be used if required.

The above represents possible typical constraints. The exact constraints to be used depend on the particular system characteristics. It is extremely important, when considering manufacturing systems modeling, to be able to tune and encompass particular system problems and requirements.

It should be emphasized that $X_{kj^\circ} > 0$ means that decision rule j° is used when the system is in state S_k and at the same time, X_{kj° is the fraction of time that the system is in state S_k.

Finally, there must also be a set of detailed balance equations to ensure that the system is stable [1]. Stable conditions add to the problem 2^{Q-1} equations of the form $BX = 0$, and one more equation to ensure that the sum of the probabilities is one. These detailed balance equations are the typical "mean-flow-out equal mean-flow-in type" equations stated in terms of the variables $\{X_{kj}\}$. A general detailed balance equation is given in the Appendix (for the interested reader). We also show an example of the detailed balance equations in the next section.

The problem in general can be written as

$$\min CX, \text{ s.t. } \begin{array}{l} DX \leq r \text{ (manufacturing service constraints)} \\ BX = 0 \\ \bar{1}X = 1 \end{array} \text{ (detailed balance equations)} \\ X \geq 0.$$

The problem is an LP and can be solved efficiently. Whenever $X_{kj} = 0$, it means that strategy j will not be used when the system is in state S_k. When $X_{kj} > 0$ for several strategies j, it means that randomized strategies are used, i.e., with probability

$$\frac{X_{kj^\circ}}{\sum_{X_{kj}>0} X_{kj}}$$

the dispatcher will use strategy j° when the system is in state S_k.

Without the set of constraints $DX \leq r$, the model above is a classical Markovian Decision problem which can be solved using Howard's iteration approach [5]. It should be emphasized that an alternative method to [5] is using LP [10]. In our model we must solve the problem using LP because of the inclusion of the set of constraints $DX \leq r$, which is the most important part of the model.

We present here a theoretical model and framework for the cooperation problem. Although excellent LP packages can be used to solve the problem, computational difficulties increase when the dimensions of the problem increase. However, when applying the model for actual cases, the state space as well as the set of decision alternatives can be greatly reduced due to 1) states with low probabilities that should be lumped and 2) according to the particular manufacturing control imposed by management some decisions at particular states that are not allowed, which further reduce the number of possibilities.

Although our model cannot explicitly include variation in service times that are due to the travel time component of the service time, a mean service time calibration [7] can be used if necessary. The principle of the method is the assumption that the mean service time can be set equal to the sum of the estimated mean service time (estimated prior to the analysis) and any related on-scene time. After the analysis, the mean travel time is a model-computed performance measure. The mean service time calibration method is described as follows.

Step 0: The mean service time of an AGV of FMS i and request j is $1/\bar{\mu}_j^i$.

Step 1: Solve the LP using $1/\bar{\mu}_j^i$ to obtain $1/\hat{\mu}_j^i$ which is the model computed mean service time $\forall\, i, j$.

Step 2: If $|1/\bar{\mu}_j^i - 1/\hat{\mu}_j^i| > \epsilon$ for at least one pair i, j set $1/\bar{\mu}_j^i \equiv 1/\hat{\mu}_j^i$ and go back to step 1; otherwise stop.

It is important to note here that based on computational experience, the mean service time calibration usually converges after few iterations (3–4 iterations).

III. Example

Consider a system G, comprised of two FMS's G_1 and G_2. Suppose there are two types of calls for service in each FMS: a) call for materials and b) call for tools. In each FMS there are three types of AGV's: type 1 AGV, which can carry only materials; type 2 AGV, which can carry only tools; and type 3 AGV, which can carry both tools and materials. We further assume that each FMS has exactly one AGV of each type. Note that the aim of this section is to illustrate the model and show cooperation strategy results. For clarity, we choose a

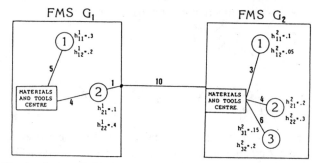

Fig. 1. System G. (Tools are resource 1 and materials are resource 2.)

small but typical example. In the previous section, we discussed other possible cases of FMS's cooperation.

The system G is depicted in Fig. 1. There are two cells in FMS 1 (G_1) and three cells in FMS 2 (G_2). The $\{h_{ij}^k\}$-probability of call type j in cell i of FMS k are given near the cells. In each FMS there is a materials/tools center where the three AGV's are located while available. The travel times on the links are in seconds. For example, the time it takes to travel from the center in FMS 1 to cell 3 in FMS 2 is (4 + 1 + 10 + 6 = 21). We remark, that if a particular route is preferred (rather than the shortest distance), then its travel time is used here. Thus this particular route is incorporated in the model, such that the resulting strategy takes it into effect. Let us assume that $\lambda^1 = 5/h$, $\lambda^2 = 4/h$, $\mu_1^1 = 6/h$ (service rate of AGV type 1 from G_1), $\mu_2^1 = 10/h$, $\mu_3^1 = 15/h$, $\mu_1^2 = 5/h$, $\mu_2^2 = 11/h$, $\mu_3^2 = 16/h$, and $P = 30$ min.

Since each AGV can be busy or free there are $2^6 = 64$ possible states. Each state is a vector of six components $\{0, 1\}$, in which the first three components describe the status of AGV type j, $j = 1, 2, 3$ in G_1, and the last three components describe the status of AGV type j, $j = 1, 2, 3$ in G_2 (see Table I).

For each state S_k, D_k, the set of strategies, can be obtained as follows. First of all, each strategy will be a vector of four components (since there are two FMS's and two types of requests for service). The first two components describe the type of cooperation if a call for service occurs in FMS 1 respectively for tools and materials. The third and fourth components describe the type of cooperation if a request for service in FMS 2 is respectively for tools and materials.

We assume in this example the following conditions regarding dispatching and cooperation (obviously other rules can easily be incorporated).

1) If a call for tools (materials) occurs in G_i, $i = 1, 2$ then first and second priorities to be dispatched are respectively AGV 1(2) and 3 in G_i (i.e., if a call for tools occurs AGV 1, if available, is dispatched, otherwise AGV 3 is dispatched).
2) If both AGV's 1(2) and 3 of G_i are busy then the following cooperation rules can be used, when a call for tools (materials) occurs in G_i:
   a) No cooperation (dispatch of a special server incurring a penalty). We number this rule by 0.

TABLE I
THE STATES

S1	0 0 0 0 0 0	S33	0 0 0 0 0 1
S2	1 0 0 0 0 0	S34	1 0 0 0 0 1
S3	0 1 0 0 0 0	S35	0 1 0 0 0 1
S4	1 1 0 0 0 0	S36	1 1 0 0 0 1
S5	0 0 1 0 0 0	S37	0 0 1 0 0 1
S6	1 0 1 0 0 0	S38	1 0 1 0 0 1
S7	0 1 1 0 0 0	S39	0 1 1 0 0 1
S8	1 1 1 0 0 0	S40	1 1 1 0 0 1
S9	0 0 0 1 0 0	S41	0 0 0 1 0 1
S10	1 0 0 1 0 0	S42	1 0 0 1 0 1
S11	0 1 0 1 0 0	S43	0 1 0 1 0 1
S12	1 1 0 1 0 0	S44	1 1 0 1 0 1
S13	0 0 1 1 0 0	S45	0 0 1 1 0 1
S14	1 0 1 1 0 0	S46	1 0 1 1 0 1
S15	0 1 1 1 0 0	S47	0 1 1 1 0 1
S16	1 1 1 1 0 0	S48	1 1 1 1 0 1
S17	0 0 0 0 1 0	S49	0 0 0 0 1 1
S18	1 0 0 0 1 0	S50	1 0 0 0 1 1
S19	0 1 0 0 1 0	S51	0 1 0 0 1 1
S20	1 1 0 0 1 0	S52	1 1 0 0 1 1
S21	0 0 1 0 1 0	S53	0 0 1 0 1 1
S22	1 0 1 0 1 0	S54	1 0 1 0 1 1
S23	0 1 1 0 1 0	S55	0 1 1 0 1 1
S24	1 1 1 0 1 0	S56	1 1 1 0 1 1
S25	0 0 0 1 1 0	S57	0 0 0 1 1 1
S26	1 0 0 1 1 0	S58	1 0 0 1 1 1
S27	0 1 0 1 1 0	S59	0 1 0 1 1 1
S28	1 1 0 1 1 0	S60	1 1 0 1 1 1
S29	0 1 1 1 1 0	S61	0 0 1 1 1 1
S30	1 0 1 1 1 0	S62	1 0 1 1 1 1
S31	0 1 1 1 1 0	S63	0 1 1 1 1 1
S32	1 1 1 1 1 0	S64	1 1 1 1 1 1

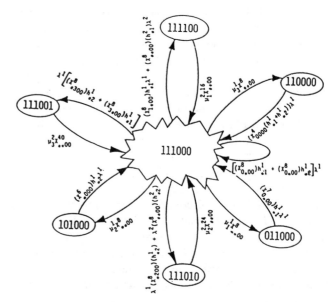

Fig. 2. Transition diagram for state $S_8 = (111000)$.

b) Dispatch AGV 1(2) from G_j, $j \neq i$, if it is available. We number this rule by 1.

c) Dispatch AGV 3 from G_j, $j \neq i$, if it is available while AGV 1(2) from G_j is busy. We number this rule by 2.

As an example of obtaining D_k consider state $S_8 = (111000)$. There are nine strategies available for this state which are (i, j, k, l), where $i \in \{0, 1, 3\}$, $j \in \{0, 2, 3\}$, and $k = l = 0$. Strategy $(0, 2, 0, 0)$, for instance, means that if a call for tools occurs in G_1 no cooperation takes place (which results in a penalty P). If a call for materials occurs in G_1 AGV type 2 from G_2 is dispatched. If a call for either tools or materials occurs in G_2 no cooperation takes place (AGV 1 will be dispatched for tools, AGV 2 will be dispatched for materials).

The flow transfer rates (where flows are products of probabilities and rates) between the states can be obtained easily. As an example Fig. 2 depicts the flow rates for $S_8 = (111000)$. The transition flow from (111000) to (111100) is $\lambda^1 X^8_{1\bullet 00} h^1_{\bullet 1} + \lambda^2 X^8_{\bullet\bullet 00} h^2_{\bullet 1}$, where for $x \bullet$ in the first component reflects a summation over $i \in \{0, 1, 3\}$ and $\bullet$ in the second place reflects a summation over $j \in \{0, 2, 3\}$. For $h \bullet$ means summation over the vertices in the appropriate FMS. This transition flow takes into account all the possibilities for AGV 1 in G_2 to become occupied (if calls for tools occur in G_1 and cooperations takes place, or if calls for tools occur in G_2).

The detailed balance equations for S_8 are therefore

$$\lambda^1 h^1_{\bullet 1}[X^8_{1\bullet 00} + X^8_{3\bullet 00}] + \lambda^1 h^1_{\bullet 2}[X^8_{\bullet 200} + X^8_{\bullet 300}]$$
$$+ \lambda^2 h^2_{\bullet 1}[X^8_{\bullet\bullet 00}] + \lambda^2 h^2_{\bullet 2}[X^8_{\bullet\bullet 00}] + \mu^1_1[X^8_{\bullet\bullet 00}]$$
$$+ \mu^1_2[X^8_{\bullet\bullet 00}] + \mu^1_3[X^8_{\bullet\bullet 00}]$$
$$= \lambda^1 h^1_{\bullet 1}[X^4_{0000} + X^7_{0\bullet 00}] + \lambda^1 h^1_{\bullet 2}[X^4_{0000} + X^6_{\bullet 000}]$$
$$+ \mu^2_1 X^{16}_{\bullet\bullet 00} + \mu^2_2 X^{24}_{\bullet\bullet 00} + \mu^2_3 X^{40}_{\bullet\bullet 00}$$

which can be written as

$$\lambda^1 h^1_{\bullet 1}[X^8_{1\bullet 00} + X^8_{3\bullet 00}] + \lambda^1 h^1_{\bullet 2}[X^8_{\bullet 200} + X^8_{\bullet 300}]$$
$$+ \lambda^2[h^2_{\bullet\bullet}][X^8_{\bullet\bullet 00}] + [\mu^1_{\bullet}][X^8_{\bullet\bullet 00}]$$
$$= \lambda^1 h^1_{\bullet 1}[X^4_{0000} + X^7_{0\bullet 00}] + \lambda^1 h^1_{\bullet 2}[X^4_{0000} + X^6_{\bullet 000}]$$
$$+ \mu^2_1 X^{16}_{\bullet\bullet 00} + \mu^2_2 X^{24}_{\bullet\bullet 00} + \mu^2_3 X^{40}_{\bullet\bullet 00}.$$

CONCLUSION

We introduced a quantitative approach to the problem of cooperation among flexible manufacturing systems. In particular we formulated and solved the cooperation among AGV's. However, using the same methodology derived here, extensions to cooperation strategies of other resources can easily be solved. In addition, the states can be defined differently (namely, taking into account the internal FMS state, such as queues for machines). The optimal policy derived in this paper is a "look forward" policy, which takes into account current system status (S) and future evolution of the system, and overall minimize response time of the system, while not increasing response time of each station if desired. A simple myopic policy usually looks at the current system status only, e.g., dispatch an AGV from FMS j (if one is available) to FMS i when all of his AGVs are busy and one is required. This simple policy does not consider future events which affect this objective function. Therefore, the simple policy is inferior to the one resulted from the computation in this paper. Note also that here the resultant policy depends on the objective function. Namely, should the manufacturing supervisor desire, the objective function can be replaced and the resulting strategy from solving the model will reflect the new objective function. For example, one model can be to minimize the system expected response time; another model will add to it such that each FMS's performance is not reduced. Thus we compensate for the initial decomposition (preallocation of resources) of the system, which is important in reducing the complexity of the real time operational control.

However, we are still limited to deal with a single objective function. Many other objective functions can be formulated as service level constraints and included in the set $DX \leq r$. For example, if certain calls for service have a very high priority (tight due dates), we can include constraints (with appropriate set of decisions) to reflect it. A very interesting future research in this regard is to study the conflicting nature of various objectives.

To measure the effect of cooperation, several criteria can be used. For example, the relative improvement is measured by

$$\frac{\text{ERT}^* - \text{ERT}^0}{\text{ERT}^0}$$

where ERT* is the optimal response time with cooperation and ERT° without cooperation (obtained by solving the same model, but considering only $X_{0000}^m \ \forall \ m$).

An important note is that hierarchical control and cooperation among subsystems is a general phenomenon in any large and complex system (large company, government administration, or military). In manufacturing, we try to understand the underlying structure of the system, model it, and—according to the model—propose optimal solution to a set of problems.

Finally, two additional directions of future research are suggested. First, instead of partitioning the AGV's, it may be interesting to investigate if it is more advantageous to pool all of them together. Second, for a particular problem a study of the special structure of the problem formulation can lead to a reduction in the state and decision space (this is in addition to some comments we made earlier in this regard).

Appendix

A general detailed balance equation for our problem is developed here. In order to simplify the analysis we assume that there are two FMS's. The number of AGV's in FMS j is Q_j, $j = 1, 2$. We also assume that there are only two types of AGV's in each FMS. The number of detailed balance equations is $2^{Q_1 + Q_2}$ (including an equation that ensures that the sum of the probabilities is 1). The decision variables are $\{X_{ijkl}\}$ where (i, j, k, l) is the strategy (i and j represent the type of cooperation respectively for AGV 1 and 2 in FMS 1; whereas k and l represent the type of cooperation respectively for AGV 1 and 2 in FMS 2).

The following detailed balance equation is for a general state S_k^1:

$$\sum_{i,j,k,l} X'_{ijkl} \left[\sum_{t=1}^{Q_1} W_t(S'_k)\mu_t^1 + \sum_{t=Q_1+1}^{Q_2} W_t(S'_k)\mu_t^2 \right.$$
$$\left. + \lambda^1 \sum_{S_k \in A^1(S'_k)} \sum_{n \in B^1(X'_{ijkl})} h^1_{\bullet n} + \lambda^2 \sum_{S_k \in A^2(S'_k)} \sum_{n \in B^2(X'_{ijkl})} h^2_{\bullet n} \right]$$
$$= \sum_{S_k \in A^1(S'_k)} \sum_{i,j,kl} X_{ijkl} \left[\sum_{t=1}^{Q_1} W_t(S_k)\mu_t^1 \right]$$
$$+ \sum_{S_k \in A^2(S'_k)} \sum_{i,j,k,l} X_{ijkl} \left[\sum_{t=Q_1+1}^{Q_2} W_t(S_k)\mu_t^2 \right]$$
$$+ \sum_{S_k \in C^1(S'_k)} \sum_{i,j,k,l} X_{ijkl} \lambda^1 \sum_{n \in D^1(X_{ijkl})} h^1_{\bullet n}$$
$$+ \sum_{S_k \in C^2(S'_k)} \sum_{i,j,k,l} X_{ijkl} \lambda^2 \sum_{n \in D^2(X_{ijkl})} h^2_{\bullet n}$$

where

$A^j(S'_k)$ the set of all states S_k that are identical to S'_k in all elements respectively for $t = 1, \cdots, Q_1$, $t = Q_1 + 1, \cdots, Q_2$, for $j = 1, 2$.

$W_t(S'_k)$ $\begin{cases} 0, & \text{if server of type } t \text{ is busy} \\ 1, & \text{if idle} \end{cases}$

$B^j(X'_{ijkl})$ the set of all calls type that cause a transition from S_k to S'_k according to X'_{ijkl}.

$C^j(S'_k)$ the set of all states S_k that are identical to S'_k in all elements respectively for $t = 1, \cdots, Q_1$, $t = Q_1 + 1, \cdots, Q_2$, for $j = 1, 2$, except for one element b_l, which is 1 in S'_k and 0 in S_k.

$D^j(X'_{ijkl})$ the set of all calls that cause a transition from S_k to S'_k according to X_{ijkl}.

References

[1] O. Berman and N. Ahituv, "Devising a cooperation policy for emergency networks," working paper WP-08-85, Faculty of Management, The University of Calgary, Calgary, AB, Canada.

[2] J. A. Buzacott, "Optimal Operating Rules for Automated Manufacturing Systems," *IEEE Trans. Automat. Contr.*, vol. 27, no. 1, Feb. 1982.

[3] C. Dupont-Gatelmand, "A survey of flexible manufacturing systems," *J. Manufacturing Syst.*, vol. 1, no. 1, pp. 1–16, 1982.

[4] J. Hatvani, "Intelligence and cooperation in heterarchic manufacturing systems," in *Proc. 16th CIRP Int. Seminar Manufact. Syst.*, 1984, pp. 1–4.

[5] R. A. Howard, *Dynamic Programming and Markov Processes*. Cambridge, MA: M.I.T., 1960.

[6] J. Kimemia and S. B. Gershwin, "An algorithm for the computer control of a flexible manufacturing system," *IIE Trans.*, pp. 353–362, Dec. 1983.

[7] R. C. Larson and A. R. Odoni, *Urban Operations Research*. Englewood Cliffs, NJ: Prentice-Hall, 1981.

[8] M. H. Lee, N. W. Hardy, and D. P. Barnes, "Error recovery in robot applications," in *Proc. Sixth British Robot Association Ann. Conf.*, 1983, pp. 217–222.

[9] O. Z. Maimon and S. Y. Nof, "Activity controller for a multiple robot assembly call," in *Contr. Manufacturing Processes and Robotics Syst.*, D. E. Hardt and W. G. Book, Eds. New York: ASME, 267–284.

[10] A. S. Mann, "Linear programming and sequential decisions," *Management Sci.*, vol. 6, pp. 259–267, 1960.

[11] O. Z. Maimon, "A multi-robot control experimental system with random parts arrival," in *Proc. IEEE Int. Conf. Robotics Automat.*, 1985, pp. 895–900.

[12] O. Z. Maimon, R. Akella, and S. B. Gershwin, "Analytic and iterative

approximation of the FMS control value function," ORSA/TIMS Joint National Meeting, Boston, MA, May 1985.

[13] S. Y. Nof, R. G. Wilhelm, and O. Z. Maimon, "Recovery planning in robot control," ORSA/TIMS Joint National Meeting, Orlando, Florida, Nov. 1983.

[14] Takenouchi, H. S. Miyamoto, and S. Seki, "A system recovering method from tracking confusion in conveyor systems," System Development Lab., Hitachi, Tokyo, Japan.

[15] P. Vrat and A. Virani, "A Cost Model for Optimal Mix of Balanced Stochastic Assembly Line and the Modular Assembly System for a Customer Oriented Production System," *International Journal of Production Research,* vol. 14, no. 4, 1976, pp. 445-463.

Oded Berman is the former Chairman of the Management Science and Information Systems Area. He has the Ph.D. degree in operations research from Massachusetts Institute of Technology and a B.A. in Economics and Statistics from Tel-Aviv University.

Dr. Berman has published articles in *Networks, Transportation Science, European Journal of Operations, INFOR, Decision Sciences, Computers and Operations, IEEE Transactions on Systems, Man and Cybernation, AIEE Transactions, Management Science, TIMS Studies in the Management Science, Operations Research, The Journal of Operations Management* and others.

Oded Z. Maimon received the B.S.I.E., B.S.M.E., and M.S. degrees in operations research from the Technion in Israel, and the Ph.D. degree from Purdue University, Lafayette, IN. His research interests are in the planning, analysis, and automatic operational control of flexible manufacturing systems in general, and robotic systems in particular.

Dr. Maimon has published in the *Transactions of the AMSE,* the *Journal of Dynamic, Systems, Measurement and Control,* the *Annals of Operations Research,* the *IIE Transactions,* the *SME Journal of Manufacturing Systems,* the *Journal of Robotic Systems,* and others.

Measuring Decision Flexibility in Production Planning

J. B. LASSERRE AND F. ROUBELLAT

Abstract—We present a measure of decision flexibility for production planning problems. The flexibility of a decision is related to the size of the choice set associated with this decision. In production planning problems this set is a convex polyhedron in an n-dimensional space. Our measure of flexibility is the volume of this set with additional information about the shape. By using a new method recently proposed we give an analytical expression of the measure of flexibility. Sensitivity of flexibility to various parameters is also given analytically as a byproduct. Experimental results on a simple inventory problem are provided. An alternative measure of flexibility for large-scale systems is also presented and discussed.

I. Introduction

IN this paper we are concerned with decision flexibility in sequential decision processes and, more particularly, in production planning problems. Typical problems that we have in mind are midterm or long-term planning problems where the material is treated as a continuous flow. Horizons vary from several months to one or two years and an elementary period of the discretized horizon can be a week or a month. Inflexibility of plans (optimal plans in particular) have been much criticized (see [1], [10], [11]). In any planning process, at each period of the horizon (which may be a rolling horizon) it is often desirable to make decisions which keep options open in an uncertain future. In production planning, uncertainty in the future mainly appears through two kinds of parameters.

- Demand is not known with certainty.
- Machine or workstation capacities are only estimated.

When a measure of uncertainty is available, one can include it in the model. Then, all the machinery available in stochastic models can be used to make a decision.

However, in many cases (in production planning or inventory control, for example) it is difficult or even impossible to get this information.

Moreover, some constraints like capacity constraints include uncertain parameters and even if they are well-defined random variables we do not know how to handle them in stochastic models. Our approach is an attempt to deal with such situations.

We consider deterministic models where the uncertain parameters have a fixed value. After the first-period decision of a plan (calculated according to this model) has been implemented, the subsequent decisions may become infeasible (due to small changes in some uncertain parameters) if decision flexibility has not been taken into account by the decision maker. It is then necessary to be able to measure the flexibility of a decision. Only a few investigations have been made in this area. Marschak and Nelson in microeconomics, Merkhofer in decision analysis, and, more recently, Hansmann, Rosenhead *et al.* in operations research

proposed definitions and sometimes measures of flexibility (see [2], [7], [8], [10], [11]).

Here, we present a measure of decision flexibility in production planning. In a way similar to Merkhofer and Rosenhead, the flexibility of a decision is related to the size of the choice set associated with this decision.

The choice set associated with the first-period decision of a plan is the set of all feasible plans starting at the beginning of the second period after the first-period decision has been implemented.

The feasibility constraints that the plans must satisfy depend on the first-period decision and on uncertain parameters. Hence, a large choice set can tolerate variations of these constraints (due to changes in the uncertain parameters) and still remain nonempty. The larger the choice set, the more flexible the first-period decision.

Clearly, the volume of this choice set provides a measure of the first-period decision flexibility.

We present a method which gives an analytical expression for this measure. Moreover, the flexibility of a decision is given analytically not only as a function of the decision variable, but also as a function of parameters such as demands, capacities, etc.

Hence, the formula can be calculated once and for all (using symbolic languages like REDUCE or MACSYMA) and will be valid for a class of problems (a problem in a class being represented by specific values given to demand and capacity parameters). A numerical version of the algorithm has also been written.

Once it is possible to measure the flexibility of a decision, the question arises of how to use this criterion to make a decision.

Generally, the decision maker is faced with the following problem. He wants to make a good decision according to some given economic criterion (there may be several criteria) but he also wants to make a flexible decision. Usually, a very flexible decision will score badly on this economic criterion and vice-versa. Hence, a tradeoff is necessary. Two approaches can be considered.

1) A supplementary constraint is added to the constraints defining the choice set. It imposes a threshold value on the economic criterion (several constraints if there are several criteria) so that plans with a bad score on this criterion are eliminated from the choice set. Then the decision maker can decide to choose the most flexible decision since flexibility includes cost considerations.

2) A classical multicriteria procedure is implemented where flexibility is one among other criteria. There exist methods for finding Pareto solutions in multiobjective optimization. In the bicriteria case see [9], for example.

As we will see later, in a production planning context, the first approach is possible only when the economic criterion is linear or piecewise linear. Simple illustrating examples are given for both approaches.

Finally, an alternative measure of flexibility is proposed for large-scale problems since the problem size is a serious limitation for the previous one. Some comments are made on these two measures.

Manuscript received October 31, 1983; revised June 26, 1984. This paper is based on a prior submission of March 29, 1983. Paper recommended by Past Associate Editor, J. Y. Luh.

The authors are with the Laboratoire d'Automatique et d'Analyse des Systèmes du C.N.R.S., Toulouse, France.

II. Decision Flexibility in Production Planning

A. Measure of Flexibility

We only consider linear production planning models. They represent a large class of problems. A feasible plan over a finite discretized horizon is a sequence of decisions admissible according to a set of linear constraints. Let us consider, for example, a simple production inventory problem with upper bounds on the production

$$x_t = x_{t-1} + u_t - d_t$$
$$x_{(0)} = x_0$$
$$0 \leq u_t \leq \overline{u_t}$$
$$0 \leq x_t \quad t = 1, \cdots, T$$

where x_t is the state variable (inventory level at the end of period t), u_t is the control (decision) variable (production in period t), d_t is the demand at period t, and $\bar{u}_t$ is the maximum production allowed in period t.

Many production planning models have similar linear equations (state equations of the system) and constraints (limited production capacity, bounds on inventories, on work force, etc., ...). Many of these models are discussed in [6].

By eliminating the state variables through the state equations (it is always possible in such models) and keeping only decision (or control) variables, an equivalent model is a set of linear inequalities $Au \leq b$ where u is the vector $(u_1, u_2, \cdots, u_T)$ and u_i is the decision vector at period i.

If we want to impose a threshold value on a linear economic criterion (to eliminate the plans with a bad score) it suffices to add a line in the matrix A.

Given a first decision u_1, the subsequent decisions $u_2, u_3, \cdots, u_T$ must satisfy the constraints

$$A^1 u^1 \leq b^1(u_1) \quad \text{where } u^1 \text{ is the vector } (u_2, \cdots, u_T)$$

and $b^1(u_1)$ is a vector which depends on u_1.

We propose to measure the flexibility of the decision u_1 by calculating the volume of the convex polyhedron defined by the set of inequalities $A^1 u^1 \leq b^1(u_1)$. Some additional information about the shape of this set will also be useful and is obtained as a byproduct of the method we give to compute the volume.

Remark: Suppose that the volume of the choice set [let us call it $\Omega(u_1)$] is zero. This means that it lies in an affine variety of dimension less than n. This means that there exists a subset I of indexes such that

$$\langle a_i^1, u^1 \rangle = b_i^1(u_1) \quad \forall i \in I, \forall u \in \Omega(u_1).$$

If among the $b_i^1(u_1)$ there exists i_o such that $b_{i_o}^1(u_1)$ is susceptible to move in the future because it contains an uncertain parameter (like demand or capacity), then the flexibility is really zero. $\Omega(u_1)$ can become empty for a small change in $b_{i_o}^1(u_1)$.

If all the $b_{i_o}^1(u_1)$ are not susceptible to move, the flexibility is not zero and the volume has to be calculated in the affine variety where $\Omega(u_1)$ lies.

B. The Volume of a Convex Polyhedron in R^n [4]

Recently, we have proposed a new method to compute the volume of a convex polyhedron in R^n defined by the linear system of inequalities $Au \leq b$ where A is an (m, n) matrix and b is an m-vector. $D = \{u: Au \leq b\}$ is a bounded convex polyhedron in an n-dimensional Euclidian space (say R^n for simplicity). Let us call $V(n, A, b)$ its volume.

In [4] we have demonstrated that

$$V(n, A, b) = (1/n) \sum_{i=1}^{m} (b_i / \|a_i\|) \, V_i(n-1, A, b) \quad (1)$$

where $\|a_i\|$ is the Euclidian norm of the vector a_i which is the ith row of A and $V_i(n-1, A, b)$ is the volume of the ith face D_i of D.

D_i is defined by

$$\langle a_i, u \rangle = b_i$$
$$\langle a_j, u \rangle \leq b_j \quad \text{for } j \neq i, j = 1, \cdots, m.$$

By elimination of a variable (using the equation $\langle a_i, u \rangle = b_i$), we obtain the projection $\tilde{D}_i$ of D_i onto an $(n-1)$-dimensional space. $\tilde{D}_i$ is defined by a new system of inequalities

$\tilde{A} y \leq \tilde{b}$ where now $\tilde{A}$ is an $(m-1, n-1)-$ matrix,

$\tilde{b}$ is an $(m-1)$-vector, and y is an $(n-1)$-vector. As before, we call $V_i(n-1, A, b)$ the volume (in an $(n-1)$-dimensional space) of D_i. We have the relationship

$$V_i(n-1, A, b) = (\|a_i\| / |\alpha|) \, V_i(n-1, \tilde{A}, \tilde{b})$$

where α is the coefficient of the eliminated variable in the constraint $\langle a_i, u \rangle = b_i$. The formula (1) can be rewritten for $V_i(n-1, \tilde{A}, \tilde{b})$ and the process is repeated $(n-1)$ times. Finally, we have to compute volumes in R defined by

$$a_i' y \leq b_i' \quad i = 1, \cdots, m-n+1 \; (y \in R).$$

The volume of this domain is given analytically by the formula

$$\max\left[0, \min_{a_i'>0} [(b_i'/a_i')] - \max_{a_i'<0} [b_i'/a_i')]\right].$$

Hence, it is possible to compute analytically $V(n, A, b)$. For more details the reader is referred to [4].

Identity (1) comes from Euler's famous theorem on homogeneous functions, and therefore $V_i(n-1, A, b)/(\|a_i\|)$ represents almost everywhere $\partial V(n, A, b)/\partial b_i$. Thus, we also get the sensitivity of $V(n, A, b)$ to b_i which will be interesting for our application.

In [4], two algorithms are given. The first one is written in the symbolic language REDUCE and gives the analytical formula. The other one is numerical algorithm written in PASCAL.

III. Application to Production Planning

We recall that the measure of the flexibility of the decision u_1 is the volume of the convex polyhedron $A^1 u^1 \leq b^1(u_1)$. Hence, by applying the results of the previous section, $V(n, A^1, b^1(u_1))$ measures the flexibility of the decision u_1. Moreover, $V_i(n-1, A^1, b^1(u_1))/\|a_i\|$ measures the sensitivity of the flexibility to the parameter $b_i^1(u_1)$. Thus, two decisions with similar flexibilities can be analyzed according to their sensitivity to some important parameters. As we shall see in an example below, we can get the sensitivity of the flexibility for a small change in certain machine capacities or demands.

Remark: Note that we get the analytical formula for a class of problems. A class is represented by the matrix A^1. A problem in this class is represented by $b^1(u_1)$ since the parameters which differentiate problems are demands and capacities (hence, affecting only the right-hand side). The computation is then made once and for all and the flexibility of a decision u_1 is then obtained by a simple evaluation of a formula. Nevertheless, we do not expect to compute analytically the flexibility formula for large-scale problems since the size of the matrix A^1 is a serious limitation (although the algorithm could be improved). The numerical version of the algorithm can handle larger size problems but not really large-scale matrices. (Evaluating the flexibility for a six-period inventory problem, A^1 being a (16, 5) matrix, required 9 s on an IBM 3033. 4 mn were necessary with the *symbolic* language for a four-period-inventory problem.) In Section IV we present another measure of flexibility for large-scale problems.

So far, we did not take any cost consideration into account. Generally, the decision maker does not want to consider plans which score badly on some economic criterion. And this criterion is generally in conflict with the flexibility criterion. The question which arises is then: how can we use both flexibility and economic criteria to make a decision or to help a decision maker?

We distinguish two cases and give an illustrating example for each case.

A. Linear Criterion

First we compute the optimal plan according to this criterion by using linear programming. C^* is the optimal cost. Then suppose $C^* + \delta \, (\delta > 0)$ is a threshold value such that plans with a score higher than $C^* + \delta$ must be eliminated. This constraint can be taken into account in the matrix A^1 so that the decision maker can choose, for example, the most flexible decision since the flexibility criterion includes cost considerations.

Example: Let us consider the simple inventory problem

$$x_t = x_{t-1} + u_t - d_t$$
$$x_{(0)} = x_0$$
$$0 \leq u_t \leq \bar{u}_t \quad t = 1, \cdots, T$$
$$0 \leq x_t$$

and suppose we want to consider the plans with a "good" score on the criterion (cost) $x_1 + x_2 + \cdots + x_T$.

Eliminating the state variables yields

$$\sum_{i=1}^{t} u_i \geq \sum_{i=1}^{t} d_i - x_0 \quad t = 1, \cdots, T.$$
$$\bar{u}_t \geq u_t \geq 0.$$

Imposing a threshold value $C^* + \delta$ is equivalent to state

$$Tu_1 + (T-1)u_2 + \cdots + u_T \leq C^* + \delta + \sum_{i=1}^{T}(T-i+1)d_i.$$

Hence, the flexibility of the decision u_1 is the volume of the domain (in R^{T-1})

$$u_2 \geq d_1 + d_2 - x_0 - u_1$$
$$u_2 + u_3 \geq d_1 + d_2 + d_3 - x_0 - u_1$$
$$\vdots$$
$$u_2 + \cdots + u_T \geq d_1 + d_2 + \cdots + d_T - x_0 - u_1$$
$$0 \leq u_t \leq \bar{u}_t \quad t = 2, \cdots, T$$

$$(T-1)u_2 + (T-2)u_3 + \cdots + u_T \leq C^* + \delta - Tu_1 + \sum_{i=1}^{T}(T-i+1)d_i.$$

Consider the following numerical example data:

$$T = 3 \quad x_0 = 0 \quad d_1 = d_2 = d_3 = 50$$
$$\bar{u}_2 = 50 \quad \bar{u}_3 = 60 \quad \bar{u}_1 = 70.$$

The optimal plan is $u_1^* = 50 = u_2^* = u_3^* \, (c^* = 0)$. $\delta = 15$.

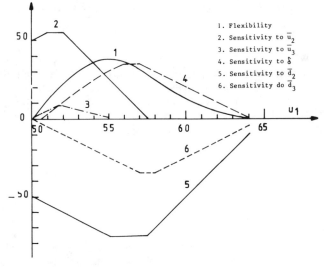

Fig. 1.

In Fig. 1 we can see the flexibility as a function of u_1. Also, sensitivity of $\bar{u}_2, \bar{u}_3, d_2, d_3, \delta$ are drawn as functions of u_1.

We can make the following observations.

1) The optimal solution according to the linear criterion is such that its first-period decision u_1^* has no flexibility. It is obvious since producing $u_1 = 50$ imposes producing $u_2 = 50$.

2) The sensitivity to the demand parameter d_i is (applying the chain rule)

$$\sum_{j=i}^{T-1} V_j(T-2, A^1, b^1(u_1))/(\|a_j\|).$$

Hence, the larger the index i, the weaker the sensitivity of $V(n, A^1, b^1(u_1))$ to d_i which means that the influence of a demand parameter at period t on the flexibility decreases when t increases.

The confirms an intuitive idea (see [6], [12], [13]).

On this example also, the sensitivity to $\bar{u}_3$ is less than the sensitivity to $\bar{u}_2$.

3) Without supplementary information, $u_1 = 55$ appears to be a flexible first-period decision because it has the largest choice set. As suggested by a referee, it would be helpful to find a means of incorporating the valuable sensitivity data with a model of parameter forecast variability.

B. Nonlinear Criterion

The approach of Section III-A is not possible any more since the criterion is not linear. A tradeoff is neccesseary between the flexibility criterion and the economic criterion. A multicriteria (in this case bicriteria) optimization method such as the one described in [9] can be implemented to help the decision maker.

Usually the decision maker is only interested in Pareto solutions (a Pareto solution is such that no other solutions with a better score on both criteria exists).

In the example below we give the Pareto solutions (see Fig. 3). For a decision u_1^0, the flexibility criterion is evaluated with the volume of the convex polyhedron, but now, without the constraint with a threshold value. The economic criterion is evaluated by solving

$$\min \, [C(u)]$$
$$\text{s.t.} \quad Au \leq b$$
$$u_1 = u_1^0$$

where solving

$$\min \; [C(u)]$$
$$\text{s.t.} \quad Au \leq b$$

is computing the cost of an optimal plan.

Example: We consider a similar example, but now the cost is as follows:

holding cost: $\quad 2x_t \quad$ if $t=1$

$\qquad\qquad\qquad x_t \quad$ if $t=2$ or 3

production cost: $\forall t \; g_t(u_t)=0 \quad$ if $u_t=0$

$\qquad\qquad\qquad 60+\sqrt{u_t} \quad$ if $u_t>0$.

Note that the criterion is nonlinear and that a setup cost is included.

The data are

$$\bar{u}_1 = 120, \; \bar{u}_2 = 100, \; \bar{u}_3 = 60$$
$$d_1 = d_2 = d_3 = 50.$$

In Fig. 2 we can see the flexibility and the economic cost as functions of u_1. The Pareto curve is plotted in Fig. 3.

An optimal solution according to the economic criterion is $u_1^* = 50$, $u_2^* = 100$, $u_3^* = 0$ with a cost of 187.07.

Comments: If we consider the Pareto curve we can observe that if the decision maker accepts a solution with a cost slightly higher than 187, then it results in an important gain in flexibility. The Pareto points for which the economic criterion is greater than 237.07 correspond to decisions u_1 ranging from 100 to 120. Those for which the economic criterion is less than 237.07 correspond to decisions u_1 ranging from 50 to 75. Hence, it is not interesting to make a decision between 75 and 100 since there is always a decision which is better according to both criteria.

With $u_1 = 99$, an optimal plan is: $u_2 = 51$, $u_3 = 0$ and the cost is 285.09. The flexibility of u_1 is 4690.

With $u_1 = 100$, an optimal plan is: $u_2 = 0$, $u_3 = 50$ and the cost is 237.07. The flexibility of u_1 is 4750.

Also, notice that, as before, other criteria can be taken into account (like sensitivity to d_2, d_3, $\bar{u}_2$, or $\bar{u}_3$ for example). The approach would be similar but more complicated.

IV. ANOTHER MEASURE OF FLEXIBILITY FOR LARGE-SCALE PROBLEMS

The measure of the flexibility presented in the previous sections is based on the volume of a convex polyhedron in an n-dimensional space. The size of the matrix A^1 is a serious limitation since the computational task increases exponentially with n. The numerical algorithm is less sensitive to n than the algorithm written in a symbolic language (which gives the analytical expression of the flexibility). But even if we use the numerical algorithm, we cannot hope to handle large-scale problems. Hence, for these problems we need another measure of flexibility.

Remark: Note that in the near future, efficient machines designed for symbolic languages will be developed since today it is a priority in the artificial intelligence research area. Hence, our algorithm's capability could be greatly improved.

A. Another Measure

As in Section II we still consider the choice set associated with u_1, i.e., the set defined by

$$A^1 u^1 \leq b^1(u_1).$$

Let us call I the set of indexes such that $b_i^1(u_1)$ [the ith component

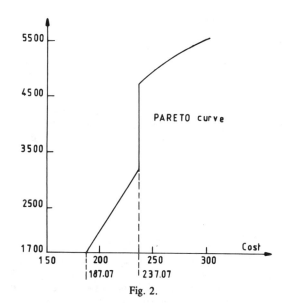

Fig. 2.

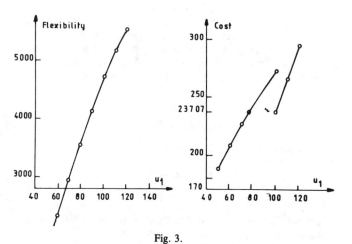

Fig. 3.

of $b^1(u_1)$] contains a parameter which is uncertain. In other words, I is the set of indexes i such that the constraint $\langle a_i^1, u \rangle \leq b_i^1(u_1)$ is susceptible to change in the future because $b_i^1(u_1)$ contains an uncertain parameter (demand, capacity, $\cdots$).

Let us define $f(u_1)$ by

$$f(u_1) = \max_u [\min_{i \in I} [d(u^1, a_i^1)]]$$

$$\text{s.t.} \quad A^1 u^1 \leq b^1(u_1)$$

where $d(u^1, a_i^1)$ is the Euclidian distance from u^1 to the hyperplane $\langle a_i^1, u^1 \rangle = b_i^1(u_1)$. If I is the set of all the indexes, $f(u_1)$ is the inner radius of the choice set (radius of the biggest hypersphere included in the choice set).

Computing $f(u_1)$ reduces to solve a linear programming problem since we have

$$f(u_1) = \max_{\delta, u^1} \delta$$

$$\langle a_i^1, u^1 \rangle + \|a_i^1\|\delta \leq b_i^1(u_1) \qquad i \in I$$

$$\langle a_j^1, u^1 \rangle \leq b_j^1(u_1) \qquad j \notin I \qquad (2)$$

Hence, we can handle large-scale problems without difficulty. In computing $f(u_1)$ we look for the point u^1 which is as "far" as possible from the set of constraints susceptible to move in the future.

Note that λ_i, the dual variable associated with the constraints

$\langle a_i^1, u^1 \rangle \le b_i^1(u_1)$, represents almost everywhere

$$\frac{\partial f(u_1)}{\partial b_i^1(u_1)}.$$

We propose to measure the flexibility of u_1 by using $f(u_1)$ and the sensitivity coefficients λ_i.

Remark: If it is possible to give some weight $\alpha_i > 0$ to the constraints (whose indexes belong to I) when some parameters are more "uncertain" or when some constraints are more important than others. $f(u_1)$ is then

$$\max_u \; [\min_i \; [\alpha_i d(u^1, a_i^1)]].$$

B. Comments

As in Section II, $f(u_1)$ is not the only information to consider. And as before, a sensitivity analysis is possible since the λ_i (dual variables) are the sensitivity coefficients for the active constraints. Moreover, it is trivial to compute the distance of a point giving the maximum in (2) to the other constraints.

This supplementary information is useful when we want to compare different decisions for which $f(u)$ are similar.

For example, suppose that we have to compare two decisions u_1^1 and u_1^2 with respective choice sets described below in Fig. 4.

The first choice set is defined by the inequalities

$$a_1 \le u_2 \le a_2$$
$$b_1 \le u_3 \le b_2.$$

The second choice set is defined by the inequalities

$$a_1' \le u_2 \le a_2'$$
$$b_1' \le u_3 \le b_2'.$$

Suppose that the only constraints susceptible to move in the future are $b_1 \le u_3 \le b_2$ and $b_2' \le u_3 \le b_2'$.

Suppose also that $b_2 - b_1 = b_2' - b_1'$.

Obviously, the most flexible decision is u_1^1. $f(u_1^1)$ is equal to $f(u_1^2)$ but the distance to the other constraints $a_1 \le u_3 \le a_2$ (or $a_1' \le u_3 \le a_2'$) is $(a_2 - a_1)/2$ for u_1^1 and $(a_2' - a_1')/2$ for u_1^2 and this permits us to conclude that u_1^1 is more flexible than u_1^2. Suppose now that $b_2 - b_1 = (b_2' - b_1') - \alpha$. ($\alpha > 0$). If α is small, can we decide that u_1^2 is more flexible than u_1^1? We do not have a definite answer to this question. It depends on the decision maker's attitude. If he is risky he will choose u_1^1 since the choice set is larger, although u_1^1 can admit a smaller variation of b_2 or b_1. But if b_2 or b_1 is a random variable with high second-order moment, then u_1^2 is a better choice.

Note that the same problem arises if we use the volume as in Section II.

Is it preferable to choose u_1^1 if the volume of its choice set is larger than the u_1^2's but also with higher sensitivity to b_2 or b_1?

In other words, what is the use of a large choice set if it can shrink strongly for a small variation of a parameter?

As we said before, flexibility is not characterized with a single value but with a vector (size of the choice set or inner radius and sensitivity coefficients) and, consequently, we do not have a total ordering of the decisions.

The first measure stresses the decisions (how many decisions will be available in the choice set). A sensitivity analysis permits us to indicate how the constraints will modify this choice set for a small variation in the right-hand side parameters. The second measure stresses the most critical constraints. Additional information on the other constraints are also available. It is difficult to say whether one is better than the other.

The example below is an illustration of the second measure. It

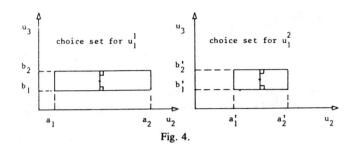

Fig. 4.

is the same example as in Section III-A. The decision set is defined by

$$u_2 \ge d_1 + d_2 - x_0 - u_1$$
$$u_2 + u_3 \ge d_1 + d_2 + d_3 - x_0 - u_1$$
$$0 \le u_2 \le \bar{u}_2$$
$$0 \le u_3 \le \bar{u}_3$$
$$2u_2 + u_3 \le C^* + \delta - 3u_1 + \sum_{i=1}^{3} (3 - i + 1) d_i.$$

Only the constraints $u_2 \ge 0$ and $u_3 \ge 0$ are not supposed to move in the future.

In Fig. 5 we can see (as in Section III-A) the different curves corresponding to

$$f(u_1) \text{ and } \frac{\partial f(u_1)}{\partial \bar{u}_2}, \frac{\partial f(u_1)}{\partial \bar{u}_3} \cdots.$$

It is worthy to note that the curve of $f(u_1)$ increases between 50 and 55 and then decreases after $u_1 = 55$, just as for the volume. Of course, $f(u_1)$ is always piecewise linear, whereas the volume is a piecewise n-polynomial function.

As before, it is also worthy to note that the sensitivity of $f(u_1)$ to d_i decreases when i increases.

In Fig. 5, by using sensitivity curves, we observe the following:

- for $u_1 0 \le u_1 < 55$ the critical constraints are

$$u_2 \le \bar{u}_2 \text{ and } u_2 \ge d_1 + d_2 - u_1$$

- for $55 \le u_1 \le 60$ the critical constraints are

$$u_2 \ge d_1 + d_2 - u_1, \; u_2 + u_3 \ge d_1 + d_2 + d_3 - u_1,$$

$$\text{and } 2u_2 + u_3 \le C^* + \delta + \sum_{i=1}^{T} (T - i + 1) d_i - 3u_1.$$

The constraint $u_3 \le \bar{u}_3$ is never critical.

CONCLUSION

Decision flexibility is an important issue for any planning methodology and, more generally, in any sequential decision process in an uncertain future. We have presented two measures of the flexibility of a decision in production planning. The first one is adapted to small size problems and can be computed analytically. The second one can be used for large-scale problems since it uses linear programming.

In both cases, a sensitivity analysis to various parameters is easy to perform. This permits us to discriminate between decisions with similar flexibility measure.

Optimizing a single economic criterion can yield a very inflexible decision and this can be dangerous (and eventually "costly") in an uncertain future. An alternative could be a

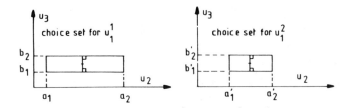

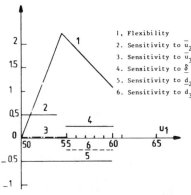

Fig. 5.

multicriteria approach where flexibility is taken into account. In this respect we hope we have made a useful contribution.

ACKNOWLEDGMENT

The authors wish to thank the area editor and the referees for the helpful comments and suggestions they have made.

REFERENCES

[1] S. Bonder, "Changing the future of operational research," in *Operational Research '78*, K. B. Haley, Ed. Amsterdam, The Netherlands: North-Holland, 1979.
[2] F. Hansmann, "On measuring flexibility," Univ. Munich, Munich Germany, Tech. Paper, 1977.
[3] L. A. Johnson and D. C. Montgomery, *Operations Research in Production Planning, Scheduling and Inventory Control*. New York: Wiley, 1974.
[4] J. B. Lasserre, "An analytical expression and an algorithm for the volume of a convex polyhedron in Rn," *J. Optimiz. Theory Appl.*, vol. 39, no. 3, pp. 363-377, 1983.
[5] ——, "Flexibilité d'une décision en planification de la production," Février,Tech. Note LAAS-SAP 82.0.14, 1982.
[6] R. A. Lundin and T. E. Morton, "Planning horizons for the dynamic lot size model: Zabel vs. protective procedures and computational results," *Oper. Res.*, vol. 23, no. 4, 1975.
[7] T. Marschak and R. Nelson, "Flexibility, uncertainty and economic theory," *Metroeconomica*, pp. 42-58, Apr., Dec. 1962.
[8] M. W. Merkhofer, "The value of information given decision flexibility," *Management Sci.*, vol. 23, no. 7, pp. 716-727, Mar. 1977.
[9] A. N. Payne and E. Polak, "An interactive rectangle elimination method for bi-objective decision making," *IEEE Trans. Automat. Contr.*, vol. AC-25, pp. 421-432, 1980.
[10] J. Rosenhead, M. Elton, and S. K. Gupta, "Robustness and optimality as criteria for strategic decisions," *Oper. Res. Quarter.*, vol. 23, pp. 413-431, 1972.
[11] J. Rosenhead, "Planning under uncertainty I: The unflexibility of methodologies," *J. Oper. Res. Soc.*, vol. 31, pp. 331-431, 1980.
[12] S. Sethi and S. Chand, "Multiple finite production rate dynamic lot size inventory models," *Oper. Res.*, vol. 29, no. 5, 1981.
[13] H. M. Wagner and T. M. Whitin, "Dynamic version of the economic lot size model," *Management Sci.*, vol. 5, pp. 89-96, 1958.

J. B. Lasserre (S'82) was born in France on May 11, 1953. He graduated from the Ecole Nationale Supérieure d'Informatique et de Mathématiques Appliquées, Grenoble, France, in 1976 and received the Docteur-ingénieur and the Docteur d'Etat degrees both from Université Paul Sabatier, Toulouse, France, in 1978 and 1984, respectively.

During the academic year 1978-1979 he was Visiting Scholar at the Department of Electrical Engineering and Computer Science, University of California at Berkeley with the support of an INRIA fellowship. Since 1980, he has been with Laboratoire d'Automatique et d'Analyse des Systèmes du C.N.R.S., Toulouse, where he is currently Chargé de Recherche. His current research interests are in control theory applied to production planning and decision flexibility.

F. Roubellat was born in France on April 5, 1941. He graduated from the Ecole Nationale Supérieure d'Electrotechnique, d'Electronique et d'Hydraulique, Toulouse, France, in 1964 and received the Docteur ès-Sciences degree from the Université Paul Sabatier, Toulouse, France, in 1969.

Since 1964, he has been with the Laboratoire d'Automatique et d'Analyse des Systèmes du C.N.R.S., Toulouse, where he is currently Maître de Recherche in charge of the Automatic Control Department. His current research interests are in production planning and workshop scheduling, with application to flexible manufacturing systems.

Optimal Scheduling for Load Balance of Two-Machine Production Lines

SADAMICHI MITSUMORI, MEMBER, IEEE

Abstract—A scheduling method for balancing work loads between a machine and the one succeeding it in a two-machine production line so as to maximize the overall production speed is proposed in this paper. The case where the production speeds are different from each other both for machines and for commodities is treated. The objective is to minimize the changeover loss under the constraints that 1) the due date must be met, 2) the queue length of commodity items must be shorter than the maximum space between the two machines, and 3) no idle time is acceptable for either machine. Both the lot size of each commodity for each production period and their production sequence is determined by the proposed method. The scheduling algorithm is a branch-and-bound procedure. In the conventional flow shop scheduling problem, the lot size of each commodity is given beforehand considering the inventory loss of the final products and the changeover loss of the production line. Only the flow sequence is determined there. Therefore, it can not perfectly balance the work loads between machines. Computational experience with the proposed method indicates that it is capable of solving problems involving 20 commodities over a planning horizon of 300 time periods. These are practical scheduling problems with a planning horizon of one week for a home appliance production line. These computation times are below one minute (using the HITAC M-180 operating at three million instructions per second (MIPS)).

I. INTRODUCTION

The number of commodities to be produced in a mass production factory tends to increase in order to meet the variety of customer demands. Therefore, one purpose of production control is to integrate large-quantity production with large-variety production.

Fig. 1(a) is an example of a problem arising from large-variety production in the case of a two-machine production line. The maximum production speed (items/s) of each commodity at each machine is given in (b). The second machine processes commodity items in the same sequence as the first machine. Thus, this production line is a flow shop.

Assume that a queue of commodity items is not acceptable between the two machines. Then, the overall production speed of each commodity on the production line is reduced to the speed of the slower machine. When the flow line produces commodities A, B, and C, their production speeds are 8, 9, and 8, respectively (Fig. 1(b)). One of the two machines is not producing at maximum speed. This means that the work loads are not balanced between the two machines.

Manuscript received September 25, 1980; revised February 26, 1981; and March 20, 1981.
The author is with the Systems Development Laboratory, Hitachi, Ltd., 1099, Ohzenji, Tama-Ku, Kawasaki, Japan.

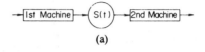

Fig. 1. Two-machine production line and example of production speeds of its machines. (a) Two-machine production line. (b) Production speeds.

The work loads are usually balanced by the following two methods. One is to regulate the capacities of the machines when changing the production commodities. This is a kind of changeover operation called the line balancing method [1]. The other is to balance the work loads between the two machines by determining a suitable production schedule.

An example of this type of scheduling problem is the n-job m-machine problem [2]. The objective of the problem is to minimize the time required to finish the given jobs. It is difficult in this method to obtain a schedule for balancing the work loads perfectly, because the second machine often has to wait for commodity items from the first machine. In this method the production lot sizes are given before the computation of scheduling. However, if the production lots and their flow sequences are decided at the same time, the work load balance can be improved. This will be explained below using the production line in Fig. 1.

The queue must have a limited allowable length in order to save production space. In order to simplify the discussion, all commodities have the same volumes. Therefore, the queue length is measured as a scalar value without discriminating between the commodities composing the queue.

Let the allowable maximum length of the queue be three items, and the initial length of the queue (at time zero) be one item of commodity A. Consider a schedule where the production line produces commodity A during time interval $[0, 1]$, commodity B during time interval $[1, 4]$, commodity A during time interval $[4, 5.5]$, $\cdots$. Fig. 2 shows the fluctuation of the queue length between the two machines when the production proceeds according to this schedule.

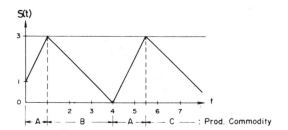

Fig. 2. Example of fluctuation of queue length formed between the two machines by a production schedule. $S(t)$: queue length formed between the two machines.

This schedule satisfies the conditions that the queue length must be shorter than three items and that the two machines must not have any idle time (waiting time). If the production line continues to produce commodity A after time interval [0, 1] and the production speed of the first machine is not reduced to the maximum production speed of the second machine, the queue length will exceed the allowable maximum length.

The conventional flow shop scheduling problem does not consider the queue length constraint. Recently, Dutta and Cunningham [3] proposed a scheduling method to sequence N jobs under this constraint. However, they assumed that all jobs have the same due date. Therefore, their method does not answer the question how to decide the lot size of each commodity.

Changeover consists of two kinds of operations. One is to regulate the capacities of machines. The other is to change the parts of machines. The loss time of the former can be eliminated by the proposed scheduling method, but the latter cannot. However since the main preparatory work for the latter has recently been carried out independently from production line work, the pauses in the production line operations are short and negligible. Therefore, the changeover loss is evaluated in terms of the cost associated with changeover operations.

II. Formulation

A. Assumptions

The two-machine production line discussed in this paper is shown in Fig. 1(a). The production line is assumed to have the following features.

1) Each machine processes one and only one item at a time.
2) The processing sequence of the items on the second machine is the same as that of the first machine.
3) The changeover loss is evaluated in terms of the cost associated with changeover operations from the discussion of I.
4) There are maximum and minimum limits to the allowable length of the queue between the machines. The minimum limit is usually zero.
5) Time t is expressed by nonnegative integers.
6) The demand and the production quantity of each commodity are normalized by the commodity production speed of the second machine. Therefore, the production quantity per unit time becomes one, irrespective of the commodity.
7) The maximum and minimum limits to the allowable length of the queue between the machines are measured in terms of the normalized production quantity defined by Assumption 6) and are otherwise independent of the type of commodities in the queue. Note that this assumption would, for example, not be satisfied when two commodities with identical machine production speed would occupy a very different amount of storage space.

B. Definition of Symbols

1) $I = \{1, 2, \cdots, n\}$: the set of commodity names (assuming that there are n kinds of commodities).
2) $[0, T]$: the scheduling horizon.
3) $D(t) = (d_1(t), d_2(t), \cdots, d_n(t))$, $t \in [0, T]$: the cumulative demand vector at time t. $d_i(t)$, $i \in I$: the cumulative demand for commodity i at time t. $D(0) = \phi$, where ϕ is a zero vector. Note that the cumulative demand is measured in terms of the time limit when production for the demand must be started at the second machine.
4) $X(t) = (x_1(t), x_2(t), \cdots, x_n(t))$, $t \in [0, T]$: the cumulative production vector at time t. $x_i(t)$, $i \in I$: the cumulative production of commodity i at time t. Note that the cumulative production is measured in terms of time when the production is started at the second machine.
5) $\{X(t), t \in [0, T]\}$: the production schedule for horizon $[0, T]$, or $\{X\}$ for brevity. $\{x_i(t), t \in [0, T]\}$: the production schedule for commodity i for horizon $[0, T]$, or $\{x_i\}$ for brevity.
6) $\Gamma = (\gamma_1, \gamma_2, \cdots, \gamma_n)$: the changeover loss coefficient vector. γ_i, $i \in I$: the loss due to changing the production commodity to commodity i. γ_i depends only on i from Assumption 3) described in Section II-A.
7) $\rho = (\rho_1, \rho_2, \cdots, \rho_n)$: the processing time difference vector. ρ_i, $i \in I$: the processing time difference of commodity i. ρ_i is defined as $p_i^{(2)} - p_i^{(1)}$, where $p_i^{(j)}$, $j = 1, 2$ is the processing time per one item of commodity i at the jth machine.
8) $S(t)$, $t \in [0, T]$: the time difference by which the production of the first machine precedes that of the second machine at time t. This time difference will be called the "float quantity."
9) M, m: the allowable maximum and minimum limits of float quantity $S(t)$.
10) $C(\{X\}) = (C_1(\{x_1\}), C_2(\{x_2\}), \cdots, C_n(\{x_n\}))$: the changeover vector for production schedule $\{X\}$. $C_i(\{x_i\})$, $i \in I$: the number of changeovers for commodity i of $\{x_i\}$.
11) $L(\{X\}) = (L_1(\{x_1\}), L_2(\{x_2\}), \cdots, L_n(\{x_n\}))$: the changeover loss vector for $\{X\}$, where $L_i(\{x_i\}) = \gamma_i C_i(\{x_i\})$.

C. Constraints

From Assumptions 1), 5), and 6) described in Section II-A, we have the following equations:

$$\Delta X(t) = X(t) - X(t-1) = e_i, \quad t \in [0, T], \quad i \in I$$
$$X(0) = X(-1) = \phi \quad (1)$$

where e_i is the ith unit vector, and ϕ is the zero vector. By adding (1) with respect to $t = 1, 2, \cdots, t$ and calculating the norm of both members, the following equation is obtained:

$$\|X(t)\| = t, \quad t \in [0, T] \quad (2)$$

where the norm $\|A\|$ of vector $A = (a_1, a_2, \cdots, a_n)$ is defined as $\|A\| = \sum_{i=1}^{n} a_i$. Equation (1) (or (2)) shows the productive capacity of the two-machine production line and may be called the "capacity constraint."

Since the demand must be met, we obtain the following inequality:

$$X(t) \geq D(t), \quad t \in [0, T] \quad (3)$$

where (3) insures

$$x_i(t) \geq d_i(t), \quad t \in [0, T], \quad i \in I. \quad (4)$$

Either inequality (3) (or (4)) is called the "due date constraint."

From (2) and (3) at time T, (5) is derived.

$$\|D(T)\| \leq T. \quad (5)$$

$T - D(T)$ is the surplus production capacity at time T. This surplus capacity may be used to produce any commodity. In other words the final production quantity A of the respective commodities may be determined arbitrarily, if A satisfies the following condition:

$$\|A\| = T, \quad A \geq D(T). \quad (6)$$

Therefore if the cumulative demand at time T is determined to satisfy

$$D(T) = A \quad (7)$$

the surplus production capacity will disappear. In the following $D(T)$ is assumed to be given to satisfy (7).

Fig. 3 shows an example of the fluctuation of the norm of cumulative production for each machine. The float quantity is also illustrated in this figure. The norm of the cumulative production of the second machine is a linear function of time t. This is because of the normalization given in Assumption 6) of Section II-A. The norm of cumulative production at the first machine is a piecewise linear function of time t. This is due to the difference between the production speed of the first machine and that of the second machine. $\tau_1(\tau_1')$ and $\tau_2(\tau_2')$ denote the time when the production commodities are changed at the first (second) machine (Assumption 2)).

We illustrate the "float quantity" using Fig. 3. BC is the number (or the length of the queue) of items found between the first and the second machines at time t (see Assumption 7)). AC is the time difference by which the production of the first machine precedes that of the second machine at time t. The former is called the "material float

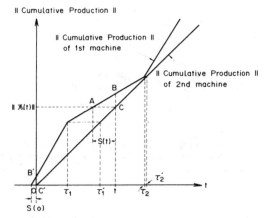

Fig. 3. Cumulative production of each commodity of each machine and float quantity.

quantity," and the latter the "time float quantity." However "float quantify" is a general term which includes both of these concepts.

Assume that the float quantity decreases and has reached value zero or allowable minimum value m for any commodity production. If the production commodity is not changed, the effective production speed of the second machine (or the overall production speed) will be reduced to the production speed of the first machine. Next assume that the float quantity increases and has reached allowable maximum value M for any commodity production. If the production commodity is not changed, the effective production speed of the first machine (or the overall production speed) will be reduced to the production speed of the second machine.

From the above discussion the overall production speed may be improved, if the production lots and their flow sequence are properly decided. However this is not always possible, for example, when $\rho_i < 0$ for all $i \in I$. In this case the maximum production speeds of one of the two machines are greater than those of the other in the production of all commodities. This does not usually occur, since, although the machine capacities are not regulated to balance the work load of each commodity, the capacities of the machines are regulated to balance the work load of the whole set of commodities. Even if this occurs, this becomes apparent in the process of finding an optimum schedule with the proposed method. Therefore a feasible schedule is assumed to exist in the following discussion.

If the material float quantity is used as the float quantity, allowable maximum value M is determined by the space between the two machines. On the other hand, if the time float quantity is used as the float quantity, the time when the next changeover must be performed can be calculated as the difference between M (or m) and time float quantity $S(t)$. In this paper the time float quantity will be used to simplify the mathematics:

$$S(t) = S(0) + \sum_{i \in I} \rho_i x_i(t), \quad t \in [0, T]. \quad (8)$$

Since the float quantity must be kept between m and M,

we obtain the following inequality:

$$m \leq S(0) + \sum_{i \in I} \rho_i x_i(t) \leq M, \qquad t \in [0, T]. \tag{9}$$

$S(0)$ is a constant which denotes the float quantity at time $t = 0$. Therefore, we can rewrite $m - S(0)$ and $M - S(0)$ as m and M, respectively, and obtain the following two inequalities:

$$m \leq \sum_{i \in I} \rho_i x_i(t) \leq M, \qquad t \in [0, T] \tag{10}$$

or

$$m \leq (\rho, X(t)) \leq M, \qquad t \in [0, T] \tag{11}$$

where (A, B) is the inner product of vectors A and B. Either (10) or (11) may be called the "float quantity constraint."

D. Definition of an Optimal Schedule

An optimal schedule for the two-machine production line must satisfy the capacity, due date, and float quantity constraints. From the above discussion the optimal schedule can be defined as follows.

Definition 1 (Optimal Schedule): Determine $\{X\}$ or $\{X(t), t \in [0, T]\}$ which satisfies

$$\min_{\{X\}} \|L(\{X\})\| \tag{12}$$

where

$$\Delta X(t) = e_i, \qquad i \in I$$
$$X(0) = \phi \tag{13}$$
$$X(t) \geq D(t) \tag{14}$$
$$m \leq (\rho, X(t)) \leq M, \qquad t \in [0, T]. \tag{15}$$

Equations (13), (14), and (15) are the "capacity constraint," the "due date constraint," and the "float quantity constraint," respectively.

The following sections give a method for obtaining an optimal schedule by the branch-and-bound method [4].

III. Mathematical Foundation

A. Feasible Schedules and Feasible Region

If the two-dimensional space in which there exists $\{x_i(t), t \in [0, T]\}$ is expressed by $X_i * T$, the $(n + 1)$-dimensional space in which there exists $\{X(t), t \in [0, T]\}$ can be expressed by $X_1 * X_2 * \cdots * X_n * T$.

Definition 2 (Feasible Schedules, Feasible Region, and Projected Feasible Regions): A feasible schedule is defined as the schedule which satisfies the capacity constraint, the due date constraint, and the float quantity constraint. A feasible schedule in the wide sense is the schedule which satisfies only the capacity constraint and the due date constraint. Feasible region $F[0, T]$ is defined as the region in which feasible schedules in the wide sense exist. Projected feasible region $F_i[0, T]$, $i \in I$ is the projection of $F[0, T]$ into subspace $X_i * T$, $i \in I$.

The above definition obviously means that feasible schedule $\{X(t), t \in [0, T]\}$ and $\{x_i(t), t \in [0, T]\}$, $i \in I$ are contained in feasible region $F[0, T]$ and projected feasible region $F_i[0, T]$, $i \in I$, respectively.

Theorem 1 (Existence Condition of Feasible Region) [5]: The necessary and sufficient condition to insure that the feasible region is not empty, i.e., $F[0, T] \neq \phi$ is

$$\|D(t)\| \leq t, \qquad t \in [0, T]. \tag{16}$$

Proof: This is proven by induction on time t. Note that all components of $D(t)$ can take only integer values from Assumptions 5) and 6). Assume that there exists a feasible schedule in the wide sense which satisfies

$$\|X(\tau)\| = \tau, \qquad X(\tau) \geq D(\tau), \qquad \tau \in [0, t].$$

Since $\|D(t + 1)\| \leq t + 1$ by (16) and $\|x(t)\| = t$, there exists feasible schedule $\{X(t), t \in [0, T]\}$ which satisfies $X(t) + e_k \geq D(t + 1)$ for any unit vector e_k. Hence, feasible schedule in the wide sense $\{X^*(\tau), \tau \in [0, t + 1]\}$ which satisfies (16) at time $t + 1$ is given by

$$X^*(\tau) = X(\tau), \qquad \tau \in [0, t]$$
$$X^*(t + 1) = X(t) + e_k.$$

The necessity of (16) is proven by contraposition. If (16) does not hold, it is obvious from the capacity constraint that the feasible schedule in the wide sense does not exist.

Theorem 2 (Projected Feasible Region): The upper boundary of projected feasible region $F_i[0, T]$ is $\alpha_i(t)$, and the lower boundary is $\beta_i(t)$, where these equations are given by

$$\alpha_i(t) = \min\left[t - \sum_{\substack{j \in I \\ j \neq i}} d_j(t), \alpha_i(t + 1)\right] \tag{17}$$

$$\beta_i(t) = \max[d_i(t), \beta_i(t + 1) - 1] \tag{18}$$

$$\alpha_i(T) = \beta_i(T) = d_i(T); \tag{19}$$

and the following equations hold:

$$\Delta \alpha_i(t) = 0 \text{ or } 1 \tag{20}$$
$$\Delta \beta_i(t) = 0 \text{ or } 1 \tag{21}$$

where $t \in [0, T]$.

Proof: From the conditions of $D(T)$ given by (6) and (7), (19) obviously holds. From (2) and (14),

$$d_i(t) \leq x_i(t) = t - \sum_{\substack{j \in I \\ j \neq i}} x_j(t) \leq t - \sum_{\substack{j \in I \\ j \neq i}} d_j(t) \tag{22}$$

and from (13),

$$x_i(t) \leq x_i(t + 1) \tag{23}$$
$$x_i(t) \geq x_i(t + 1) - 1. \tag{24}$$

Let $X(t) = (x_1(t), x_2(t), \cdots, x_n(t))$ be a constant vector which satisfies $x_i(t) = t - \sum_{j \in I}^{j \neq i} x_j(t)$, $x_k(t) = d_k(t)$, $k \neq i$, $k \in I$. Then, there exists a feasible schedule in the wide sense by which cumulative demand $X(t)$ is produced until time t. This $X(t)$ satisfies $\|X(t)\| = t$. Hence, there exists a case that $x_i(t) = t - \sum_{j \in I}^{j \neq i} d_j(t)$ in (22). From this result and (23), (17) is concluded to be the upper bound of the

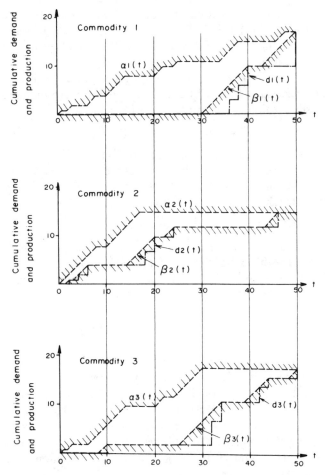

Fig. 4. Example of projected feasible region. --- $\alpha_i(t)$ or $\beta_i(t)$. -·- $d_i(t)$. Region surrounded by hashed lines is projected feasible region.

projected feasible region at time t. Equation (18) is proven in a similar manner.

Next (20) and (21) will be proven. From (16) and (17) it is obvious that $\Delta \alpha_i(t) \geq 0$. If $\Delta \alpha_i(t) > 0$, from (17)

$$\alpha_i(t-1) = t - 1 - \sum_{\substack{j \in I \\ j \neq i}} d_j(t-1) \leq \alpha_i(t) \leq t - \sum_{\substack{j \in I \\ j \neq i}} d_j(t).$$

Hence,

$$\Delta \alpha_i(t) = \alpha_i(t) - \alpha_i(t-1)$$
$$\leq t - \sum_{\substack{j \in I \\ j \neq i}} d_j(t) - \left\{ t - 1 - \sum_{\substack{j \in I \\ j \neq i}} d_j(t-1) \right\}$$
$$= 1 - \sum_{\substack{j \in I \\ j \neq i}} \Delta d_j(t)$$
$$\leq 1$$

and (20) is obvious. Equation (21) can be deduced from (18). An example of the projected feasible region is shown in Fig. 4.

B. Restrictions of Feasible Schedules

We have mainly discussed feasible schedules in the wide sense in the preceding section. In the following section we will discuss feasible schedules which satisfy the float quantity constraint as well as the capacity constraint and the due date constraint. However it is difficult to obtain an optimal schedule in the set of all feasible schedules. Therefore we consider a subset of the set of feasible schedules which will be defined in Definition 3, and obtain an optimal schedule from the subset. The feasible schedules contained in the subset are called feasible schedules in the narrow sense.

Definition 3 (Feasible Schedule in the Narrow Sense): A feasible schedule in the narrow sense is feasible schedule $\{X(t), t \in [0, T]\}$ satisfying the following condition: if a changeover of the production is determined to be accomplished at time $\tau_0 \in [0, T]$, time $\tau(\varepsilon[0, \tau_0])$ of the preceding changeover is determined so as to maximize production quantity $\tau_0 - \tau$ of a commodity at time interval $[\tau, \tau_0]$. In the following discussion a feasible schedule in the narrow sense will be called a feasible schedule.

C. Local Feasible Schedules

A feasible schedule must satisfy the constraints (13)–(15). However, if feasible schedule $\{x_i(t), t \in [0, T]\}, i \in I$ is investigated within projected feasible region $F_i[0, T]$, it is found to be a local feasible schedule satisfying the following definition. Therefore the numbers of changeover for optimal local feasible schedules can be used for the computation of a lower bound on the number of changeovers for feasible schedules.

Definition 4 (Local Feasible Schedule): When projected feasible region $F_i[0, T] (\neq \phi)$ and cumulative production b_i to be produced until time t are given, $\{x_i(\tau), \tau \in [0, T]\}$ satisfying

$$\{x_i(\tau), \tau \in [0, T]\} \in F_i[0, T]$$
$$\Delta x_i(\tau) = 0 \text{ or } 1, \quad \tau \in [0, T]$$
$$x_i(t) = b_i$$
$$|\rho_i| \cdot (\tau_q - \tau_p) \leq M - m, \text{ if } \Delta x_i(\tau) = 1, \tau \in [\tau_p, \tau_q]$$

is defined as the local feasible schedule for b_i to be produced until time t. Note that if $t = T$, then $b_i = d_i(T)$. A local feasible schedule is not guaranteed to satisfy the constraints of (13)–(15) for even only one commodity.

Definition 5 (Optimal Local Feasible Schedule): An optimal local feasible schedule is a local feasible schedule $\{x_i^L\}$ which minimizes the number of changeovers $C_i(\{x_i^L\})$.

Theorem 3 (Lower Bound of the Number of Changeovers for an Optimal Local Feasible Schedule): A lower bound on the number of the changeovers for $\{x_i^L\}$ by which cumulative production b_i is produced until time t ($\leq T$) in $F_i[0, T] (\neq \phi)$ is obtained by the following algorithm.

Algorithm:

1) $C_i := 0, \tau_0 := t, x_i^L(\tau_0) := b_i,$

2) $x_i^L(\tau) := x_i^L(\tau_0), \quad \tau \in [\tau_\alpha, \tau_0]$, where
$$\tau_\alpha = \max[\tau / \alpha_i(\tau - 1) < x_i^L(\tau_0), \quad \tau \leq \tau_0], \alpha_i(-1) = 0.$$

3) If $\tau_\alpha = 0$, terminate,

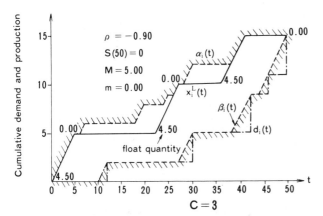

Fig. 5. Example of optimal local feasible schedule. —$x_i^L(t)$. --- $\alpha_i(t)$ or $\beta_i(t)$ ·-·- $d_i(t)$. Region surrounded by hashed lines is projected feasible region.

4) $x_i^L(\tau) := x_i^L(\tau + 1) - 1, \quad \tau \in [\tau_\beta, \tau_\alpha]$, where

$$\tau_\beta = \max[\tau/\beta_i(\tau - 1) > x_i^L(\tau_\alpha) - (\tau_\alpha - \tau + 1),$$

$$|\rho_i| \cdot (\tau_\alpha - \tau) \leq M - m, \quad \tau < \tau_\alpha], \beta_i(-1) = 0,$$

5) $C_i := C_i + 1$,

6) If $x_i^L(\tau_\beta) - \tau_\beta \neq 0$, terminate,

7) $\tau_0 := \tau_\beta$; go to 2).

Proof: From Theorem 2, since $\Delta\alpha_i(t) = 0$ or 1, the border of $F_i[0, T]$ is a monotone nondecreasing function with respect to τ. Therefore it is easily understood that if the local feasible schedule is determined so as to maximize the production quantity for each commodity in a production period, it is optimal. Since it is the objective of this algorithm to obtain a local feasible schedule by which cumulative production b_i is produced until time t, a local feasible schedule must be determined from time t and downward. For these reasons, in order to maximize the production quantity for each period, the production starting time is determined to be as early as possible in step 2), and the duration of the production is maximized in step 4). If $x_i^L(\tau_\beta) = \tau_\beta \neq 0$ holds in step 6), this algorithm fails to find an optimal local feasible schedule. However, C_i is obviously the lower bound to be obtained.

Fig. 5 shows the optimal local feasible schedule and the number of changeovers which have been obtained by this algorithm. The following relation holds between the number of changeovers of an optimal feasible schedule and that of a local feasible schedule.

Theorem 4 (Lower Bound on the Number of Changeovers): If $\{x_i^0(\tau), \tau \in [0, t_H]\}$ (or $\{x_x^0\}$), $i \in I$ is the optimal feasible schedule in feasible region $F[0, T]$ by which the cumulative production quantity $H(= (h_1, h_2, \cdots, h_n)), m \leq (\rho, H) \leq M$ by (15)) is produced until time t_H, and $\{x_i^L(\tau), \tau \in [0, t_H]\}$ (or $\{x_i^L\}$), $i \in I$ is the local feasible schedule in projected feasible region $F_i[0, T]$ by which cumulative production quantity h_i is produced until time t_H, the following inequality is derived

$$C_i(\{x_i^0\}) \geq \min_{\{x_i^L\} \in F_i[0,T], x_i^L(t_H) = h_i} C_i(\{x_i^L\}). \quad (25)$$

IV. Formulation of the Scheduling Algorithm by the Branch-and-Bound Method

A. Partition of Set of Feasible Schedules into Subsets

The branch-and-bound method necessitates partitioning of the set of feasible schedules into subsets. The following is the definition of the subsets used in the proposed method.

Definition 6 (Subset of Feasible Schedules): Subset $[H]$ ($H \neq 0, m \leq (\rho, H) \leq M$ by (15)) of feasible schedules (Definition 3) in feasible region $F[0, T]$ is one which satisfies the following conditions.

For all $\{X(t), t \in [0, T]\} \in [H]$,

1) $X(t_H) = H, 0 \leq t_H \leq T, t_H = \|H\|$,
2) $X(\tau), \tau \in [t_H, T]$ is determined. The schedule within $[t_H, T]$ previously given is expressed as $S[H]$.

Next, we will discuss the method for partitioning the set of feasible schedules into subsets in the branch-and-bound algorithm.

Theorem 5 (Method for Partitioning the Subset of Feasible Schedules): Subset $[H]$ ($H \neq 0, m \leq (\rho, H) \leq M$ by (15)) of feasible schedules in feasible region F is partitioned into $[G_p], p \in I^*$ which satisfies the following conditions:

$$G_p = H - k_p e_p, \quad p \in I^* \quad (26)$$

where

$$k_p = \max[k/H - ke_p \in F, m \leq (\rho, H - ke_p) \leq M, k \geq 0] \quad (27)$$

$$I^* = \{p/p \in I, k_p > 0\}. \quad (28)$$

Hence,

$$S[G_p] = (k_p)_p \wedge S[H]. \quad (29)$$

The right side of (29) shows that quantity k_p of commodity p is produced just before $S[H]$. If $k_p = 0$ holds for all $p \in I$, it is obvious from (15) that no feasible schedule exists.

Proof: Since schedules considered here are feasible schedules in the narrow sense (Definition 3), any $[G_p^*]$ which satisfies $G_p^* = H - ke_p, 0 < k < k_p, p \in I^*$ is not in the subset of feasible schedules in the narrow sense with regards to Definition 3, (27) and (28). Therefore, $[G_p], p \in I^*$ is a decomposition of $[H]$. And obviously, $[G_p], p \in I^*$ is a subset given in Definition 6, for each $p \in I^*$.

B. Equation for Calculating the Lower Bound

Theorem 6 (Equation for Calculating Lower Bound): Lower bound $LB([H])$ of the objective function (12) for subset $[H]$ ($\neq \phi$), where $H = (h_1, h_2, \cdots, h_n)$, in feasible region $F[0, T]$ is obtained by the following problem:

$$\min_{p \in I^*}\left[\gamma_p\left\{\min_{\{x_p^L\} \in F_p, x_p^L(t_H - k_p) = h_p - k_p} C_p(\{x_p^L\}) + 1\right\}\right.$$
$$\left. + \sum_{\substack{i \in I \\ i \neq p}} \gamma_i \min_{\{x_i^L\} \in F_i, x_i^L(t_H) = h_i} C_i(\{x_i^L\})\right] + \|L(S[H])\| \quad (30)$$

where I^*, k_p, t_H are given in Theorem 5 and Definition 6. F_i, $i \in I$ is the projected feasible region of $F[0, T]$, and $\{x_i^L\}$, $i \in I$ is the local feasible schedule in F_i.

Proof: Consider a feasible schedule by which the cumulative production quantity H is achieved at time $t_H(= \|H\|)$, and assume that p is the production commodity in the production period just preceding time t_H. In this case the production quantity in this production period is k_p from Theorem 5. Let N_H be the number of changeovers in the optimal local feasible schedule by which cumulative production quantity h_p is achieved at time t_H. And let N'_H be the same for the optimal local feasible schedule by which the cumulative production quantity $h_p - k_p$ is achieved until time $t_H - k_p$. Hence N_H is larger than N'_H by one. Therefore, N_H is

$$\min_{\{x_p^L\} \in F_p,\, x_p^L(t_H - k_p) = h_p - k_p} C_i\left(\{x_p^L\}\right) + 1.$$

A lower bound on the number of changeovers for the optimal local feasible schedule for commodity $i(\neq p)$ is given as

$$\min_{\{x_i^L\} \in F_i,\, x_i^L(t_H) = h_i} C_i\left(\{x_i^L\}\right).$$

Commodity p can be selected arbitrarily. Therefore, from Theorem 4, the lower bound on the number of changeovers for the optimal schedule by which H is achieved at time t_H is calculated as the first term of (30). The lower bound on the objective function for $[H]$ is obtained by adding a changeover loss corresponding to $S[H]$ to the first term. Since the calculation method for obtaining min $C_i(\{x_i^L\})$ is given by Theorem 3, the lower bound can be obtained from Theorem 6.

V. Scheduling Algorithm

Fig. 6 shows an outline of the scheduling algorithm formulated by the branch-and-bound method. Feasible regions $F_i[0, T]$, $i \in I$ are computed in block 2 in Fig. 6, since they are commonly used when calculating the lower bound for each subset of feasible schedules. Let $\mathscr{F}$ denote the family of the subsets which are obtained at each stage of the computation process. Therefore, $\mathscr{F}$ contains only $[H]$, where $H = D(T)$, at the beginning of the computation process, as shown in block 3.

In block 4 subset $[H]$ is partitioned into subsets $[G_p]$, $p \in I^*$ by the procedure given in Theorem 5. Therefore, the subset $[H]$ in $\mathscr{F}$ is substituted by $[G_p]$, $p \in I^*$ in block 4. In block 5 the next subset is selected which is to be partitioned in block 4. However, if H is a zero vector, the computation process is terminated. Then $S[H]$ is an optimal schedule.

VI. Example of Calculation

Example 1

Number of commodities: $n = 3$.
Scheduling horizon: [0, 50].

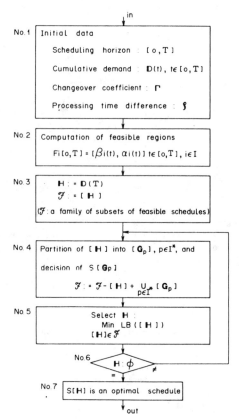

Fig. 6. Flowchart of optimal scheduling algorithm.

Cumulative demand: $D(t) = (d_1(t), d_2(t), d_3(t))$, $t \in [0, 50]$ as shown in Fig. 7.
Processing time difference: $\rho = (\rho_1, \rho_2, \rho_3) = (-0.90, 0.50, 0.30)$.
Changeover loss coefficient: $\Gamma = (\gamma_1, \gamma_2, \gamma_3) = (2, 1, 1)$.
Initial value of float quantity: $S(0) = 0$.
Allowable maximum and minimum values of float quantity: $M = 5$, $m = 0$.

The process of finding an optimal schedule by the branch-and-bound method is shown in Fig. 8, and the optimal schedule is shown in Fig. 7. $[(p, q, r)]$ in Fig. 8 shows $[H]$ in Definition 6, when $H = (p, q, r)$. The numbers over and under the leaf expressing subset $[H]$ are the values of its float quantity and $LB([H])$, respectively.

In Fig. 7 an optimal schedule $X_i(t)$, $t \in [0, T]$, $i \in I$ is shown as a solid line in $F_i[0, T]$, $i \in I$. $F_i[0, T]$, $i \in I$ is the region including the border surrounded by hashed lines. The numbers attached to the solid lines are the float quantity values at the production times. The slanting parts of the solid line in $F_i[0, T]$ express the production time period of the commodity. Fig. 7 shows that the production line produces one and only one commodity item at a time.

Example 2

Number of commodities: $n = 3$.
Scheduling horizon: [0, 50].
Cumulative demand: $D(t) = (d_1(t), d_2(t), d_3(t))$, $t \in [0, 50]$ as shown in Fig. 9.
Processing time difference: $\rho = (\rho_1, \rho_2, \rho_3) = (-0.95, 0.60, 0.40)$.

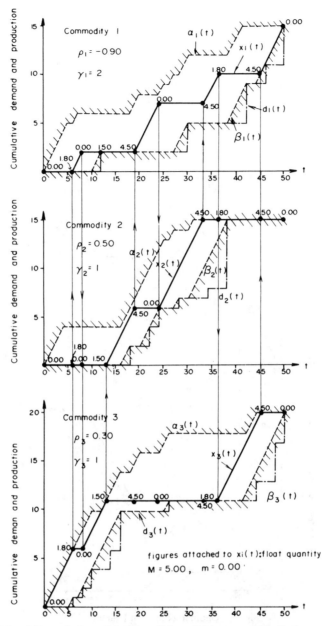

Fig. 7. Optimal schedule of Example 1.—$x_i(t)$. --- $\alpha_i(t)$ or $\beta_i(t)$ · -·- $d_i(t)$. Region surrounded by hashed lines is projected feasible region.

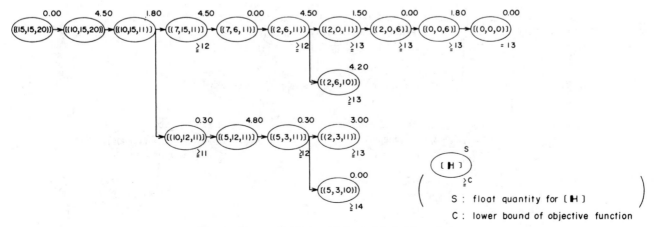

Fig. 8. Process deciding optimal schedule for Example 1.

347

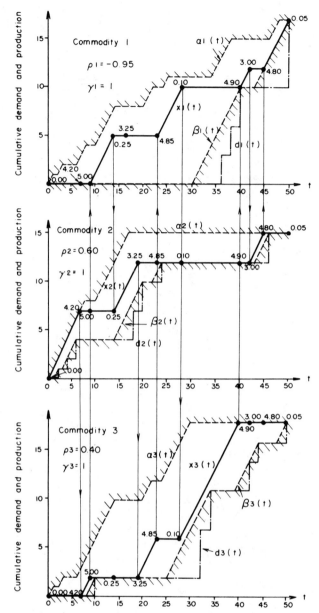

Fig. 9. Optimal schedule of Example 2.

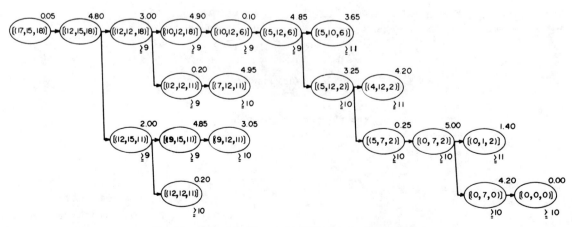

Fig. 10. Process deciding optimal schedule for Example 2.

Changeover loss coefficient: $\Gamma = (\gamma_1, \gamma_2, \gamma_3) = (1, 1, 1)$.
Initial value of float quantity: $S(0) = 0$.
Allowable maximum and minimum values of float quantity: $M = 5, m = 0$.

The process of finding an optimal schedule is shown in Fig. 10, and the optimal schedule is shown in Fig. 9. Computational experience with this proposed method indicates it is capable of solving problems involving 20 commodities over a planning horizon of 300 time periods. These are practical scheduling problems with a planning horizon of one week for a home appliance production line. These computation times are below one minute (using the HITAC M-180 operating at 3 MIPS).

The computation time is considered to increase rapidly with increases in the size of the problem. However the computational experience indicates that this increase is not so large. The optimal schedules for both Examples 1 and 2 are obtained by the shortest searching path. This is because the region to be searched for an optimal feasible schedule is restricted by the feasible region and the float quantity constraint. Therefore the broader the feasible region, or the larger the allowable maximum limit of the float quantity, the larger the computation time will become.

VII. Conclusion

This paper presented a scheduling method for balancing work loads between a machine and the one succeeding it in a two-machine production line so as to maximize the production speed. This paper treats the case where the production speeds of both machines and commodities are different from each other. The objective is to minimize the changeover loss under the constraints that 1) the demand must be met, 2) the buffer storage must be lower than an upper limit, and 3) idle time is not acceptable for either machine. The schedule obtained by the proposed scheduling method gives the lot size of each commodity for each production period, and its flow sequence.

A scheduling algorithm was formulated using branch-and-bound method. The geometrical characteristics of the region where the feasible schedules exist were used to calculate the lower bound of the objective function for the subset of feasible schedules. This method was proposed for the first time in the author's previous paper [5].

Acknowledgment

The author would like to thank General Manager T. Miura, Chief Researcher T. Mitsumaki, and Senior Researcher K. Mitome of Systems Development Laboratory, Hitachi Ltd., for interesting and helpful discussion, and the referees whose comments have greatly improved the presentation of the material contained in this paper.

References

[1] E. J. Ignall, "A review of assembly-line balancing," *J., Ind. Eng.*, vol. 16, no. 4, 1965.
[2] R. W. Conway, W. L. Maxwell, and L. W. Miller, *Theory of Scheduling*. Reading, MA: Addison-Wesley, 1967.
[3] S. K. Dutta and A. A. Cunningham, "Sequencing two-machine flow-shops with finite intermediate storage," *Management Sci.*, vol. 21, no. 9, pp. 989–996, 1975.
[4] R. E. Lawler and D. E. Wood, "Branch-and-bound method: A survey," *Oper. Res.*, vol. 14, pp. 699–719, 1966.
[5] S. Mitsumori, "Optimum production scheduling of multicommodity in flow line," *IEEE Trans. Syst., Man, Cybern.*, vol. 2, no. 4, pp. 486–493, 1972.

Concurrent Routing, Sequencing, and Setups for a Two-Machine Flexible Manufacturing Cell

ENG-JOO LEE, STUDENT MEMBER, IEEE, AND PITU B. MIRCHANDANI, SENIOR MEMBER, IEEE

Abstract—An approach proposed for scheduling of Flexible Manufacturing Systems (FMS) is to sequentially solve three related subproblems, commonly referred to as the "FMS loading," "job routing," and "operation sequencing" problems. However, the use of this approach results in assigning a fixed set of resources for each machine prior to sequencing, and thus may unnecessarily restrict the capability of the flexible manufacturing system, especially if the machines are very versatile.

In this paper, we present an approach which allows in-process "machine loading" through magazine setups, and which concurrently routes and sequences the jobs on the versatile machines.

To illustrate the concurrent approach, a specific two-versatile-machine flowshop scheduling problem, referred to as 2-VFSP, is defined and studied in detail. Theoretical results show that the optimal schedule for 2-VFSP need not have more than two in-process magazine setups, giving rise to the three possible scheduling configurations. Further, it is proven that to obtain the optimal schedule is an NP-complete problem.

A heuristic is developed and tested on sets of random joblists, with different setup times corresponding to varying degrees of machine versatility. Results indicate that schedules which incorporate in-process magazine setup(s) may be more desirable when the setup time is small. Thus the use of concurrent scheduling approach which allows in-process magazine setups may be especially useful for systems with very versatile machines.

I. INTRODUCTION

THE ADVANCEMENTS in integrated automation, computer, and machine tool technology have led to the development of machines which are versatile. For example, a modern numerically controlled (NC) machine can automatically perform many different operation types, and can now replace several conventional machines. Such versatile machines are especially attractive to the mid-volume production sector of the metal-working industry where "flexibility" has become an important and desirable feature [15]. When these NC machines are further automated and integrated by advanced computer technology, a new mode of manufacturing emerges in the form of a Flexible Manufacturing System (FMS) [4].

In particular, a Flexible Manufacturing Cell (FMC) is a class of FMS's consisting of two or more versatile machines (such as machining centers) with a material handling system that can transport partpieces to the machines within the cell quickly (such as a robot). The operational design of FMC's are becoming increasingly important especially since recent surveys [4], [16] indicate that both the users of flexible automation and machine tool makers have begun emphasizing a step-by-step approach, starting with small-scale flexible manufacturing cells.

An important aspect of the operational design of an FMC involves scheduling. In this paper, we shall study a specific two-versatile-machine FMC. Section II explains the "concurrent strategy" undertaken in developing the scheduling algorithm. This is followed by a section on notation and definitions. Some theoretical results and heuristics for the problems are reported in Section IV. Schedules of sets of random joblists are developed using the heuristics. Finally, Section V concludes with a discussion on this scheduling strategy, and some possible research extensions.

II. THE "CONCURRENT" SCHEDULING STRATEGY

Briefly, scheduling includes the basic decisions of routing (machines-to-a-partpiece allocation) and sequencing (partpieces-to-a-machine assignment). In the traditional dedicated-machine environment, each machine can only perform one operation type, and, therefore, the technological constraints of the processing needs dictate the route for each partpiece. Consequently, the corresponding operational scheduling problem reduces to only the sequencing of the partpieces.

In the FMC versatile-machine environment, since each machine can perform more than one operation type, a particular operation of a partpiece may be performed by more than one machine. This enables each partpiece to have several alternative routings through the machines. Thus the corresponding operational scheduling problem involves both routing and sequencing.

In essence, the significant difference between the dedicated-machine environment and the versatile-machine environment lies in that the only scheduling concern in the former case depends solely on the characteristics of the processing requirements of the partpieces such as job precedence and operations ordering. Scheduling in the latter case, however, requires the consideration of the capabilities (versatility) and constraints (tool magazine capacity limitation) of the machines in addition to the processing requirements of the partpieces.

When the total number of tools required to process all the partpieces does not exceed the magazine capacity of each machine, then each partpiece can basically be processed on any of the machines. If the partpiece requires a tool that is resident in the tool magazine, then a tool exchange occurs

---

Manuscript received December 17, 1986; revised September 4, 1987. This work was partially supported by IBM under Grant on Manufacturing Education, and by the FMS Program at RPI sponsored by ALCOA, GE, GM, KODAK, and RCA. Part of the material in this paper was presented at the 1986 IEEE International Conference on Robotics and Automation, San Francisco, CA, April 7-10, 1986.

The authors are with the Electrical, Computer, and Systems Engineering Department, Rensselaer Polytechnic Institute, Troy, NY 12180.

IEEE Log Number 8718906.

between the machine spindle and the magazine. This tool exchange, sometimes also referred to as the spindle-setup, is performed automatically with negligible setup time, and this is what basically constitutes the versatility feature of the machine.

However, in most practical applications, the total number of tools exceeds the magazine capacity. In this case, three possible situations may occur when a partpiece arrives at the machine. The first two situations occur when the required tool to process the partpiece 1) is resident on the spindle, or 2) is available on the tool magazine. In these situations, either 1) no setup, or 2) an automatic tool exchange takes place.

If, however, 3) the required tool is not resident in the magazine, then another type of setup may be required. This setup, referred to as a *magazine setup*, involves the replacement of some or all of the tools in the magazine. In scheduling a set of partpieces, the number of times that magazine setups are performed depends on the tool allocation and the tool distribution strategies employed by the tool management system (TMS). In addition, the amount of time needed to perform each magazine setup will depend on the FMC and TMS technologies of the system; it may range from a negligibly short time interval to a significant amount of time.

By allowing in-process magazine setups to take place, the versatility of the machines in an FMC can be utilized to their fullest potential, and at the same time, a greater number of part types may be processed during a production period due to the increased system flexibility. In the concurrent scheduling strategy developed in this paper, the scheduling decisions involving routing and sequencing are performed concurrently and are integrated with decisions on if and when any magazine setups are to be performed.

In the literature, only a few research studies related to this type of magazine setups have been reported. Hankins and Rovito [7] conducted computer simulations and reported performance results for two different magazine setup strategies. They concluded that a comprehensive tool management system is essential to the success of an FMS.

In her investigation of automated tooling for FMS's, ElMaraghy [5] concluded that, irrespective of the selected level of automation, productivity can be maximized and idle time minimized through the increased utilization of machines by ensuring the availability of the necessary tools when needed—using in-process magazine setups.

Tang and Denardo [13], [14] considered the problem of finding the sequence in which to process a joblist on a single machine and the tools to load on the machine, if necessary, before each job is processed. The two optimality criteria considered were the minimization of the total number of tool switches, and the minimization of the total number of instants at which tools are switched. The procedures developed for each optimality criterion were also extended to the case with m machines in sequence.

III. NOTATION AND DEFINITIONS

For each physical partpiece is associated a "logical" entity referred to as its *job*, which describes its processing needs. The processing activity requiring a single tool is referred to as a *tool-task*. A dedicated machine can perform only a single tool-task, while a versatile machine can perform a number of different tool-tasks depending on the number of tools its tool magazine can hold. The maximum number of distinct tool-tasks that machine i can perform without the need for additional magazine setup corresponds to the size of its tool magazine capacity C_i. Thus for a total number of N distinct tool-tasks among the jobs—and hence, N distinct tools required—and a tool magazine capacity of size C_{max} among the machines, at least $\lceil N/C_{max} \rceil$ different tool-sets need to be loaded.

An *operation* consists of one or more tool-tasks. An operation that may be performed by tool-set j is referred to as an *operation type* j. To the machines, each operation type corresponds to one of its possible feasible tool-sets that can be loaded on its magazine.

As an illustration, suppose three types of partpieces, A, B, and C, are to be scheduled in a production cycle, requiring, respectively, tool sets $\{1, 2, 3\}$, $\{4, 5, 6\}$, and $\{1, 5, 6\}$ (note $N = 6$). Suppose we have a single versatile machine with $C_{max} = 4$. If operation type X corresponds to tool set $\{1, 2, 3, 4\}$ and operation type Y to $\{1, 2, 5, 6\}$, then part A requires only operation type X, part C requires only operation type Y, and part B requires both operation types. Thus scheduling in our context is viewed in terms of partpieces, each partpiece being defined as a job comprised of operations, and each operation belonging to an operation type.

IV. THE 2-VFSP PROBLEM

A scheduling problem can be described in terms of its shop characteristics, its job characteristics, and the optimality criterion with which the evaluation of each schedule can be made (see, e.g., [3]). The specific 2-VFSP problem studied here is described below using this framework.

Shop Characteristics

- The shop consists of two identical machines, P and Q; each with a tool magazine of equal *finite* capacity C.
- Each machine can perform *all* operations of the two different operation types, X and Y.
- Each machine can perform *only one* operation at a time.
- Each operation type corresponds to C tools.
- The *setup* operation, S_{XY} (S_{YX}), changes the tool-set for operation type X (Y) with the tool-set for operation type Y (X). Both setup operations are assumed to take the same amount of time to perform; the time may be negligible, or finite.
- A setup operation must be performed when the machine needs to perform the other operation type.
- The transfer or transport time of a partpiece from one machine to the other is *negligible*.

Job Characteristics

- Each partpiece requires a job composed of one operation of type X, followed by one operation of type Y.
- The operations cannot be performed simultaneously on a partpiece.
- The type X operation *must* be completed before its type

Y operation may commence. This corresponds to an *operation precedence constraint*.
- All partpieces are *immediately* available for processing once production begins.

Optimality Criterion
- To *minimize* the makespan of the schedule for the given joblist.

Clearly, this problem resembles the well-known conventional two-machine flowshop problem [9] except that it has versatile machines rather than dedicated machines. In the dedicated machine environment, a flowshop is defined in terms of workstations (or machines): the routing of each job through the workstations is the same. Thus a sequencing of the jobs defines a schedule for the flowshop. For the versatile machine environment, even when the operation order is the same for each of the jobs (type X operation before type Y operation), jobs may have different routings (some jobs may be done on a single machine, others may be done on both machines; see, for example, Fig. 1). In this context, we define a flowshop in terms of operation types: the operations precedence order of each job is the same. It is for this reason that we refer to this problem as the two-versatile-machine flowshop scheduling problem (2-VFSP).

A. Some Theoretical Results

We present here two theoretical results specific to our 2-VFSP problem. The first, Theorem 1 and its corollary, shows that the optimal schedule requires at most two setups—each one on a different machine. The second, Theorem 2, states that the 2-VFSP problem belongs to the class of "difficult" problems known as NP-complete [6].

Theorem 1: For the 2-VFSP, an optimal schedule *exists* with at most two setups, of which at most one is in each machine and both are of the S_{XY} type.

Proof: Without loss of generality, we first show that all S_{YX} setups in a machine, if any exist, can be removed in a feasible schedule to obtain another feasible schedule with only one S_{XY} setup on this machine *without* increasing the makespan.

Consider a feasible schedule that has an S_{YX} type setup on, say, machine P. Thus there is a set of consecutive jobs whose type Y operations are scheduled immediately before this setup. Since the schedule is feasible for the given precedence constraints, the type X operation of each of these jobs must have been scheduled to be completed before any of the type Y operations commence. (These type X operations may have been scheduled on either of the machines.) Therefore, if the type Y operations are rescheduled to be processed at a later time, the resultant schedule will still remain feasible.

Similarly, if the set of consecutive type X operations that are scheduled immediately after this S_{YX} setup are rescheduled to be processed at an earlier time, the schedule will remain feasible *without* any increase in the makespan.

Hence, if these two sets of operations are interchanged so that the set of type X operations are scheduled before the set of

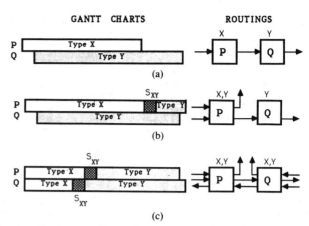

Fig. 1. The three possible configurations of the optimal schedule and the corresponding routing patterns for the partpieces. (Doubly cross-hatched region represents setup operation.) (a) *Zero-setup* schedule. (b) *One-setup* schedule. (c) *Two-setup* schedule.

type Y operations, the resultant new schedule remains feasible. However, the setups immediately before and after this S_{YX} setup, if any exist, are of S_{XY} type. Thus the interchange will make these S_{XY} setups redundant and eliminate their need, but will require an S_{XY} in place of the S_{YX} setup.

Clearly, therefore, any feasible schedule with S_{YX} types of setups can undergo a series of such interchanges until at most one setup, an S_{XY} setup, results on machine P.

The "interchange" process done on one machine is performed independently without affecting the original schedule of the other machine. Thus the same "interchange" process can be performed on machine Q, if there exist any S_{YX} setups, until machine Q will also have at most one setup, an S_{XY} setup. Therefore, if we have an optimal schedule with more than one setup on either machine, we can perform these "interchanges" until we have at most one S_{XY} setup on each machine and still remain optimal. This completes the proof. ∎

When the setup time is nonzero, we have a stronger result, namely Corollary 1 below, which states that an optimal schedule will *not* have more than two necessary setups. We state this corollary without proof.

Corollary 1: For the 2-VFSP, if the setup time is nonzero, then an optimal schedule is one which has no more than two setups, both being of the S_{XY} type, and of which at most one is on each machine.

Theorem 1 and Corollary 1 provide valuable information concerning the possible configuration of the optimal schedule, and its corresponding routing pattern. Note that the optimal solution of the 2-VFSP provides both the sequencing and the routing decisions concurrently. In all, there are three possible optimal schedule configurations—with zero, one, or two setups. Fig. 1 depicts these configurations with the corresponding routing patterns.

The existence of three possible configurations of the optimal schedule suggest an approach to solve the 2-VFSP which involves solving a subproblem corresponding to each configuration. These subproblems shall be referred to as *zero-setup*, *one-setup*, and *two-setup* problems, respectively. The first subproblem corresponds to the conventional two-machine flowshop problem which is solved optimally by the polynomial-time algorithm of Johnson [9]. However, for the other two subproblems—the *one-setup* and *two-setup* problems—there are no known algorithms.

The next theoretical result, Theorem 2, states that 2-VFSP belongs to the class of problems known as NP-complete, such as the traveling-salesman problem, which has defied attempts for exact solutions, short of some type of enumeration of all feasible solutions. In essence, it is suspected that no polynomial-time algorithm exists which exactly solves all problem instances of an NP-complete problem; in the unlikely event that a polynomial-time algorithm exists for an NP-complete problem, then all problems in this class are solvable in polynomial time.

Theorem 2: 2-VFSP is NP-complete.

Proof: To prove Theorem 2, it suffices to show that at least one of the subproblems is NP-complete and that each subproblem of 2-VFSP is either polynomially solvable or NP-complete.

We first prove that the *two-setup* problem is NP-complete. This subproblem may be formulated as a decision problem as follows:

TWO-SETUP:

Given n jobs, each job i requiring integer operation times of c_i and c_{n+i} (with c_i preceding c_{n+i}), corresponding to the first and the second operation type, respectively, and setup time of t_s, is there a schedule such that for $S \subseteq \{1, 2, \cdots, 2n\}$, we have

$$\sum_{j \in S} c_j + t_s = \sum_{j \notin S} c_j + t_s \ ?$$

Clearly, *TWO-SETUP* is in NP since a nondeterministic algorithm need only to guess a subset S of $\{1, 2, \cdots, 2n\}$ and check in polynomial time that the total processing time of the operations in S is the same as that for the rest of the operations, and that the precedence constraint has not been violated.

Now consider the following well-known NP-complete problem:

0-1 KNAPSACK:

Given integers c_j, $j = 1, \cdots, n$ and K; is there a subset $S \subseteq \{1, 2, \cdots, n\}$ such that

$$\sum_{j \in S} c_j = K ?$$

There exists a polynomial transformation for a problem instance from *0-1 KNAPSACK* to *TWO-SETUP*, as follows:

Given any instance of $(n - 2)$ integers $c_1, \cdots, c_{n-2}$, and K of *0-1 KNAPSACK*, we construct the following instance of *TWO-SETUP* with:

i) a setup time of $t_s \geq 0$;
ii) processing times for type X operations of n jobs, being $c_1, \cdots, c_{n-1}, c_n$, with $c_{n-1} = c_n = 0$;
iii) processing times for type Y operations being $c_{n+1} = c_{n+2} = \cdots = c_{2n-2} = 0$, $c_{2n-1} = 3M - 2K$, and $c_{2n} = 2M$ where

$$M = \sum_{j=1}^{n} c_j > K.$$

We claim that there exists a subset $S \subseteq \{1, 2, \cdots, n-2\}$ with

$$\sum_{j \in S} c_j = K$$

if and only if there exists an $E \subseteq \{1, 2, \cdots, 2n\}$ such that

$$\sum_{j \in E} c_j + t_s = \sum_{j \notin E} c_j + t_s.$$

(If:) Clearly, since c_{2n-1} and c_{2n} adds up to

$$5M - 2K > \sum_{j=1}^{n} c_j$$

they must be separated and each performed on a different machine. Hence, since $c_n = 0$, we have

$$\sum_{j \in S} c_j + t_s + c_{2n-1} = \sum_{\substack{j \notin S \\ j \neq 2n-1, 2n}} c_j + t_s + c_{2n}.$$

Since

$$\sum_{\substack{j \notin S \\ j \neq 2n-1, 2n}} c_j = M - \sum_{j \in S} c_j$$

it follows directly from arithmetic that

$$\sum_{j \in S} c_j = K.$$

(Only if:) Suppose that

$$\sum_{j \in S} c_j = K$$

for some $S \subseteq \{1, 2, \cdots, n-2\}$. Then it follows immediately that

$$\sum_{j \in S} c_j + t_s + c_{2n-1} = \sum_{\substack{j \notin S \\ j \neq 2n-1, 2n}} c_j + t_s + c_{2n}.$$

Next, we need to show that the precedence constraint has not been violated. Clearly, since $c_{n-1} = c_n = c_{n+1} = \cdots = c_{2n-2} = 0$, no operation pair c_i and c_{n+i} violates that constraint.

This proves that *TWO-SETUP*, and hence, the *two-setup* problem, is NP-complete.

In a similar fashion, the *one-setup* problem can be formulated as a decision problem, and proved to be NP-complete.

Finally, since the *zero-setup* problem is polynomially solvable, it follows that each subproblem of 2-VFSP is either polynomially solvable or NP-complete. Hence 2-VFSP is NP-complete. ∎

Obviously, the results of Theorem 2 suggest that it is impractical to develop exact algorithms to solve the *one-setup* and *two-setup* problems optimally. Thus heuristics that solve these NP-complete subproblems of 2-VFSP suboptimally have been developed and described in the next subsection.

B. Heuristics for 2-VFSP

As mentioned before, there already exists a known (Johnson's) algorithm that optimally solves the *zero-setup* subproblem. In this section, we describe two heuristics, ONE-SETUP and TWO-SETUP developed for the *one-setup,* and the *two-setup* subproblems, respectively.

The One-Setup Case

One may make several important observations on the characteristics of the *one-setup* schedule. Fig. 2 shows a *one-setup* schedule with machine P as the *setup* machine. In this schedule, there are essentially two classes of jobs. The first class consists of "2-machine" jobs, each using *both* machines to process it—in this case, type X operation is performed on machine P, and type Y on machine Q. The second class consists of "1-machine" jobs each having both of its operations processed on one machine, the *setup machine*. In this case, machine P is the setup machine.

One-setup scheduling can thus be viewed as the simultaneous scheduling of these two job classes so as to minimize the makespan. Clearly, due to the operation precedence constraint on each job, any schedule obtained by sequencing the "2-machine" jobs will inevitably incur idle periods I_P and I_Q for machines P and Q, respectively. The *optimal* schedule, therefore, essentially "fits" the set of "1-machine" jobs (plus the setup operation) into the idle period of the setup machine. Therefore, to obtain the optimal *one-setup* schedule, it is necessary to identify both the setup machine and the two job classes.

We next observe that Johnson's (*zero-setup*) schedule consists of only "2-machine" jobs; and it minimizes not only makespan, but also the idle periods on each machine. The *one-setup* scheduling heuristic in essence *improves* on the *zero-setup* schedule by shortening the makespan by "moving" some operations from one machine (with the shorter idle period) to the other machine.

Note that there may be cases when the *one-setup* schedule for a joblist becomes undesirable. This occurs when the sum of the setup time and the smallest processing time among the jobs performed by the nonsetup machine is greater than the idle period of the setup machine in the *zero-setup* schedule. In this case, a setup on either of the machines would, instead, increase the makespan of the *zero-setup* schedule. In the algorithm presented below, we include a check to ensure that the determination of the *one-setup* schedule terminates when this case occurs.

ONE-SETUP (Heuristic 1)

Step 1 (To select the setup machine):

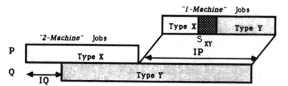

Fig. 2. *One-setup* schedule with P as a setup machine.

Obtain the *zero-setup* schedule using the Johnson's algorithm. Let I_P be the total idle time for machine P and I_Q for machine Q. The setup machine, say M, is selected to correspond to the machine with the larger total idle time; that is, $I_M = \max(I_P, I_Q)$.

Termination Check. Let $b = \max\{\min(c_i, i = 1, \cdots, n), \min(c_j, j = n + 1, \cdots, 2n)\}$. If $t_s > (I_M - b)$, stop (a *one-setup* schedule cannot be optimal), otherwise, go to step 2.

Step 2 (To identify the set of 1-machine jobs):

Let T denote the "desired" total processing time of the operations reassigned from the nonsetup machine to the setup machine, where $T = (I_M - t_s)/2$.

From the initial *zero-setup* schedule in step 1, select[1] the operations on the nonsetup machine that have a combined total processing time "closest" to T. The jobs corresponding to the selected operations will be the 1-machine jobs; the rest form the 2-machine jobs.

Step 3 ("Fitting" the 1-machine jobs in the setup machine):

Obtain the partial *zero-setup* schedule for the 2-machine jobs using the Johnson's algorithm. Then, fit the 1-machine jobs into the partial schedule by sequencing them in the idle period of the setup machine so that their type X operations are processed first, followed by a setup, and then their type Y operations.

The Two-Setup Case

In the *two-setup* schedule, generally there are *three* job classes assigned on each machine; a set of 1-machine jobs, a subset of 2-machine jobs having type X operation on this machine, and the complementary subset of 2-machine jobs having type Y operation on this machine.

The scheduling heuristic for the *two-setup* case involves two steps. First, an initial schedule consisting *only* of 1-machine jobs for each machine is determined using a partitioning heuristic. The second step is a refinement that tries to improve on the initial schedule by balancing the workloads on the machines. This is achieved by identifying a subset of the 1-machine jobs which will give a more balanced workload if their operations are distributed so that each machine performs one operation—thus transforming them into 2-machine jobs. (A 0-1 KNAPSACK heuristic may be employed in this refinement step.)

TWO-SETUP (Heuristic 2)

Step 1 (Partition the 1-machine jobs):

Treat each job i as a 1-machine job with an integer value, $d_i = (c_i + c_{n+i})$, corresponding to the combined processing times of both its operations.

---
[1] We can use a 0-1 KNAPSACK heuristic for this selection, with $K = T$ (see, e.g., [8]).

Partition these n integers (on the two machines) so as to minimize

$$\left| \sum_{j \in S} d_j - \sum_{j \notin S} d_j \right|$$

with $S \subseteq \{1, 2, \cdots, n\}$, using the following *first-fit decreasing* heuristic for the bin-packing problem (see, e.g., [8]).

a) Sort the jobs such that $d_{[1]} \geq d_{[2]} \cdots \geq d_{[n]}$, where $[i]$ denotes the job number of the ith largest combined processing times in $\{d_1, d_2, \cdots, d_n\}$.

b) If all jobs have been assigned, then stop. Otherwise, assign the current largest job to the machine with the *smaller* interim total processing time. Repeat this until every job has been assigned.

Step 2 (Refinement to balance workload):
Let c_0 be the smallest processing time for all the operations, where $c_0 = \min \{c_i; i = 1, \cdots, 2n\}$.

Let T_P and T_Q be the total processing times of machines P and Q, respectively, and denote their difference by D, where $D = |T_P - T_Q|$.

a) Let machine M be the machine with maximum total processing time $T_M = \max(T_P, T_Q)$.
If there are *no* more 1-machine jobs on machine M, then stop.
Otherwise
  i) If $D < 2c$, then stop.
  ii) *Otherwise*, from the 1-machine jobs on machine M, identify the job j which has one of its operation times being "closet" to $D/2$.
  If $|c_j - D/2| < |c_{n+j} - D/2|$, then classify job j as a 2-machine job by distributing its operations so that its type Y operation remains on the same machine, but its type X operation is reassigned to the other machine.
  If $|c_j - D/2| > |c_{n+j} - D/2|$, then split job j so that its type X operation remains on the same machine, but its type Y operation is reassigned to the other machine. (Break ties arbitrarily.)

b) Update the values of T_P, T_Q, and D. Go to step 2a).

C. An Illustrative Example

We will illustrate the above heuristics with an example. Suppose we need to schedule five jobs, A to E, with processing times as shown in Table I. Let the setup time be 2 time units.

The *zero-setup* schedule gives a makespan of 26 time units, while the *one-setup* and *two-setup* schedules obtained from the above heuristics give makespans of 23 and 24 time units, respectively. Fig. 3 depicts these schedules with the sequence of the jobs at each machine, and the corresponding routing patterns for each job.

Briefly, the *one-setup* schedule was obtained (using Heuristic 1) as follows:

Step 1:
The *zero-setup* schedule gives the values of the two idle

TABLE I
JOB PROCESSING TIMES FOR THE ILLUSTRATIVE EXAMPLE

Operation Type	Job Name					Total
	A	B	C	D	E	
X	3	5	7	8	2	25
Y	4	2	1	5	6	18
Total	7	7	8	13	8	

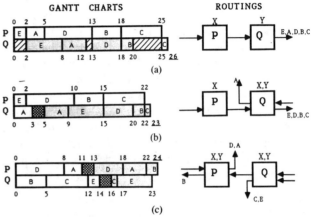

Fig. 3. The schedules and their corresponding routing patterns for the three subproblems of the illustrative example. (a) *Zero-setup* schedule. (b) *One-setup* schedule. (c) *Two-setup* schedule.

time periods, with $I_P = 1$ and $I_Q = 8$. Hence, machine Q is identified as the setup machine, and $I_M = I_Q = 8$.

Step 2:
Now, $T = I_M - t_s/2 = 3$. From those operations on the *non-setup* machine (P), we select job A as a (in this case, the only) 1-machine job because its type X operation time ($A_X = 3$) is "closest" to T.

Step 3:
Implementing the Johnson's algorithm for the 2-machine jobs $\{B, C, D, E\}$, we obtain the partial schedule which has an idle time of 8 units on the setup-machine Q. Rescheduling the operations on machine Q so that all existing idle time occurs before job E (the first job in Johnson's schedule) results in 9 time units becoming available to perform both operations of job A (the 1-machine job) plus the setup.

Thus fitting in job A on machine Q gives the final *one-setup* schedule of Fig. 3(b) with a makespan of 23 time units.

In the two-setup case, Heuristic 2 was used as follows:
Step 1:
Using the first-fit decreasing heuristic on the combined processing time for each job results in a schedule with jobs $\{D, A\}$ on machine P and jobs $\{B, C, E\}$ on machine Q.

Step 2:
Now, $c_0 = C_Y = 1$, and the difference between the total processing times of machines P and Q, $D = |T_P - T_Q| = 3$.

a) Machine Q is the machine with the larger total processing time and all the jobs are 1-machine jobs.
  i) Clearly, $D = 3 > 2c_0$.
  ii) Here we select job B which has its type Y operation of 2 time units being closest to $D/2 = 1.5$ time units.

We then split job B so that its type X operation remains on machine Q, but its type Y operation is reassigned to machine P.

b) The updated values for the machines are $T_P = 24$, $T_Q = 23$, and $D = 1$.

Going to step 2a, we now have $D < 2c$, so we stop.

The final *two-setup* schedule is shown in Fig. 3(c) with a makespan of 24 time units.

Specific to the above example, we remark that an exhaustive enumeration of all feasible solutions reveals that each of the schedules obtained in Fig. 3 has the shortest makespan corresponding to the *zero-setup, one-setup,* and *two-setup* cases, respectively. Among these, the *one-setup* schedule has the shortest makespan and hence is the optimal solution to this instance of 2-VFSP problem. The optimal routings of the jobs to the two machines are also given in Fig. 3(b).

Table II shows the improvements achieved when in-process magazine setups are allowed. The machine utilization ratio is defined as the ratio of machine utilization time to makespan. Observe that, in this example, both the average utilization increases and the makespan decreases when in-process magazine setups are allowed.

D. Some Empirical Tests

Johnson's (*zero-setup*) algorithm and the two heuristics were computer coded in Fortran 77 for empirical testing. Ten different joblists were randomly generated. The values for the "number of jobs" for each joblist, and for the processing times of the two types of operations for each job, were also randomly generated. These corresponding values ranged from 5 to 10 (for the number of jobs in a joblist), and from 1 to 10 (for the processing times).

These ten joblists were each tested with different values of setup times so as to correspond to varying degrees of machine versatility. Setup times of 1, $\cdots$, 8, 10, and 12 time units were considered. For each value of setup time and each joblist, the three schedules corresponding to the *zero-setup, one-setup,* and *two-setup* schedules were generated. Among these three schedules, the one with the minimum makespan was designated the "best" schedule.

Table III summarizes the results of our tests. For each value of the setup time that corresponds to a row in Table III, ten joblists were used. Thus the entries in the column for "best" schedule are from ten different joblists. The results reveal that the use of setups during scheduling may or may not be desirable depending on the magnitude of the setup time. When the setup time is small, the schedules that incorporate setup(s) tend to have shorter makespan than the corresponding *zero-setup* schedules. As the setup time increases, the reverse occurs; schedules without setups tend to be the "best." This observation is in line with one's intuition that "as setup time increases, incorporating setups in the schedule becomes less and less desirable."

A joblist parameter available prior to determining any schedule corresponds to the "difference between the two total processing times of type X and type Y operations" which we shall denote by D_{XY}. Our preliminary findings also suggest an interesting relationship between this joblist parameter and the number of setups in the "best" schedule for that joblist. Table IV illustrates this relationship.

In Table IV, each row corresponds to one of the ten joblists tested. The entries in the column for "best" schedule are the results of testing of ten different setup times for each joblist. In the *zero-setup* schedules obtained from Johnson's algorithm, we note that the maximum among the total idle periods of the two machines, IM, is directly correlated with D_{XY}. Since it is essentially this idle time period of the *zero-setup* schedule that is used to improve the makespan via setups, the schedules with setups appear to give better results when D_{XY} is large. This can be observed in Table IV, where we can see a gradual decrease in the number of times the *zero-setup* schedule is the "best," when D_{XY}, and hence IM, increases.

E. Some Further Remarks on the 2-VFSP

Note that we solved the 2-VFSP by solving all the three subproblems. It may sometimes be possible to solve the 2-VFSP without necessarily solving all three subproblems by using some useful bounds associated with the *one-setup* and *two-setup* schedules. We outline such a procedure below.

The lower bound for each of these schedules can be calculated from the joblist data as follows. For the *one-setup* schedule, this corresponds to L_1, where $L_1 = \lceil (T_P + T_Q + t_s + t_{min})/2 \rceil$; where $t_{min} = \min \{c_i, i = 1, \cdots, 2n\}$. And for

TABLE II
IMPROVEMENTS ACHIEVED WHEN IN-PROCESS SETUPS ARE ALLOWED

	Schedule Type (Makespan)					
	Zero-Setup (26)		One-Setup (23)		Two-Setup (24)	
Machine	P	Q	P	Q	P	Q
Machine utilization time	25	18	22	21	22	21
Utilization ratio	0.96	0.69	0.96	0.91	0.92	0.88
Average utilization	0.83		0.94		0.90	
Improvement in average utilization	(NA)		10.8%		6.9%	
Makespan improvement	(NA)		11.5%		7.7%	

TABLE III
SUMMARY OF THE EMPIRICAL TESTS

Setup Time	Number of Times the "Best" Schedule Has		
	Zero Setup	One Setup	Two Setups
1	0	5	5
2	0	6	4
3	2	6	2
4	4	5	1
5	5	4	1
6	5	4	1
7	5	4	1
8	7	3	0
10	9	1	0
12	9	1	0

TABLE IV
RELATIONSHIP BETWEEN D_{XY} AND THE NUMBER OF SETUPS USED IN THE "BEST" SCHEDULE

Difference D_{XY}	Idle Time IM	Number of Times the "Best" Schedule Has		
		Zero Setup	One Setup	Two Setups
0	3	8	0	2
1	4	8	0	2
5	6	7	3	0
7	8	7	3	0
7	8	6	4	0
9	10	3	7	0
11	12	2	6	2
11	12	2	8	0
12	12	3	0	7
21	23	0	8	2

the *two-setup* schedule, this corresponds to L_2, where $L_2 = \lceil (T_P + T_Q + 2t_s)/2 \rceil$.

The procedure first determines the *zero-setup* schedule and uses this makespan, M_0, to compare with the two lower bounds, L_1 and L_2. If $M_0 \leq \min \{L_1, L_2\}$, then the *zero-setup* schedule is optimal; otherwise, the *one-setup* schedule needs to be determined next. Then the *one-setup* makespan M_1 is compared with M_0 and L_2. If $M_1 < M_0$ and $M_1 \leq L_2$, then the *one-setup* schedule is optimal; otherwise, the *two-setup* schedule also needs to be determined. Thus with this checking of lower bounds, it is possible to solve the 2-VFSP without always solving the three subproblems together.

Finally, we remark that although in our empirical tests we have assumed finite setup times, new advancement in machine-tool technology and tool management systems may make it possible for magazine setups to become automated resulting in negligible setup times. For this case, in general, the optimal schedule is one that incorporates setup(s).

V. CONCLUSIONS

In this paper, we introduced a *concurrent* approach to scheduling a 2-machine FMC. The approach utilizes the machine versatility feature to integrate the tool management system into the scheduling system by allowing in-process "machine loading" through magazine setups. In essence, it solves both the routing and sequencing subproblems concurrently. To illustrate the utility of the concurrent approach, we defined in detail a specific two-versatile-machine flowshop scheduling problem, referred to as 2-VFSP.

We distinguish two types of setups, the tool exchange (or the spindle-setup) which is automatic, and the magazine setup which may or may not be automatic. The concurrent strategy adopted allows both types of setups to take place during a production cycle and solves both the routing and sequencing problems concurrently.

An alternative *sequential* strategy may be employed. This strategy allows only tool exchanges to take place and does not permit any in-process magazine setups. Here, production is scheduled so that no operation is assigned to a machine that does not have the tools (required for the operation) resident in the magazine. Such type of strategies has been reported for the scheduling of large-scale FMS's by Stecke [11], Kusiak [10], Berrada and Stecke [2], and Ammons *et al.* [1] among others, where the "loading," "routing," and "sequencing" problems are solved sequentially.

If the sequential strategy is employed in the 2-VFSP, then each machine would have been dedicated with only the tool-set of a particular operation type. This, in essence, transforms the 2-VFSP into the conventional 2-(dedicated) machine flowshop problem solvable by Johnson's algorithm—thereby giving only the *zero-setup* schedule. However, we remark that the "loading" problem for deciding on which tools to load onto the magazine becomes the critical problem in this sequential strategy.

In the theoretical study of 2-VFSP, we showed that an optimal schedule exists which has at most two in-process magazine setups—one on each machine. This gives rise to three possible configurations for an optimal schedule having zero, one, or two setups, respectively. However, we proved that to obtain the optimal schedule is an NP-complete problem. We, therefore, developed a heuristic approach with which the subproblems of determining the *one-setup* and *two-setup* schedules are solved approximately.

The heuristic was tested on sets of random joblists, for different values of setup times corresponding to varying degrees of machine versatility. Results showed that in-process magazine setups may be desirable especially if the setup time is small; by appropriately scheduling the consequent in-process magazine setups, the makespan of the resultant schedule may be shorter than the Johnson's (*zero-setup*) schedule.

Finally, although we have applied the concurrent approach to problems involving only two machines, we have gained valuable insights. The results for the 2-VFSP suggest that concurrent routing and sequencing more effectively captures the available machine versatility for FMC's consisting of versatile machines, and the *zero-setup* schedule obtained using the sequential scheduling strategy may not be optimal in such cases. This points to the need for further research on the concurrent approach for scheduling FMC's with a larger number of machines, that may incorporate appropriate in-process magazine setups when necessary.

REFERENCES

[1] J. C. Ammons, C. B. Lofgren, and L. F. McGinnis, "A large scale work station loading problem," in *Proc. 1st ORSA/TIMS Special Interest Conf. on FMS* (Michigan), pp. 249-256, 1984.

[2] M. Berrada and K. E. Stecke, "A branch-and-bound approach for machine loading," in *Proc. 1st ORSA/TIMS Special Interest Conf. on FMS* (Michigan), pp. 256-271, 1984.
[3] R. W. Conway, W. L. Maxwell, and L. W. Miller, *Theory of Scheduling*. Reading, MA: Addison-Wesley, 1967.
[4] J. S. Edghill and A. Davies, "Flexible manufacturing systems—The myth and reality," *Int. J. Adv. Manuf. Technol.*, vol. 1, no. 1, pp. 37-54, 1985.
[5] H. A. ElMaraghy, "Automated tool management in flexible manufacturing," *J. Manuf. Syst.*, vol. 4, no. 1, pp. 1-13, 1985.
[6] M. R. Garey and D. S. Johnson, *Computers and Intractability: A Guide to the Theory of NP-Completeness*. San Francisco, CA: Freeman, 1979.
[7] S. L. Hankins and V. P. Rovito, "A comparison of two tool allocation and distribution strategies for FMS," in *Proc. 1st ORSAS/TIMS Special Interest Conf. on FMS* (Michigan), pp. 272-277, 1984.
[8] T. C. Hu, *Combinatorial Algorithms*. New York, NY Addison-Wesley, 1982.
[9] S. M. Johnson, "Optimal two- and three-stage production schedules with setup times included," *Naval Res. Logistics Quart.*, vol. 1, pp. 61-68, 1954.
[10] A. Kusiak, "Loading models in flexible manufacturing systems," in *Proc. 7th Int. Conf. on Production Research* (Windsor, Ont., Canada), pp. 641-649, 1983.
[11] K. E. Stecke, "Formulation and solution of nonlinear integer production planning problems for flexible manufacturing systems," *Manag. Sci.*, vol. 29, pp. 273-288, 1983.
[12] ——, "Design, planning, scheduling, and control problems of flexible manufacturing systems," in *Proc. 1st ORSA/TIMS Special Interest Conf. on FMS* (Michigan), pp. 1-7, 1984.
[13] C. S. Tang and E. V. Denardo, "Models arising from a flexible manufacturing machine—Part I: Minimization of the number of tool switches," Working Paper No. 341, Graduate School of Management, UCLA, CA, 1986.
[14] ——, "Models arising from a flexible manufacturing machine—Part II: Minimization of the number of switching instants," Working Paper No. 342, Graduate School of Management, UCLA, CA, 1986.
[15] D. M. Zelenovic, "Flexibility: A condition for effective production systems," *Int. J. Production Res.*, vol. 20, pp. 318-337, 1982.
[16] J. Zygmont, "Flexible manufacturing systems: Curing the cure-all," *High Technology*, pp. 22-27, Oct. 1986.

Eng-Joo Lee (S'86) received the B.S. and M. Eng. degrees in computer and systems engineering from Rennsselaer Polytechnic Institute (RPI), Troy, NY, in 1984 and 1985, respectively. He is currently a doctoral candidate at RPI.

Since 1984, he has been a Research Assistant in the Information and Decision Systems Laboratory at RPI. Currently he is involved with the industry-sponsored FMS Program within the Center for Manufacturing Productivity and Technology Transfer at RPI. His research interests include combinatorial optimization and the application of operations research and systems engineering to robotics and automation.

Mr. Lee is a member of Eta Kappa Nu, Tau Beta Pi, ORSA, and ACM.

Pitu B. Mirchandani (M'60-SM'82) received the B.S. and M.S. degrees from the University of California at Los Angeles in 1966 and 1967, respectively, and the Sc.D. degree in operations research from Massachusetts Institute of Technology, Cambridge, in 1975.

He is currently a Professor in the Electrical, Computer, and Systems Engineering Department at Rensselaer Polytechnic Institute, Troy, NY. His research interests include optimization of stochastic systems, network analysis, combinatorial optimization, and the application of operations research and systems engineering to logistics, transportation, manufacturing, automation, and decision support systems. He is the co-author of *Location on Networks: Theory and Algorithms*. He is an associate editor for *Transportation Science*.

Dr. Mirchandani is a member of ORSA, TIMS, IIE, ACM, and the Mathematical Programming Society.

Selection of Process Plans in Automated Manufacturing Systems

ANDREW KUSIAK AND GERD FINKE

Abstract—Most of the planning models for automated manufacturing systems are based on the assumption that for each part there is only one process plan available. In this paper, we take the more realistic point of view that for each part a number of different process plans are generated, each of which may require specific types of tools and auxiliary devices such as fixtures, grippers, and feeders. A model for the selection of a set of process plans with the minimum corresponding manufacturing cost and minimal number of tools and auxiliary devices is formulated. The developed model for m parts is equivalent to the problem of finding the maximum clique in an m-partite graph with the minimum corresponding cost. Heuristic algorithms and numerical results are discussed.

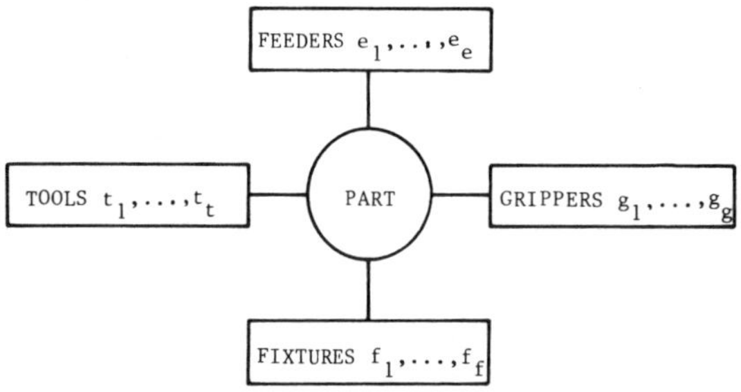

I. INTRODUCTION

AUTOMATED manufacturing systems possess a number of distinct features. One of these features relates to the number of auxiliary devices used such as (see Fig. 1):

fixtures for holding parts to be machined or assembled
grippers for handling parts
feeders for presenting parts.

Of course, parts are machined or assembled using tools whose design tends to be different from the design used in the classical manufacturing systems. The reason for that is due to the automated handling of tools. Typically, for a part to be manufactured in an automated manufacturing system, a number of different process plans are generated. Each process plan specifies requirements for tools, auxiliary devices, as well as operations (for example, machining operations) to be performed at a given cost. The process and production planners face the problem of selecting from the set of process plans a subset of process plans with the minimum number of fixtures, grippers, feeders, and tools and minimum corresponding total cost.

There are at least three reasons for solving the process plan selection problem:

a) reduction of the production cost,
b) limited capacity of tool magazines,
c) reduction of the number of types of auxiliary devices.

While the first two reasons seem obvious, the last requires some comments. It may appear that the reduction of the

Manuscript received February 17, 1987; revised October 23, 1987. Part of the material in this paper was presented at the IEEE International Conference on Robotics and Automation, San Francisco, CA, April 7–10, 1986.
A. Kusiak is with the Department of Industrial and Management Engineering, The University of Iowa, Iowa City, IA 52242.
G. Finke is with the Department of Applied Mathematics, Technical University of Nova Scotia, Halifax, NS, Canada B3J 2X4.
IEEE Log Number 8820158.

Fig. 1. Auxiliary devices and tools associated with a part.

number of different types of tools and auxiliary devices decreases the scheduling flexibility (see, for example, [1], [2]), however, it has an opposite effect. A manufacturing system with a limited number of tools and auxiliary devices is easier to schedule than a large system. Since in such a system the bottleneck tools and devices are relatively easy to determine, if required, the scheduling flexibility can be easily increased by using multiple tools and auxiliary devices. This option is always less costly than the development of unique tools and devices.

To simplify the presentation of the process plan selection problem in the subsequent sections of this paper, for a given part, rather than all three auxiliary devices, only one at a time is considered; namely, a fixture. In fact this complies with practice, where frequently

- fixtures are used to hold prismatic parts in machining systems,
- grippers are used for handling rotational parts in machining and assembly systems,
- feeders are used for presenting small parts for assembly.

In Section II of this paper the process planning problem is defined. Graph-theoretical and integer programming formulation of the process selection problem are presented in Section III. To solve the integer programming formulation, two heuristic algorithms are developed in Section IV. The two heuristics and computational results are presented in Section V. Final conclusions are drawn in Section VI.

II. THE PROCESS PLANNING PROBLEM

Before the process plan selection problem will be formulated, the process planning problem is briefly discussed.

There have been at least two books written on the process planning problem (see [3] and [4]). Extensive review papers have been published by Srinivasan and Liu [6], Weill [7], and Requicha and Vandenbrande [5].

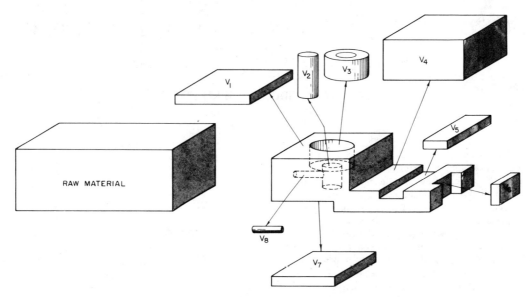

Fig. 2. A typical part with volumes $v_1, \cdots, v_8$ to be removed.

Although frequently in literature, and also in this paper, the process planning is discussed in the context of machining systems, all considerations presented hold for assembly systems as well.

To define a process plan the following notation is adopted:

- $V_l = (v_{l_1}, \cdots, v_{l_n})$ is a set of material volumes to be removed in one setup, i.e., without resetting a part on the currently used fixture or changing the fixture.
- V is the set of all volumes to be removed for a given part. Note that

$$V = \bigcup_{l=1}^{m} V_l$$

where m is the total number of setups.
- $T_l = (t_{l_1}, \cdots, t_{l_k})$ is a set of tools for removing V_l.
- f_l is a fixture for holding a part while removing V_l.

Based on the above notation, a process plan P_i is defined as a set of 3-tuples

$$P_i = \{(V_1; T_1; f_1), \cdots, (V_m; T_m; f_m)\}.$$

To illustrate this definition consider the following example.

Example 1

Given the three-dimensional part in Fig. 2 with material volumes $v_1, \cdots, v_8$ to be removed, generate sample process plans.

For the part in Fig. 2 one can generate the following two process plans:

$$P_1 = \{(v_1, v_4, v_5, v_6; t_1, t_2; f_1), (v_7; t_3; f_2),$$
$$(v_2, v_3, v_8; t_4, t_5, t_6; f_3)\}$$

$$P_2 = \{(v_4, v_5, v_6; t_1, t_7; f_4),$$
$$(v_1, v_7, v_2, v_3, v_8; t_1, t_4, t_5, t_6; f_4)\}. \bullet$$

The corresponding costs of removing material volumes $v_1, \cdots, v_8$ are c_1 and c_2, where usually $c_1 \neq c_2$.

Define for each process plan P_i, $i \in N$, the following incidence column vector:

$$x_i = [x_{1i}, \cdots, x_{ai}, x_{bi}, \cdots, x_{ci}]^T$$

where

$$x_{ti} = \begin{cases} 1, & \text{if a tool } t \text{ is used in } P_i, t \in \{1, \cdots, a\} \\ 0, & \text{otherwise} \end{cases}$$

$$x_{fi} = \begin{cases} 1, & \text{if a fixture } f \text{ is used in } P_i, f \in \{b, \cdots, c\} \\ 0, & \text{otherwise.} \end{cases}$$

For any two process plans P_i and P_j define the weighted Hamming distance

$$d_{ij} = \sum_{q=1}^{c} w_q \delta(x_{qi}, x_{qj}), \quad \text{for all } i \text{ and } j \qquad (1)$$

where

$$\delta(x_{qi}, x_{qj}) = \begin{cases} 1, & \text{if } x_{qi} \neq x_{qj} \\ 0, & \text{otherwise} \end{cases}$$

and w_q is the weight coefficient of the tool (or auxiliary device) q. The weighted Hamming distance d_{ij} measures dissimilarity between process plans P_i and P_j. One can note that assuming $w_q = 1$ for all q, the weighted Hamming distance (1) becomes the Hamming distance.

Due to the different importance of each tool (auxiliary device) the Hamming distance has been modified by introducing the weight coefficient w_q for each tool (auxiliary device) q. For example, the weight assigned to a fixture has typically a higher value than the weight assigned to a tool. A way to determine the value of the weight w_q, for all q, is to set it proportional to the cost of the tool (auxiliary device) q.

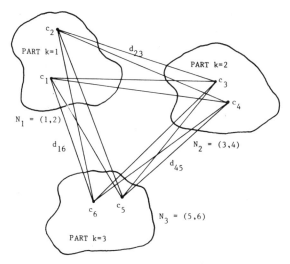

Fig. 3. A 3-partite graph.

III. FORMULATION OF THE PROCESS PLAN SELECTION PROBLEM

In this section two formulations of the process plan selection problem are presented:

1) a graph-theoretical formulation, and
2) an integer programming formulation.

A. Graph-Theoretical Formulation

In order to present a graph-theoretical formulation of the process plan selection problem, the following notation is introduced:

$K = \{1, 2, \cdots, m\}$ is the set of parts to be manufactured.

N_k is the set of process plans for a part k, $\forall k \in K$.

N is the set of all process plans (vertices), where

$$\bigcup_{k \in K} N_k = N.$$

A is the set of arcs connecting process plans (vertices) from set N_k to set N_l, $(k, l) \in K \times K$ and $k \neq l$.

d_{ij} is the weighted Hamming distance, $\forall (i, j) \in A$.

c_i is the cost of process plan P_i, $\forall i \in N$.

One can note that from the definition of N_k, $\forall k \in K$, and the set A, the resulting graph $G = (N, A)$ is a complete m-partite graph.

The process plan selection problem considers all sets $S = \{S_1, S_2, \cdots, S_m\}$ which contain one representative node S_k from each N_k.

Denote by $A^s \subseteq A$ the set of all (nondirected) arcs between the points of S. The objective is then to minimize the total cost

$$\sum_{(i,j) \in A^s} d_{ij} + \sum_{i \in S} c_i.$$

Briefly stating, the process plan selection model is to find the minimum cost maximum clique (S, A^s) in the m-partite graph (N, A).

An example of the 3-partite graph, corresponding to an $m = 3$ part problem for which three process plans are selected, is shown in Fig. 3.

B. Integer Programming Formulation

In addition to the notation introduced in the previous subsection, define two decision variables:

$$x_i = \begin{cases} 1, & \text{if process plan } i \text{ has been selected} \\ 0, & \text{otherwise} \end{cases}$$

$$y_{ij} = \begin{cases} 1, & \text{if process plans } i \text{ and } j \text{ have been selected} \\ 0, & \text{otherwise.} \end{cases}$$

The process plan selection problem is to choose, for each set of process plans (a part), only one representative in such a way that the sum of the distances among the selected process plans and the sum of their costs is minimized.

(IP) $$\min \frac{1}{2} \sum_{(i,j) \in A} d_{ij} y_{ij} + \sum_{i \in N} c_i x_i \qquad (2)$$

so that

$$\sum_{i \in N_k} x_i = 1, \quad k \in K \qquad (3)$$

$$x_i + x_j - 1 \leq y_{ij}, \quad (i, j) \in A \qquad (4)$$

$$x_i = 0, 1, \quad i \in N \qquad (5)$$

$$y_{ij} = 0, 1, \quad (i, j) \in A. \qquad (6)$$

Constraint (3) ensures that for each part exactly one process plan is selected. Consistency is imposed by constraint (4). Constraints (5) and (6) ensure integrality.

Note that since from the definition $y_{ij} = y_{ji}$ and the term $d_{ij} y_{ij} = d_{ji} y_{ji}$ appears twice in (2), a factor 1/2 has been introduced. One can see that due to the nature of constraint (5), the integrality of constraint (6) is redundant. Relaxing integrality of this constraint, problem (IP) is transformed into the following mixed integer programming problem:

(MIP) $$\min \frac{1}{2} \sum_{(i,j) \in A} d_{ij} y_{ij} + \sum_{i \in N} c_i x_i \qquad (7)$$

so that

(3), (4), (5)

$$y_{ij} \geq 0, \quad (i, j) \in A. \qquad (8)$$

IV. HEURISTIC ALGORITHMS

Practical process plan selection problems cannot be solved exactly in the form (IP) or (MIP) because of high computational complexity. Instead, two quick and efficient heuristic algorithms are developed.

A. Construction Algorithm A1

Consider a given sequence of the process plan sets $N_1, N_2, \cdots, N_m$. We want to construct the set $S = \{S_i\}$ following this sequence, i.e., first $S_1 \in N_1$ is chosen, then $S_2 \in N_2$, $\cdots$, and, finally, $S_m \in N_m$.

Let $S^{(k)} = \{S_1, S_2, \cdots, S_k\}$ denote the selected elements at stage k. The selection of process plans (nodes) is done as follows:

Step 0: Set $S^{(0)} = \phi$ and $k = 1$.
Step 1: [Selection of $S_k \in N_k$].
Determine the cost

$$C_i = c_i + \sum_{s \in S^{(k-1)}} d_{is}, \quad \forall i \in N_k.$$

Let S_k be the process plan with the minimum corresponding cost.
Step 2: Set $S^{(k)} = S^{(k-1)} + \{S_k\}$.
If $k = m$, exit ($S^{(m)} = S$ is the final set of process plans);
else set $k = k + 1$ and go to Step 1.

In Step 1, the best possible element $S_k \in N_k$ is chosen which leaves all previous elements $S^{(k-1)}$ fixed. Initially, $S_1 \in N_1$, is simply a process plan with the minimum corresponding cost $c_i, i \in N_1$. In general, the minimum cost represents exactly the expansion cost for a clique in the k-partite graph spanned by $N_1, N_2, \cdots, N_k$.

B. Exchange Algorithm A2

Algorithm A1 constructs a complete set S of process plans. An attempt is made to improve S by exchanging process plans within each group N_k, starting with N_1, N_2, etc.

Step 0: Set a counter $c = 0$ and $k = 1$.
Step 1: [Exchange in N_k].
Determine the minimum cost

$$C_i' = c_i + \sum_{s \in S - \{S_k\}} d_{is}, \quad \forall i \in N_k.$$

If the minimum is attained for an element $T_k \neq S_k$, exchange this pair, set $S = S - \{S_k\} + \{T_k\}$, and set $c = c + 1$.
Step 2: If $k = m$, stop if $c = 0$ and repeat from Step 0 if $c > 0$. If $k < m$, set $k = k + 1$ and go to Step 1.

Whenever an exchange leads to an improvement, the new element is adopted and the remaining sets N_k are checked. As long as one or more exchanges occurs during one full pass of algorithm A1, all sets N_k are searched again. At the end, a stable set of process plans is found. Algorithms A1 and A2 are based on the preselected sequence $N_1, N_2, \cdots, N_m$. This sequence is, in fact, arbitrary and may be replaced by $N_{\pi(1)}$, $N_{\pi(2)}, \cdots, N_{\pi(m)}$ for any permutation π of $1, 2, \cdots, m$. The solution procedure for the process plan selection problem consists therefore of the following three phases:

Phase 1: Generate a random permutation of process plan sets.
Phase 2: Apply algorithm A1 to obtain an initial set S of process plans.
Phase 3: Improve S using algorithm A2.

Several runs of the solution procedure should be performed to avoid low-quality local minima. Application of the developed procedure for solving the process plan selection problem is illustrated in Example 2.

Example 2

Assume the following data:

a) a set of parts $K = \{1, 2, 3, 4\}$,
b) sets of process plans $N_1 = \{1, 2\}$, $N_2 = \{3, 4, 5\}$, $N_3 = \{6, 7\}$, $N_4 = \{8, 9, 10\}$,
c) the incidence matrix for 4 parts (10 process plans).

$$\text{tools} \begin{cases} t_1 \\ t_2 \\ t_3 \\ t_4 \\ t_5 \end{cases} \text{fixtures} \begin{cases} f_1 \\ f_2 \\ f_3 \end{cases} \begin{bmatrix} \overbrace{1 \quad \;\; 1 \;\; 1}^{\text{part 1}} & \overbrace{1 \;\; 1 \;\; 1}^{\text{part 2}} & \overbrace{\;\;\;\; 1}^{\text{part 3}} & \overbrace{\;\;\;\;}^{\text{part 4}} \\ \end{bmatrix} \quad (9)$$

(incidence matrix with columns $P_1\ P_2\ P_3\ P_4\ P_5\ P_6\ P_7\ P_8\ P_9\ P_{10}$)

d) a vector of process plan costs

$$c_j = [5.8, 9.4, 11.6, 5.7, 3.4, 4.3, 5.1, 6.4, 5.2, 5.3],$$

e) a vector of weights

$$w_i = [1, 1, 1, 1, 1, 1, 1, 1]^T.$$

For the incidence matrix c) and weight vector e) calculate the matrix $D = [d_{ij}]$ of Hamming distances

$$D = \begin{bmatrix} \infty & \infty & 2 & 1 & 3 & 3 & 1 & 4 & 1 & 4 \\ & \infty & 5 & 6 & 4 & 4 & 8 & 5 & 6 & 5 \\ & & \infty & \infty & \infty & 1 & 3 & 4 & 3 & 4 \\ & & & \infty & \infty & 2 & 2 & 5 & 1 & 5 \\ & & & & \infty & 6 & 4 & 3 & 2 & 5 \\ & & & & & \infty & \infty & 5 & 4 & 3 \\ & & & & & & \infty & 3 & 2 & 3 \\ & & & & & & & \infty & \infty & \infty \\ & & & & & & & & \infty & \infty \\ & & & & & & & & & \infty \end{bmatrix}. \quad (10)$$

One can note that in matrix (10) for entries $(i, j) \in A$, $d_{ij} \geq 0$ and for entries $(i, j) \notin A$, $d_{ij} = \infty$.

For the above data the process plan selection problem is solved using the three-phase solution procedure.

Phase 1. A random permutation of process plan sets

$$N_{\pi(1)} = \{N_1, N_2, N_3, N_4\} \text{ is generated.}$$

Phase 2. Algorithm A1 is applied to obtain the initial set of process plans (nodes) S.
  Step 0: $S^{(0)} = \phi$, $k = 1$.
  Step 1: Since

  $$\sum_{s \in \phi} d_{is} = 0$$

  the costs c_i, $\forall i \in N_1$, are compared, i.e., $c_1 = 5.8$, $c_2 = 9.4$; hence $S_1 = 1$.
  Step 2: $S^{(1)} = \{1\}$, $k = 2$.
  Step 1: Compare

  $$c_i + \sum_{s \in S^{(1)}} d_{is}, \quad \forall i \in N_2$$

  i.e., $c_3 + d_{31} = 11.6 + 2 = 13.6$, $c_4 + d_{41} = 5.7 + 1 = 6.7$, $c_5 + d_{51} = 3.4 + 3 = 6.4$. The minimum cost defines $S_2 = 5$.
  Step 2: $S^{(2)} = \{1, 5\}$, $k = 3$.
  Step 1: Compare

  $$c_i + \sum_{s \in S^{(2)}} d_{is}, \quad \forall i \in N_3$$

  i.e., $c_6 + d_{61} + d_{65} = 13.3$, $c_7 + d_{71} + d_{75} = 10.1$; hence $S_3 = 7$.
  Step 2: $S^{(3)} = \{1, 5, 7\}$, $k = 4$.
  Step 1: Compare

  $$c_i + \sum_{s \in S^{(3)}} d_{is}, \quad \forall i \in N_4$$

  i.e., $c_8 + d_{81} + d_{85} + d_{87} = 16.4$, $c_9 + d_{91} + d_{95} + d_{97} = 10.2$, $c_{10} + d_{10.1} + d_{10.5} + d_{10.7} = 17.3$; hence $S_4 = 9$.
  Step 2: The selected set of process plans $S = S^{(4)} = \{1, 5, 7, 9\}$ with the corresponding total cost $c_1 + c_5 + c_7 + c_9 + d_{15} + d_{17} + d_{19} + d_{57} + d_{59} + d_{79} = 32.5$.

Phase 3. Algorithm A2 is used to improve the set of process plan S.
  Step 0: Set $c = 0$ and $k = 1$.
  Step 1: Compare

  $$c_i + \sum_{s \in \{5,7,9\}} d_{is}, \quad \forall i \in N_1$$

  i.e., $c_1 + d_{15} + d_{17} + d_{19} = 10.8$, $c_2 + d_{25} + d_{27} + d_{29} = 27.4$. The set S remains unchanged.
  Step 2: Set $k = 2$.
  Step 1: Compare $c_3 + d_{31} + d_{37} + d_{39} = 19.6$, $c_4 + d_{41} + d_{47} + d_{49} = 9.7$, $c_5 + d_{51} + d_{57} + d_{59} = 12.4$. Thus $T_2 = 4$ which replaces $S_2 = 5$.
  A new set $S = \{1, 4, 7, 9\}$ with a corresponding total cost 29.8 is constructed.
  Set $c = 1$.

The algorithm continues for $k = 3$ and $k = 4$, without changing the current set S. Since $c > 0$, the set S has been altered, and the algorithm restarts from Step 0 with the current set S. Again, no changes take place since the set S is optimal.

Here algorithms A1, A2 use the sequence of process plan sets N_1, N_2, N_3, N_4, i.e., the identity permutation $\pi(i) = i$, $i = 1, 2, 3, 4$. The heuristics may be based on any of the $m! = 24$ arrangements of the process plan sets. It is interesting to note that there are $2 \times 3 \times 2 \times 3 = 36$ possible ways of selection sets S of process plans. The optimal solution is represented by the following set of process plans:

$$\{P_1, P_4, P_7, P_9\}$$

with a corresponding total cost 29.8. One may use any of the $m! = 24$ permutations to initiate the heuristics. It is interesting to note that in five of these permutations, algorithm A1 already produces the optimal solution. However, for all permutations, algorithm A2 terminates with the optimal solution (which, of course, cannot be expected, in general, for large-scale problems). As one can see from matrix (9) the above set of process plans requires the subset $\{t_1, t_3, t_4\}$ of tools and a fixture f_1. As compared to the worst case solution, our solution results in the savings of the tools $\{t_2, t_5\}$ and fixtures $\{f_2, f_3\}$.

V. Computational Results

In order to analyze the performance of algorithms A1 and A2, random problems of various sizes (n, m) were generated. Here $n = |N|$ is the total number of process plans and $m = |K|$ is the total number of parts. The costs c_i, for each $i \in N$, were uniform in interval $(0, 10)$ and the Hamming distances were random integers in $[1, 10]$. Table I summarizes the results obtained.

Algorithm A1 is very efficient, its computation time $c(A1)$ is 3 to 9 times less than the time $c(A2)$ of algorithm A2 for the range of problem sizes considered. Nevertheless, the total run time $c(A1) + c(A2)$ is only about 0.5 s per random permutation for the largest problem of size (100, 40) that has been solved. The improvement of solution quality from 3 to 8 percent seems to make the application of algorithm A2 worthwhile.

The last two columns of Table I demonstrate the performance of the two heuristics. A total of 20 random permutations were generated. The "First Occurrence" column lists the average number of runs required to detect, for the first time, the best solution of the 20 repetitions. As indicated in the last column of Table I, the best solution occurs rather frequently in the first run of the solution procedure. Also most of the other solutions are good quality ones. Their values are only about 1 percent off the value of the best solution. Although several runs are certainly advisable, a small number of reruns (five to ten) seems sufficient.

In order to analyze the influence of parameter fluctuations, more extreme random problems were generated. The costs c_i were uniformly distributed real numbers in $(0, 100)$ and the Hamming distances were random integers in $[1, 100]$. The finding in Table II are similar to Table I with no increase in the computation time. Here, the repeated occurrence of the best solution is even more remarkable because of the wider intervals for costs and distances. All best solutions found

TABLE I
PERFORMANCE OF ALGORITHMS A1 AND A2
(Average Values for 5 Instances of Each Problem Solved.)

Size (n, m)	Comparison $c(A2)/c(A1)$	Improvement by A2(%)	Run Time* (s)	Best Solution (20 runs)	
				First Occurrence	Frequency
(25, 5)	3.3	8.2	0.014	2.6	10.2
(50, 10)	4.8	7.1	0.055	1.8	12.5
(50, 15)	5.9	4.1	0.100	1.2	8.0
(100, 20)	6.6	4.2	0.280	5.6	3.8
(100, 25)	7.7	5.3	0.380	2.2	7.2
(100, 30)	8.6	3.9	0.460	7.2	5.0
(100, 40)	8.9	2.8	0.570	2.2	10.6

* CDC CYBER 170-720.

TABLE II
PERFORMANCE FOR MODIFIED DATA
(Average Values for 5 Problems Solved.)

Size (n, m)	Best Solution (20 runs)	
	First Occurrence	Frequency
(100, 20)	5.4	2.6
(100, 25)	3.4	4.8
(100, 30)	5.8	4.4
(100, 40)	2.4	6.6

appear to be optimal or very close to optimality. They could not be further improved by even a large number of additional runs.

VI. CONCLUSIONS

The process plan selection model can be applied in two phases of manufacturing:

1) design
2) planning (management).

Process plans in large-scale manufacturing systems are designed by a number of different process planners. Each of them creates requirements for tools and auxiliary devices such as fixtures, grippers, and feeders. Solving the process plan selection model in the design phase may result in significant savings due to the reduction of the total number of tools and auxiliary devices. For example, a single fixture may cost more than $5000.

Planning of automated systems is known for its high computational complexity. Since the process plan selection model reduces the overall number of tools and auxiliary devices, it simplifies the planning problem.

Although the process plan selection model was found to be a complex combinatorial problem, two efficient heuristic algorithms were developed. Computational experience has confirmed that solutions generated by the two heuristics are close to optimality.

REFERENCES

[1] A. Kusiak, "Flexible manufacturing systems: A structural approach," *Int. J. Prod. Res.,* vol. 23, pp. 1057–1073, 1985.
[2] A. Kusiak, A. Vannelli, and K. R. Kumar, "Grouping problem in scheduling flexible manufacturing systems," *Robotica,* vol. 3, pp. 245–252, 1985.
[3] T. C. Chang and R. Wysk, *An Introduction to Automated Process Planning Systems.* Englewood Cliffs, NJ: Prentice-Hall, 1985.
[4] G. Halevi, *The Role of Computers in Manufacturing Processes.* New York, NY: Wiley, 1980.
[5] A. A. G. Requicha and J. Vandenbrande, "Automated systems for process planning and part programming," in A. Kusiak, Ed, *Artificial Intelligence: Implications for Computer Integrated Manufacture.* New York, NY: Springer-Verlag, 1988.
[6] R. Srinivasan and C. R. Liu, "Evolutionary trends in generative process planning," in U. Rembold and R. Dillmann, Eds, *Methods and Tools for Computer Integrated Manufacturing.* New York, NY: Springer-Verlag, 1983, pp. 179–193.
[7] R. D. Weill, "Present tendencies in computer aided process planning," in *Proc. PROLMAT Conf.,* pp. 155–168, 1985.

Andrew Kusiak received the B.Sc. degree in precision engineering and the M.Sc. degree in mechanical engineering, both from the Warsaw Technical University, Warsaw, Poland, and the Ph.D. degree in operations research from the Polish Academy of Sciences, Warsaw.

He is a Professor and Chairman of the Department of Industrial and Management Engineering at the University of Iowa, Iowa City. His primary research interests are in artificial intelligence, operations research, and manufacturing systems. He is a member of the editorial board of the *International Journal of Advanced Manufacturing Technology, International Journal of Computer Integrated Manufacturing, International Journal of Production Research,* and *Robotica.* He is also the editor of two book series, Artificial Intelligence in Industry (IFS and Springer, New York) and Applied Artificial Intelligence (Taylor & Francis, UK).

Gerd Finke received the M.Sc. and Ph.D. degrees in mathematics from the University of Kiel, Kiel, West Germany in 1969.

After graduation he joined the Faculty of Engineering of the Technical University of Nova Scotia, Halifax, Canada, where he is currently Professor in the Departments of Applied Mathematics and Industrial Engineering. His main interests are in combinatorial optimization and network design.

Chapter 7: Additional Readings in Manufacturing Systems Engineering

7.1: Introduction

This chapter surveys additional literature related to the modeling and control of automated manufacturing systems. It takes a brief look at artificial intelligence in manufacturing, automatic error recovery for manufacturing systems, modeling and simulation tools, applications, and research issues.

The chapter is intended to serve as suggestions for further reading. Some of the topics represent relatively well developed areas (modeling and simulation), and others are emerging and fruitful areas of research (automated error recovery).

7.2: Artificial Intelligence in Manufacturing

The shop floor is a dynamic environment where the unexpected continuously occurs, forcing changes to planned activities. Hence, plans must be altered during the manufacturing process. These decisions are being automated with the help of some techniques from artificial intelligence. Bourne and Fox [1] present a good overview and introduction to the issues and concepts that are needed to achieve autonomous manufacturing.

Autonomous manufacturing refers to the complete automation of decision making on the shop floor. This includes process planning, process selection, process sequencing, shop level scheduling, and monitoring and control. These decisions are both *predictive* and *reactive*. Planning and scheduling are predictive decisions while reacting to errors in a manufacturing cell falls into the second category.

The problem of scheduling orders in a job-shop raises a number of issues of interest to the artificial intelligence community including [1]

- Knowledge representation semantics for organization modeling
- Extending knowledge representation techniques to include the variety of constraints found in the scheduling domain
- Integrating constraints into the search process
- Relaxing constraints when conflict occurs
- Analyzing the interaction between constraints to diagnose poor solutions

The scheduling problem can be viewed from a constraint directed reasoning perspective. Constraint-directed reasoning is a heuristic search technique in which domain knowledge is represented as constraints that bound and guide the search. For example, in scheduling, the next step of an operation can be viewed as a precedence constraint and the due date for an order as the goal constraint.

Process selection may also be automated by means of constraint-directed reasoning techniques and knowledge-based systems. Knowledge-based systems can also be used for sequencing [4–6].

Handling the changes that occur on the shop floor is the major goal of these software systems. One technique for determining the effect of change focuses on goal-directed rule-based processing. For each category of error, a set of rules is defined that specifies the procedure to be followed. Another approach embeds dependency links in the shop model that define how the information was derived and what would have to be done to re-derive it. The specification of the derivation procedure defines the work required to correct the situation. In [1], the authors present two systems that are based on some of these ideas.

7.2.1: ISIS

ISIS is a constraint-directed reasoning system that addresses the problem of how to construct accurate, timely, realizable schedules, and manage their use in job-shop environments. The system constructs schedules by performing a hierarchical, constraint-directed search in the space of alternative schedules. The search is divided into four levels: order selection, capacity analysis, resource analysis, and resource assignment. ISIS starts with a null schedule and alternative partial schedules are generated either forward from the start date or backward from the due date. The system provides [1]

- A knowledge-representation language for modeling organizations and their constraints
- Hierarchical, constraint-directed scheduling of orders
- Analytic and generative constraint relaxation
- Techniques for the diagnosis of poor schedules

7.2.2: Transcell

Transcell (transportable cell) is a system designed to manage a wide variety of machines, built from different technologies, all of which communicate in different languages. It is a generic operating and database system that reacts to this overwhelming problem. More details can be found in [1].

7.3: Automated Error Recovery

In recent years, artificial intelligence techniques have been explored for automating error recovery in manufacturing systems [7,8]. We define *automatic error recovery* as the process by which a system returns itself to normal operation after conditions exist that generally hinder the intended purpose of the system. For example, consider a vibrating bowl feeder that feeds screws to an assembly station [9]. The feeder automatically clears jammed parts by ejecting them with air jets or by using an air-driven plunger to push the screws back into the bowl. In either case, the jam is detected and action is taken to clear the jam. Hence, the feeder is capable of automatic error recovery. Without some form of automatic error recovery the average uninterrupted running time would go from 80 minutes to a brief 2.02 minutes [9]!

A common approach to error recovery is to hard-code routines that will handle every conceivable error. This results in enormous programs in which the error recovery code comprises 80–90 percent of the workstation code [10,11]. An alternate approach is to incorporate machine intelligence to supplement such hard-coded programs.

7.3.1: Approaches to Automated Error Recovery

Even if 80–90 percent of one's effort is dedicated to writing error recovery routines for an uncertain environment, then even that great effort will be inadequate. As a result, researchers have begun to look for ways to reduce the effort involved in developing these automatic error recovery routines.

7.3.2: Zero-Based Error State Programming

This term is used to describe the approach taken by Williams, Rogers, and Upton [12]. They have developed a cell covered with binary sensors. The bit string produced by the sensors uniquely identifies the present state of the system.

Their system starts with no task code whatsoever, hence the name zero-base. Error states are defined as situations where the controller has no prescribed action. Therefore, when the system is initialized, every state is an error state.

The system handles error states by asking the operator to select a correct action for task continuation. The system then stores the operators's response and repeats the action whenever that state is encountered again. In this approach, the programming process and the generation of error recovery routines are automated.

7.3.3: Frame-Based Approaches

Frames are a relational data structure that have been applied to the problem of automatic-error recovery [13–15]. Each frame may be considered as a node in a network of relations. The top levels contain fixed information while the lower levels contain slots that must be filled by specific instance or data. The frames allow a contextual explanation of all of the work-cell activities, thus forming a substrate for the essential diagnosis of work-cell-task errors, which ultimately leads to the recovery procedure.

The frame slots determine what actions are to be performed on the environment and what actions are to be received from the environment. In their system [13-15], developing a work-cell program begins with a list of available tasks and a user selecting, from the list, the desired tasks that will enable the job to be accomplished. From their frame point of view, the user is basically building a job frame that points to the various task frames. Action frames may then point to either efferent frames and/or perception frames. Efferent frames point to command frames, which point to interface frames, which actually make something happen. Perception frames are associated with sensor frames, which are associated with interface frames, which actually collect data from the sensors.

7.3.4: Failure Reason Analysis

Srinivas has developed an approach [16,17] for handling error recovery in robots. Srinivas considers a task to be a sequence of transformations that changes the world from an initial state S(0), into a goal state S(g). Along the way, any particular action A(j) transforms the present state of the system S(n) into the next state S(n+1) that eventually leads to the goal state S(g). When an error is detected, it means that the system has recognized that action A(j) has not produced the expected next state S(n+1) but instead has resulted in some failure state S(f).

Srinivas [17] points out that traditional planning systems treat the failure state S(f) as though it were some arbitrary state and respond to the failure by trying to develop a plan that will move the system to the "nearest" intermediate subgoal. He observes that such an approach does not take advantage of the information implicit in the reason for the failure. Working along these lines, Srinivas' failure reason analysis starts by asking "Why did A(j) fail?" Was it due to some operational or informational error? Was some precondition for the action not met? Or, did the action fail to meet a desired constraint? The idea is that once the cause of the error has been found, the solution is close at hand.

Srinivas' system has been implemented on a JPL robot, and basic success of the technique has been demonstrated.

7.3.5: Semantic Extraction and Failure Reason Analysis

Semantics and the operation's intent are an important part of Gini's system [18–21] for error interpretation and error recovery. Their system is based on Srinivas' failure reason analysis technique and uses semantic information to help constrain the search for the possible causes of an error.

This system starts with a manipulator-level task program. A preprocessor operates on the task program, by applying pattern-matching techniques to extract the program's semantics. This process generates an augmented program, (AP). The AP processor executes the AP, monitors the execution of the robot program, and maintains a history of the robots activities called the event trace.

In all of these systems, the problem centers around the ability to acquire new knowledge. There are at least two learning techniques that may be helpful in extending the a priori knowledge bases of the methods: observation learning and failure driven learning. Each of these techniques have their origin in explanation-based learning.

7.3.6: Explanation-Based Learning Systems [22]

These revolve around the development of an explanation, sometimes called a justification or proof, that indicates the information, features, and inference rules that were needed to solve a sample problem. This is done by constructing a causally complete representation of the scenario. Any crucial information missing from the instance must be inferred and the causal relations between components must be discovered and made explicit. Unlike most current understanding systems, the understanding process of an explanation based learning system, is unique in that it must maintain data dependency links connecting each representation with all of the inference rules from the domain model used to justify the scenario during the understanding process, including causal information, goal enablements, planning information, and so forth. The amalgam of all the data dependency links in the understood representation is called the inference justification network.

One key problem for a learner, be it either human or computer, is to identify which features of a particular event should be expected to appear in future events. Explanation-based learning solves this problem, in that the generalized scenario retains only the features that were necessary to produce the explanation, within the current domain theory. The relevant features are then precisely those features that enable the causal and intentional relationships to be inferred. With this background, we can now consider the following two applications of explanation-based learning.

Observation learning: Segre and DeJong are trying to replace robot teaching with robot learning by working on a robot planning system that can acquire planning concepts through observation [23]. They have pursued this goal, with some success, by building on work done in explanation based learning.

In their system, when a new structure is to be built the user must first specify a goal for the process; the goal includes a functional specification of the assembly. Then via low-level gripper primitive commands, the system observes the user driving a conventional arm to task completion. For the observation to be useful, the system starts with some degree of background knowledge that enables it to understand the observed sequence of low-level gripper primitives.

The justification analyzer constructs a causally complete explanation of how the observed gripper commands achieved the specified goal. Besides developing the causal links interrelating the input actions, the explanation includes a set of constraints crucial to successful task completion. The generalizer applies domain-specific background knowledge to the explanation, by generating a generalized task scheme that does not reflect the peculiarities of the observed problem solving episode.

The newly created task scheme is placed in a library that does not distinguish between built-in and acquired schemata. The library of schemata is indexed by the goals that the various schema achieve. Future interaction, can then invoke the planner, which when presented with a new situation, will select a schema from the library and formulate its preconditions as a series of subgoals to be achieved.

Failure driven learning: Pazzani has applied failure-driven learning to the development of an expert system for the diagnosis of failures in the attitude control of the DSCS-III satellite [24]. The attitude control expert system (ACES) integrates model-based and heuristic-based diagnosis.

The initial diagnostic heuristics are simple, often definitional in nature. The heuristics examine atypical features and hypothesize potential faults. Since the initial heuristics are so simple, they will often hypothesize faults that will be denied by the device models—this is called a hypothesis failure. When a hypothesis failure occurs, the reason for the improper diagnosis is noted, and the heuristic that suggested the nonexistent fault is revised so the hypothesis will not be proposed again, in future similar cases.

Pazzani notes that failure-driven learning dictates two important facets of learning: When to learn—when a hypothesis failure occurs; and, What to learn—features that distinguish a fault in one component from faults in other components. What is not specified is how to learn, and it is here that he calls on the concepts of explanation-based learning. Through the techniques of explanation-based learning, ACES learns how to avoid a hypothesis failure after just one example. This is done by finding the most general reason for the hypothesis failure and then modifying the accusing heuristic.

The work [24] is fully implemented in a combination of Lisp and Prolog on a Symbolics 3600. He has conducted some experiments with the system, and the results are impressive. He has found that his naive expert system after learning experience was able to out perform an expert system built by using rules based on a human's expert advice.

7.4: Modeling and Simulation

Simulation is a widely used tool for designing manufacturing systems, for evaluating scheduling algorithms, and for computing performance measures [25-50]. The motivation for using simulation stems from the fact that manufacturing systems are very expensive to design. Once they are implemented, it is sometimes costly to make changes. The operation of manufacturing systems is not so easy to predict because of contention of service, limited waiting areas, parallel operations, multiple interactions, and complex decision mechanisms [48]. Performance modeling, although a combination of simulation and analysis, is one approach to the design of automated manufacturing systems.

Suri et al. have developed MANUPLAN [28-30], which uses queueing models to do a rough-cut dynamic analysis of the manufacturing system. It calculates resource utilization, work-in-process, and lead time for parts and resources. Manufacturing system parameters can be chosen with this software to refine system performance. Recently [30], MANUPLAN has been integrated into a software system for the rapid design and analysis of manufacturing systems. Their approach integrates four modeling methods: spreadsheets (Lotus 1-2-3), queueing models (MANUPLAN II), simulation models (SIMAN), and animation (CINEMA). This approach covers the entire design process from the rough-cut analysis to the final design. An application to a metal part manufacturing cell consisting of multiple machines for five machining operations has been reported [30]. The model development time was less than one day.

Simulation software, based on conventional packages such as SLAM [31], MAP/1 [27], and GPSS [47], has also been developed. Others are based directly on queueing networks such as SPEEDS [41] and RESQ [48].

The research queueing package (RESQ) is a tool developed by IBM to construct and solve models of systems with jobs contending for service from many resources. RESQ uses modeling elements to represent the system. These consist of service centers, customers, nodes, and chains. The models are analyzed by using two possible solution methods. An analytic solution solves equations to determine the performance measures. Simulation can also be used to create a statistical experiment, which observes the behavior of the model and that generates the performance measures from these observations. RESQ has been used to analyze and compare four work-in-process policies on an assembly line: (1) a push system, (2) a pull system, (3) a transfer line, and (4) a closed loop system.

Hardware simulations are also being developed to represent manufacturing systems [33,35,36]. They are an alternative to existing analytic and numerical simulation methods, by using special-purpose hardware simulators that exhibit a direct correspondence between the movement of electronic pulses through a digital circuit and the movement of pieces in a production network. Hardware modules for machines and buffers have been developed. The main advantage of this aproach has been demonstrated by a hardware simulator that has achieved high simulation rates and good statistical performance. In addition, these simulators can represent production rates, random failures and repairs, and finite buffer capacities. Hardware simulators present the potential for inexpensive, reconfigurable simulators that can model a variety of production networks [36]. Comparisons have been made with the two-machine one-buffer transfer line of Gershwin and Berman [Chapter 2].

Physical simulators offer still another approach [34]. They are based on actual scaled-down equipment that has been configured to model a flexible production line. Scaled-down versions of mills, lathes, robots, conveyors, and automatic storage and retrieval systems can all be computer controlled through a personal computer. These models are ideal tools for educational purposes, and they are also efficient and inexpensive ways to analyze and test actual systems prior to their construction [34]. Scale models can also be used as on-line simulators to answer "what if?" questions for the actual system under different operating policies.

Several applications and case studies that use simulation have appeared in the literature [25,31,37–39,48].

7.5: Applications

There is a lot of discussion in this area on getting started, or converting to flexible and automated manufacturing systems [51,53]. Specific applications have been reported in the aircraft industry [52], the integrated circuits area [55–58], and for in-process storage systems [60]. Also, Wittrock [54] presents some results for scheduling algorithms in flexible flow lines at IBM.

It should also be pointed out that many other applications have appeared throughout this tutorial and are not repeated here (e.g., Petri net controllers, transfer lines (Ho), printed-circuit-card-assembly lines (Akella, Gershwin, and Choong)).

7.6: Research Issues

Wemmerlov and Hyer [64] and Sinha and Hollier [65] present two thorough reviews of future directions in cellular manufacturing. Also, Cassidy et al. [Chapter 1, Ref. 28] review the research needs in manufacturing systems.

Finally, the policy of just-in-time (JIT) manufacturing [61–63] is receiving considerable attention as a result of its success in Japan. It is noteworthy that a review paper [63] has specifically addressed the JIT strategy for small manufacturers. It is very likely that the future will concentrate on automation, integration, and flexibility in manufacturing for small business.

7.7: References

7.7.1: Knowledge Based Systems for Manufacturing

1. D.A. Bourne and M.S. Fox, "Autonomous Manufacturing: Automating the Job-Shop," *IEEE Computer*, Sept. 1984, pp. 76–86.
2. C. Fellenstein, C.O. Green, L.M. Palmer, A. Walker, and D.J. Wyler, "A Prototype Manufacturing Knowledge Base in Syllog," *IBM Journal of Research and Development,* Vol. 24, No. 4, July 1985, pp. 413–421.
3. A.C. Kak, K.L. Boyer, C.H. Chen, R.J. Safranek, and H.S. Yang, "A Knowledge-Based Robotics Assembly Cell," *IEEE Expert*, Spring 1986, pp. 63–83.
4. A. Kusiak, "FMS Scheduling: A Crucial Element in an Expert System Control Structure," *Proceedings of the 1986 IEEE International Conference on Robotics and Automation*, San Francisco, Calif., April 1986, pp. 653–668.
5. J. Erschler and P. Esquirol, "Decision-Aid in Job Shop Scheduling: A Knowledge Based Approach," *Proceedings of the 1986 IEEE International Conference on Robotics and Automation*, San Francisco, Calif., April 1986, pp. 1651–1656.
6. E. Bensana, M. Correge, G. Bel, and D. Dubois, "An Expert System Approach to Industrial Job-Shop Scheduling," *Proceedings of the 1986 IEEE International Conference on Robotics and Automation*, San Francisco, Calif. April 1986, pp. 1645–1650.

7.7.2: Automated Error Recovery

7. P.J. Fielding, F. DiCesare, G. Goldbogen, and A. Desrochers. "Intelligent Automated Error Recovery in Manufacturing Workstations," *Proceedings of the 1987 IEEE International Symposium on Intelligent Control*, IEEE Computer Society Press, Washington, D.C., 1987, pp. 280–285.
8. P.J. Fielding, F. DiCesare, and G. Goldbogen, "Error Recovery in Automated Manufacturing through the Augmentation of Programmed Processes," *Journal of Robotic Systems*, Vol. 5, No. 4, 1988, pp. 337–362.
9. H.P. Wiendahl and W. Ziersch, "Increasing the Availability of Assembly Systems," *Assembly Automation*, Vol. 5, No. 4, 1985, pp. 217–224.
10. A.L. Giacobbe, "Diskette Labeling and Packaging System Features Sophisticated Robot Handling," *Robotics Today*, Vol. 6, No. 2, 1984, pp. 73–75.
11. M. Gini and R. Smith "Reliable Real-time Robot Operation Employing Intelligent Forward Recovery," *Technical Report TR 85-30*, University of Minnesota, Minneapolis, Minn., Sept. 1985.
12. D.J. Williams, P. Rogers, and D.M. Upton, "Programming and Recovery in Cells for Factory Automation," *The International Journal of Advanced Manufacturing Technology*, Vol. 1, No. 2, 1986, pp. 37–47.
13. M.H. Lee, D.P. Barnes, and N.W. Hardy, "Research into Error Recovery for Sensory Robots," *Sensor Review*, Oct. 1985, pp. 194–197.
14. M.H. Lee, D.P. Barnes, and N.W. Hardy, "Error Recovery in Robot Applications," *Proceedings of the 6th British Robot Associations Conference*, May 1983, pp. 217–222.
15. N.W. Hardy, M.H. Lee, and D.P. Barnes, "Knowledge Engineering in Robot Control," *Proceedings Expert Systems 83*, 1983, pp. 70–77.
16. S. Srinivas, "Error Recovery in Robots through Failure Reason Analysis," *AFIPS Conference Proceedings, 47, 1978 National Computer Conference*, AFIPS Press, Reston, Va., June 1978, pp. 275–282.
17. S. Srinivas, "Error Recovery in Robot Systems," Ph.D. Thesis, California Institute of Technology, Pasadena, Calif. 1977.
18. R.E. Smith and M. Gini, "Robot Tracking and Control Issues in an Intelligent Error Recovery System," *Proceedings of the 1986 IEEE International Conference on Robotics and Automation*, IEEE Computer Society Press, Washington, D.C., 1986, pp. 1070–1075.
19. M. Gini, R. Doshi, S. Garver, M. Gluch, R. Smith, and I. Zualkernan, "Symbolic Reasoning as a Basis for Automatic Error Recovery in Robots," *Technical Report TR 85-24*, University of Minnesota, Minneapolis, Minn., July 1985.
20. M. Gini and G. Gini, "Towards Automatic Error Recovery in Robot Programs," *Proceedings of the 8th International Joint Conference on Artifical Intelligence*, Karlsruhe, West Germany, Aug. 1983, pp. 821–823.
21. M. Gini, R. Doshi, M. Gluch, R. Smith, and I. Zualkernan, "The Role of Knowledge in the Architecture of a Robust Robot Control," *Proceedings of the 1985 IEEE International Conference on Robotics and Automation*, IEEE Computer Society Press, Washington, D.C., March 1985, pp. 561–567.
22. M.J. Pazzani, "Explanation Based Learning for Knowledge-Based Systems," *Proceedings of the Knowledge Acquisition for Knowledge-Based Systems Workshop*, Banff, Alberta, Canada, Nov. 1986, pp. 34-0—34-17.
23. A.M. Serge and G. DeJong, "Explanation-Based Manipulator Learning: Acquisition of Planning Ability through Observation," *Proceedings of the 1985 IEEE International Conference on Robotics and Automation*, St. Louis, Mo., March 1985, pp. 555–560.

24. M.J. Pazzani, "Refining the Knowledge Base of a Diagnostic Expert System: An Application of Failure Driven Learning," *Proceedings of the Fifth National Conference on Artificial Intelligence*, Morgan Kaufman Publishers, Los Altos, Calif., Aug. 1986, pp. 1029–1035.

7.7.3: Modeling and Simulation

25. R. Akella, J. Bevans, and Y. Choong, "Simulation of a Flexible Manufacturing System," *Laboratory for Information and Decision Systems Report No. LIDS-P-1435*, M.I.T., Cambridge, Mass., March 1985.

26. L. Pun, G. Doumeingts, and A. Bourely, "The GRAI Approach to the Structural Design of Flexible Manufacturing Systems," *International Journal of Production Research*, Vol. 23, No. 6, 1985, pp. 1197–1215.

27. L. Rolston, "Modeling Flexible Manufacturing Systems with MAP/1," *Proceedings of the First ORSA/TIMS Conference on Flexible Manufacturing Systems*, Ann Arbor, Mich., 1984, pp. 199–204.

28. R. Suri and G.W. Diehl, "Quick and Easy Manufacturing Systems Analysis Using MANUPLAN," *Proc. Spring IIE Conference*, Institute of Industrial Engineers, Dallas, Texas, pp. 195–205, May 1986.

29. R. Suri and G.W. Diehl, "MANUPLAN: A Precursor to Simulation for Complex Manufacturing Systems," *Proc. Winter Simulation Conference*, IEEE Press, New York, 1985.

30. K.R. Anderson, G.W. Diehl, M. Shimizu, and R. Suri, "Integrating Spreadsheets, System Modeling, and Animation for Rapid Computer-Aided Design of Manufacturing Systems," *Proceedings of the Sixth Annual Conference on University Programs in Computer-Aided Engineering, Design and Manufacturing*, Georgia Institute of Technology, Atlanta, Ga., June 1988, pp. 22–27.

31. X.L. Chang, R.S. Sullivan, and J.R. Wilson, "Using SLAM to Design the Material Handling System of a Flexible Manufacturing System," *International Journal of Production Research*, Vol. 24, No. 1, 1986, pp. 15–26.

32. R.F. Garzia, M.R. Garzia, and B.P. Ziegler, "Discrete-Event Simulation," *IEEE Spectrum*, Dec. 1986, pp. 32–36.

33. H. D'Angelo, M. Caramanis, S. Finger, A. Mavretic, Y.A. Phillis, and E. Ramsden, "Event Driven Model of an Unreliable Production Line with Storage," *Proceedings of the 24th IEEE Conference on Decision and Control*, Ft. Lauderdale, Fla., Dec. 1986, pp. 1694–1698.

34. B. Khoshnevis and A. Kiran, "A FMS Physical Simulator," *Journal of Manufacturing Systems*, Vol. 5, No. 1, 1986, pp. 65–68.

35. E. Ramsden, M. Ruane, H. D'Angelo, M. Caramanis, and A. Mavretic, "A Digital Production System Simulation Engine," *Proceedings of the 1986 IEEE International Conference on Systems, Man, and Cybernetics*, Atlanta, Ga., Oct. 1986, pp. 698–702.

36. E. Ramsden, M. Ruane, H. D'Angelo, and A. Mavretic, "The Factory on an Integrated Circuit Chip," *Proceedings of the 1985 IEEE International Conference on Cybernetics and Society*, Tucson, Ariz., Nov. 1986.

37. D.W. Andrews, J.E. Stoner, and D.H. Withers, "A Discrete-Event Simulation of a Computer-Controlled Materials Distribution System," *IBM TR 08.111*, Lexington, Ky., **Note:** IBM Research Reports may be obtained from: IBM T.J. Watson Research Center, Distribution Services F-11 Stormytown, P.O. Box 218, Yorktown Heights, N.Y. 10598.

38. J. Pasquier, "High Level Simulation of Flexible Card Assembly Lines," *IBM Research Report RC 10881*, IBM, Yorktown Heights, N.Y.

39. B. Dietrich and B. Marsh, "An Application of a Hybrid Approach to Modeling a Flexible Manufacturing System," *IBM Research Report RC 10887*, IBM, Yorktown Heights, N.Y.

40. H. Engelke et al. "Structured Modeling of Manufacturing Processes," *Proceedings Annual Simulation Symposium*, Tampa, Fla., 1983, pp. 55–68.

41. Y.C. Ho (Ed.), "SPEEDS: A New Technique for the Analysis and Optimization of Queuing Networks," *TR 675*, Division of Applied Sciences, Harvard University, Cambridge, Mass., Feb. 1983.

42. A.M. Law and W.D. Kelton, *Simulation Modeling and Analysis*, McGraw-Hill, New York, 1982.

43. W.L. Maxwell and R.C. Wilson, "Dynamic Network Flow Modelling of Fixed Path Material Handling Systems," *AIIE Transactions*, Vol. 13, No. 1, 1981, pp. 12–21.

44. Y.V. Reddy and M.S. Fox, "KBS—An Artificial Intelligence Approach to Flexible Simulation," *CMU-RI-TR-82-1*, The Robotics Institute, Carnegie Melon University, Pittsburgh, Penn., 1982.

45. C.L. Beck and B.H. Krogh, "Models for Simulation and Discrete Control of Manufacturing Systems," *Proceedings of the 1986 IEEE International Conference on Robotics and Automation*, San Francisco, Calif., April 1986, pp. 305–310.

46. S.C. Mathewson, "The Application of Program Generation Software and Its Extensions to Discrete Event

Simulation Modeling," *IIE Transactions*, Vol. 15, No. 1, March 1984, pp. 3–18.

47. T.J. Schriber, "The Use of GPSS/H in Modeling a Typical Flexible Manufacturing System," *Proceedings of the First ORSA/TIMS Conference on Flexible Manufacturing Systems*, Ann Arbor, Mich., 1984, pp. 168–182.

48. W.M. Chow, E.A. MacNair, and C.H. Sauer, "Analysis of Manufacturing Systems by the Research Queueing Package," *IBM Journal of Research and Development*, Vol. 29, No. 4, July 1985, pp. 330–342.

49. H. Engelke, J. Grotrian, C. Scheuing, A. Schmackpfeffer, W. Schwarz, B. Solf, and J. Tomann, "Integrated Manufacturing Modeling System," *IBM Journal of Research and Development*, Vol. 29, No. 4, July 1985, pp. 343–355.

50. A.W. Naylor and M.C. Maletz, "The Manufacturing Game: A Formal Approach to Manufacturing Software," *IEEE Transactions on Systems, Man, and Cybernetics*, Vol. SMC-16, No. 3, May/June 1986, pp. 321–334.

7.7.4: Applications

51. H.J. Warnecke, R. Steinhilper, and H.P. Roth, "Developments and Planning for FMS—Requirements, Examples, and Experiences," *International Journal of Production Research*, Vol. 24, No. 4, 1986, pp. 763–772.

52. G. Handke, "Computer Integrated and Automated Manufacturing Systems in Aircraft Manufacturing," *International Journal of Production Research*, Vol. 24, No. 4, 1986, pp. 811–823.

53. A.S. Carrie, E. Adhami, A. Stephens, and I.C. Murdoch, "Introducing a Flexible Manufacturing System," *International Journal of Production Research*, Vol. 22, No. 6, 1984, pp. 907–916.

54. R.J. Wittrock, "Scheduling Algorithms for Flexible Flow Lines," *IBM Journal of Research and Development*, Vol. 24, No. 4, July 1985, pp. 401–412.

55. D.L. Krause and D.A. Locy, "Hybrid Integrated Circuit Manufacturing Process as Controlled by Shop Information Systems," *IEEE Transactions on Components, Hybrids, and Manufacturing Technology*, Vol. CHMT-3, No. 3, Sept. 1980, pp. 345–353.

56. E.G. Smith, "Automatic Control of Large-Scale Integrated Circuit Fabrication Processes—Process Control Algorithms," *IEEE Transactions on Components, Hybrids, and Manufacturing Technology*, Vol. CHMT-3, No. 3, Sept. 1980, pp. 331–338.

57. A.M. Spence and D.J. Welter, "Capacity Planning of a Photolithography Work Cell in a Wafer Manufacturing Line," *Proceedings of the 1987 IEEE International Conference on Robotics and Automation*, Raleigh, N.C., pp. 702–708.

58. C. Lozinski and S.B. Gershwin, "Dynamic Production Scheduling in Computer-Aided Fabrication of Integrated Circuits," *Proceedings of the 1986 IEEE International Conference on Robotics and Automation*, San Francisco, Calif., April 1986, pp. 660–663.

59. J.A. Buzacott and D. Cheng, "Quality Modelling and Quality Control of a Manufacturing System," *Proceedings of the 23rd IEEE Conference on Decision and Control*, Las Vegas, Nev., Dec. 1984, pp. 226–229.

60. T. Liu, D. Scott, H. Romanowitz, R. Innes, and D. Chin, "An In-Process-Storage System Case Study," *Proceedings of the 1986 IEEE International Conference on Robotics and Automation*, San Francisco, Calif., April 1986, pp. 497–503.

7.7.5: Just-In-Time Manufacturing

61. P.R. Philipoom, L.P. Rees, B.W. Taylor, III, and P.Y. Huang, "An Investigation of the Factor Influencing the Number of Kanbans Required in the Implementation of the JIT Technique with Kanbans," *International Journal of Production Research*, Vol. 25, No. 3, 1987, pp. 457–472.

62. M. Ebrahimpour and R.J. Schonberger, "The Japanese Just-In-Time/Total Quality Control Production System: Potential for Developing Countries," *International Journal of Production Research*, Vol. 23, No. 3, 1984, pp. 421–430.

63. B.J. Finch and J.F. Cox, "An Examination of Just-In-Time Management for the Small Manufacturer: With an Illustration," *International Journal of Production Research*, Vol. 24, No. 2, 1986, pp. 329–342.

7.7.6: Research Issues

64. U. Wemmerlov and N.L. Hyer, "Research Issues in Cellular Manufacturing," *International Journal of Production Research*, Vol. 25, No. 3, 1987, pp. 413–431.

65. R.K. Sinha and R.H. Hollier, "A Review of Production Control Problems in Cellular Manufacturing," *International Journal of Production Research*, Vol. 22, No. 5, 1984, pp. 773–789.

Author Index

Agerwala, T., 252
Akella, R., 129, 145
Baptiste, P., 283
Berman, O., 74, 327
Bruno, G., 96
Buzacott, J.A., 84, 104
Cao, X.R., 230
Cassandras, C., 206, 225
Chien, T.T., 177
Choong, Y., 129
Conterno, R., 96
Dato, M.A., 96
Davis, R.P., 60
Eyler, M.A., 177
Favrel, J., 283
Finke, G., 359
Gershwin, S.B., 35, 74, 119, 129
Gruver, W.A., 162
Haruna, K., 301
Hildebrant, R.R., 35
Ho, G.S., 273
Ho, Y.C., 177, 201, 204, 206, 225, 230
Hyung, L.-K., 283
Kamath, M., 262
Kennedy, W.J., Jr., 60
Kera, K., 290

Kimemia, J., 119
Komoda, N., 290, 301
Kubo, T., 290
Kumar, P.R., 145, 319
Kusiak, A., 359
Lasserre, J.B., 334
Lee, E.-J., 350
Lin, W., 319
Maimon, O., 327
Matsumoto, K., 301
Menga, G., 96
Merchant, M.E., 12
Mirchandani, P.B., 350
Mitsumori, S., 340
Mitter, S.K., 35
Molloy, M.K., 268
Murata, T., 301
Narasimhan, S.L., 162
Ramamoorthy, C.V., 273
Roubellat, F., 334
Shanthikumar, J.G., 84
Sharifnia, A., 156
Suri, R., 35
Tanimoto, H., 16
Viswanadham, N., 262

IEEE Computer Society Press

Publications Activities Board

Vice President: Duncan Lawrie, University of Illinois
James Aylor, University of Virginia
P. Bruce Berra, Syracuse University
Jon T. Butler, US Naval Postgraduate School
Tom Cain, University of Pittsburgh
Michael Evangelist, MCC
Eugene Falken, IEEE Computer Society Press
Lansing Hatfield, Lawrence Livermore National Laboratory
Ronald G. Hoelzeman, University of Pittsburgh
Ez Nahouraii, IBM
Guylaine Pollock, Sandia National Laboratories
Charles B. Silio, University of Maryland
Ronald D. Williams, University of Virginia

Editor-in-Chief: Ez Nahouraii, IBM
Editors: Jon T. Butler, US Naval Postgraduate School
Garry R. Kampen, Seattle University
Krishna Kavi, University of Texas, Arlington
Arnold C. Meltzer, George Washington University
Frederick R. Petry, Tulane University
Charles Richter, MCC
Sol Shatz, The University of Illinois, Chicago
Kit Tham, Mentor Graphics Corporation
Rao Vemuri, University of California, Davis

T. Michael Elliott, Executive Director
Eugene Falken, Publisher
Margaret J. Brown, Managing Editor
(Tutorials and Monographs)
Denise Felix, Production Coordinator (Reprint Collections)
Janet Harward, Promotions Production Manager

Submission of proposals: For guidelines on preparing CS Press Books, write Editor-in-Chief, IEEE Computer Society, 1730 Massachusetts Avenue, N.W., Washington, DC 20036-1903 (telephone 202-371-1012).

Offices of the IEEE Computer Society

Headquarters Office
1730 Massachusetts Avenue, N.W.
Washington, DC 20036-1903
Phone: (202)371-1012
Telex: 7108250437 IEEE COMPSO

Publications Office
10662 Los Vaqueros Circle
Los Alamitos, CA 90720
Membership and General Information: (714)821-8380
Publications Orders: (800)272-6657

European Office
13, Avenue de l'Aquilon
B-1200 Brussels, Belgium
Phone: 32 (2) 770-21-96 Telex: 25387 AWALB

Asian Office
Ooshima Building
2-19-1 Minami-Aoyama, Minato-ku
Tokyo 107, Japan

IEEE Computer Society Press Publications

Monographs: A monograph is a collection of original material assembled as a coherent package. It is typically a treatise on a small area of learning and may include the collection of knowledge gathered over the lifetime of the authors.

Tutorials: A tutorial is a collection of original materials prepared by the editors and reprints of the best articles published in a subject area. They must contain at least five percent original materials (15 to 20 percent original materials is recommended).

Reprint Books: A reprint book is a collection of reprints that are divided into sections with a preface, table of contents, and section introductions that discuss the reprints and why they were selected. It contains less than five percent original material.

Technology Series: The technology series is a collection of anthologies of reprints each with a narrow focus of a subset on a particular discipline.

Purpose

The IEEE Computer Society advances the theory and practice of computer science and engineering, promotes the exchange of technical information among 97,000 members worldwide, and provides a wide range of services to members and nonmembers.

Membership

Members receive the acclaimed monthly magazine *Computer*, discounts, and opportunities to serve (all activities are led by volunteer members). Membership is open to all IEEE members, affiliate society members, and others seriously interested in the computer field.

Publications and Activities

Computer. An authoritative, easy-to-read magazine containing tutorial and in-depth articles on topics across the computer field, plus news, conferences, calendar, interviews, and new products.

Periodicals. The society publishes six magazines and four research transactions. Refer to membership application or request information as noted above.

Conference Proceedings, Tutorial Texts, Standards Documents. The Computer Society Press publishes more than 100 titles every year.

Standards Working Groups. Over 100 of these groups produce IEEE standards used throughout the industrial world.

Technical Committees. Over 30 TCs publish newsletters, provide interaction with peers in specialty areas, and directly influence standards, conferences, and education.

Conferences/Education. The society holds about 100 conferences each year and sponsors many educational activities, including computing science accreditation.

Chapters. Regular and student chapters worldwide provide the opportunity to interact with colleagues, hear technical experts, and serve the local professional community.

Ombudsman

Members experiencing problems — magazine delivery, membership status, or unresolved complaints — may write to the ombudsman at the Publications Office.

Other IEEE Computer Society Press Texts

Monographs

Integrating Design and Test: Using CAE Tools for ATE Programming:
Written by K.P. Parker
(ISBN 0-8186-8788-6 (case)); 160 pages

JSP and JSD: The Jackson Approach to Software Development (Second Edition)
Written by J.R. Cameron
(ISBN 0-8186-8858-0 (case)); 560 pages

National Computer Policies
Written by Ben G. Matley and Thomas A. McDannold
(ISBN 0-8186-8784-3 (case)); 192 pages

Physical Level Interfaces and Protocols
Written by Uyless Black
(ISBN 0-8186-8824-6 (case)); 240 pages

Protecting Your Proprietary Rights in the Computer and High Technology Industries
Written by Tobey B. Marzouk, Esq.
(ISBN 0-8186-8754-1 (case)); 224 pages

Tutorials

Ada Programming Language
Edited by S.H. Saib and R.E. Fritz
(ISBN 0-8186-0456-5); 548 pages

Advanced Computer Architecture
Edited by D.P. Agrawal
(ISBN 0-8186-0667-3); 400 pages

Advanced Microprocessors and High-Level Language Computer Architectures
Edited by V. Milutinovic
(ISBN 0-8186-0623-1); 608 pages

Communication and Networking Protocols
Edited by S.S. Lam
(ISBN 0-8186-0582-0); 500 pages

Computer Architecture
Edited by D.D. Gajski, V.M. Milutinovic, H.J. Siegel, and B.P. Furht
(ISBN 0-8186-0704-1); 602 pages

Computer Communications: Architectures, Protocols and Standards (Second Edition)
Edited by William Stallings
(ISBN 0-8186-0790-4); 448 pages

Computer Grahics (2nd Edition)
Edited by J.C. Beatty and K.S. Booth
(ISBN 0-8186-0425-5); 576 pages

Computer Graphics Hardware: Image Generation and Display
Edited by H.K. Reghbati and A.Y.C. Lee
(ISBN 0-8186-0753-X); 384 pages

Computer Grahics: Image Synthesis
Edited by Kenneth Joy, Max Nelson, Charles Grant, and Lansing Hatfield
(ISBN 0-8186-8854-8 (case)); 384 pages

Computer and Network Security
Edited by M.D. Abrams and H.J. Podell
(ISBN 0-8186-0756-4); 448 pages

Computer Networks (4th Edition)
Edited by M.D. Abrams and I.W. Cotton
(ISBN 0-8186-0568-5); 512 pages

Computer Text Recognition and Error Correction
Edited by S.N. Srihari
(ISBN 0-8186-0579-0); 364 pages

Computers for Artificial Intelligence Applications
Edited by B. Wah and G.-J. Li
(ISBN 0-8186-0706-8); 656 pages

Database Management
Edited by J.A. Larson
(ISBN 0-8186-0714-9); 448 pages

Digital Image Processing and Analysis: Volume 1: Digital Image Processing
Edited by R. Chellappa and A.A. Sawchuk
(ISBN 0-8186-0665-7); 736 pages

Digital Image Processing and Analysis: Volume 2: Digital Image Analysis
Edited by R. Chellappa and A.A. Sawchuk
(ISBN 0-8186-0666-5); 670 pages

Digital Private Branch Exchanges (PBXs)
Edited by E.R. Coover
(ISBN 0-8186-0829-3); 400 pages

Distributed Control (2nd Edition)
Edited by R.E. Larson, P.L. McEntire, and J.G. O'Reilly
(ISBN 0-8186-0451-4); 382 pages

Distributed Database Management
Edited by J.A. Larson and S. Rahimi
(ISBN 0-8186-0575-8); 580 pages

Distributed-Software Engineering
Edited by S.M. Shatz and J.-P. Wang
(ISBN 0-8186-8856-4 (case)); 304 pages

DSP-Based Testing of Analog and Mixed-Signal Circuits
Edited by M. Mahoney
(ISBN 0-8186-0785-8); 272 pages

End User Facilities in the 1980's
Edited by J.A. Larson
(ISBN 0-8186-0449-2); 526 pages

Fault-Tolerant Computing
Edited by V.P. Nelson and B.D. Carroll
(ISBN 0-8186-0677-0 (paper) 0-8186-8667-4 (case)); 432 pages

Gallium Arsenide Computer Design
Edited by V.M. Milutinovic and D.A. Fura
(ISBN 0-8184-0795-5); 368 pages

Human Factors in Software Development (Second Edition)
Edited by B. Curtis
(ISBN 0-8186-0577-4); 736 pages

Integrated Services Digital Networks (ISDN) (Second Edition)
Edited by W. Stallings
(ISBN 0-8186-0823-4); 404 pages

(Continued on inside back cover)

For Further Information:

IEEE Computer Society, 10662 Los Vaqueros Circle, Los Alamitos, CA 90720

IEEE Computer Society, 13, Avenue de l'Aquilon, 2, B-1200 Brussels, BELGIUM

IEEE Computer Society, Ooshima Building, 2-19-1 Minami-Aoyama, Minato-ku, Tokyo 107, JAPAN

Interconnection Networks for Parallel and Distributed Processing
Edited by C.-l. Wu and T.-y. Feng
(ISBN 0-8186-0574-X); 500 pages

Local Network Equipment
Edited by H.A. Freeman and K.J. Thurber
(ISBN 0-8186-0605-3); 384 pages

Local Network Technology (3rd Edition)
Edited by W. Stallings
(ISBN 0-8186-0825-0); 512 pages

Microprogramming and Firmware Engineering
Edited by V. Milutinovic
(ISBN 0-8186-0839-0); 416 pages

Modern Design and Analysis of Discrete-Event Computer Simulations
Edited by E.J. Dudewicz and Z. Karian
(ISBN 0-8186-0597-9); 486 pages

New Paradigms for Software Development
Edited by William Agresti
(ISBN 0-8186-0707-6); 304 pages

Object-Oriented Computing—Volume 1: Concepts
Edited by Gerald E. Peterson
(ISBN 0-8186-0821-8); 214 pages

Object-Oriented Computing—Volume 2: Implementations
Edited by Gerald E. Peterson
(ISBN 0-8186-0822-6); 324 pages

Office Automation Systems (Second Edition)
Edited by H.A. Freeman and K.J. Thurber
(ISBN 0-8186-0711-4); 320 pages

Parallel Architectures for Database Systems
Edited by A.R. Hurson, L.L. Miller, and S.H. Pakzad
(ISBN 0-8186-8838-6 (case)); 478 pages

Programming Productivity: Issues for the Eighties (Second Edition)
Edited by C. Jones
(ISBN 0-8186-0681-9); 472 pages

Recent Advances in Distributed Data Base Management
Edited by C. Mohan
(ISBN 0-8186-0571-5); 500 pages

Reduced Instruction Set Computers
Edited by W. Stallings
(ISBN 0-8186-0713-0); 384 pages

Reliable Distributed System Software
Edited by J.A. Stankovic
(ISBN 0-8186-0570-7); 400 pages

Robotics Tutorial (2nd Edition)
Edited by C.S.G. Lee, R.C. Gonzalez, and K.S. Fu
(ISBN 0-8186-0658-4); 630 pages

Software Design Techniques (4th Edition)
Edited by P. Freeman and A.I. Wasserman
(ISBN 0-8186-0514-0); 736 pages

Software Engineering Project Management
Edited by R. Thayer
(ISBN 0-8186-0751-3); 512 pages

Software Maintenance
Edited by G. Parikh and N. Zvegintzov
(ISBN 0-8186-0002-0); 360 pages

Software Management (3rd Edition)
Edited by D.J. Reifer
(ISBN 0-8186-0678-9); 526 pages

Software-Oriented Computer Architecture
Edited by E. Fernandez and T. Lang
(ISBN 0-8186-0708-4); 376 pages

Software Quality Assurance: A Practical Approach
Edited by T.S. Chow
(ISBN 0-8186-0569-3); 506 pages

Software Restructuring
Edited by R.S. Arnold
(ISBN 0-8186-0680-0); 376 pages

Software Reusability
Edited by Peter Freeman
(ISBN 0-8186-0750-5); 304 pages—

Software Reuse: Emerging Technology
Edited by Will Tracz
(ISBN 0-8186-0846-3); 392 pages

Structured Testing
Edited by T.J. McCabe
(ISBN 0-8186-0452-2); 160 pages

Test Generation for VLSI Chips
Edited by V.D. Agrawal and S.C. Seth
(ISBN 0-8186-8786-X (case)); 416 pages

VLSI Technologies: Through the 80s and Beyond
Edited by D.J. McGreivy and K.A. Pickar
(ISBN 0-8186-0424-7); 346 pages

Reprint Collections

Selected Reprints: Dataflow and Reduction Architectures
Edited by S.S. Thakkar
(ISBN 0-8186-0759-9); 460 pages

Selected Reprints on Logic Design for Testability
Edited by C.C. Timoc
(ISBN 0-8186-0573-1); 324 pages

Selected Reprints: Microprocessors and Microcomputers (3rd Edition)
Edited by J.T. Cain
(ISBN 0-8186-0585-5); 386 pages

Selected Reprints in Software (3rd Edition)
Edited by M.V. Zelkowitz
(ISBN 0-8186-0789-0); 400 pages

Selected Reprints on VLSI Technologies and Computer Graphics
Edited by H. Fuchs
(ISBN 0-8186-0491-3); 490 pages

Technology Series

Artificial Neural Networks: Theoretical Concepts
Edited by V. Vemuri
(ISBN 0-8186-0855-2); 160 pages

Computer-Aided Software Engineering (CASE)
Edited by E.J. Chikofsky
(ISBN 0-8186-1917-1); 132 pages